Chipote, P., Mgxekwa, B., & Godza, P. (2014). "Impact of financial liberalization on economic growth: A case study of South Africa", *Mediterranean Journal of Social Sciences*, 5(23), pp. 1–8.

Chiwira, O., Bakwena, M., Mupimpila, C., & Tlhalefang, J. B. (2016). "Integration, inclusion, development in the financial sector and economic growth Nexus in SADC: Empirical review", *British Journal of Economics, Management & Trade,* 11(4), pp. 1–15.

Dandume, M. Y. (2014). "Financial sector development, economic growth and poverty reduction: New evidence from Nigeria"/Finansalsektordegelisim, ekonomikbuyume, yoksullukoranindusurulmesi: NijeryaOrnegi. *CankiriKaratekinUniversitesiIktisadiveIdariBilimlerFakultesiDergisi*, 4(2), pp. 1–21.

Evans, D. S., & Jovanovic, B. (1989). "An estimated model of entrepreneurial choice under liquidity constraints", *The Journal of Political Economy*, 1(1), pp. 808–827.

Fowowe, B., & Abidoye, B. (2013). "The effect of financial development on poverty and inequality in African countries", *The Manchester School*, 81(4), pp. 562–585.

Galor, O., & Moav, O. (2004). "From physical to human capital accumulation: Inequality and the process of development", *Review of Economic Studies*, 71(4), pp. 1001–1026.

Galor, O., & Zeira, J. (1993). "Income distribution and macroeconomics", *Review of Economics,* 60(1), pp. 35–52.

Greenwood, J., & Jovanovic, B. (1990). "Financial development, growth and the distribution of income", *Journal of Political Economy,* 98(5), pp. 1076–1107.

Holtz-Eakin, D., Joulfaian, D., & Rosen, H. S. (1994). "Sticking it out: Entrepreneurial survival and liquidity constraints", *Journal of Political Economy*, 102(1), pp. 53–75.

Jalilian, H., & Kirkpatrick, C. (2005). "Does financial development contribute to poverty reduction?", *Journal of Development Studies*, 41(4), pp. 636–656.

Keho, Y. (2017). "Financial development and poverty reduction: Evidence from selected African countries", *International Journal of Financial Research*, 8(4), pp. 90–98.

Koenker, R., & Bassett, Jr. G. (1978). "Regression quantiles", *Econometrica*, 46(1), pp. 33– 50.

Koenker, R., & Hallock, F. K. (2001). "Quantile regression", *Journal of Economic Perspectives*, 15(4), pp. 143–156.

Majid, M. S. A., Dewi, S., & Kassim, S. H. (2019). "Does financial development reduce poverty? Empirical evidence from Indonesia", *Journal of the Knowledge Economy*, 10(3), pp. 1019–1036.

Nwani, S. E, & Osuji, E. (2020). "Poverty in Sub-Saharan Africa: The dynamics of population, energy consumption and misery index", *International Journal of Management, Economics and Social Sciences,* 9(4), pp. 247–270.

Odhiambo, N. M. (2010). "Is financial development a spur to poverty reduction? Kenya's experience", *Journal of Economic Studies*, 37(3), pp. 343–353.

Odhiambo, N. M. (2014). "Financial systems and economic growth in South Africa: A dynamic complementarity test", *International Review of Applied Economics*, 28(1), pp. 83–101.

Ofori, I. K., Armah, M. K., Taale, F., & Ofori, P. E. (2021). "Addressing the severity and intensity of poverty in Sub-Saharan Africa: How relevant is the ICT and financial development pathway?" *Heliyon,* 7(10), e08156.

Orji, A., Aguegboh, E., & Anthony-Orji, O. I. (2015). "Real sector output and financial liberalisation in Nigeria", *Journal of Infrastructure Development,* 7(2), pp. 136–150.

Rashid, A., & Intartaglia, M. (2017). Financial development–does it lessen poverty? *Journal of Economic Studies*, 44(1), pp. 69–86.

Tchamyou, V. S. (2019). "The role of information sharing in modulating the effect of financial access on inequality", *Journal of African Business*, 20(3), pp. 317–338.

Tchamyou, V. S., & Asongu, S. A. (2017a). "Information sharing and financial sector development in Africa", *Journal of African Business*, 18(1), pp. 24–49.

Tchamyou, S. A., & Asongu, S. A. (2017b). "Conditional market timing in the mutual fund industry", *Research in International Business and Finance,* 42(December), pp. 1355–1366.

Tchamyou, V.S., Erreygers, G., & Cassimon, D. (2019). "Inequality, ICT and financial access in Africa", *Technological Forecasting and Social Change*, 139(February), pp. 169–184.

Tsaurai, K. (2020). "Financial development-poverty reduction nexus in BRICS: A panel data analysis approach", *Applied Econometrics and International Development*, 20(2), pp. 18–32.

Uddin, G. S., Shahbaz, M., Arouri, M., & Teulon, F. (2014). "Financial development and poverty reduction nexus: A cointegration and causality analysis in Bangladesh", *Economic Modelling*, 36(C), pp. 405–412.

Zahonogo, P. (2017). "Financial development and poverty in developing countries: evidence from Sub-Saharan Africa", *International Journal of Economics and Finance*, 9(1), pp. 211–220.

Appendices

Appendix 17.1 Definitions and Sources of Variables

Variables	*Definitions*	*Sources*
Poverty headcount	Poverty headcount ratio at national poverty lines (% of population)	WDI (World Bank)
Severity of poverty	"Poverty severity, which measures the degree of inequality among the poor by putting more weight on the position of the poorest". Squared of poverty gap index	Generated
Financial institutions depth index	"The Financial Institutions Depth (FID) Index, which compiles data on bank credit to the private sector, pension fund assets, mutual fund assets, and insurance premiums (life and non-life) as percentages of GDP".	Findex (World Bank)
Financial institutions access index	"The Financial Institutions Access (FIA) Index, which compiles data on the number of bank branches and the number of automatic teller machines (ATMs) per 100,000 adults"	Findex (World Bank)
Financial institutions efficiency index	"The Financial Institutions Efficiency (FIE) Index, which compiles data on the banking sector's net interest margin, the lending–deposits spread, the ratios of non-interest income to total income and overhead costs to total assets, and the returns on assets and equity".	Findex (World Bank)
Inflation	Inflation, consumer prices (annual %)	WDI (World Bank)
Foreign aid	Net Official Development Assistance received (% of GNI)	WDI (World Bank)
Government expenditure	General government final consumption expenditure (% of GDP)	WDI (World Bank)
Economic growth	GDP growth (annual %)	WDI (World Bank)
Foreign investment	Foreign direct investment, net inflows (% of GDP)	WDI (World Bank)
Income inequality (Gini)	"The Gini coefficient is a measurement of the income distribution of a country's residents".	WDI (World Bank)
Remittances	Remittance inflows (%GDP)	WDI (World Bank)
Trade	Trade is the sum of exports and imports of goods and services measured as a share of gross domestic product.	WDI (World Bank)

GDP: gross domestic product. GNI: Gross National Income. WDI: World Development Indicators. IMF: International Monetary Fund.

Appendix 17.2 Summary Statistics

	Mean	*SD*	*Min*	*Max*	*Obs*
Poverty headcount	48.215	14.055	7.900	73.200	1,680
Severity of poverty	16.529	22.480	0.000	169.299	1,681
Financial institutions depth	0.097	0.147	0.000	0.880	1,680
Financial institutions access	0.077	0.128	0.000	0.880	1,680
Financial institutions efficiency	0.494	0.199	0.000	0.990	1,680
Inflation	32.026	593.191	−13.056	23773.13	1,680
Foreign aid	11.345	11.527	−0.250	94.946	1,680
Government expenditure	5.353	25.868	−17.463	565.538	1,680
GDP growth	3.635	5.173	−50.248	35.224	1,680
Foreign direct investment	2.938	6.456	−28.624	103.337	1,680
Inequality (Gini)	53.250	19.829	0.000	86.832	1,680
Remittances	4.385	17.842	0.000	235.924	1,680
Trade openness	67.240	35.588	6.320	311.354	1,680

SD: standard deviation. Min: minimum. Max: maximum.

Appendix 17.3 Correlation Matrix (Uniform Sample Size: 1,680)

	PovHC	*SoPov*	*FID*	*FIA*	*FIE*	*Infl*	*NODA*	*Gov.*	*GDPg*	*FDI*	*Gini*	*Remit*	*Trade*
PovHC	1.000												
SoPov	0.071	1.000											
FID	−0.069	−0.207	1.000										
FIA	−0.264	−0.283	0.412	1.000									
FIE	−0.338	−0.146	0.312	0.305	1.000								
Infl	0.055	0.066	−0.025	−0.022	0.001	1.000							
NODA	0.375	0.084	−0.251	−0.164	−0.246	−0.013	1.000						
Gov.	−0.044	−0.023	0.036	0.018	0.073	−0.095	0.092	1.000					
GDPg	−0.111	−0.036	0.001	0.029	0.069	−0.062	−0.017	0.146	1.000				
FDI	0.004	−0.050	0.058	0.196	−0.010	−0.017	0.069	0.031	0.081	1.000			
Gini	0.120	0.139	0.001	−0.156	−0.034	0.012	0.097	0.017	0.005	−0.094	1.000		
Remit	0.082	−0.046	0.111	−0.013	−0.052	−0.009	0.034	0.088	0.031	0.014	0.044	1.000	
Trade	−0.146	−0.054	0.255	0.380	0.005	−0.028	−0.056	0.083	0.059	0.308	−0.040	0.305	1.000

PovHC: poverty headcount. SoPov: severity of poverty. FID: financial institutions depth. FIA: financial institutions access. FIE: financial institutions efficiency. Infl: inflation. NODA: foreign aid. Gov: government expenditure. GDPg: gross domestic product growth. FDI: foreign direct investment. Gini: the Gini coefficient. Remit: remittances.

Index

Note: **Bold** page numbers refer to tables; *italic* page numbers refer to figures and page numbers followed by "n" denote endnotes.

Printed in the United States
by Baker & Taylor Publisher Services

T0388997

Hegemony and Resistance around the Iranian Nuclear Programme

The Iranian nuclear crisis is a proxy arena for competing visions about the functioning of international relations.

This book is the first to provide comprehensive and comparative analyses to conceptualise the interaction between 'hegemonic structures' and those actors resisting them using the Iranian nuclear case as an illustration. It analyses the foreign policies of China, Russia and Turkey towards the Iranian nuclear programme and thereby answers the question to what extent these policies are indicative of a security culture that resists hegemony. Based on 70 elite interviews with experts and decision-makers closely involved with the Iranian nuclear file, it analyses resistance to hegemony across its ideational, material and institutional framework conditions. The cases examined show how 'compliance' on the part of China, Russia and Turkey with parts of US approaches to the Iranian nuclear conflict has been selective and how US policy preferences in the Iran dossier have been resisted on other occasions. As such, the Iran nuclear case serves as an illustration to shed light on the contemporaneous interaction of the forces of consent and coercion in international politics.

This book will be of key interest to scholars, students and practitioners in International Relations, Security Studies and Foreign Policy Analysis.

Moritz Pieper is Lecturer in International Relations at the University of Salford, UK. He has been a visiting Research Fellow at China Foreign Affairs University (CFAU), Beijing, the German Institute for International and Security Affairs, Brussels and at the School of Oriental and African Studies (SOAS), London.

Routledge Studies on Challenges, Crises and Dissent in World Politics

Series editors

Karoline Postel-Vinay

Centre for International Studies, and Research (CERI) France

and

Nadine Godehardt

German Institute for International and Security Affairs, Germany

This new series focuses on challenges, crises and dissent in world politics and the major political issues that have surfaced in recent years. It welcomes a wide range of theoretical and methodological approaches including critical and postmodern studies, and aims to improve our present understanding of global order through the exploration of major challenges to inter/national and regional governability, the effects of nationalism, extremism, weak leadership and the emergence of new actors in international politics.

1 **Empires of Remorse**
Narrative, Postcolonialism and Apologies for Colonial Atrocity
Tom Bentley

2 **Crisis and Institutional Change in Regional Integration**
Edited by Sabine Saurugger and Fabien Terpan

3 **Violent Non-State Actors**
From Anarchists to Jihadists
Ersel Aydinli

4 **Power-Sharing**
Empirical and Normative Challenges
Edited by Allison McCulloch and John McGarry

5 **Hegemony and Resistance around the Iranian Nuclear Programme**
Analysing Chinese, Russian, and Turkish Foreign Policies
Moritz Pieper

Hegemony and Resistance around the Iranian Nuclear Programme

Analysing Chinese, Russian and Turkish Foreign Policies

Moritz Pieper

LONDON AND NEW YORK

First published 2017
by Routledge
2 Park Square, Milton Park, Abingdon, Oxon OX14 4RN

and by Routledge
711 Third Avenue, New York, NY 10017

Routledge is an imprint of the Taylor & Francis Group, an informa business

British Library Cataloguing in Publication Data
A catalogue record for this book is available from the British Library

Library of Congress Cataloging in Publication Data
Names: Pieper, Moritz, author.
Title: Hegemony and resistance around the Iranian nuclear programme : analysing Chinese, Russian, and Turkish foreign policies / Moritz Pieper.
Description: Abingdon, Oxon ; New York, NY : Routledge, 2017. |
Series: Routledge studies on challenges, crises and dissent in world politics ; 5 | Includes bibliographical references and index.
Identifiers: LCCN 2016049107| ISBN 9781138205666 (hardback) |
ISBN 9781315466378 (ebook)
Subjects: LCSH: Nuclear nonproliferation–Iran. | Nuclear nonproliferation–International cooperation. | Nuclear nonproliferation–Government policy–China. | Nuclear nonproliferation–Government policy–Russia (Federation) | Nuclear nonproliferation–Government policy–Turkey. | Hegemony.
Classification: LCC JZ5675 .P54 2017 | DDC 327.1/7470955–dc23
LC record available at https://lccn.loc.gov/2016049107

ISBN: 978-1-138-20566-6 (hbk)
ISBN: 978-1-315-46637-8 (ebk)

Typeset in Times New Roman
by Wearset Ltd, Boldon, Tyne and Wear

In loving memory of Clemens Pieper

Contents

Acknowledgements

Research for this book let me travel around the globe – from Turkey, Russia and the United States, to China and all over Europe. Managing such a project would not have been possible without the help and support of many people. First and foremost, I want to thank Dr Tom Casier, who has been my primary advisor and mentor in all matters related to my doctoral dissertation out of which this book has emerged. His enduring faith in my research and persistent encouragement was as motivating as it was inspiring, and his perceptive yet congenial supervision made the journey a very enjoyable one. I equally wish to thank Dr Tugba Basaran, whose sharp comments helped to crystallise many thoughts in the making and flesh out the unspoken word behind the lines. Throughout this journey at the University of Kent's Brussels campus, I was fortunate to have been surrounded by many fantastic colleagues who have enhanced my research experience in so many ways. I thank above all Octavius Pinkard, who has travelled with me around the globe and back to Bonnefooi in the course of a PhD programme that has turned academic interactions into a friendship for a lifetime. I also wish to thank John Heieck, Teresa Cabrita, Shahriar Sharei, Tomislava Penkova, Shubranshu Mishra, Klaudia Tani, Richa Kumar, Inez Summers and Michael Sewell for making BSIS such a pleasant place to work.

My international research trips would not have been possible without the financial support of a number of organisations. Primarily, I was honoured to have received a 50th Anniversary Doctoral Scholarship from the University of Kent. This privilege gave me the financial independence to embark on this project in the first place. In addition, external travel grants were made available by the Centre for East European Language Based Area Studies (CEELBAS), the Association for the Study of the Middle East and Africa (ASMEA), the International Studies Association (ISA), the SETA Foundation and the Santander Travel Bursary Scheme.

I thank the China Foreign Affairs University in Beijing for my affiliation as a Visiting Researcher in April and May 2014 and the Friedrich-Ebert Foundation in Shanghai and Beijing for suggestions for interviewees. I am also grateful to Bek, Tomasz, Maria, Natasha, Cécile, Alex, Nais, Jean-Thomas, John, Fazal and Alice for a warm welcome in China and for helping me out with Chinese language issues. Useful suggestions and help in setting up interviews in Turkey came from the Friedrich-Ebert Foundation in Brussels and Istanbul as well as from the German Marshall Fund in Ankara. For help with logistics and practical arrangements in Turkey, I am very grateful to Sacid Örengul and Taner Zorbay. I am also indebted to a number of government officials, for helping me establish contacts for interviews in Brussels, Turkey, Russia, China and the US. They have asked to remain anonymous, but they know who they are, and I am very grateful for their help in 'getting the snowball rolling'.

The German Institute for International and Security Affairs (SWP) has given me the opportunity to spend some most productive months at their Brussels office as a Visiting Research Fellow from January to April 2015. I am very thankful for the warm welcome I received from Dr Dušan Reljić and his team. I am also pleased to acknowledge the support of the School of Oriental and African Studies (SOAS) in London and Prof. Arshin Adib-Moghaddam in particular for providing resources and facilities during my stay as a Visiting Scholar in May 2015. I equally thank the Higher School of Economics in Moscow, my Russian *alma mater*, for receiving me as a Visiting Researcher in June 2016. Sincere thanks must also go to the Directorate of Politics and Contemporary History at the University of Salford, my current port of call, for providing a congenial and stimulating surrounding. I especially wish to thank Dr Iván Garcia and Prof. Alaric Searle for their encouragement and support during my work on this project.

Partial results and sections of this book benefitted from conferences at the Russian Academy of Sciences, Moscow; the University of Tartu; the Middle East Technical University, Ankara; the University of Leeds; the Middle East Studies Association's 47th Annual Meeting, New Orleans; the National Press Club, Washington; the International Studies Association's 55th Annual Convention, Toronto; the University of West Bohemia, Plzeň; and the University of Deusto, Bilbao. An earlier version of Chapter 2 was published as a book chapter by the SETA Foundation in 2015; an earlier version of Chapter 3 was published in *International Politics*, vol. 52, no. 5; and an earlier version of Chapter 5 was published in the *Asian Journal of Peacebuilding*, vol. 2, no. 1. Formulations in the introduction, Chapter 1 and Chapter 5 are derived, in part, from an article published in *European Security*, vol. 26, no. 1. I thank the editors for granting the permission to re-use the material here.

At Routledge, I thank Dr Andrew Taylor, Sophie Iddamalgoda, as well as the two series editors Dr Nadine Godehardt and Dr Karoline Postel-Vinay for the smooth publishing process. For their engagement with and constructive comments on this work, I am grateful to Dr Peter Ferdinand, Dr Albena Azmanova and Dr Amanda Klekowski von Koppenfels. All remaining errors are my own.

Finally, my heartfelt gratitude will always be to my parents. Without their love and unwavering support, this book could not have been possible. And it is to them that it is dedicated.

Moritz Pieper
Manchester, March 2017

Transliteration and translation

For all transliterations from the Chinese to the Latin alphabet, the standard *Pinyin* system (without diacritic markers) has been used for all proper names and translations (e.g. Xi Jinping, Zhuhai Zhenrong, taoguang yanghui).

Special characters of the modern Turkish alphabet have been used for all Turkish proper names, authors and terms (e.g. Orta doğu, Erdoğan, Uluslararası İlişkiler).

For transliteration from Russian, the British standard version has been used, thus ю becomes 'yu', я becomes 'ya', ъ is apostrophised, etc. Proper names have been anglicised (e.g. Sergei, Andrei and Alexander instead of Aleksander) – except when in another author's citation. Translations from Russian are the author's, except where indicated otherwise.

While there is no unified system for the transliteration of Farsi, the romanisation of Farsi names and titles has largely followed the Library of Congress system (e.g. Ahmadinejad, Rouhani, Diplomasi-ye Hastehi).

Acronyms

ABC	Atomic, Biological, Chemical (weapons)
AEOI	Atomic Energy Organisation of Iran
AKP	*Adalet ve Kalkınma Partisi* (Justice and Development Party)
BBC	British Broadcasting Corporation
BMD	ballistic missile defense
BRICS	acronym referring to Brazil, Russia, India, China, South Africa
CCP	Chinese Communist Party
CENESS	Center for Energy and Security Studies
CFAU	China Foreign Affairs University
CFSP	Common Foreign and Security Policy (of the European Union)
CICIR	China Institute of Contemporary International Relations
CIIS	China Institute of International Studies
CIISS	China Institute for International Strategic Studies
CIS	Commonwealth of Independent States
CISADA	Comprehensive Iran Sanctions, Accountability, and Divestment Act
CITIC	China International Trust and Investment Corporation
CNOOC	China National Offshore Oil Corporation
CNPC	China National Petroleum Corporation
CRS	Congressional Research Service
CTBT	Comprehensive Nuclear-Test-Ban Treaty
E3	the 'European Three' (Germany, France, United Kingdom)
E3 + 3	the 'E3' plus China, Russia, United States
ECFR	European Council on Foreign Relations
EEAS	European External Action Service
EU	European Union
FFI	Foreign Financial Institution
GCC	Gulf Cooperation Council
HEU	Highly Enriched Uranium
HST	Hegemonic Stability Theory
IAEA	International Atomic Energy Agency
ILSA	Iran–Libya Sanctions Act

IMEMO	*institut mirovoj ekonomiki i meshdunarodnikh otnoshenii rossijskoi akademii nauk* (Institute of World Economy and International Relations, Russian Academy of Sciences)
IR	International Relations
IRGC	Iranian Revolutionary Guard Corps
IRISL	Islamic Republic of Iran Shipping Lines
ISA	Iran Sanctions Act
JCPOA	Joint Comprehensive Plan of Action (of 14 July 2015)
JPOA	Joint Plan of Action (of 23 November 2013)
LEU	Low Enriched Uranium
LNG	Liquefied Natural Gas
MD	Missile Defense
MFA	Ministry of Foreign Affairs
Minatom	Russian Ministry of Atomic Energy (until 2004)
MİT	Milli İstihbarat Teskilati (Turkish Intelligence Agency)
MoU	Memorandum of Understanding
MTCR	Missile Technology Control Regime
NAM	Non-Aligned Movement
NATO	North Atlantic Treaty Organisation
NDAA	National Defense Authorization Act
NNWS	Non-nuclear Weapon State
NPT	Nuclear Non-Proliferation Treaty
NSG	Nuclear Suppliers Group
NWFZ	Nuclear weapons-free zone
NWS	Nuclear Weapon State
OFAC	Office of Foreign Asset Control
OPEC	Organisation of Petroleum-Exporting Countries
OSCE	Organisation for Security and Cooperation in Europe
P5+1	the five permanent UN Security Council members plus Germany
PJAK	*Partiya Jiyana Azad a Kurdistanê* (Party of Free Life of Kurdistan)
PKK	*Partiya Karkerên Kurdistanê* (Kurdistan Worker's Party)
PLA	People's Liberation Army
PMD	Possible Military Dimension
PUK	Patriotic Union of Kurdistan
Rosatom	Russian Ministry of Atomic Energy (since 2004)
SIIS	Shanghai Institute for International Studies
SIPRI	Stockholm Peace Research Institute
START	Strategic Arms Reduction Treaty
SWIFT	Society for Worldwide International Financial Telecommunication
TANAP	Trans-Anatolian Pipeline
TAP	Trans-Adriatic Pipeline
TPAO	Turkish Petroleum Company
TRR	Tehran Research Reactor

UK	United Kingdom
UN	United Nations
UNSC	United Nations Security Council
UNSCR	United Nations Security Council Resolution
US	United States
USSR	Union of Soviet Socialist Republics
WTO	World Trade Organisation
WWII	World War Two

Introduction

The Iran question is the question of our age.

(Turkish diplomat in conversation with the author,
Washington, 14 February 2014)

In 2002, an Iranian exile opposition group revealed the existence of nuclear facilities in Iran that had been undeclared to the International Atomic Energy Agency (IAEA). With a uranium enrichment site at Natanz and a heavy-water reactor under construction at Arak suddenly in the spotlight of international attention, a heated debate ensued whether Iran was in breach of its Safeguards Agreement under the nuclear non-proliferation treaty (NPT). Iran had acceded to the NPT in 1968, ratified the treaty in 1970 and signed a Safeguards Agreement in 1974. With the discovery of facilities that could be used to produce weapons-grade fissile material, speculations about a 'possible military dimension' of Iran's nuclear programme flared up. The 'E3' (Germany, France, the United Kingdom) started to negotiate with Iran to de-escalate what was soon turning into a delicate political conflict, and were joined in late 2004 by the EU's High Representative for Foreign Affairs and Security Policy Javier Solana. Under the E3's negotiation efforts, nuclear talks initially made progress from the stalled 2003 Sa'dabad negotiations and culminated in the November 2004 Paris agreement, which outlined broad-based European-Iranian cooperation in a number of issue areas, including comprehensive cooperation in the nuclear, technological and economic field in return for 'objective guarantees' in the exclusively peaceful nature of the Iranian nuclear programme. Developments as from 2005, most notably the election of Mahmoud Ahmadinejad as president of Iran in June, brought this momentum to a halt. Despite European efforts to prevent this from happening, the Iranian nuclear file eventually was referred from the IAEA Board of Governors to the United Nations Security Council (UNSC) according to Art. XII.C of the IAEA Statute after Iran had resumed previously suspended uranium enrichment in August 2005. The file was taken to the UNSC in 2006, where the negotiation format was enlarged to the E3+3, or P5+1 (the five permanent UNSC member states plus Germany). The United States, China and Russia had now joined the negotiations. When Iran was found in non-compliance of first

UN resolutions and exhibiting insufficient transparency with the IAEA, international sanctions were imposed soon after. Years of missed opportunities, stonewalling, and tactical deceptions on different sides followed (Porter 2014).

As a reflection of the fruitlessness of diplomatic engagements, the UN Security Council adopted six resolutions on Iran's nuclear programme between 2006 and 2010, of which four imposed international sanctions that included trade and financial restrictions, travel bans and asset freezes against Iranian individuals, as well as arms embargoes (UN 2006, 2007, 2008, 2010). This was followed by US unilateral and EU sanctions, imposed between 2010 and 2012, that had a noticeable impact on the Iranian economy. New US financial sanctions targeted and sanctioned foreign entities' engagement in the Iranian energy sector (US Department of State 2011, 2013), while the EU imposed an oil and gas embargo in 2012, banned European companies from providing insurance for oil or petrochemical shipments from Iran, banned transactions between European and Iranian banks, and froze assets of the Iranian Central Bank held within the EU (EU Council 2012a, 2012b, 2012c). The extent to which these measures led the Iranian government to modify negotiation positions was hotly debated between sanctions advocates and their critics. In 2013, however, Hassan Rouhani, who had been chief nuclear negotiator in the early years of the nuclear crisis until 2005, was elected president partly due to a campaign run on the promise to improve the economy. A markedly more constructive Iranian approach to negotiations was met with a willingness on the US side to engage in direct US–Iranian talks. This was a diplomatic novelty since the break in diplomatic relations following the Iranian revolution in 1979. The US and Iranian administrations had already engaged in secret-channel diplomacy via Oman prior to the election.

Only months later, on 24 November 2013, a first interim agreement between the P5+1 and Iran was reached in Geneva. Besides first tangible confidence-building measures on both sides such as limited, temporary sanctions relief for Iran and the halting by Iran of uranium enrichment above 5 per cent, this 'Joint Plan of Action' gave all delegations a much-needed breathing space to work on a comprehensive final accord. A political framework agreement was reached in Lausanne in April 2015. Three months later, the signing of a 'Joint Comprehensive Plan of Action' (JCPOA) on 14 July 2015 marked the end of long marathon negotiations (EEAS 2015a). Under the JCPOA, Iran commits itself to reduce its centrifuge numbers, restructure facilities such as the underground facility at Fordow and the heavy-water reactor at Arak, reduce its enriched uranium inventory to 300 kilograms and limit the enrichment level to 3.67 per cent in accordance with the standard requirement for civilian nuclear power plants. In return, all UN, EU, and US nuclear-related sanctions on Iran began to be lifted on 16 January 2016 after the confirmation by the IAEA that Iran had fulfilled agreed commitments under the nuclear accord (IAEA 2016).[1] The US and EU administrations initiated the lifting of sanctions specified in Annex V of the JCPOA, while relevant sanctions provisions in UN Security Council Resolutions (UNSCR) 1696, 1737, 1803, 1835, 1929 and 2224 were terminated.[2]

16 January 2016 thus marked the 'Implementation Day' – in the agreement's parlance – of the JCPOA to resolve the stand-off on the controversial Iranian nuclear programme that has bedeviled international politics ever since its discovery in 2002. This in itself is a historic success of international diplomacy to prevent war in the Middle East, and the first successful precedent for the rewinding of UN sanctions imposed over proliferation charges without resorting to military means or regime change plans. The verified rollback of nuclear capabilities is also rare in arms-control history (Fitzpatrick 2016).

The stakes could not be higher. Iran's international standing is altered in a process where international trade, political and security relations with Iran will no longer have to take place under the dangling Damocles sword of the nuclear conflict. Governments and companies alike express an interest in entering the Iranian market and renewing commercial relations with a country that had been securitised for decades of 'institutionalised enmity' (Parsi 2012: 240). The path is cleared for an altered engagement of Iran with the world. But the Iran nuclear file cannot be laid to rest yet. Until 2025 ('Termination Day'), Iran will be subject to a special monitoring and inspection regime and only then be treated as any other non-nuclear weapon state (NNWS) under the NPT. A UN weapons embargo will be lifted only after five years from Implementation Day, ballistic missile sanctions after eight years. And whilst nuclear-related sanctions began to be lifted in January 2016, the persistence of non-nuclear-related US financial sanctions will continue to have extraterritorial effects on foreign entities (so-called 'secondary sanctions'). In addition, the persistence of US primary sanctions on US entities also extends to foreign subsidiaries of US companies. This opens up a range of legal grey zones where non-US companies will have to effectively firewall their Iran businesses from any links to the US. Banks are reluctant to lend credits for Iran businesses because of the persistence of what Iran complains is a 'psychological sanctions architecture'.[3] The EU and the US administration have both published lengthy guidelines and information notes to help the private sector navigate the complex world of Iran sanctions regimes as the JCPOA has entered its implementation phase (EEAS 2016; US Department of the Treasury 2016). Structural constraints continue to complicate the 'fundamental shift' that the JCPOA aspires (EEAS 2015b). To complicate matters further, domestic politics especially in Iran and in the United States have the potential to endanger the implementation of this complex agreement. The fact that this is a Plan of Action, not an international treaty, makes it legally and politically more vulnerable to attempts by domestic critics, or future administrations, to undermine or even rewind pledges made.

The final contours of Iran's nuclear status will have far-reaching implications for the future of the NPT, global security governance and the regional security architecture. 'This deal is not only a political agreement', Iranian president Rouhani stated in his remarks at the 71st UN General Assembly in September 2016, 'It also represents a creative approach and method for constructive interaction with a view to peacefully resolving crises and challenges' (Rouhani 2016). The solution of the Iran nuclear crisis on 'Termination Day' of the agreement will shape the future working relationship between 'the West' and Iran. But on an equally fundamental and understudied dimension, the more than decade-old Iranian nuclear conflict served as a battle ground between 'the modernised' and the 'modernising' world (Patrikarakos 2012: 30), and between hegemonic powers and norm-shapers in the making.

It is that crucial nexus to which this book will direct its attention. This research project analyses Chinese, Russian, and Turkish foreign policies towards the Iranian nuclear programme. Underlying this research is the question how and where their foreign policies interact with mechanisms dominated and perpetuated by another actor whose involvement, for a number of reasons, is critical for any resolution of the nuclear stand-off with Iran – namely the United States. Not least because of traumatised US–Iranian relations and the centrality of the US in Iranian foreign policy discourse, Washington holds considerable sway over Iran's nuclear future, and the strong US–Iranian bilateral negotiation track under their respective foreign ministers John Kerry and Javad Zarif testifies to this. But also on a structural level, the omnipresence of US financial power in international governance and the extent to which this particular leverage shapes policy formulations of other actors creates what in this book will be called 'hegemonic structures'. These structures have met criticism and outright rejection by a range of actors. Russia, China and Turkey have voiced principled scepticism about the use of sanctions in international politics, but have accepted and approved of sanctions mandated by the UN Security Council. Sanctions adopted by national governments like the US or regional organisations like the EU, however, have been rejected and partially circumvented in areas where they affected third country activities in Iran. Yet, China, Russia and Turkey have all shown a level of receptiveness to US pressure and have, to varying degrees, preemptively taken policies to avoid financial sanctions for having 'violated' US sanctions lists. On the diplomatic and technical level, all three states under investigation here have introduced proposals to solve the Iranian nuclear conflict, and have thus served as challengers of and complements to 'Western' positions on Iran at the same time. This confronts us with the puzzle of how to conceptualise the interaction between China, Russia and Turkey on the one side and US-dominated power structures on the other.

While Chinese, Russian, and Turkish Iran policies, respectively, as well as their foreign policies towards the controversial nuclear programme of Iran have been analysed before, no comparative analyses at book length have been produced thus far that use the Iranian nuclear case as an illustration to conceptualise the interaction between 'hegemonic structures' and those actors resisting them. The significance of China, Russia and Turkey for the success of the nuclear talks, their involvement in international security governance, and their intricate status between what scholars have labelled the 'West' and 'rising' or 'emerging' new power centres will be analysed in the chapters that follow. In this effort, the research will be guided by the central question to what extent Russian, Chinese and Turkish foreign policies in the Iranian nuclear conflict are indicative of a security culture that resists hegemony. Dividing lines that emerged at negotiations indicated that the implementation of any nuclear agreement would be fraught with intricate legal, institutional, and profoundly normative disagreements. Throughout the nuclear talks, as well as those periods when negotiations broke down, technical questions about nuclear physics have been held hostage to political narratives on different sides. China's and Russia's involvement in the P5+1 format from 2006 onwards was crucial in dispelling the impression that the Iranian nuclear conflict in essence was a stand-off between Iran and 'the West'. Yet, within the P5+1 and between the P5+1 and external mediators (such as Turkey), diverging views crystallised about the best approach to resolve the nuclear crisis. Especially the debate over the imposition of sanctions became a thorny issue that stood emblematic for wider questions of legitimacy in international governance, hegemonic politics and conceptions of World Order. The Iranian nuclear case, in a sense, serves as a laboratory to examine fundamental questions about international relations that will continue to reverberate long after the Iranian nuclear file will be closed. Between Iran and the other participants in the JCPOA, the long schedules of implementation will ensure that the Iranian nuclear case will continue to be high on the world political agenda well beyond 2016. Against this background, the timeframe of this book's analysis is roughly set from the onset of the Iranian nuclear conflict in 2003 up until the early aftermath of the JCPOA's 'Implementation Day' in January 2016.

Outline of the book

The structure of this book is as follows. A first chapter will outline and justify the conceptual and theoretical framework that guides the analysis. It will argue how neo-Gramscian scholarship embedded in the scholarly literature on 'norm dynamics' is a fruitful angle to the research project. Such a conceptual approach has never been applied to analyses of the Iranian nuclear crisis. This first chapter will make sense of and elucidate terms such as 'security culture', 'resistance' and 'hegemony', and situate the theoretical approach taken in the scholarly literature. Within the latter, the roles of actors said to be 'rising' or 'emerging' have been analysed and predicted, but have remained under-studied in work at the interstice of security, International Relations and Area Studies. The chapter will therefore situate the approach taken here within the body of work that attempts to elucidate the relationship between Western and 'non-Western' powers. Chinese, Russian, and Turkish Iran policies will be presented as three in-depth case studies to illustrate the workings of resistance and hegemony in international politics.

Chapters 2, 3 and 4 present the empirical case studies of this book. Chapter 2 analyses Turkish foreign policy towards Iran's nuclear programme, and shows how Turkish Iran policies are torn between resistance to US approaches because of ideational and material disagreements and accommodation with US positions because of institutional framework conditions. Turkey's NATO membership and a shared neighbourhood with Iran in particular will be portrayed as influential factors that have led many analysts to situate Turkey between different geostrategic and political 'camps'. It was this 'in-between-ness' coupled with a Turkish drive for more pro-active international diplomacy that led Turkey, together with Brazil, to negotiate the first-ever agreement in 2010 that got an Iranian approval. With a momentum for the adoption of sanctions at the UN Security Council under way, however, this episode turned into an exemplary case for failed expectations and the workings of structural path dependency. It will be shown how Turkish foreign policy towards the Iranian nuclear programme was formulated before and after its active period of involvement in 2009–2010 both on a discursive and a behavioural level. The chapter will thereby reflect on what this tells us about the role of actors external to the P5 + 1 negotiation format in the Iran case, but also about the role of 'new power centres' in international security governance at large.

Chapter 3 analyses Russian Iran policies and thereby introduces the second in-depth case study of this book. It will process-trace Russia's positioning in the Iranian nuclear negotiations, especially when the case was referred to the UN Security Council in 2006. Russia's history as a nuclear partner of Iran on the civilian nuclear energy market has provided Moscow with a unique 'entrance point' into understanding Iranian positions, both on a technical and a normative level. This also gave Russia the technical expertise and political credibility to position itself as a self-proclaimed constructive intermediary between Iran and other parties in the talks. Russian contracts to build Iran's only nuclear power plant at Bushehr and Russia's oftentimes vocal criticisms of its Western dialogue partners, however, have not meant an automatic alignment with Iran. The chapter will disentangle reasons for Russian and Iranian mutual frustrations with each other as well as tactical approximations in Iran's nuclear stand-off, and thereby also illustrate how Russia became an invaluable partner of the West in addressing technical proliferation-related aspects of Iran's nuclear file. Doing so, the chapter outlines the different ideational, material, and institutional factors impacting Russia's foreign policy towards the Iranian nuclear programme.

Chapter 4 presents an analysis of Chinese foreign policy towards the Iranian nuclear programme. It will be shown how China, similarly to Russia, sought to balance between upholding good relations with Iran, braking an American push for sanctions, and displaying a willingness to accommodate US concerns at other times. Especially the intensive commercial nature of Sino-Iranian relations at a time when Western companies had largely withdrawn from the Iranian market has meant that Beijing was potentially very vulnerable to US financial sanctions imposed on third country entities trading with Iran. Manoeuvring between this conundrum, the outside perception of Chinese Iran policies and a desire to be seen as 'rising peacefully' became a political tightrope walk. It will be analysed why China criticised Iran sanctions, yet did not veto their adoption in the UN Security Council. Finally, China's role and participation in the Iran talks will be contextualised in the scholarly and policy debates about the nature and future direction of Chinese foreign policy. Questioning projections of 'clashes' between the West and non-Western powers, the chapter shows how the Iran case is an example of a more nuanced balance between contestation of dominant normative paradigms and accommodation on a behavioural level with those very structures that Beijing has criticised on a discursive level.

Chapter 5 comparatively analyses the research findings of these three in-depth case studies and draws conclusions on the extent to which Chinese, Russian and Turkish Iran policies are indicative of a security culture that resists hegemony, before the last chapter concludes the book with a synthesis and an outline of possible further areas of research. This book tells the story of how the Iranian nuclear crisis has morphed into a contestation over international justice, the power of dissent in international security governance, and the relationship between hegemony, resistance and World Order.

Notes

1 As specified in paragraphs 15.1–15.11 of Annex V of the JCPOA.

2 However, some restrictions related to nuclear-, conventional arms-, and ballistic missile-related activities apply under UNSCR 2231 of 20 July 2015, which has endorsed the JCPOA. See US Department of the Treasury (2016: 2). Legal terminologies differ with regard to the lifting of sanctions, and have subsequently left room for debate as to their political implications. The EU has *terminated* relevant provisions of Council Regulation No 267/2012, *suspended* some provisions of Council Decision 2010/413/CFSP and *amended* others. The US has *ceased the application of* sanctions of relevant legislative Acts and *terminated* relevant executive orders. See Annex V of the JCPOA.

3 Author's interview with Iranian diplomat, Berlin, 29 August 2016.

References

EEAS. 2015a. *Joint Comprehensive Plan of Action*. Vienna, 14 July 2015. Available at https://eeas.europa.eu/statements-eeas/docs/iran_agreement/iran_joint-comprehensive-plan-of-action_en.pdf (accessed 17 September 2016).

EEAS. 2015b. Joint statement by EU High Representative Federica Mogherini and Iranian Foreign Minister Javad Zarif Vienna, *Press statement*, 14 July. Available at www.eeas.europa.eu/statements-eeas/2015/150714_01_en.htm (accessed 13 August 2015).

EEAS. 2016. Information Note on EU sanctions to be lifted under the Joint Comprehensive Plan of Action (JCPOA). Brussels, 23 January 2016. Available at http://collections.internetmemory.org/haeu/content/20160313172652/http://eeas.europa.eu/top_stories/pdf/iran_implementation/information_note_eu_sanctions_jcpoa_en.pdf (accessed 17 September 2016).

EU Council. 2012a. *Council Decision 2012/35/CFSP of 23 January 2012 Amending Decision 2010/413/CFSP concerning restrictive measures against Iran*. Available at http://eur-lex.europa.eu/legal-content/EN/TXT/?uri=CELEX%3A32012D0035 (accessed 16 September 2016).

EU Council. 2012b. *Council Regulation No 267/2012 of 23 March 2012 Concerning Restrictive Measures against Iran and Repealing Regulation (EU) 961/2010*. Available at http://eur-lex.europa.eu/legal-content/EN/TXT/?uri=uriserv:OJ.L_.2012.088.01.0001.01.ENG&toc=OJ:L:2012:088:TOC (accessed 16 September 2016).

EU Council. 2012c. *Council Decision 2012/635/CFSP of 15 October 2012 Amending Council Decision 2010/413/CFSP Concerning Restrictive Measures against Iran*. Available at http://eur-lex.europa.eu/LexUriServ/LexUriServ.do?uri=OJ:L:2012:282:0058:0069:EN:PDF (accessed 16 September 2016).

Fitzpatrick, Mark. 2016. The Iran Deal Shows its Worth. Politics and Strategy, *The Survival Editors' Blog*, 18 January 2016. Available at www.iiss.org/en/politics%20

and%20strategy/blogsections/2016-d1f9/january-c129/the-iran-deal-shows-its-worth-ded4 (accessed 2 March 2016).

IAEA. 2016. Verification and Monitoring in the Islamic Republic of Iran in Light of United Nations Security Council Resolution 2231 (2015). *IAEA board report*, 16 January 2016. Available at www.iaea.org/sites/default/files/gov-inf-2016-1.pdf (accessed 6 September 2016).

Parsi, Trita. 2012. *A Single Roll of the Dice: Obama's Diplomacy with Iran*. New Haven, CT: Yale University Press.

Patrikarakos, David. 2012. *Nuclear Iran. The Birth of an Atomic State*. London, New York: I.B. Tauris.

Porter, Gareth. 2014. *Manufactured Crisis. The Untold Story of the Iran Nuclear Scare*. Charlottesville: Just World Books.

Rouhani, Hassan. 2016. *Statement at the General Debate of the General Assembly of the United Nations*, 22 September 2016. Available at https://gadebate.un.org/sites/default/files/gastatements/71/71_IR_en.pdf (accessed 23 September 2016).

UN. 2006. *Resolution 1737 (2006), adopted by the Security Council at its 5612th Meeting, on 23 December 2006*. Available at www.un.org/ga/search/view_doc.asp?symbol=S/RES/1737%282006%29 (accessed 26 September 2016).

UN. 2007. *Resolution 1747 (2007), Adopted by the Security Council at its 5647th Meeting on 24 March 2007*. Available at www.iaea.org/sites/default/files/unsc_res1747-2007.pdf (accessed 26 September 2016).

UN. 2008. *Resolution 1803 (2008), adopted by the Security Council at its 5848th meeting, on 3 March 2008*. Available at www.un.org/ga/search/view_doc.asp?symbol=S/RES/1803%282008%29 (accessed 26 September 2016).

UN. 2010. *Resolution 1929 (2010), Adopted by the Security Council at its 6335th Meeting, on 9 June 2010*. Available at www.iaea.org/sites/default/files/unsc_res1929-2010.pdf (accessed 26 September 2016).

US Department of State. 2011. *Fact Sheet: Comprehensive Iran Sanctions, Accountability, and Divestment Act (CISADA)*, 23 May 2011. Available at www.state.gov/e/eb/esc/iransanctions/docs/160710.htm (accessed 26 September 2016).

US Department of State. 2013. *Fact Sheet: Iran Freedom and Counter-Proliferation Act of 2012*, 23 April 2013. Available at www.state.gov/documents/organization/208111.pdf (accessed 25 September 2016).

US Department of the Treasury. 2016. *Guidance Relating to the Lifting of Certain U.S. sanctions pursuant to the Joint Comprehensive Plan of Action on Implementation Day*. 16 January.

1 Security discourse on Iran

The power to construct international relations

The Iranian nuclear crisis is a proxy arena for diverging projections of World Order. Different approaches to the Iranian nuclear crisis not only reveal different understandings of the nuclear non-proliferation regime. They also unravel diverging views on how international politics should function at large. Before this book delves into the question how China, Russia and Turkey have positioned themselves in these fundamental debates about how the world works, a few conceptual parentheses are in order. If we want to understand how the Iranian nuclear conflict has morphed into a battle ground for a deep-seated malaise in international politics, we need to understand first the normative aspect in all talk about 'security' and 'insecurity'. Structures of power condition hierarchies of agency.[1] It is this premise that captures the logic of the concept of 'securitisation' as developed by the so-called 'Copenhagen School' (Buzan, Waever and de Wilde 1998; Buzan and Waever 2003). According to this concept, being in a position to 'securitise' an issue or an actor means having the authority to define what counts as 'exceptional' measures to deal with the perceived threat (Buzan *et al.* 1998: 25). The securitisation of Iran's nuclear dossier, as will be seen, offers illuminating examples. Here, the rendition of enmity is order-constituting: The power to label other actors 'rogues' reconfirms prevalent power structures (see also Homolar 2011; Hoyt 2000; Senn 2009). If such a discursive practice of threat construction is to have an effect, there needs to be a relevant audience that lends confidence in authority (Williams 2003). It necessitates a power standing of the securitising agent and the acceptance thereof by said audience. It will be shown in the chapters that follow how a discourse that sought to 'securitise' an Iranian nuclear programme as a threat to international peace and security was met with both acceptance and resistance. The ensuing political controversy was revelatory for the relation between power and audience that the 'Copenhagen School' aims to dissect. Yet, rather than staying within the canon developed around the political ontology of state practice, analyses of power and security discourses will benefit from 'politicizing security', as advocated by the 'Aberystwyth School' (Bilgin 2013: 103; Booth 2005). Instead of 'desecuritisation' as discursive de-escalation, such an angle helps understand *the political* in policy divergences surrounding the Iranian nuclear crisis. Similarly, Iranian rhetoric about 'prestige', perceived leadership of the Islamic and non-aligned world, Third Worldism and 'unfairness' in international politics (Barzegar 2012: 233) all rely on political assumptions about power and interaction effects. Power relations and implicit value systems are audience-specific (Balzacq 2010), and each audience will re-define the meaning of *justice* (see Azmanova 2012; Gehring 1994: 366; Müller 2013; Welch 1993).

1 Analysing international norm dynamics

In their oft-cited work on norm acceptance and norm dynamics, Finnemore and Sikkink (1998) argue that 'norms evolve in a patterned "life cycle" and that different behavioural logics dominate different segments of the life cycle' (888). After an initial promotion of norms by what they call 'norm entrepreneurs', a process of socialisation sets in which they label a 'norm cascade', in which a sufficiently critical mass of actors accepts and adopts that particular norm (whereby a 'tipping point' is reached). In a last stage, actors internalise norms, rendering compliance automatic and thus creating a new 'logic of appropriateness', in the words of March and Olsen (1998: 949; see also March and Olsen 1989). The 'norm life cycle' presents a framework for the emergence of norms in international politics, and sheds light on the mechanisms of change, albeit in a somewhat schematic fashion that fails to systematically account for contingent power relations. The 'classical' norm literature that draws on Finnemore's and Sikkink's work proceeds from a conception of relatively static stages (see also Florini 1996; Johnstone 2007; Keck and Sikkink 1998; Nadelmann 1990). Alexander Wendt (1999) speaks of 'Culture' as a self-fulfilling prophecy (184ff.), and thereby makes the important observation that the expectation of 'appropriate behavior' (culture) structures how agents behave – itself a precondition for the preservation or alteration of culture. Norms become a 'conventionally' accepted standard of appropriate behaviour. 'Consensus becomes common sense, and common sense structures our thoughts', Ali Ansari (2006: 5) writes in *Confronting Iran* and cautions: 'conclusions are reached on what we choose to see. And often what we choose to see supports our preconceptions, even if they are misconceptions' (82). Constructivism's focus on socialisation through interaction, in other words, 'reveal(s) a great deal of trust in the replication and self-sustainment of identities', as Terhalle puts it (2011: 349). Questioning such dominant normative structures is an exercise in implicitly advocating alternative norms, or at least a modified norm understanding. It is the latter prospect, however, that has been largely neglected in the 'classical' scholarly norm literature. As Wunderlich (2014; see also Wunderlich *et al.* 2013) convincingly demonstrates, this has been due to a research bias towards a unidirectional understanding of norm diffusion. Agency-based analyses of norm dynamics have tended to conflate norm diffusion with the promotion of 'positive', i.e. Western, liberal norms ('altruistic norm advocacy', Wunderlich 2014: 85). In reaction to this strand of literature, new studies have been written on changes in international norms as being essentially dispute-driven (Stiles and Sandholtz 2009: 323f.), on the contestation of norms (Bob 2012; Krook and True 2012;

Kersbergen and Verbeek 2007; Sandholtz 2007; Wiener 2008), and on the link between norms and power structures (Adler-Nissen 2014; Epstein 2012a, 2012b; Towns 2012). These works thus have contributed to the advancement of a 'new' generation or a 'second wave' of norm literature (see Cortell and Davis 2000; Wunderlich 2013) that conceives norms as essentially contested narratives. Wunderlich (2014) reverses the directionality of Finnemore's and Sikkink's original 'norm entrepreneurship' model and asks to what extent 'norm violators' (such as Iran) can be conceived of as 'roguish' norm entrepreneurs and contribute to a norm 'renovation' (88–89). Similarly, Bloomfield and Scott (2017) explore the politics of resistance of what they call 'norm antipreneurs'. Such scholarship weaves together norm dynamics, contingent power structures and issues of international legitimacy. It helps us understand 'how rising powers are socialized into the (current) order, while at the same time reshaping it when they enter' (Terhalle 2011: 345). The social construction of identities and 'order' is intimately linked with hegemonic legitimacy. The link between 'hegemonic discourse' and the structures of power has also been emphasised in the works of Laclau (1988) and Laclau and Mouffe (1985).[2]

Recurring but essentially contested concepts such as 'the international community' to which the nuclear programme of Iran allegedly presents a threat call for a critical re-evaluation of taken-for-granted assumptions about the conditions of legitimacy in international politics. Here, challenging questions arise about the conditions for acceptance of and admission to an 'international community'[3] that may give rise to critiques of the functioning of the international system as a reflection of selective norms of certain dominant states (see also Epstein 2012a; Widmaier and Park 2012). Chomsky (2016) singles out the US State Department's use of the term 'international community' as 'a technical term referring to the United States and whoever happens to agree with it' (51). Katzenstein (1996) similarly remarks that 'World society carries standardised oppositional ideologies that are usually selective reifications of elements of dominant world ideology' (48); of 'restricted subsets of global society', in the words of Nincic (2005: 11).

The NPT regime as a politico-legal framework underlying all discussions surrounding Iran's nuclear crisis is a case in point for such latent power structures and conceptions of world order. Dividing its signatories into those in possession of nuclear weapons before 1 January 1967 and non-nuclear weapon (NNW) states that are not allowed to acquire nuclear weapons according to Art. II, the treaty effectively imposed an arbitrary freezing of the political status quo. At the same time, in Art. IV and V, it granted NNWS the right to civilian nuclear energy. For years of negotiations in the Iranian nuclear conflict, the US had upheld the suspicion of Iran's nuclear programme potentially having a military dimension, whereas Teheran adamantly upholds its legitimate 'right to enrichment' for civilian nuclear purposes in accordance wit Art. IV of the NPT.[4] When the US insisted on a 'complete cessation' of an Iranian nuclear fuel cycle, as was the US's initial position in 2003–2004, it was not only violating the spirit and letter of the NPT,[5] but it also forcefully demonstrated how international law can be held victim to political narratives: The first steps toward an Iranian nuclear programme were laid with an agreement reached in 1957 between the US and Iran ruled by Shah Reza Pahlevi. Under the auspices of the US 'Atoms for Peace' programme, the US provided Iran with a nuclear reactor and Highly Enriched Uranium (HEU), followed by US support for Iran to acquire a reprocessing facility for plutonium extraction (Oborne and Morrison 2013: 38f.).[6] After the Iranian revolution in 1979 swept away a state that had been a regional ally of the US, Washington cancelled its contracts and nuclear agreements with Iran. In the 2000s, an Iranian nuclear programme was being securitised[7] whose initial stepping stones were laid with the help of the US (Porter 2014: 23–38).[8] In an Iranian reading, the US's approach to the Iranian nuclear programme in this early stage of the Iranian nuclear conflict was the expression of Western arrogance, imperialist attitudes and nuclear double standards (Moshirzadeh 2007: 524; Mousavian 2012: 468). This also has created a political fault line between the developed and the developing, the modernised and the modernising world (Patrikarakos 2012: 30). In his memoirs, former IAEA Secretary General Mohamed ElBaradei writes of an 'asymmetry of the global security system' (172) that he detects in the perpetuation of the existing global nuclear framework conditions, as emphasised by the 'linkage between nuclear proliferation and the sluggish pace of disarmament' (254). This 'discrimination between the haves and have-nots' (Hurrell 2006: 11) and the non-recognition of Iran's 'inalienable right' to nuclear technology for peaceful purposes by key actors in the Iranian nuclear dossier[9] was a recurrent theme underlying all negotiation efforts with Iran (Wunderlich *et al.* 2013: 263–272). This nuclear discrimination purposely built into the NPT became the bone of contention for much wider 'fairness' issues between the 'developing world' and 'Western domination', as framed by Iranian argumentation.

The Iranian nuclear crisis needs to be understood not only as a politico-legal challenge for the NPT regime, but also as a proxy issue for a wider re-rethinking about world order and 'nuclear hypocrisy' (Oborne and Morrison 2013: 32f.). This book sheds light on how China, Turkey and Russia positioned and position themselves toward these essentially *normative* questions and what this tells us about underlying conceptions of security governance, the non-proliferation regime and world order at large. China and Russia are both nuclear weapon states (NWSs) themselves and permanent members of the UNSC. A binary distinction between nuclear 'haves' and 'have-nots' thus fails to account for differences in foreign policy approaches to Iran between these two states and the West. While China, Turkey and Russia are driven by myriad factors and motivations in pursuing their respective Iran policies, the case studies in this book will carve out the existence of common themes of diplomacy and foreign policy principles on which their positions converge and on which they potentially differ from 'Western' approaches. While this is not to argue that China, Russia and Turkey act in a concerted manner to accelerate the demise of prevalent governance structures, it will be shown in this research how their respective foreign policies share commonalities in terms of normative conceptions of how international relations should be governed.

1.1 Conceptualising security cultures and resistance to hegemony

Drawing on Katzenstein's (1996) definition of 'culture' as

> a set of evaluative standards (such as norms and values) and a set of cognitive standards (such as rules and models) that define what social actors exist in a system, how they operate, and how they relate to one another,
>
> (21)

a security culture is understood as a set of evaluative and cognitive standards in the security governance realm. It is the yardstick with which states assess legitimate means and ends in security policies. While 'evaluative standards' in the form of norms and values will be further elaborated upon in the following section, the use of 'cognitive standards' understood as 'rules and models' as taken from Katzenstein's definition may require further clarification at this point. If 'norms and values' are concrete convictions and conceptions, 'rules and models' relate to the broader macro-structure that regulates the way these norms and values are communicated, applied, or changed. An example is the UN system setting rules and models by way of its institutional make-up and treaty stipulations (decision-making procedures, legal proceedings, constitution of UN bodies) that serves as a broader frame for the channelling of concrete norms and values that account for the former's political content and are often at the heart of debate among its member states (such as sovereignty, non-interference). Rules and models are structures underlying norms and values. If China, Russia and Turkey regard US unilateral Iran sanctions as illegitimate, but accept international (i.e. UN-backed) Iran sanctions, they reveal their own understanding of certain *norms* in international politics, while demonstrating an adherence to the *rules* of the UN family.

Such a definition of security cultures assumes that the recognition, shaping and evaluation of security cultures is necessarily relational: The assessment of legitimacy here presupposes interaction effects, as alternative discourses aiming at the modification of prevalent norms make no sense in an isolated system (Wendt 1999: 158). Both Western states and those under investigation in this research project (China, Russia, Turkey) have reproached each other at times of interpreting and consequently circumventing or even blatantly ignoring international treaty provisions according to their political convenience. Examples here include the reproach brought forward against Western countries that the insistence on a complete shutdown of Iran's nuclear programme is a violation of the spirit and the letter of Art. IV NPT, or the reproach that the export of certain products and technologies to Iran violate the spirit and the letter of pertinent UNSC sanctions resolutions on Iran – whose legality in turn was questioned by Iran. These examples forcefully make the point that what is deemed 'appropriate behaviour' in international politics is always the outcome of an intersubjective evaluation and presupposes what Wendt (1999) calls *Culture* as 'socially shared knowledge' (141). The acceptance or rejection of a foreign policy behaviour as appropriate is the expression of *security cultures*. 'Security' becomes situational and is intimately and inextricably linked with images of 'Self' and 'Other'.

A further conceptual element of a security culture between hegemony and resistance to be clarified here is that of 'hegemony', 'counter-hegemony' and 'resistance'. Aware of scholarly contributions sceptical of the possibility of an application of Antonio Gramsci's concept of 'hegemony' to international relations as the expansion of a concept originally conceived to apply to the nation-state (Femia 2005; Germain and Kenny 1998; Burnham 1991; Bellamy 1990; Worth 2009), this book moves away from Gramsci's focus on a dominant social class, but contends that the idea of hegemony can nevertheless be usefully applied to the international arena (see also Rupert 1995).[10] Moving away from the strong actor-centricity of hegemony as understood in the hegemonic stability theory (HST) and the 'benign' liberal variant of hegemony (Gilpin 1981; Ikenberry 2001: 47, 2011; Keohane 1984; Kindleberger 1973, 1981), this book draws on Gramsci's understanding of hegemony as developed in his *Prison Notebooks* and conceives of hegemony as political, social and economic structures enabling dominance in international politics (Gramsci 1971: 171–172). Here, a dominant class forms a relationship with subaltern classes that is characterised both by consent and coercion (Gramsci 1971: 55–60, 415–425).[11] An actor may lead single-handedly for a while, but a long-term survival of any 'system' needs to be built on consent (Cox 1981, 1983; Worth 2015). Hegemony is therefore different from imperialism as the latter is understood as 'accumulation by dispossession' (Luxemburg 2003). The latter, exploitative conception features more prominently in World systems theory (Wallerstein 1979).

A Gramscian notion of hegemony proceeds from the importance of the *Making of Culture* as key to the crafting of Order (Fiori 2013: 130). To the extent that cultural codes formulated by dominant actors are accepted throughout society and become *common sense*, hegemonic order is always based on the dual foundation of consent *and* coercion – just as a centaur is always half man, half beast, in the analogy Gramsci borrowed from Machiavelli (Cox 1993: 52; Gramsci 1971: 169–170). In the same logic, ideas alone never create lasting order. Instead, theory (ideas) and practice are inextricably interrelated and inter-penetrate each other. Gramsci's therefore necessarily is a circular understanding of how hegemony comes into being, and how it changes gradually. Such an ensemble of and arrangement between hegemonic structures and the wider society represents a 'historic bloc'. In a neo-Gramscian understanding of hegemony in International Relations as developed, e.g. in Cox's seminal works on *World Order* (Cox 1996: 131), the prevalence of dominant structures that are accepted and sustained by a sufficiently large number of other actors can also constitute such a 'historic bloc'. Order here relies on a 'multitude of agencies' (Worth 2015: 173). To the extent that other states act upon, sustain and reinforce US dominant structures in the social, economic and political sphere, US hegemony post-1945 has brought about a historic bloc in a Gramscian understanding that is being upheld by the vast majority of states in the Western hemisphere.

A security culture is understood as hegemonic if shaped by a dominant actor that holds sufficient power so as to induce adaption and acceptance thereof by other actors. Such an understanding draws on Gramsci's conception of *cultural* hegemony (Service 2007: 139–140), but adds an under-studied dimension of the security and foreign policy realm and thus moves away from the strong association of neo-Gramscianism with the discipline of International Political Economy (see also Worth 2011: 386–390). Rather than focusing on narrowly materialist conceptions of hegemony, it is argued here that the *perception of legitimacy* is crucial for the preservation or alteration of hegemonic structures.[12]

By implication, a security culture is counter-hegemonic if it confronts or challenges a prevailing hegemonic framework and the normative pull toward socialisation with it. The latter, drawing on a definition by Finnemore and Sikkink (1998), is 'the dominant mechanism of a norm cascade – the mechanism through which norm leaders persuade others to adhere' (902). 'Gramsci's concern', Schwarzmantel (2009) writes, 'lay in challenging the dominant ideas or hegemonic concepts, and forming a new historic bloc or constellation of social forces to create an alternative set of ideas' (8). This 'withering away of one bloc and the formation of another', Worth (2015) adds, 'happens over time as ideas, material circumstances and overriding hegemonic forces challenge the prevailing order and replace it with another' (20). Resistance to normative structures underlying hegemony lies at the root of every attempt to challenge consensual relationships, to resist a dominant 'historic bloc'. If hegemony is based on consent and coercion, questioning prevalent practice and contesting norms touches the foundational pillars of any hegemonic consensus.

It is at this point that an important conceptual differentiation between 'resistance' and 'counter-hegemony' needs to be made. 'A counter-hegemony would consist of a coherent view of an alternative world order, backed by a concentration of power sufficient to maintain a challenge to core countries', Robert Cox writes (1981: 150). The notion of 'counter-hegemony' thus implies a drive towards power transitions. While it may be possible conceptually to think of 'counter-hegemony', the latter is unlikely to exist in practice. Even examples coming to mind such as North Korea, which withdrew from the NPT and eventually tested atomic bombs, cannot serve as an empirical example of a phenomenon we could term 'counter-hegemony'. The setting up of such essentially reductionist dichotomies would not be able to capture a more fine-grained picture of how states position themselves towards hegemony, and would be immediately vulnerable to criticism as to how one could judge whether China, Russia, or Turkey are moving more or less towards the extreme end of such a spectrum. Their track record of engagement with Western dialogue partners, mutual collaboration in diplomatic forums as well as approval of sanctions resolutions passed on Iran render talk of 'counter-hegemony' meaningless from the outset. It is also worth noting that Gramsci himself never used the term 'counter-hegemony'.

The ideas of consent and coercion developed in Gramscian thought therefore transcend a dichotomous juxtaposition of 'domination' and 'resistance'.

Resistance is never a 'totalizing act' (Jones 2006: 76), but always a qualified form of disagreement with a part of, or even a number of, hegemonic policies. Resistance and hegemonic power are locked in a constant process of engagement with each other. Contrary to the idea of 'counter-hegemony', resistance is therefore conceptualised here as a qualified form of disagreement with hegemony, with established power constellations. It thus borrows more from Gramsci's concept of a longer-term 'war of position' in which meanings and values gradually become contested, rather than from his idea of a 'war of manoeuvre', which Steve Jones (2006) likens to an 'all-out frontal attack' (31).

It is therefore suggested here that 'resistance' to hegemony captures more accurately the conceptual continuum that eschews distortionary simplifications and allows for empirically more qualified analyses of foreign policies. Conceptually and empirically, 'counter-hegemony' entertains the idea that one hegemony will eventually be replaced by another. Yet, contestation need not mean endeavours to topple the hegemonic system, but can be a first step towards reforming it. Instead of an *anti*-American world, we should think of *post*-American variants (Acharya 2014; Bilgin 2008; Hart and Jones 2011; Zakaria 2011). Another helpful distinction here may be that of radicalism (negation of the prevalent order) versus resistance (qualified disagreement with parts of that order).[13] Moreover, the unlikeliness of 'counter-hegemony' on a state level may be explained by the threefold distinction of hegemony made by Gramsci: since a hegemonic world order implies predominance in the social, economic and political sphere, the attempt to resist hegemony in the security and political sphere may be cushioned by a state's dependence on and integration in the international economic order that is predicated upon the US-inspired neoliberal world order (the economic sphere). It is precisely in this context that an analysis of Chinese, Russian and Turkish Iran policies becomes an examination of the friction between consent and coercion, between resistance and hegemony.

Lastly, both in an IR realist as well as in a critical understanding, 'hegemony' has been treated largely as a state-centric concept. The analysis presented in this book, however, is more sympathetic to a Coxian understanding of hegemony in that it focusses on the material, ideational and institutional conditions that are underpinning hegemonic structures (Cox 1996: 97f., 135f.; Gill 1993, 2003; Pijl 1984; Rupert 1995, 2000). The interplay between these factors helps us understand the positioning of China, Turkey and Russia towards hegemonic structures and sheds light on what might at first sight seem to be ambiguous policies towards a monolithic hegemonic pole. While a security culture in a hegemonic understanding implies an expected adherence to the norms of this hegemonic power structure (as the set of standards by which behaviour is considered 'appropriate'), the understanding of a security culture in a Chinese, Russian and Turkish reading experiences a significant shift of emphasis in that security is understood to be security *from* such a normative hegemony. As will be shown, this explains the discursive re-iteration of the foreign policy norms of sovereignty and non-interference. This is not to imply that China, Russia and Turkey act in a concerted manner to craft a security culture of their own understanding in an attempt to openly challenge 'the system'.[14] Instead of assuming strategic convergence in their foreign policy behaviour, the case studies in this book analyse adherence to a common security culture discursively, while their respective material and institutional interests may be different or may even be conflicting with each other.

2 Whose hegemony?

It has been argued so far that a security culture is understood as hegemonic if shaped by dominant power structures that hold sufficient power so as to induce adaption and acceptance thereof by other actors. 'Hegemonic strategies,' Leverett and Leverett (2013) write, 'are inherently expansionist: a state uses military, political, and economic power not just to defend its interests but to bend others into accommodating them' (332). The US here is deemed a single most dominant actor that, as of yet, remains unmatched in its potential to influence other actors through such hegemonic structures (see also Cox 2001, 2007; Hart and Jones 2011; Krahmann 2005: 533; Wicht 2002: 77; Patel and Hansmeyer 2009; Quinn 2011; Stokes 2014; Wilkinson 1999: 142). The workings of this 'influence' will be elucidated throughout the case studies in the chapters that follow. Discussing reports of the telecommunications company Apple entering the Iranian market, a State Department official remarked in an interview that the 'motivation is to provide Iranians with technology that is in line with broader US foreign policy goals, (with) the post-WWII idea to maintain [*sic*] global capitalism'.[15] As telling as such a statement is for the US importance attached to the maintenance of a global neo-liberal ideology, it also stands in striking contrast to US efforts to regulate third countries' trade relations with Iran for the sake of upholding a narrative of enmity carefully constructed for almost four decades since Iran's Islamic Revolution in 1979.

The US unilateral sanctions regime that was put in place during the Iranian nuclear stand-off is arguably the ultimate expression of power that draws on hegemonic structures: the US imposed punitive measures onto other states that do business with Iran (Lohmann 2013) – an 'imperial extension of American power and [...] (a) sheer effrontery by which America sought to impose its political position', as Ali Ansari (2006: 144) puts it. The first effort to sanction third country entities engaging in business with Iran was enacted with the 1992 Iran–Iraq Nonproliferation Act that prohibited the transfer of goods or technologies that could facilitate the development of atomic, biological or chemical (ABC) weapons (Takeyh and Maloney 2011: 1301). This sector-specific sanctions regime was significantly expanded with the Iran–Libya Sanctions Act (ILSA) of 1996, which was modified in 2001 and renamed the Iran Sanctions Act in 2006 (Leverett and Leverett 2013: 310). Beyond the initial proliferation dimension, the ILSA imposed sanctions on third country entities investing more than US$40 million in the development of petroleum resources in Iran. This extension of the scope and applicability of Iran sanctions signified a sea change in US Iran policies in the way these gradually served to create a comprehensive regime that was to be adhered to on a global basis. From mid-2005, the US thus discovered the leverage power of financial sanctions. Rather than the State Department, the more relevant institutional actor in sanctions matters became the Treasury Department, and in particular the Office of Foreign Assets Control (OFAC) therein. Based on Executive Order 13382, the US anticipated that foreign companies would forfeit business with Iran because of their fear of losing access to the US capital market (Taylor 2010: 68). To that end, the US Treasury Department also maintains a list of 'Foreign Sanctions Evaders'. A US Treasury Department FAQ website notes:

> By cutting off access to the U.S. marketplace and financial system to such sanctions evaders, Executive Order 13608 provides Treasury with a powerful tool to prevent and deter such behavior and to hold such persons accountable and to convince them to change their behavior.
>
> (US Department of the Treasury 2016)

The extraterritorial application of these US unilateral sanctions was introduced with the justification of 'Iranian sponsoring of international terrorism', and thus acquired an explicit non-nuclear dimension (Lohmann 2015). Unilateral US 'secondary' sanctions[16] were considerably expanded with the 2010 Comprehensive Iran Sanctions, Accountability, and Divestment Act (CISADA) sanctioning purchases of Iranian oil as well as business transactions with the Iranian Central Bank,[17] the 2012 Iran Threat Reduction and Syria Human Rights Act, the 2013 National Defense Authorization Act, and the 2012 Iran Freedom and Counter-Proliferation Act (US Department of State 2013). Such a process of 'extraterritorialising' US legislation and enforcing political conceptions onto other states through compliance under the threat of economic costs operates with financial coercion on the basis of the US predominance in the global trade, financial and economic system. While international and unilateral nuclear-related Iran sanctions began to be lifted from 16 January 2016, the persistence of non-nuclear-related US Iran sanctions that had been imposed on Iranian individuals and entities because of human rights violations or 'sponsorship of terrorism' will continue to complicate third country entities' activities in Iran. The same holds for the continued application of sanctions for 'foreign subsidiaries' of US companies, which proves to be a challenging legal grey zone in practice (IISS 2016).

'The reason Airbus needed a license to build new airliners for Iran's civil aviation [following the JCPOA implementation] was that 10 per cent of Airbus is of US origin, so their Iran contracts would fall under US financial sanctions', an EEAS official explained to me, and quipped: 'But of course the US lifted the ban on Iranian pistachios, because apparently they didn't manage to grow good pistachios in California.'[18]

The former high-level US officials Leverett and Leverett (2013) even hold that US secondary sanctions are 'almost certainly' illegal under WTO regulations, but that no one has litigated the question so far (280).[19] The Helms-Burton and D'Amato-Kennedy Acts, upon which CISADA was modelled, have been considered unlawful under international law (Dupont 2012: 4; Lowe 1997). Yet, '[t]he fear of reputational costs', Giumelli and Ivan (2013) write, 'has led banks to adopt cautious behavior in order to avoid paying the costs of defying UN, EU and especially US financial bans' (15) – no matter their formal (un)lawfulness. The observation that 'psychology' and 'reputation' have led companies to 'over-comply' with sanctions lists for fear of possible (unintended) violations was confirmed by several officials interviewed for this study.[20] 'Reputational costs' here underlines the perceived need to adhere to such a sanctions regime out of fear of future retributions and forcefully ties in with points made above on the relational aspect of politics and the normative pull towards dominant discourses. 'Companies simply have no risk appetite for controversial business. They are cautious because of the financial terror', a barrister working with EU sanctions cases explained in an interview.[21]

Such US unilateral sanctions policies, in addition, were met with a transatlantic sanctions consensus after 2010 (Pieper 2016). Differences between the US and the E3 over approaches to the Iranian nuclear crisis initially ran deeply. These ranged from the referral of the case to the UNSC, opposed by the E3 at first but advocated by the Bush administration, the use of force, and the use of sanctions to pressure Iran (Meier 2013). In the early phase of nuclear diplomacy, at least until the November 2004 agreement, the Europeans' negotiating credibility lay in their potential to block the Americans from pushing the case to be referred to the Security Council and in serving as a 'human shield' against possible American or Israeli unilateral attacks on Iranian nuclear facilities (Porter 2014: 141). The creation of the E3+3 format introduced a new momentum to the talks, and from 2010 onwards, the EU imposed hard-hitting economic sanctions that went beyond the sanctions regime imposed by the UNSC. With this decision at the latest, the EU ceased to be a balancer against US positions, as seen by Tehran. Former Iranian official Mousavian writes in his 2014 book:

> One of the harshest blows to the Iranian financial system came with the US Congress threatening to place sanctions on the Belgian-based Society for Worldwide International Financial Telecommunication (SWIFT) unless they cut ties with all Iranian banks [...]. Unsurprisingly, the EU yielded to US threats and consequently cut off the Iranian Central Bank from the international financial system.
>
> (38)

Asked about the imposition of EU sanctions in addition to UN and US sanctions, a European diplomat thus explains the rationale in an interview as follows:

> Europe was also under serious pressure. But we are part of the 3+3 format. The Americans like to impose sanctions, but have little trade with Iran. So they are looking at their partners to put pressure on Iran. It was only logical that we followed suit.[22]

'The most effective sanctions are those that are not written', another European Iran desk officer explained, 'there is a moral suasion; an international mood where you could not have any cooperation with Iran because of US pressure. You could not even have dared to export spaghetti to Iran.'[23] Such 'moral suasion' propels internalisation: It is important to note that key EU member states had adapted their positions in early 2012 to allow for the imposition of far-reaching European sanctions modeled on American ones. Alcaro and Tabrizi trace such a level of cohesion 'back to European alignment with US policy rather than to EU–US policy convergence', referring to 'alignment' as a conscious decision to 'follow the US lead' (2014: 18). While in November 2011, the UK had been the only EU member state to unilaterally impose an oil embargo on Iran (Tabrizi and Santini 2012: 2), such a step became a policy consensus half a year later: In July 2012, the EU Foreign Affairs Council adopted an EU oil embargo on Iran. A gas embargo followed suit in October 2012. Further sanctions included a ban on the trade of diamonds and precious metals, the cut-off of Iran from the Belgian-based SWIFT system, the banning of European companies from providing insurance for oil or petrochemical shipments from Iran, the banning of transactions between European and Iranian banks, and the freezing of assets of the Iranian Central Bank held within the EU.[24] These were far-reaching measures just short of a blanket trade embargo. It is important to note that these were deliberate EU policy decisions. Dupont argues that the lawfulness of these 'restrictive measures' (in the EU's own terminology) is dubious under international law, as they are countermeasures taken by a regional organisation that go beyond UNSC sanctions taken pursuant to Chapter VII (Dupont 2012). With these EU Council decisions, European sanctions policies were on a par with US policy preferences. 'We made a sacrifice once', an EEAS official who was involved in the nuclear negotiations explained in an interview, 'we had a lot of trade with Iran, but we voluntarily scaled it back and imposed an oil embargo. Such things are political commitments'.[25]

While the caveat should be added that EU and US positions are not to be conflated into a general 'Western bloc' approach, the 2012 round of EU sanctions has generated a transatlantic sanctions consensus that confronted Iran with an unprecedented economic embargo situation. The present analysis will shed light on Chinese, Russian and Turkish efforts to position themselves in opposition to a security culture involving a US-inspired system of pressure on Iran whose legitimacy is being called into question by a security culture advocated by China, Russia and Turkey. In doing so, however, this study rejects a simplistic and easily-adoptable anti-Americanism underlying some of the literature on US dominance that sometimes is at the boundary of analysis and advocacy.[26] Scholars setting out to understand 'non-Western' perspectives need to be careful not to uncritically adopt an essentialised critique of 'Western' global hegemony. Occidentalism is as much an intellectual trap to be avoided as Orientalism (Adib-Moghaddam 2011; Bilgin 2004; Howard 1995; Puchala 1997; Said 2003).[27] Especially the example of EU sanctions nuances the idea of US hegemony. What is being argued here is that EU sanctions acted upon structures as outlined above in a way that saw the strengthening of a hegemonic consent (see Pieper 2016).

3 Discourse and behaviour between *doublespeak* and 'compliance': understanding Chinese, Russian and Turkish Iran policies

An analysis of changing and conflicting conceptions of security cultures as an expression of diverging interests and identities of different actors effectively becomes an investigation into alternative normative narratives of international relations at large. Resistance to hegemony, so the power transition paradigm, can lead to counter-hegemonic struggles by which new 'power centres' ultimately create a system of international relations crafted according to their own political conceptions of legitimacy (Gilpin 1981; Organski 1968; Organski and Kugler 1980). Hegemonic transition theories and power transition theories have focused on the extent to which emerging new power poles replace existing dominant power structures to create new models of governance.[28] This literature has regained much attention in the context of 'emerging powers' and in the debate about who the architects of the future global order are. Here, one needs to be careful not to read the existence or emergence of a non-Western bloc alternative, acting as a monolith of resistance to the Westernisation of discourses, into any voices of dissent. While Chinese, Russian and Turkish foreign policies toward the Iranian nuclear programme breathe the ambition to partially 'de-Westernise' security cultures and discourses toward Iran, as will be shown in the chapters that follow, we must not over-theorise on indications of 'counter-hegemonic forces' struggling to topple the prevailing power system. Analyses of powers that 'rise' or 'emerge' in international relations typically consider the changing international economic system in a multipolar world and the shift of power equations that growing economies of these new power centres bring along. The BRICs-label, coined in 2001 by Goldman Sachs, is but the most prominent example of a plethora of acronyms to identify groups of 'emerging' powers (Goldman Sachs 2001). Analyses of newly emerging powerful states in the context of systemic leadership contestation have brought the literature on hegemonic transition theory and power transition to renewed scholarly attention, albeit steeped in a very state-centric ontology (Bevir and Gaskarth 2015: 75). Two fundamental flaws bedevil predictions of coming power transitions between a group of like-minded states and established powers: first, they proceed from a teleological reading of political developments. The history of power transition in the twenty-first century cannot be written in hindsight yet. And it certainly is no linear development. And second, it proceeds from an ontology of security that stems from established practices of dominant power centres. What came to be termed globalisation may be perceived as American hegemony elsewhere. And 'terrorism' is a relational term by default. These are labels that *produce* reality, and it is argued here that norm contestation over these, and other, terms need not necessarily lie at the basis of 'power shifts', but can be a first step towards norm renovation. The 'power transition' projection, in addition, is complicated by domestic foreign policy debates, and hence differing ideas on scope, goals and means of what it means to be a 'rising' or 'global' power. As Serfaty (2012/2013) observes: 'Russia, China and India have more interest in the United States and the EU than in each other, though each for different reasons' (33; see also Brütsch and Papa 2013). China, Russia, and other countries necessarily have 'competing international identities that try to satisfy a variety of international (and domestic) constituencies', (36–37) Shambaugh and Xiao write in their chapter in Nau's and Ollapally's edited volume on *Worldviews of Aspiring Powers* (2012). The case studies that follow show how such an observation applies to China, Russia, and Turkey, and the penultimate chapter sheds light on the implications thereof for the level of their 'resistance to hegemony' in a comparative fashion. Importantly for this study, the role of such 'emerging' powers in international security governance has remained understudied (see also Biersteker 2015: 61). Adherence to an alternative security culture is not assumed to imply a concerted act of states to challenge 'core countries' representing a hegemonic security culture. This book aims to differentiate such a categorical depiction of systemic power transitions by investigating the extent to which Chinese, Russian and Turkish foreign policies toward the Iranian nuclear programme stand indicative of alternative security cultures toward Iran in a 'process of power de-concentration' (Tessman and Wolfe 2011: 218) in which dominant power structures have not been replaced by alternative governance structures (yet). This accounts for more continuous and nuanced power shifts in world politics and holds that 'resistance is immanent to power' (Adib-Moghaddam 2014: 91).

At this juncture, this book makes a distinction between a discursive and a behavioural level in foreign policy behaviour to introduce a two-level model to better capture such qualified resistance to hegemony: An advocacy for a non-hegemonic security culture on a discursive level can be paralleled by compliance with a hegemonic security culture on a behavioural level. A seeming discrepancy between both levels can thus be observed. The possible variation in norm compliance described here is visualised by a two-level model to capture 'resistance to hegemony', as shown in Table 1.1.

Normative divergence in conjunction with rules divergence would be 'counter-hegemony', because it rejects hegemony on both a discursive and a behavioural level. This scenario is not represented in the table because the case studies that follow aim to make sense of a perceived discrepancy between both levels. In the same logic, it excludes the occurrence of normative convergence with rules convergence, because this scenario represents perfect adherence to

Table 1.1 Two-level model to conceptualise 'resistance to hegemony'

	Discursive level	*Behavioural level*
Adherence to security culture	Advocacy for non-hegemonic security culture	Compliance with a US-inspired hegemonic security culture
Degree of resistance to hegemony	*Normative* divergence from hegemony	*Rules* convergence with hegemony

hegemonic policies. Neither of these two scenarios applies to the Iran policies of China, Russia and Turkey, so the present study is interested in shedding light on 'ambivalent' cases, where the discursive level need not be a function of the behavioural level, where a state's foreign policy displays incoherence, in other words. A state can advocate for a non-hegemonic security culture discursively, but still comply with a hegemonic security on a behavioural level. 'Norms' relate to the discursive level as this research proceeds from the assumption that actors convey, talk about, and refer to norms as 'evaluative standards', to take up Katzenstein's terminology (1996: 21). Discourse differs where evaluations presuppose diverging norms. 'Rules', then, relate to the behavioural level in the way they condition and structure actors' 'cognitive standards' (ibid.). Partial acceptance of hegemonic structures on a behavioural level even when conveying normative divergence on a discursive level may be predicated upon a level of perceived political and material dependence on the US. The latter observation and the extent to which it can be discerned in the case studies under investigation here will be qualified in the empirical chapters for Chinese, Russian and Turkish Iran policies, respectively.

On a theoretical note, however, it is felt that an important parenthesis on the compatibility of materialist motivations with the overall theoretical framework as laid out above should be inserted at this point. Rather than subscribing to either a purely neorealist analysis where social mediation and ideational factors in foreign policy are treated as epiphenomenal or to an exclusively ('thick') social constructivist angle where the assumption of 'ideas all the way down' does not allow for material forces to play a prominent role, this book is sympathetic to the theoretical position where material predispositions as well as the aspect of social mediation are recognised as playing a role in a state's foreign policy decision-making. The idea of social mediation cannot exist in 'a material vacuum' either, in the words of Kowert and Legro (1996: 490–491). Adopting a purely materialist or purely ideational framework to analyse politics and policies, in other words, would make no sense if social and material factors co-determine each other (Katzenstein and Okawara 2001/2002; Katzenstein and Sil 2004; Nau 2002; Sørensen 2008).[29]

It is this conception of correlational complementarity that allows for an analysis of divergences between a discursive and a behavioural level. The critique may be brought forward that discourse in itself already constitutes behaviour. This reading has been emphasised by Habermas' (1981) writings on speech acts as communicative action. But if we are to accept that interests can be material as well as social, it would be an ontological fallacy to hold that discourse always is an empirical act. Beyond semantic hair-splitting, it is perhaps instructive to think of the concept of 'doublespeak'.[30] According to this concept, what an actor says does not always correspond with how he acts. An actor, in other words, can fall short of acting upon the discourse he conveys. It is in this understanding that this book makes a distinction between a discursive and a behavioural level of foreign policy to arrive at a more comprehensive understanding of an actor's security culture.

The understanding and perception of a state's standing in 'the international system' may lead states to adopt foreign policies that they would not otherwise adopt if the system were such that interaction effects did not matter; if states were to regard themselves as isolated actors, in other words. Based on an understanding of the US's power position and its ability to enforce unilateral sanctions on third countries' companies, for example, China, Russia and Turkey may adopt policies that are not beneficial economically (e.g. reduction of Iranian oil imports). The pursuance of materialist foreign policy objectives thus takes place *at least next to* ideational motivations; or, more often than not, as *the outcome of* implicit foreign policy identity perceptions and conceptions about legitimacy. As social context always (co)determines a state's foreign policy behaviour, 'interests' are ideational as well as material. The advocacy for resistance to hegemony in Chinese, Russian and Turkish security cultures can therefore be accompanied by an awareness of material dependence that makes these states (temporarily) accept foreign policy decisions as the expression of norms of hegemonic structures (such as intrusive sanctions regimes). In the long-term, however, their resistance to hegemonic security cultures may reflect on their aspiration to bring about a non-hegemonic understanding of *norms* regulating international politics and a more equitable adherence to the *rules* of the UN system.

4 Methods, case selection and data analysis

In analysing different approaches to the Iranian nuclear programme and with the aim to disentangle underlying conceptions of security cultures, analyses of Chinese, Russian and Turkish foreign policies serve as investigations into Iran policies that are different from 'the West'. Departing from the guiding research puzzle underlying this study, analyses of the foreign policies of China, Russia and Turkey serve as case studies to illustrate different approaches to the Iranian nuclear programme and the existence or possible emergence of security cultures resisting hegemony. While I believe there is much value in adopting similar frameworks as outlined above in analyses of international diplomacy and security at large, the focus of this work lies on a comparison of Chinese, Russian and Turkish policies toward the Iranian nuclear programme from 2003–2016. The research design adopts an integrative comparative case study design that combines analyses of Chinese, Russian and Turkish foreign policies toward the Iranian nuclear programme on a within-case level, respectively, with an eventual cross-case comparison (Rohlfing 2012; see also Collier 1993; Lieberson 1991; Ragin 1987). Research methods on the within-case level were process-tracing, qualitative data analysis as well as qualitative expert interviewing. Before the remainder of the chapter will elaborate on each of these, a rationale for the comparison of the cases chosen must be given first.

4.1 Case selection

This study analyses Chinese, Russian and Turkish Iran policies as three case studies of 'non-Western' foreign policies towards a contested issue area. A focus often placed in the scholarly literature on foreign policies of states like China, Turkey and Russia is their perceived inherent 'non-Western-ness'.[31] Underlying many analyses of their foreign policies is a bifurcation into two thinkable scenarios: either these states socialise into a Western-dominated world, including its governance structures, or they will herald a power transition and 'emerge' as challengers to this Western-dominated world.[32] Acknowledging that the 'emerging power' label has been criticised on substantial and conceptual grounds, I aim to shed light on the working relationship between powers that have created, crafted and sustained the prevailing international governance architecture, and those powers that will, one way or the other, play an important role in the gradual modification thereof. With structural imbalances built into a system that was defined by the powerful position of certain Western states, 'rising powers' will have an interest in advocating security cultures that challenge such power asymmetries short of articulating radical alternatives (Narlikar 2013: 575). As much as 'hegemony' and 'resistance' are never absolute and in a continuous process of interaction, an analysis of Chinese, Russian and Turkish respective foreign policies towards Iran's nuclear programme becomes an examination of *non-Western* foreign policies in processes of interaction and inter-penetration between allegedly exclusive camps.

At the same time, the US is acknowledged as the prevalent hegemonic point of reference whose preservation depends on the acceptance thereof by other states. This is the consent that actors give – tacitly or explicitly – to established power structures without which the latter could well come into being, but could not be sustained for long. Such an acceptance presupposes a level of dependence that lets actors subscribe to dominant normative frameworks. Turkey's political dependence on the US already is a more formalised relation due to (NATO) alliance structures and Cold War historical legacies (Baran and Lesser 2009; Fuller 2008; Giragosian 2008; Lesser 1992; Mastney and Nation 1996). But also Russia's and China's foreign policies have widely been analysed as balancing acts between independent policies and foreign policy stances that were more accommodating to the US and arguably crafted in reaction to or anticipation of US perceptions of Chinese and Russian policies. Russia's quest for (a new) international identity after the collapse of the Soviet Union until the present day underwent distinct phases of re-orientation that always involved a re-balancing of relations with the US (see Adomeit 2013; Belopolsky 2009: 14–28; Casier 2006; Mankoff 2009; Sakwa 2002, 2012; Shakleina 2013: 166–174; Stent 2014; Trenin 2001, 2006; Tsygankov 2007).

And also China's economic 'rise' essentially was an opening-up to and adherence to an international economic system that was created and dominated by the US after 1945 (Breslin 2013; Jacques 2012; Johnston 2003; Johnston and Ross 1999; Jones and Breslin 2015; Ross and Zhu 2008; Shambaugh 2000, 2005; Yan 2006; Zhao 2008, 2013). This opening-up of China to a system determined to a large extent by the United States was ushered in under Deng Xiaoping's leadership as from 1978, but preconditioned on the re-opening of US–Chinese relations under Nixon and Mao in 1972 (Kissinger

2011). The fact that China, Russia and Turkey, albeit grudgingly, accept and comply with parts of an elaborate system of 'extraterritorialised' US legislation suggests a substantial level of political and financial dependence on the US on the part of these countries. This observation lends itself to a research puzzle for an examination of the friction between the advocacy for security cultures that resist this hegemony and the factual adherence to prevailing hegemonic structures. The case studies that follow will analyse how and why China, Russia and Turkey have had to balance such competing positions, and how 'political dependence' on the US need not exclude resistance to US policies on other occasions. The same holds true for the case countries' interest in relations with Iran. The chapters that follow will illustrate how these partially centrifugal forces oftentimes present a policy dilemma for China, Russia and Turkey, and how this played out in their positions on the Iranian nuclear file.

Finally, the interest in the three cases chosen derives from the importance assigned to these countries' weight in the Iranian nuclear dossier. China and Russia are permanent UNSC members and have therefore been part of the P5+1 format. Their inclusion has become inevitable with the referral of the Iran file from the IAEA to the UNSC in 2006 at the latest, where Chapter VII sanctions resolutions require the consent of Moscow and Beijing. Turkey, however, is not an obvious candidate in this comparison. At least until 2011, Turkey was often portrayed as having the potential to act as a 'facilitator' of talks, being embedded in Western strategic security cultures, and, at the same time, a geographic neighbour of Iran that, throughout history, has learnt the necessity to maintain good-neighbourly relations with Iran. It is this position that led analysts to see Turkey in a bridge-building function conducive to diplomatic de-escalation (Davutoğlu 2013; Gürzel 2012; Gürzel and Ersoy 2012; Kibaroğlu and Caglar 2008; Kibaroğlu 2009; Önis and Yılmaz 2009; Ülgen 2012; Üstün 2010). 'Mediation' in diplomacy has been defined in different ways, and its preconditions and required components have been discussed controversially (Bercovitch 1992: 8; Blake and Mouton 1985: 15; Kleiboer 1996, 1998; Moore 2003: 8). Underlying many analyses of the effectiveness of mediation in different conflict situations is the view that mediators should be impartial, be in a position to influence the conflicting parties' perceptions or behaviour and (as a precondition for the latter) have credibility as a mediating party (Brookmore 1980). Bercovitch and Schneider (2000), in their review of the mediation literature, speak of 'leverage, power potential, and influence' (149). For a while, Turkey acted as a facilitator of talks and conduit of messages between Iran and the West. With the phase of remarkable shuttle diplomacy in 2009 and 2010 that led to the signing of the Tehran declaration in May 2010, however, Turkey had moved beyond a mere facilitating role and had directly engaged in mediatory diplomatic efforts (Fitzpatrick 2010; Kibaroğlu 2010; Leverett and Leverett 2013: 361f.; Ozkan 2010; Parsi 2012: 172f.; Pieper 2015). The next chapter will explore the reasons for Turkey's Iran diplomacy, and provide answers for the dramatic failure of the 2010 initiative.

4.2 Methods and data analysis

The primary research methods comprise process-tracing and qualitative data analysis of policy documents (primary sources, e.g. declassified government documents, UN and IAEA documents, press releases, transcripts of speeches), memoirs of decision-makers, policy briefs and the scholarly literature. Press releases and statements, in government parlance, belong to the category of public relations and therefore always have to be read as socially mediated language with a purpose. Documents of international organisations, like minutes of Security Council meetings, will be used to document when and how Chinese, Russian and Turkish officials have voiced objections, concerns or approval. And memoirs of former officials intimately involved in the Iranian nuclear file will be drawn upon throughout the chapters that follow, as these are highly informative accounts of decision-makers that often write from the convenient position of elder statesmen who do not have to mince their words because of the secretive nature of both the dossier and their official position with all caveats on open information that ensue.

In this endeavour, process-tracing is used as a research method on the within-case level (Beach and Pedersen 2012; Hall 2006; Mahoney 2000: 412). In the present research, process-tracing as a method looks at the way that China, Russia and Turkey have positioned themselves toward the Iranian nuclear programme through an analysis of concrete foreign policy decisions by way of constant comparison of the data used (see Savin-Baden and Major 2013: 436–437). As such, process-tracing is used here in its outcome-oriented variant that aims to understand decisional processes (Beach and Pedersen 2013: 3). A reconstruction of the way that Chinese, Russian and Turkish foreign policies toward the Iranian nuclear programme have been crafted relies on the collection of observations that we can find in policy documents, briefs, legislative documents and through conducting qualitative interviews.

For this purpose, semi-structured elite interviews were conducted as a method of empirical data generation. This research project benefited from face-to-face interviews that have been conducted with delegates from the respective nuclear negotiation teams and the respective Iran desks of China, Russia, Germany, the UK, France, the US, and the EEAS. Travelling between Beijing, Moscow, Brussels, Berlin, Vienna, London and Washington, I have spoken to representatives from all P5+1 state parties as well as Iran. For the purpose of this project's research focus, contacts with the Chinese, Russian and Turkish foreign policy establishments were the most relevant. Current and former desk officers have been identified in relevant departments in charge of Iran policies, security policy and non-proliferation, as well as members of the delegations at the nuclear talks. Besides such key officials in the respective foreign ministries in Ankara, Beijing and Moscow, interviews have also been conducted with diplomats at their permanent representations to the IAEA in Vienna. Their insights and professional involvement with the subject matter added an important dimension of expertise at the governmental level. Interviewees (especially officials) function as norm-carriers in this regard (see also Foyle 1997; Young and Schafer 1998). State agents 'anthropomorphize the state' (Li 2010: 356). Such encounters not only become an invaluable methodological tool in qualitative social science research into norms in international politics, but are also an important means of exchange between the policy community and the academy.

Besides these governmental foreign policy actors, research institutes and think tanks in the three case countries inform foreign policy, 'float' policy ideas and thereby partially also shape the decision-making process. The most influential Chinese think tanks in the realm of international relations are think tanks that are officially affiliated with the State Council and the foreign ministry such as the China Institute of International Studies (CIIS) and the China Institutes of Contemporary International Relations (CICIR), the latter of which enjoys close ties with the ministry for state security (Downs 2004: 28).[33] The Shanghai Institute for International Studies (SIIS) also regularly briefs different ministries of the central government as well as the Foreign Affairs Office of the Party Central Committee. These institutes act as foreign policy consultants and craft conceptual policy recommendations for the Chinese government and thus have a considerable impact on the Chinese foreign policy decision-making process (see also Lanteigne 2009: 29). To a lesser extent, the same holds true for Russian think tanks. Russian Iran policy is shaped by the foreign ministry, and important strategic decisions are made by the president who, according to the Russian constitution, holds ultimate authority in foreign policy matters. Yet, a number of Russian think tanks provide a considerable amount of expertise and consultancy related to nuclear non-proliferation and energy policy. These include primarily the CENESS (Center for Energy and Security Studies) and the PIR Center (the Russian Center for Policy Studies), the latter of which has close ties to the Russian government and is regularly briefing the foreign ministry and the defense ministry on foreign policy, nuclear proliferation and arms control. The Turkish think tank scene is somewhat less pronounced in terms of foreign policy consultancy. Instead, renowned experts at Turkish universities brief and consult their government, and it is with a number of these that interviews were conducted on Turkish Iran policies (at Middle East Technical University, Hacettepe University, Kadir Has University, IPEK University). Other experts included a range of policy analysts, consultants and experts at think tanks and universities in Brussels, Moscow, Ankara, Berlin, Vienna, London, Istanbul, Washington, Beijing and Shanghai. In total, 70 elite interviews with experts, consultants and decision-makers were conducted for this research project in 10 countries.

Notes

1 Such an understanding draws on the central tenet of co-determination between agent and structure that has defined and shaped much of social constructivist scholarship. See Katzenstein (1996); Kubálková, Onuf and Kowert (1998); Onuf (1989); Wendt (1987, 1999); See also Adler (1997); Kratochwil (2001); Lapid (1989); Lapid and Kratochwil (1996); Ruggie (1998).

2 Albeit in an explicit poststructuralist reading, where discourse and practice are constitutive of each other at different levels throughout society. This book, however, makes a distinction between discourse and foreign policy behaviour, as will be outlined in section 5 below.
3 Hedley Bull's work (1977, 1995) focusses on bridging the assumed dichotomy between 'power' and 'norms' and has theorised on a 'third conception of an international society' (Bull 1966: 79). While this is a valuable starting point, it departs from a conception of norm promotion that itself favours those states in the most powerful position to convey narratives of 'appropriate behaviour'. It is this notion of uni-directional norm diffusion that a 'second-generation' literature on norm dynamics has critically addressed.
4 Hassan Rouhani (2011) describes in his memoirs of his time as chief nuclear negotiator that Iran's right to enrich was an objective he sought to achieve during negotiations with the EU-3 (61, 666). The US, and other Western governments and experts, hold that a 'right to enrich' is not given by the NPT. Since 'enrichment' is not explicitly mentioned in the treaty provisions, another position taken by some states is that 'enrichment' is an implicit right. In her political memoirs, former Secretary of State Hillary Clinton's account of the Iranian nuclear talks implicitly insinuates Iranian military intentions (Clinton 2014: 416–446).
5 Iran was one of the first states to sign the NPT in 1968 and ratify it in 1970.
6 Iran's nuclear infancy under the 'Atoms for peace' programme constituted its 'unsophisticated socialization into the nuclear world', as Homayounvash (2016: 1) writes.
7 Jason Jones (2011) has analysed how the Iranian nuclear issue has become securitised in US media and policy discourse and how this discourse impacted favourably on the Bush administration's policy strategising on Iran. See also Porter (2014).
8 For a more detailed and concise discussion of US–Iran nuclear cooperation in the 1960s and 1970s, see Kibaroğlu (2007); Patrikarakos (2012); Ansari (2006).
9 The initial US insistence on a complete cessation of Iranian nuclear activities 'led to more distrust toward the West in Iran and disillusionment with disarmament treaties' from an Iranian perspective (Mousavian 2012: 451).
10 See Gilpin (1987), Joseph (2002), Keohane (1984) for realist analyses and usages of the concept; Ferguson (2003), Ikenberry (2004) for liberalist analyses (the 'benign hegemon' theory); Clark (2009a, 2009b) for an English school approach to hegemony; Arrighi (1993), Cox (1981), Rupert (1995), Worth (2009) for neo-Gramscian analyses of 'hegemony'.
11 This concept of hegemony needs to be understood in the context of their time. Written for revolutionary Marxist politics, especially Lenin's conception of hegemony was almost interchangeable with 'political leadership' (Cox 1993: 50; Service 2000: 170–171), but was also understood as a function of the working class (Worth 2015: 65). Gramsci acknowledged having drawn on Lenin in theorising on 'hegemony' (1971: 381). In a sense, a widespread acceptance of a Gramscian application of 'hegemony' in IR (and thus outside of Marxist circles) is somewhat surprising and has, over time, acquired an academically instrumental meaning of its own and thus moved away intellectually from its progenitor under the wider umbrella of Critical Theories. Jones (2006) writes of a 'Gramsci industry' that has become tailored to different disciplines (see also Hobsbawm 2011: 316).
12 This is an argument also made by Ian Clark, discussing the concept of 'hegemonic succession' with a view to China–US relations (2011). See also Clark (2005) on Legitimacy in 'International Society'; Rapkin and Braaten's (2009) conceptualisation of hegemonic legitimacy; Reus-Smit (2014) on 'Power, Legitimacy, and Order'; and Kitchen (2009: 84–85) on 'hegemonic transition'. Keohane (1984) rightly notes that 'theories of hegemony should seek not only to analyse dominant powers' decisions to engage in rule-making and rule-enforcement, but also to explore why secondary states defer to the leadership of the hegemony'. Such theories should 'account for the legitimacy of hegemonic regimes and for the coexistence of cooperation' (39).

13 Also see Adib-Moghaddam (2014: 116–118) on this differentiation.
14 As would be an underlying assumption of a 'power transition' theory.
15 Author's interview, Washington, 30 October 2014.
16 Called 'secondary' because they do not stop at sanctioning the target state directly, but also aim to punish third country entities' dealings with the target state.
17 CISADA also included sanctions against the sale of Iranian caviar, carpets and pistachios, which had previously been exempted under the Clinton administration.
18 Author's interview with EEAS official, Brussels, 15 April 2016.
19 In order to pre-empt formal counter action by EU member states in the WTO, the US government seemed to have found a *modus operandi* over the application of the ILSA. To this effect, it is worth quoting at length from Kenneth Katzman's (2006) CRS Report for Congress:

> Traditionally skeptical of economic sanctions as a policy tool, the European Union states opposed ILSA as an extraterritorial application of U.S. law. The EU countries threatened formal counter-action in the World Trade Organization (WTO), and in April 1997, the United States and the EU formally agreed to try to avoid a trade confrontation over ILSA (and a separate 'Helms-Burton' Cuba sanctions law, P.L. 104–114). The agreement contributed to a decision by the Clinton administration to waive ILSA sanctions on the first project determined to be in violation: a $2 billion contract, signed in September 1997, for Total SA of France and its minority partners, Gazprom of Russia and Petronas of Malaysia to develop phases 2 and 3 of the 25-phase South Pars gas field. The Administration announced the 'national interest' waiver (Section 9(c) of ILSA) on May 18, 1998, after the EU pledged to increase cooperation with the United States on non-proliferation and counter-terrorism. The announcement indicated that EU firms would likely receive waivers for future projects that were similar.

Since 2010, however, US President Obama enforced US unilateral sanctions also against European companies by way of executive orders (Lohmann 2015).
20 Author's interview with European Iran desk officer, Berlin, 14 November 2014; author's interview with State Department official, Washington, 30 October 2014; author's interview with Turkish diplomat, Washington, 30 October 2014; author's interview with Iranian diplomat, Berlin, 29 August 2016; author's interview with Dr Aniseh Tabrizi, Research Fellow at RUSI, London, 15 September 2016.
21 Author's interview at Brick Court Chambers, London, 1 April 2016.
22 Author's interview with European diplomat, Brussels, 6 February 2015.
23 Author's interview with European Iran desk officer, Rome, 8 August 2016.
24 See Council Decision 2012/35/CFSP amending Decision 2010/413/CFSP, 23 January 2012; Council Regulation 267/2012, 23 March 2012; Council Decision 2012/635/CFSP amending Decision 2010/413/CFSP, 15 October 2012.
25 Author's interview with EEAS official, Brussels, 15 April 2016.
26 This is true if we think of the underlying emancipatory appeal of, e.g. neo-Marxist theorising. Scharzmantel (2009) cautions against a 'straightforward and relatively banal idea of US hegemony in the post-Cold War world' (6) and calls for a differentiated application of neo-Gramscian thoughts in International Relations.
27 Policy disagreements about World Order need to be reflected in the way we study them. Bilgin (2008) therefore has made the case for more cross-disciplinary analysis that transcends the mutually reinforcing worldviews between scholars and foreign policy practitioners (10) as well as what she calls, drawing on Edward Said (1978), 'epistemological Orientalism' prevalent in hegemonic disciplinary discourses in the academy (14).
28 Seminal examples discussing global power shifts and prospects for international cooperation are DiCicco and Levy (2003); Gilpin (1981); Jones (2011); Kagan (2002, 2012); Keohane (1984); Kugler and Lemke (1996); Kupchan (2012); Organski

(1968); Organski and Kugler (1980). See also Chan (2004b); Clark (2014); Kitchen (2009: 82–87); Lebow and Valentino (2009); Lemke (1997); Vezirgiannidou (2013).

29 See also Neal's chapter on 'empiricism without positivism' (2013).

30 The term is often traced back to George Orwell's novel *Nineteen Eighty-Four*, published in 1949. While it does not appear in the book as such, the term is conceptually close to that of 'doublethink', which is used in the book.

31 Pertinent to much theorising on Turkish foreign policy, for example, is the reading that Turkey represents a 'bridge' between the Western world and a region and culture that is not Western. In this context, Barry Buzan *et al.* have coined the label 'Westernistic' to characterise states like Turkey or Japan that breathe the ambition to be perceived as a, if not Western country, then at least as being close to its political cultures (Buzan and Diez 1999: 49; Buzan and Segal 1998).

32 Analyses of power transitions due to the emergence of new 'power poles' include Armijo (2007); Brawley (2007); Hancock (2007); Hurrell (2006); Laidi (2012, 2014); Lieber (2014); MacFarlane (2006); Morris (2011); Nel (2010); Patrick (2010); Phillips (2008); Rynning and Ringsmore (2008); Sakwa (2012); Schiffer and Shorr (2009); Serfaty (2011); Sotero and Armijo (2007); Subacchi (2008); Vezirgiannidou (2013).

33 CIIS is one of the most influential civilian foreign policy research institutes and counts as 'the research arm of the MFA' (Downs 2004: 28). See also Glaser and Saunders (2002: 597–616); Shambaugh (2002: 583–585). CICIR is the 'primary civilian intelligence organ and has direct access to the Politburo Standing Committee' (Downs 2004: 28).

References

Acharya, Amitav. 2014. *The End of American World Order*. Cambridge: Polity Press.

Adib-Moghaddam, Arshin. 2011. *A Metahistory of the Clash of Civilisations. Us and Them Beyond Orientalism*. New York: Columbia University Press.

Adib-Moghaddam, Arshin. 2014. *On the Arab Revolts and the Iranian Revolution. Power and Resistance Today*. London: Bloomsbury.

Adler, Emanuel. 1997. Seizing the Middle Ground: Constructivism in World Politics, *European Journal of International Relations* vol. 3, no. 3: 319–363.

Adler-Nissen, Rebecca. 2014. Stigma Management in International Relations: Transgressive Identities, Norms and Order in International Society. *International Organisation* vol. 68, no. 1: 143–176.

Adomeit, Hannes. 2013. Fehler im Betriebssystem. Die russisch-amerikanischen Beziehungen. *Osteuropa* vol. 63, no. 9: 57–78.

Alcaro, Ricardo and Tabrizi, Aniseh Bassiri. 2014. Europe and Iran's Nuclear Issue: The Labours and Sorrows of a Supporting Actor. *The International Spectator* vol. 49, no. 3: 14–20.

Ansari, Ali. 2006. *Confronting Iran. The Failure of American Foreign Policy and the Next Great Conflict in the Middle East*. New York: Basic Books.

Armijo, Leslie Elliott. 2007. The BRIC Countries (Brazil, Russia, India, and China) as Analytical Category: Mirage or Insight? *Asian Perspective* vol. 31, no. 4: 7–42.

Arrighi, Giovanni. 1993. The Three Hegemonies of Historical Capitalism. In: Gill, S. (ed.), *Gramsci, Historical Materialism and International Relations*. Cambridge: Cambridge University Press, pp. 148–185.

Azmanova, Albena. 2012. *The Scandal of Reason. A Critical Theory of Political Judgment*. New York: Columbia University Press.

Balzacq, Thierry. 2010. *Securitization Theory: How Security Problems Emerge and Dissolve*. New York: Routledge.

Baran, Zeyno and Lesser, Ian O. 2009. Turkey's Identity and Strategy: A Game of Three-Dimensional Chess. In: Michael Schiffer and David Shorr (eds), *Powers and Principles. International Leadership in a Shrinking World*. Plymouth: Lexington Books, pp. 197–224.

Barzegar, Kayhan. 2012. Iran's Nuclear Program. In: Kamrava, Mehran (ed.), *The Nuclear Question in the Middle East*. New York: Columbia University Press.

Beach, Derek and Pedersen, Rasmus Brun. 2012. *Process Tracing Methods*. Ann Arbor: University of Michigan Press.

Bellamy, Richard Paul. 1990. Gramsci, Croce and the Italian Political Tradition. *History of Political Thought* vol. 11, no. 2: 313–337.

Belopolsky, Helen. 2009. *Russia and the Challengers. Russian Alignment with China, Iran, and Iraq in the Unipolar Era*. Basingstoke: Palgrave Macmillan.

Bercovitch, Jakob. 1992. *Mediation in International Relations: Multiple Approaches to Conflict Management*. New York: St. Martin's Press.

Bercovitch, Jacob and Schneider, Gerald. 2000. Who Mediates? The Political Economy of International Conflict Management. *Journal of Peace Research* vol. 37, no. 2: 145–165.

Bevir, Mark and Gaskarth, Jamie. 2015. Global Governance and the BRICs: Ideas, Actors, and Governance Practices. In: Gaskarth, Jamie (ed.), *Rising Powers, Global Governance and Global Ethics*. Abingdon and New York: Routledge, pp. 74–96.

Biersteker, Thomas and Moret, Erica. 2015. Rising Powers and Reform of the Practices of International Security Institutions. In: Gaskarth, Jamie (ed.), *Rising Powers, Global Governance and Global Ethics*. Abingdon and New York: Routledge, pp. 57–73.

Bilgin, Pinar. 2004. Whose Middle East? Geopolitical Inventions and Practices of Security. *International Relations* vol. 18, no. 1: 17–33.

Bilgin, Pinar. 2008. Thinking Past 'Western' IR? *Third World Quarterly* vol. 29, no. 1: 5–23.

Bilgin, Pinar. 2013. Critical Theory. In: Williams, Paul D. (ed.), *Security Studies. An Introduction*. Abingdon and New York: Routledge, pp. 93–106.

Blake, R.A. and Mouton, J.S. 1985. *Solving Costly Organizational Conflicts*. San Francisco: Josse-Bass.

Bloomfield, Alan and Scott, Shirley V. 2017. *Norm Antipreneurs and the Politics of Resistance to Global Normative Change*. Abingdon: Routledge.

Bob, Clifford. 2012. *The Global Right Wing and the Clash of World Politics. Cambridge Studies in Contentious Politics*. New York: Cambridge University Press.

Booth, Ken. 2005. Critical Explorations. In: Booth, Ken (ed.), *Critical Security Studies and World Politics*. Boulder, CO: Lynne Rienner, pp. 1–25.

Brawley, Mark R. 2007. Building Blocks or a BRIC Wall? Fitting U.S. Foreign Policy to the Shifting Distribution of Power. *Asian Perspective* vol. 31, no. 4: 151–175.

Breslin, Shaun. 2013. China and the Global Order: Signaling Threat or Friendship? *International Affairs* vol. 3: 615–634.

Brookmore, David A. and Sistrink, Frank. 1980. The Effects of Perceived Ability and Impartiality of Mediator and Time Pressure on Negotiation. *Journal of Conflict Resolution* vol. 24, no. 2: 311–327.

Brütsch, Christian and Papa, Mihaela. 2013. Deconstructing the BRICS: Bargaining Coalition, Imagined Community, or Geopolitical Fad? *The Chinese Journal of International Politics* vol. 6: 299–327.

Bull, Hedley. 1977. *The Anarchical Society: A Study of Order in World Politics*. New York: Columbia University Press.

Bull, Hedley. 1995 [1966]. Society and Anarchy in International Relations. In: Der Derian (ed.), *International Theory: Critical Investigations*, Basingstoke: Macmillan, pp. 75–93.

Burnham, Peter. 1991. Neo-Gramscian Hegemony and International Order. *Capital and Class* vol. 45: 73–95.

Buzan, Barry and Diez, Thomas. 1999. The European Union and Turkey. *Survival* vol. 41, no. 1: 41–57.

Buzan, Barry and Segal, Gerald. 1998. A Western Theme. *Prospect* February 1998: 18–23.

Buzan, Barry and Waever, Ole. 2003. *Regions and Powers. The Structure of International Society*. Cambridge: Cambridge University Press.

Buzan, Barry, Waever, Ole and de Wilde, Jaap. 1998. *Security. A New Framework for Analysis*. Boulder, CO and London: Lynne Rienner.

Casier, Tom. 2006. Putin's Policy Towards the West: Reflections on the Nature of Russian Foreign Policy. *International Politics* vol. 43, no. 3: 384–401.

Chan, Steve. 2004. Can't Get No Satisfaction? The Recognition of Revisionist States. *International Relations of the Asia-Pacific* vol. 4: 207–238.

Chomsky, Noam. 2016. *Who Rules the World?* London: Hamish Hamilton.

Clark, Ian. 2005. *Legitimacy in International Society*. Oxford: Oxford University Press.

Clark, Ian. 2009a. Towards an English School Theory of Hegemony. *European Journal of International Relations* vol. 15, no. 2: 203–228.

Clark, Ian. 2009b. Bringing Hegemony Back In. *International Affairs* vol. 85, no. 1: 23–26.

Clark, Ian. 2011. China and the United States: A Succession of Hegemonies? *International Affairs* vol. 87, no. 1: 13–28.

Clark, Ian. 2014. International Society and China: The Power of Norms and the Norms of Power. *The Chinese Journal of International Politics* vol. 7, no. 3: 315–340.

Clinton, Hillary Rodham. 2014. *Hard Choices*. New York: Simon & Schuster.

Collier, David. 1993. The Comparative Method. In: Finifter, A.W. (ed.), *Political Science: The State of the Discipline II*. Washington, D.C.: American Political Science Association, pp. 105–119.

Cox, Michael, 2001. Whatever Happened to American Decline? International Relations and the New United States Hegemony. *New Political Economy* vol. 6, no. 3: 311–340.

Cox, Michael, 2007. Is the United States in Decline – Again? An Essay. *International Affairs* vol. 83, no. 4: 643–653.

Cox, Robert W. 1981. Social Forces, States and World Orders: Beyond International Relations Theory. *Millennium: Journal of International Studies* vol. 10, no. 2: 126–155.

Cox, Robert W. 1983. Gramsci, Hegemony and International Relations: An Essay in Method. *Millenium* vol. 12, no. 2: 162–175.

Cox, Robert W. 1993. Gramsci, Hegemony and International Relations: An Essay in Method. In: Gill, Stephen (ed.), *Gramsci, Historical Materialism and International Relations*. Cambridge: Cambridge University Press, pp. 49–66.

Cox, Robert W. 1996. *Approaches to World Order*. Cambridge: Cambridge University Press.

Davutoğlu, Ahmet. 2013. Turkey's Mediation: Critical Reflections from the Field. *Middle East Policy* vol. 20, no. 1: 83–90.

DiCicco, Jonathan M. and Levy, Jack S. 2003. The Power Transition Research Program: A Lakatosian Analysis. In: Elman, Colin and Elman, Miriam (eds), *Progress in International Relations Theory: Appraising the Field*. Cambridge, MA: MIT Press, pp. 109–159.

Downs, Erica S. 2004. The Chinese Energy Security Debate. *The China Quarterly* vol. 177: 21–41.

Dupont, Pierre-Emmanuel. 2012. Countermeasures and Collective Security: The Case of the EU Sanctions Against Iran. *Journal of Conflict and Security Law* vol. 17, no. 3: 301–336.

ElBaradei, Mohamed. 2012. *The Age of Deception. Nuclear Diplomacy in Treacherous Times*. New York: Bloomsburg Publishing Plc.

Epstein, Charlotte. 2012a. Stop Telling Us How to Behave: Socialization or Infantilization? *International Studies Perspectives* vol. 13, no. 2: 135–45.

Epstein, Charlotte. 2012b. Symposium: Interrogating the Use of Norms in International Relations: An Introduction. *International Studies Perspectives* vol. 13, no. 2: 121–122.

Femia, Joseph. 2005. Gramsci, Machiavelli, and International Relations. *Political Quarterly* vol. 76, no. 3: 341–349.

Ferguson, Neill. 2003. Hegemony or Empire? *Foreign Affairs* vol. 82, no. 5: 154–162.

Finnemore, Marta and Sikkink, Kathryn. 1998. International Norm Dynamics and Political Change. *International Organization* 52, no. 4: 887–917.

Fiori, Giuseppe. 2013 (1979). *Das Leben des Antonio Gramsci. Eine Biographie*, trans. Heimbucher, Renate and Schoop, Susanne. Berlin: Rotbuch.

Fitzpatrick, Mark. 2010. Iran: The Fragile Promises of the Fuel-Swap Plan. *Survival* vol. 52, no. 3: 67–94.

Florini, Ann. 1996. The Evolution of International Norms. *International Studies Quarterly* vol. 40, no. 3: 363–389.

Foyle, Douglas. 1997. Public Opinion and Foreign Policy: Elite Beliefs as a Mediating Variable. *International Studies Quarterly* vol. 41, no. 1: 141–169.

Fuller, Graham E. 2008. *The New Turkish Republic: Turkey as a Pivotal State in the Muslim World*. Washington, D.C.: United States Institute of Peace.

Gehring, Thomas. 1994. *Dynamic International Regimes: Institutions for International Environmental Governance*. Frankfurt am Main: Lang.

Germain, Randall and Kenny, Michael. 1998. Engaging Gramsci: International Relations Theory and the New Gramscians. *Review of International Studies* vol. 24, no. 1: 3–21.

Gilpin, Robert. 1981. *War and Change in World Politics*. Cambridge: Cambridge University Press.

Gilpin, Robert. 1987. *The Political Economy of International Relations*. Princeton, NJ: Princeton University Press.

Gill, Stephen (ed.) 1993. *Gramsci, Historical Materialism and International Relations*. Cambridge: Cambridge University Press.

Giragosian, Richard. 2008. Redefining Turkey's Strategic Orientation. *Turkish Policy Quarterly* vol. 6, no. 4: 33–40.

Giumelli, Francesco and Ivan, Paul. 2013. *The Effectiveness of EU Sanctions. An Analysis of Iran, Belarus, Syria and Myanmar (Burma)*. EPC Issue Paper No. 76.

Glaser, Bonnie S. and Saunders, Philip C. 2002. Chinese Civilian Foreign Policy Research Institutes: Evolving Roles and Increasing Influence. *The China Quarterly* vol. 171: 597–616.

Goldman Sachs. 2001. Building Better Global Economic BRICs. *Global Economics Paper No. 66*. Available at [last accessed 7 August 2014]: www.goldmansachs.com/our-thinking/archive/archive-pdfs/build-better-brics.pdf.

Gramsci, Antonio. 1971. *Selections from the Prison Notebooks*, ed. and trans. Hoare, Q. London: Lawrence & Wishart.

Gürzel, Aylin. 2012. Turkey's Role in Defusing the Iranian Nuclear Issue. *The Washington Quarterly* vol. 35: 141–152.

Gürzel, Aylin and Ersoy, Eyüp. 2012. Turkey and Iran's Nuclear Program. *Middle East Policy*, vol. 19, no. 1: 37–50.

Habermas, Jürgen. 1981. *Theorie des kommunikativen Handelns (Bd. 1: Handlungsrationalität und gesellschaftliche Rationalisierung; Bd. 2: Zur Kritik der funktionalistischen Vernunft)*. Frankfurt am Main: Suhrkamp.

Hall, Peter A. 2006. Systematic Process Analysis: When and How to Use It. *European Management Review* vol. 3: 24–31.

Hancock, Kathleen J. 2007. Russia: Great Power Image Versus Economic Reality. *Asian Perspective* vol. 31, no. 4: 71–98.

Hart, Andrew F. and Jones, Bruce D. 2011. How Do Rising Powers Rise? *Survival* vol. 52, no. 6: 63–88.

Hobsbawm, Eric. 2011. *How to Change the World. Tales of Marx and Marxism*. London: Abacus.

Homayounvash, Mohammad. 2016. *Iran and the Nuclear Question. History and Evolutionary Trajectory*. London and New York: Routledge.

Homolar, Alexandra. 2011. Rebels without a Conscience: The Evolution of the Rogue States Narrative in US Security Policy. *European Journal of International Relations* vol. 17, no. 4: 705–727.

Howard, Rhoda E. 1995. Occidentalism, Human Rights, and the Obligations of Western Scholars. *Canadian Journal of African Studies* vol. 29, no. 1: 110–126.

Hoyt, Paul D. 2000. The 'Rogue State' Image in American Foreign Policy. *Global Society* vol. 14, no. 2: 297–310.

Hurrell, Andrew. 2006. Hegemony, Liberalism and Global Order: What Space for Would-be Great Powers? *International Affairs* vol. 82, no. 1: 1–19.

Ikenberry, John G. 2001. *After Victory. Institutions, Strategic Restraint, and the Rebuilding of Order after Major Wars*. Princeton and Oxford: Princeton University Press.

Ikenberry, John G. 2004. Liberalism and Empire: Logics of Order in the American Unipolar Age. *Review of International Studies* vol. 30: 609–630.

Jacques, Martin. 2012. *When China Rules the World*. 2nd edition. London: Penguin Books.

Johnston, Alastair Iain, and Ross, Robert (eds). 1999. *Engaging China: The Management of an Emerging Power*. London: Routledge.

Johnston, Alastair Iain. 2003. Is China a Status Quo Power? *International Security* vol. 27: 5–56.

Johnstone, Ian. 2007. The Secretary-General as Norm Entrepreneur: In: Chesterman, Simon (ed.), *Secretary or General? The UN Secretary-General in World Politics*. Cambridge: Cambridge University Press, pp. 123–138.

Jones, Bruce. 2011. Beyond Blocs: *The West, Rising Powers and Interest-Based International Cooperation*. Policy Analysis Brief, The Stanley Foundation.

Jones, Catherine and Breslin, Shaun. 2015. China in East Asia. Confusion on the Horizon? In: Gaskarth, Jamie (ed.), *Rising Powers, Global Governance and Global Ethics*. Abingdon and New York: Routledge, pp. 115–132.

Joseph, Jonathan. 2002. *Hegemony: A Realist Analysis*. London: Routledge.

Kagan, Robert. 2002. *Paradise and Power: America and Europe in the New World Order*. London: Atlantic.

Kagan, Robert. 2012. *The World America Made*. New York: Knopf.

Katzenstein, Peter. J. (ed.). 1996. *The Culture of National Security. Norms and Identity in World Politics*. New York: Columbia University Press.

Katzman, Kenneth. 2006. The Iran–Libya Sanctions Act (ILSA). *CRS Report for Congress*, 26 April 2006. Available at [last accessed 27 September 2014]: http://fpc.state.gov/documents/organization/66441.pdf.

Keck, Margaret E. and Sikkink, Kathryn. 1998. *Activists Beyond Borders: Advocacy Networks in International Politics*. Ithaca, NY: Cornell University Press.

Keohane, Robert O. 1984. *After Hegemony: Cooperation and Discord in the World Political Economy*. Princeton: Princeton University Press.

Kersbergen, Kees van and Verbeek, Bertjan. 2007. The Politics of International Norms. *European Journal of International Relations* vol. 13, no. 2: 217–238.

Kibaroğlu, Mustafa. 2007. Iran's Nuclear Ambitions from a Historical Perspective and the Attitude of the West. *Middle Eastern Studies* vol. 43, no. 2: 223–245.

Kibaroğlu, Mustafa. 2010. The Iranian Quagmire: How to Move Forward. Position: Resuscitate the Nuclear Swap Deal. *Bulletin of the Atomic Scientists*: 1–7.

Kindleberger, Charles P. 1973. *The World in Depression 1929–1939*. Berkeley: University of California Press.

Kissinger, Henry A. 2011. *On China*. New York: Penguin.

Kitchen, Nick. 2009. Hegemonic Transition and US Foreign Policy. In: Parmar, Inderjeet, Miller, Linda B. and Ledwidge, Mark (eds), *Obama and the World. New Directions in US Foreign Policy*. London and New York: Routledge, pp. 80–92.

Kowert, Paul and Legro, Jeffrey. 1996. Norms, Identity and Their Limits: A Theoretical Reprise. In: Katzenstein, Peter J. (ed.), *The Culture of National Security*. New York: Columbia University Press, pp. 451–497.

Kleiboer, Marieke. 1996. Understanding Success and Failure of International Mediation. *The Journal of Conflict Resolution* vol. 40, no. 2: 360–389.

Kleiboer, Marieke. 1998. *The Multiple Realities of International Mediation*. Boulder, CO: Lynne Rienner.

Krahmann, Elke. 2005. American Hegemony or Global Governance? Competing Visions of International Security. *International Studies Review* vol. 7: 531–545.

Kratochwil, Friedrich. 2001. Constructivism as an Approach to Interdisciplinary Study. In: Fierke, Karin M. and Jorgensen, Knud Erik (eds), *Constructing International Relations. The Next Generation*. New York: M.E. Sharpe, pp. 13–35.

Krook, Mona Lena and True, Jacqui. 2012. Rethinking the Life Cycles of International Norms: The United Nations and the Global Promotion of Gender Equality. *European Journal of International Relations* vol. 18, no. 1: 103–127.

Kubálková, Vendulka, Onuf, Nicholas and Kowert, Paul (eds). 1998. *International Relations in a Constructed World*. New York and London: M.E. Sharpe, pp. 3–24.

Kugler, Jacek and Lemke, Douglas. 1996. *Parity and War: Evaluations and Extensions of the War Ledger*. Ann Arbor: University of Michigan Press.

Kupchan, Charles. 2012. *No One's World: The West, the Rising Rest and the Coming Global Turn*. Oxford: Oxford University Press.

Laclau, Ernesto. 1988. Metaphor and Social Antagonisms. In: Nelson, Cary and Grossberg, Lawrence (eds), *Marxism and the Interpretation of Culture*. Basingstoke: Macmillan Education, pp. 249–257.

Laclau, Ernesto and Mouffe, Chantal. 1985. *Hegemony and Socialist Strategy: Towards a Radical Democratic Politics*. London: Verso.

Laidi, Zaki. 2012. BRICS: Sovereignty Power and Weakness. *International Politics* vol. 49: 614–632.

Laidi, Zaki. 2014. Towards a Post-hegemonic World: The Multipolar Threat to the Multilateral Order. *International Politics* vol. 51: 350–365.

Lanteigne, Marc. 2009. *Chinese Foreign Policy. An Introduction*. New York: Routledge.

Lapid, Yosef. 1989. The Third Debate: On the Prospects of International Theory in a Post-Positivist Era. *International Studies Quarterly* vol. 33, no. 3: 235–254.

Lapid, Yosef and Kratochwil, Friedrich (eds). 1996. *The Return of Culture and Identity in IR Theory*. Boulder, CO and London: Lynne Rienner.

Lebow, Richard Ned and Valentino, Benjamin. 2009. Lost in Transition: A Critical Analysis of Power Transition Theory. *International Relations* vol. 23, no. 3: 389–410.

Lemke, Douglas. 1997. The Continuation of History: Power Transition Theory and the End of the Cold War. *Journal of Peace Research* vol. 34, no. 1: 23–26.

Lesser, Ian O. 1992. *Bridge or Barrier? Turkey and the West After the Cold War*. Santa Monica: RAND Corporation.

Leverett, Flynt and Leverett, Hillary Mann. 2013. *Going to Tehran. Why the United States Must Come to Terms with the Islamic Republic of Iran*. New York: Metropolitan Books.

Li, Xiaojun. 2010. Social Rewards and Socialization Effects: An Alternative Explanation for the Motivation Behind China's Participation in International Institutions. *The Chinese Journal of International Politics* vol. 3: 347–377.

Lieber, Robert J. 2014. The Rise of the BRICS and American Primacy. *International Politics* vol. 51: 137–154.

Lieberson, Stanley. 1991. Small N's and Big Conclusions: An Examination of the Reasoning in Comparative Studies Based on a Small Number of Cases. *Social Forces* vol. 70, no. 2: 307–320.

Lohmann, Sascha. 2013. Unilaterale US-Sanktionen gegen Iran. *SWP-Aktuell* vol. 63: 1–8.

Lohmann, Sascha. 2015. Zwang zur Zusammenarbeit. *SWP-Aktuell* vol. 54: 1–8.

Luxemburg, Rosa. 2003 (1913). *The Accumulation of Capital*. London: Routledge.

MacFarlane, S. Neil. 2006. The 'R' in BRICs: Is Russia an Emerging Power? *International Affairs* vol. 82, no. 1: 41–57.

Mahoney, James. 2000. Strategies of Causal Inference in Small-N Analysis. *Sociological Methods & Research* vol. 28, no. 4: 387–424.

Mankoff, Jeffrey. 2009. *Russian Foreign Policy. The Return of Great Power Politics*. Plymouth: Rowman & Littlefield.

March, James G. and Olsen, Johan P. 1989. *Rediscovering Institutions: The Organizational Basis of Politics*. New York: Free Press.

March, James G. and Olsen, Johan P. 1998. The Institutional Dynamics of International Political Orders. *International Organization* vol. 52, no. 4: 943–969.

Mastney, Vojtech and Nation, Craig R. 1996. *Turkey Between East and West. New Challenges for a Rising Regional Power*. Boulder, CO: Westview Press.

Meier, Oliver. 2013. European Efforts to Solve the Conflict over Iran's Nuclear Programme: How Has the European Union Performed? *EU Non-Proliferation Consortium. Non-Proliferation Papers, No. 27*, February 2013. Available at www.sipri.org/sites/default/files/EUNPC_no-27.pdf (accessed 28 September 2016).

Moore, Christopher W. 2003. *The Mediation Process*. San Francisco: Josse-Bass.

Morris, Ian. 2011. *Why The West Rules – For Now. The Patterns of History and What they Reveal about the Future*. London: Profile Books.

Moshirzadeh, Homeira. 2007. Discursive Foundations of Iran's Nuclear Policy. *Security Dialogue* vol. 38, no. 4: 521–543.

Mousavian, Seyed Hossein. 2012. *The Iranian Nuclear Crisis. A Memoir*. Washington: Carnegie Endowment for International Peace.

Mousavian, Seyed Hossein (with Shahir Shahidsaless). 2014. *Iran and the United States. An Insider's View on the Failed Past and the Road to Peace*. New York, London: Bloomsbury.

Müller, Harald. 2013. Introduction. Where It All Began. In: Müller, Harald and Wunderlich, Carmen (eds), *Norm Dynamics in Multilateral Arms Control: Interests, Conflicts, and Justice*. Athens: University of Georgia Press, pp. 1–19.

Nadelmann, Ethan A. 1990. Global Prohibition Regimes: The Evolution of Norms in International Society. *International Organization* vol. 44, no. 4: 479–526.

Narlikar Amrita. 2013. Introduction. Negotiating the Rise of New Powers. *International Affairs* vol. 89, no. 3: 561–576.

Nau, Henry R. 2012. Introduction: Domestic Voices of Aspiring Powers. In: Nau, Henry R. and Ollapally, Deepa M. (eds), *Worldviews of Aspiring Powers. Domestic Foreign Policy Debates in China, India, Iran, Japan, and Russia*. New York: Oxford University Press, pp. 3–35.

Neal, Andrew W. 2013. Empiricism without Positivism. King Lear and Critical Security Studies. In: Salter, Mark B. and Mutlu, Can E. (eds), *Research Methods in Critical Security Studies*. New York: Routledge, pp. 42–45.

Nel, Philip. 2010. Redistribution and Recognition: What Emerging Regional Powers Want. *Review of International Studies* vol. 36: 951–974.

Nincic, Miroslav. 2005. *Renegade Regimes: Confronting Deviant Behavior in World Politics*. New York: Columbia University Press.

Oborne, Peter and Morrison, David. 2013. *A Dangerous Delusion. Why the West Is Wrong about Nuclear Iran*. London: Elliot and Thompson.

Onuf, Nicholas G. 1989. *World of Our Making: Rules and Rule in Social Theory and International Relations*. Columbia: University of South Carolina Press.

Organski, Abramo Fimo Kenneth. 1968. *World Politics*. New York: Knopf.

Organski, Abramo Fimo Kenneth and Kugler, Jacek. 1980. *The War Ledger*. Chicago: University of Chicago Press.

Ozkan, Mehmet. 2010. Turkey-Brazil Involvement in Iranian Nuclear Issue: What Is the Big Deal? *Strategic Analysis* vol. 35, no. 1: 26–30.

Parsi, Trita. 2012. *A Single Roll of the Dice: Obama's Diplomacy with Iran*. New Haven: Yale University Press.

Patel, Ketan and Hansmeyer, Christian. 2009. American Power, Patterns of Rise and Decline. In: Parmar, Inderjeet, Miller, Linda B. and Ledwidge, Mark (eds), *Obama and the World. New Directions in US Foreign Policy*. London and New York: Routledge, pp. 275–288.

Patrick, Stewart. 2010. Irresponsible Stakeholders: The Difficulty of Integrating Rising Powers. *Foreign Affairs* vol. 89, no. 6: 44–53.

Patrikarakos, David. 2012. *Nuclear Iran. The Birth of an Atomic State*. London, New York: I.B. Tauris.

Phillips, Lauren M. 2008. *International Relations in 2030: The Transformative Power of Large Developing Countries. Discussion Paper 3/2008*. Bonn: Deutsches Institut für Entwicklungspolitik.

Pieper, Moritz. 2015. Turkey's Iran Policy: A Case of Dual Strategic Hedging. In: Kanat, Kilic Bugra, Tekelioglu, Ahmet Selim and Ustun, Kadir (eds), *Politics and Foreign Policy in Turkey. Historical and Contemporary Perspectives*. Ankara: SETA Publications, pp. 107–130.

Pieper, Moritz. 2016. The Transatlantic Dialogue on Iran: The European Subaltern and Hegemonic Constraints in the Implementation of the 2015 Nuclear Agreement with Iran. *European Security* vol. 26, no. 1: 99–119.

Pijl, Kees van der. 1984. *The Making of an Atlantic Ruling Class*. London: Verso.

Porter, Gareth. 2014. *Manufactured Crisis. The Untold Story of the Iran Nuclear Scare*. Charlottesville: Just World Books.

Puchala, Donald. 1997. Some Non-Western Perspectives on International Relations. *Journal of Peace Research* vol. 34, no. 2: 129–134.

Quinn, Adam. 2011. The Art of Declining Politely: Obama's Prudent Presidency and the Waning of American Power. *International Affairs* vol. 87, no. 4: 803–824.

Ragin, Charles C. 1987. *The Comparative Method: Moving Beyond Qualitative and Quantitative Strategies*. Berkeley and Los Angeles: University of California Press.

Reus-Smit, Christian. 2014. Power, Legitimacy, and Order. *The Chinese Journal of International Politics* vol. 7, no. 3: 341–359.

Rapkin, David P. and Braaten, Dan. 2009. Conceptualising Hegemonic Legitimacy. *Review of International Studies* vol. 35: 113–149.

Rohlfing, Ingo. 2012. *Case Studies and Causal Inference: An Integrative Framework*. Basingstoke: Palgrave Macmillan.

Ross, Robert S. and Zhu, Feng (eds) 2008. *China's Ascent: Power, Security, and the Future of International Politics*. Ithaca, NY: Cornell University Press.

Rouhani, Hassan. 2011. *Amniyat Melli va Diplomasi-ye Hastehi Iran (National Security and Nuclear Diplomacy)*. Tehran: Center for Strategic Research.

Ruggie, John Gerard. 1998. What Makes the World Hang Together? Neo-utilitarianism and the Social Constructivist Challenge. *International Organization* vol. 52, no. 4: 855–885.

Rupert, Mark. 1995. *Producing Hegemony: The Politics of Mass Production and American Global Power*. Cambridge: Cambridge University Press.

Rupert, Mark. 2000. *Ideologies of Globalization*. London: Routledge.

Rynning, Sten and Ringsmore, Jens. 2008. Why Are Revisionist States Revisionist? Reviving Classical Realism as an Approach to Understanding International Change. *International Politics* vol. 45, no. 1: 19–39.

Sandholtz, Wayne. 2007. *Prohibiting Plunder: How Norms Change*. New York: Oxford University Press.

Said, Edward. 2003 [1978]. *Orientalism*. London: Penguin Books.

Sakwa, Richard. 2002. *Russian Politics and Society*. 3th edition. London and New York: Routledge.

Sakwa, Richard. 2012. The Problem of 'the International' in Russian Identity Formation. *International Politics* vol. 49, no. 4: 449–465.

Savin-Baden, Maggi and Major, Claire Howell. 2013. *Qualitative Research. The Essential Guide to Theory and Practice*. London and New York: Routledge.

Schiffer, Michael and Shorr, David. 2009. *Powers and Principles. International Leadership in a Shrinking World*. Plymouth: Lexington Books.

Schwarzmantel, John. 2009. Introduction. Gramsci in His Time and in Ours. In: McNally, Mark and Schwarzmantel, John (eds), *Gramsci and Global Politics: Hegemony and Resistance*. London and New York: Routledge, pp. 1–16.

Senn, Martin. 2009. *Wolves in the Woods. The Rogue State Concept from a Constructivist Perspective*. Baden-Baden: Nomos.

Serfaty, Simon. 2011. Moving into a Post-Western World. *The Washington Quarterly* vol. 34, no. 2: 7–23.

Serfaty, Simon. 2012/2013. The West in a World Recast. *Survival* vol. 54, no. 6: 29–40.

Service, Robert. 2000. *Lenin. A Biography*. London: Pan Books.

Service, Robert. 2007. *Comrades. Communism: A World History*. London: Pan Books.

Shakleina, Tatiana. 2013. Russia in the New Distribution of Power. In: Nadkarni, Vidya and Noonan, Norma (eds), *Emerging Powers in a Comparative Perspective. The Political and Economic Rise of the BRIC Countries*. London: Bloomsbury Academic, pp. 163–187.

Shambaugh, David. 2000. Sino-American Strategic Relations: From Partners to Competition. *Survival* vol. 42, no. 1: 97–115.
Shambaugh, David. 2002. China's International Relations Think-tanks: Evolving Structure and Process. *The China Quarterly* vol. 171: 575–596.
Shambaugh, David. 2005. Introduction: The Rise of China and Asia's New Dynamics. In: Shambaugh, David (ed.), *Power Shift: China and Asia's New Dynamics*. Berkeley: University of California Press, pp. 1–22.
Shambaugh, David and Xiao, Ren. 2012. China: The Conflicted Rising Power. In: Nau, Henry R. and Ollapally, Deepa M. (eds), *Worldviews of Aspiring Powers. Domestic Foreign Policy Debates in China, India, Iran, Japan, and Russia*. New York: Oxford University Press, pp. 36–72.
Sørensen, Georg. 2008. The Case for Combining Material Forces and Ideas in the Study of IR. *European Journal of International Relations* vol. 14, no. 5: 5–32.
Sotero, Paulo and Armijo, Leslie Elliott. 2007. Brazil: To Be or Not to Be a BRIC? *Asian Perspective* vol. 31, no. 4: 43–70.
Stent, Angela E. 2014. *The Limits of Partnership. U.S.–Russian Relations in the Twenty-First Century*. Princeton and Oxford: Princeton University Press.
Stiles, Kendall and Sandholtz, Wayne. 2009. *International Norms and Cycles of Change*. New York: Oxford University Press.
Stokes, Doug. 2014. Achilles' Deal: Dollar Decline and US Grand Strategy after the Crisis. *Review of International Political Economy*, vol. 21, no. 5: 1071–1094.
Subacchi, Paola. 2008. New Power Centres and New Power Brokers: Are They Shaping a New Economic Order? *International Affairs* vol. 84, no. 3: 485–498.
Tabrizi, Aniseh Bassiri and Santini, Ruth Hanau. 2012. EU Sanctions against Iran: New Wine in Old Bottles? *ISPI Analysis*, No. 97, March 2012.
Takeyh, Ray and Maloney, Suzanne. 2011. The Self-limiting Success of Iran Sanctions. *International Affairs* vol. 87, no. 6: 1287–1312.
Taylor, Brendan. 2010 *Sanctions as Grand Strategy*. Adelphi series, International Institute for Strategic Studies. Abingdon: Routledge.
Terhalle, Maximilian. 2011. Reciprocal Socialization: Rising Powers and the West. *International Studies Perspectives* vol. 12: 341–361.
Tessman, Brock and Wolfe, Wojtek. 2011. Great Powers and Strategic Hedging: The Case of Chinese Energy Security Strategy. *International Studies Review* vol. 13: 214–240.
Towns, Ann E. 2012. Norms and Social Hierarchies: Understanding International Policy Diffusion 'from Below'. *International Organization* vol. 66, no. 2: 179–209.
Trenin, Dmitri. 2001. *The End of Eurasia. Russia on the Border Between Geopolitics and Globalization*. Washington, D.C.: Carnegie Endowment for International Peace.
Trenin, Dmitri. 2006. Russia Leaves the West. *Foreign Affairs* vol. 85, no. 4: 87–96.
Tsygankov, Andrei P. 2007. Finding a Civilizational Idea: 'West', 'Eurasia', and 'Euro-East' in Russia's Foreign Policy. *Geopolitics* vol. 12, no. 3: 375–399.
US Department of State. 2013. *Fact Sheet: Iran Freedom and Counter-Proliferation Act of 2012*, 23 April 2013. Available at www.state.gov/documents/organization/208111.pdf (accessed 25 September 2016).
Ülgen, Sinan. 2012. *Turkey and the Bomb*. Washington, D.C.: Carnegie Endowment for International Peace.
Üstün, Kadir. 2010. Turkey's Iran Policy: Between Diplomacy and Sanctions. *Insight Turkey* vol. 12, no. 3: 19–26.
Vezirgiannidou, Sevasti-Eleni. 2013. The United States and Rising Powers in a Post-hegemonic Global Order. *International Affairs* vol. 89, no. 3: 635–651.

Wallerstein, Imanuel. 1979. *The Capitalist World Economy*. Cambridge: Cambridge University Press.
Welch, David E. 1993. *Justice and the Genesis of War*. Cambridge: Cambridge University Press.
Wendt, Alexander. 1987. The Agent-Structure Problem in International Relations Theory. *International Organization* vol. 41, no. 3: 335–70.
Wendt, Alexander. 1999. *Social Theory of International Politics*. Cambridge: Cambridge University Press.
Widmaier, Wesley W. and Park, Susan. 2012. Differences Beyond Theory: Structural, Strategic and Sentimental Approaches to Normative Change. *International Studies Perspective* vol. 13, no. 2: 123–134.
Wicht, Bernard. 2002. *Guerre et Hégémonie. L'éclairage de la longue durée*. Paris: Georg Editeur.
Wiener, Antje. 2008. *The Invisible Constitution of Politics: Contested Norms and International Encounters*. Cambridge and New York: Cambridge University Press.
Williams, Michael C. 2003. Words, Images, Enemies: Securitization and International Politics. *International Studies Quarterly* vol. 47: 511–531.
Wilkinson, David. 1999. Unipolarity without Hegemony. *International Studies Review* vol. 1, no. 2: 141–172.
Worth, Owen. 2009. Beyond World Order and Transnational Classes. The (Re)application of Gramsci in Global Politics. In: McNally, Mark and Schwarzmantel, John (eds), *Gramsci and Global Politics: Hegemony and Resistance*. London and New York: Routledge, pp. 19–31.
Worth, Owen. 2011. Recasting Gramsci in International Politics. *Review of International Studies* vol. 37: 373–392.
Worth, Owen. 2015. *Rethinking Hegemony*. London: Palgrave Macmillan.
Wunderlich, Carmen. 2013. Theoretical Approaches in Norm Dynamics. In: Müller, Harald and Wunderlich, Carmen (eds), *Norm Dynamics in Multilateral Arms Control: Interests, Conflicts, and Justice*. Athens: University of Georgia Press, pp. 20–47.
Wunderlich, Carmen. 2014. A 'Rogue' Gone Norm Entrepreneurial? Iran within the Nuclear Nonproliferation Regime. In: Wagner, Wolfgang, Werner, Wouter and Onderco, Michal (eds), *Deviance in International Relations. 'Rogue States' and International Security*. Basingstoke: Palgrave Macmillan, pp. 83–104.
Yan, Xuetong. 2006. The Rise of China and its Power Status. *The Chinese Journal of International Politics* vol. 1: 5–33.
Young, Michael D. and Schafer, Mark. 1998. Is There Method in Our Madness? Ways of Assessing Cognition in International Relations. *Mershon International Studies Review* vol. 42: 63–96.
Zakaria, Fareed. 2011. *The Post-American World. And the Rise of the Rest*. London: Penguin Books.
Zhao, Suisheng. 2008. *China–U.S. Relations Transformed. Perspectives and Strategic Interactions*. New York: Routledge.
Zhao, Suisheng. 2013. China: A Reluctant Global Power in the Search for its Rightful Place. In: Nadkarni, Vidya and Noonan, Norma (eds), *Emerging Powers in a Comparative Perspective. The Political and Economic Rise of the BRIC Countries*. London: Bloomsbury Academic, pp. 101–130.

2 Turkish foreign policy towards the Iranian nuclear programme

1 'Strategic depth' and the reproach of nuclear double standards: discursive divergence from US policies after 2003

Turkey signed the NPT in 1969 and ratified it in 1980. Turkey also ratified the NPT's Additional Protocol in 2000, is a member of the Nuclear Suppliers Group (NSG),[1] and has been supportive of the Middle East nuclear weapons-free zone initiative. Ankara was thus always a steadfast supporter of international nuclear non-proliferation efforts and endorses the IAEA's safeguards regime and verification mechanisms to ensure non-diversion of nuclear material to military purposes.[2] Yet, with the revelation of a uranium enrichment site at Natanz and a heavy-water reactor under construction at Arak, Turkey has been very cautious not to join Western scepticism on Iran (Udum 2012: 103f.). As a foreign ministry official formulates in an interview: 'We shouldn't judge on the basis of assumptions'.[3] These facilities were hitherto undeclared to the IAEA. A controversy thus ensued as to whether Iran had been in breach of its 1974 Safeguards Agreements which stipulated transparency about all nuclear facilities. While Iran argued that it was only required to declare their existence six months prior to feeding nuclear fuel into them,[4] word of a 'possible military dimension' started to make the rounds in Western diplomatic communities. The heavy-water reactor at Arak was disconcerting news as it opened up the possibility of Iran accumulating weapons-grade plutonium.

Statements coming from the Turkish foreign ministry and then-prime minister Erdoğan similarly underscored the Turkish insistence on the upholding of Iranian sovereignty and the principle of non-interference,[5] which Ankara saw as threatened by an emerging US securitisation of the Iranian nuclear programme when the Bush administration was eager to take Iran's case from the IAEA to the UN Security Council (Porter 2014: 145). Turkey refrained from assuming Iranian military intentions. Murat Mercan, Chairman of the Foreign Affairs Committee of the Turkish Grand National Assembly from 2007 to 2011, summarises Turkey's Iran policy as follows:

> We advocate diplomatic and economic engagement of Iran rather than isolationist policies as a more effective way to address the challenges that we are

> facing in the region. We will continue to encourage all our counterparts to take a conciliatory approach in order to better tackle the problems in the Middle East [...]. Any interference from the outside world will have a boomerang effect and will be counter-productive. Therefore, the international community should refrain from any attempt to interfere in Iran to the detriment of the social and political fiber of Iran.
>
> (Mercan 2009: 18, 19)

It is this conception that brought the Turkish position occasionally closer to China's and Russia's with regard to a joint hesitant approach to the use of sanctions on Iran and with regard to a shared conviction of the foreign policy principle of 'non-interference'. Turkey is committed to the nuclear non-proliferation regime and would not want to see a nuclear Iran emerge in the region that would also challenge Turkey from a regional power perspective. Yet, Turkey made a clear differentiation between legitimate non-proliferation concerns and what it perceives as unhelpful pressure on a country that has legitimate rights to nuclear energy. Erdoğan criticised what he saw as an 'unfair' and one-sided focus of the West on a possible military dimension of the Iranian programme, while Israel's de facto nuclear weapons were ignored (Hunter 2010: 166). Iran, so the message goes, should not be the sole focus of nuclear security considerations. In a 2007 interview, when asked whether he shared Western fears about an Iranian nuclear bomb, Erdoğan replied with a snide remark: 'We are against nuclear weapons, regardless of whether they are in the hands of Iran or Israel or any Western country. But obviously some states are allowed to have weapons of mass destruction while others are not.'[6] Besides this emphasis on an equitable application of nuclear non-proliferation efforts, Turkey was also well aware of the regionally destabilising effect that a war with Iran would have entailed, if the diplomatic engagement was to break down. Notwithstanding the collapsing bilateral trade, Turkey would most likely have received an influx of Iranian refugees crossing the Turkish–Iranian border, be drawn into a regional proxy war or even become the target of Iranian counterstrikes (Larrabee and Nader 2013: 26; Ülgen 2011: 159). Given the stakes, Erdoğan did not mince his words, calling the prospect of a war with Iran 'crazy' (in Parsi 2012: 180).

With this foreign policy discourse, Erdoğan's government was in line with its regional foreign policy formula of 'zero problems with neighbors' that had been formulated by foreign minister Ahmet Davutoğlu. This concept was aimed to convey Turkey's striving for good relations with its neighbours, including Syria, Iraq and Iran in the course of a general foreign policy shift toward a stronger regional commitment.[7] The formulation of the foreign policy doctrines of 'zero problems with neighbors' and 'Strategic Depth' (Davutoğlu 2001) has inspired categorisations of Turkish Middle Eastern policies as that of a middle power (Sandal 2014), of a 'pivotal' state in the region (Fuller 2008; Kazan 2005; Larrabee and Lesser 2003: 2–3; Lewis 2006; Winrow 2003), and of a 'bridge' or 'facilitator' in conflict resolution (Altunışık 2008; Gürzel 2012; Gürzel and Ersoy 2012; Ülgen 2012; Üstün 2010). Condemning Iran over its Iranian nuclear programme and exerting pressure on its neighbour, in this reading, would have run counter to the idea of 'Zero problems with neighbours' and 'Strategic Depth' in the region. Such a policy also entailed a more critical stance towards Western policies in the region, as illustrated by geopolitical overlaps of interests between Iran and Turkey in the run-up to the looming US-led Iraq war in 2003. The US administration was struck by the refusal of the Turkish Grand Assembly on 1 March 2003 to approve of US troop deployments on Turkish soil for combat operations in neighbouring Iraq (Aykan 2007; Gunter 2005). Turkey instead focused on the 'Neighbouring Countries of Iraq Platform' initiative whose priority was to safeguard Iraqi territorial integrity.

As Turkey adopted a more US-critical foreign policy discourse publicly, it also became more critical of Israeli foreign policy and the Israeli settlement policy. Turkish officials and especially Prime Minister Erdoğan started to publicly criticise Israel on harsh terms (Barkey 2013: 146; Gul 2011; Kibaroğlu 2012: 87; Kibaroğlu and Caglar 2008: 63; Oktav 2007: 89). Turkish criticism of and alienation from the US and Israel went hand in hand with a warming of relations with US-defiant countries like Iran. The furtherance of good-neighbourly relations led to the signing of a memorandum of understanding on the transport of Iranian gas to Europe via Turkey between Iran's petroleum minister Kazem Vaziri-Hamaneh and Turkish energy minister Hilmi Guler. At their meeting, they also addressed the Turkish development of Iran's South Pars gas field (Hiro 2009: 389). In this context, a foreign policy rhetoric overly critical of Iran's lack of transparency with the IAEA would have sent mixed signals in a phase of Turkish–Iranian rapprochement in line with Davutoğlu's new regional policy.

When the Iranian nuclear file eventually was taken from the IAEA to the UNSC in 2006 and the imposition of chapter VII sanctions became imminent, Turkey criticised the use of sanctions as a political instrument and emphasised that sanctions can complement diplomacy, but should never be an end in themselves.[8] This position is to be explained by an upholding of the principle of 'non-interference' and the Turkish understanding and conviction of sovereignty that lies at the basis of its scepticism of Western pressure against a country with legitimate nuclear rights as an NPT member, as explained by a high-ranking Turkish diplomat in an interview.[9] In addition, it has been pointed out that the experience of destabilising sanctions in neighbouring Iraq is another factor in Turkey's perception of sanctions as a counterproductive instrument and another driving force for Ankara's insistence on the need to find political solutions (Larrabee and Nader 2013: 27; Üstün 2010: 20). A briefing on 'Iran's Nuclear Program – The Turkish Perspectives', issued by the Turkish foreign ministry in June 2010, formulates: 'For decades, Turkey has borne the burden of tragic events unfolding in its vicinity. Adverse economic and political implications of the sanctions against Iraq in the 1990s bear testament to this fact' (Turkish foreign ministry 2010). At the same time, the scepticism toward sanctions carries the Turkish awareness that the weakening of its neighbour Iran will impact on the Turkish economy due to close trade links. 'We explain to our US partners: If you want to hurt Iran, don't hurt us at the same time', a Turkish diplomat explained in an interview.[10] Foreign minister Davutoğlu publicly stated that Turkey was against sanctions with a view to concerns over possible constraints on regional trade that new sanctions regimes might entail (Raphaeli 2010). It will be the subject of a later section to analyse the effect of such material factors on Turkey's Iran policy and its stance on sanctions.

In all Turkey's discursive positioning on the Iranian nuclear crisis, Turkey's NATO membership functioned as the proverbial elephant in the room. Turkey's public discourse on Iran oftentimes displayed a level of friction between evolving national security perceptions and NATO commitments and, closely related to the latter, solidarity with its US ally. In 2010, the Turkish Security Council approved of the removal of Iran and Syria from the list of countries posing a security threat to Turkey and instead explicitly named Israel as a destabilising force that could potentially trigger a regional arms race (Lutz 2010; Vahedi 2010). Military officials and politicians, such as retired General and former secretary-general of the National Security Council Tuncay Kilinc, openly started raising the question whether Turkey should withdraw from NATO and rather engage in other regional organisations (Baran and Lesser 2009: 210; Kibaroğlu and Caglar 2008: 68;). The 'Missile Defense' episode of 2011 therefore introduced a major irritant into Turkish–Iranian relations. Presented as a missile defense shield with a missile-defense radar to be stationed in Kürecik, Turkey, the US introduced its new ballistic missile defense (BMD) plans as being directed against potential missile attacks coming from Iran (Barkey 2013: 154). These plans triggered a phase of irritation between Turkey and Iran, as the Iranian government was concerned that its neighbour Turkey agreed to the deployment of the radar on its soil (Gürzel 2012: 148; Kibaroğlu 2013: 228).[11] This move was preceded by the Turkish reluctance to specifically name Iran as a threat to the Alliance – because of the sensitivity of this issue in bilateral relations and out of an understanding that such a listing would give reason to Tehran to advance its missile capabilities as counter-balancing measures (Kibaroğlu 2013: 232; Ülgen 2012: 10).[12] In Ömer Taşpınar's assessment, the Turkish decision to host the radar system was evidence of Turkey having become a proponent of the containment of Iran – something that boosted Turkey's image in American eyes (Taşpınar *et al.* 2012: 10).[13] Under the pressure of what he calls a 'double gravity predicament', Philip Robins (2013) therefore writes that Ankara chose 'its global strategic relationship with the United States ahead of any region-based considerations' (394). Turkish divergence from US foreign policy on a discursive level with regards to the Iranian nuclear programme, especially in the latter half of the 2000s, was thus oftentimes paralleled by more mixed signals on a behavioural level, as the Missile Defense episode demonstrated. A later section will return to this question of behavioural convergence. Suffice to note at this point that the 'MD' episode described above occurred after Turkey's failed mediatory efforts in the Iranian nuclear crisis in 2009–2010, and thus during a period of tacit re-alignment with American positions.

In its foreign policy discourse on the emerging Iranian nuclear crisis, however, Turkey conveyed a normative divergence from the US government advocating for sanctions and from an approach employing pressure on Iran over the latter's lack of transparency. While international sanctions, backed by UNSC resolutions, were accepted by Turkey, Ankara views EU and US unilateral sanctions as unhelpful measures that have an adverse effect on the Turkish economy. They are also seen as undermining dialogue efforts with Iran and strengthening Iranian hardliner positions.[14] Turkish officials had repeatedly stressed their discontent with 'unhelpful' pressure on Iran in the form of sanctions for the latter's noncompliance with the IAEA and UNSC stipulations. In October 2009, weeks after adoption of UNSCR 1887 on strengthening the NPT regime with a Turkish vote in favour, Erdoğan underlined Iran's right to nuclear technology for peaceful purposes and criticised Western one-sided pressure on Tehran for suspected illicit nuclear activities and the fact that Israel's possession of nuclear weapons was dealt with as a political taboo at the same time (Seufert 2012: 26).[15] One month later, Turkey abstained from condemning Iran in the IAEA Board of Governors (ibid.). Without a permanent UNSC membership, Turkey's insistence on the IAEA as the main body to legitimately watch over the non-proliferation regime,[16] Turkish suspicion against UN-backed pressure on Iran therefore also carries an element of awareness that the transfer from Vienna to New York deprives Turkey of an instrument to make its voice heard and takes the issue to the international arena of Great Power politics. 'The world is bigger than five', president Erdoğan revealingly formulated his general discontent of the structure of the Security Council at the 69th UN General Assembly in 2014, and again in 2016.[17] Then-prime minister Davutoğlu strikes a similar cord when he publicly complains that the P5 are deciding on other countries' behalf.[18] 'The Turkish policy regarding sanctions is a microcosm for Turkish nuclear diplomacy in general', Sinan Ülgen (2012: 7) therefore writes. Tellingly, a Turkish foreign ministry official remarked in an interview that 'we use the same talking points on both international and unilateral sanctions'.[19] If the same diplomatic language applies to all sanctions, regardless of the institutional basis for their adoption, so the message, the use of sanctions as a political instrument is viewed with *categorical* scepticism by the Turkish government.

2 Turkey as a facilitator in the Iranian nuclear dossier

The period of Turkey's pro-active shuttle diplomacy in the Iranian nuclear dossier in 2009 and 2010 is an illuminating period to analyse for two main reasons. First, Turkey lacks a permanent UNSC seat. Analyses of China's and Russia's Iran policies thus proceed from a somewhat different starting point structurally and institutionally. While Russia and China are both permanent Security Council members holding veto power and have a more authoritative say over the adoption of such sanctions, Turkey does not need to formally adopt them, absent a permanent UNSC seat. Turkey's active diplomatic involvement in the Iranian nuclear dossier in 2009–2010, however, allows us to analyse its Iran policies from a position of altered institutional stakes that approximated Turkey's role to the P5+1. Second, analysing Turkey's foreign policy in and the rationale for this active shuttle diplomacy allows for a concrete analysis of the behavioural dimension of Turkish Iran policy in line with the two-level model introduced in Chapter 1. It allows us, in other words, to pin-point Turkey's *deeds* in diplomacy, next to the discursive edifice as worked out in the preceding section. This behavioural level also plays out in the way sanctions regimes are either complied with or circumvented. Sanctions implementation directly and concretely translates security cultures into trade restrictions to be communicated to companies.

The fact that Turkey is a NATO member and a regional neighbour of Iran inspired and informed theorisations of Turkey as a potential facilitator of talks and conduit of messages between Iran and the West, rather than as a fully-fledged mediator (see Fuller 2008; Giragosian 2008; Gürzel 2012; Gürzel and Ersoy 2012; Kibaroğlu 2009; Kibaroğlu and Caglar 2008; Önis 2009; Ülgen 2012).[20] The idea of facilitation is distinct from mediation in that a facilitating third party should not interfere in the process (Burton 1969; Fisher 1972). The perception of Turkish facilitation was given life with the choice of Turkey as a venue for negotiations between the E3 and their Iranian counterparts, and between the extended format of P5+1 and Iran. On Turkey's mediatory potential, a Turkish foreign ministry official remarked: 'We are ready to host, but we never invited. We only provided logistical support when we were approached. We never imposed ourselves as mediators.'[21] Talks between the Secretary of Iran's Supreme National Security Council Ali Larijani and EU High Representative Javier Solana in April 2007 in Ankara and P5+1 negotiations at the political directors-level in Istanbul on several occasions, the latest having taken place in April 2012, were the reason for the perception of Turkey as an impartial host and venue for negotiations (Önis and Yilmaz 2009: 19).

Yet, Erdoğan's government did not leave it to the role of facilitator and started pronouncing an interest in a more proactive diplomatic involvement in the Iranian nuclear dossier. 'We are ready to be the mediator', Erdoğan stated the Turkish interest in getting involved in the nuclear file in November 2008, and continued: 'I do believe we could be very useful' (in Parsi 2012: 145). 'Turkey has reoriented its foreign policy by means of an active, multidimensional and visionary framework. Mediation is an integral part of this policy', former foreign minister Davutoğlu writes in praiseful words (2013: 90). While Turkish interlocutors tend to emphasise Turkey's natural role as a facilitator and a more candid dialogue atmosphere than the P5+1 have with Iran,[22] other (mainly Western) insiders and experts are more critical. Critiques of the perception of Turkey acting as a conflict mediator range from rectifying that Turkey's role may rather be comparable to that of Switzerland providing 'good services' (i.e. more as a conduit of messages)[23] all the way to the assertion that Turkey does not play any role whatsoever and is the most overrated actor in the nuclear dossier.[24] Switzerland, a common comparison, possesses credible channels that Turkey does not have because of Turkey's questionable neutrality and power political interests in the region. The idea of Turkish mediation has also been questioned out of a practical understanding that enlarging the format makes coordination more difficult. 'When you multilateralise', Ali Vaez from the International Crisis Group puts it, 'you make it [the format] a dysfunctional mechanism for diplomacy. It takes away capital for innovation because you are more likely to be on lowest common ground.'[25]

2.1 Brazilian–Turkish mediation and the 2010 Tehran declaration

Notwithstanding these different perspectives on the idea of Turkish mediation, Turkey, together with Brazil, started negotiating as a mediator between Iran and the P5+1 as from early 2010. Turkey's entry into the Iranian nuclear negotiations as a mediator is to be contextualised against the background of the unsuccessful attempts in 2009 to negotiate a nuclear fuel swap deal, a proposal by the 'Vienna group' (consisting of the US, France, Russia, and the IAEA) in which Iran would have agreed to send three-quarters of its Low Enriched Uranium (LEU) out of the country to be refined in Russia, and would have in turn received the nuclear fuel needed for the Tehran Research Reactor (TRR) in the form of nuclear fuel rods from France. Even though the US administration was hesitant to accept Turkey as a mediator at first, largely due to Turkey's new regional assertiveness and Erdoğan's occasional anti-Israel rhetoric (Parsi 2012: 181f.), the US State Department conveyed already in early 2009 the US appreciation of any Turkish efforts to help alleviate tensions over the Iranian nuclear case (ibid.). As the nuclear fuel swap proposal of the Vienna group lost momentum in late 2009, Brazil and Turkey seized their chance of diplomatic initiative – separate, at first, then by way of a coordinated shuttle diplomacy from January 2010 onwards. Both, in addition, had become non-permanent members of the UN Security Council in January 2009 – a fact that conveniently bolstered their political weight as mediators in line with Davutoğlu's vision of 'rhythmic diplomacy' (Ilgit and Ozkececi-Taner 2014).

With the aim to propose a diplomatic initiative to ease tensions, Turkish foreign minister Davutoğlu, together with his Brazilian and Iranian counterparts, negotiated the Tehran declaration in May 2010, a declaration that guaranteed Iran the right to use nuclear energy for the Tehran Research Reactor (TRR) and specified the transport of 1,200 kg of LEU from Iran to Turkish soil to be held in escrow. This was an idea that, in its details, was similar to the earlier proposal of a nuclear swap deal by the Vienna group in late 2009, but conveyed a conception of non-Western power constellations in the search for a solution to the Iranian nuclear crisis. Iran hailed the establishment of a new world order, aligning itself with the South American regional leader Brazil and Turkey, an influential regional actor in the Near East (see Savyon 2010). Turkey, however, understood this declaration as an attempt to engage Iran and let diplomacy work (Kibaroğlu 2010: 4f.; see also Santos 2011).[26] Turkey's mediation in the Iran file served at least two purposes strategically. It assured the US of Turkey's commitment to international diplomacy that was in line with Western security and non-proliferation priorities. This understanding stemmed from the communication between the US and Turkish diplomatic teams. Second, Turkey's mediation conveyed a prioritisation of political dialogue and a rejection of sanctions as a counterproductive means of pressuring the Iranians at the same time. Seizing a moment where other diplomatic initiatives on Iran were scarce, the Turkish–Brazilian diplomacy was conceived to at least allow for more 'breathing space' before sanctions would become an option again. Coupled with the Turkish awareness that Turkey's geographic location and cultural proximity makes Ankara a more natural interlocutor with Tehran than any Western negotiator (Cetinsaya 2003),[27] Turkish foreign policy activism in the Iran nuclear file in 2010 also served to demonstrate to the Iranians that Turkey would not abuse their trust by simply advocating a Western agenda (Parsi 2012: 189f.).

As confirmed by a Turkish diplomat who was directly involved in these negotiations, the US was briefed on the plans for the TRR deal, but did not think that Turkey would 'deliver on this', so they had already 'gotten China and Russia on board for UNSC sanctions'.[28] Then-Secretary of State Clinton complacently remarked in a testimony to the Senate Foreign Relations Committee on 18 May 2010, referring to the impending sanctions momentum in the Security Council: 'I guess that tells us all we need to know about the deal Brazil and Turkey tried to work out with Iran' (Hunter 2010: 154). While the deal actually did meet a lot of US points content-wise,[29] the 'timing was bad', the Turkish diplomat intimately involved in the negotiations explained in an interview.[30] And on a global power political level, there was a certain astonishment on the US side as to 'who these upstarts [i.e. Brazil and Turkey] are'.[31] The reading that Turkey had seized its chance to play on the stage of international diplomacy usually reserved for 'Great Powers' and was thus boxing above its weight is a political one that has much to do with the perpetuation of certain narratives *after the fact*. A former E3 delegation member put it in an interview:

> Turkey is not a mediator, but part of the problem. The Tehran declaration was an unfortunate story. It was a good attempt, but it came too late. This often happens in international politics. Remember the 2003 Swiss initiative. It simply was not wanted politically by the Bush administration at the time. It was bad timing.[32]

Ironically, the Tehran declaration had gotten the US backing through a letter that president Obama had sent in April 2010 to Brazilian president Lula in which he welcomed a Brazilian–Turkish diplomatic initiative (Mousavian 2012: 383; Parsi 2012: 187). The fact that Iran reacted positively to the proposed fuel swap deal confronted the US with a dilemma: Either to allow a diplomatic break-through in the nuclear dossier, or to go ahead with a new round of international sanctions. They opted for the latter. Justifying her government's decision, then-US Secretary of State Hillary Clinton laconically writes in her memoir that Turkey's '"Zero Problems with Neighbors" foreign policy [...] made Turkey overeager to accept an inadequate diplomatic agreement with its neighbour, Iran, that would have done little to address the international community's concerns about Tehran's nuclear program' (Clinton 2014: 217). If anything, this passage is a prime example of a selective reading of facts for political reasons.

Against the background of the Turkish conviction that the Tehran declaration was the first to have secured the approval of Iran and could have served as an important confidence-building measure,[33] Turkey voted against UN Security Council Resolution 1929 on 9 June 2010, along with Brazil. Lebanon abstained. The Turkish vote against this latest round of international sanctions was to be read as a frustrated reaction to the impatience of Western powers only one month after the Turkish–Brazilian diplomatic initiative, as was also formulated explicitly in Erdoğan's letter to the leaders of 26 countries before the adoption of the resolution in the Security Council (Mousavian 2012: 382). Before voting against this resolution in the Security Council, Turkey's UN representative Ertuğrul Apakan expressed his government's frustration with the negative reception of the Tehran declaration: 'Sufficient time and space should be allowed for its implementation. We are deeply concerned that the adoption of sanctions would negatively affect the momentum created by the declaration and the overall diplomatic process' (UN 2010b: 3–4).

As NATO member Turkey had never voted against the American position since 1952, this constituted a watershed in American–Turkish relations and resulted in a public diplomatic fall-out between the Turkish and the American side (Hiro 2009: 426; Leverett and Leverett 2013: 363; Parsi 2012: 193f.). The sentiment of irritation and frustration was also publicly demonstrated by an op-ed that Brazilian foreign minister Amorim and his Turkish counterpart Davutoğlu had published in the *International Herald Tribune* in late May 2010 (Davutoğlu and Amorim 2010). After this experience of a failed diplomatic initiative, Turkey started coordinating closer with the Americans in the Iranian nuclear dossier[34] – out of a desire to avoid such miscommunication in the future, but arguably also out of a frustrated understanding that its pro-active role in 2010 did not meet much enthusiasm.[35] 'Turkey's mediator role ended with the No-vote against resolution 1929 in 2010', Soli Özel puts it.[36] From the role of mediator, Turkey retreated to the role of 'facilitator' again, with Istanbul hosting the January 2011 P5+1 talks with Iran. Other back channels (Oman) were used thereafter, and the strong bilateral US–Iran negotiation track after the election of Rouhani in 2013 had brought a new momentum for diplomacy to the nuclear talks effectively that rendered Turkish mediation obsolete. The political framework agreement reached in Lausanne on 2 April 2015 as well as the nuclear agreement reached on 14 July 2015 were applauded by Turkey. Telling for the Turkish emphasis of missed opportunities in 2010 was the reference to the Tehran declaration in Turkey's press release one day after the successful conclusion of the talks in Lausanne in April 2015: 'Turkey has actively supported the processes for a peaceful solution through dialogue and has contributed to them including through finalising [sic] of the Tehran Joint Declaration in 2010' (Turkish foreign ministry 2015). As Mustafa Akyol (2015) holds, the Turkish government's memory of seeing its own deal fall apart five years ago partially explained the relative public silence in Turkey on the talks between Iran and the P5+1 in early 2015. The conclusion of the JCPOA in July 2015, however, was welcomed as a historic milestone by Turkey. 'We always favoured diplomacy', a Turkish official stated in an interview, 'our initiative in 2010, together with Brazil, was an indication of that. We welcome the deal because it is compatible with what we have always said.'[37]

Turkey's motivation for the 2010 Tehran declaration was a mixture of the Turkish insistence on a resolution of the nuclear conflict through dialogue and engagement and a pragmatic attempt to engage Iran in order to reduce tensions. Turkey pursued a pro-active diplomatic role in an attempt to reduce those tensions that, as Ankara felt, were only sustained and increased by international sanctions. While the Turkish government believed to enjoy the backing of the US administration, however, it was left in the belief that it did not intentionally resist a US foreign policy line: The US backing for Turkey's diplomatic initiative made its negotiators believe they were adhering to an approach approved by the US government. In the sequence of events, it was the failure of the West to honour and act upon the Tehran declaration and the subsequent adoption of sanctions that led to Turkey's public positioning against the US official position. To be faced with the dilemma of choosing between sanctions and acting upon a diplomatic break-through negotiated outside of the P5+1 framework only becomes a dilemma when one understands the institutionalised nature of the impending sanctions opportunity. The adoption of UNSCR 1929 seized a 'momentum' (because it was understood on the US side that Russia and China would not veto the upcoming sanctions resolution), and it kept US domestic critics of Iran diplomacy in Congress happy. Choosing sanctions over diplomacy was thus the expression of a structural path dependency. The fact that Turkey retreated from its mediatory role thereafter and, as confirmed by Turkish officials, even communicated and coordinated its policies more closely with the American administration in the aftermath of this short episode of behavioural divergence, evidences Turkey's desire to convey 'solidarity with the US' and 'comply' with US positions. Despite a discursively underscored normative divergence, behavioural convergence with US positions eventually (and noticeably quickly) took place after the short-lived episode of a public alienation.

3 Between 'gold-for-gas' and sanctions waivers: Turkey's balancing act regarding unilateral Iran sanctions

The fact that the Turkish Iran diplomacy in 2010 ended up being portrayed as a dilemma between diplomacy and sanctions and that diplomacy came too late in a Western reading drove a wedge between the US and Turkey and severely angered the Turkish side – not least because Turkey was sure to enjoy the backing of the Obama administration and because of prime minister Erdoğan's personally invested capital in this episode of Iran diplomacy. It also underlined once more very clearly the Turkish position that international sanctions can complement diplomacy, but always complicate political negotiations and should never be an end in themselves.[38] The institutional nature of sanctions is relevant here: UN-mandated international sanctions are seen with scepticism, but are not rejected categorically by Turkey. Unilateral sanctions, however, that are adopted by actors outside of UN structures, pertain to a security culture that Turkey was not supportive of, as this section will show. It is primarily the predicament of being dependent on Iranian energy economically that explains Turkey's sometimes ambivalent stance on US unilateral sanctions. Especially given Ankara's frustration with the lack of credit for its active Iran diplomacy and the adoption of sanctions resolution 1929, it has therefore been speculated by US officials that Turkey might offset the effect of sanctions on Iran by promoting trade relations with Iran and thereby undermine Western sanctioning efforts. Besides the energy dimension which will be elucidated in the following paragraphs, Iran is economically relevant for Turkey as a transit for Turkish trucks heading to Central Asia, for the Turkish tourism industry, and for Turkish companies' plans to invest in Iranian construction and service sectors (Larrabee and Nader 2013: 32; Ülgen 2011: 158). It was also against this backdrop that Turkey welcomed the JCPOA from an economic perspective. 'If the nuclear deal makes the West and Iran friends, it will help our economic relationship gain pace', a Turkish official remarked in an interview, referring to the lifting of sanctions in January 2016 on 'Implementation Day' of the JCPOA.[39]

What was central to US allegations of 'sanctions-busting' while nuclear-related US sanctions against non-US entities were still in place was the perception of Turkey becoming a financial lifeline at a time when US authorities were stepping up their efforts to isolate Iran from international payment transfers. When Dubai started imposing restrictions on transit goods destined for Iran, the Turkish foreign economic relations board announced its willingness to step in and provide Turkish ports for those shipments (Raphaeli 2010). This led to US concerns about Turkish subterfuges to continue with trade relations that would undermine US efforts to dry up financial lifelines to Iran. The most publicly discussed annoyance to the US in this regard has been Turkey's 'gold-for-gas' trade with Iran, in which Turkey had been exporting gold to Iran via Dubai as an indirect payment for Iranian natural gas deliveries and as a means to circumvent sanctions regimes that were aimed to cut off Iran from the international banking system (Kandemir 2013). Relevant US sanctions legislation in this regard was the 2012 'Iran Threat Reduction and Syria Human Rights Act' that essentially was meant to cut off Iran from international payment in US dollars for its oil sales. Turkey had therefore paid Iran in Turkish Lira, held in Halkbank accounts, with which Iran in turn had been buying gold from Turkey. When the US tightened its control on the sales of precious metals in the beginning of 2013, a motivation explicitly was to stop the alleged Turkish attempt to find loopholes in the US sanctions regime (ibid.; Daly 2013). As then-US Treasury Department undersecretary for terrorism and financial intelligence David Cohen told the House of Representative's Foreign Relations Committee in May 2013: 'I can assure you we are looking very carefully at evidence that anyone outside of Iran is selling gold to the government of Iran' (ibid.). And in spite of international optimism regarding the implementation of the Joint Plan of Action as part of the interim nuclear agreement struck between Iran and the P5 + 1 in November 2013 in Geneva, the US Treasury Department added Turkish businesses and individuals to a list of violators of US Iran sanctions in early February 2014 (Gladstone 2014).

In frank words that are indicative of the Turkish government's perception of the legitimacy of unilateral sanctions, a foreign ministry official commented on the 'gold-for-gas' controversy in an interview as follows: 'if the criticism is that we undermine unilateral sanctions, then I'm sorry: Nobody is at the centerpiece of the world. The US is not making laws for the world community.'[40] The same official sums up Turkey's stance on the sanctions question as follows:

> Concerning companies' exports [of potentially sanctioned goods], they always ask for approval by the foreign ministry. If sanctions are mandated by the UN Security Council, we say no. If companies, however, are affected by unilateral sanctions, it is their decision. In the resolution text, a 'restrictive/limited approach' is mentioned. We don't intervene in companies' dealings.

The distinction here is that of unilaterally adopted sanctions that run counter to *norms* Turkey subscribed to as governing international relations (sovereignty, non-interference), versus UN-adopted international sanctions that adhere to a *rules*-based international order. Legitimate 'institutions' in a Turkish understanding are those of the UN system that legitimise the penetration of sovereign rights, if deemed necessary and if so decided by the Security Council. But Turkey resisted the nature of US sanctions enforcement in its appeal to regulate world trade relations ('The US is not making laws for the world community').

However, Turkish officials explain that Turkey has to strike a balance between showing solidarity with the US, disagreeing with the latter's sanctions policies, and the imperatives of geography.[41] Official interviewees confirmed that US officials have regularly asked Turkey to reduce their oil imports from Iran.[42] While Turkey shows receptiveness to the idea by discussing import reductions and diversification plans,[43] Turkey is regarding neighbouring Iran as an inevitable trading partner. 'Iran simply has the best oil', as formulated by a foreign ministry official.[44] Iran is Turkey's second largest oil and gas supplier after Russia. This is expected to remain the case, despite the increased oil output levels by Iran after the lifting of sanctions in January 2016, and despite the temporary deterioration in Turkish–Russian relations following the shooting down of a Russian military jet by Turkey in November 2015.[45] Turkey is importing around 20 per cent of its gas and around 40 per cent of its oil from Iran, a 'US report required by section 505 (a) of the Iran Threat Reduction and Syria Human Rights Act of 2012' noted (US Energy Information Administration 2012: 6; 2013: 2). The report goes on to explore alternatives for Turkish gas supplies that would allow Turkey to cut down on its energy imports from Iran. The Turkish dependence on oil and gas from Iran was an issue of permanent contention in Turkish–American relations with a view to strengthening the effect of unilateral US Iran sanctions.[46] In March 2012, Turkey announced for the first time a planned reduction of oil imports from Iran by 20 per cent and engaged in talks with Libya and Saudi Arabia to make up for the resulting shortage (see Habertürk 2012).[47] A 20 per cent Iranian oil import reduction is the amount required to qualify for US 'sanctions waivers' as outlined in section 1245 of the US National Defense Authorization Act (NDAA) (US Department of State 2013). Turkey was subsequently granted a 'sanctions waiver' on 11 June 2012 which was renewed for another 180 days on 7 December 2012 (Daly 2013). 'We have lowered our energy imports from Iran to escape financial sanctions. The US gave us this leeway, because they understood our challenge', a Turkish official explained in an interview.[48]

This receptiveness to US sanctions was a remarkable development in Turkey's approach to unilateral US sanctions – Ankara had previously always stated that it saw itself bound by UN sanctions only, as analysed above. Enforcement of unilateral US and EU sanctions had always been left up to the Turkish private sector (Ülgen 2012: 9). The Turkish foreign ministry 'informs private companies of the impact of new sanctions', as formulated repeatedly by foreign ministry officials interviewed in this project.[49] While leaving sanctions enforcement up to the private sector can be read as a convenient way to guard a political level of passivity on the part of the Turkish government as concerns pressure on Iran, it is worth pointing out that the oil and gas company BOTAŞ Petroleum Pipeline Corporation is wholly state-owned since 1995. Turkey's oil refiner Tüpraş is also government-owned, and a Turkish diplomat confirms in an interview that 'upon request by the US, Tüpraş has decreased its oil imports from Iran'.[50] Compliance with unilateral US energy sanctions in order to qualify for 'sanctions waivers' therefore is a deliberate act of demonstrating solidarity with the US on a governmental level. As confirmed by a Turkish foreign ministry official working on Turkish–Iranian bilateral issues, cutting down on oil trade with Iran, as in 2012, is 'a ministerial decision'.[51] US concerns, the same official went on to state, 'are taken into account in some sectors' (ibid.). In a similar vein, a Turkish diplomat to the US points out that the Turkish–Iranian trade volume fell from $22 billion in 2012 to $18 billion in 2013 because of US unilateral sanctions. Scaling back trade with Iran was, the same diplomat asserts, a 'governmental decision', and adds: 'You have to do that. The United States has such a power in the world economy that your companies suffer otherwise'.[52] Another Turkish official thus explains the rationale for bowing to US financial power: 'They print the dollars, you know.' The power of US financial sanctions stems from the ubiquity of the US dollar as the international reserve currency – and the global payment method in dollars for resource commodities like oil and gas.

Another instance of Turkey reacting to US pressure was its decision in 2008 to cancel the agreement with Iran for the Turkish Petroleum Company (TPAO) to invest US$5.5 billion and operate in the South Pars oil fields (Kardas 2010; Larrabee and Nader 2013: 31). Such an investment would have 'violated' the US Iran Sanctions Act (ISA) that foresees sanctions against foreign companies investing more than US$20 million in the Iranian energy sector. The Turkish renouncement of the investment signalled that Turkish investment in the Iranian oil infrastructure was sacrificed for a politically higher-valued alignment with US interests concerning Iran.

A publicly demonstrated normative divergence from US power structures, and ambivalent cases such as the 'gold-for-gas' deal discussed above, did not prevent Turkey from demonstrating a behavioural convergence with the very same power structures. While emphasising Turkey's conviction that 'non-interference' in Iran's domestic affairs should guide the search for political solutions to the Iranian nuclear crisis, Ankara underscores the importance of solidarity with its US ally and even shows compliance with US unilateral sanctions. 'Now under the JCPOA, the import of Iranian oil and gas is no longer prohibited. Of course we welcome that', a Turkish official explains in an interview, referring to the lifting of US nuclear-related secondary sanctions on 'Implementation Day' that now renders such balancing acts between energy import policies and compliance with sanctions regimes obsolete.[53] '[T]he United States has ceased efforts to reduce Iran's crude oil sales, including limitations on the quantities of Iranian crude oil sold and the nations that can purchase Iranian crude oil', a US Treasury Department guidance note reads (2016: 13).

Other Turkish investment projects in the Iranian market, however, may still be put on hold because companies are cautious not to fall under US financial sanctions that stay in place because of 'sponsorship of terrorism' charges and human rights violations. Moreover, the 'foreign subsidiary' classification of US companies may prove a challenging task to prove in sanctions avoidance: Companies with links to the US will have to 'firewall' their business in Iran between domestic and foreign operations to avoid being caught by US financial sanctions. And 'probably almost all Turkish companies have some US component like email servers or software', as a Turkish diplomat acknowledges in an interview.[54] Given the legal uncertainties pertaining to this regulation, companies may continue to be overly cautious in the Iran business, even if they qualify for a general license.

4 Turkey's Iran policy between ideology, geostrategy and alliance management

The history of Turkish–Iranian relations is a history of a balancing dynamic between competition and cooperation. Their mutual border has not shifted since the 1639 Treaty of Qasr-e Shirin. Both countries have managed to balance geopolitical differences, diverging ideological outlooks and converging economic interests. Likewise, Turkish foreign policy is a multidirectional one that has to reconcile distinctive regional policies with an embeddedness in international alliance structures. In practice, such structural constraints find expression in compliance with but criticism of Iran sanctions, energy trade with Iran with the simultaneous planning of import reductions and supply-side diversifications, and US-critical discourses with a concurrent commitment to NATO alliance structures and security identities. Turkey's Iran policy after 2002 often was a balancing act between geostrategic pragmatism, ideational foreign policy projections and Western alliance management. This balancing act helps explain the research finding of a mixed rules convergence with hegemonic structures which concluded the previous section. As administrations in Western capitals following the 2010 Tehran declaration publicly started questioning Turkey's stance in the Iranian nuclear file and reproaching Ankara with having been outsmarted by Iranian deception (Parsi 2012: 195f.),[55] Turkey's abandonment of a mediator role thereafter and 'closer communication with the Americans'[56] as a result can be read as a desire to avoid getting their hands burnt by playing a pro-active role that is not wanted politically by the P5+1 format – an eventual rules convergence, in other words, with those policies that Ankara had tried to overcome with its mediation in the first place.

In addition, Turkey's power to defuse the Iranian nuclear crisis and its ability to mediate between the P5+1 parties and Iran was further eroded in the wake of power shifts in the region starting with the outbreak of what has come to be coined the 'Arab Spring' that saw Iran and Turkey oppose each other with fundamentally differing perceptions of regional order – as epitomised by their diametrically opposed interests especially in the Syrian civil war (Barkey 2013; Kibaroğlu 2012; Larrabee and Nader 2013: 8f.; Pieper 2013; Taşpınar 2012). While Iran remains one of Assad's latest steadfast allies in the region,[57] Turkey has positioned itself increasingly as more outspoken against any future prospects of Assad holding power in Syria. After an initial Turkish support for Assad, Ankara has officially called the Assad regime an illegitimate one, with Erdoğan even calling on Assad to step down (Taşpınar 2012: 137). In harsh terms, Prime Minister Davutoğlu criticised the Western failure to find common ground in opposing Assad. With the so-called 'Islamic State' becoming a much bigger security challenge on the agenda of Western policy circles, Davutoğlu lamented calculations in Western capitals that Assad could eventually be considered a lesser evil and, in the face of the terrorist challenge posed by 'the IS', 'be seen as the good guy and stay in Damascus'.[58]

Such resentment over wider regional policies spilled over into Iran's perception of Turkey's role in the nuclear crisis. The reason Istanbul was no longer wanted as a venue by Iran for the February and April 2013 nuclear negotiations (which took place in Almaty, Kazakhstan) lay in differences of the two countries' positions toward Syria.[59] Only a few years after Iranian receptiveness to Turkish diplomatic activism, Turkey was sidelined as a credible broker in the Iran file by the course of events (Alcaro 2013: 98). On the one hand, Turkey started siding more pro-actively with political actors in the wake of societal upheavals throughout the Arab world that destroyed what was left of Turkey's perception as a neutral broker (Seufert 2014).[60] On the other, Turkey and Iran face each other in grim opposition concerning geostrategic conceptions in the Middle East. It is quite telling that in 2013 and 2014, Oman acted as a conduit of messages between the US and Iran[61] – a role that, five years ago, was Turkey's. Instead of 'zero problems with neighbours', Turkey now faced zero neighbours without problems.

It was in this context of Turkish–Iranian diverging agendas for regional policies that president Erdoğan's outburst against Iran following the conclusion of the political framework agreement on Iran's nuclear programme on 2 April

2015 had to be understood: While foreign minister Çavuşoğlu was one of the first to applaud the outcome of the nuclear talks, Erdoğan warned of a boost to Iran's role in the region and criticised Iran's policies in Yemen, Syria, and Iraq (Pamuk 2015). Such statements were not linked to the negotiated outcome on the nuclear file per se, but were aimed at mending fences with Saudi Arabia and a wider 'Sunni bloc' of Arab states in the region (Arslan 2015). Saudi–Turkish relations had suffered after Turkey had taken sides for the Muslim Brotherhood in Egypt, which the Kingdom in Riyadh considers a terrorist organisation. Iran-critical statements coming from Turkey are music in the ears of the Saudi kingdom, which is a regional rival of Iran and likes to see Iran's role in the region diminished. As Ülgen holds, closer Saudi–Turkish relations are beneficial for Turkey's regional agenda, as they help confront common challenges, including 'the Islamic State', and the pro-Iranian Houthi militia in Yemen.[62] In reaction to Turkey's public support for Saudi policies in Yemen, 65 Iranian parliamentarians had demanded a cancellation of Erdoğan's planned visit to Iran in early April 2015 (Bar'el 2015). The visit eventually did take place despite the preceding war of words. Turkey's Iran policies, as such considerations demonstrate, are often the outcome of partially conflicting policy priorities on a number of often diverse issues (security, economic) and on different levels (global, regional).

Fences in sour Turkish–Iranian relations began to be mended after Hassan Rouhani was elected president in Iran in June 2013. Rouhani's pledges to follow policies of 'prudence and moderation' enabled a tentative Turkish–Iranian rapprochement. In an interview with Al-Jazeera on 12 February 2014, Erdoğan called Iran and Turkey 'strategic partners', and explicitly underlined that purchasing oil and natural gas is part and parcel of such a partnership.[63] This was followed by a visit Erdoğan paid to Iran on 28–29 January 2014, where he signed trade deals in the hope of increasing energy ties – much to the dislike of the US administration, which again must have seen the potential of its sanctions regime being undermined (Hafezi 2014). The visit was reciprocated in June 2014 by a visit of Rouhani to Turkey – the first visit of an Iranian president to Turkey since 1996. The interest in expanding bilateral relations, including increases in annual trade volumes, was also underlined at Iranian foreign minister Zarif's visit to Turkey in March 2016 (Turkish foreign ministry 2016). The lifting of nuclear-related Iran sanctions as part of the JCPOA's implementation is seen as an opportunity to expand and deepen Turkish–Iranian trade relations. Turkey has no active role in the agreement's implementation, but welcomes the positive effects that a post-sanctions environment can bring about.[64] With the deepening of economic relations facilitated in a post-sanctions environment, Turkey and Iran announced their intention to increase their mutual trade volume to US$30 billion (Fars News 2016). Despite disagreements over their respective regional policies, the Turkish and Iranian government regard each other as significant trade partners and, indeed, important anchors of stability in a region that has been troubled by virulent political changes. This helps explain the fact that the Iranian government was the first to condemn the coup attempt against the Turkish government of 15 July 2016 (Hashem 2016).

Finally, when conceptualising clashes between Turkish Iran policies and Western governments' policies towards the region, it might be worth pointing out that a tendency for public defiance of US policies was stronger in the prime minister's office under Erdoğan's leadership than in the foreign ministry.[65] Not known for mincing his words, Erdoğan was a prime minister who displays more consistency between a discursive and a behavioural level of resistance to hegemony – and retains that characteristic in the president's office. Barkey (2013) even states that Erdoğan 'has [at times] articulated positions that mirror Tehran's' (149). And among experts and Turkey watchers, it has been an open secret that within the Turkish cabinet, influential actors take a more pro-Iran stance than others. The former head of the Turkish intelligence agency (Milli İstihbarat Teskilati, MİT), Hakan Fidan, was renowned as one of the most influential pro-Iran actors and a close confidant of then-Prime Minister Erdoğan. Experts had long speculated that Turkey's Iran policy constituted a certain *domaine réservé* for the former MİT-head, where he had enjoyed a disproportionately big leeway,[66] even to the point where he had been alleged to pass on intelligence information to the Iranian government (Entous and Parkinson 2013). The impression that otherwise objectionable interactions with Iran might have enjoyed the backing from high government officials was nurtured further when the state prosecutor's investigations had 'exposed serious connections between Iranian agents and senior government officials,' Turkish media wrote in February 2015 (Today's Zaman 2015).[67] Tellingly, the investigations were brought to a halt by the Turkish government.[68]

While difficult to verify, it thus cannot be excluded that personal politics and even 'networks' with vested interests have factored in as a variable to influence Turkey's overall foreign policy towards Iran. Turkish diplomats in turn, in interviews for this research project and elsewhere, tend to emphasise the importance of solidarity with the US as an important regional and global ally – even when Turkish compliance with certain US-inspired policies, as the sanctions question has shown, proves to be 'costly'.[69] At times the object of tense discussions between US and Turkish officials, Turkey's regional policies are bound to partially conflict with the US and other NATO partners. Public disagreements about Turkey's domestic politics have played up these differences in regional policies. Here, a growing Western-Turkish alienation in policies towards Syria, Iraq and Iran has overlapped with Western concerns about Turkey's societal and domestic policies, especially in the aftermath of the averted coup attempt in July 2016.

5 Conclusion

From the outset of the Iranian nuclear crisis in 2002, Turkey was cautious with criticism of the Iranian government and did not follow public positions at the time that emphasised Iran's violations of its IAEA Safeguards Agreement. Ankara reiterated the importance of the Westphalian principles of 'sovereignty' and 'non-interference' to govern international relations and emphasised Iran's rights to civilian nuclear energy as an NPT signatory. Turkey's public positioning on the emerging nuclear crisis coincided with a regional foreign policy re-orientation that sought a deepening of relations with Turkey's immediate neighbourhood. This new foreign policy outlook was epitomised by the two concepts of 'Strategic Depth' and 'zero problems with neighbours', albeit more so for public relations purposes than for operational policy implementation.

But while implementing what they saw as a 'multidirectional' foreign policy, Turkey's NATO membership gives it an inroad into Western security perspectives at the same time. This position lent Ankara credibility in the form of mediatory potential and enabled Turkey to facilitate talks between the P5 + 1 and Iran between 2007–2011, and to mediate in a period of shuttle diplomacy in 2009–2010. Absent a permanent UN Security Council seat and external to the P5 + 1 format for nuclear negotiations with Iran, Turkey's activism in the Iran nuclear file gave it 'leverage' and thus approximated its role to that of other, institutionally more established, actors, in Iran diplomacy. It was an exercise of Great Power diplomacy and demonstration of Turkey's foreign policy self-perception and portrayal, and was an attempt to avert the imposition of a new round of sanctions.

The May 2010 Tehran declaration to defuse the Iranian nuclear crisis was an intriguing precedent of a non-Western attempt to ease tensions and make use of Turkey's political capital in Tehran. Yet, the mediatory efforts by Brazil and Turkey did not encounter much appreciation in Western capitals. The episode stood indicative of Turkey's insistence on the need to resolve the Iranian nuclear crisis through political dialogue and diplomacy on a discursive level and resist self-perpetuating hegemonic power structures on a behavioural level that had hitherto obstructed meaningful attempts at crisis diplomacy. Turkey's short mediatory role ended with the passing of UNSCR 1929 in June 2010, an outcome that infuriated Ankara because it illustrated the Western insincerity in finding a genuine solution to the Iran crisis, in a Turkish reading. Turkey's criticism of Western impatience and the Turkish vote against UNSCR 1929 in June 2010 forcefully made this point also on a behavioural level. Voting against this sanctions resolution as a non-permanent Security Council member at the time, Turkey drove home a strong message of resistance. Turkey's retreat to the role of facilitator and closer alignment with US positions thereafter, however, underlined again Ankara's official re-iteration of the importance of solidarity with its US ally. Resistance to US positions on a behavioural level, in other words, never was a consistent policy imperative for Turkey.

The same conclusion can be drawn from the examination of Turkey's position on international and unilateral Iran sanctions presented in this chapter. Turkey's diplomacy on the sanctions issue was a balancing line between disagreeing with a securitisation of the Iranian nuclear file and resulting sanctions regimes on the one hand and still complying with Western sanctions efforts, on the other. Here, a distinction between the institutional nature of sanctions has been made: The Turkish government has implemented provisions of UN-mandated Iran sanctions, but reiterated not to be bound by unilateral sanctions. Imposed outside of UN structures, these are seen as circumventing the *rules* of the UN system.

Turkey has consistently been criticising the imposition of sanctions before and after the 2010 Tehran declaration also because of economically adverse effects on the significant Turkish–Iranian trade relations. The government 'informs' the Turkish private sector of the impact and consequences of US unilateral sanctions, so the official stance. Resistance to financial coercion in the form of extraterritorialised US legislation took place on a behavioural level by Turkish evasion mechanisms. Such circumventions, like the controversial 'gold-for-gas' deal, never were consistently outright counter-hegemonic policies by design. As has been shown, Turkey showed responsiveness to US allegations of 'sanctions-busting', and eventually complied with US unilateral sanctions regimes despite Turkey's high dependence on Iranian oil and gas supplies. As a consequence, Turkey qualified for US 'sanctions waivers' because of a reduced level of Iranian oil imports. Turkish Iran policies are the outcome of a complex calculation of economic pragmatism, regional foreign policy imperatives, international alliance structures and a politico-cultural sensibility for Iranian resentment to Western pressure. This chapter has shown how a public advocacy for a non-hegemonic security culture can be paralleled by a foreign policy behaviour that falls short of acting upon this discourse. A basic adherence to US power structures eventually prevailed over the possibility of an outright resistance to hegemony.

Disagreements between Iran and Turkey after 2011 in regional policies, and here especially over approaches to the war in Syria, saw Turkey's potential as a facilitator in the nuclear talks decrease. While the election of Rouhani has enabled a gradual rapprochement between Ankara and Tehran also in regional matters, the Turkish government has watched from the sidelines as the final elements for a nuclear accord fell into place in 2015. Turkey also plays no role in the implementation of the JCPOA, but benefits from the opportunities that the lifting of economic and financial sanctions has brought for Turkish trade relations with Iran. Turkish Iran policy, in other words, no longer has to balance between pursuing a foreign policy seen as defiant by Western sanctions-monitors and appeasing to US decision-makers.

Notes

1 The NSG, created in 1974, is a group of nuclear supplier countries that sets guidelines for nuclear and nuclear-related exports, aiming to strengthen the global non-proliferation regime.
2 Author's interview with high-ranking Turkish diplomat to the IAEA, over phone, 23 August 2013.
3 Author's interview with Turkish foreign ministry official, Ankara, 20 August 2014.
4 Modified Code 3.1 to the IAEA's Subsidiary Arrangement stipulates that signatories ought to inform the agency about the construction of new nuclear facilities – a provision that Iran did not see itself bound to. Following the JCPOA in July 2015, however, Iran agreed to comply with modified Code 3.1.
5 The importance of these principles has been emphasised again by Turkish prime minister Davutoğlu at a keynote speech in Brussels, organised by Friends of Europe, 15 January 2015.

6 SPIEGEL Interview with Turkish prime minister Recep Tayyip Erdoğan. Available at www.spiegel.de/international/europe/spiegel-interview-with-turkish-prime-minister-recep-tayyip-erdogan-if-the-eu-doesn-t-want-us-they-should-say-it-now-a-477448.html (accessed 24 March 2015).

7 Another implicit motivation in Turkish–Iranian relations likely was the fathoming of a Turkish–Iranian approximation of their policies towards 'the Kurdish question' (Aras 2001: 109–111; Scholl-Latour 2001: 177). Especially after PKK's leader Abdullah Öcalan's expulsion from Syria and ensuing arrest by Turkish authorities, Turkey started suspecting Iran more than Syria of providing sanctuary for PKK fighters (Olson 2004: 48). At the same time, Iran feared that Turkey could play the 'Azeri card' and stir up separatist sentiments in the Iranian Northern provinces with a strong Azeri population (ibid., 11). And with PJAK, an offshoot of the PKK, waging an armed struggle against Iranian authorities since 2004, Iran may have had a stronger incentive to coordinate 'Kurdish' policies with Turkey. The war in Syria from 2011 has damaged this dynamic. Iran's support for the Patriotic Union of Kurdistan (PUK) brings geopolitical tensions with the Turkish government (Rafizadeh 2016: 110).

8 Author's interview with high-ranking Turkish diplomat, Ankara, 17 June 2013; author's interview with Turkish diplomat, Turkish embassy Beirut, 14 August 2014.

9 Author's interview with high-ranking Turkish diplomat to the IAEA, by phone, 23 August 2013.

10 Author's interview, Ankara, 17 June 2013. Also formulated almost identically by another Turkish foreign ministry official at the Turkish embassy in Washington, author's interview, Washington, 14 February 2014.

11 Even though the Iran nuclear issue is not on the official policy planning agenda of NATO, it is always a topical issue, as a NATO official explained in an interview. Author's interview, Brussels, 29 January 2015. Tertrais (2010/2011) writes that a nuclear Iran would 'undoubtedly have an impact on NATO's internal nuclear debate. A possible outcome of these deliberations would be for NATO nuclear weapons to "move South"; the weapons would be maintained in Italy and Turkey' (47). In the remainder of his article, he develops scenarios for Turkey's possible reaction to a nuclear Iran. With Turkey being a NATO member, Turkey's policy planning in this regard has far-reaching implications for the alliance (55–56).

12 According to a Turkish high-ranking diplomat, the Turkish agreement to the radar system was also tied to the insistence that it should not be specified against which potential threats the defense shield was targeted. Author's interview, Brussels, 29 May 2013. Acknowledging Iranian concerns, another Turkish foreign ministry official conceded: 'Of course, they don't like rockets on their borders'. Author's interview, Ankara, 17 June 2013.

13 It has also been argued elsewhere that Turkey's 'nuclearisation', when faced with a possible Iranian nuclear bomb, would only be prevented by the credibility of NATO and US security guarantees (Baran and Lesser 2009: 213; Perkovich and Ülgen 2015b; Udum 2007; Ülgen 2012: 23). In the context of debates about the likeliness of a regional arms race with a nuclear armed Iran, Turkey would abstain from striving to counterbalance as long as it can effectively rely on NATO's 'extended deterrence' clause and Article V of the NATO Charter (Author's interview with Sinan Ülgen, visiting scholar Carnegie Europe, Brussels, 6 May 2013; Author's interview with Dr Şebnem Udum, expert at Hacettepe University, Ankara, 18 June 2013). Others argue that even NATO's 'nuclear umbrella' would not prevent Turkey from nuclearising when faced with a nuclear Iran, as this would severely unsettle the regional balance to Turkey's disadvantage (Author's interview with Dr Meliha Altunışık, expert at Middle East Technical University, Ankara, 14 June 2013; Author's interview with Prof. Dr Hüseyin Bağcı, expert on Turkish foreign policy, head of department for International Relations, Middle East Technical University, Ankara, 18 June 2013). See also Perkovich's and Ülgen's edited book (2015a) on Turkey's Nuclear Future.

14 Author's interview with high-ranking Turkish diplomat, Ankara, 17 June 2013; author's interview with high-ranking Turkish diplomat, Ankara, 20 August 2014.
15 As also reiterated by a Turkish foreign ministry official. Author's interview, Ankara, 20 August 2014.
16 Author's interview with high-ranking Turkish diplomat to the IAEA, by phone, 23 August 2013.
17 Erdoğan's Addresses to the 69th UN General Assembly. Available at www.tccb.gov.tr/news/397/91138/president-erdogan-addrebes-the-un-general-abembly.html (accessed 8 January 2015); Erdoğan's Addresses to the 71th UN General Assembly. Available at www.tccb.gov.tr/en/news/542/52361/president-erdogan-addresses-the-un-general-assembly.html (accessed 25 September 2016). Elaborating on this statement in responses to his address at the World Economic Forum in 2014, he said: '(The Security Council) should not be a platform where only five countries have a say. Such a system is not fair at all. This world is not a slave of these five permanent countries.' Summary of the transcript available at www.tccb.gov.tr/news/397/91194/president-erdogan-reiterates-turkeys-stance-in-regional-ibues.html (accessed 24 March 2015).
18 As he did at a 'Policy Spotlight' keynote speech in Brussels, organised by Friends of Europe, 15 January 2015.
19 Author's interview with Turkish foreign ministry official, Ankara, 20 August 2014.
20 In this context, see also former Turkish foreign minister Davutoğlu's article 'Turkey's Mediation: Critical Reflections from the Field' (2013).
21 Author's interview, Ankara, 20 August 2014.
22 Author's interview with high-ranking Turkish diplomat, Ankara, 17 June 2013; author's interview with Prof. Dr Hüseyin Bağcı, Ankara, 18 June 2013; author's interview with Turkish foreign ministry official, Ankara, 20 August 2014.
23 Author's interview with EEAS official, Brussels, 4 June 2013.
24 Author's interview with Dr Walter Posch, senior Iran analyst, SWP, Berlin, 25 June 2013; author's interview with high-ranking European diplomat to the IAEA, Vienna, 13 August 2013.
25 Author's interview with Dr Ali Vaez, via Skype, 25 July 2013. Similar points were made by Soli Özel, foreign affairs editor at Habertürk and lecturer at Kadir Has University, author's interview, Istanbul, 7 September 2013.
26 As also stated by a Turkish foreign ministry official. Author's interview, Ankara, 20 August 2014.
27 Author's interview with Turkish high-ranking Turkish foreign ministry official, Ankara, 17 June 2013. As another Turkish official puts it: 'Iranian and Turkish markets are compatible, we have cultural affinities, we even eat the same kind of kebabs.' Author's interview, London, 30 March 2016.
28 Author's interview with Turkish high-ranking diplomat, Brussels, 29 May 2013. In her memoir, Hillary Clinton (2014) admits that 'throughout the spring of 2010 we worked *aggressively* to round up the votes' (427) [for UNSC sanctions, emphasis added].
29 The US State Department and other critics of the 2010 proposed swap deal argued that the Iranian decision for a 20 per cent uranium enrichment was left unaddressed, that Iran's uranium stockpile had increased since the latest Vienna group proposal of October 2009 ('time has overtaken the original proposal'), and that the Tehran declaration did not address Iranian past defiance of UNSC resolutions (see US Department of State 2010). See also Fitzpatrick (2010: 79).
30 Author's interview with Turkish high-ranking diplomat, Brussels, 29 May 2013.
31 Ibid.; point also made by Soli Özel, Foreign affairs editor at Habertürk and lecturer at Kadir Has University, author's interview, Istanbul, 7 September 2013. The US take on the proposed swap deal at the time was that 'this is not Turkey's league to play in', he put it. In a similar vein, Hillary Clinton (2014) describes Turkey and Brazil as

'prime examples of the "emerging powers" whose rapid economic growth was fueling big ambitions for regional and global clout' (430). The implicitly dismissive tone becomes even more discernible as she goes on to say that 'They also happened to have two confident leaders in Luiz Inácio Lula da Silva of Brazil and Recep Tayyip Erdoğan of Turkey, both of whom considered themselves to be men of action able to bend history to their will' (ibid.).

32 Author's interview with former E3 delegation member, Brussels, 29 January 2015. His reference to the 'Swiss initiative' refers to the 'Guldimann memorandum' of 2003. In it, Iran offered the US government an unprecedented level of cooperation on all issues of concern, full transparency with the IAEA and guarantees for the peaceful nature of Iran's nuclear programme, and a 'roadmap' detailing proposals for negotiations to break the 'wall of silence' between Iran and the US. The Swiss ambassador to Tehran, Tim Guldimann, transmitted the memorandum to the US State department. In the White House, however, it fell on deaf ears.
33 Author's interview with high-ranking Turkish diplomat to the IAEA, by phone, 23 August 2013.
34 Author's interview with Turkish high-ranking diplomat, Brussels, 29 May 2013; 'the US has learnt to understand Turkey better since 2010', as one Turkish foreign ministry official put it, Interview, Ankara, 17 June 2013.
35 Author's interview with Dr Meliha Altunışık, Ankara, 14 June 2013.
36 Foreign affairs editor at Habertürk and lecturer at Kadir Has University, author's interview, Istanbul, 7 September 2013.
37 Author's interview with Turkish official, London, 30 March 2016.
38 Author's interview with high-ranking Turkish diplomat, Ankara, 17 June 2013.
39 Author's interview, London, 30 March 2016.
40 Author's interview with Turkish foreign ministry official, Ankara, 20 August 2014.
41 Author's interview with Turkish high-ranking Turkish foreign ministry official, Ankara, 17 June 2013; author's interview with Turkish diplomat to the US, Washington, 14 February 2014.
42 Author's interview with Turkish high-ranking Turkish foreign ministry official, Ankara, 17 June 2013.
43 Supply diversification plans included ideas to increase oil imports from Northern Iraq. Leaving aside political and logistical problems, the interest was reciprocated from the Iraqi side. Interview with Dr Meliha Altunışık, Ankara, 14 June 2013. Author's interview with Turkish diplomat to the US, 14 February 2014; author's interview with Turkish foreign ministry official, Ankara, 20 August 2014.
44 Ibid.
45 Author's interview with Turkish diplomat, Brussels, 15 April 2016.
46 Author's interview with high-ranking Turkish foreign ministry official, Ankara, 17 June 2013.
47 These talks with Libya and Saudi Arabia, however, 'came to nothing', as confirmed by a Turkish foreign ministry official. Author's interview, Washington, 14 February 2014.
48 Author's interview with Turkish official, London, 30 March 2016.
49 Author's Interviews with Turkish diplomat, Ankara, 17 June 2013; Washington, 14 February 2014; Beirut, 14 August 2014, Ankara, 20 August 2014.
50 Author's interview with Turkish diplomat to the US, Washington, 30 October 2014.
51 Author's interview with Turkish foreign ministry official, Ankara, 20 August 2014. Another foreign ministry official describes the government's role in Tüpraş' trade with Iran as 'advisory'. Author's interview, Ankara, 20 August 2014.
52 Author's interview with Turkish diplomat to the US, Washington, 30 October 2014.
53 Author's interview, London, 30 March 2016.
54 Author's interview, Turkish delegation to the European Union, Brussels, 15 April 2016.
55 Turkey has even been branded 'Iran's lawyer' (Stein and Bleek 2012: 137).

56 Author's interview with Turkish high-ranking diplomat, Brussels, 29 May 2013.
57 President Rouhani, in June 2015, let it be known that Iran will support the Syrian government 'until the end of the road' (Reuters 2015).
58 Ahmet Davutoğlu, keynote speech in Brussels, organised by Friends of Europe, 15 January 2015.
59 As confirmed by a Turkish high-ranking diplomat. Author's interview, Brussels, 29 March 2013.
60 The taking of sides for Sunni Muslim forces in the region was perceived as biased, and culminated in Turkey's pro-Muslim Brotherhood position on Egypt, the uncompromising public support for Sunni anti-Assad actors in Syria (even if radical) and markedly sharper tones in rhetoric against Israel. 'Erdoğan's government departed too quickly from its own version of a foreign policy oriented at win–win results and overhastily plunged into a foreign policy determined by confessional identities and interests,' Günther Seufert (2014: 34) attests.
61 Author's interview with EEAS official, Brussels, 20 March 2014.
62 Ibid. On other issues, Turkish–Saudi convergence of interests is not always given. On the Syrian conflicts, both Turkey and Saudi Arabia share their enmity with Bashar al-Assad. However, Turkey and Qatar support opposition groups that are clashing with those backed by Saudi Arabia.
63 'Erdoğan: Turkey's role in the Middle East,' Talk to Al Jazeera, 12 February 2014. Available at www.aljazeera.com/programmes/talktojazeera/2014/02/erdogan-turkey-role-middle-east-201421282950445312.html (accessed 20 February 2014).
64 Author's interview with Turkish official, London, 30 March 2016; author's interview with Turkish diplomat, Brussels, 15 April 2016.
65 Erdoğan's term as prime minister ended in 2014. On 12 August 2014, he was elected as the 12th Turkish president, and as the first president to be elected by popular vote.
66 Author's interview with Dr Mark Hibbs, non-proliferation expert at the Carnegie Endowment for International Peace, Berlin, 16 March 2015; author's interview with Gökhan Bacık, expert at Ipek University, Ankara, 9 September 2013. See also Hibbs (2015).
67 Zaman is unlikely to produce such critical reports in the future, following its nationalisation in March 2016.
68 This had been made possible due to new legislation enabling a stronger political interference in the judiciary that was introduced in 2014, and had resulted in international concern over the politicisation of the Turkish judiciary. Among other things, the new law gives the government great clout in the nomination of judges and prosecutors. See Pamuk (2014).
69 Author's interview with Turkish diplomat to the US, Washington, 14 February 2014.

References

Akyol, Mustafa. 2015. Why Doesn't Turkey Speak up on Iran Nuclear Issue? *Turkey Pulse, Al monitor*, 26 February 2015. Available at www.al-monitor.com/pulse/originals/2015/02/turkey-geneva-nuclear-iran-un.html# (accessed 12 May 2015).

Alcaro, Riccardo. 2013. Friends and Foes of a United States-Iran Nuclear Agreement. *Turkish Policy Quarterly* vol. 12, no. 3: 93–101.

Altunışık, Meliha Benli. 2008. The Possibilities and Limits of Turkey's Soft Power in the Middle East. *Insight Turkey* vol. 10, no. 2: 41–54.

Aras, Bülent. 2001. Turkish Foreign Policy towards Iran: Ideology and Foreign Policy in Flux. *Journal of Third World Studies* vol. 18, no. 1: 105–124.

Arslan, Deniz. 2015. Iran Nuclear Talks May Pose Opportunities and challenges for Turkey. *Today's Zaman*, 1 April 2015. Available at www.todayszaman.com/diplomacy_iran-nuclear-talks-may-pose-opportunities-and-challenges-for-turkey_376894.html?mkt_tok=

3RkMMJWWfF9wsRoivqjOZKXonjHpfsX64u8rUKCg38431UFwdcjKPmjr1YcFTcp0aPyQAgobGp5I5FEIQ7XYTLB2t60MWA%3D%3D (accessed 7 April 2015).

Aykan, Mahmut Bali. 2007. A Retrospective Analysis of Turkey-United States Relations in the Wake of the US War in Iraq in March 2003. In: Guney, Nursin Atesoglu (ed.), *Contentious Issues of Security and the Future of Turkey*. Hampshire: Ashgate, pp. 51–70.

Baran, Zeyno and Lesser, Ian O. 2009. Turkey's Identity and Strategy: A Game of Three-Dimensional Chess. In: Michael Schiffer and David Shorr (eds), *Powers and Principles. International Leadership in a Shrinking World*. Plymouth: Lexington Books, pp. 197–224.

Bar'el, Zvi. 2015. Erdogan in Tehran: Turkey Wants to Dance at Every Mideast Wedding. *Haaretz*, 8 April 2015. Available at www.haaretz.com/news/middle-east/1.650923 (accessed 9 May 2015).

Barkey, Henri J. 2013. Turkish–Iranian Competition after the Arab Spring. *Survival* vol. 54, no. 6: 139–162.

Burton, John W. 1969. *Conflict and Communication: The Use of Controlled Communication in International Relations*. New York: The Free Press.

Cetinsaya, Gökhan. 2003. Essential Friends and Natural Enemies: The Historic Roots of Turkish–Iranian Relations. *Middle East Review of International Affairs* vol. 7, no. 3: 116–132.

Clinton, Hillary Rodham. 2014. *Hard Choices*. New York: Simon & Schuster.

Daly, John. 2013. How Far Will Turkey Go in Supporting Sanctions Against Iran? *The Turkey Analyst* vol. 6, no. 13. Available at www.turkeyanalyst.org/publications/turkey-analyst-articles/item/48-how-far-will-turkey-go-in-supporting-sanctions-against-iran?.html (accessed 8 March 2014).

Davutoğlu, Ahmet. 2001. *Stratejik derinlik: Türkiye'nin uluslararası konumu* (Strategic Depth: Turkish Foreign Policy). Istanbul: Küre Yayınları.

Davutoğlu, Ahmet. 2013. Turkey's Mediation: Critical Reflections from the Field. *Middle East Policy* vol. 20, no. 1: 83–90.

Davutoğlu, Ahmet and Amorim, Ceslo 2010. Giving Diplomacy a Chance. *International Herald Tribune*, 26 May 2010. Available at www.nytimes.com/2010/05/27/opinion/27iht-eddavutoglu.html?_r=0 (accessed 29 July 2014).

Entous, Adam and Parkinson, Joe. 2013. Turkey's Spymaster Plots Own Course on Syria. *The Wall Street Journal*, 10 October 2013. Available at www.wsj.com/news/articles/SB10001424052702303643304579107373585228330 (accessed 15 April 2015).

Fars News. 2016. *Iran, Turkey to Boost Trade Exchanges to $30bln*. Fars News Agency, 17 June 2016. Available at http://en.farsnews.com/newstext.aspx?nn=13950328000682 (accessed 28 September 2016).

Fisher, Ronald J. 1972. Third Party Consultation: A Method for the Study and Resolution of Conflict. *Journal of Conflict Resolution* vol. 16: 67–94.

Fitzpatrick, Mark. 2010. Iran: The Fragile Promises of the Fuel-Swap Plan. *Survival* vol. 52, no. 3: 67–94.

Fuller, Graham E. 2008. *The New Turkish Republic: Turkey as a Pivotal State in the Muslim World*. Washington, D.C.: United States Institute of Peace.

Giragosian, Richard. 2008. Redefining Turkey's Strategic Orientation. *Turkish Policy Quarterly* vol. 6, no. 4: 33–40.

Gladstone, Rick. 2014. US Issues Penalties Tied to Iran Sanctions. *International New York Times*, 8–9 February 2014.

Gul, Abdullah. 2011. Israel a Burden on its Allies, Turkish President Gül Says. *Today's Zaman* November 23. Available at www.todayszaman.com/news-263676-israel-a-burden-on-its-allies-turkish-president-gul-says.html (accessed 29 April 2013).

Gunter, Michael M. 2005. The US–Turkish Alliance in Disarray. *World Affairs* vol. 167, no. 3: 113–123.

Gürzel, Aylin. 2012. Turkey's Role in Defusing the Iranian Nuclear Issue. *The Washington Quarterly* vol. 35: 141–152.

Gürzel, Aylin and Ersoy, Eyüp. 2012. Turkey and Iran's Nuclear Program. *Middle East Policy*, vol. 19, no. 1: 37–50.

Habertürk. 2012. Turkey to Cut Iran Oil Imports. *Habertürk* March 30. Available at www.haberturk.com/general/haber/729605-turkey-to-cut-iran-oil-imports (accessed 1 April 2013).

Hafezi, Parisa. 2014. Turkey's Erdogan Visits Iran to Improve Ties After Split over Syria. *Reuters*, 29 January 2014. Available at www.reuters.com/article/2014/01/29/us-iran-turkey-erdogan-idUSBREA0S11T20140129 (accessed 7 August 2014).

Hashem, Ali. 2016. Why Iran Stood with Erdogan. *Al monitor, Iran pulse*, 19 July 2016. Available at www.al-monitor.com/pulse/originals/2016/07/iran-reactions-turkey-coup-attempt-zarif-erdogan.html (accessed 28 September 2016).

Hibbs, Mark. 2015. Turkey's Interests and Tanideh. *Arms Control Wonk*, 23 February 2015. Available at http://hibbs.armscontrolwonk.com/archive/3117/turkeys-interests-and-tanideh?mkt_tok=3RkMMJWWfF9wsRolsq7JZKXonjHpfsX64u8rUKCg38431UFwdcjKPmjr1YAJRct0aPyQAgobGp5I5FEIQ7XYTLB2t60MWA%3D%3D (accessed 15 April 2015).

Hiro, Dilip. 2009. *Inside Central Asia. A Political and Cultural History of Uzbekistan, Turkmenistan, Kazakhstan, Kyrgyzstan, Tajikistan, Turkey, and Iran*. New York: Overlook Duckworth.

Hunter, Robert E. 2010. Rethinking Iran. *Survival* vol. 52, no. 5: 135–156.

Ilgit, Asli and Ozkececi-Taner, Binnur. 2014. Turkey at the United Nations Security Council: 'Rhythmic Diplomacy' and a Quest for Global Influence. *Mediterranean Politics* vol. 19, no. 2: 183–202.

Kandemir, Asli. 2013. Exclusive: Turkey to Iran Gold Trade Wiped out by New US Sanction. *Reuters*, 15 February 2013. Accessed June 10, 2013. www.reuters.com/article/2013/02/15/us-iran-turkey-sanctions-idUSBRE91E0IN20130215.

Kardas, Saban. 2010. Turkish–Iranian Energy Cooperation in the Shadow of US Sanctions on Iran. *Eurasia Daily Monitor* vol. 4, no. 144. 27 July 2010. Accessed 22 September 2013. www.jamestown.org/single/?no_cache=1&tx_ttnews[tt_news]=36672#.Uj88XD-971U.

Kazan, Isil. 2005. Turkey: Where Geopolitics Still Matters. *Contemporary Security Policy* vol. 26, no. 3: 588–604.

Kibaroğlu, Mustafa. 2009. Turkish Perspectives on Iran's Nuclearization. *EurasiaCritic* April: 48–54.

Kibaroğlu, Mustafa. 2010. The Iranian Quagmire: How to Move Forward. Position: Resuscitate the Nuclear Swap Deal. *Bulletin of the Atomic Scientists*: 1–7.

Kibaroğlu, Mustafa. 2012. What Went Wrong with the 'Zero Problems with Neighbors' Doctrine? *Turkish Policy Quarterly* vol. 11, no. 3: 85–93.

Kibaroğlu, Mustafa and Caglar, Boris. 2008. Implications of a Nuclear Iran for Turkey. *Middle East Policy* vol. 15: 59–80.

Larrabee, Stephen F. and Lesser, Ian O. 2003. *Turkish Foreign Policy in an Age of Uncertainty*. Santa Monica: RAND Corporation.

Larrabee, Stephen F. and Nader, Alireza. 2013. *Turkish–Iranian Relations in a Changing Middle East*. Santa Monica: RAND Corporation.

Leverett, Flynt and Leverett, Hillary Mann. 2013. *Going to Tehran. Why the United States Must Come to Terms with the Islamic Republic of Iran*. New York: Metropolitan Books.

Lewis, Jonathan Eric. 2006. Replace Turkey as a Strategic Partner? *The Middle East Quarterly* vol. 13, no. 2: 45–52.

Lutz, Meris. 2010. Turkey: Ankara Adds Israel to List of Strategic Security Threats. *Free Republic* November 1. Accessed 20 April 2013. www.freerepublic.com/focus/f-news/2619192/posts.

Mercan, Murat. 2009. Turkish Foreign Policy and Iran. *Turkish Policy Quarterly* vol. 8: 13–19.

Mousavian, Seyed Hossein. 2012. *The Iranian Nuclear Crisis. A Memoir*. Washington: Carnegie Endowment for International Peace.

Oktav, Ozden Zeynep. 2007. The Limits of Change: Turkey, Iran, Syria. In: Guney, Nursin Atesoglu (ed.), *Contentious Issues of Security and the Future of Turkey*. Hampshire: Ashgate, pp. 85–98.

Olson, Robert W. 2004. *Turkey-Iran Relations 1979–2004: Revolution, Ideology, War, Coups, and Geopolitics*. Costa Mesa: Mazda Publishers.

Önis, Ziya. 2009. Between Europeanization and Euro-Asianism: Foreign Policy Activism in Turkey During the AKP Era. *Turkish Studies* vol. 10, no. 1: 7–24.

Önis, Ziya and Yılmaz, Şuhnaz. 2009. Between Europeanization and Euro-Asianism: Foreign Policy Activism in Turkey during the AKP Era. *Turkish Studies* vol. 10: 7–24.

Pamuk, Humeyra. 2014. Turkey's President Approves Law Tightening Grip on Judiciary. *Reuters*, 26 February 2014. Available at http://uk.reuters.com/article/2014/02/26/uk-turkey-judiciary-idUKBREA1P1MF20140226 (accessed 15 April 2015).

Pamuk, Humeyra. 2015. Turkey's Erdogan Says Can't Tolerate Iran Bid to Dominate Middle East. *Reuters*, 26 March 2015. Available at www.reuters.com/article/us-yemen-security-turkey-idUSKBN0MM2N820150326 (accessed 1 September 2016).

Parsi, Trita. 2012. *A Single Roll of the Dice: Obama's Diplomacy with Iran*. Yale: Yale University Press.

Perkovich, George and Ülgen, Sinan (eds). 2015a. *Turkey's Nuclear Future*. Washington: Carnegie Endowment for International Peace.

Perkovich, George and Ülgen, Sinan (eds). 2015b. Why Turkey Won't Go Nuclear. *Project Syndicate*, 10 April 2015. Available at www.project-syndicate.org/commentary/turkey-iran-nuclear-proliferation-by-george-perkovich-and-sinan-ulgen-2015-04?mkt_tok=3RkMMJWWfF9wsRoivKrLZKXonjHpfsX64u8rUKCg38431UFwdcjKPmjr1YcERMB0aPyQAgobGp5I5FEIQ7XYTLB2t60MWA%3D%3D (accessed 15 April 2015).

Pieper, Moritz. 2013. Turkish Foreign Policy toward the Iranian Nuclear Programme: In Search of a New Middle East Order after the Arab Spring and the Syrian Civil War. *Alternatives – Turkish Journal of International Relations* vol. 11, no. 3: 81–92.

Porter, Gareth. 2014. *Manufactured Crisis. The Untold Story of the Iran Nuclear Scare*. Charlottesville: Just World Books.

Rafizadeh, Majid. 2016. Odd Bedfellows: Turkey and Iran. *Turkish Policy Quarterly* vol. 14, no. 4: 109–115.

Raphaeli, Nimrod. 2010. Turkey Throws Iran a Safety Net. *Inquiry & Analysis Series Report* No. 629, 3 August 2010: Middle East Media Research Institute.

Reuters. 2015. Iran's Rouhani Vows to back Syria 'Until the End of the Road'. *Reuters*, 2 June 2015. Available at http://uk.reuters.com/article/2015/06/02/uk-mideast-crisis-syria-iran-idUKKBN0OI0UN20150602 (accessed 8 June 2015).

Robins, Philip. 2013. Turkey's 'Double Gravity' Predicament: The Foreign Policy of a Newly Activist Power. *International Affairs* vol. 89, no. 2: 381–397.

Santos Vieria de Jesus, Diego. 2011. Building Trust and Flexibility: A Brazilian View of the Fuel Swap with Iran. *The Washington Quarterly* vol. 34: 61–75.

Savyon, Ayelet. 2010. The Iran–Turkey–Brazil Nuclear Agreement: In the Iranian Perception, a New World Order Led By Iran. Middle East Media Research Institute: *Inquiry & Analysis Series Report* no. 610, May 17, 2010.

Scholl-Latour, Peter. 2001. *Allahs Schatten über Atatürk. Die Türkei in der Zerreißprobe*. München: Goldmann.

Seufert, Günter. 2012. Foreign Policy and Self-Image. The Societal Basis of Strategy Shifts in Turkey. *SWP Research Paper*. Berlin: Stiftung Wissenschaft und Politik. German Institute for International and Security Affairs.

Seufert, Günter. 2014. Türkei. In: *Mächte und Milizen. Wer sind die wichtigsten Akteure des Nahen Ostens? Und was wollen sie?* Internationale Politik Nr. 5, September/October, pp. 32–34.

Stein, Aaron and Bleek, Philipp C. 2012. Turkish–Iranian Relations: From 'Friends with Benefits' to 'It's Complicated'. *Insight Turkey* vol. 14, no. 4: 137–150.

Taşpınar, Ömer. 2012. Turkey's Strategic Vision and Syria. *The Washington Quarterly* vol. 35: 127–140.

Taşpınar, Ömer, Malley, Robert and Sadjadpour, Karim. 2012. Symposium: Israel, Turkey and Iran in the Changing Arab World. *Middle East Policy* vol. 19, no. 1: 1–24.

Tertrais, Bruno. 2010/2011. A Nuclear Iran and NATO. *Survival* vol. 52, no. 6: 45–62.

Today's Zaman. 2015. Prosecutor Illegally Sought Opinion of Government on Evidence in Iran-backed Terror Probe. *Article*, 10 February 2015. Available at http://mobile.todayszaman.com/national_prosecutor-illegally-sought-opinion-of-government-on-evidence-in-iran-backed-terror-probe_372288.html (accessed 15 April 2015).

Turkish foreign ministry. 2010. Iran's Nuclear Program. The Turkish Perspective. June 2010, *Powerpoint Briefing*, Available at http://losangeles.cg.mfa.gov.tr/images/TemsilcilikOzel/b15prjzpaplfysi5nz31z4qmIrans%20Nuclear%20Program.pdf (accessed 6 May 2015).

Turkish foreign ministry. 2015. No: 102, 3 April 2015, Press Release Regarding the Agreement on the Nuclear Program of Iran. *Press Release*. Available at www.mfa.gov.tr/no_-102_-3-april-2015_-press-release-regarding-the-agreement-on-the-nuclear-program-of-iran.en.mfa (accessed 6 May 2015).

Turkish foreign ministry. 2016. Visit of the Foreign Minister of Iran to Turkey. *Press statement*, 19 March 2016. Available at www.mfa.gov.tr/visit-of-the-foreign-minister-of-iran-to-turkey.en.mfa (accessed 23 June 2016).

Udum, Şebnem. 2007. Turkey's Non-nuclear Weapons Status. A Theoretical Assessment. *Journal on Science and World Affairs* vol. 3, no. 2: 51–59.

Udum, Şebnem. 2012. Türkiye'nin İran Nükleer Meselesindeki Siyaseti (Turkish Policy on the Iranian Nuclear Issue). *Ortadoğuz Analiz* vol. 43, no. 4: 98–107.

UN. 2010. Provisional Summary, 6335th Security Council Meeting. *S/PV. 6335*. Available at www.un.org/en/ga/search/view_doc.asp?symbol=S/PV.6335 (accessed 13 October 2014).

US Department of State. 2010. Background Briefing on Nuclear Nonproliferation Efforts with Regard to Iran and the Brazil/Turkey Agreement. *Special Briefing*, Senior Administration Officials. Washington, May 28, 2010. Available at www.state.gov/r/pa/prs/ps/2010/05/142375.htm (accessed 19 April 2014).

US Department of State. 2013. Regarding Significant Reductions of Iranian Crude Oil Purchases. *Remarks*, 29 November 2013. Available at www.state.gov/secretary/remarks/2013/11/218131.htm (accessed 17 May 2014).

US Department of the Treasury. 2016. *Guidance Relating to the Lifting of Certain US Sanctions Pursuant to the Joint Comprehensive Plan of Action on Implementation Day*. 16 January.

US Energy Information Administration. 2012. Natural Gas Exports from Iran. *Independent Statistics & Analysis* October 2012. Available at www.eia.gov/analysis/requests/ngexports_iran/pdf/full.pdf (accessed 10 April 2013).

US Energy Information Administration. 2013. Turkey. *Analysis Briefs* February 1. Available at www.eia.gov/countries/analysisbriefs/Turkey/turkey.pdf (accessed 10 April 2013).

Ülgen, Sinan. 2011. The Security Dimension of Turkey's Nuclear Program: Nuclear Diplomacy and Non-Proliferation Policies. *Center for Economic and Foreign Policy Studies (EDAM)*, 9 December 2011. Available at www.edam.org.tr/EDAMNukleer/section5.pdf (accessed 23 January 2013).

Ülgen, Sinan. 2012. *Turkey and the Bomb*. Washington, D.C.: Carnegie Endowment for International Peace.

Üstün, Kadir. 2010. Turkey's Iran Policy: Between Diplomacy and Sanctions. *Insight Turkey* vol. 12, no. 3: 19–26.

Vahedi, Elias. 2010. Turkey Removes Iran from 'Red Book' Threat List. *Iran Review*, 5 September 2010. Available at www.iranreview.org/content/Documents/Turkey_Removes_Iran_from_%E2%80%9CRed_Book%E2%80%9D_Threat_List.htm (accessed 28 August 2014).

Winrow, Gareth M. 2003. Pivotal State or Energy Supplicant? Domestic Structure, External Actors, and Turkish Policy in the Caucasus. *Middle East Journal* vol. 57, no. 1: 76–92.

3 Russian foreign policy towards the Iranian nuclear programme

1 Bushehr as burden and leverage: Russian reactions to the 2002 nuclear revelations[1]

With the revelation of a uranium enrichment facility at Natanz and a heavy-water plant under construction at Arak, Russian nuclear cooperation with Iran suddenly appeared in a disconcerting light (Lata and Khlopkov 2003).[2] Russo-Iranian nuclear cooperation had started in September 1994 when a protocol was signed between the Russian Atomic Energy Minister, Viktor Mikhaylov, and the president of the Atomic Energy Organisation of Iran (AEOI), Reza Amrollahi, in which the Russians expressed their willingness to complete the 1,000-MW power reactor at Bushehr worth US$800 million. Bushehr was Iran's only nuclear power plant project that had been started by German Kraftwerk Union AG in 1970 (Yurtaev 2005: 107), but was abandoned in the wake of the Islamic Revolution (Orlov and Vinnikov 2005: 50). Unable to get nuclear technology from its former European partners that had cooperated with Iran in the starting phases of the Iranian nuclear programme under the Shah in the 1960s and 1970s,[3] Iran had turned to China and the USSR (with Russia succeeding the latter). As from the mid-1990s and despite US pressure, Russia had become Iran's nuclear partner (see also Freedman 2000; Sarukhanyan 2006: 88–108).

Against the backdrop of the uranium fuel sales for the construction of Bushehr, Putin appeared pugnacious and downplayed the revelations of a covert Iranian nuclear programme, calling non-proliferation concerns a 'means of squeezing Russian companies out of the Iranian market' in 2003 (Parker 2009: 221).[4] Such a statement neatly captures the Russian *zeitgeist* at the time on the nexus between non-proliferation and legitimate nuclear cooperation that continued to underwrite Russian foreign policy in the Iranian nuclear dossier for the years to come: Russian economic benefits had to be weighed against political and security concerns of technology sales to Iran that might be of a dual-use nature. Russia's official position thus indicated that Moscow did not share the American state of alert and was apprehensive for early signs of an emerging securitisation of the Iranian nuclear issue.[5] As a Russian diplomat formulated in an interview: 'Concerning Iranian nuclear intentions, we are not so hysterical as the Americans'.[6] It was already at this early juncture in the Iranian nuclear crisis that different security conceptions towards Iran's nuclear programme between the US and Russia became apparent.

Terms such as the 'pursuance of national interests' therefore have to be understood as relational concepts: What reads as an act of defiance of hegemonic powers can be an act of necessary resistance against instrumental politicisations in a non-hegemonic reading. In the context of Russia's interests increasingly clashing with those of the US over the looming war in Iraq in 2003, Moscow saw no reason to comply with US pressure and renounce its commercial ties with Tehran. Tellingly, the abrogation of the 1995 Gore–Chernomyrdin Commission under president Putin in 1999 (Antonenko 2001) had effectively put an end to US–Russian consultations concerning arms and technology transfers to Iran. This binational commission with a number of working groups had been established with a view to mutual consultations on a range of policies of mutual concern (Stent 2014: 18). But Putin's administration did not turn its back on the US. Having officially announced a 'strategic partnership' with the US and endorsed the NATO–Russia Council in 2002 (Conrad 2011: 45), Putin continued defending the Bushehr project and thereby indirectly sat on the fence when it came to judging the security implications of an Iranian nuclear programme in a US reading. The United States was trying hard to end Russia's nuclear cooperation with Iran (Belopolsky 2009: 101–107; Porter 2014: 106). In a Russian reading, Bushehr was a legitimate civilian nuclear power plant, unconnected to any hitherto covert uranium enrichment facilities (Aras and Ozbay 2006: 134).[7] Linking the light-water reactor Bushehr to proliferation risks was, indeed, inconsistent with IAEA Safeguards (Porter 2014: 107). Recurrence to 'legitimate' projects therefore conveyed a sense of self-determination and independence from an alarmist rhetoric about nuclear Iran: In the context of a looming securitisation of Iran's nuclear dossier in the first half of the 2000s, Russian recurrence to 'legitimacy' in international politics and external economic policies aimed to position Moscow against the outlining US approach to Iran's nuclear programme. These admonitions of the early 2000s were later rehearsed when progress in diplomacy became more tangible. Russian official statements after the conclusion of the political framework agreement on 2 April 2015, negotiated in Lausanne between the P5 + 1 and Iran, served as a reminder that Russia had supported Iran's right to peaceful nuclear energy all along (Russian foreign ministry 2015). Annex III of the JCPOA details conditions for civil nuclear cooperation with Iran, which are regarded as separate from proliferation concerns addressed in the JCPOA itself. 'This separation was very important for us', a member of the Russian delegation to the nuclear talks asserted in an interview.[8]

From the beginning of the nuclear stand-off in 2002, Putin had repeatedly emphasised the Iranian right to nuclear power (Mousavian 2012: 163f.; Putin 2003). This, as well as the track record of nuclear cooperation between Russia and Iran made Russia the logical candidate in Iran's search for allies as international public opinion turned against Iran and as Western governments grew more impatient with the Iranian lack of cooperation and transparency while talks over the nuclear programme proceeded. That Iran stayed quiet on Russia's Chechnya policy could be seen as a 'quid pro quo' on the side of the Iranians for Russian economic and technological assistance and for Russian 'protection' of Iran from international pressure at least in the first few years of the Iran dossier following 2002, as observers have noted.[9] In a broader 'Eastern bloc approach' embraced especially with the coming into office of president Ahmadinejad in 2005, Iran reached out to Russia, China and the Non-Aligned Movement, hoping to find an international coalition supportive of Iran. Especially the support of China and Russia, two permanent UNSC members, was deemed crucial in resisting US pressure (Mousavian 2012: 84, 141). This was the case in the run-up to the first IAEA Board of Governors meeting in 2003 dealing with Iran, and in the course of the following years when referral of the Iranian nuclear file from the IAEA to the UNSC still might have been prevented. Ideologically inflated as a 'looking to the East' policy with the advent of Ahmadinejad as president and Ali Larijani as chief nuclear negotiator, Iran was trying to garner support of these states in order to build a broad anti-US coalition to break the format of Iran facing Western negotiation partners over its nuclear programme that met increasingly fierce opposition (Mousavian 2012: 190f.). Therewith, however, Iran was misinterpreting Russian intentions: Following Iran's resumption of nuclear enrichment activities in August 2005 after a period of temporary suspension (Jafarzadeh 2007: 159), testimony to the failure of nuclear negotiations between Iran and the E3, the file was referred to the UNSC in 2006. Of the 35 members of the IAEA Board of Governors, 27 endorsed the board resolution, of which Russia was one. The Russian endorsement became possible after a reference to 'international peace and security' had been omitted. An earlier resolution had still contained the reference and was therefore vetoed by Russia (Fitzpatrick 2006: 21). The Iran nuclear case now had been transferred from Vienna to New York. The format for negotiations was extended to the P5+1, or 'Group of Six', as it was referred to in Russian government and media parlance (*Gruppa Shesti*). Russia, hesitant to join the negotiations at first (International Crisis Group 2006: 14), was forced to take a stance by now at the latest by nature of its permanent Security Council membership. Negotiations with Iran now required policy coordination with China and Russia. The extension of the format, however, now also had brought the US to the negotiation table, whose absence had been a structural deficit in the earlier E3–Iran talks.

In P5+1 meetings, Russia found itself in a camp with China arguing for a less pressuring approach to Iran than Western governments were pushing for and argued against the adoption of a UNSC resolution (Mousavian 2012: 235f.; Patrikarakos 2012: 224). In an attempt to broker a political solution to the crisis, Russia proposed a plan in 2006 by which Iran would have to transfer its enrichment programme onto Russian soil while still benefitting from its output.[10] The idea of such a transfer was quickly rejected by the Iranians. This decision signalled to Moscow that Iran would not accept indefinite reliance on Russia in the field of nuclear technology and constituted a watershed both for Russia's perception of the Iranian goals and for US–Russian cooperation over the Iranian nuclear file: Not only did this episode prove the 'total failure of the "looking to the East" policy', it 'opened a new chapter in the nuclear standoff in which Russia began to move closer to the West', as Houssein Mousavian (2012) writes in his memoirs (256–257). And after the failure of renewed P5 + 1 initiatives to reach a politically acceptable compromise in the following months, Russia did not make use of its veto right and approved of UNSC resolution 1696 in July 2006 under chapter VII (article 40), which made the suspension of uranium enrichment mandatory.

Increasingly aware of Iranian delaying techniques and against the background of the rejection of 'the Russian plan', Russia voted for UNSC Resolution 1737 in December 2006, approving for the first time the imposition of chapter VII sanctions on Iran.[11] The sanctions regime was intensified and reaffirmed by UNSC Resolution 1747 in March 2007, followed by Resolution 1803 in March 2008 and Resolution 1835 in September the same year. While Russia aimed at averting or at least slowing down international pressure on Iran, it aimed at using the Bushehr project as a leverage to use in Iran's nuclear conflict at the same time. This was evidenced by the constant pushing back of the date of completion of the Bushehr power plant, which, on the surface of it, was attributed to 'technical' issues (Katz 2010: 64; 2012: 58), but was also read as a Russian sensitivity to US concerns – or at least of then-minister of Atomic Energy Alexander Rumyantsev (see also Nizameddin 2013: 266; Stent 2014: 150–151). It equally prolonged the Iranian dependence on Russian technology.[12] Therme (2012) therefore writes that Russia was 'instrumentalising' nuclear cooperation with Iran in order to both keep Iran dependent on Russian know-how and to hinder an Iranian 'nuclear success' (190).[13] While this strategy can be read as a rational commercial calculation, it also served to show responsiveness to US security perceptions at the same time and fulfilled a double purpose for Russia in this sense. Russia's public statements against unhelpful pressure on Iran entailed advocacy for a security culture that resisted US positions, while Moscow still managed to steer a course that was avoiding outright rejection of US policies.

2 Russia's position on Iran sanctions

2.1 Rule convergence: Russia's approach to UNSC sanctions on Iran

In accordance with the Russian hesitance when it comes to international sanctions against Iran, Moscow has always reiterated the importance of dialogue and diplomacy, rejecting a military solution to the crisis and calling on Iran to comply with the IAEA. In one of his newspaper articles published prior to the 2012 presidential elections, Putin wrote that a military attack on Iran would have 'catastrophic' consequences (Putin 2012). In addition to braking the sanctions track, Moscow has thus (in tandem with China) worked toward weakening their impact by watering down provisions contained in the UNSC resolution drafts (Kuchins and Weitz 2009: 176).[14] 'Sanctions', President Putin reminded his audience at the Valdai Club in October 2014, 'are already undermining the foundations of world trade, the WTO rules, and the principle of inviolability of private property' (Putin 2014).

Moscow's eventual support for pressure and sanctions on Iran was the result of an ambivalent position: Russia appeared to heed US concerns about the Iranian nuclear activities and, unofficially, made sure that it would remain the exclusive provider of nuclear fuel for Iran by slowing down Iran's nuclear advances. At the same time, it angered the Iranians and shattered any illusion that Russia was a reliable ally and would always protect Tehran from Western pressure. In Tehran, the impression was fuelled that the Iranian nuclear programme constituted a 'bargaining chip' for Moscow and that 'Russia is intentionally stalling in dealing with Iran to wring concessions from the United States' (Mousavian 2012: 93). In his Third-Worldist rhetoric, Ahmadinejad's 'overreliance on Russia' even became a cause for domestic criticism within Iranian policy circles (ibid.: 320). With Russia approving of successive rounds of UNSC-backed international sanctions, Iran had learnt the hard way that it could not rely on Russia as a diplomatic shield. But the disillusionment was mutual: Also in Russia, official voices began to worry that 'Tehran had […] outsmarted Moscow by using Russia's diplomatic screen to advance Iranian goals that were inimical to Russia's own security interests' (Parker 2009: 249). 'The Iranians are quite skilled at wasting our time', a Russian official involved in the nuclear talks quipped in an interview.[15] And also Russian foreign policy heavyweight Yevgeny Primakov shared the assessment that Iran was playing for time (2016: 230). Russian–Iranian nuclear cooperation thus by no means implied an automatic lenience with Iran in the nuclear talks. Russia's history of nuclear partnership with Iran was fraught with mutual frustration and occasional public accusations. A former European ambassador to both Moscow and Tehran formulates in an interview: 'The grievances and mistrust between Russia and Iran are many, from the Treaty of Turkmenchai until today'.[16]

Russia's support for UN sanctions under chapter VII thus has to be seen in this context of Russian scepticism regarding Iranian intentions and of wanting to be seen as a constructive partner for its Western interlocutors. The calculation was a mixture of geostrategic as well as global power political considerations, as will also be discussed in the remainder of this chapter. Remarks delivered by Russia's UN representative, Vladimir Churkin, in Security Council sessions that passed the sanctions resolutions, conveyed a balance between the cautious admonition of Iran's failures to address international concerns about its nuclear file and a principled reservation regarding the use of sanctions (UN 2006: 2–3; UN 2007: 10–11; UN 2010: 8–9).

Russia's approval for sanctions also was a reaction to political circumstances at the time that would have made resistance to sanctions difficult to sell politically. This was the case with the revelation of the existence of Iran's second uranium enrichment facility in Fordow near Qom in September 2009, hitherto unknown to the IAEA,[17] and also to Russian intelligence services (Parsi 2012: 126; Stent 2014: 232–233). Russia was taken by surprise and therefore angered by the Iranian lack of transparency, but was also not pleased by the fact that

Western intelligence sources had not been shared with Moscow (ibid.).[18] Another undercurrent was the fact that Iran had rejected the Vienna group's proposal in 2009 that had centred around Russia as a key actor in the fuel swap deal.[19] The latest sanctions regime against Iran was approved by the UN Security Council in June 2010 with Resolution 1929. Russia's vote for sanctions therefore also has to be seen in the context of this political momentum, where Russia's frustration with the Iranian lack of cooperation was one factor in the calculation and where a veto in the UNSC would have constituted an outright rejection of (not only Western) security concerns regarding Iran's lack of transparency, as demonstrated again with the revelation of the Fordow facility. In addition, the publicly upheld threat of unilateral Israeli airstrikes against Iranian nuclear facilities likely was another incentive for Russia to agree to the imposition of international sanctions which would temporarily disarm the argument for military operations (Porter 2014: 271).

Russia's approval of international sanctions arguably also was linked to concessions offered by the US administration in exchange for Moscow's consent in the sanctions question. In what has been described as 'horse-trading' taking place between the US and Russia,[20] the controversial Missile Defense (MD) episode in US–Russian relations became interlinked with Iran sanctions in the UN Security Council whereby the Obama administration offered concessions in the MD plans (renouncement of the missile defence shield deployment in Poland and the Czech Republic) and would be guaranteed Russia's cooperation in the Iran nuclear file in return, i.e. Russian green light for a new round of international sanctions on Iran (Kuchins and Weitz 2009: 168; Patrikarakos 2012: 256). Landler (2016) writes that a secret letter to that effect by President Obama to President Medvedev was hand delivered by then-NSC official (and later US ambassador to Russia) Michael McFaul and under-secretary of state William Burns in February 2009 (270). President Medvedev's public reaction to this secret letter about an Iran-missile shield bargain indicated that Moscow was not happy to publicly discuss the matter (France24 2009). The fact that Moscow did not reconsider its support for sanctions in June 2010 after the surprising Brazilian–Turkish diplomatic break-through in May 2010 and Tehran's unexpected approval of the Tehran declaration, however, goes to show that Russia had been promised too many important concessions by the US, as Trita Parsi contends (2012: 196). Russia even suspended the planned sale of its S-300 long-range air-defense system to Iran. Medvedev issued presidential decree 1154 in October 2010 to that effect. The cancellation of the S-300 contract was a major annoyance for Iran with which Russia squandered a good deal of its 'leverage power' over Tehran.[21] Esfandiary and Fitzpatrick (2011) even suggest that Russia 'went beyond the strict reading of the UN sanctions by cancelling (the S-300 contract) [...]. This was a decision that may have had the most significant impact on Iran of any national measure' (145) – a decision that has even been likened to Russia's own unilateral sanctions on Iran (see Kozhanov 2015b).[22] The importance of the Iran issue and the acknowledged necessity to work with the Russians on Iran has been an important (if not the most important) motivation behind the US–Russian 'reset' policy in 2009, as confirmed by former Secretary of State Clinton in her memoir (Clinton 2014: 235; see also Deyermond 2013; Fitzpatrick 2010: 71; Kozhanov 2015b; Parsi 2012: 94).[23]

2.2 Norm divergence: Russia's position on unilateral Iran sanctions

Russia was and is sceptical of the use of sanctions as a means of pressuring Iran into compliance. Throughout the years of nuclear diplomacy, Russia's permanent representative to the IAEA Vladimir Voronkov re-iterated that the removal of all sanctions should be the result of the IAEA's clarification of all remaining questions on Iran's nuclear programme (Fars News Agency 2014). In contrast to Russia's grudging acceptance of international sanctions, however, unilateral sanctions as imposed by the US and the EU, it is being reiterated from the Russian foreign ministry, are not seen as legitimate instruments of international politics (Medvedev 2010; Reuters 2010; Russian foreign ministry 2012b; Sheridan 2009).[24] 'We view unilateral sanctions as illegal', a Russian foreign ministry official working on the Iranian nuclear dossier puts it in an interview.[25] Next to the finding that unilateral sanctions have 'only brought a disrupture of the E3+3 dynamics' (ibid.), such a frank statement conveys conceptions of legitimacy in international politics. Sanctions, imposed unilaterally by the US and, after 2010, also by the EU, are viewed as breaching a normative framework that should govern international relations. 'Another risk to world peace and stability is presented by attempts to manage crises through unilateral sanctions and other coercive measures, including armed aggression, outside the framework of the UN Security Council,' the official Russian Foreign Policy Concept formulates (Russian foreign ministry 2013a). 'Rules and models' of the UN system, in Katzenstein's terminology (1996: 21) to understand 'culture', are not adhered to if sanctions are adopted outside of the UN Security Council, in a Russian understanding. Reacting to the EU decision to impose an oil embargo on Iran, effective from 1 July 2012, Foreign Minister Lavrov publicly deplored what he described as unilateral steps designed to 'punish Iranian stubbornness' (Russian foreign ministry 2012a). He emphasised that Russia regarded such steps as a 'deeply faulty line' and reiterated the importance of political dialogue instead of punitive measures. The EU was criticised for the fruitlessness of an excessive tilt toward a punitive sanctions track which is regarded as 'unhelpful' by Russia and is said to complicate the search for common policy positions within the P5+1, as emphasised by a Russian diplomat in an interview.[26]

Presented publicly as motivated by the adverse effect sanctions have on diplomacy, Russia's rejection of unilateral sanctions and its hesitancy to use international sanctions is also to be explained by the adverse material effect these have on Russian companies: In the context of a growing anti-Iran climate in Western policy circles as the nuclear dossier was dragging, the US criticised Russian exports of weapons and defensive systems and explicitly started sanctioning Russian firms for conducting such business with Iran. The aircraft manufacturer Sukhoi and arms exporter Rosoboronexport were sanctioned in 2006,

2007 and 2008 (Belopolsky 2009: 127; Defense Industry Daily 2006).[27] The US thereby was aiming at hampering what was perceived as a 'cynical' Russian two-track policy in which Moscow was officially committed to the international sanctions regime but not supporting the spirit of it (Patrikarakos 2012: 228). Russian weapons deliveries, however, were outlawed on a multilateral basis in 2010 by the weapons embargo adopted through UNSCR 1929. But Western sanctions also hampered Russian trade with Iran in other areas, as a Russian diplomat remarked in an interview: 'Our trade volume (with Iran) is not as high as it used to be because of unilateral sanctions. This is because of globalised trade: The EU SWIFT sanctions made bank communication with Iran more difficult'.[28]

Russia, as demonstrated by its language on sanctions, is motivated by a normative understanding of their (il)legitimacy, but is equally motivated by the material dimension of the effect of sanctions. This finding differentiates Russia's stance on the sanctions regime: While Moscow criticises the political effects of sanctions, its acceptance of the latter appears to be selective and dependent on the US position, the impact of sanctions on Russia, and the nature of the sanctions adopted (unilateral or international). This is an important point to retain for an examination of a Russian security culture as resisting hegemony and its constituent material, ideational and institutional underpinnings (Cox 1996: 97f.; 135f.). A later section will come back to this point. An intermediate conclusion suggests that Russia's foreign policy discourse bears indications of a security culture that resists hegemony: Russia emphasises legitimate rights of Iran, legitimate Russian commercial ties with Iran, counterproductive sanctions pressure and illegal Western unilateral sanctions and therewith emphasises the sovereignty of Iran to be upheld. The impression on a behavioural level is more mixed: Russia was slowing down the sanctions track, but eventually adopted and complied with international sanctions, yet was blacklisted for trade with Iran that ostensibly contravened sanctions regimes. The next section introduces a Russian discourse of constructive mediation as an additional important element that feeds into Russia's security culture on Iran.

3 Russia in the Iranian nuclear talks: the notion of 'constructive mediation' and the role of technical intermediary

Since the 1990s, Russia had to reconcile security perceptions with legitimate commercial aspects of Russo-Iranian relations. As much as this had been a source for disagreement between Russia and its Western counterparts, it had demonstrated Russia's defense of a security culture that would resist securitising discourses, which were presenting Iran's nuclear programme as an inherent threat and basing policies on allegations. In the second half of the 2000s, however, decisions by the Iranian leadership had alienated Moscow and contributed their part in bringing Moscow to agree to the imposition of sanctions. President Medvedev's decision not to deliver the S-300 defense system to Iran, that would have allowed the interception of long-range missiles, was a political tilt toward a course more accommodating to Western security concerns. Medvedev's presidency was characterised by a wider understanding of foreign policy as a 'modernisation resource' that nurtured the impression of the possibilities for closer cooperation with Moscow in the Iran dossier, but also in the US–Russia dialogue in general. Such a course impacted on the negotiations on the New-START treaty about the reduction of strategic nuclear weapons between the US and Russia, the granting of transit routes for US and NATO troop supplies to Afghanistan, and a new round of UN sanctions on Iran in 2010.

The depiction of Russia's role in the Iran dossier as that of an obstructionist veto player indulgent with the Iranians would therefore be a fallacy. In its official diplomacy, Russia's emphasis on the need to find a political solution to the nuclear crisis through dialogue was complemented by concrete technical suggestions to achieve the former. Proposals such as the creation of an international fuel centre on Russian soil by president Putin are a case in point (Diakov 2007: 135f.). Former IAEA Secretary-General Mohamed ElBaradei notes in his memoirs in this context:

> Contrary to allegations made at times by the West, Putin strongly opposed Iran's acquisition of nuclear weapons and questioned its need for nuclear enrichment capability; but he concurred that Iran should be offered attractive assistance, including nuclear technology, and he supported an international guarantee of reactor fuel supply. Putin also put forward an idea for an international repository for spent fuel, which I applauded.
>
> (ElBaradei 2012: 137)

As nuclear proliferation expert Mark Hibbs holds, a Russo-Iranian nuclear cooperation agreement[29] would have achieved to embed Iran's nuclear activities within a transnationally controllable and verifiably peaceful framework and would inevitably align Russia closer with Western powers in emphasising the need for full Iranian cooperation with the IAEA (Hibbs 2013). The first international nuclear fuel reserve bank in Angarsk, Russia, that Russia had built together with the IAEA is an example of an attempt of such a multilateralisation of the nuclear fuel cycle (Meier 2014: 22). After the failed attempts in 2009 and 2010 to negotiate fuel swap deals, however, it seemed doubtful whether the scenario of a multilateralisation of the fuel cycle to contain Iranian domestic enrichment was still an option on the negotiation table thereafter, and reportedly also was a 'sticking point' in negotiations in early 2015 (Richter and Mostaghim 2015). The nuclear agreement reached in July 2015, however, foresees transnational control mechanisms through which Russia is responsible for the shipment of enriched uranium out of Iran (Ryabkov in: CENESS 2015). The JCPOA detailed that Iran was to maintain a stockpile of enriched uranium of no more than 300 kg. During the nuclear talks, Russia became the recipient of the excess uranium to be shipped out of Iran.[30] Russia's state-owned nuclear energy agency Rosatom, in addition, took on the task of aiding Iran in the conversion of the uranium enrichment facility at Fordow into a production centre for medical isotopes and continues to supply the nuclear fuel for civilian nuclear energy usage.[31] The reconstruction of Fordow is based on a Russian–Iranian bilateral agreement that is the outcome of an initiative of the Russian delegation at the nuclear talks (Khlopkov 2016: 1). To this end, President Putin signed Decree Nr. 567 on 23 November 2015, lifting restrictions on the supply of technology and equipment to Iran that is needed for the reconstruction of the Fordow facility.

Such involvement of Russia as a key actor on a technical level is testament to the importance that Russia had in crucial phases of nuclear diplomacy with Iran. Russian officials stress that Moscow has introduced several constructive proposals throughout the process, some of which are known (like Lavrov's 'step-by-step' plan in 2011 or the proposal for an international fuel consortium in 2006), while others are unknown to the public and were circulated within the P5+1 format.[32] Reiterating the importance of constructive dialogue and Moscow's contribution to that end, Russian foreign ministry officials noted the similarity between the 'reciprocal approach' underlying the proposal discussed in Geneva in November 2013 and Lavrov's earlier step-by-step proposal.[33] Russia was thus even being ascribed the role of an intermediary and facilitator of talks between Iran and the West by some observers (Aras and Ozbay 2006: 139).[34] In line with Russia's desire to be perceived as a responsible global power, such interpretations reflect on Moscow's willingness to be seen as a cooperative and pragmatic dialogue partner in the Iranian nuclear file. Seyed Hossein Mousavian (2012) even writes that it was a strategic mistake of the West not to have given the Russian 'step-by-step' plan more consideration (457).

Next to such concrete technical proposals, Russia's public stance oftentimes gave Moscow the role of an admonisher against a lack of cooperation on both the Western and the Iranian side. A Russian delegation member to the nuclear talks puts it the following way: 'In [the nuclear talks in] Vienna, one side was heated up, the other was ice cold. We managed to make a nice room temperature. We thus see ourselves much in the role of a mediator.'[35] Another Russian diplomat formulates: 'Russia tried to play the role of mediator between the US and Iran. Our role was to facilitate, to reconcile, to narrow differences.'[36] The unprecedented progress in nuclear talks following the Iranian presidential elections in June 2013 that led to a historic first interim deal between Iran and the P5+1 were therefore publicly welcomed and appreciated by the Russian foreign ministry, as was the progress made in Vienna, Lausanne and Geneva leading up to the successful negotiation of the JCPOA, which was seen in no small part due to Russia's active contributions.[37]

This should not necessarily be attributed to an ideational convergence of security cultures. 'Rather than "norms" and "public goods",' Kuchins and Weitz (2009) remark in this context, 'Russian leaders and political analysts frame Russia's terms of international cooperation as realpolitik bargains and "trade-offs" of interests' (168). If 'multilateralism' in international relations is understood as legitimacy by way of international socialisation, then the Russian reference to multilateralism is at least surprising in the face of Russia's political discourse on contested international issues that is marked much more by geopolitical conceptions and 'spheres of interest'.[38] Despite policy-specific Russian proposals, Russia does not capitalise on such instances of Russian-led multilateral approaches. Russia's approach to multilateral cooperation, in other words, remains ad-hoc and compartmentalised. The understanding of 'trade-offs', in a positive reading, allows selective cooperation on some issues, even when conflicting interests prevail in others. The aspect of convergence of security cultures was cast in doubt even more strongly by the outbreak of the 'Ukraine crisis' and the ensuing deterioration in relations between Russia and the West.

4 The impact of the Ukraine crisis on Russian Iran policies

As a catalyst for an unprecedented deterioration in relations between Russia and the West, the Ukraine crisis has affected policy coordination in almost all areas – with the notable exception of the Iran nuclear talks, where a constructive level of collaboration between Russia and the West had remained intact. Here, a partial discrepancy between a discursive and a behavioural level quickly became apparent. At a time when policy coordination with Russia was suspended in most other formats (like the G-8 group or the NATO-Russia Council), Russian official rhetoric alluded to the possibility that the Russian government could recalibrate its position in the Iran nuclear talks as a reaction to Western pressure on Moscow over its policies towards the Ukraine conflict. The impression that Moscow was flirting with the idea of using the Iran nuclear talks as a vehicle for obstructionism was nurtured when on 19 March 2014, Russian deputy foreign minister Sergei Ryabkov stated that Russia could reconsider its position on the Iranian nuclear dossier in the context of Western sanctions discussions directed against Russia in the wake of Russia's annexation of Crimea in March 2014. This was an indication for the impression that Russia occasionally has used the Iran nuclear talks as a 'bargaining chip' to get concessions in other issue areas (Fitzpatrick 2014). Importantly, as a member of the European negotiation team pointed out in an interview, Ryabkov did not hint at a Crimea-Iran connection until after the round of talks at the time in Vienna had ended, indicating that his statement was intended in an audience-specific (and not substantive) context.[39] President Putin's Valdai speech in October 2014 was pregnant with such discursive warnings and was the strongest high-level outburst of what can be read as an outright counter-hegemonic positioning in the wake of a Russian–Western resentment in 2013–2014. The speech has been dubbed 'Munich II', referring to Putin's speech at the 2007 Munich Security Conference, in which he accused the US of undermining global stability, provoking a new arms race, and toppling foreign governments and warned of NATO eastward expansion and unilateralism. Yet, the 'Ukraine crisis' arguably took the Russian resistance-rhetoric to new levels. Speaking of the 'unilateral diktat' and 'dictatorship over people and countries' of a 'self-proclaimed leader', Putin stated in his 2014 Valdai speech:

> It does not matter who takes the place of the center of evil in American propaganda, the USSR's old place as the main adversary. It could be Iran, as a country seeking to acquire nuclear technology, China, as the world's biggest economy, or Russia, as a nuclear superpower.
>
> (Putin 2014)

Asked about Russia's cooperation on the Iranian nuclear file, Putin vaguely stated that 'external conditions might force us to re-consider some of our positions'. In an interview in March 2015, deputy foreign minister Ryabkov (2015) was playing on a similar rhetoric, stating that Russia reserves itself a 'maximum of manoeuvrability'.

Despite these public warnings and gloomy rhetoric on a discursive level, however, retaliatory moves affecting the Iranian nuclear talks did not materialise on a behavioural level.[40] Officials and experts shared the assessment that Russian hints at a change of position in the Iranian nuclear talks remained symbolic politics, but were not followed by substantive policies.[41] Occasional irritations about Russia's positioning were nurtured by reports in the summer of 2014 about a Russian–Iranian oil-for-goods barter agreement worth US$1.5 billion. According to the terms of the agreement, Russia would be importing 500,000 barrels of Iranian oil per day in exchange for Russian equipment and goods exported to Iran (Arbatov 2014; Saul and Hafezi 2014).[42] Given that Russia is an oil-exporting country itself, the import of additional Iranian oil would not make sense in an economic thinking, but was read as a favourable macroeconomic aid for the Iranians suffering from the financial effects of existing sanctions regimes.[43] Just why this would be in Russia's interest lacked intuitive obviousness.[44]

Short of actual economic diversification options, such negotiations can put up a smokescreen as a reaction to Western economic isolation attempts, so the assessment outside of Russia.[45] Against the backdrop of the imposition of Western sanctions on Russia and the downgrading of Russian creditworthiness by rating agencies such as Moody's, Fitch, and Standard & Poor's, the attractiveness of Russia as an investment target has decreased.[46] While Russia dismisses such downgrading as a 'political decision', the Russian economy has experienced an intensified capital flight. But the economic alienation is mutual: The Russian government has shown a tendency of economic alienation from US-inspired financial and economic instruments – in addition to the level of political resentment, and in addition and reaction to Western attempts to isolate Russia economically. Examples are the Putin administration's announcement to substitute embargoed manufactured goods from the West by domestic produces; indirect taxes and direct product bans; and relevant changes in the customs legislation (Libman 2014).

In all this maelstrom of a deteriorating political climate between Russia on the one hand and the EU and the US on the other in the wake of the Ukraine crisis, the negotiations on the Iranian nuclear file have remained as a rare policy domain where constructive cooperation on a working level continued. This finding was indicative of the high importance that Moscow had attached to a political solution of the simmering Iranian nuclear conflict (Meier and Pieper 2015).[47] The following section suggests a number of reasons why that is so.

5 *Derzhavnichestvo* in practice: Russia between status quo politics and resistance in the JCPOA implementation

A peculiar combination of factors lets Russia resist US policies, while on other occasions, pressure on Iran and the upholding of tensions surrounding Iran's nuclear programme was supported by Russia during the phases of negotiations. In what follows, the chapter disentangles this seeming variation in Russia's foreign policy line by following the two-level model of a discursive and a behavioural dimension of Russia's Iran policy as introduced earlier with regard to Russian interests in the implementation of the JCPOA. A number of material factors will therefore be analysed to complement the preceding analysis of Russia's discourse and role perception.

A first factor in Russia's support for continued pressure on Iran during the JCPOA implementation is Russia's comparative advantage on the European gas and oil market. Sceptics have pointed out that this is the strongest counterargument for Moscow to be genuinely interested in a long-term solution to Iran's nuclear crisis.[48] Should Iran's final nuclear status be settled as a result of the successful implementation of the nuclear agreement, a partial normalisation of relations between Iran and the West would ensue. The substantial interest on the part of the business community to re-enter the Iranian market indicates that this could happen long before the eventual 'Termination Day' of the JCPOA. As a result, Russia could, in the mid- to long term, be faced with the emergence of a competitor on the European energy market.[49] Beyond the official welcoming statements from the Russian government, the public reaction to the nuclear agreement in Russia therefore was not enthusiastic.[50] Russia's current near-monopoly position on the European gas and oil market, so the reading, would be endangered. The scenario of a sudden Iranian oil and gas competitor, however, falls short of accounting for more nuanced market structures: Even before the imposition of the EU's oil embargo against Iran in 2012, Iran provided a stable 6 per cent of the EU's oil imports (Eurostat 2012); Russia's share is around 30 per cent. Russian officials are thus relaxed about the prospect of Iran becoming a rival on the European oil market any time soon – even after the lifting of Iran sanctions.[51] Russian officials emphasise that a transition to fully-normalised trade relations between Iranian and Western companies 'won't happen in a day', and that there are merely normal market dynamics at play that Russia does not have to fear.[52] Instead, so the official assertion, Russia would stand to gain from developing cooperation with Iran in other spheres, including investments in the Iranian oil and gas production (ibid.).

Russia's share on the European gas market lies at 30 per cent. Even though Iran holds the world's second largest gas reserves, it lacks production and transportation structures. Russia also knows that the existing pipeline structure benefits Russian gas interests, while pipelines from Iran to Europe do not exist and would have to be built.[53] Even alternative projects like the Trans-Anatolian Pipeline (TANAP) and the Trans-Adriatic Pipeline (TAP) would transport natural gas from Azerbaijan and thus circumvent Iran. Yet, the rapid deterioration in relations between Russia and the EU in the course of the Ukraine crisis has sped up Europe's efforts to diversify its energy sources away from Russia (ECFR 2015; European Commission 2015).[54] But also Russia is seeking to diversify its oil and gas customers and increasingly seeks to export LNG to the Asian market (Westphal 2014). A Russian diplomat acknowledges in an interview that 'in the mid-term, Iranian oil could put some pressure on the Russian economy, depending on oil consumption levels and price fluctuations. In the long term, however, our priority is to make our economy less dependent on oil and gas'.[55] After all nuclear-related sanctions began to be lifted in January 2016, Iran re-started exporting oil to the European Union. This, in addition to the historically low oil price of US$30–40 per barrel in 2014 and 2015, exerted a degree of pressure on the Russian economy. Russia, not an OPEC member, thus has an interest in 'freezing' the oil price at a global level together with OPEC members – a prospect of no interest to Iran. Tellingly, Iran abstained from an OPEC meeting in April 2016 which had precisely this point on the agenda (Gamal and Lawler 2016). While the prospect of Iran becoming an important alternative energy supplier for Europe is still unclear, it cannot be excluded that political dynamics have the potential to shake Russia's position on the European energy market (see also Sasnal and Secrieru 2015). In any event, the JCPOA has taken away the economic justification for the Russian–Iranian oil-for-goods agreement as discussed in the summer of 2014.

A second factor explains why it is advantageous for Russia if international investments still face hindrances on the Iranian market despite the lifting of sanctions as part of the JCPOA: Russia would be in a position to monopolise certain economic sectors on that market. Cooperation in the nuclear technology area (Bushehr) has been the flagship of Russian–Iranian economic cooperation, despite mutual accusations and frustrations as described in section 2 above. On 11 November 2014, Rosatom announced its intention to construct eight additional nuclear power units in Iran (TASS 2014). Spent nuclear fuel will be returned to Russia for reprocessing and storage, while Russia will produce and deliver the nuclear fuel. It is not always clear, however, whether the furtherance of nuclear technology cooperation is a purely commercially driven project (by Rosatom), or whether Rosatom's Iran projects are not at least partially co-decided by Russia's political leadership.[56] 'There is a strong perception in Moscow', Arbatov and Sazhin (2016) write, 'of the indispensable importance of Russia's military and civilian nuclear energy for its global status, defense, economic development, and foreign policy' (10).

It is this sometimes ambivalent business culture where the dividing lines between government and the corporate world become blurred that has led some to speak of 'Russia, Inc.', with Vladimir Putin as its CEO (Hill and Gaddy 2013: 201; Stent 2014: 180–181). Experts from Rosatom were involved in the E3+3 nuclear negotiations with Iran.[57] 'They do a good technical job', a Russian diplomat said in an interview, 'the reconstruction of the Fordow facility is now one of their obligations under the JCPOA. This is a commercial operation for Rosatom, but of course we [the Russian foreign ministry] are in close contact with them'[58] and 'all sensitive activities of Rosatom are under supervision of the government'.[59] In the Iran nuclear talks, 'Rosatom was not a participant, it was a tool', as the well-connected Russian journalist Fjodor Lukyanov points out.[60] Given that Rosatom emerged from the restructured Ministry of Atomic Energy in 2004, this is perhaps not surprising. The official Rosatom website states that 'It [Rosatom] is responsible for meeting Russia's commitments in the nuclear industry with a specific focus on the international nuclear non-proliferation effort'.[61] With the reconstruction of the Fordow facility, Anton Khlopkov writes, 'commercial mechanisms are being used to achieve non-proliferation goals' (2016: 2–3). Here, Rosatom also assumes a responsibility for safety and security standards in transferring sensitive materials and technology. In the Iran case, this is at the border between the technical and the political given that Iran has not signed the Convention on Nuclear Safety, is not a party to the Convention on the Physical Protection of Nuclear Material or the Convention for the Suppression of Acts of Nuclear Terrorism. Beyond the commercial operations of Rosatom, the Russian government could use its leverage to lobby Iran to sign such agreements (Baklitsky and Weitz 2016: 14).

Western apprehensions about business with Iran could guarantee Russia an advantageous market position that results from the absence of Western competition. A normalisation of Iran's relations with the West could change this equation, and the interest on the part of European companies to re-enter the Iranian market following the lifting of nuclear sanctions in January 2016 is a clear indication for a possibly stronger future competition over bidding projects in Iran. At the same time, Russian companies theoretically might be in an advantageous position as long as US sanctions on 'foreign subsidiaries' of US companies continue to be applied. European companies are likely to be affected more by this regulation than Russian ones. 'But if you do transactions in US dollars, you are affected,' a Russian diplomat explains. 'In times of globalisation, companies are linked, so we all feel the effects of US sanctions.'[62] However, the Russian–Iranian trade volume does not account for a big share on either side's external trade balance: Russia's trade with Israel almost reaches numbers comparable to Russian–Iranian trade, despite the fact that Israel's population is 10 times smaller than Iran's (Sazhin 2010). The Iranian–Turkish trade volume is roughly eight times higher than the Russian–Iranian one; in 2014, Iran's trade volume with China was even 30 times higher than the one with Russia. Iran's economy is heavily dependent on oil exports to China, while the latter was happy to fill the void left behind by the absence of Western companies and is proactive entering new contracts following the conclusion of the JCPOA.[63] Yet, as Kozhanov (2015b) points out, what ails the prospect of Russia's stronger economic activity in Iran is not the competition with China, but Russia's structural economic and technological problems. The equipment and technology that Iran is in need of, in other words, Russia lacks itself. 'There are probably not many Russian companies that provide goods and technology that Iran needs. These are probably more European and American companies in the first place', a Russian foreign ministry official admits in an interview.[64] Russian politicians have pointed out on several occasions that the weapons market is an area Russia has an interest in capitalising on again. While the UN weapons embargo in UNSCR 1929 has been lifted, a five-year ban on conventional weapons deliveries was written into the JCPOA. And while Russian officials make reference to the possibility of a UN authorisation prior to 2020 for the sale of weapons,[65] Russian experts are sceptical whether the US would not veto the delivery thereof in the Security Council.[66]

Third, and arguably the most important reason from a global power and prestige perspective, Russia's self-understanding of being an unavoidable global power player enters into the calculation about the direction Russia's Iran policy ought to take. Russia, as a permanent UNSC member, wants to be understood as a state among equals. Russian scepticism voiced during the nuclear talks in Lausanne in March and April 2015 and in the negotiations leading up to the JCPOA over a 'snap back' provision that would automatically re-impose sanctions on Iran if the latter was found in non-compliance with the agreement was indicative in this regard: Moscow's concerns hinted at the dilution of its veto power that a Security Council authority over sanctions matters entails. An EEAS official who was present at the negotiations explained in an interview:

> Russia didn't want any one state to [be able to] notify the Security Council and trigger the automatic re-imposition of sanctions. Now with the Joint Commission, there is a pre-Security Council process. You multilateralise the issue. This makes it legitimate. And it was also a face-saving element for the Iranians.[67]

The JCPOA established a 'Joint Commission' consisting of the E3+3 and Iran 'to monitor the implementation of this JCPOA' (Preamble ix). If a consultation among JCPOA participants about a possible breach of the agreement by any one participant fails to resolve the issue, this issue can be referred to the Joint Commission, which operates according to an elaborate dispute resolution mechanism (Para 36 JCPOA). The re-imposition 'of the old UN Security Council resolutions' (Para 37) has been kept as an eventuality, but only as the result of a lengthy institutional review process (UN 2015: Para 11–13). As a Russian diplomat explained in an interview: 'This was a mechanism introduced by our Western partners who wanted to retain leverage over Iran [...]. They want Iran to see this Damocles sword hanging over them'.[68] He continued: 'How the snap-back would work in practice is a mystery to everyone. If companies are already there [in Iran], how will you unwind complex commercial relations?'[69]

Russian criticism of the modalities of such a mechanism for the re-imposition of sanctions and its acceptance of the compromise found was illustrative of the understanding analysed in this chapter of Russia's approach to international cooperation at large. Drawing on Sakwa's concept of 'neo-revisionism', it is understood here that Russia's working with international organisations of the UN system does not constitute an appeal by Moscow to fundamentally challenge the system of international governance, but to partially revise its functioning (Sakwa 2011, 2015: 28). Viatcheslav Morozov strikes a similar chord, arguing that Russia 'does not challenge the western-dominated world order in any radical way; rather, it claims a legitimate voice in the debate how this world order must evolve' (2013: 19). This observation ties in with the distinction made in Chapter 1 between 'rules and models' as cognitive standards versus 'norms and values' as evaluative standards (Katzenstein 1996: 21) and gives a preliminary answer to the question whether Russian foreign policy on Iran indicates a security culture that resists hegemony. While Russia supports and adheres to the 'rules' and basic functioning of the UN system, its disagreement with other UN members and US power structures reveals a different *normative* understanding of what is deemed legitimate in international relations. Putin's 2014 Valdai speech underlined this distinction as follows: 'We must clearly identify where unilateral actions end and we need to apply multilateral mechanisms' (Putin 2014). The concept of 'neo-revisionism' thus captures the striving for more equitable and thus non-hegemonic international relations, while falling short of outright opposition to hegemony. 'As the existing order is visibly crumbling', Dmitri Trenin (2009) writes in a commentary to Kuchins' and Weitz's chapter in *Powers and Principles* (2009), 'Moscow wants to be present at the creation of its replacement' (189), and therewith echoes their analysis of 'Russia's Place in an Unsettled Order', in which 'an international system of global American hegemony [is] evaporating and being replaced by genuine multipolarity' (166). Elsewhere in Putin's 2014 Valdai speech, the root cause of the new ice age between Russia and the West was therefore formulated as follows: 'The Cold War ended, but it did not end with the signing of a peace treaty with clear and transparent agreements on respecting existing rules or creating new *rules and standards*' (emphasis added). While 'norms' can differ, so the message, a rules-based arrangement between Russia and the West should have ensured an equitable co-existence.[70]

In practice, this should entail a Western recognition and acknowledgement of Russia's status as a Great Power (*derzhavnichestvo*), in a Russian understanding. Russia is thus very aware of its power to veto new sanctions in the UNSC. Combined with deliberations about the state of US–Russian relations, Russia's Iran and Middle East policy can tip the scales in a process either towards greater consultation with Russia or towards international isolation of Russia (Katz 2008). Russia's Great Power status is understood as the recognition of Russia on an equal footing with the United States, and as being in a position to work towards the multipolarisation of international relations – away from what is perceived as a US-centric world order. Russia's foreign policy towards the Syrian civil war is a case in point for the implications of Russian resistance to US-inspired power structures, where not only Russia's steadfast support for the ruling Assad government, but also its military incursion from September 2015 have rendered Russia an unavoidable dialogue partner for the West – and generated substantial levels of frustration over the lack of common approaches to end the bloodshed.[71] In its support for Assad, Russian and Iranian regional interests are converging, while Western and Russian ones had been further drifting apart.[72] 'For Iran, it is an issue of regional balance of power and its own security, and for Russia it is an issue of upholding certain principles of international order and rejection of US pressure', Lukyanov (2014) reflects on this geopolitical convergence of interests between Russia and Iran over Syria. In a Western reading, Russia's foreign policy is thus perceived at times as rebellious, obstinate, and disruptive at worst. In a Russian reading, there is cause for a general distrust of US commitments to keep issue-specific agreements like the Iran deal separate from disagreements in other policy areas. The persistence of US unilateral sanctions and the imposition of new sanctions because of Iranian ballistic missile tests only a day after 'Implementation Day' of the JCPOA are two examples of what is seen as a politicisation of a sectoral agreement.[73] 'The JCPOA is a nuclear agreement. It should not be taken to express disagreements about Iranian regional policies', a foreign ministry official remarked in an interview.[74]

Russian policies towards Ukraine, Syria or Iran reflect the understanding that Russia's voice cannot be overlooked in world politics. The understanding that the foreign policy principles of 'territorial integrity' and 'sovereignty' should govern international politics also has to be understood in this context of the idea that states should treat each other on equal terms. The official Russian Foreign Policy Concept breathes this ambition to 'democratise' international relations (Russian foreign ministry 2013a). 'Democratisation' of international relations would thus accurately characterise Russia's understanding of a desirable security culture to govern international politics. In the reading that 'democratisation' entails the deconstruction of power hierarchies, this is an endeavour explicitly questioning hegemony. If international relations are 'democratic', existing power asymmetries are smoothed out, eliminating hegemonic structures by definition. Such rhetoric underscores an advocacy for a non-hegemonic security culture.

Throughout the decade-old complexities of international politics surrounding the Iranian nuclear case, Russia has alternated in alienating both 'the West' and Iran: Russia is not shying away from resuscitating at a later moment in time potentially controversial deals that had been temporarily halted due to US pressure and (unfavourable) international attention. The much-discussed S-300 deal is a case in point: Frozen under US pressure in 2009 by the Medvedev administration and suspended after the adoption of UNSCR 1929 in 2010, the Putin administration resuscitated the sale in 2015 after the diplomatically encouraging signals emanating from Lausanne and Vienna have provided a favourable window of opportunity to do so. The impression thus occasionally prevailed that while Russia was purporting to propose plans in the P5+1/E3+3 format (the Russian plan, the Lavrov plan), a plethora of Russian commercial interests, Russian energy politics and its global role understanding obstructed a long-term solution to the Iranian nuclear crisis, as formulated, e.g. by former high-ranking European diplomats involved with the Iran file.[75] A former European ambassador stated in an interview: 'Russia creates problems in order to sell solutions.'[76] A high-ranking Swiss diplomat puts it more pointedly: 'Nobody trusts the Russians'.[77]

Russia's foreign policy towards the Iranian nuclear dossier is torn between the public advocacy for more 'democratic' international relations and a security culture that understands security as security *from* hegemonic frameworks. On a behavioural level, the political dependence on the US makes Russia follow a partially accommodating course on other occasions. Dmitri Trenin (2014) speaks of a 'compartmentalised environment' defining US–Russian relations and a pragmatic approach to specific issue areas in which both cooperation and disagreement is possible at the same time.[78] As Angela Stent (2014) has analysed concisely, relations between Russia and the West after 1991, and between Russia and the US in particular, were always characterised by periods of cooperation on issues of common concern and periods of tensions and outright confrontation at the same time. That such remarkable progress and professionalism in the Iran nuclear talks leading up to the JCPOA in the summer of 2015 was possible while an unprecedented crisis of trust between Russia and the West over events in Ukraine and the 'common neighbourhood' and resentments over Western sanctions against Russia froze all other cooperation, is probably the starkest testament to this observation.

6 Conclusion

Russian foreign policy toward Iran and its nuclear programme has to be seen in the context of Russia's political relations with 'the West' in general, and with the US in particular. Russo-Iranian commercial cooperation in the civilian nuclear sphere was a legacy that impacted on Russia's reaction to the 2002 revelations of hitherto covert nuclear facilities in Iran. Russian foreign policy in the nascent Iranian nuclear crisis made a distinction between purely commercial and legitimate nuclear technology usage (Bushehr), and a security political dimension of the Iranian nuclear programme ('Western allegations of military intentions remain unproven'). Here, Russia was also driven by an understanding that US policies towards Iran's file were essentially politicised during the Bush administration. When the Iranian nuclear file was referred to the UNSC in 2006, Russia was slowing down the sanctions track, but eventually accepted and approved of international sanctions. This was seen as a lesser evil because the perceived alternative to sanctions could have been war with Iran, especially at a time when voices calling for unilateral air strikes grew louder. The approval of sanctions also had to be read as a frustration with Iran's rejection of Russian technical proposals to de-escalate the tensions in 2006. Especially the revelation of the Fordow facility in 2009 set the stage for a momentum towards a P5 compromise on a new sanctions round in 2010. Here, working through UN channels to shape the outcome of sanctions was seen as the best way to exert influence over the Iran file and prevent worse policies from materialising, while Russian public rhetoric conveyed a scepticism of the use of sanctions all along. Moscow's striving to preserve Great Power status in what Russian decision-makers refer to as a 'polycentric' world explains its advocacy for a security culture that breathes the ambition to 'democratise' international relations and resist US pressure. Russia and its Western counterparts in the P5+1 framework occasionally have appeared to be standing on two opposite ends of the spectrum of political instruments when it came to approaching the Iranian nuclear file. This was all the more true when Russia was vocal about its criticism of Western attempts from 2011 onwards to isolate Iran economically. As this chapter has shown, however, taking such disagreements as signs of an unalterable freezing into mutually opposed camps and portraying Russia as a cumbersome veto player in the UNSC does not do justice to much more complex foreign policy positions that have to bridge official discourse(s) with largely material, global power political and security motivations.

As has been discussed, an occasionally remarkable degree of pragmatic cooperation with the US on the Iran file has been observed – not only at the peak of the Obama–Medvedev 'reset' policy, but also in diplomatic negotiations in the P5+1 format. An outspoken criticism of Western unilateral sanctions policies was paralleled by a desire to be perceived as a constructive player in the Iranian nuclear dossier. A number of Russian initiatives (Putin's 2006 proposal for the multilateralisation of the nuclear fuel cycle, the 2011 'step-by-step' plan, the creation of a fuel consortium on Russian soil) are indicative of this Russian willingness to make proposals on a technical level. The constructive Russian contribution to the Iran nuclear talks in spite of the freezing of policy coordination with the West in most other areas in the wake of the 'Ukraine crisis' from 2014 was testament to a compartmentalisation of foreign policy that allowed progress on specific issue areas. In the Iran talks, Russia helped to resolve a number of key issues, including the reconstruction of the Fordow enrichment plant and the shipment of excessive quantities of low-enriched uranium out of Iran. This contribution helped pave the way to the successful negotiation of the Joint Comprehensive Plan of Action in 2015. Moscow's subsequent approval of UNSCR 2231 that endorsed the JCPOA meant that Russia gave its full backing to the solution of one of the most ardent international security problems, despite arguments that it might stand to lose economically from an altered international standing of Iran that could result from this.

Russia's Iran policy is an illustration of a state's foreign policy that challenges hegemonic structures, but works within and through the system of international governance as the best means to check on such power structures. It is an example of a friction between contestation and accommodation, between resistance and consent. Russia's search for a foreign policy identity, like that of any other state, is an iterative process as the outcome of the state's international context, its self-understanding, and the perception thereof by other actors (see also Ziegler 2012). This finding explains the seeming variation in Russia's Iran policy, where the advocacy for a security culture that resists US-inspired power structures does not always coincide with divergence from these very structures on a behavioural level.

Notes

1 An earlier version of this chapter was published as Pieper, Moritz (2015). Between the Democratisation of International Relations and Status Quo Politics: Russia's Foreign Policy towards the Iranian Nuclear Programme. *International Politics* vol. 52, no. 5: 567–588.

2 Seyed Hossein Mousavian (2012) writes that Russia and Iran had been negotiating 'secretly for cooperation on an enrichment facility in Natanz and heavy-water reactors in Arak, but Russia halted this cooperation under pressure from the United States in the late 1990s' (55).
3 Most notably Germany and France.
4 Russia is contracted to deliver fuel for the Bushehr plant until at least 2021 (Meier 2014: 19), but has agreed to provide fuel for the plant's lifetime in an additional protocol.
5 In the 1990s and early 2000s, the US portrayed the Bushehr power plant as a problem, while not recognising the real proliferation risk that lay in non-declared nuclear facilities. 'These were dilettantes in the White House that did not understand the nuclear fuel cycle', Carnegie nuclear proliferation expert Dr Mark Hibbs commented in an interview with the author, 18 March 2015. The US scepticism of Russian nuclear activities in Iran was reversed a decade later: In nuclear talks with Iran in early 2015, the US delegation was supportive of the idea to include the Russian atomic industry as a technical actor for a final nuclear settlement of the Iranian nuclear programme.
6 Author's interview with Russian diplomat to the US, Washington, 31 October 2014.
7 Also re-iterated by a Russian foreign ministry official. Author's interview, Moscow, 18 April 2013.
8 Author's interview at the Russian permanent mission to the IAEA, Vienna, 18 August 2016.
9 Iran exhibited a 'quietist narrative towards Russia's hegemonic posture [...] in Chechnya and the north Caucasus', as Dorraj and Entessar (2013: 8) put it. Adib-Moghaddam (2010) lists Iran's silence on Russia's Chechnya policy as an example of Iranian '"eclectic" pragmatism' (74). See also Freedman (2007: 199); Stent (2014: 32). A caveat on sectarianism in political Islam should sound a note of caution on Iran's ability to stir up tensions in this region, however: While Iran is a majoritarian Shia country, Muslims in the Caucasus are predominantly Sunni. As in Afghanistan, both Iran and Russia have an interest in staunching Sunni extremism in the region. This point was also made by Dr Vladimir Sazhin, Author's interview, Moscow, 17 April 2013. The rise of fundamentalist Sunni groups like the so-called 'Islamic State' has underscored this convergence even more.
10 Referred to as 'the Russian plan'.
11 Evegeny Primakov writes that the Iranian rejection of Putin's proposal to produce nuclear fuel for Iran on Russian soil was unfortunate. This plan, he held, 'could have been a way out of a delicate position' (Primakov 2016: 230).
12 With Bushehr, Moscow held leverage over Iran as far as fuel and the technical operation of the plant was concerned. Fuel fabrication and insertion is a technically difficult process and is best carried out by the actual producer of the plant. In addition, Iran needed the Russian technicians to operate the plant, as was also evidenced by the informal prolongation of the initially contracted two-year period during which Russian technicians were supposed to work in Bushehr. In this sense, through its nuclear technology cooperation, Moscow had channels through which it was able to make its voice heard in Tehran. Author's interview with Dr Anton Khlopkov, Director of CENESS, Moscow, 17 April 2013.
13 While Russian officials cited Iranian payment delays as reasons for the delay in Bushehr going 'online', Iran was referring to technical deficiencies on Russia's end. Author's interview with Dr Anton Khlopkov, Moscow, 17 April 2013.
14 As also confirmed by a German foreign ministry official, conversation with author, Berlin, 4 February 2013.
15 Author's interview, Moscow, 18 April 2013.
16 The treaty of Turmenchai was an agreement between Persia and the Russian empire and concluded the Russo-Persian war (1826–1828). In it, Iran's Caucasian territories were ceded to Russia.
17 Therewith also supposedly breaching IAEA modified code 3.1, which stipulates the acknowledgement of new facilities already as from their planning phase (IAEA 2011).

18 Obama's national security advisor General James Jones revealed the existence of the Qom facilities to Medvedev's national security advisor Sergei Prikhodko in September 2009 (Stent 2014: 232–233). In her memoir, Hillary Clinton (2014) recalls how the 'steely Lavrov' appeared 'flustered and at a loss of words' when the US informed Russia on the Qom facility. She attributes Russia's noticeable change in rhetoric to this revelation (425). Fitzpatrick (2010) writes that 'the Russians were not amused that they had to hear about the Fardow plant from the Americans' (71).
19 As confirmed by a former Russian diplomat posted to Tehran. Conversation with author, Washington, 1 November 2014.
20 Author's interview with Dr Ali Vaez, chief Iran analyst, International Crisis Group, via Skype, 25 July 2013.
21 A politically more tactical alternative might have been the temporary (unlimited) freezing of the contract, instead of having it completely cancelled altogether. Author's interview with Dr Anton Khlopkov, Moscow, 17 April 2013.
22 While imposing a weapons embargo, paragraph 7 of UNSCR 1929 makes no mention of surface-to-air systems. The S-300 is such a system, so on technical grounds, Russia could have delivered the system to Iran. Iran has thus filed a lawsuit before the International Arbitration Court in Geneva against Russia for cancelling the S-300 contract. Reportedly, the S-300 deal was subject of discussion during Russian defense minister Sergei Shoigu's meeting with his Iranian counterpart Hossein Dehqan in Tehran in January 2015 (Fars News Agency 2015). And on 13 April 2015, two weeks after the conclusion of the political framework agreement with Iran in Lausanne, President Putin officially cancelled the suspension and paved the way for an eventual delivery (Kozhanov 2015a). The delivery began in April 2016 and was publicly criticised by the US (Ria Novosti 2016).
23 Dr Alexei Arbatov, head of the Center for International Security at IMEMO, asserts that the S-300 decision was the 'peak' of the US–Russian reset policy. Author's interview, Moscow, 13 November 2013.
24 This point was also persistently made in all interviews with Russian officials carried out for this research. Russian ambassador to the EU Vladimir Chizhov also reiterated this position at a keynote speech on 2 February 2015; University of Kent's EU–Russia conference, Brussels.
25 Author's interview with Russian foreign ministry official, Moscow, 18 April 2013.
26 Author's interview with Russian foreign ministry official, Moscow, 18 April 2013.
27 Along with Rosoboronexport, the Bush Administration also sanctioned the Tula Instrument-Making Design Bureau and the Kolomna Machine-Building Design Bureau in 2007 (Belopolsky 2009: 127). The sanctions against Sukhoi, imposed because of alleged violations of the Iran Nonproliferation Act of 2000 were lifted again in November 2006 (Ria Novosti 2007).
28 Author's interview with Russian diplomat, Brussels, 19 March 2015.
29 In the form of an international fuel bank or with Russia serving as destination for the return of Iranian spent fuel rods.
30 Author's interview with Andrey Baklitsky, Moscow, 10 June 2016; author's interview with Dr Alexei Arbatov, Moscow, 21 June 2016. The removal of the excess LEU took place in late December 2015. 'The LEU shipment was a Safeguards nightmare,' a Russian diplomat at the Russian permanent mission to the IAEA recalls in an interview. 'The Iranians had multiple stockpiles all over the country with different safety standards. Taking care of this gave us many sleepless nights'. Author's interview, Vienna, 18 August 2016.
31 Author's interview with Andrey Baklitsky, Moscow, 10 June 2016. See also paragraph 7.1 of JCPOA Annex III on Civil Nuclear Cooperation.
32 Author's interview with Russian foreign ministry official, Moscow, 18 April 2013.
33 Author's interview with Russian foreign ministry official, Moscow, 12 November 2013; Author's interview with Russian diplomat to the US, Washington, 31 October 2014.

34 Author's interview with Dr Anton Khlopkov, Moscow, 17 April 2013; conversation with German diplomat, Moscow, 18 April 2013.
35 Author's interview at the Russian mission to the IAEA, Vienna, 18 August 2016.
36 Author's interview, London, 1 April 2016.
37 Author's interview with Russian foreign ministry official, Moscow, 12 November 2013, Author's interview with Russian diplomat, Brussels, 15 April 2016.
38 This point was made by Prof. Andrei Makarychev. Panel discussion, University of Kent's EU–Russia conference, Brussels, 3 February 2015.
39 Author's interview with EEAS official, Brussels, 20 March 2014.
40 Asked about Ryabkov's statement about a possible Ukraine–Iran issue linkage, a Russian diplomat remarked in an interview: 'Yes, but that was in March [of 2014]. These were emotional times'. Author's interview, Washington, 31 October 2014.
41 It should not come as a surprise that Russian diplomats emphasise this point. But also non-Russian officials and experts have confirmed this assessment: Author's interview with Dr Mark Fitzpatrick, London, 11 July 2014; with Dina Esfandiary, London, 11 July 2014; with European diplomat, Berlin, 14 November 2014; with former E3 delegation member, Brussels, 29 January 2015; with European diplomat, Brussels, 6 February 2015; EEAS official, Brussels, 24 February 2015; with E3 official, London, 10 March 2015; with Dr Mark Hibbs, Berlin, 18 March 2015 with EEAS official, Brussels, 15 April 2016.
42 Russian exports to Iran are mainly metals, food, and machinery.
43 A European diplomat also questioned the desirability of the agreement on the Iranian side as follows: 'What does Russia have that Iran really wants?'. Author's interview, Brussels, 6 February 2015.
44 See Kozhanov (2015b) on the 'revitalization of Russian–Iranian relations' and the impact of tensions with the West thereon. Due to a clause in the Joint Plan of Action of November 2013 negotiated between Iran and the P5+1, the implementation of such an oil-for-goods deal would also be in contravention of the JPoA. The latter 'allows Iran to continue exporting a total of 1 million barrels a day of oil to six countries: China, India, Japan, South Korea, Taiwan and Turkey'. Since Russia was not an oil costumer of Iran at the time the JPoA was negotiated, it was not included in this provision.
45 Author's interview with European Iran desk officer, Berlin, 14 November 2014. See also Kozhanov (2015b).
46 In the wake of Russia's annexation of Crimea, Russia has been downgraded to one notch above 'junk' status first. In January 2015, Standard & Poor's downgraded Russia's foreign currency credit rating to junk status, thereby placing it below investment grade (Andrianovna and Galouchko 2015).
47 Andrey Baklitsky also held that decisions of such strategic importance are generally taken by the president's office. The decision 'not to blow everything up' and go ahead with negotiations thus was a top-level political decision. Author's interview, Moscow, 10 June 2016.
48 Author's Interview with high-ranking Swiss diplomat, Berlin, 26 August 2013. Similar points were also made by Dr Walter Posch, senior Iran analyst at SWP, author's interview, Berlin, 25 June 2013.
49 Author's interview with European diplomat, Iran desk, Berlin, 14 November 2014; Author's interview with Swiss high-ranking diplomat, Berlin, 26 August 2013; author's interview with Dr Walter Posch, Berlin, 25 June 2013; author's interview with former German ambassador to Iran, Berlin, 26 August 2013; author's interview with European diplomat, Brussels, 6 February 2015. See also Mironova (2015).
50 Author's interview with Fjodor Lukyanov, Moscow, 14 June 2016.
51 Author's interview with Russian diplomat, Brussels, 19 March 2015. Email correspondence with Russian diplomat to the EU, 12 March 2015; author's interview with Russian diplomat, Brussels, 7 October 2014.
52 Russian official at the permanent mission of the Russian federation to the EU, email correspondence with author, 13 March 2014.

53 A Russian diplomat put it in an interview as follows:

> (The) EU has a strong dependence on Russian gas. Iran would not have such a strong input once sanctions are lifted. Maybe energy-starved Southeast-Asian countries would jump on that opportunity, but (the) Russian energy export is also fixed because of pipelines going to Europe. Nabucco is not moving forward, pipelines need time to be built, it takes a while.
>
> (Author's interview, Brussels, 19 March 2015)

54 Fjodor Lukyanov points out a paradox here: Since Ukraine makes a profit from the transit of Russian gas to Europe, European attempts to decrease its gas dependence on Russia over the Ukraine crisis would harm Ukraine economically. Author's interview, Moscow, 14 June 2016.

55 Author's interview, London, 1 April 2016.

56 Proliferation expert Dr Mark Hibbs held the latter view. Rosatom's economic incentive to invest in Iran, which is chronically short of money, can be questioned. If it was not for a political signal of commercial autonomy (that the Kremlin is determined to send), Rosatom's board might be better advised to invest in more financially stable markets abroad, Dr Hibbs held. Author's interview, Berlin, 18 March 2015. Traditionally, Minatom, and since 2004 Rosatom, have had an important role in driving Russian Iran policy, as Belopolsky (2009) holds: 'Given the economic rewards to be reaped in these markets, Minatom Ministers freelanced, developing policy towards these states [China, Iran, Iraq] which was not coordinated or necessarily in line with state policy' (58). Yet, Fjodor Lukyanov argues that of all the Russian state companies, Rosatom is the least politicised, unlike Gasprom, which has been used for political goals before. Author's interview, Moscow, 14 June 2016. Another Russian nonproliferation expert stated that 'Rosatom does not do charity work for the sake of nonproliferation' and is interested in commercial returns. Author's interview with Dr Alexei Arbatov, Moscow, 21 June 2016.

57 Author's interview with Russian diplomat, Brussels, 15 April 2016; author's interview with Russian foreign ministry official and former member of the UN Panel of Experts on Iran, Moscow, 14 June 2016.

58 Author's interview with Russian diplomat, Brussels, 15 April 2016.

59 Author's interview with Russian foreign ministry official and former member of the UN Panel of Experts on Iran, Moscow, 14 June 2016.

60 Author's interview, Moscow, 14 June 2016.

61 www.rosatom.ru/en/global-presence/international-relations/ (accessed 17 June 2016).

62 Author's interview, Brussels, 15 April 2016.

63 On this, see the following chapter on China's foreign policy toward Iran.

64 Author's interview, Moscow, 14 June 2016.

65 Author's interview with Russian diplomat, Brussels, 15 April 2016; author's interview with Russian foreign ministry official, Moscow, 14 June 2016.

66 Author's interview with Andrey Baklitsky, Moscow, 10 June 2016.

67 Author's interview, Brussels, 15 April 2016.

68 Author's interview, Brussels, 15 April 2016.

69 Author's interview, Brussels, 15 April 2016.

70 Western responses to this reading usually hold that the Russian reference to mutual respect and equality in diplomacy often is upheld as a disguise to breach international obligations.

71 Fjodor Lukyanov made the same point in an interview. 'One motivation behind the Syrian intervention was to broaden the framework of cooperation and communication', he said. Author's interview, Moscow, 14 June 2016.

72 For useful analyses of Russian policies in the unfolding Syrian crisis, see Allison 2013; Averre and Davies 2015; Charap 2013; Trenin 2013.

73 In Annex B of the JCPOA:

> Iran *is called upon* not to undertake any activity related to ballistic missiles designed to be capable of delivering nuclear weapons, including launches using such ballistic missile technology, until the date eight years after the JCPOA Adoption Day or until the date on which the IAEA submits a report confirming the Broader Conclusion, whichever is earlier.
>
> (Emphasis added)

74 Author's interview with Russian foreign ministry official and former member of the UN Panel of Experts on Iran, Moscow, 14 June 2016.
75 Ibid.; author's interview with Dr Walter Posch, Berlin, 25 June 2013; author's interview with former German ambassador to Iran, Berlin, 26 August 2013; author's interview with high-ranking Swiss diplomat, Berlin, 26 August 2013. See also Mousavian (2012: 397f.).
76 Author's interview with former European ambassador, London, 15 September 2016.
77 Author's interview with high-ranking Swiss diplomat, Berlin, 26 August 2013.
78 An E3 official speaks of 'pockets of cooperation' between Russia and the West. Author's interview, London, 10 March 2015.

References

Adib-Moghaddam, Arshin. 2010. *Iran in World Politics. The Question of the Islamic Republic*. New York: Columbia University Press.

Allison, Roy. 2013. Russia and Syria: Explaining Alignment with a Regime in Crisis. *International Affairs* vol. 89, no. 4: 795–823.

Andrianovna, Anna and Galouchko, Ksenia. 2015. S&P Cuts Russia's Rating to Junk; Sanctions and Oil Slump Hammer Ruble. *Bloomberg News*, 27 January 2015. Available at www.bloomberg.com/news/2015-01-26/russia-credit-rating-cut-to-junk-by-s-p-for-first-time-in-decade.html (accessed 27 January 2015).

Antonenko, Oksana. 2001. Russia's Military Involvement in the Middle East. *Middle East Review of International Affairs*, vol. 5, no. 1: 1–14.

Aras, Bülent and Ozbay, Fatih. 2006. Dances with Wolves: Russia, Iran and the Nuclear Issue. *Middle East Policy* vol. 13, no. 4: 132–147.

Arbatov, Alexei. 2014. Iran, Russia, and the Ukrainian Crisis. *The National Interest*, 17 July 2014. Available at http://nationalinterest.org/blog/the-buzz/iran-russia-the-ukrainian-crisis-10902?page=3 (accessed 20 October 2014).

Arbatov, Alexei and Sazhin, Vladimir. 2016. The Nuclear Deal With Iran: The Final Step or a New Stage? *Carnegie Article*, 20 April 2016. Available at http://carnegie.ru/2016/04/20/nuclear-deal-with-iran-final-step-or-new-stage/ixc4?mkt_tok=eyJpIjoiWVRNNE1UbGlPRFZpTTJGaSIsInQiOiJTOFpReDM3Q1c5MFZtUGhqZzJWXC94Njd6MHB0TFJRN2ZjTzV4YWpTNUl0V0tXT2pBbmxHUDFrVEJNa3pranpwXC9DRDBoWFg2cWZBU29FMDg1NjI5cERRTTl2NkZXejhIeWp2YitVM1BKczZNPSJ9 (accessed 27 June 2016).

Averre, Derek and Davies, Lance. 2015. Russia, Humanitarian Intervention and the Responsibility to Protect: the Case of Syria. *International Affairs* vol. 91, no. 4: 813–834.

Baklitsky, Andrey and Weitz, Richard. 2016. The Iranian Deal: Opportunities and Obstacles for Russian–US Cooperation. *Valdai Papers* no. 46, April 2016.

Belopolsky, Helen. 2009. *Russia and the Challengers. Russian Alignment with China, Iran, and Iraq in the Unipolar Era*. Basingstoke: Palgrave Macmillan.

Center for Energy and Security Studies. 2015. The Iran Nuclear Deal: Russia's Interests and Prospects for Implementation. 14 August. *Transcript of a Meeting with Russian*

Deputy Foreign Minister Sergey Ryabkov. Available at: http://ceness-russia.org/data/page/p1494_1.pdf?mkt_tok=3RkMMJWWfF9wsRogvazKZKXonjHpfsX64u8rUKCg38431UFwdcjKPmjr1YUDScd0aPyQAgobGp5I5FEIQ7XYTLB2t60MWA%3D%3D (accessed 2 September 2015).

Charap, Samuel. 2013. Russia, Syria and the Doctrine of Intervention. *Survival* vol. 55, no. 1: 35–41.

Clinton, Hillary Rodham. 2014. *Hard Choices*. New York: Simon & Schuster.

Conrad, Matthias. 2011. *NATO-Russia Relations under Putin. Emergence and Decay of a Security Community? An Analysis of the Russian Discourse on NATO (2000–2008)* Wien/Münster: LIT Verlag.

Cox, Robert. 1996. *Approaches to World Order*. Cambridge: Cambridge University Press.

Defense Industry Daily. 2006. US Ban on Russian Defense Firms Raises the Stakes. *Defense Industry Daily*, 11 August 2006. Available at www.defenseindustrydaily.com/us-ban-on-russian-defense-firms-raises-the-stakes-02522/ (accessed 17 October 2013).

Deyermond, Ruth. 2013. Assessing the Reset: Successes and Failures in the Obama Administration's Russia Policy, 2009–2012. *European Security* vol. 22, no. 4: 500–523.

Diakov, Anatoli. 2007. The Nuclear Fuel Cycle. In: Arbatov, Alexei (ed.), *At the Nuclear Threshold. The Lessons of North Korea and Iran for the Nuclear Non-Proliferation Regime*. Moscow: Carnegie Endowment for International Peace, pp. 127–141.

Dorraj, Manochehr and Entessar, Nader. 2013. Iran's Northern Exposure: Foreign Policy Challenges in Eurasia. Georgetown University School of Foreign Service in Qatar, Center for International and Regional Studies, *Occasional Paper* No. 13.

ElBaradei, Mohamed. 2012. *The Age of Deception. Nuclear Diplomacy in Treacherous Times*. New York: Bloomsburg.

Esfandiary, Dina and Fitzpatrick, Mark. 2011. Sanctions on Iran: Defining and Enabling 'Success'. *Survival* vol. 53, no. 5: 143–156.

European Commission. 2015. Energy Union Package. Communication from the European Commission to the European Parliament, the Council, the European Economic and Social Committee, the Committee of the Regions and the European Investment Bank: A Framework Strategy for a Resilient Energy Union with a Forward-Looking Climate Change Policy. *European Commission Communication* COM(2015) 80, 25 February 2015. Available at http://ec.europa.eu/priorities/energy-union/docs/energyunion_en.pdf (accessed 13 April 2015).

European Council on Foreign Relations. 2015. Europe's Alternatives to Russian Gas. *Article*, 9 April 2015. Available at www.ecfr.eu/article/commentary_europes_alternatives_to_russian_gas311666 (accessed 13 April 2015).

Eurostat. 2012. EU Oil Imports, by Country of Origin. *Statistics Explained*, 19 July 2012. Available at http://epp.eurostat.ec.europa.eu/statistics_explained/index.php?title=File:EU_Oil_imports,_by_country_of_origin_(in_%25).png&filetimestamp=20120719133014 (accessed 16 October 2013).

Fars News Agency. 2014. Russian Diplomat: IAEA Report on Iran Should Result in Annulment of All Sanctions. *Fars News Agency*, 5 February 2014. Available at http://english.farsnews.com/newstext.aspx?nn=13921116001172 (accessed 22 February 2014).

Fars News Agency. 2015. Iran, Russia Agree on Delivery of S300 Defense System. *Fars News Agency*, 20 January 2015. Available at http://english.farsnews.com/newstext.aspx?nn=13931030000940&mkt_tok=3RkMMJWWfF9wsRoluavIZKXonjHpfsX64u8rUKCg38431UFwdcjKPmjr1YACTMd0aPyQAgobGp5I5FEIQ7XYTLB2t60MWA%3D%3D (accessed 27 January 2015).

Fitzpatrick, Mark. 2006. Iran and North Korea: The Proliferation Nexus. *Survival* vol. 48, no. 1: 61–80.

Fitzpatrick, Mark. 2010. Iran: The Fragile Promises of the Fuel-Swap Plan. *Survival* vol. 52, no. 3: 67–94.

Fitzpatrick, Mark. 2014. The Ukraine Crisis and Nuclear Order. *Survival* vol. 56, no. 4: 81–90.

France24. Moscow Rebuffs Obama's Secret Iran-missile Shield Deal. *France24*, 4 March 2009 [last accessed 19 October 2013]: www.france24.com/en/20090303-moscow-rebuffs-secret-obama-deal-medvedev-missile-shield-iran.

Freedman, Robert O. 2000. Russian–Iranian Relations in the 1990s. *Middle East Review of International Affairs* vol. 4, no. 2: 65–78.

Freedman, Robert O. 2007. Russia, Iran and the Nuclear Question: The Putin Record. In: Kanet, Robert E. (ed.), *Russia. Re-Emerging Great Power*. Basingstoke: Palgrave Macmillan, pp. 195–221.

Gamal, Rania El and Lawler, Alex. 2016. With or without Iran, Oil Producers to Meet in April on Output Deal. *Reuters Markets*, 16 March 2016. Available at www.reuters.com/article/us-oil-opec-talks-idUSKCN0WI0Q2 (accessed 9 June 2016).

Hibbs, Mark. 2013. An Iran Deal Buy-in for Russia? *Carnegie Article*, 3 January 2013. Available at http://carnegieeurope.eu/2013/01/03/iran-deal-buy-in-for-russia/ez4g (accessed 11 March 2013).

Hill, Fiona and Gaddy, Clifford G. 2013. *Mr. Putin. Operative in the Kremlin*. Washington: Brookings Institution Press.

IAEA. 2011. Implementation of the NPT Safeguards Agreement and Relevant Provisions of Security Council Resolutions in the Islamic Republic of Iran. *IAEA board report,* 24 May 2011. Available at www.iaea.org/Publications/Documents/Board/2011/gov2011-29.pdf (accessed 24 February 2014).

International Crisis Group. 2006. Iran: Is There a Way Out of the Nuclear Impasse? *Crisis Group Middle East Report* no. 51, 23 February 2006.

Jafarzadeh, Alireza. 2007. *The Iran Threat. President Ahmadinejad and the Coming Nuclear Crisis*. New York: Palgrave Macmillan.

Katz, Mark. 2008. Russian–Iranian Relations in the Ahmadinejad Era. *Middle East Journal* vol. 62, no. 2: 202–216.

Katz, Mark. 2010. Russia–Iranian Relations in the Obama Era. *Middle East Policy* vol. 17, no. 2: 62–69.

Katz, Mark. 2012. Russia and Iran. *Middle East Policy* vol. 19, no. 3: 54–64.

Katzenstein, Peter. J. (ed.). 1996. *The Culture of National Security. Norms and Identity in World Politics*. New York: Columbia University Press.

Khlopkov, Anton. 2016. JCPOA: Early Results. A View from Russia. 14 July 2016. Available at https://csis-prod.s3.amazonaws.com/s3fs-public/160714_Anton_Khlopkov_JCPOA.pdf (accessed 31 August 2016).

Kozhanov, Nikolai. 2015a. Tshem obernetsya reshenie Rossii nostavlyat‘ Iranu S-300 (Russia's S-300 Sale to Iran: An Expected Surprise). *Carnegie Commentary*, 15 April 2015. Available at http://carnegie.ru/2015/04/15/ru-59780/i73t (accessed 22 April 2015).

Kozhanov, Nikolai. 2015b. Understanding the Revitalization of Russian–Iranian Relations. *Carnegie Paper*, 5 May 2015. Available at http://carnegie.ru/2015/05/05/understanding-revitalization-of-russian-iranian-relations/i86n?mkt_tok=3RkMMJWWfF9wsRojuq%2FOZKXonjHpfsX64u8rUKCg38431UFwdcjKPmjr1YcJRcF0aPyQAgobGp5I5FEIQ7XYTLB2t60MWA%3D%3D (accessed 13 May 2015).

Kuchins, Andrew and Weitz, Richard. 2009. Russia's Place in an Unsettled Order: Calculations in the Kremlin. In: Schiffer, Michael and Shorr, David (eds), *Powers and Principles. International Leadership in a Shrinking World*. Plymouth: Lexington Books, pp. 165–196.

Landler, Mark. 2016. *Alter Egos. Hillary Clinton, Barack Obama, and the Twilight Struggle over American Power*. London: WH Allen.

Lata, Vasily and Khlopkov, Anton. 2003. Iran's Missile and Nuclear Challenge: A Conundrum for Russia. *PIR Center report*. Available at www.pircenter.org/report/lata_05-08-2003.pdf (accessed 28 January 2013).

Libman, Alexander. 2014. Aussenwirtschaftlicher Protektionismus in Russland. *SWP-Aktuell* 69, November.

Lukyanov, Fyodor. 2014. Russia Plays the Iran Card. *Al-monitor*, 17 January 2014. Available at www.almonitor.com/pulse/originals/2014/01/geneva-ii-russia-syria-diplomacy.html?utm_source=dlvr.it&utm_medium=twitter&goback=%2Egde_1491617_member_5830606003795693568#%21 (accessed 20 January 2014).

Medvedev, Dmitry. 2010. In: Russia's Medvedev Criticizes Iran Sanctions: Report. *Reuters*, 17 June 2010. Available at www.reuters.com/article/2010/06/18/us-nuclear-iran-medvedev-idUSTRE65H0M720100618www.reuters.com/article/2010/06/18/us-nuclear-iran-medvedev-idUSTRE65H0M720100618 (accessed 19 October 2013).

Meier, Oliver. 2014. In der Krise liegt die Chance. Der Atomkonflikt mit Iran und seine Auswirkungen auf das nukleare Nicht-verbreitungsregime. *SWP research paper*, October 2014. Berlin: SWP.

Meier, Oliver and Pieper, Moritz. 2015. Russland und der Atomkonflikt mit Iran. Kontinuitäten und Brüche bei den russischen Interessen im Zeichen der Ukrainekrise. *SWP-Aktuell* 38, April 2015: 1–4.

Mironova, Irina. 2015. Rossia i Iran na mirovykh gasovykh rynkakh: budushaya konkurentsia neisbeshna? (Russia and Iran on the global gas market: is future competition inevitable?) PIR Center Analysis. *Index Bezopasnosti* Nr. 112 (Security Index). Available at http://pircenter.org/media/content/files/13/14301343850.pdf (accessed 28 May 2015).

Morozov, Viatcheslav. 2013. Subaltern Empire? *Problems of Post-Communism* vol. 60, no. 6: 16–28

Mousavian, Seyed Hossein. 2012. *The Iranian Nuclear Crisis. A Memoir*. Washington: Carnegie Endowment for International Peace.

Nizameddin, Talal. 2013. *Putin's New Order in the Middle East*. London: Hurst.

Orlov, Vladimir A. and Vinnikov, Alexander. 2005. The Great Guessing Game: Russia and the Iranian Nuclear Issue. *The Washington Quarterly* vol. 28, no. 2: 49–66.

Parker, John W. 2009. *Persian Dreams. Moscow and Tehran Since the Fall of the Shah*. Washington, D.C.: Potomac Books.

Parsi, Trita. 2012. *A Single Roll of the Dice: Obama's Diplomacy with Iran*. New Haven: Yale University Press.

Patrikarakos, David. 2012. *Nuclear Iran. The Birth of an Atomic State*. London, New York: I.B. Tauris.

Porter, Gareth. 2014. *Manufactured Crisis. The Untold Story of the Iran Nuclear Scare*. Charlottesville: Just World Books.

Primakov, Yevgeny. 2016. *Mir bez Rossii? K chemu vedet politicheskaya blizorukost' (A World Without Russia? Where Political Shortsightedness Leads us)*. Moscow: Centrpoligraf.

Putin, Vladimir. 2003. Interv'iu Prezidenta Rossii V.V. Putina Amerikanskim telekanalam. Novo-Ogoreva. September 20, 2003 Soobshschenie Press Sluzhby Prezidenta Rossijskoj Federatsii. MID RF DIP, Informatsionnij Biulleten' (Interview of American TV channels with Russian president V. Putin). *Bulletin of the Press Service of the Russian Federation*, 24 September 2003.

Putin, Vladimir. 2012. Rossiia i meniaiushchiisia mir (Russia and the Changing World). *Moskovskie novosti*, Article, 27 February 2012. Available at www.mn.ru/politics/78738 (accessed 9 June 2016).
Putin, Vladimir. 2014. Speech at the Meeting of the Valdai International Discussion Club. *Kremlin Transcript*. Available at http://eng.kremlin.ru/news/23137 (accessed 26 January 2015).
Reuters. 2010. Russia Warns U.S. against Unilateral Iran Sanctions. *Reuters*, 13 May 2010. Available at www.reuters.com/article/2010/05/13/us-russia-iran-us-idUSTRE64C1SU 20100513 (accessed 19 October 2013).
Ria Novosti. 2007. Moscow to Press U.S. for Sanction Cancellation – FM Lavrov. *Ria Novosti* 3 February 2007. Available at http://en.ria.ru/world/20070203/60158997.html (accessed 17 October 2013).
Ria Novosti. 2016. SShA vystupaiut protiv prodazhi Rossiei Iranu kompleksov S-300 (USA protests against Russian S-300 sale to Iran). *Ria Novosti*, 18 April 2016. Available at http://ria.ru/world/20160418/1414660614.html (accessed 9 June 2016).
Richter, Paul and Mostaghim, Ramin. 2015. Iran Shifts on Key Issue in Nuclear Talks as Deadline Looms. *Los Angeles Times*, 30 March 2015. Available at www.latimes.com/world/europe/la-fg-iran-nuclear-20150330-story.html (accessed 2 April 2015).
Russian foreign ministry. 2012a. Kommentarii Departamenta Informatsii i Pechati MID Rossii v sviazi s vvedeniem evropejskim soiuzom novykh sanktsij v otnoshenii Irana (Statement by the Russian foreign ministry press department regarding the new EU sanctions against Iran), *Press statement*, 23 November 2012. Available at www.mid.ru/bdomp/ns-rasia.nsf/1083b7937ae580ae432569e7004199c2/c32577ca00174586442579 8e005323a7!OpenDocument, accessed 1 March 2012 (accessed 2 February 2014).
Russian foreign ministry. 2012b. Kommentarii Departamenta Informatsii i Pechati MID Rossii na vopros agenstva 'Interfaks' otnositel'no vozmozhnosti provedenia 'Shesterkoj' novoj vstrechi s uchastiem Irana (Statement by the Russian foreign ministry press department on the question of 'Interfax' agency on the possibility of the new P5+1 meeting with Iran). *Press statement*, 17 October 2012. Available at www.mid.ru/bdomp/ns-rasia.nsf/1083b7937ae580ae432569e7004199c2/c32577ca0017458644257a 9a00618ef9!OpenDocument (accessed 2 February 2014).
Russian foreign ministry. 2013a. *Concept of the Foreign Policy of the Russian Federation*. Available at www.mid.ru/bdomp/ns-osndoc.nsf/1e5f0de28fe77fdcc32575d900298676/86 9c9d2b87ad8014c32575d9002b1c38!OpenDocument (accessed 15 January 2014).
Russian foreign ministry. 2015. Zaiavlenie MID Rossii po itogam peregovorov v Lozanne ministrov inostrannykh del 'Gruppy shesti' i Irana po voprosu ob uregulirovanii situatsii vokrug iranskoi iadernoi programmy (Foreign ministry statement on the foreign minister-level talks in Lausanne between the 'Group of Six' and Iran concerning the resolution of the situation surrounding the Iranian nuclear programme). *Statement*, 2 April 2015. Available at http://mid.ru/brp_4.nsf/newsline/1172F6446B05FF5C43257E 1B0068E88B (accessed 4 May 2015).
Ryabkov, Sergei. 2015. Interview with deputy foreign minister Sergei Ryabkov, *Russia today*, 2 March 2015. Available at http://mid.ru/BDOMP/Brp_4.nsf/arh/7A81B5F85B AFCAEC43257DFC00537D23?OpenDocument (accessed 23 March 2015).
Sakwa, Richard. 2011. Russia and Europe: Whose Society? *Journal of European Integration* vol. 33, no. 2: 197–214.
Sakwa, Richard. 2015. *Frontline Ukraine. Crisis in the Borderlands*. London: I.B. Tauris.
Sarukhanyan, Sevak. 2006. *Rossia i Iran. 10 let yaderonovo cotrudnitshestva* (Noravank. Russia and Iran. 10 years of nuclear cooperation). Yerevan: Nautshno-obrasovatelnij fond.

Sasnal, Patrycja and Secrieru, Stanislav. 2015. Out of the Comfort Zone: Russia and the Nuclear Deal with Iran. *Strategic File*, No. 11 (74), June 2015. Warsaw: The Polish Institute of International Affairs.

Saul, Jonathan and Hafezi, Parisa. 2014. Exclusive: Iran, Russia Negotiating Big Oil-for-Goods Deal. *Reuters*, 10 January 2014. Available at www.reuters.com/article/2014/01/10/us-iran-russia-oil-idUSBREA090DK20140110 (accessed 20 October 2014).

Sazhin, Vladimir. 2010. What Kind of Iran Is Advantageous to Russia? *Vremya Novostey* 5 March 2010.

Sheridan, Mary Beth. 2009. Russia Not Budging on Iran Sanctions. *Washington Post*, 14 October 2009. Available at http://articles.washingtonpost.com/2009-10-14/world/36795978_1_sanctions-and-threats-secret-nuclear-facility-iranian-nuclear-program (accessed 19 October 2013).

Stent, Angela E. 2014. *The Limits of Partnership. U.S.–Russian Relations in the Twenty-First Century*. Princeton and Oxford: Princeton University Press.

Therme, Clément. 2012. *Les Relations entre Téhéran et Moscou depuis 1979*. Paris: Les Presses Universitaires de France.

Trenin, Dmitri. 2009. Reaction to: Kuchins, Andrew and Weitz, Richard. 2009. Russia's Place in an Unsettled Order: Calculations in the Kremlin. In: Schiffer, Michael and Shorr, David (eds), *Powers and Principles. International Leadership in a Shrinking World*. Plymouth: Lexington Books, pp. 188–192.

Trenin, Dmitri. 2013. The Mythical Alliance. Russia's Syria Policy. *The Carnegie Papers*, February 2013. Moscow: Carnegie Endowment for International Peace.

Trenin, Dmitri. 2014. Russia: Sealing the New Quality of Its Foreign Policy. *Eurasia Outlook*, 20 January 2014. Available at http://carnegie.ru/eurasiaoutlook/?fa=54244 (accessed 22 February 2014).

UN. 2006. Provisional Summary, 5612th Security Council Meeting, *S/PV.5612*. Available at www.un.org/en/ga/search/view_doc.asp?symbol=S/PV.5612 (accessed 13 October 2014).

UN. 2007. Provisional Summary, 5647th Security Council Meeting. *S/PV.5647*. Available at www.un.org/en/ga/search/view_doc.asp?symbol=S/PV.5647 (accessed 13 October 2014).

UN. 2010. Provisional Summary, 6335th Security Council Meeting. *S/PV. 6335*. Available at www.un.org/en/ga/search/view_doc.asp?symbol=S/PV.6335 (accessed 13 October 2014).

UN. 2015. *Resolution 2231 (2015), Adopted by the Security Council at its 7488th Meeting, on 20 July 2015*. Available at www.un.org/ga/search/view_doc.asp?symbol=S/RES/2231(2015) (accessed 23 September 2016).

Westphal, Kirsten. 2014. Russland und Europas Energiemix. Krim-Krise: Abhängigkeiten und Strategien. *Energlobe*, 20 March 2014. Available at http://energlobe.de/politik/krim-krise-abhaengigkeiten-und-strategien-russland-und-europas-energiemix (accessed 13 April 2015).

Yurtaev, Vladimir. 2005. Bushehr. Russia and Iran's Meeting Point. *Russian Analytica* vol. 4, April 2005: 99–114.

Ziegler, Charles E. 2012. Conceptualizing Sovereignty in Russian Foreign Policy: Realist and Constructivist Perspectives. *International Politics* vol. 49: 400–417.

4 Chinese foreign policy towards the Iranian nuclear programme

1 'Win–win' and Peaceful Co-existence: Chinese discourse in Iran's nuclear file

At a time when all eyes of the international and certainly Chinese security and non-proliferation community were on the North Korean nuclear case in 2002, the revelation by an Iranian exile opposition group of the existence of clandestine Iranian nuclear facilities hit the news (IAEA 2003). China's reaction was reserved: While stressing Iran's obligation to prove the exclusively peaceful character of its nuclear programme, China was repeatedly emphasising Iran's legitimate right to peaceful nuclear energy under Article IV of the NPT and was critical of pressure on Tehran because of non-proven proliferation concerns. China's public discourse underlined Iranian sovereign rights and cautioned against measures that would interfere with Iran's domestic structures (Dorraj and Currier 2008; Garver 2011: 81f.; International Crisis Group 2010; Mazza 2011; Nourafchan 2010: 39; Pieper 2013; Swaine 2010: 6f.; Yuan 2006).[1]

This rhetoric was not new. The upholding of the principle of non-intervention and sovereignty is a recurring key Chinese foreign policy conception that influences the formulation of Chinese foreign policy and diplomacy since the 1950s. Much of China's foreign policy conceptualisation dates back to the Maoist doctrine of the Five Principles of Peaceful Co-existence articulated at the 1955 Bandung conference.[2] These principles comprise mutual respect for 'territorial integrity and sovereignty, mutual non-aggression, mutual non-interference in domestic affairs, equality and mutual benefit, and peaceful co-existence' (Lanteigne 2009: 11, 46). While it is not difficult to recognise the Westphalian emphasis on the centrality of state sovereignty in international relations in these principles, Chinese foreign policy thinking is complemented by the more anti-hegemonic tone of the 'four no's', namely 'no hegemony, no power politics, no military alliances, and no arms racing' (ibid.). John Garver (2006a) therefore calls China's Five Principles of Peaceful Co-existence an 'implicitly antihegemonist code of behavior' (47). This public diplomacy was complemented by calls for a 'common endeavor to promote democracy in international relations', as formulated in 2007 by then-president Hu Jintao at the CCP's 17th National Congress (Hu 2007: 59). Reminiscent of the Russian plea for more democratic international relations formulated in the Russian Foreign Policy Concept (see previous chapter), China's understanding of democracy *between states* can therefore be translated as a refusal of hegemonic politics.[3] This understanding aims at the levelling of power asymmetries in international relations. If international relations are 'democratised', states become equals, just as voters nominally are in democratic political systems on the domestic level. The advocacy for 'democratic' international relations, in other words, checks hegemonic ambitions because it contests power hierarchies. Jones and Breslin (2015) thus write:

> [China] attempts[s] to work on behalf of other developing and rising powers to democratize the global order and undermine the power of the West. The official promotion of this supposedly anti-normative global agenda thus becomes, in a strange way, a normative agenda in its own right.
>
> (117)

The guiding principles of 'territorial integrity' and the mutual respect for sovereignty continued to serve as a discursive framework for Chinese foreign policy and recurred in official talking points. It therefore also repeatedly resonated in China's approach to the Iranian nuclear file. China thus has always insisted on a political dialogue (as opposed to sanctions) as the only way forward to solve the nuclear crisis (Calabrese 2006: 10; Garver 2011: 81f.; Mazza 2011; Nourafchan 2010: 39; Swaine 2010: 6f.; Yuan 2006). In a rare statement conveying the urgency of the subject matter and the importance of political negotiations, the Chinese delegation to the P5+1 warned of 'wasted time', should the talks in April 2015 on a political framework agreement fail. 'If the negotiations are stuck, all previous efforts will be wasted', the statement went (Reuters 2015a). Chinese political talking points on the Iran talks usually were of a reserved and vague nature, underlining that a comprehensive agreement should be 'equitable, balanced and win–win [*sic*] to all at an early date' (Hua 2015).

China, however, was critical of what it perceived as double standards in nuclear diplomacy, with Iran being harshly criticised for its lack of transparency, while the West remained silent on nuclear activities of non-NPT members such as Israel, Pakistan and India.[4] This testified to what China criticised as 'nuclear favouritism' (International Crisis Group 2010: 4). On passing UNSCR 1887 in 2009 on nuclear non-proliferation and nuclear disarmament, Chinese president Hu Jintao's remarks in the Security Council did not once make reference to Iran, but instead warned against 'double standards' in the non-proliferation regime (UN 2009: 11–12). Tong Zhao (2015) makes the argument in this context that the Chinese interest in strengthening the nuclear non-proliferation regime is also conditioned by its security interests in Southeast Asia: A final settlement of Iran's nuclear conflict could serve as a model to be applied to other non-nuclear weapon states under the NPT. The virtual nuclear capability of Japan, Zhao holds, was a concern that factored into China's overall interest in seeing the Iranian nuclear conflict resolved with a robust IAEA verification regime. The nuclear programme of Vietnam and the possibility of a Philippine one were also cited in this line of argumentation. China's position here seems to be cautiously optimistic that a diplomatic solution of the Iranian nuclear crisis could serve as a model for the resolution of other proliferation problems (North Korea included). This stance differs from the Russian one in that Moscow adamantly insisted that the solution of the Iranian nuclear crisis only applies to Iran and is not a model to be replicated elsewhere (Arbatov and Sazhin 2016: 9). During the nuclear talks, Russia seems to have won that debate. Paragraph 11 of the JCPOA, as well as paragraph 27 of UNSCR 2231 now state that the agreement's provisions and measures 'should not be considered as setting precedents for any other state' (EEAS 2015a: Para xi).

It is also worth pointing out that Chinese interviewees in this research project mentioned the idea of a nuclear weapons-free zone (NWFZ) in the Middle East when discussing the Iranian nuclear issue.[5] Such a zone, however, would be meaningless without the inclusion of Israel. Reference to the NWFZ project can therefore be read as an expression of dissatisfaction with selective nuclear non-proliferation policies in the Middle East. China's argumentation here echoes that of Iran (Wunderlich *et al.* 2013: 270). A further Chinese criticism was targeted at the heavy bias towards non-proliferation efforts on the part of the Western nuclear powers, while the unwillingness on the part of nuclear powers to effectively engage in nuclear disarmament was uncovered as hypocrisy and a lack of credibility (International Crisis Group 2010: 4).[6] With this discourse, China's position resembles those of the NAM more than any other nuclear-weapon state (NWS).[7] Fey *et al.* (2013) call this a position of 'solidarity *with*, and a distanced and privileged position *toward*, the NAM' (187, emphasis in the original). In her study of the NAM and non-proliferation, Yew (2011) finds little principled convergence between national positions on non-proliferation issues and NAM positions, but carves out an associational value of Third World solidarity in the face of perceived unfairness and disproportionality in global proliferation dynamics. By implication, the NAM has come to be seen as a platform for states like Iran to cultivate an anti-Americanism in a multilateral setting and to act as the 'Avenger of the Dispossessed' (Wunderlich *et al.* 2013: 266; Wunderlich 2014: 94; Yew 2011: 9–10).

The similarity between Chinese foreign policy discourse and NAM's emphasis on nuclear disproportionality had to be understood not only against the background of China's publicly formulated foreign policy norms, but also against the background of China's 'complicity' in the set-up of controversial Iranian nuclear facilities. China had passed on sensitive nuclear technology supplies to Pakistan and Iran in the 1980s and 1990s that were at odds with the efforts of the West at the time to consolidate the nuclear non-proliferation regime. Beijing provided a nuclear reactor for the Isfahan Nuclear Technology Centre; signed a memorandum whereby China committed itself to train Iranian scientists and engineers; shared knowledge for the design of nuclear facilities needed for uranium conversion and directly contributed to the building of a uranium conversion facility in Isfahan and heavy water production plants (Djallil 2011: 236; Garver 2006a: 139–165; Patrikarakos 2012: 122, 135–137). The main controversy concerning

Chinese contributions to Iranian nuclear technology was the sale of natural uranium in 1991 – a sale that the IAEA did not know of and that was uncovered only in 2003 at the Jabr Ibn Hayan Multipurpose Laboratories at the Tehran Nuclear Research Center. This finding would later resurface in the charges against Iranian violations of its Comprehensive Safeguards Agreement (ElBaradei 2012: 117; Mousavian 2012: 54; Patrikarakos 2012: 157).[8]

In the latter half of the 1990s, China abandoned its close cooperation with Iran in the nuclear realm and complied more strictly with international non-proliferation regimes. The outside perception of Chinese assistance to an Iranian nuclear programme might spoil China's intention to be seen as a 'responsible Global Power', American interlocutors explained to their Chinese counterparts (Garver 2006a: 201–236; Medeiros 2007: 62–63). During the 1990s, China therefore signed up to relevant treaties and agreements on nuclear non-proliferation. In 1989, Beijing concluded a voluntary safeguards agreement with the IAEA for IAEA safeguards to be applied in China, signed the NPT in 1992 and the Comprehensive Nuclear-Test-Ban Treaty in 1996. China also joined the Zangger Committee and the Nuclear Suppliers Group in 1992 and was supportive of the fissile material reduction treaty (Malik 2000; Nourafchan 2011: 42).[9] With these framework policy shifts, discourse and behaviour followed suit. '[W]hen China joined the Nuclear Suppliers Group in 1992', former Iranian nuclear spokesperson Hossein Mousavian (2012) recalls, 'it ceased nuclear cooperation with Iran under American pressure' (54). And in 2003, China published a remarkable White Paper on Non-Proliferation Policy in which it committed itself to 'continue to take an active part in international non-proliferation endeavours, and exert great efforts to maintain and strengthen the existing non-proliferation international law system within the UN framework'.[10] Working through the international non-proliferation regimes in place has strengthened China's standing in international negotiations on Iran as well. Rather than simply resisting 'Western hegemonic' Iran policies, it provided Beijing with a legal footing in its positioning on the Iran file. This helped both keep extralegal policy options (unilateral airstrikes) of other actors in check and strengthen multilateral non-proliferation efforts. '[M]aterial considerations alone do not justify the decision to join these regimes', Li Xiaojun (2010) underlines, and suggests that socialising effects had replaced 'instrumental calculations' (349). By choosing to sign on to these treaties and agreements, China was thus formally set on a path that allowed a compatibility of interests with the West regarding Iran – regional stability, non-proliferation, and compliance with global institutions like the IAEA. Yet, in all its dealings with Iran in the nuclear infrastructure domain, China had never given the same sense of urgency and concern to Iranian nuclear ambitions as Western governments had. Suspicions about loose Chinese enforcements of its nuclear non-proliferation commitments, in addition, persisted (Malik 2000: 455). Beijing's reserved reaction to the discovery of hitherto undeclared Iranian nuclear facilities in 2002 mirrored this divergence between China and the West. China resisted the emerging securitisation of the Iranian nuclear case and warned against the danger of PMD allegations.

China's official 'Five Principles for a Comprehensive Solution of the Iranian Nuclear Issue', outlining China's understanding of the necessary approach to a successful closure of the Iranian nuclear file and set forth by Deputy foreign minister Li Baodong in 2013, hark back to the tone and spirit of the 'Five Principles of Peaceful Co-existence' referred to above: Reiterating the importance of dialogue, respect for Iranian legitimate rights, reciprocity, good will, and a holistic approach to security, China's government implicitly rejects punitive policies and explicitly calls for a gradual lifting of all unilateral and multilateral sanctions imposed on Iran (Chinese foreign ministry 2014; Hua 2014). This is a diplomatic language that takes up the 'Five Principles' and applies them to the Iranian nuclear talks. The call to act with restraint and 'on the basis of mutual respect' was a stance that had conflicted with efforts to impose international sanctions on Iran just three years before, when China did not veto UNSCR resolution 1929 in June 2010. The next section dissects reasons for such a decision.

2 China's position on Iran sanctions

2.1 'Delay and weaken': China's position on international Iran sanctions

The effect of sanctions fundamentally goes against the spirit of the Five Principles of Peaceful Co-existence as outlined above. China regards sanctions as a concrete expression of an intrusive approach breaching the principle of non-interference in the internal affairs of sovereign countries. When Iran's nuclear file was referred to the Security Council in 2006, E3 + 1 sanctions resolution initiatives on Iran therefore were delayed by China and the content of the resolutions significantly 'diluted' by Chinese amendments in what has been described as a 'delay-and-weaken strategy' (International Crisis Group 2010: 12; see also Mousavian 2012: 236). In braking the imposition of sanctions and watering down their content, the Chinese government was thus not only resisting US policies, but actively working against them. Instead of assuming straightforward resistance to US hegemony as the prime foreign policy motivation, however, the eventual adoption of sanctions in the UN Security Council bespeaks a Chinese desire to maintain the 'responsible stakeholder' imagery (Chan 1999: 146), whereby the Chinese government essentially succumbed to pressure from the US rather than resisting it. In addition, the security argument that nuclear powers are reluctant to see new nuclear powers enter their exclusive club helps understand why international sanctions were, at a minimum, slowing down Iranian progress on its nuclear fuel-cycle activities. The aggressive Iranian rhetoric of the Ahmadinejad administration after 2005, moreover, was not conducive to furthering Iranian interests in seeing weaker, rather than stronger, international sanctions adopted. It is thus instructive to see Chinese dilution attempts of sanctions resolutions as an effort to water them down without emptying them.

In pursuing this strategy in sanctions negotiations, however, cooperation with Russia was crucial, as China sees 'isolation in the Security Council as something to be strictly avoided' (International Crisis Group 2010: 15; see also Ferdinand 2013: 17; Wuthnow 2010: 66).[11] Before P5+1 meetings, the Russian and Chinese negotiation teams convened to agree on joint approaches concerning the proposal of amendments of sanctions resolution texts (as did the E3+1, i.e. the E3+the US, dialogue partners).[12] In practice, Chinese–Russian joint efforts consistently managed to water down the initial resolution's provisions, with China proposing what was called 'amendments' (which in practice were deletions of complete passages) to certain paragraphs, while Russia was proposing its own 'amendments' (read: deletions) to the other remaining paragraphs.[13] In a context where the momentum for sanctions increasingly gained traction in 2006 after the referral of Iran's nuclear case to the UNSC, China also supported the 'Russian plan' in 2006 to transfer uranium enrichment to Russian soil (Mousavian 2012: 235). The realisation of this plan would have defused tensions and disrupted the momentum for sanctions in the Security Council. After Iran's rejection of the plan, the lack of Chinese high-level participation in P5+1 meetings in New York at the time was another political signal conveying China's unhappiness with sanctions (Mousavian 2012: 365). Iran, for its part, knew how to play on differences in policy priorities within the P5+1 format. In late 2009, when the idea of a fuel-swap plan was declared dead politically by most other actors, China kept arguing its case. Fitzpatrick (2010) writes: 'Iran fed China's position by engaging various interlocutors as potential intermediaries. Needing Beijing's support, or at least acquiescence, for a new UN sanctions resolution, Washington kept the door officially open for as long as it could' (77).

Yet, having worked against their imposition in the UNSC, China eventually approved of sanctions on Iran. Interviewees in China, both from government-consulting institutes as well as former officials, confirm that this was due to a realisation that Iran did not cooperate transparently enough with the IAEA, but that it also conveyed a Chinese willingness to demonstrate a cooperative spirit to the Americans.[14] It is the effect of Sino-American relations on China's Iran policy that largely explains China's voting for UN sanctions resolutions. Even though Iran sanctions entail negative effects on Sino-Iranian commercial relations (a subject of section 3) and run counter to China's principled opposition to the intrusive effect of sanctions, China eventually supported them. In this context, Garver (2011) writes of a 'Dual Game' that China is playing in Iran. In other words, China displayed a behavioural convergence with policies Beijing was criticising from a normative point of view. This is important to underline as it nuances a relatively straightforward attempt to check US hegemony. While the imposition of international sanctions on Iran was pushed and lobbied for by the US government, China could have stopped such efforts with a Security Council veto. Knowing that the actual use of this veto right would have reputational political consequences, however, China has thus traditionally favoured 'the practice of abstention and acquiescence' (Wuthnow 2010: 63). Eventual support for UNSC sanctions on Iran therefore is, in Katzenstein's terminology, compliance with 'rules and models' (of the UN system), while China's public statements indicate disagreement with the 'norms and values' underlying these sanctions policies. China's position here is congruent with that of Russia and Turkey as analysed in the two preceding chapters.

But was China complying with the Iran sanctions regimes that the Chinese government itself had voted for? Resolution 1929 (2010) 'calls upon states to *take appropriate measures*' [UN 2010, emphasis added] to restrict Iranian financial capacities and 'decides that all States shall [...] *exercise vigilance* when doing business with entities incorporated in Iran or subject to Iran's jurisdiction, including those of the IRGC and IRISL'[15] (ibid., emphasis added). Such a vague wording, coupled with the absence of UN enforcement mechanisms, would still allow China to continue commercial interactions that other parties might regard as violating sanctions provisions. Charges by the US Treasury Department pertaining to China's compliance with UNSCR 1929, for example, were that IRISL used front companies and transferred ownership to companies in Hong Kong in order to conceal its identity (US Treasury Department 2011). The US administration also hinted at China's continued business with Iranian financial institutions and warned China's biggest banks not to accept transfers from IRISL's insurer Moallem (Rubenfeld 2011). Replies by Chinese authorities made reference to insufficient intelligence capabilities and the difficulty in identifying front companies and deceptive practices from legitimate operators (Broadhead 2011).[16] Etel Solingen (2012) therefore argues that 'China's compliance with multilateral sanctions has been selective, reluctant, and intermittent, often relying on linguistic and behavioral contortions to justify inconsistencies' (333). Asked about the charges that IRISL still lay anchor in Hong Kong, a Chinese diplomat replied in an interview that

> Chinese customs officials pay close attention to the implementation of relevant UN Security Council resolutions. We also have our internal laws. But sometimes some parties want more than what is stated in the resolution provisions. We obey the Security Council resolutions, but we don't want to go much beyond that.[17]

This is sometimes referred to by sanctions scholars as the 'floor' versus 'ceiling' debate (Biersteker and Moret 2015: 70): China (and Russia) regard UN sanctions not as the base for additional restrictive measures (floor), but as a legitimate limit of what they see as acceptable (ceiling).

Arguably, an important factor also was the political momentum at the time of adoption of sanctions resolutions. Even though China has been calling for patience with Iran, political framework conditions made Beijing approve of sanctions when diplomatic soothing strategies would not work anymore against the background of a publicly growing discontent with Iran's lack of transparency. A Chinese veto would have constituted an outright rejection of the public display of Western security concerns in such a case – and international isolation was something Beijing was keen to avoid. This was the case with resolution 1737 in 2006, when Iran had removed IAEA seals from its enrichment facilities in order to re-start uranium enrichment instead of suspending it as stipulated in the preceding resolution 1696 (IAEA 2006; UN 2006a); with resolution 1803 in 2008 when Iran further refused to suspend heavy-water related activities (UN 2008a); and with resolution 1929 in 2010, which was adopted after the revelation of yet another (hitherto unknown) enrichment facility near Qom in autumn 2009 (UN 2010a). However, as Gareth Porter (2014) argues, it was especially the publicly repeated threats of Israeli airstrikes on Iranian facilities and the ensuing risk of war that likely was the bigger incentive for China (and Russia) to agree to economic sanctions, rather than any Iranian decisions in its file with the IAEA (271).

In all these cases, moreover, China gave what Wuthnow (2010) calls a 'qualified acceptance' (74): China's consent to the imposition of sanctions was accompanied by remarks conveying equivocation, as becomes clear from the minutes of relevant Security Council meetings and statements before or following the vote. On passing UNSCR 1737, China's UN representative Wang Guangya stated that the sanctions were considered 'limited and reversible', and called on 'all parties concerned' to 'practice restraint' (UN 2006b: 7–8). These talking points reappeared in China's remark on passing UNSCR 1747, in which Wang also urged all parties (in a somewhat awkward phraseology) to 'creatively seek to resume negotiations' (UN 2007: 12) and also deemed it important to point out that UNSCR 1747 did not alter the exemption provisions of UNSCR 1737 (ibid.: 11). Reiterating the reversibility of sanctions, Wang's statement after voting for UNSCR 1803 also called upon all parties to 'give full play to initiative and creativity and demonstrate determination and sincerity in resuming negotiations' (UN 2008b: 17). President Hu Jintao's remarks following China's vote for UNSCR 1887 in 2009, as noted in the previous section, did not refer to Iran once, and instead expanded on the need for a strong global non-proliferation regime in general (UN 2009: 11–12). And in justifying its vote for the much-discussed UNSCR 1929 in 2010, China's representative Li Baodong again reverted to earlier talking points about reversible sanctions, the 'incremental' and 'clearly targeted' nature of sanctions adopted, and nebulously stated that 'Security Council unity is essential to resolving the Iranian nuclear issue' (UN 2010b: 10–11). The latter statement reads especially contradictory in view of China's strong discourse against sanctions on Iran on other occasions as analysed in the previous section.

Finally, China's support for Iran sanctions at the UN level had to be seen in the context of China's interest in a stable Middle East. Diluting sanctions and eventually adopting them may not be such a contradictory policy as it seems on the face of it: China's 'delay-and-weaken' policy takes out the most restrictive elements of sanctions provisions, while the adoption of sanctions takes the wind out of the sails of those actors that advocated military strikes on Iran. In this understanding, China's work through the UN (including the adoption of sanctions, if necessary) is a way to check US power. The adoption of sanctions can thus also be seen as the lesser evil that, at a minimum, gives breathing space to counteract more confrontational actors. A regional destabilisation through the outbreak of an open military conflict would have severely endangered and disrupted Chinese oil supplies and commercial interests in the Iranian market, the impact of which will be discussed further below. In addition, not being proactive itself, but waiting for Western initiatives to de-escalate the nuclear crisis, China conveniently followed a strategy of maintaining its market position in Iran while benefitting politically from ('free riding' on) Western diplomatic efforts.[18]

2.2 Norm divergence: China's position on unilateral Iran sanctions

Irrespective of oft-repeated suspicions about Chinese irregular compliance with international sanctions, the Chinese government publicly pledges its commitment to the institutional structure of the UN. China has worked through the mechanisms of the UN family and displayed an acceptance of its legitimacy to govern international relations since Beijing took the permanent membership in the Security Council from Taipei in 1971 (Zhang 1998: 73–91). While international sanctions adopted by the United Nations are thus adopted with the support of China as a permanent UNSC member, the Chinese government opposes any unilateral sanctions. This formulation is repeated nearly verbatim across the range of interviewees – be they government consultants, Chinese academics, or Chinese officials.[19] Unilateral sanctions are seen as the extension of domestic law onto other sovereign states and thus as breaching international law (Hong 2016). China's rhetoric against unilateral sanctions imposed on Iran has consistently conveyed China's 'principled opposition' (Garver 2006a: 66) against what is seen as the expression of 'arrogant hegemonism' (ibid.: 84).

But China's opposition to such sanctions was never merely an ideological conviction. China's dislike for an extraterritorialised US legislation also had to be understood in the context of Chinese companies having been sanctioned unilaterally by the US for interactions with Iran that were seen as undermining US efforts at changing Iranian behaviour. Chinese companies have been sanctioned by the US because of missile and technology supplies that assisted Iran's ballistic missile programme,[20] and because of Chinese nuclear technology transfers to Iran that potentially assisted the setting-up of Iranian *unsafeguarded* facilities, as specified in the first section of this chapter (thus in breach of IAEA stipulations). The Iran–Libya Sanctions Act (ILSA) of 1996 furthermore stipulated that investments made in the Iranian energy sector exceeding 20 million US dollar in one year are sanctionable activities (Katzman 2006). The 'Comprehensive Iran Sanctions, Accountability, and Divestment Act' (CISADA) of 2010 also includes the provision of refined petroleum products to Iran as breaches of that act. The Chinese company Zhuhai Zhenrong was sanctioned under CISADA in 2012 (US Department of State 2012). The Chinese Kunlun bank, together with the Iraqi Elaf Islamic Bank, was the first foreign bank to be sanctioned under CISADA in July 2012 for having financed oil-related businesses with Iranian banks that had been blacklisted before (Lohmann 2015: 5). As a result, the US Treasury Department has 'prohibited U.S. banks from opening or maintaining correspondent accounts or payable-through accounts in the United States for Bank of Kunlun – effectively cutting off Bank of Kunlun's direct access to the U.S. financial system' (US Department of the Treasury 2016).

The effect of Chinese entities being listed carries a significant labelling effect that the Chinese government cannot ignore. Insisting on the illegality of American pressure and the freedom of Chinese trade relations lies on one side of the spectrum, showing a cooperative spirit and 'investigating' issues of concern to the US on the other. This holds for 'controversial' investment projects as well as for oil trade, which had become an area the US had identified as a leverage with which to dry up financial lifelines to Iran. The US administration is aware of the Chinese prioritisation of Sino-US relations over Sino-Iran relations, as a State Department official confirmed in an interview: 'China really values its relationship with the US. It values it above that with Iran'.[21] Acknowledging precisely this conundrum, the US has been encouraging Arab oil exporters (like Saudi Arabia) 'to boost oil exports to China in an attempt to decrease reliance on Iranian oil and secure agreement to sanctions' (International Crisis Group 2010: 14). Another idea on the part of the State Department was to ask Saudi Arabia to sell its oil to China at a lower price (Kemenade 2010: 109). And in her efforts to round up support for UNSCR 1929 in 2010, Secretary of State Clinton told Chinese State Councillor Dai Bingguo that if China supports the resolution and 'reduces its commercial ties to Iran, we could help it find other sources of energy', as she writes in her memoir (Clinton 2014: 432).[22]

The extent to which the Chinese government was happy to accept such US-brokered supply alternatives is not publicly documented. At a minimum, however, Beijing showed receptiveness to US demands to decrease its purchases of Iranian oil (Tsukimori and Goswami 2013). Such a policy both serves to respond to US perceptions of China's Iran policies and to qualify for the US sanctions 'waivers' granted to those countries that 'significantly reduce' their import of Iranian oil as outlined in section 1245 of the US National Defense Authorization Act (NDAA) (US Department of State 2013). A 20 per cent reduction of Iranian oil imports is considered a 'significant reduction' according to this US sanctions legislation. Such a reduction qualifies for a waiver granted to states that would otherwise be sanctioned by US authorities (Lohmann 2013: 4). The vague formulation, however, suggests that the granting of such waivers comes with a considerable political leeway. Supply diversification talks, in this sense, may also serve as 'proof' that a decrease of dependence on Iranian oil is envisaged. China's reduction of Iranian oil imports and plans to diversify its oil suppliers can, against the background of US sanctions legislation, be read as a direct response to American policy concerns and serves to demonstrate Chinese cooperation to their American counterparts.[23] China repeatedly qualified for these periodic waivers since 2012.[24] Opaque business structures may be another way of evading sanctions. 'Chinese are very smart', an Iranian diplomat said in an interview, 'they establish fake companies just to work with Iran'.[25] Oil-for-goods barter agreements were another way for China to evade sanctions and for Iran to receive products in times of Western embargoes (Slavin 2011).

Yet, a caveat on the interaction between the Chinese government and state-owned companies should be inserted here: A former Chinese diplomat to Iran explained in an interview that Chinese companies have an interest in enlarging their market access and profits abroad, while the Chinese government has an interest in enlarging its 'soft power' abroad.[26] Yet, this can also serve as a convenient argument for China's government to 'save face' both vis-à-vis its domestic business lobbies as well as vis-à-vis its American counterparts. To the latter, a reduction in Chinese–Iranian oil trade appears as a sign of governmental cooperation in pressuring Iran, while the government can fall back on the 'effects of markets' argument for the former.[27] Corporate interests need not be Chinese 'national interests', but the tightly interwoven dependencies between these two makes it difficult to assess who in China decides on sanctions 'compliance'.[28] Tellingly, a Chinese diplomat to the US offered a glimpse of political frankness when asked about the effect of EU and US unilateral sanctions by saying: 'Even if we don't admit it, EU and US sanctions do work. Chinese companies should take care. Both the government and companies have to comply.'[29]

A final element to keep in mind when analysing China's position on sanctions is their effect on Chinese investments in Iran in sectors not affected by sanctions, but abandoned by the West for political reasons: During the times of international and unilateral sanctions imposed on Iran, China made use of the economic vacuum created by the Western self-imposed embargo situation and sold its products that were unavailable to Iran otherwise. An Iranian diplomat puts it in an interview as follows: 'Whether it was the Tehran–Mashhad railway connection, the Tehran–Isfahan railway connection, or the Tehran metro, China completed infrastructure projects, China sent its products when Iran couldn't get material from Germany during the sanctions phase. China filled that vacuum.'[30] China therefore has benefited from the embargo situation created by Western Iran sanctions, as it has enlarged the Chinese share in the Iranian market, despite persistent Iranian complaints about the quality of Chinese goods.[31] Chinese government officials, however, are quick to deny claims that Western sanctions are an opportunity for China, arguing that market competition is 'just normal' and that China would not be worse off with the prospect of increased Western investments in Iran.[32] The next section will shed light on the importance of Chinese material interests in Iran for overall Chinese Iran policies, and analyse to what extent Chinese Iran policies indeed have to adapt to a new business environment in the wake of the implementation of the JCPOA.

3 Chinese–Iranian economic relations and the impact of the JCPOA

China is exporting capital goods, engineering services and arms to Iran. Chinese corporations have invested in non-hydrocarbon sectors: joint ventures have been created;[33] Chinese companies have been investing in Iranian infrastructure projects; China signed agreements on cooperation in industrial and mining sectors, China's largest steel factory developer is building plants in the Yazd province, and the China International Trust and Investment Corporation (CITIC), together with Chinese Norinco, was contracted for the completion of the Tehran metro system (Calabrese 2006: 6–9).[34] China has built bridges, railways, dams and tunnels in Iran, and has invested in upstream operations in Iranian oil and natural gas fields such as the Azadegan and Yadavaran fields (Harold and Nader 2012: 10–12). This web of economic activities is crucial to bear in mind when wanting to understand China's stakes in Iran. China is primarily importing oil from Iran. At the same time, China is shipping some of its own refined oil into northern Iran, as Iran – despite its oil wealth – does not have sufficient refining capacities.[35] As from 2009, China had become Iran's most significant foreign trade partner. The Chinese–Iranian trade volume is extensive. In 2014, it reached a record high of US$52 billion (compared to a Russian–Iranian trade volume of only US$1.6 billion and a trade volume of US$10 billion between Iran and the EU-28 combined). In 2015, it somewhat dropped to US$34 billion – still 26 times the size of the Russian–Iranian trade volume.[36]

The importance of oil shipments in Chinese–Iranian economic relations was underlined by a number of major oil deals that have tied the two countries' economies together even more closely, cementing Iran's position as one of China's biggest oil suppliers. China also became a key stakeholder and one of the largest investors in the Iranian oil industry and is active developing and exploring oil fields in Iran (Derakhshi 2009; International Crisis Group 2010: Annex B; Shen 2006: 61). This high dependency explained why China was unwilling to cut back on oil imports from Iran when asked to do so by the US. The lifting of nuclear-related Iran sanctions as part of the implementation of the JCPOA has 'legalised' the Chinese–Iranian oil trade: US unilateral sanctions penalising third country business with the Iranian oil economy do not apply anymore. 'Significant reductions' of Chinese oil imports from Iran in order to pre-empt US financial penalties under the 2010 CISADA are thus no longer needed (US Department of the Treasury 2016: 13). Equally important for China's import of Iranian crude oil is the US waiver of sanctions against 'non-US persons who own, operate, or control, or insure a vessel used to transport crude oil from Iran to another country' as part of the JCPOA (ibid.: 12). China is now also free to do transactions with Iran's shipping companies and port operators (ibid.: 17).

Beijing's involvement in the Iranian economy and especially in the oil sector is to be explained by China's interest in the stability of oil supplies ever since China became a net oil importer in 1993. Seen in the context of China's rise as an 'emerging' global power, this need for stable oil supplies becomes a crucial determinant in China's Iran policy ('resource diplomacy', see Lanteigne 2009: 51f.). As China's 'rise' since Deng Xiaoping's reforms primarily meant a 'rise' in the economic sphere, the Chinese government sees the necessity for political stability in the Middle East through the lens of economic supply stability.[37] While China's main oil supplier is Saudi Arabia,[38] Iran usually ranges second. Viewed through the prism of Sino-US relations, Saudi Arabia was a less controversial oil supplier because of the special alliance relationship between Washington and Riyadh. US security guarantees to Saudi Arabia thus also translated into energy security guarantees for China. Given shifts in US–Saudi relations and the prospect of a changing security and economic environment that the lifting of Iran sanctions entails, Saudi Arabia's value from a Chinese perspective might decrease (Garver 2016: 7). What likely remains is China's sensitivity to the US in its conduct of external economic relations in this region.

An illustrative material factor for this sensitivity is the American presence in the Malacca Strait, through which most of Chinese oil supplies from Iran are shipped.[39] It has been argued that China conveniently 'free rides' on US protection of sea lanes (Downs 2004: 32). In a liberal reading, this finding would suggest that benign hegemony provides global public goods (see also Ikenberry 2008; Ikenberry 2011: 7). This 'functional logic of liberal hegemony' has been maintained throughout the Obama administration (Rudolf 2016). That China stands to benefit from these public goods nuances the idea of 'resistance' to US power across the board. At a minimum, the Chinese dependence on this maritime bottleneck explains why a deterioration in Sino-US bilateral relations is not desirable for China already out of important logistical reasons (Goldstein 2011: 96). In 2003, former president Hu Jintao therefore even spoke of a 'Malacca dilemma' that China faces in case of a potential blockage (from either terrorists or other states) of such an essential lifeline for the Chinese economy that connects East Asia with the Middle East. Hinting at the presence of the US in that waterway, Hu noted that 'certain powers have all along encroached on and tried to control navigation through the strait' (in: Lanteigne 2009: 86). In an acknowledgement of such a logistical vulnerability, China has invested in the construction and development of the port of Gwadar in Southwest Pakistan as well as in the planning of overland transportation lines like the Karakoum Highway between Pakistan and Western China. Such transmission belts would allow the supply of Iranian oil from Pakistan to mainland China through Gwadar (Markey 2014: 10).[40] This diversification of supply lines aims to secure the stability of oil supplies from the Middle East via overland routes and testifies a Chinese awareness of its current dependence on American benevolence.

Other economic projects that aim to link inter-regional trade relations already count on an altered standing of Iran in the wake of the implementation of the JCPOA that would also facilitate trade in the absence of nuclear-related sanctions: Chinese projects like the so-called 'One Belt, One Road' initiative comprising the Maritime Silk Road and the Silk Road Economic Belt are conceptualised as economic corridors through the Eurasian continent (Godehardt 2014). With this project, Chinese external trade and foreign policy planning has envisioned a joint policy towards central Asia, the Middle East, the Black Sea region and the Caucasus. A stable Middle East becomes a precondition for the implementation of such a project. Given the central geographical location of Iran, the removal of trade complications as sanctions regimes have been lifted from Iran is good news for China. The 'One Belt, One Road' project has therefore been highlighted as a Chinese long-term priority on the occasion of President Xi Jingping's state visit to Iran in January 2016 (Chinese foreign ministry 2016b).

The lifting of Iran sanctions may also see Chinese material interests in Iran clash with Russian ones. In the past, both China and Russia have invested in the Iranian nuclear infrastructure and were main arms suppliers to Iran (Gill 1998,

1999). Chinese attempts to restart arms deals with Iran will likely face Russian competition. Russia had made it clear that it was planning to capitalise on the eventual lifting of arms embargoes from Iran (Saunders 2015). While the UN weapons embargo will only be lifted five years after 'Implementation Day', states can apply for a UN authorisation already before this date. John Garver suggests in this context that 'a number of Chinese military technologies would be attractive to Iran, such as anti-ship cruise missiles, long distance air-to-air missiles and sea mines' (Garver 2016: 5). On the nuclear energy market, however, Russia had surpassed China as the most important nuclear sponsor of Iran in the 1990s, as evidenced not least by Rosatom's construction of Iran's only nuclear power plant at Bushehr (see previous chapter). Yet, China's proposal for the conversion of Iran's Arak reactor into a nuclear research centre at the nuclear talks has been accepted as the most reasonable one among other alternatives.[41] The conversion is carried out by an international consortium, with China in the lead in the 'Working Group' on the Arak conversion. The Russian–Iranian nuclear cooperation thus ceases to be an exclusive one, with Chinese nuclear expertise now also actively involved in the conversion of the heavy-water reactor at Arak as part of the implementation of the JCPOA. The removal of the calandria (or core reactor) at the Arak facility in January 2016 was a first step towards its reconstruction, which is intended to cut off the route to a plutonium bomb (Arbatov and Sazhin 2016: 5). Importantly, the conversion of the Arak facility is a proliferation-related task carried out under the JCPOA. But it cannot be excluded that Chinese companies in the future become involved on the Iranian market in sectors hitherto dominated by Russia.

In these deliberations of a post-JCPOA positioning, it has to be kept in mind that China's foreign policy is partially motivated and informed by economic interest groups. China's state-owned business corporations and oil traders enjoy a powerful position in lobbying the Chinese government. Decision-makers in Beijing have to carefully weigh the pursuance of commercial interests with the perception of China's foreign policy on the part of other major stakeholders. While major investment projects may nurture the impression of extensive Chinese–Iranian technology transfers, part of these did not materialise yet or remain in the planning phase. This sometimes is the outcome of a Chinese behaviour not to endanger the 'responsible Great Power' image that a pursuance of commercial contracts in outright disagreement with US policies would entail. The 'extraterritorialisation' of US legislation by way of unilateral sanctions affecting third country companies served as another, more material deterrent. It was also a political deterrent because the listing of Chinese entities is a publicly visible and therefore undesirable labelling, as analysed in the previous section. However, it should also not be forgotten that the non-materialisation of Chinese investment projects in Iran may be the result of political decisions in Tehran. And in view of a politicised environment when doing business in Iran, the Chinese government may decide to sign legally non-binding MoU that allow for political manoeuvrability. Such commitments are easier to revoke in case of unfavourable political framework conditions. The JCPOA has made these calculations easier for the Chinese government. With the explicit aim to open a 'new chapter' in relations with Iran (EEAS 2015b), the JCPOA offers the framework conditions for an intensified Chinese business presence in Iran. President Xi Jinping was the first head of state to visit Iran following the conclusion of the nuclear negotiations. It was also the first visit by a Chinese president to Iran in 14 years. On his visit in January 2016, Xi signed 17 agreements in energy, trade and industry sectors, indicating an eagerness to expand trade relations with Iran (Glenn 2016). While China was the main beneficiary of Western embargoes on Iran, the lifting of nuclear-related sanctions as part of the implementation of the JCPOA entails the possibility of stronger future competition for Chinese companies on the Iranian market. But – leaving aside Iranian complaints about the quality of Chinese products – China may have a structural advantage here: The 'foreign subsidiary' component in US sanctions that continue to apply regardless of the lifting of nuclear-related Iran sanctions is likely to affect European companies more than Chinese ones. In this context, Xi's visit can be read as an early positioning in a shifting business climate, as does his re-assurance to Supreme Leader Ali Khamenei that China will always be a 'reliable cooperative partner' of Iran (Chinese foreign ministry 2016b). 'Currently, China–Iran relations are embracing a new development opportunity', a foreign ministry announcement reads (Chinese foreign ministry 2016a). Whether this means that Chinese policy towards Iran will be a more assertive one both vis-à-vis Western critics and economic competitors is doubtful. China's diplomacy on Iran continues to be a cautious one, as the next section discusses on the basis of the preceding analysis of China's foreign policy discourse, its position on sanctions policies, and the material dimension of China–Iran relations.

4 Chinese Iran diplomacy between triangulation and resistance

Deng Xiaoping had outlined a pragmatic doctrine that should accompany China's modernisation process as from 1978 ('hide our capabilities and bide our time (*taoguang yanghui*); be good at maintaining a low profile; and never claim leadership'; in: Jacques 2012: 590).[42] During Hu Jintao's presidency, Marc Lanteigne (2009) holds, China's doctrine of 'biding its time' was being replaced by exercises of great power diplomacy (*daguo zhanlue*, 21). In characteristically vague wording, President Hu had announced in 2009 that China should 'continuously keep a low profile *and proactively get some things done*' (Chen and Pu 2014: 178, emphasis added). The foreign policy mantra under president Xi Jinping shifted from *taoguang yanghui* to *fenfa youwei*, which translates to striving for achievement (Tong Zhao 2015).[43] The 18th Party Congress report (of November 2012) stressed the concept of China's 'peaceful development', but equally warned, in a quite explicit language reminiscent of the 'four no's' in Chinese foreign policy thinking, of the danger of 'hegemonism', 'power politics', and 'neo-interventionism' (Chatham House 2013). This rhetoric was echoed by then-party general secretary Xi Jinping in January 2013 and was seen as a timely positioning against the backdrop of the NATO intervention in Libya, attempted UNSC resolutions on Syria and sabre-rattling over Iran.

An additional discursive undercurrent in Chinese–Iranian relations since the 1980s and 1990s was that of a 'civilisational' level of Third World solidarity and opposition to 'US hegemony' (Burman 2009: 26–27, 159f.; Garver 2006a: 3–28).[44] Statements would commonly refer to the fact that both countries had to cede territory, partially lost sovereignty and suffered national humiliations at the hands of Western imperialism (Harold and Nader 2012: 2). China's flirtation with such an overtly anti-hegemonic rhetoric was toned down in the later part of the 1990s, as was discussed above. China was cautious not to let its principled normative divergence from US hegemony (and ensuing sympathy with Iranian anti-hegemonism) go out of hand and endanger Chinese relations with the US. The latter's benevolence was crucial for China's path of economic modernisation ushered in under Deng Xiaoping which essentially required China's acceptance into a US-dominated capitalist system. Beijing's desire to portray itself as a 'responsible Great Power' (*fuzeren de daguo*; Chan 1999: 146) and to convey the image of China's 'peaceful rise' indicated a discursive willingness not to endanger the US's acceptance of China as an equal power on the world scene. The concept of 'peaceful rise' as introduced in a 2003 White Paper was even replaced by the more harmonious-sounding concept of 'peaceful development' (Glaser and Medeiros 2007). If China 'develops peacefully', so the narrative, it does not challenge a prevailing status quo. As a consequence, China's biting swipes at US 'arrogance' and 'imperialism' became fewer, its foreign policy discourse more cautious.[45]

Analyses of Chinese foreign policy tend to be couched in terms of status quo versus revisionist policies.[46] Against this background, the analysis of China's Iran policies presented in this chapter has shown how China's foreign policy can display elements of resistance to hegemony as well as accommodation with the very same power structures China criticises on a discursive level. China's position on the Iranian nuclear programme bespeaks a diplomatic tightrope walk in which Chinese governments had to 'triangulate their various interests with Washington and Tehran' without wanting to choose between the two (Shen 2006: 63). China's act of triangulating the effects of Sino-US relations on Sino-Iran relations and vice versa emerged as a continuing pattern in China's relations between these two countries after the Islamic revolution in 1979 and during the Iran–Iraq war. But in practice, commercial interests in Iran were weighed against US pressure and the perception of Chinese foreign policy in Washington. It often was the anticipation of possible effects on relations to the US that determined how far China was willing to go in courting Iran by publicly lending support. Anti-hegemonic sympathies with Iran always were balanced against the importance China attached to the Sino-US relationship and the possible detrimental effect on China's economic modernisation path that an alienation from Washington could have entailed.

With regard to the sanctions question, Beijing therefore knew that its voting pattern in the UNSC was a positioning with far-reaching political implications in one way or the other. Voting for UNSC resolutions imposing sanctions on Iran lets Beijing lose political capital in Iran, while voting against them would have alienated those powers pushing for a tougher stance toward Iran and the 'sanctions sponsors', most notably the US. Abstaining from a vote might have been a way for China to circumvent this dilemma, but does not do justice to Chinese claims to being seen as an influential power taking responsibility on issues of global security.[47] A member of the Chinese delegation to the nuclear talks explained in an interview that the Chinese delegation was engaging in mediation between the US and the Iranian position in P5+1 negotiations with Iran.[48] And a former Chinese diplomat to Iran even remarked in an interview that China was pressing Iran to compromise on the question of the uranium enrichment level – a crucial point of contention in P5+1 talks with Iran.[49] It is important to note, however, that this is Chinese official rhetoric. As former Iranian officials recall, China was even more at the periphery of the negotiations than Russia – at least during the earlier years of nuclear diplomacy (Mousavian 2012: 182, 264). Asked about the dynamic and relationship between Russia and China in the Iran diplomacy, both current and former officials and experts working on Iran, however, confirm that the impression prevailed that China was 'following Russia's lead' in the Security Council for the years of nuclear diplomacy.[50] A former E3 delegation member formulates:

> At times, China was hiding behind Russia, and on other occasions, it was the other way around. China had Russia build up a counter-position to the West. China generally has an observing and restrained role. China is tackling problems with a long-term view. They don't want to fall out with the Americans.[51]

This is an observation that emphasises the perception that China did not want to be seen at the forefront of diplomatic activity. This puts China into an intricate position. Its institutional weight as a permanent Security Council member makes it too 'big' a state to hide behind rhetorical contortions and act in the shadow of a publicly more assertive Russia. In an interview, the Russian political analyst Fjodor Lukyanov here mentions the Libyan case as an example where China, despite its policy preferences, decided not to block UNSC resolution 1973 because Russia abstained and 'China does not feel comfortable being alone'.[52] In deliberating key foreign policy interests and their portrayal to the world, it can thus be assumed that China holds views that differ from the Russian government's, but are subordinated to a higher-valued goal of preventing diplomatic isolation – a finding that dispels the fallacy of assuming a Chinese–Russian 'bloc position'. Naturally, a stronger Chinese dissociation from Russian positions in public posturing would come at the expense of convenient 'covers'. Whether Chinese positions in the future will clash more openly with American ones can be doubted on the basis of the cautious characteristics of the Chinese foreign policy machinery carved out in this chapter.

In addition, China's growing economic weight, both in the Middle East and globally, forces China to make choices and adopt public positions in line with its geopolitical influence. Close economic ties with Iran and a perception of sanctions as an interference into the domestic politics of sovereign states on the one hand needed to be reconciled with the desire to be perceived as a 'responsible Great Power' that is actively supporting and endorsing nuclear non-proliferation efforts on the other hand. The latter meant an eventual Chinese endorsement of sanctions resolutions against Iran, even though such a policy went against the Chinese public discourse on how to approach the Iranian nuclear issue. While this is not to argue that China shared an American approach on an ideational basis, this factual observation has been qualified throughout this chapter: A number of economic, security, and political reasons explain why China's interests were best served by working through the P5 + 1 format. The latter has meant working with and responding to influential American policy preferences, and entailed a means for China to exercise institutional control over them. China's 'rise' will continue to spur debates and attempts at predicting the future direction of its foreign policy. As this chapter has shown, however, the complex picture of Chinese foreign policy in the Iranian nuclear file eschews an easy categorisation of either a status quo or a revisionist power.

5 Conclusion

Re-iterating that China regards sanctions as a violation of the principle of non-interference and an infringement of Iran's sovereignty, Chinese decision-makers made it clear that attempts to punish Iran for its lack of cooperation and transparency over its nuclear programme ran against China's understanding of an ideational framework that should govern international politics as soon as Iran's nuclear case was referred to the UNSC in 2006. Nothing captures this normative divergence more clearly than China's recurrence to its 'Five Principles of Peaceful Co-existence' as guiding principles of foreign policy, which represent an explicitly anti-hegemonic posture when read against China 'four no's' in public diplomacy. And ultimately, an additional discursive undercurrent in Chinese–Iran relations is the aspect of 'civilisational' solidarity that captures the joint historical experience of humiliation at the hands of Western imperialism. On this basis, there was an obvious public diplomacy utility for Iran of 'keeping China as a friendly voice in the negotiations', as one E3 official put it.[53] However, this must not be confused with an ideological stylisation of an 'Eastern bloc' policy as pursued by the Ahmadinejad administration. Such a policy did not meet much enthusiasm on the Chinese side (Mousavian 2012: 84, 141). Mousavian formulates: 'China's foreign policy is based on development advancement, while Ahmadinejad's foreign policy is based on ideological advancement' (401). And in a frank judgment about his country's foreign policy, former Iranian ambassador to China, Javad Mansouri, described the 'looking to the East' policy a mistake (in: Mousavian 2012: 443). The recurring Chinese rhetoric about 'civilisational solidarity', it thus appears, does not correspond with actual policy preferences on either side. For all its ideological posturing, Iran is much more interested in the West. It has the youngest population in the region, is craving for recognition on an equal footing with Western powers, and needs Western technology and products more than Chinese ones. Chinese–Iranian cooperation in bilateral relations is mostly limited to economic (energy) issues.

On a larger plane, the Iran nuclear dossier served as a microcosm to analyse Chinese conceptions about how international relations are to be governed. The Iran sanctions issue here has offered fitting illustrations of the workings of resistance and accommodation. Conveying a serious discontent with international sanctions as legitimate instruments to pressure Iran, China has engaged in a 'delay and weaken' strategy with regards to sanctions resolutions in the UNSC. Factors influencing China's Iran policy and its ultimate approval of international sanctions include US flexibility towards China (exempting China from unilateral sanctions, introducing China to alternative oil suppliers), the character of Iranian noncompliance with the IAEA, and international market (dis)incentives. The first is a political aspect and is subject to behind-the-doors bargaining between Chinese and American counterparts in the administration. The second is a security argument: China is a nuclear-weapon state and has an in-built desire not to see the number of nuclear-weapon states grow. Such a development could potentially entice other states to follow suit and unleash a dangerous proliferation dynamic. It also endangers stability conditions in the Middle East and beyond. The aspect of market disincentives relates to the argument put forward by Chinese officials that market dynamics are out of the hands of the central government. Companies decide to stay in or out of Iran business faced with sanctions, so the argument. The fact that Chinese companies operating in Iran are big state-owned ones, however, casts a certain damp on this line of reasoning. This nurtured the point often made by non-Chinese experts and officials that China actually stood to benefit from Western embargoes against Iran.

Lastly, this chapter has analysed material factors that influence China's Iran policy. These are particularly relevant in the context of China's energy needs. The Chinese government has an interest in good economic relations and in securing energy supplies from Iran's huge oil and gas fields. Iran, in turn, imports part of its refined petrol from China due to its own limited refining capacities. Iranian–Chinese bilateral trade is intensive, and China is indicating a strong interest in deepening trade relations following the lifting of international and unilateral Iran sanctions on 16 January 2016. The increased interest in the Iranian market on the part of Western companies might usher in a more competitive business environment for Chinese companies in Iran. Uncertainties linger, however, concerning the continued application of US unilateral financial sanctions, and the future implementation of the JCPOA. On non-proliferation aspects of the JCPOA, China's involvement in the reconstruction of the Iranian nuclear facility at Arak as a technical element of the nuclear accord means that China is a participant co-responsible for its dutiful implementation, and indicates the Chinese government's commitment to a successful closure of the Iranian nuclear file. Beijing's P5 membership also makes it a crucial member of the

'Joint Commission' set up as a dispute resolution mechanism. China's foreign policy discourse on the Iranian nuclear crisis has always been the expression of a security culture that, at its core, propagated stability for politico-economic reasons alongside a cautious criticism of sanctions policies that were seen as eroding that stability. China's active involvement in the implementation of the JCPOA as well as its pronounced interest in making use of economic opportunities that may arise on the Iranian market are indications for an understanding in Beijing that the successful implementation of the nuclear agreement would be the best guarantee for the endurance of a diplomatic compromise that serves Chinese interests on a material, institutional and ideational level.

Notes

1 These talking points were also given in an interview with a former Chinese diplomat to Iran. Author's interview, Beijing, 22 April 2014.

2 This conference, held in Indonesia, saw the gathering of many heads of states of 'developing countries' and was perceived by the Chinese leadership under chairman Mao Zedong as the expression of an anti-imperialist momentum. Public references to the 'Bandung Spirit' throughout the 1960s underlined the importance China attached to ties with the developing world, especially at a time when Chinese–Soviet relations increasingly faltered in the wake of the Mao-Krushchev rift (Lanteigne 2009: 133).

3 On a strategic level, the fear that interventionist policies can lead to precedents affecting China's delicate relations with Taiwan and Tibet has been analysed as another underlying factor for China's steadfast stance for 'non-interference' (Wu 2010: 295).

4 Author's Interview with former Chinese diplomat to Iran, Beijing, 22 April 2014.

5 Author's Interview with former Chinese diplomat to Iran, Beijing, 22 April 2014; author's Interview with Li Xin, China Institute of contemporary international relations, Beijing, 24 April 2014. See also foreign minister Wang Yi's link between the implementation of the agreement between Iran and the P5+1 and the establishment of a nuclear weapons-free zone in the Middle East (Wang 2014). The NWFZ in the Middle East is called for in the 1995 Resolution on the Middle East, and was linked to the indefinite extension of the NPT the same year at the NPT Review Conference (Potter and Mukhatzhanova 2012: 38–39).

6 Justified as this reproach might be, it is somewhat misplaced as coming from nuclear-armed China. With this position, however, China is targeting the US and Russia as possessors of the world's largest nuclear arsenals (see also Fey *et al.* 2013: 182–183).

7 The position of the NAM in the Iranian nuclear dossier was based on an 'emphasis on the multiculturalism and challenge against the nuclear disarmament and non-proliferation, on the one hand, and non-compromising position on inalienable and non-discriminatory right for peaceful uses of nuclear energy' (Soltanieh, in Mousavian 2012: 471). On NAM positions on non-proliferation and disarmament, see Braveboy-Wagner (2009: 24); Litwak (2014: 58–59); Potter and Mukhatzhanova (2012); Yew (2011).

8 The failure to report the sale was mentioned in the IAEA's June 2003 report on the implementation of the NPT safeguards agreement in Iran (IAEA 2003: 2).

9 China, however, has not yet ratified the CTBT.

10 White Paper on China's Non-Proliferation Policy Published. Available at www.china.org.cn/english/2003/Dec/81312.htm (accessed 2 October 2014).

11 This wording was used almost verbatim by a European diplomat in conversation with the author, Moscow, 18 April 2013.

12 European Iran desk officer, conversation with author, Berlin, 4 February 2013.

13 European Iran desk officer, conversation with author, Berlin, 4 February 2013. A comprehensive analysis of a joint Chinese–Russian negotiation behaviour is beyond the scope of this chapter. Suffice to recall at this point that inferring from such pre-negotiations the existence of a united Chinese–Russian 'bloc' confronting the West would be an analytical fallacy. The previous chapter has analysed Russia's positions on Iran in the Security Council as well as Russia's respective foreign policies that are distinct from China's. The next chapter will draw on the conclusions reached in the two chapters on Russian and Chinese Iran policies in synthesising the research findings from a comparative perspective. The following chapter will therefore also elaborate on the positional dynamics between China and Russia in the UN Security Council.

14 Author's interview with former Chinese diplomat to Iran, Beijing, 22 April 2014; author's interview with Dr Jin Lianxiang, senior research fellow at the Shanghai Institute for International Studies, Shanghai, 16 April 2014; Author's interview with Dr Su Hao, Beijing, 6 May 2014.

15 UNSCR 1929 significantly expanded sanctions on the Iranian financial sector, prohibiting the establishment of Iranian banks abroad and froze foreign account assets of Iranian entities. It also for the first time directly identified the Iranian Revolutionary Guards Corps (IRGC) as a target of sanctions and imposed sanctions on the Iranian transport
sector (Iran Air Cargo and the Islamic Republic of Iran Shipping Lines – IRISL).

16 Under international law, in addition, it is hardly possible to prevent 'flag hopping' because legal requirements of ownership and manning of ships remain vague. The applicable 1986 Registration Convention has never entered into force (Sohn *et al.* 2010: 52). If Iranian cargo vessels thus change flags, third parties cannot contest. Sanctioning Iranian cargo traffic therefore involves an active enforcement on the part of port states.

17 Author's interview with Chinese diplomat to the US, Washington, 31 October 2014.

18 EEAS official, author's interview, Brussels, 13 March 2013; author's interview with high-ranking Swiss diplomat, Berlin, 26 August 2013; author's interview with former E3 delegation member, Brussels, 29 January 2015; author's interview with Dr Ali Vaez, Crisis Group, senior Iran analyst at the International Crisis Group, via Skype, 25 July 2013; author's interview with Dr Dina Esfandiary, expert at the International Institute for Strategic Studies, London, 11 July 2014. However, Carnegie scholar Tong Zhao (2015) writes that

> For its part, Beijing is interested in shedding the label of free rider and more actively contributing to peace and stability in the region. But China won't be able to complete the process overnight. Beijing has generally preferred to take an indirect and gradualist approach.

19 Author's interview with Chinese foreign ministry official, Beijing, 18 April 2014; author's interview with former Chinese diplomat to Iran, Beijing, 22 April 2014; author's interview with Chinese diplomat to the US, Washington, 31 October 2014; author's interview with Dr Li Xin, China Institutes of contemporary international relations, Beijing, 24 April 2014.

20 Between China and the US, China's compliance with the MTCR became a constant issue of contention. China recognised Annex I that lists which missiles and related equipment are banned (Category 1 missiles), but not Annex II, which lists dual-use 'missile-related technology and items' whose export is to be controlled by an export licensing system on a case-by-case basis (Garver 2006a: 212–213). The US decision in 2003 to ban Norinco from the US market for several years was likely the most outspoken and decisively public sanctions effort in this regard. Norinco, formally China North Industries Corporation, is a Chinese manufacturing company that is known, i.a. for its defense products.

21 Author's interview, Washington, 30 October 2014.
22 President Obama also reportedly had a phone conversation with then-president Hu Jintao precisely on China's position on Iran sanctions in the run-up to UNSCR 1929 (Zhao 2013: 117), while US ambassador to the UN Susan Rice lobbied in New York (Landler 2016: 245).
23 Author's interview with former Chinese diplomat to Iran, Beijing, 22 April 2014; author's interview with Dr Jin Lianxiang, senior research fellow at the Shanghai Institute for International Studies, Shanghai, 16 April 2014.
24 Author's interview with Dr Jin Lianxiang, Shanghai, 16 April 2014.
25 Author's interview with Iranian diplomat, Berlin, 29 August 2016.
26 Author's interview with former Chinese diplomat to Iran, Beijing, 22 April 2014. On this, see also Zhao (2015).
27 In an insightful article, Downs (2004) writes that the top leadership positions in the major Chinese (state-owned) oil companies like CNPC, Sinopec and CNOOC are appointed by the Central Committee of the CCP. This gives company heads direct access to and political clout over the Chinese leadership, in addition to the state's 'fiscal dependence' on these companies (25).
28 Much in the same logic of referring to the complications of international markets, this reading would also explain why the Chinese government had conducted most of its material assistance to Iran, both in the military domain and in the provision of 'dual-use technology', e.g. nuclear infrastructure, covertly or semi-covertly (Garver 2006a: 125). When China began to supply arms to Iran (and Iraq) during the Iran–Iraq war, the government relied on third country intermediaries, 'thus allowing Beijing to claim with narrow accuracy that *China* had not sold weapons to Iran' (ibid.: 81). This 'cover' effect that Garver calls a 'policy of plausible denial' (ibid.: 194) would allow China to present its dealings with Iran in one light or the other, depending on the intended audience. Malik (2000) adds that 'complex, family-connected networks (*Guanxi*) operate across military organizations, government ministries, and nominal civilian corporations' and 'can be unresponsive to admonitions from the MFA' (461).
29 Author's interview, Washington, 31 October 2014.
30 Author's interview with Iranian diplomat, Berlin, 29 August 2016.
31 Dr Ali Vaez, chief Iran analyst with the International Crisis Group, therefore succinctly characterised the Sino-Iranian economic relationship as an 'oil for junk' exchange. Author's interview via Skype, 25 July 2013. In a similar fashion, a US State Department official stated in an interview that China and India 'have been taking advantage of (Western) sanctions. But the quality is crap.' Author's interview, Washington, 30 October 2014. The idea of China taking advantage of Western embargoes against Iran has been expressed by a number of interviewees (author's interview with EEAS official, Brussels, 24 February 2015; author's interview with former Chinese diplomat to Iran, Beijing, 22 April 2014; author's interview with European diplomat, Brussels, 6 February 2015).
32 Author's interview with Chinese foreign ministry official, Beijing, 18 April 2014; author's interview with former Chinese diplomat to Iran, Beijing, 22 April 2014; author's interview with Chinese diplomat to the US, Washington, 31 October 2014.
33 In 2007 and 2008 respectively, a Sino-Iranian joint venture of automobile companies was created (between the Chinese company Chery and the Iranian company Majmoeh Mazi Toos and between Chinese LiFan and the Iranian KMC Company). See International Crisis Group (2010: 7).
34 See also Tehran Metro website (2013), http://tehran-metro.com/featured/iran-seeks-2bn-from-china-to-complete-tehran-metro; author's interview with former Chinese diplomat to Iran, Beijing, 22 April 2014.
35 E.g. oil from the China Petroleum National Corporation-led consortium in Kazakhstan. This ties China, Iran and Kazakhstan together in an economic triangle (Hiro 2009: 387).

36 These are official UN trade numbers retrieved from the Statistics Division of the Department of Economics and Social Affairs. Available at http://comtrade.un.org/data/.

37 Author's interview with Dr Jin Lianxiang, senior research fellow at the Shanghai Institute for International Studies, Shanghai, 16 April 2014.

38 In 2011, around 20 per cent of Chinese crude oil imports came from Saudi Arabia, while imports from Iran accounted for 11 per cent of Chinese overall crude oil imports. Angola also ranks as one of China's biggest oil suppliers. See International Energy Agency (2012: 6).

39 The Malacca Strait is a maritime strait between Indonesia, Malaysia and Singapore, linking the Indian to the Pacific Ocean and the South China Sea. See International Energy Agency (2012). Burman (2009) further names the Lombok Strait and the Strait of Macassar, 'each of which could "fairly" easily be blockaded by the US Navy' (117).

40 See also Garver (2006b) on China's development of overland transportation links with Central Asia. This infrastructure expansion, he contends, is a projection of Chinese economic interests in need of logistical protection.

41 Author's interview with Russian nuclear delegation member, Russian permanent mission to the IAEA, Vienna, 18 August 2016; author's interview with French nuclear delegation member, French permanent mission to the IAEA, Vienna, 17 August 2017; author's interview with Iranian diplomat, Berlin, 29 August 2016. While a 'basic agreement' between China and Iran about the reconstruction of Arak has been reached, the first Iranian progress report on the implementation of the JCPOA mentions that the main company in charge of the redesign is Iranian (www.farsnews.com/newstext.php?nn=13950129001029). Ariane Tabatabai (2016) writes that this may have been intended as a 'way to try to get a better deal with Beijing'. A 'Working Group', with China in the lead, is composed of P5+1 participants and facilitates the Arak conversion. The latter is carried out by an International Partnership, composed of the Working Group and Iran. The project manager of the actual operation is Iran.

42 Shambaugh and Xiao (2012) attribute the scholarly obsession with 'concealing leadership' to a translation error and contrast several possible translations of Deng's dictum *taoguang yanghui, bu dang tou, zousuo zuowei* (40–41).

43 See Lampton's (2014) study for insights into questions of strategic leadership from Deng Xiaoping to Xi Jinping. Zhao (2013: 114–120) contrasts Chinese 'Core Interests' with 'Global Power Responsibility' and shows how Chinese decision-makers were traditionally torn between bilateralism and multilateralism to address major foreign policy issues. A comprehensive and insightful overview over Chinese domestic foreign policy debates has been written by Shambaugh and Xiao (2012). While domestic debates about China's foreign policy role abound,

> government officials in the Foreign Ministry and Central Committee Foreign Affairs Office are pragmatically centered between these two schools (Major Powers and Global South), but they must respond to Nativist and Realist voices in society, the military, and the Communist Party.
>
> (67)

44 The aspect of 'civilisational familiarity' in Chinese–Iranian relations was also mentioned by Dr Su Hao, expert at China Foreign Affairs University. Author's interview at China Foreign Affairs University, Beijing, 6 May 2014.

45 Yet, Shambaugh and Xiao (2012) assert: 'anti-hegemony (*fan ba*) remains the *sine qua non* of the Chinese worldview and foreign policy' (47).

46 For readings of a Chinese 'revisionist' foreign policy, see Chin and Thakur (2010); Christensen (2011); Deng (2006); Economy (2010); Feng (2009); Friedberg (2005, 2011); Halper (2010); He and Feng (2012); Holslag (2011); Jacques (2009); Kissinger (2011: 487f.); S. Chan (2004); Shirk (2007); Swaine (2011). For analyses

of a more cooperative Chinese foreign policy, see G. Chan (2006); Gill (2007); Ikenberry (2008); Johnston (2003, 2004, 2013); Kang (2003, 2007); Liang (2007); Medeiros and Fravel (2003). For 'pragmatic' approaches, see Breslin (2013); Da (2010); Foot (2006; 2009/2010); Geeraerts and Holslag (2007); Hung (2016); Kastner and Saunders (2012); Wang (2011); Yan (2006); Zhao (2006). In this context, see also the 'power transition theory': Levy (2008). Zhu Liqun's study *China's Foreign Policy Debates* (2010) offers an effective overview of these and other major debates on China's foreign policy from a Chinese perspective. For an insightful overview of Chinese domestic foreign policy debates, see Shambaugh and Xiao (2012: 36–72).

47 Zhao (2013) fittingly writes of a 'balance between taking a broad great power responsibility and focusing on its narrowly defined core interests to play down its pretense of being a global power' (121).

48 Author's interview with Chinese foreign ministry official, Beijing, 18 April 2014.

49 Author's interview with former Chinese diplomat to Iran, Beijing, 22 April 2014.

50 Author's interviews with Dr Mark Fitzpatrick, International Institute for Strategic Studies London, 11 July 2014; with Dr Dina Esfandiary, International Institute for Strategic Studies, London, 11 July 2014; with EEAS official, Brussels, 13 March 2013; with EEAS official, Brussels, 4 June 2013; with former German ambassador to Iran, Berlin, 26 August 2013; with EEAS official, Brussels, 20 March 2014; with European diplomat, Brussels, 18 March 2014; with former Chinese diplomat to Iran, Beijing, 22 April 2014; with E3 official, London, 10 March 2015; with Fjodor Lukyanov, Moscow, 14 June 2016. A European diplomat described the Chinese negotiation technique as 'not completely passive, but rather low-level', and continued: 'China is not so creative. They [*sic*] help where they can, but display a "strategic patience"'. Author's interview, Brussels, 6 February 2015.

51 Author's interview with former E3 delegation member, Brussels, 29 January 2015.

52 Author's interview, Moscow, 14 June 2016.

53 Author's interview with E3 official, London, 10 March 2015.

References

Arbatov, Alexei and Sazhin, Vladimir. 2016. The Nuclear Deal With Iran: The Final Step or a New Stage? *Carnegie Article*, 20 April 2016. Available at http://carnegie.ru/2016/04/20/nuclear-deal-with-iran-final-step-or-new-stage/ixc4?mkt_tok=eyJpIjoiWVRNNE1UbGlPRFZpTTJGaSIsInQiOiJTOFpReDM3Q1c5MFZtUGhqZzJWXC94Njd6MHB0TFJRN2ZjTzV4YWpTNUl0V0tXT2pBbmxHUDFrVEJNa3pranpwXC9DRDBoWFg2cWZBU29FMDg1NjI5cERRTTl2NkZXejhIeWp2YitVM1BKczZNPSJ9 (accessed 27 June 2016).

Biersteker, Thomas and Moret, Erica. 2015. Rising Powers and Reform of the Practices of International Security Institutions. In: Gaskarth, Jamie (ed.), *Rising Powers, Global Governance and Global Ethics*. Abingdon and New York: Routledge, pp. 57–73.

Braveboy-Wagner, Jaqueline Anne. 2009. *Institutions of the Global South*. New York: Routledge.

Breslin, Shaun. 2013. China and the Global Order: Signaling Threat or Friendship? *International Affairs* vol. 3: 615–634.

Broadhead, Ivan. 2011. Hong Kong Shipping Under Scrutiny for Iran Links. *Voice of America* 5 October 2011. Available at www.voanews.com/content/hong-kong-shipping-under-scrutiny-for-iran-links-131213804/146250.html (accessed 13 October 2014).

Burman, Edward. 2009. *China and Iran. Parallel History, Future Threat?* Stroud: The History Press.

Calabrese, John. 2006. China and Iran: Mismatched Partners. *Jamestown Occasional Papers*. Available at www.jamestown.org/docs/Jamestown-ChinaIranMismatch.pdf (accessed 10 February 2013).

Chan, Gerald. 1999. *Chinese Perspectives on International Relations: A Framework for Analysis*. London: MacMillan.

Chan, Gerald. 2006. *China's Compliance in Global Affairs. Trade, Arms Control, Environmental Protection, Human Rights*. Singapore: World Scientific.

Chan, Steve. 2004. Can't Get No Satisfaction? The Recognition of Revisionist States. *International Relations of the Asia-Pacific* vol. 4: 207–238.

Chatham House. 2013. China's New Leadership: Approaches to International Affairs. *Asia Meeting Summary*, 7 March 2013.

Chen, Dingding and Pu, Xiaoyu. 2014. Correspondence with Alastair Iain Johnston. Letter to the Editors. *International Security* vol. 38, no. 3: 176–183.

Chin, Gregory and Thakur, Ramesh. 2010. Will China Change the Rules of Global Order? *The Washington Quarterly* vol. 33: 119–138.

Chinese foreign ministry. 2014. *China's Five Principles for a Comprehensive Solution of the Iranian Nuclear Issue*. 19 February 2014. Available at www.fmprc.gov.cn/mfa_eng/wjb_663304/zzjg_663340/jks_665232/jkxw_665234/t1129941.shtml (accessed 1 May 2014).

Chinese foreign ministry. 2016a. Xi Jinping Arrives in Tehran for State Visit to Iran. *Foreign Ministry Communication*, 23 January 2016. Available at www.fmprc.gov.cn/mfa_eng/zxxx_662805/t1335158.shtml (accessed 22 June 2016).

Chinese foreign ministry. 2016b. Xi Jinping Meets with Supreme Leader Ali Khamenei of Iran. *Foreign Ministry Communication*, 24 January 2016. Available at www.fmprc.gov.cn/mfa_eng/topics_665678/xjpdstajyljxgsfw/t1335153.shtml (accessed 22 June 2016).

Christensen, Thomas. 2011. The Advantages of an Assertive China: Responding to Beijing's Abrasive Diplomacy. *Foreign Affairs* vol. 90, no. 2: 54–67.

Da, Wei. 2010. Has China Become 'Tough'? *China Security* vol. 6, no. 3: 97–104.

Deng, Yong. 2006. Reputation and the Security Dilemma: China Reacts to the China Threat Theory. In: Alastair Johnston and Robert Ross (eds), *New Directions in the Study of China's Foreign Policy*. Stanford, CA: Stanford University Press, pp. 186–214.

Derakhshi, Reza. 2009. Sinopec in $6.5 Billion Iran Refinery Deal: Iranian Media. *Reuters*, 25 November 2009. Available at www.reuters.com/article/2009/11/25/us-iran-china-refineries-idUSTRE5AO20C20091125 (accessed 25 March 2015).

Djallil, Lounnas. 2011. China and the Iranian Nuclear Crisis: Between Ambiguities and Interests. *European Journal of East Asian Studies* vol. 10: 227–253.

Dorraj, Manochehr and Currier, Carrie. 2008. Lubricated with Oil: Iran–China Relations in a Changing World. *Middle East Policy*, vol. 15, no. 2: 66–80.

Downs, Erica S. 2004. The Chinese Energy Security Debate. *The China Quarterly* vol. 177: 21–41.

EEAS. 2015a. *Joint Comprehensive Plan of Action*. Vienna, 14 July 2015. Available at https://eeas.europa.eu/statements-eeas/docs/iran_agreement/iran_joint-comprehensive-plan-of-action_en.pdf (accessed 17 September 2016).

EEAS. 2015b. Joint statement by EU High Representative Federica Mogherini and Iranian Foreign Minister Javad Zarif Vienna, *Press statement*, 14 July. Available at www.eeas.europa.eu/statements-eeas/2015/150714_01_en.htm (accessed 13 August 2015).

Economy, Elizabeth C. 2010. The Game Changer: Coping with China's Foreign Policy Revolution. *Foreign Affairs* vol. 89, no. 9: 142–152.

ElBaradei, Mohamed. 2012. *The Age of Deception. Nuclear Diplomacy in Treacherous Times*. New York: Bloomsburg.

Feng, Huiyun. 2009. Is China a Revisionist Power? *The Chinese Journal of International Politics* vol. 2: 313–334.

Ferdinand, Peter. 2013. The Positions of Russia and China at the UN Security Council in the Light of Recent Crises. *Briefing Paper*, March 2013. Directorate-General for External Policies of the European Union.

Fey, Marco, Hellmann, Andrea, Klinke, Friederike, Plümmer, Franziska and Rauch, Carsten. 2013. Established and Rising Great Powers: The United States, Russia, China, and India. In: Müller, Harald and Wunderlich, Carmen (eds), *Norm Dynamics in Multilateral Arms Control: Interests, Conflicts, and Justice*. Athens: University of Georgia Press, pp. 163–206.

Foot, Rosemary. 2006. Chinese Strategies in a US-Hegemonic Global Order: Accommodating and Hedging. *International Affairs* vol. 82, no. 1: 77–94.

Foot, Rosemary. 2009/2010. China and the United States: Between Cold and Warm Peace. *Survival* vol. 51, no. 6: 123–146.

Fitzpatrick, Mark. 2010. Iran: The Fragile Promises of the Fuel-Swap Plan. *Survival* vol. 52, no. 3: 67–94.

Friedberg, Aaron L. 2005. The Future of US–China Relations: Is Conflict Inevitable? *International Security* vol. 30, no. 2: 7–45.

Friedberg, Aaron L. 2011. *A Contest for Supremacy: China, America, and the Struggle for Mastery in Asia*. London and New York: W.W. Norton.

Garver, John. 2006a. *China and Iran. Ancient Partners in a Post-imperial World*. Seattle: University of Washington Press.

Garver, John. 2006b. Development of China's Overland Transportation Links with Central, South-west and South Asia. *The China Quarterly* vol. 185: 1–22.

Garver, John. 2011. Is China Playing a Dual Game in Iran? *The Washington Quarterly* vol. 34, no. 1: 75–88.

Garver, John. 2016. China and Iran: An Emerging Partnership Post-Sanctions. *MEI Policy Focus* 3.

Geeraerts, Gustaaf and Holslag, Jonathan. 2007. The 'Pandragon'. China's Dual Diplomatic Identity. *BICCS Asia Papers* vol. 2, no. 1: 1–15.

Gill, Bates. 1998. Chinese Arms Exports to Iran. *Middle East Review of International Affairs*, vol. 2, no. 2: 55–70.

Gill, Bates. 1999. Chinese Arms Exports to Iran. In: Kumaraswamy, P.R. (ed.), *China and the Middle East. The Quest for Influence*. New Delhi: Sage, pp. 117–141.

Gill, Bates. 2007. *Rising Star: China's New Security Diplomacy*. Washington, DC: Brookings Institution Press.

Glaser, Bonnie S. and Medeiros, Evan S. 2007. The Changing Ecology of Foreign Policy-making in China: the Ascension and Demise of the Theory of 'Peaceful Rise'. *China Quarterly* vol. 190: 291–310.

Glenn, Cameron. 2016. Economic Trends: January 2016. *The Iran Primer*, 1 February 2016. Available at http://iranprimer.usip.org/blog/2016/feb/01/economic-trends-january-2016 (accessed 27 September 2016).

Godehardt, Nadine. 2014. China's 'neue' Seidenstrasseninitiative. *SWP-Studie*, June. Available at www.swp-berlin.org/fileadmin/contents/products/studien/2014_S09_gdh.pdf (accessed 24 March 2015).

Goldstein, Lyle J. 2011. Resetting the US–China Security Relationship. *Survival* vol. 53, no. 2: 89–116.

Halper, Stefan. 2010. *The Beijing Consensus: How China's Authoritarian Model Will Dominate the Twenty-First Century*. New York: Basic Books.

Harold, Scott and Nader, Alireza. 2012. China and Iran. Economic, Political, and Military Relations. *RAND Occasional Paper*. Santa Monica: RAND Corporation. Available at www.rand.org/pubs/occasional_papers/OP351.html (accessed 6 September 2016).

He, Hai and Feng, Huiyun. 2012. Debating China's Assertiveness: Taking China's Power and Interests Seriously. *International Politics* vol. 49, no. 5: 633–644.

Hiro, Dilip. 2009. *Inside Central Asia. A Political and Cultural History of Uzbekistan, Turkmenistan, Kazakhstan, Kyrgyzstan, Tajikistan, Turkey, and Iran*. New York: Overlook Duckworth.

Holslag, Jonathan. 2011. *Trapped Giants. China's Troubled Military Rise*. Abingdon: Routledge, Adelphi Books.

Hua, Chunying. 2014. Foreign Ministry Spokesperson Hua Chunying's Regular Press Conference on February 21, 2014. *MFA Spokesperson's Remarks*. Available at www.chinamission.be/eng/fyrth/t1131024.htm (accessed 23 April 2014).

Hua, Chunying. 2015. Foreign Minister Wang Yi to Attend Foreign Ministers' Meeting on the Iranian Nuclear Issue. *MFA Spokesperson's Announcement*, 2 July 2015. Available at www.fmprc.gov.cn/mfa_eng/wjdt_665385/wsrc_665395/t1278070.shtml (accessed 21 June 2016).

Hu, Jintao. 2007. *Documents of the 17th National Congress of the Communist Party of China*. Beijing: Foreign Languages Press.

Hung, Ho-fung. 2016. *The China Boom. Why China Will Not Rule the World*. New York: Columbia University Press.

IAEA. 2003. Implementation of the NPT Safeguards Agreement in the Islamic Republic of Iran. *IAEA Board Report*. Available at www.iaea.org/Publications/Documents/Board/2003/gov2003-40.pdf (accessed 20 December 2012).

IAEA. 2006. Implementation of the NPT safeguards agreement in the Islamic Republic of Iran. *IAEA Board Report*. Available at www.iaea.org/Publications/Documents/Board/2006/gov2006-53.pdf (accessed 20 December 2012).

Ikenberry, John G. 2008. The Rise of China and the Future of the West. Can the Liberal System Survive? *Foreign Affairs* vol. 87. No. 1: 23–37.

Ikenberry, John G. 2011. *Liberal Leviathan: The Origins, Crisis, and Transformation of the American World Order*. Princeton: Princeton University Press.

International Energy Agency. 2012. World Oil Choke Points. *Analysis Briefs*. Available at www.eia.gov/countries/regions-topics.cfm?fips=wotc&trk=p3 (accessed 19 February 2013).

International Crisis Group. 2010. The Iran Nuclear Issue: The View from Beijing. *Asia Briefing* Nr. 100, 17 February 2010.

Jacques, Martin. 2012. *When China Rules the World*. 2nd edn. London: Penguin.

Johnston, Alastair Iain. 2003. Is China a Status Quo Power? *International Security* vol. 27: 5–56.

Johnston, Alastair Iain. 2004. Beijing's Security Behavior in the Asia-Pacific: Is China a Dissatisfied Power? In: Suh, J.J., Katzenstein, Peter J. and Carlson, Allen (eds), *Rethinking Security in East Asia: Identity, Power, and Efficiency*. Stanford, CA: Stanford University Press, pp. 34–96.

Johnston, Alastair Iain. 2013. How New and Assertive Is China's New Assertiveness? *International Security* vol. 37, no. 4: 7–48.

Jones, Catherine and Breslin, Shaun. 2015. China in East Asia. Confusion on the Horizon? In: Gaskarth, Jamie (ed.), *Rising Powers, Global Governance and Global Ethics*. Abingdon and New York: Routledge, pp. 115–132.

Kang, David C. 2003. Getting Asia Wrong: the Need for New Analytical Frameworks. *International Security* vol. 27, no. 4: 57–85.
Kang, David C. 2007. China Rising: *Peace, Power, and Order in East Asia*. New York: Columbia University Press.
Kastner, Scott L. and Saunders, Phillip C. 2012. Is China a Status Quo or Revisionist State? Leadership Travel as an Empirical Indicator of Foreign Policy Priorities. *International Studies Quarterly* vol. 56, no. 1: 163–77.
Katzenstein, Peter. J. (ed.). 1996. *The Culture of National Security. Norms and Identity in World Politics*. New York: Columbia University Press.
Katzman, Kenneth. 2006. The Iran–Libya Sanctions Act (ILSA). *CRS Report for Congress*, 26 April 2006. Available at http://fpc.state.gov/documents/organization/66441.pdf (accessed 27 September 2014).
Kemenade, Willem van. 2010. China vs. the Western Campaign for Iran Sanctions. *The Washington Quarterly*, vol. 33, no. 3: 99–114.
Kissinger, Henry A. 2011. *On China*. New York: Penguin.
Lampton, David M. 2014. *Following the Leader. Ruling China, From Deng Xiaoping to Xi Jinping*. Berkeley: University of California Press.
Landler, Mark. 2016. *Alter Egos. Hillary Clinton, Barack Obama, and the Twilight Struggle over American Power*. London: WH Allen.
Lanteigne, Marc. 2009. *Chinese Foreign Policy. An Introduction*. New York: Routledge.
Levy, Jack S. 2008. Power Transition Theory and the Rise of China. In: Ross, Robert S. and Feng, Zhu (eds), *China's Ascent: Power, Security, and the Future of International Politics*. Ithaca, NY: Cornell University Press, pp. 11–33.
Li, Xiaojun. 2010. Social Rewards and Socialization Effects: An Alternative Explanation for the Motivation Behind China's Participation in International Institutions. *The Chinese Journal of International Politics* vol. 3: 347–377.
Liang, Wei. 2007. China: Globalization and the Emergence of a New Status Quo Power? *Asian Perspective* vol. 31, no. 4: 125–149.
Litwak, Robert. 2014. *Iran's Nuclear Chess: Calculating America's Moves*. Washington, D.C.: Woodrow Wilson Center Press.
Liqun, Zhu. 2010. China's Foreign Policy Debates. *Chaillot Papers*, September 2010. Paris: European Union Institute for Security Studies.
Lohmann, Sascha. 2013. Unilaterale US-Sanktionen gegen Iran. *SWP-Aktuell* vol. 63: 1–8.
Lohmann, Sascha. 2015. Zwang zur Zusammenarbeit. *SWP-Aktuell* vol. 54: 1–8.
Malik, J. Mohan. 2000. China and the Nuclear Non-Proliferation Regime. *Contemporary Southeast Asia* vol. 22, no. 3: 445- 478.
Markey, Daniel S. 2014. Reorienting U.S. Pakistan Strategy. From Af-Pak to Asia. *Council Special Report* No. 28, January. New York: Council on Foreign Relations.
Mazza, Michael. 2011. China–Iran Ties: Assessment and Implications for U.S. Policy. *AEI Iran Tracker*, 21 April 2011. Available at www.irantracker.org/analysis/michael-mazza-china-iran-ties-assessment-and-implications-us-policy-april-21-2011 (accessed 11 February 2013).
Medeiros, Evan S. 2007. *Reluctant Restraint: The Evolution of China's Nonproliferation Policies and Practices, 1980–2004*. Stanford: Stanford University Press.
Medeiros, Evan S. and Fravel, Taylor M. 2003. China's New Diplomacy. *Foreign Affairs* vol. 82, no. 6: 22–35.
Mousavian, Seyed Hossein. 2012. *The Iranian Nuclear Crisis. A Memoir*. Washington: Carnegie Endowment for International Peace.

Nourafchan, Nicolo. 2010. Constructive Partner or Menacing Threat? Analyzing China's Role in the Iranian Nuclear Program. *Asian Security* vol. 6, no. 1: 28–50.

Patrikarakos, David. 2012. *Nuclear Iran. The Birth of an Atomic State*. London, New York: I.B. Tauris.

Pieper, Moritz. 2013. Dragon Dance or Panda Trot? China's Position towards the Iranian Nuclear Programme and Its Perception of EU Unilateral Iran Sanctions. *European Journal of East Asian Studies* vol. 12: 295–316.

Porter, Gareth. 2014. *Manufactured Crisis. The Untold Story of the Iran Nuclear Scare*. Charlottesville: Just World Books.

Potter, William C. and Mukhatzhanova, Gaukhar. 2012. *Nuclear Politics and the Non-aligned Movement: Principles vs. Pragmatism*. Abingdon, New York: Routledge.

Rubenfeld, Samuel. 2011. US Threatens Sanctions on Chinese Banks over Iran. *The Wall Street Journal*, 28 September 2011. Available at http://blogs.wsj.com/corruption-currents/2011/09/28/us-threatens-sanctions-on-chinese-banks-over-iran/ (accessed 13 October 2014).

Rudolf, Peter. 2016. Liberale Hegemonie und Aussenpolitik unter Barack Obama. *SWP Comments* 56, August 2016.

Saunders, Paul J. 2015. Russia Eyes Iranian Arms Deal after Lausanne. *Al-monitor*, 6 April 2015. Available at www.al-monitor.com/pulse/originals/2015/04/iran-s300-sale.html (accessed 18 April 2015).

Shambaugh, David and Xiao, Ren. 2012. China: The Conflicted Rising Power. In: Nau, Henry R. and Ollapally, Deepa M. (eds), *Worldviews of Aspiring Powers. Domestic Foreign Policy Debates in China, India, Iran, Japan, and Russia*. New York: Oxford University Press, pp. 36–72.

Shen, Dingli. 2006. Iran's Nuclear Ambitions Test China's Wisdom. *Washington Quarterly* vol. 29, No. 2: 56–57.

Shirk, Susan L. 2007. *China Fragile Superpower: How China's Internal Politics Could Derail Its Peaceful Rise*. Oxford: Oxford University Press.

Slavin, Barbara. 2011. Iran Turns to China, Barter to Survive Sanctions. *Atlantic Council: Iran Task Force*. Available at www.acus.org/files/publication_pdfs/403/111011_ACUS_IranChina.PDF (accessed 11 February 2013).

Sohn, Lous B., Juras, Kristen Gustafson, Noyes, John E. and Franckx, Erik. 2010. *Law of the Sea in a Nutshell*. 2nd edn. St. Paul: West Nutshell Series.

Solingen, Etel. 2012. Ten Dilemmas in Nonproliferation Statecraft. In: Solingen, Etel (ed.), *Sanctions, Statecraft, and Nuclear Proliferation*. Cambridge: Cambridge University Press.

Swaine, Michael D. 2010. Beijing's Tightrope Walk on Iran. *China Leadership Monitor nr. 33*, 28 June. Available at www.hoover.org/publications/china-leadership-monitor/article/35436 (accessed 11 February 2013).

Tabatabai, Ariane. 2016. Iran Issues First Progress Report on Nuclear Deal. *Bulletin of the Atomic Scientists, Column*, 19 April 2016. Available at http://thebulletin.org/iran-issues-first-progress-report-nuclear-deal9350?mkt_tok=eyJpIjoiWVRNNE1UbGlPRFZpTTJGaSIsInQiOiJTOFpReDM3Q1c5MFZtUGhqZzJWXC94Njd6MHB0TFJRN2ZjTzV4YWpTNUl0V0tXT2pBbmxHUDFrVEJNa3pranpwXC9DRDBoWFg2cWZBU29FMDg1NjI5cERRTTl2NkZXejhIeWp2YitVM1BKczZNPSJ9 (accessed 27 June 2016).

Tsukimori, Osamu and Goswami, Manash. 2013. Exclusive: Iran Oil Exports to Plunge, No Dividend Yet from Easing Tensions. *Reuters*. Available at www.reuters.com/article/2013/10/25/us-iran-oil-asia-exclusive-idUSBRE99O0GF20131025 (accessed 11 November 2013).

UN. 2006a. Security Council Imposes Sanctions on Iran for Failure to Halt Uranium Enrichment, Unanimously Adopting Resolution 1737. *Resolution 1737*. Available at www.un.org/News/Press/docs//2006/sc8928.doc.htm (accessed 21 December 2012).

UN. 2006b. Provisional Summary, 5612th Security Council Meeting, *S/PV.5612*. Available at www.un.org/en/ga/search/view_doc.asp?symbol=S/PV.5612 (accessed 13 October 2014).

UN. 2007. Provisional Summary, 5647th Security Council Meeting. *S/PV.5647*. Available at www.un.org/en/ga/search/view_doc.asp?symbol=S/PV.5647 (accessed 13 October 2014).

UN. 2008a. Security Council Tightens Restriction on Iran's Proliferation-sensitive Nuclear Activities, Increases Vigilance over Iranian Banks, Has States Inspect Cargo. *Security Council Resolution 1803*. Available at www.un.org/News/Press/docs/2008/sc9268.doc.htm (accessed 21 December 2012).

UN. 2008b. Provisional Summary, 5848th Security Council Meeting. *S/PV.5848*. Available at www.un.org/en/ga/search/view_doc.asp?symbol=S/PV.5848 (accessed 13 October 2014).

UN. 2009. Provisional Summary, 6191st Security Council Meeting. *S/PV.6191*. Available at www.un.org/en/ga/search/view_doc.asp?symbol=S/PV.6191 (accessed 13 October 2014).

UN. 2010a. Security Council Imposes Additional Sanctions on Iran, Voting 12 in Favour to 2 Against, with 1 Abstention. *Security Council Resolution 1929*. Available at www.un.org/News/Press/docs/2010/sc9948.doc.htm (accessed 21 December 2012).

UN. 2010b. Provisional Summary, 6335th Security Council Meeting. *S/PV. 6335*. Available at www.un.org/en/ga/search/view_doc.asp?symbol=S/PV.6335 (accessed 13 October 2014).

UN. *Resolution 2231 (2015), Adopted by the Security Council at its 7488th Meeting, on 20 July 2015*. Available at www.un.org/ga/search/view_doc.asp?symbol=S/RES/2231 (2015) (accessed 27 September 2016).

US Department of State. 2012. Three Companies Sanctioned Under the Amended Iran Sanctions Act. *Media Note*, 12 January. Available at www.state.gov/r/pa/prs/ps/2012/01/180552.htm (accessed 4 May 2014).

US Department of State. 2013. Regarding Significant Reductions of Iranian Crude Oil Purchases. *Remarks*, 29 November. Available at www.state.gov/secretary/remarks/2013/11/218131.htm (accessed 17 May 2014).

US Department of the Treasury. 2011. Treasury Designates Ten Shipping Companies, Three Individuals Affiliated with Iran's National Shipping Line. *Press Release* 20 June. Available at www.treasury.gov/press-center/press-releases/Pages/tg1212.aspx (accessed 13 October 2014).

US Department of the Treasury. 2016. OFAC FAQs: Iran Sanctions. *US Department of the Treasury Resource Center*. Available at www.treasury.gov/resource-center/faqs/Sanctions/Pages/faq_iran.aspx#fse (accessed 28 September 2016).

Wang, Jisi. 2011. China's Search for a Grand Strategy: a Rising Great Power Finds its Way. *Foreign Affairs* vol. 90, no. 3: 68–79.

Wang, Yi. 2014. Wang Yi: The Agreement Between the P5+1 and Iran Should Be Properly Implemented. *Press Cut*, 9 January. Ministry of Foreign Affairs of the People's Republic of China. Available at www.fmprc.gov.cn/mfa_eng/wjdt_665385/zyjh_665391/t1116503.shtml (accessed 23 April 2014).

Wu, Xinbo. 2010. Four Contradictions Constraining China's Foreign Policy Behavior. *Journal of Contemporary China* vol. 10, no. 27: 293–301.

Wunderlich, Carmen. 2014. A 'Rogue' Gone Norm Entrepreneurial? Iran within the Nuclear Nonproliferation Regime. In: Wagner, Wolfgang, Werner, Wouter and Onderco, Michal (eds), *Deviance in International Relations. 'Rogue States' and International Security*. Basingstoke: Palgrave Macmillan, pp. 83–104.

Wunderlich, Carmen, Hellmann, Andrea, Müller, Daniel, Reuter, Judith, and Schmidt, Hans-Joachim. 2013. Non-Aligned Reformers and Revolutionaries: Egypt, South

Africa, Iran and North Korea. In: Müller, Harald and Wunderlich, Carmen (eds), *Norm Dynamics in Multilateral Arms Control: Interests, Conflicts, and Justice*. Athens: University of Georgia Press, pp. 246–295.

Wuthnow, Joel. 2010. China and Cooperation in UN Security Council Deliberations. *The Chinese Journal of International Politics* vol. 3: 55–77.

Yan, Xuetong. 2006. The Rise of China and its Power Status. *The Chinese Journal of International Politics* vol. 1: 5–33.

Yew, Yvonne. 2011. *Diplomacy and Nuclear Non-Proliferation: Navigating the Non-Aligned Movement*. Discussion Paper 7, 2011. Cambridge, MA: Belfer Center for Science and International Affairs.

Yuan, Jing-Dong. 2006. China and the Iranian Nuclear Crisis. *Jamestown Foundation: China Brief*, 1 February. Available at www.jamestown.org/programs/chinabrief/single/?tx_ttnews%5Btt_news%5D=3926&tx_ttnews%5BbackPid%5D=196&no_cache=1 (accessed 24 January 2013).

Zhang, Yonjing. 1998. *China in International Society since 1949: Alienation and Beyond*. New York: St. Martin's Press.

Zhao, Suisheng. 2006. China's Pragmatic Nationalism: Is it Manageable? *The Washington Quarterly*, vol. 29, no. 1: 131–144.

Zhao, Suisheng. 2013. China: A Reluctant Global Power in the Search for its Rightful Place. In: Nadkarni, Vidya and Noonan, Norma (eds), *Emerging Powers in a Comparative Perspective. The Political and Economic Rise of the BRIC Countries*. London: Bloomsbury Academic, pp. 101–130.

Zhao, Tong. 2015. China and the Iranian Nuclear Negotiations. *Carnegie Article*, 2 February 2015. Available at http://carnegietsinghua.org/2015/02/02/china-and-iranian-nuclear-negotiations (accessed 24 March 2015).

5 Chinese, Russian and Turkish policies in the Iranian nuclear dossier

1 Contesting hegemony: normative opposition to extra-UN instruments

When Iranian nuclear facilities were uncovered in 2002, hitherto undeclared to the IAEA and therefore suspected to be in breach of Iran's Safeguards Agreement, it did not take long for discursive dividing lines to emerge. Russian, Chinese and Turkish foreign policies towards the controversial Iranian nuclear programme displayed a security culture that differed from a US-inspired security culture. The Bush administration was pursuing an assertive foreign policy line on Iran, going so far as to threaten Iran with an attack and regime change: The incorporation of Iran into the infamous 'Axis of Evil' was a discursive escalation (see Bush 2002). This was astonishing given that Iran had substantially cooperated with the US on defeating the Taliban in Afghanistan in the wake of the 9/11 attacks (Parsi 2007: 228–229). Against the background of the US-led military action in neighbouring Iraq in 2003, observers thus feared that Iran could be next on the list for regime change. The proliferation concerns of Iran's nuclear programme, in this context, could have provided a convenient political pretext. China, Russia and Turkey reacted with caution and were more hesitant to assume Iranian intentions on the basis of contested proliferation concerns. They held that pressure was not conducive to achieving greater cooperation by the Iranians concerning their nuclear file with the IAEA, and publicly reiterated that diplomacy was of the utmost importance for working towards a political solution to the emerging nuclear crisis. In their foreign policy discourse, they contested a reading of the NPT and of Iranian rights as an NPT member that was perceived as politicised. In Articles II and III of the NPT, non-nuclear weapon states commit themselves not to acquire nuclear weapons, while Articles IV and V make mention of an 'inalienable right' (a formula that Iran continuously referred to) of all parties to use nuclear technology for peaceful purposes.[1] The crux of the Iranian nuclear crisis for many years lay in whether non-nuclear weapon states (like Iran) have a 'right to enrich'. China, Russia, Turkey, and other non-Western actors like India, Brazil, South Africa and the non-aligned movement regarded the US's initial insistence on the denial of that right as a hegemonic exercise, as 'an effort to rewrite the NPT unilaterally', as former senior director for Middle East affairs on the US National Security Council Flynt Leverett (2013) formulates (260).[2] Placing restrictions on nuclear fuel cycle developments is perceived as encroachments on sovereignty, and it is here that Chinese, Russian and Turkish rhetoric becomes congruent with Iranian argumentation. The challenge in the stand-off over Iran's nuclear programme, a Turkish foreign ministry official reflects in an interview, is to 'strengthen the NPT regime without infringing sovereign rights'.[3]

'Compliance' with international law is selective and situational (on this, see also Litwak 2012: 34–39). Contrary to North Korea's walk away from the NPT in 2003, the Iran case is different in nature.[4] Rather than weakening the treaty regime through another withdrawal from the treaty, the Iranian case was challenging the regime *from within*. The impasse in diplomatic progress towards a settlement of Iran's nuclear status soon made it clear that the legal contestation on the surface gave way to a much deeper-seated malaise in international power constellations. In their positions towards the Iranian nuclear programme, China, Russia and Turkey implicitly started to question the legitimacy of US-dominated power structures to decide over legality and illegality in international relations. They began to resist hegemony. In this, their foreign policy discourse has occasionally shown more appreciation for the Iranian perspective according to which the West aims to deprive Iran of technology that it has a legal right to use. Chinese, Russian and Turkish public diplomacy can be read as a discursive attempt at desecuritisation of the Iranian nuclear file. China and Russia put the brakes on sanctions efforts in the UNSC, watered down resolutions and condemned pressure on Iran, and showed a stronger public anti-US posturing and anti-sanctions stance than Turkey did.[5] This diplomacy was also meant as an act of de-escalation of what was felt to be an emotionally charged, politicised discourse emanating especially from Washington. The Nuclear Non-Proliferation Treaty, as Patrikarakos (2012: 30) aptly writes, created a political faultline between the developed and the developing, the modernised and the modernising world (see also Guldimann 2007: 171; Perthes 2008).

Applying the UAE model, where the United Arab Emirates have voluntarily renounced enrichment and reprocessing capabilities in the 123 Agreement on Peaceful Civilian Nuclear Energy between the UAE and the US was never an option in the Iran case due to deep-seated Iranian perceptions of prestige and justice claims. The stand-off over 'enrichment' lay at the basis of years of semantic hairsplitting as to how to interpret the relevant treaty provisions. International treaty law was instrumentalised by diametrically opposed factions. The contested nature of the dispute over Iranian nuclear rights, in this context, also casts doubts on the 'legality' of UN Security Council resolutions passed under Article 40 calling on Iran to manadatorily suspend enrichment (Dupont 2012: 3; UN 2006, 2008, 2010). Because it is here that the crux of the international stand-off over Iran's nuclear programme lies, such provisions can, and have been, read as an act of hegemonic instrumentalisation of the UN system. 'Legality and 'illegality' of UNSC resolutions is highly contentious in the legal literature. The Security Council is empowered under the constituent treaty (i.e. the UN Charter) to pass resolutions binding on all members of the organisation (see Art. 24(1); Art. 25). With these institutional provisions, SC resolutions are binding even when in conflict with another treaty – such as the NPT – *provided that* resolutions are passed as decisions pursuant to Chapter VII (see the supremacy clause of Art. 103).[6] Iran has rejected the legality of Security Council Resolutions demanding that it halts uranium enrichment. Then-permanent representative of Iran to the UN Javad Zarif (now foreign minister) called these demands in UNSCR 1696 'illegal' and 'unwarranted', and depicts the Security Council as an 'instrument of pressure' and as the wrong venue for discussing the issue which should be dealt with by the IAEA (see Zarif 2006; see also Pirseyedi 2013: 158). It has also been argued that the Security Council has

> never determined that Iran's nuclear program was a 'threat to the peace, breach of peace, or act of aggression', in the meaning of Article 39 of the UN Charter, while such determination is a prerequisite for adoption of measures under Chapter VII.
>
> (Dupont 2012: 4; see also Tzanakopoulos 2011: 60)

Here, the legality of UNSC resolutions is questioned on procedural grounds. That China and Russia did not derail the sanctions momentum against this background therefore has to be seen as a political decision, as the outcome, it has been argued, of an interaction with structural hegemony.

A Coxian understanding of hegemony proceeds from an analysis of its underlying material, ideational, and institutional structures. According to Cox, hegemony is 'based on a coherent conjunction or fit between a configuration of material power, the prevalent collective image of world order (including certain norms) and a set of institutions which administer the order with a certain semblance of universality' (1981: 139). The cases presented in the preceding chapters have shown how these structural levels of hegemony apply to Chinese, Russian and Turkish contestations of a US-inspired security culture on the Iranian nuclear programme. As argued in the preceding chapters, the United States was the first to establish a form of hegemony that corresponds closest with a Gramscian understanding of the term. To the extent that other states act upon, sustain and reinforce dominant structures in the social, economic and political sphere, US hegemony has brought about a 'historic bloc' in a Gramscian understanding that is being upheld by the vast majority of states in the Western hemisphere. The explicit acceptance by Western governments of increasingly comprehensive US sanctions regimes on Iran with extraterritorial effect was testament to the creation of such a consent. Hard-hitting EU sanctions on Iran from 2010 onwards testified to the creation of a transatlantic consensus that had formed around a transnational Iran sanctions architecture (Alcaro and Tabrizi 2014). Political consent was complemented with, and conditioned by, coercion in the form of powerful financial penalties that will be addressed further below in this chapter.

Suffice to recall that a distinction between norms and rules was made in Chapter 1 of this book. Drawing on Katzenstein's (1996) definition of culture as

> a set of evaluative standards (such as norms and values) and a set of cognitive standards (such as rules and models) that define what social actors exist in a system, how they operate, and how they relate to one another,
>
> (ibid.: 21)

'norms and values' are here understood as concrete convictions and conceptions, while 'rules and models' relate to the broader macro-structure that regulates the way these norms and values are communicated, applied, or changed. All three states investigated here advocate an adherence to the institutional framework of the UN system as embodying the underlying rules and models of international politics. Unilateral sanctions regimes, however, circumvent these rules and models. In what became contested as an essential exercise of hegemonic power, the US started imposing intrusive sanctions not only on a target country (Iran), but also onto third countries that are engaged in business with Iran (so-called 'secondary' sanctions). Unilateral sanctions thus propose 'rules and models' that run counter to those of the United Nations because they relinquish UN mandates for their adoption – a fact which implicitly undermines 'rules and models' of multilateral decision-making in international organisations (Armstrong *et al.* 2004). Contesting the legitimacy of extraterritorialised US legislative action, therefore, becomes a normative divergence from hegemony. The cases under investigation here have resisted hegemony – on the basis of the institutional nature of sanctions.

Referring to the extent of Chinese and Russian convergence of interests at the nuclear talks, a Russian delegation member thus stated that 'we usually have the same lines. We also don't have unilateral sanctions, so naturally, Iran considers us more of a partner than the West.'[7] This perception was confirmed by an Iranian diplomat.[8] Chinese, Russian and Turkish conceptions of legitimacy in this regard pertain to a desirable security culture that resists established power constellations reserving the right to 'make [...] laws for the world community' outside of UN structures, as a Turkish official put it in an interview.[9] The respective language (diplomatic talking points) used on unilateral Iran sanctions by Chinese, Russian, and Turkish officials is quite revealing in this regard. In interviews conducted for this research project, Turkish diplomats described such sanctions as 'unhelpful' or 'counterproductive', while the discursive divergence from such policies went noticeably further in statements by Chinese and Russian officials. The former 'oppose' such measures, while Russian officials even described them as 'illegal'.[10] Thus, 'sovereignty' and 'non-interference' are norms that should govern international relations; they are the ideational underpinning of what China, Russia and Turkey understand as a 'democratic system of international relations'. Rejecting 'hegemonism' and speaking on the Russian and Chinese alignment of positions on this subject, Russian foreign minister Lavrov stated at his appearance in front of the UN General Assembly in September 2016 that

> the principles of sovereign equality of States and non-interference in the internal affairs should become a measure of decency and legitimacy of any

member of the world community, especially if it claims to have privileged positions in the international affairs.

(Lavrov 2016)

In this context, interference is not only understood as the physical intrusion into the territory of another state, as in the case of a military invasion, but equally captures the intrusive effect that a comprehensive sanctions regime can have on a country. The rendition of enmity is an order-constituting exercise because it presupposes an acceptance thereof by a relevant audience (Balzacq 2010; Williams 2003: 514). Being able to exert control over transgressions thus means being truly sovereign (see Agamben 2002: 25; Schmitt 1993: 19). If China, Russia and Turkey, therefore, reiterate legitimate Iranian rights to develop nuclear energy, they implicitly raise questions of sovereignty in that they question the non-granting of such rights on the part of other dominant actors.

2 Material disagreements with Iran sanctions regimes: barter, circumvention and sanctions compliance

The US has woven a web of truly intrusive Iran sanctions regimes composed of different sub-regimes for different charges (human rights abuses, terrorism charges, nuclear-related) that has sought the country's international isolation. The use of political discourse to capture this degree of 'intrusion' is most revealing and needs no further explanatory commentary when then-US Secretary of State Clinton called for 'crippling' sanctions.[11] More relevant for our understanding of transnational coercion, 'secondary sanctions' have had an intrusive effect on third countries because of undesired interactions with the sanctioned entities, and continue to do so after the JCPOA entered into force. 'Iran remains designated as a state sponsor of terrorism [...] and the JCPOA does not alter that designation', a joint Treasury and State Department guidance on the lifting of Iran sanctions reads very clearly (US Department of the Treasury 2016: 42). The US narrative of Iran has woven a dense tapestry of enemy projection that is very difficult to unravel – and the continued persistence of non-nuclear-related unilateral US Iran sanctions in spite of the lifting of most international and unilateral nuclear sanctions in January 2016 is a seminal case in point. The US will retain secondary sanctions targeting third parties 'for dealings with Iranian persons on our SDN list, including those designated under our terrorism, counter-proliferation, missile, and human rights authorities'. Such parties would 'put themselves at risk of being cut off from the US financial system. This includes foreign financial institutions, who would risk losing their correspondent account with US banks' (White House 2015). Complicating matters for companies eager to invest in Iran is the extent to which Iranian companies control relevant parts of the economy that will continue to fall into the above categories. Examples here are banks listed for carrying out terrorism-related transactions; construction, trading and transport companies tied to the IRGC; or telecommunication companies (Kagan 2015). Richard Nephew, former Principal Deputy Coordinator for

Sanctions Policy at the US State Department, therefore advises companies to 'develop clear force majeure clauses and other schemes to insure themselves against the risk of immediate sanctions reapplication' (Nephew 2015: 3). As Mark Fitzpatrick, former US State Department official, remarks: '[The] absence of reliable banking channels means other businesses will have no way to repatriate profits' (Fitzpatrick 2016). Companies and banks are requested to exercise 'due diligence' when engaging with Iranian entities and remain cautious about exploring new opportunities. Because of the opaque nature of business networks in Iran, detecting front companies and avoiding US financial penalties is a big hurdle for companies and obstructs the opening of the 'new chapter' in relations with Iran that the JCPOA aspires to.

The power of the US dollar as international reserve currency secures the US an additional, and most effective, tool to extend its grip on the international movement of services, goods, persons, and money that can fairly be described as financial interventionism. Any transaction in US dollars has to go through US capital markets, and trade with and investments in Iran are prohibited in US dollars, forcing business partners to rely on alternative currencies or risk sanctions such as financial penalties from the US Treasury Department or the loss of US operating licences. US financial threats (secondary sanctions, threats to exclude trading partners from US financial institutions) served to *coerce* other actors into acceptance of hegemonic policies. 'We present our partners with a choice: You either trade with the Bank of Iran or you trade with the Bank of America', former US National Security Council official and Woodrow Wilson scholar Robert Litwak bluntly stated in an interview.[12] On the basis of the US dominance in the global financial system, US legislation is being extraterritorialised. If a Foreign Financial Institution (FFI) trades in US dollars, it becomes subject to US jurisdiction again. The sheer size of the US capital market and the fact that many of the biggest clearance banks operate in New York thus has introduced a powerful structural leverage to coerce both governments and private sector actors into compliance. Violations of US unilateral financial sanctions carry high penalties, and the fines levied against the French bank BNP Paribas in 2014 for having executed money transfers of companies, including Iranian ones, that are *on US sanctions lists*, was but the most recent and publicised example among a plethora of others (Lohmann 2014: 6). The US Treasury imposed penalty payments on the British banks Lloyds, HSBC, Standard Chartered, Barclays, against Dutch ING and ABN Amro, and against the Royal Bank of Scotland and Swiss UBS (ibid.; Taylor 2010: 71). The declared intention was to change the behaviour of third parties, and the position of the US dollar as international reserve currency served as the leverage tool to coerce other actors into succumbing to US policy preferences.

These secondary sanctions have affected China, Russia and Turkey in different sectors of the economy, respectively. As the previous three chapters have shown, their stakes in Iran sanctions are different, and so are their stakes in resisting not only the ideational, but also the material dimension of these sanctions. Arms trade constituted a bigger part of the Russian–Iranian trade volume than it did in the Chinese–Iranian trade volume.[13] Determined to restrict Russian arms exports to Iran already before the imposition of UN embargoes, the US sanctioned the Russian aircraft manufacturer Sukhoi and arms exporter Rosoboronexport in 2006 and 2008 (Defense Industry Daily 2006).[14] Both Russian and Chinese weapons deliveries, however, had to be suspended by the UN arms embargo in UNSCR 1929 adopted in 2010. On balance, though, Russian 'compliance' with different Iran sanctions was 'easier' economically because of the relatively negligible Russian–Iranian trade volume than it was for China. China has bigger stakes in trade with Iran and in energy relations in particular that make it hard to renounce on Iranian oil imports. The attempt to dry up financial lifelines to Iran by gradually hampering its access to the financial markets as well as the US prohibitions to invest significantly in the Iranian economy thus proved to be more of a challenge to China than to Russia. Dependence on the US-dominated world economy explains behavioural deviation from foreign policies that would normally advocate resistance. The Iran–Libya Sanctions Act (ILSA) of 1996 stipulated that investments made in the Iranian energy sector exceeding US$20 million in one year are deemed sanctionable activities (Katzman 2006). The 'Comprehensive Iran Sanctions, Accountability, and Divestment Act' (CISADA) of 2010 then also included the provision of refined petroleum products to Iran as breaches of that act, and the Chinese company Zhuhai Zhenrong was sanctioned under the CISADA in 2012 (US Department of State 2012). Beijing showed a receptiveness to direct and indirect pressure emanating from Washington: As much as China is dependent on Iranian crude oil supplies, policy-makers in Beijing are cautious not to overstep the mark set out by US Iran sanctions. The Chinese reduction of Iranian oil imports in order to qualify for US sanctions waivers against Chinese companies (Lohmann 2013: 4) was a forceful case in point for an adherence to a US normative framework out of economic considerations. China's increased sensitivity to US sanctions in the energy field was explained by the need for oil imports, while Russia is endowed with its own natural resources. The lifting of nuclear-related Iran sanctions as part of the implementation of the JCPOA has 'legalised' the Chinese–Iranian oil trade. US unilateral sanctions penalising third country business with the Iranian oil economy (such as those contained in CISADA) have been made inapplicable.

Turkey's economic dependence on Iran also primarily lies in the energy sector. With roughly 20 per cent of gas imports and 40 per cent of oil imports coming from Iran, Turkey's government had to find a balance in continuing trade relations with Iran, and wanting to show receptiveness to US concerns. Turkey constantly found itself in need of justification for its imports of Iranian oil which Washington saw as counterproductive to US efforts to convince its allies of its economic isolation policies vis-à-vis Iran. But while showing receptiveness to US intentions as evidenced by reducing the levels of oil imports from Iran, Turkey did not cut off these most cost-effective energy imports from neighboring Iran. Geography and economic considerations partially trump perceptions of solidarity with US policies. Turkey's 'gold-for-gas' trade with Iran in which

Turkey had been exporting gold to Iran via Dubai as an indirect payment for Iranian natural gas deliveries was perceived in Washington as an attempt to circumvent US financial sanctions on Iran, and Turkish businesses were on the watch list of US legislators and sanctions-enforcement agencies (Daly 2013).[15] Turkey complied with Iran sanctions, even though this compliance was 'costly', as formulated by a Turkish foreign ministry official.[16] The lifting of Iran sanctions under the JCPOA therefore is welcome news for Turkey, as it facilitates cross-border trade with its neighbour Iran.

Here, the important caveat should be added that a state's *national* position on sanctions is complicated by the domestic arena. The interstice of and interaction between different economic and political policy prioritisations can lead to frictions between various domestic actors in the formulation of foreign policy – a fact so complacently dismissed by IR 'realist' schools of thought. It should not come as a surprise to Foreign Policy analysts that a state's 'Iran policy' is the aggregate outcome of a debate between various domestic actors, often with partially clashing agendas. Governmental politics is often path-dependent in respective institutional contexts (Welch 1992: 116) and enriched by varied domestic foreign policy debates (Nau 2012). Perhaps nowhere was this more noticeable than in Russia during Yeltsin's administration (1991–1999). In a climate of administrative chaos and economic decline during the early Yeltsin years, foreign policy with regard to nuclear export control regimes oscillated as competing voices within Russia rendered Moscow's Iran policy ambiguous at best (Orlov 2010).[17] A non-transparent commercial climate became the hallmark as the old Soviet structures disintegrated. While the Ministry of Atomic Energy (MinAtom), headed by Viktor Mikhaylov, approved of nuclear technology sales to Iran[18] without consulting President Yeltsin, Moscow officially committed itself through the Gore-Chernomyrdin Commission to consult the Americans about arms and technology transfers to Iran of a dual-use nature (Antonenko 2001).[19] And analyses of Russia's contemporary Iran diplomacy, as Chapter 3 has shown, have to take into account inter-ministerial (and within ministries, inter-departmental) differences in approaches and shifting priorities between the Kremlin, the foreign ministry, and the Ministry of Atomic Energy (known as RosAtom since 2004). Turkey's Iran policy is torn between conflicting priorities of the prime minister's and president's office, the foreign ministry, the ministry of economics, and commercial (albeit state-owned) actors such as the oil and gas company BOTAŞ Petroleum Pipeline Corporation, and the oil refiner Tüpraş. In addition, the Turkish governmental position is complicated by the presence of cabinet members with more 'pro-Iran' sympathies than some of their colleagues. As for the Chinese case, Chapter 4 has demonstrated that there is a considerable degree of ambivalence over the relation between the Chinese government and companies such as Norinco, the China National Petroleum Corporation (CNPC) or the China National Offshore Oil Corporation (CNOOC). Tellingly, companies were sanctioned unilaterally by the US administration, as mentioned above (e.g. the Chinese company Zhuhai Zhenrong or Russia's aircraft manufacturer Sukhoi and arms exporter Rosoboronexport), while the respective diplomatic machinery (primarily, but not exclusively, the foreign ministry) pledged cooperation with their US counterparts. Such institutional (read: bureaucratic and ministerial) structures often provide a level playing field for domestic actors to act out disagreements over approaches to Iran and US hegemony. The states under investigation here (China, Russia, Turkey) are thus by no means to be understood as respective monolithic blocs, and the preceding case studies have process-traced the outcomes of domestic decision-making processes and the occasional plethora of voices on the formulation of an overall governmental 'Iran policy'.

On an aggregate level, however, the Russian 'oil-for-goods' scheme that was reported on in the summer of 2014, Chinese 'goods for oil', and Turkish 'gold for gas' barter schemes all speak one language: The isolation of Iran over its controversial nuclear programme has been resisted commercially in what Khanna (2014) calls a 'break between commercial and nuclear diplomacy'. Faced with US and EU financial sanctions that complicate payment modalities before the implementation of the JCPOA, China and Turkey reverted to barter agreements (much to the dismay of the Iranians). Writing on the US practice of sanctioning third country entities for their interactions with Iran, Leverett and Leverett (2013) thus remark: 'American policy is now incentivising emerging powers to develop alternatives to established, US-dominated mechanisms for conducting, financing, and settling international transactions. As Washington continues on this course, it will hasten the shift of economic power from West to East' (282). Such alternative payment mechanisms were also seen in the form of Chinese and Turkish payments to Iran in currencies that circumvent the 2010 CISADA and the 2012 Iran Threat Reduction Act (like Turkish Lira or Chinese renminbi) or the use of front companies for 'sanctions-busting' purposes. These are examples of circumvention mechanisms in the short run. In the long run, however, they may serve as indicators that new economic power centres can form to challenge the US financial monopoly (see also Ferdinand 2013: 18). Any alternative mechanisms aimed at undermining the 'dollar hegemony' threaten to erode the financial infrastructure that constitutes the financial basis for the US's powerful position in international affairs (Eichengreen 2011; Hung 2016: 115–144; Stokes 2014). While this is not to suggest that such alternative schemes have the potential to replace the latter, they at least indicate seeds of resistance to financial omnipotence.

The commercialisation of relations between Iran and other states that criticise the use of economic and financial power for political leverage has been a recurring discursive theme in the Ahmadinejad administration. The 'looking to the East' policy has been the explicit attempt to diversify Iranian trade relations in order to better resist US pressure (Mousavian 2012: 190f.). A subject of domestic criticism, and in the realisation of its relative fruitlessness, this policy objective has eventually been dropped. But the Iranian reference to vocabulary like that of a 'resistance economy' in 2014 (Khajepour 2014) bears a striking resemblance to earlier such attempts of commercial resistance (at least discursively) to denote not only a political disagreement with US-dominated world order that is part of the Islamic Republic's foundational narrative, but also an economic model that 'resists' Western economic power. As such, it has been a strong illustration of 'counter-hegemony'. Whereas this is more propagandistic rhetoric than an actual policy preference on the Iranian side (Barzegar 2014), China, Russia and Turkey cannot but arrange themselves with global financial and economic structures that they depend on – not only economically more so than Iran, but also politically, as the preceding chapters have shown.[20]

Does this finding imply that resistance to US policies in the Iranian nuclear conflict purely stemmed from an economic motivation? To the extent that trade relations with Iran constituted a significant part of the respective trade volume, this material argument held considerable explanatory power. And it is here that economic sanctions have their biggest leverage. But the preceding chapters have also shown how resistance to intrusive policies such as the adoption of unilateral financial sanctions was driven by an ideational opposition to the claim to be in a position to regulate third country policies. It was the latter aspect that met resistance on normative grounds because it resembled most closely the idea of hegemony with an overarching cultural, economic, and political control. It is the construction of norms that influences material imperatives pertaining to adherence or rejection of hegemonic structures, and vice versa. This observation harks back to the reciprocal effects between ideational, material and institutional underpinnings of power structures in a Coxian understanding of hegemony. Rather than singling out one dominant side of the equation, it is the joint effect of these factors that needs to be taken into account for an understanding of Chinese, Russian and Turkish policies in relation to Iran. The cases analysed in the preceding chapters have shown how resistance to power structures relates to the interaction effects between these power structures and perceptions of legitimacy in international relations at large (Clark 2011; Rapkin and Braaten 2009; Reus-Smit 2014).

Finally, however, it is important to stress that differences in political prioritisations between the states investigated here persist and account for varying degrees of 'resistance to hegemony' beyond observed patterns of similarity in discourse and behaviour, as carved out above. The three previous chapters have given ample empirical evidence of the multilayered nature of Chinese, Russian and Turkish foreign policies towards Iran. There have repeatedly been arguments for Chinese and Russian interests in a protracted stalemate of an internationally isolated Iran because of economic advantages that the Western embargo situation entailed. Chinese–Iranian trade ties are substantial (aggregate bilateral trade flows reached a record height of US$52 billion in 2014), whereas Russian–Iranian trade ties are comparatively negligible (bilateral trade worth US$3.6 billion constituted a high in 2010, which fell to a trade flow of US$1.3 billion in 2015).[21] While the Russian–Iranian nuclear cooperation was the flagship of their bilateral relations during the times of international Iran sanctions, China capitalised on the sale of manufactured goods and engineering services and was dependent on the import of Iranian crude oil. This dependence, however, made China vulnerable to US financial sanctions targeting the Iranian energy sector. Chinese–Iranian trade ties have now even been made easier with the lifting of Western nuclear Iran sanctions as part of the implementation of the JCPOA. The lifting of nuclear-related Iran sanctions eases Chinese payments for Iranian oil deliveries as Iran is allowed into the SWIFT banking system again, and as investments in the Iranian oil industry are no longer subject to financial sanctions.

3 The de-Westernisation of Iran discourses

'Rationality' and 'responsibility' in the international system are inherently subjective notions. Besides material motivations as described above, compliance with US sanctions arguably also had to do with the notion of 'responsible stakeholder'.[22] Having become de facto nuclear partners of Iran in the 1990s, both China and Russia had halted cooperation under pressure from the United States, as the previous chapters have shown.[23] This Chinese and Russian receptiveness to American pressure has also been noticed with disgruntlement in Iran and is being recalled by Iranian officials (see Mousavian 2012: 54–55). 'After all', Hossein Mousavian (2012) writes, 'although Russia, China, and the Non-Aligned Movement exercised considerable clout in international diplomacy, they could not be relied on as a dynamic coalition leading the way toward a resolution of Iran's nuclear issue' (86). The depiction that China and Russia have effectively functioned as Iran's diplomatic covers and spoiled efforts to exert pressure on Iran is thus too short-sighted and is disproved by their track record of cooperation with the US. 'Responsible stakeholder' is a term as relational as it is normative. Biersteker and Erica (2015) note that talk of responsibility is correlated with the reinforcement of established practices, and works out a 'rather condescending approach' in the perpetuation of that label (58). Responsibility is attributed to actors that subscribe to norms of a power bloc, irresponsibility to those that disregard them.

The discursive usage of these terms, therefore, presupposes intersubjectively shared meanings. The narrative of an irrational, irresponsible Iranian leadership is a powerful example for the instrumentalisation of a discourse on logic. States that do not share hegemonic values or political structures are labeled irresponsible, unreasonable, renegade counter-poles. As was carved out in Chapter 1, a critical reading of international politics needs to ask who is 'revolting' against an existing order and who gets to act as 'custodians of the seals of international approval and disapproval' (Claude 1966: 371–372). Since the Islamic Revolution propelled an anti-American regime to power in 1979, Iran has been positioning itself in opposition to policies crafted by the West, and the United States in particular. Both sides (Iran and the United States) have used this official rhetoric of mutual stigmatisation for political reasons for the last three decades. The 'axis of resistance' rhetoric, used to denote resistance to Israel and American presence in the Middle East, is illustrative of such a discursive construction of competing worldviews (Posch 2013: 27). The social construction of statist identities for public relations purposes is often simplistic and even dualistic when combined with the means for foreign policy portrayal of an 'enemy'. What can be a self-proclaimed axis of resistance may be an Axis of Evil in an antagonistic discourse. The language on 'rogue states' flouting international norms not because of their behaviour but because of the *nature of their regime* (Litwak 2008: 92; 2012: 9–19; 2014: 17–33) is juxtaposed on the other side by the language on 'global arrogance', denoting an all-encompassing, almost Orwellian, reach of a roughshod superpower. Regimes of exception and truth are produced and reproduced through powerful narratives. The 'state sponsor of terrorism' label used by US administrations to discard the 'rationality' of Iran's government is another indicator for the political nature of a mutual demonisation, of labels that *produce* reality: Successive US governments have deliberately supported regimes and governments that, by any account, have to be considered 'terrorist'. Historical examples abound. '"[T]errorism" as a noun and "terroristic" as an adjective', as Adib-Moghaddam (2014) therefore concisely puts it, 'are the terminological surface effect of discursive representations' (167). Foreign policy discourse and state identities thus always have to be understood in a relational context.

In their advocacy for the non-interference of external actors in the fabric of a third country, states like China and Russia advance such an essentially *relational* understanding of security. Publicly rejecting the idea of interventionism, they mould security cultures that counteract the attempt to 'internationalise' norms (Bevir and Gaskarth 2015: 76). China's 'four no's', discussed in Chapter 4, are an attempt to provide a negative interpretation for what was crafted as 'universal values' by Western states, and to portray them as destabilising to China's 'national' culture. The nationalist underbelly (more or less pronounced at different times in the last two decades) of Russia's official discourse in the search for a post-Soviet identity is another such example where 'counter-cultures' take state identities and societal orders to a level that, by design, shuts out 'the international' (Sakwa 2012). The observed discrepancy between official rhetoric and the undermining of the principle of 'non-interference' on other occasions is an intriguing one. Publicly pledging for equitable international relations and the non-interference in other country's domestic affairs, Russia and China make self-defined exceptions (Ferguson 2015: 33). Russia's terminology of a 'near abroad' is most indicative in this regard, and Russian semi-covert interference in politics in neighboring Ukraine has only been the most emblematic of examples. And contested territorial disputes in Southeast Asia testify to China's partially particularistic definitions of sovereignty. While a thorough examination of such cases is beyond the scope of this book, suffice to retain here that 'non-interference' and an understanding of 'national interests' are highly contested terms and applied inconsistently by China and Russia themselves. While the preceding chapters have cast light on varying domestic motivations in Russian, Chinese, and Turkish foreign policies, further research is needed to illuminate the nexus between foreign policies and the internal determinants of 'international norms' discourses. In their aggregate effect on the international level, however, as the preceding case studies have shown, norm contestation becomes an act of resistance to hegemony. In this context, the academic debate about 'declining' powers should not be misconstrued as exclusively encompassing 'systemic' power transitions (mostly framed in rationalist terminology), but should be inclusive of the fragmentation of *interpretative frameworks* to World Order.

Russia's and China's advocacy for more 'democratic' international relations are indicative of an effort to advance the ideational dimension of a security culture that eliminates power asymmetries in international relations, and hence 'democratises' them. Speeches by Chinese high-ranking officials on endeavors to 'promote democracy in international relations' (Hu 2007: 59) and the idea of democratic international relations as put forward in the official Russian Foreign Policy Concept (2013) and as repeated continuously by Russian officials are seminal cases in point. And when Turkish president Erdoğan reminds the assembled delegates at the UN General Assembly that 'the world is bigger than five', he is breathing the same spirit and invokes the ambition to decentralise hierarchical relations.[24] Turkey's non-inclusion in the UN Security Council, however, gives such statements coming from the Turkish head of state another dimension absent in the Chinese and Russian rhetoric. Turkey here directly challenges institutionalised power such as that of the permanent UNSC members. China's and Russia's reference to 'democratic' international relations thus have to be understood as a critique of the asymmetrical power disposal in international relations *beyond* such concrete institutionalised forms of power. China and Russia base their global power understanding to a large extent on their P5 membership and would not want modifications to its format, yet voice criticisms about US dominance in international security governance. This mirrors the distinction made between the 'international system' (which does not always serve US interests) and US dominance in it. The call for 'democratic' international relations reflects George Orwell's famous dictum from *Animal Farm* that 'All animals are equal. But some are more equal than others'. Russian foreign ministry Lavrov himself even references Orwell in an article published in the June 2016 edition of *Russia in Global Affairs*:

> All talk of 'revisionism' does not stand up to scrutiny and is essentially based on the primitive logic that only Washington can call the tune in international affairs today. This logic suggests that the principle stated by George Orwell years ago that all are equal but some are more equal than others seems to have been adopted at the international level.
>
> (Lavrov 2016a: 17)

He took up his reference to Orwell's *Animal Farm* again in his remarks at the General Debate in the UN General Assembly in September 2016 (Lavrov 2016b). Leaving aside his not-so-subtle swipes against 'Washington', his conception of dominance in international relations is one that becomes congruent with the Chinese and Turkish emphasis on the need to de-concentrate power: Despite formal equity, contingent power relations between states have created asymmetrical hierarchies. Depending on political stances and orientations, these more powerful states can be labeled 'primes inter pares', superpowers, or hegemons. The 'democratisation' of international relations, in this reasoning, would bring about the leveling of such power relations.[25] This is thus not to be confused with the translation of democracy as a societal model for international emulation.[26] Quite on the contrary, Chinese and Russian calls for 'democratic' international politics presuppose the adherence to rules on an equitable basis internationally, while domestic rules can and should stay distinct, and while norm conceptions can differ. Chinese, Russian and Turkish warnings of the counterproductive effect of punitive policies that undermine diplomacy are indicative of the same intent. China, Russia and Turkey have publicly advocated for a security culture that rejects politics of aggression vis-à-vis Iran and therewith have sought to 'de-Westernise' discourse on Iran. In not vetoing sanctions resolutions, however, their behaviour fell short of acting upon their own discourse. Their acceptance of the rules of the UN system demonstrates their adherence to the institutional dimension of the broader macro-structure through which international politics are channeled, conveyed and communicated, as shown in the previous section. The Chinese and Russian approval of international sanctions in the UNSC is a case in point for their understanding that working through UN structures provides a means of checking and co-formulating such punitive measures as international sanctions. China's and Russia's sanctions policies were instrumental and served to prevent other (military) policy choices that were seen as a worse course than sanctions. In addition, a 'key factor' for Russian and Chinese votes in favour of UNSC resolutions imposing sanctions, Peter Ferdinand writes, 'is maintaining the reputation of the UN and its agencies for impartial assessment of controversial evidence' (2013: 4).

4 'Compliance' with 'international norms' or with hegemonic structures?

A discrepancy between discursive and behavioural levels in Chinese, Russian and Turkish Iran policies can be observed, as the previous section has concluded. Advocacy for an explicitly *non*-hegemonic security culture discursively here can be paralleled by compliance with parts of a US-inspired security culture on a behavioural level. Grudging acceptance of a US-inspired sanctions regime on Iran allows China, Russia and Turkey collectively to join the camp of 'responsible' states because they adhere to material structures put in place for politico-ideological reasons. On a behavioural level, China and Russia here also demonstrated partial compliance with US demands, as the examples cited above have demonstrated. A normative disagreement with US power structures, therefore, did not entail an all-out rejection of US policies towards Iran. Normative divergence and rules convergence take place concurrently. Variation in norm compliance on the part of the cases examined here has to be seen in the context of their respective bilateral ties with the United States, the perception of their foreign policies toward Tehran and elsewhere, and their stakes in avoiding a confrontation with Iran. These stakes are, as has been shown, both material and ideational. As analysed in the previous chapter, China heavily depends on Iranian crude oil imports, which explains its indirect disregard for certain (unilateral) Iran sanctions regimes. The Chinese eventual approval of multilateral sanctions (and therewith compliance with US demands) was attributed to the higher value Beijing attaches to the US than to Iran.[27] Turkey, as analysed in Chapter 2, did not face the dilemma of formally having to approve of UN sanctions on Iran, yet still emphasises its alliance solidarity with the US. By way of institutional imperatives (NATO), Turkey already has a stronger 'built-in compliance' with the US government, even though this is subject to Turkey's sometimes markedly different 'national interests', as not only the case study presented in Chapter 2, but also diverging approaches to transnational Islamism and other regional political and security issues have demonstrated. And of the three cases presented in this book, Russia is the one where publicly articulated 'norms and values' are in starkest contrast to US preferences. Tensions over the future of Ukraine loom large in attempts to explain the deterioration in relations between the US and Russia, and between the West and Russia in general, but should not be seen as sole explanatory factors. Despite allusions to the contrary, Russia has not allowed this general state of frosty relations to affect the Iran nuclear talks, and has acted with a professionalism that was indicative of the high importance attached in Moscow to the negotiation of the JCPOA (see Meier and Pieper 2015). The fact that the Russian government has not 'spoiled' the Iran nuclear talks is indicative of its understanding carved out in Chapter 3 that the closure of Iran's nuclear file is in Russia's own security interest, and that working through the P5 + 1 format is legitimised by UNSC resolutions. Constructive collaboration in this format is thus not to be misconstrued as 'compliance' with US interests, but underscores the Russian (and Chinese) approval of multilateral institutional structures to approach Iran's nuclear programme. In Katzenstein's distinction, 'rules and models' that have the power to *constrain* hegemonic projects institutionally are adhered to, while 'norms and values' can differ. Both China and Russia have eventually approved of UN-mandated Iran sanctions because they were negotiated through UN structures. The latter lend multilateral approaches legitimacy, which has elsewhere been defined as 'compliance that is not coercive' (Clark 2005: 12; Hurd 1999). 'Compliance' with policies is therefore not to be misconstrued as 'compliance' with 'norms' of an ill-defined 'international community'. Norm conceptions differ within the international system. Asking who is 'complying' with what norms and for what purpose necessarily brings up the question of contingent power constellations. Clark formulates: 'power suffuses legitimacy, but does not empty it of normative content' (2005: 4). Russian president Putin underlined this point in his 2014 Valdai speech in stark anti-hegemonic terms:

> In a situation where you had domination by one country and its allies, or its satellites rather, the search for global solutions often turned into an attempt to impose their own universal recipes. This group's ambitions grew so big that they started presenting their policies they put together in their corridors of power as the view of the entire international community. But this is not the case.
>
> (Putin 2014)

'Compliance' is thus a terminology that suggests hierarchical relationships. The case studies examined here have shown how 'compliance' on the part of China, Russia and Turkey with approaches to the Iranian nuclear conflict to which they object from an ideational standpoint has been selective, and how US policy preferences in the Iran dossier have been resisted on other occasions. The aggregate result from an analytical point of view, then, is the erosion of the concept of 'compliance', and a relational take on the legitimacy of policies with a global impact. Suffice it to note that this is not to imply that the Iran policies of China, Russia and Turkey are crafted in a joint effort to challenge the US-dominated system of governance. Interviewees confirm that consultations between their governments are made, but equally with all parties involved.[28] A Turkish diplomat acknowledges that, for instance, 'our [the Turkish] position is closer to the Russian discourse than to the Western one',[29] a Russian diplomat states that 'our [the Russian and the Chinese] positions are close (and) we coordinate positions and have similar goals',[30] and a Chinese diplomat affirms that 'we [the Chinese] are closer to the Russians than to the other four [P5+1 member states]'.[31] Yet, their foreign policies are seen as separate tracks. The extent of diplomatic engagement and contestation differs as well: Turkey was less involved institutionally than the UNSC permanent members Russia and China, and was engaged in diplomatic mediation between Iran and the established negotiating format of the P5+1. This position allowed Turkey to be more critical of the imposition of multilateral sanctions, which it did not have to decide on itself (in the absence of a Security Council seat). Its non-permanent Security Council membership, however, gave Turkey the ability to exert institutional influence on procedural matters, even though its veto against UNSCR 1929 did not prevent the adoption of sanctions (at least nine out of 15 votes in favour are required). And while Turkey's mediation and diplomacy might have facilitated the dialogue between Iran and the P5+1, there is 'no particular role for Turkey' during the implementation of the JCPOA, as Turkish officials remarked in interviews.[32]

But also between the two non-Western P5+1 states, China and Russia, there are nuances in official diplomatic contestation that distinguish their positions from each other, as has become clear by the preceding two in-depth case studies. An impression prevails that China is 'hiding behind' a more assertive Russian diplomatic positioning.[33] Joint 'resistance' to dominant power structures on a global level therefore does not exclude the possibility of competing power distributions on regional levels. And resistance to global hegemony can be paralleled by ambitions to construct regional hegemony, resulting in overlapping tapestries of global, regional and inter-regional security orders (see also Acharya 2014: 20; 101–105; Buzan and Wæver 2003; Worth 2015: 177). Russia may be considered a regional hegemon in the post-Soviet space just as China may be seen as a regional hegemon in East Asia. Rather than theorising on counter-hegemonic bloc movements at the global level, it is therefore arguably a more insightful endeavor to analyse the respective interactions of foreign policies with dominant power structures and to examine their collective effect on the crafting of a security culture that resists those structures – even though these foreign policies have their different motivations, constraints and preconditions. It is no wonder, then, that some Western observers of Russian, Chinese or Turkish foreign policies cannot but conclude that their policies appear ambiguous. States that sit on fences because they are torn between different security cultures, strategic circles, or geographic crossroads are bound to pursue multidirectional foreign policies. To the outside observer, these policies occasionally appear incoherent, opportunistic at best and politically unfaithful at worst. Status quo actors, as dominant forces in this order, have an inherent interest in domesticating alternative initiatives (Cox 1996: 130). This research has shown how an investigation into Chinese, Russian and Turkish policies toward Iran can offer more nuanced understandings of foreign policies between such exclusive camps. Exclusive and essentially simplistic categorisations are a recurring mantra of foreign policy projections. This does not translate into an automatic stand-off between 'the West' and 'the Rest', and it is to this question that the final section now turns its attention.[34]

5 Contesting hegemony and moving into a post-American world

Iran's nuclear crisis was never only about nuclear physics. Its significance can also not be fully grasped by an exclusive public international law perspective, which is inevitably doomed to reproduce the circular semantic analyses that have bedeviled international negotiations since 2003. Iran's nuclear crisis was about perceived hegemonic politics and about a conflict whose resolution has far-reaching implications for the dialectic between dominant power centres and actors that resist them. Henry Kissinger, in his 2014 book *World Order*, is clear about who he thinks is on the right side of history: 'Though couched in terms of technical and scientific capabilities, the issue is at heart about international order – about the ability of the international community to enforce its demands against sophisticated forms of rejection' (159). Beyond such binary conceptions of acceptance or rejection of a generic 'international community', however, the gradual implementation of the JCPOA and its ramifications for Iran's standing in the world will affect perceptions of world order in a process where the US role as a shaper of world hegemony is diminishing.

At the heart of the Cold War over Iran's nuclear programme lie over three decades of traumatised US–Iranian relations. Technical solutions to end this nuclear crisis have been proposed, discussed and rejected for more than a decade before a nuclear agreement was eventually negotiated in July 2015 (Arms Control Association 2014). The replacement of the confrontational rhetoric of the Ahmadinjad administration by a more conciliatory tone under Rouhani's leadership has allowed for a historically constructive dialogue with Iran. But also personnel changes on the US side let diplomacy gain pace after US Secretary Clinton, a proponent of sanctions, had been replaced by John Kerry in February 2013. The qualitatively new phase in US–Iran talks in the decade-old nuclear crisis was marked by a first interim agreement on Iran's nuclear programme in

November 2013, a political framework agreement in April 2015, followed by a comprehensive nuclear agreement three months later. New political realities coin their own terminologies. Against the backdrop of the unprecedented progress at the nuclear talks from the summer of 2013 onwards, a Russian foreign ministry official remarked in an interview that the US position had embraced an approach that Moscow had already been advocating for years. The American referral to a 'step-by-step' approach, the same official held, was in fact harking back to diplomatic language that had been used by Russian diplomats more than three years ago.[35] Dmitri Trenin (2014) similarly writes that 'the US adoption of a gradualist approach toward Iran that Russia had long favored resulted in a breakthrough on the Iranian nuclear issue.' The same wording was repeated nearly verbatim to me by a Turkish official, but comparing a 'new' US approach with longstanding Turkish diplomatic preferences.[36]

The Obama administration's subtle reframing of the challenge emanating from Iran (from 'rogue' to 'outlier': Litwak 2014: 17–33) already upon entering office signaled a shift from purely unilateral approaches towards Iran in expectance of followership (during the Bush 43 administration) to one that embeds US foreign policy in a multilateral framework and an *international* norm discourse. During the early years of nuclear diplomacy with Iran, China, Russia, Turkey and others have resisted a language that could be conceived as a pretext for military actions against Iran (Litwak 2014: 28). The inclusion of China and Russia in the P5+1 negotiation format was vital in pre-empting the impression of 'Western powers' attempting to negotiate away Iranian rights. Their inclusion also ensured that China's and Russia's interests and negotiation positions were taken into consideration. The principle of 'P5 unity' served to prevent Iran from playing off different P5 positions against each other. But it also served as an insurance to Moscow and Beijing that their positions formed an inevitable part in the consultation process. A French delegation member to the nuclear talks recalls:

> We first discuss within the E3+1, then we go to Russia and China, who add their own views, then we have meetings with Iran. Russia, for example, would tell us at the E3+3 level that we [the E3+1] are going too far with a certain proposition, but that they wouldn't say so at the talks with Iran in order to keep the 'P5 unity'.[37]

Rather than assuming a coherent bloc challenge to US hegemony, the cases analysed here have therefore disentangled qualified forms of disagreement instead of all-out resistance. The notion of a coherent bloc challenge is further distinguished by the fact that Chinese, Russian, and Turkish Iran policies have different rationales, respectively, and are often in conflict with each other. The lifting of sanctions as part of the implementation of the JCPOA may entail future economic competition between Chinese and Russian companies in some sectors on the Iranian market, as the previous chapter has argued. Chinese companies may enter the Iranian market in sectors where Russian ones were hitherto dominant. The Chinese assistance in the conversion of the Arak heavy-water reactor in Iran was an early indicator for a new actor on the Iranian nuclear infrastructure market that had previously been dominated by Russian companies. Whilst the conversion of the Arak facility is a proliferation-related task carried out under the JCPOA, it cannot be excluded that Chinese companies develop an interest in bidding for commercial projects in sectors hitherto dominated by Russia. As for Turkey, Russian–Iranian trade volumes are surpassed by a significantly higher Russian–Turkish trade volume.[38] Even after the lifting of Iran sanctions, Turkey might therefore constitute a competitor for Iran in expanding economic relations with Russia. While Iranian trade preferences lie elsewhere than with Russia, the point is that China's, Russia's and Turkey's bilateral relations with each other complicate the idea of a unified bloc challenge to 'the West' that their Iran policies constitute.

Conceptions of security cultures change gradually. As such, this study has offered illustrative case studies that break down the schematic framework of 'systemic' analyses of power shifts in areas often misleadingly labeled 'global governance'. The decline of hegemony and 'the rise of the rest' are too often portrayed in dichotomous terms. Even a flag-waving critical scholar such as Cox, in his 1996 *Approaches to World Order*, suggests that the 'task of changing world order begins with the long, laborious effort to build new historic blocs within national boundaries' (140–141). Here, I slightly deviate from his conception in conceiving international order as the outcome of the reciprocal interaction between state-centric and international factors. The preceding analysis has shown how international diplomacy (nuclear talks) can lead to regimes and agreements (JCPOA) that in turn will impact on 'historic blocs within national boundaries' (rather than strictly the other way around). Such an approach becomes especially relevant when actors meet in the international arena that do not share the same views on how international politics should work. Understanding foreign policies that are not 'Western' and do not necessarily share the same normative framework – but still work with the rules of the system instead of working to overthrow them – requires a more differentiated perspective on the dynamics of gradual power shifts. The scholarly debate about 'China's rise' is a seminal example here. Underlying analyses of a 'rising China' is an implicit assumption that a more pronounced Chinese foreign policy in international relations could come at the expense of US power. The 'international system' would experience a shift from the 'Washington consensus' to a 'Beijing consensus' (Ramo 2004). The 'China Threat' school (employed by both realist and liberal commentators) formulates most emblematically that 'China's rise' is a negative development (Gertz 2000). Such scholarly interpretations are misconstrued. In his detailed analysis of China's contemporary economic development, Ho-fung Hung (2016) argues how both romanticised and alarming accounts of China's 'rise' as the harbinger of a subversive moment in world affairs are exaggerated. China's politico-economic benefits from the US-centred neoliberal global order, he argues, help explain why China perpetuates that order. He attributes this phenomenon largely to the 'hegemonic role of the dollar' (119) as the most widely used reserve currency and international transaction currency. As long as China's economy retains an export-oriented focus and holds massive amounts of US Treasury bonds, speculations of a Chinese interest in subverting the 'global order' are misplaced.[39]

The co-existence of established and 'emerging' powers will inevitably determine the design of the future world order. Amitav Acharya makes this point in his 2014 *The End of American World Order*. While not joining the chorus of voices pointing towards 'US decline', his work argues that

> [a]ny reconstituted American hegemony has to change a lot, and accommodate, rather than co-opt, other forces and drivers, including the emerging powers and regional groups. It has to adapt to a new multilateralism that is less beholden to American power and purposes.
>
> (4)

It is also in this context that Richard Sakwa (2011) has coined the concept of 'neo-revisionism' to make sense of foreign policies that do not directly question or challenge the essence of the international system (as revisionist states would do), but indirectly aim to revise its functioning (see also Sakwa 2015: 30–35). In a similar vein, Simon Serfaty (2011: 18) describes China and Russia as 'prudent revisionist powers', and Peter Ferdinand writes that neither China nor Russia 'is trying to overthrow [...] globalization', but 'may be able to bend international trends in directions that are more advantageous to them' (2007: 680). Such observations aim to avoid over-theorising about foreign policies that do not coincide with that of the 'system leader' as expressions of the advancement of alternative norms in international security governance (see also Breslin 2009, 2013; Clark 2011: 25–26). 'Every international order must sooner or later face the impact of two tendencies challenging its cohesion: either a redefinition of legitimacy or a significant shift in the balance of power' (365), Kissinger (2014) writes towards the end of his lengthy historical treatise on World Order. His proposed solution to such challenges is wise statesmanship, which naturally, he thinks, falls to American leadership (367–371). Beyond such complacent US-centrism, this book has offered in-depth case studies to illustrate the more nuanced workings of resistance and cooperation between 'Western' and 'non-Western' actors. Torn between resistance and accommodation under the centrifugal forces of commercial, institutional and normative constraints, the foreign policies of China, Turkey and Russia in the Iranian nuclear dossier cannot but appear as 'essentially ambiguous' when filtered through a Western lens.[40]

The US leadership role is declining, but an entirely alternative international order is not yet in sight. The distinction between 'power' and 'leadership' is one that is also made by proponents of a liberal hegemonic order. 'Power', it has been suggested throughout this book, must not be solely understood in terms of material capabilities. Patel and Hansmeyer (2009) thus note that talk of US decline,[41] 'while catchy in its pithiness, [...] does not exactly lend itself to precision' (275), and suggest a clearer distinction between power and leadership across multiple policy domains and 'areas of impact' (ibid.). The projection that any of the so-called 'rising powers' will overtake the United States militarily remains illusory (Cox 2001, 2007; Hart and Jones 2011; Quinn 2011; Stokes 2014). And economically, the projected overtake by China in absolute GDP numbers is not to be equated with economic pre-eminence, as the US will continue to lead in terms of personal wealth (GDP per capita) and innovation ahead of China (Patel and Hansmeyer 2009: 279).[42] While the economic output of the US measured in terms of percentage share of world GDP has been declining since the 1950s (Lundestad 2013: 292), and its current account and fiscal deficits have dramatically worsened, it remains the world's single biggest economic and capital market as well as military power, and the role of the US dollar as the international lead currency continues to cement this position. Even a notorious critic of 'US hegemony' as Noam Chomsky writes in his 2016 *Who Rules the World?*: 'While the United States remains the most powerful state in the world, nevertheless, global power is continuing to diversify, and the United States is increasingly unable to impose its will. But decline has many dimensions and complexities' (66).

Against this background, the cases analysed in this book have shown how US financial sanctions continue to exercise powerful leverage. The US will continue to play a preponderant role in international relations. A lot of the dividing lines in the scholarly debate about 'decline' tend to be shrouded in a neo-realist terminology adopted from International Political Economy. What is often left out of the picture is the link between international leadership and perceptions of legitimacy (see also Clark 2005; Schirm 2010).[43] What has thus been suggested here is that a new dynamic of norm diffusion has set in, whereby dominant interpretations of World Order are being challenged by non-American actors. It is this long-term development of re-balanced power relationships that Acharya (2014) termed the 'end of American World Order', which might be followed by 'a concert among the established and emerging powers and a network of predominantly regionalized orders' (9). He further contends: 'They do make the world less American-centric, but far from heralding a global fragmentation or the rise of regional hegemonies, these regional worlds could be an essential foundation for sustaining a multiplex world order in the twenty-first century' (105). Further research is needed at the interstice of IR scholarship and Area Studies expertise to examine this relationship beyond the accommodation-confrontation spectrum.

As long as the US-dominated historic bloc exists, China, Russia and Turkey will work with and through an international architecture in place that guarantees the best means to contain and check US influence. This does not mean that these states will accept the same norms. Despite disagreements with the United States at the macro level, China, Russia and Turkey do not appear as revisionist states constituting a bloc challenge to American dominance. An acknowledgement thereof allows for the debunking of confrontational policy rhetoric and for more nuanced research on post-hegemonic power shifts. It also transcends the unhelpful divide between 'norm-setter' and 'norm-taker', and helps us to reflect more accurately on the future co-existence between former hegemonic and emerging powers. The policy task, then, is to accept diverging national cultures, but still craft rules and models that allow actors to coordinate policies that allow for peaceful co-existence.[44]

6 Conclusion

This chapter has picked up the twofold distinction between a discursive and a behavioural level that has been worked out in Chapter 1 and applied to the three empirical case studies throughout this book: While Chinese, Russian and Turkish officials publicly advocate for an adherence to a security culture that emphasises compliance with the norms of 'sovereignty' and 'non-interference', their level of perceived material and political dependence on the United States has prompted them to follow foreign policies that still complied with US policy preferences on some occasions and resisted them on others. The US unilateral sanctions regime and compliance with an extraterritorialised US legislation is the most prominent case in point. Since the discovery of hitherto covert Iranian nuclear facilities in 2002, Russia has emphasised the Iranian right to use peaceful nuclear energy and, until the referral of the Iranian nuclear file to the UNSC and the adoption of first Security Council resolutions, has largely shielded Tehran from international pressure – as has China. While Western governments observed Russian–Iranian nuclear technology cooperation with a watchful eye, Moscow continued making a distinction between legitimate commercial projects and an alleged military dimension of Iran's nuclear programme. Likewise, the history of Chinese supplies of sensitive nuclear technology to Iran, its commercial exploitation of Western embargoes on Iran and its dense ties with the Iranian oil economy, were seen as undermining Western attempts to increase international pressure on Iran. Turkey presents itself as a US ally in the region and is committed to NATO alliance structures. Materially, its location as a geographic neighbor of Iran and economic imperatives, however, imposed constraints on Turkey that led Ankara enter into friction with US attempts to isolate Iran economically. All three states were also motivated by a principled scepticism of what they regarded as a biased and politicised reading of the NPT regime. Chinese, Russian and Turkish foreign policies toward the Iranian nuclear programme have therefore breathed the ambition to 'de-Westernise' security cultures and discourses toward Iran. Here, the chapter has argued against over-theorisations of indications for supposedly counter-hegemonic forces struggling to topple the prevailing power system. Respective foreign policy motivations are diverse, and the respective commonalities in foreign policy positions between China, Russia and Turkey have been juxtaposed to their many differences and even clashes. The 'de-Westernisation' of Iran discourses is therefore not to be confused with a joint endeavor to create a counter-hegemonic bloc opposing US leadership. To the contrary, their relative dependence on the United States has led China, Russia and Turkey to follow foreign policies that they resisted on a discursive level. Tellingly, China and Turkey have reduced their oil imports from Iran in order to qualify for US sanctions exemptions.

The acceptance of UN-backed international sanctions explains a convergence of rules that are still accepted as governing international relations at large. It explains an adherence to 'rules and models' that have the power to *constrain* hegemonic policies because they have to be channeled through *multilateral* institutions. Working through UN mechanisms was seen as the best way to contain unilateral military options and shape the outcome of UN sanctions resolutions. Doing so, Russia and China have expressed disagreements with its Western dialogue partners in the P5+1 format. Turkey shares with China and Russia an aversion for Western pressure on Iran. Emphasising that sanctions are counterproductive political tools to force Tehran to the negotiating table, the Turkish government engaged in a phase of proactive mediation in 2009 and 2010. Even though it secured the historic first Iranian agreement to a proposed deal in May 2010, this episode of Turkish mediation ended as a policy failure when UNSCR 1929 was adopted only a month later. The subsequent retreat from an active role as mediator – while not shying away from venting its disappointment – shelved the idea of outside mediation in Iran's nuclear file. It underscored that a resolution to Iran's nuclear crisis had to be found within the P5+1 format – under framework conditions that were not yet given in 2010, but only three years later.

With different prioritisations and conceptions of legitimacy by the different actors involved becoming manifest, the Iranian nuclear crisis was not only a battlefield for the survival of the NPT regime, but a debate about differing conceptions of world order and security governance. Chinese, Russian and Turkish foreign policies toward the Iranian nuclear programme, as analysed in this chapter, stand indicative of alternative security cultures toward Iran in a 'process of power de-concentration' (Tessman and Wolfe 2011: 218) in which dominant power structures have not been replaced by alternative governance structures (yet). As much as 'historic blocs' are sustained through the complementary forces of consent and coercion, hegemony – conceptually and empirically – can never be absolute. By implication, it is impossible to conceive of its contrary extreme – counter-hegemony – in the absolute. 'A counter-hegemony would consist of a coherent view of an alternative world order, backed by a concentration of power sufficient to maintain a challenge to core countries', Robert Cox formulates (1981: 150), and thus importantly implies that 'counter-hegemony' denotes a drive towards power transitions. The conceptual distinction between an all-out confrontation (what in Gramscian terms resembles a 'war of manoeuvre' most closely) and an issue-specific contestation of hegemonic structures (a 'war of position' in Gramsci's thought) has been shown to apply to the Iran policies of China, Russia and Turkey. 'Resistance' has been conceptualised as a qualified disagreement with parts of the hegemonic order, and it was shown how such resistance takes place across ideational, material and institutional dimensions of such an order. Such a finding sheds light on the dynamics of international power shifts that will, one way or another, determine international politics and the co-existence between hegemonic powers and norm-shapers in the making.

Notes

1 Now-president Hassan Rouhani (2011) writes in his memoirs on his time as chief nuclear negotiator that 'any government that accepts long-term suspension or stopping enrichment is doomed to collapse' (61, 666). In a similar fashion, Hossein Mousavian (2014) writes that Iran would never accept denial of access to enrichment under the NPT (138). This 'right' has been likened in importance to the nationalisation of the Iranian oil industry in 1951 under Prime Minister Mossadegh (cf. Parasiliti 2012: 35). Possibly aware of the fruitlessness of maximalist positions, governments may change the language used in negotiations (without affecting the substance). For instance, a European Iran desk officer calls the 'right to enrich' question a pseudo debate, and instead states that what is relevant is 'factual on-the-ground' enrichment. Author's interview, Berlin, 14 November 2014.

2 The Obama administration's tacit recognition of Iranian uranium enrichment has to be read as a realisation of the political untenability of the Bush administration's insistence on 'zero enrichment'. 'We lost that battle', former US National Security Council official Dr Robert Litwak acknowledged in an interview. Author's interview, Washington, 31 October 2014. And former high-level US diplomat William Burns formulates:

> The reality is that the Iranians have developed over the course of the last decade or more the know-how to enrich, they know their way around basic enrichment technology, and you can't wish that away, you can't dismantle it away, you can't bomb it away [...]. I understand the argument for no enrichment, but I just think that train left the station.
>
> (Burns and Glasser 2015)

3 Author's interview, Ankara, 20 August 2014.

4 Attempts at political negotiations to end the first North Korean nuclear crisis that culminated in the 1994 Agreed Framework did not produce reliable guarantees to ensure that North Korea was not using nuclear material to manufacture a nuclear weapon. A second crisis in 2002 led to the expulsion of IAEA inspectors from the country, and the eventual withdrawal of North Korea as a state party to the NPT in January 2003. Ensuing UN sanctions and several rounds of six-party talks (between the US, Russia, China, Japan, North Korea and South Korea) could not prevent North Korea's march to a nuclear weapon. The testing of missiles and underground nuclear tests in 2006 drove the final nail in that coffin (see Bulychev and Vorontsov 2007: 13–29). Contrary to North Korea, Iran does not claim to have nuclear weapons ambitions and wants its nuclear programme to be accepted as legal and legitimate. Effective comparisons between the North Korean and the Iranian nuclear case are Fitzpatrick (2006) and Litwak (2008).

5 As also confirmed by Dr Dina Esfandiary, International Institute for Strategic Studies, Author's interview, London, 11 July 2014.

6 I wish to thank John Heieck for clarifying this point.

7 Author's interview at the Russian permanent mission to the IAEA, Vienna, 18 August 2016.

8 Author's interview with Iranian diplomat, Berlin, 29 August 2016.

9 Author's interview with Turkish foreign ministry official, Ankara, 20 August 2014.

10 Author's interview with Russian foreign ministry official, Moscow, 18 April 2013; author's interview with Russian diplomat to the EU, Brussels, 7 October 2014; author's interview with Chinese foreign ministry official, Beijing, 18 April 2014.

11 As former US Secretary of State Hillary Clinton has done 'for years', as she writes in her memoirs. She adds with barely concealed pride how the US assembled a 'coalition' to adopt them, and that 'Bibi Netanyahu told me he liked the phrase so much that he had adopted it as his own' (Clinton 2014: 441).

12 Author's interview, Washington, 31 October 2014.

13 See the arms trade databases at SIPRI (http://armstrade.sipri.org/armstrade/page/trade_register.php). Subsequent UNSC resolutions, especially UNSCR 1929, have imposed arms embargoes that impede Russian arms sales to Iran. Asked about Russia's compliance with Iran arms embargoes, a Russian official in October 2014 states in an interview that Russia's arms export control regimes are now even stricter than the US and UN ones. Author's interview, Washington, 31 October 2014. Under the JCPOA, the weapons embargo will be lifted five years after 'Implementation Day'. However, states can apply for a UN authorisation for the delivery of arms already before the end of this five-year period.

14 The sanctions against Sukhoi, imposed because of alleged violations of the Iran Nonproliferation Act of 2000, were lifted again in November 2006 (Ria Novosti 2007).

15 Lohmann (2014) writes:

> US sanctions enforcement agencies are primarily those authorities that monitor the implementation of US financial sanctions. The Bureau of Industry and Security in the US Department of Commerce monitors the export of dual-use goods, the Directorate of Defense Trade Controls in the State Department monitors the sale of military hardware, and the Office of Terrorism and Financial Intelligence (and here especially the Office of Foreign Asset Control, OFAC) in the US Finance Department monitors compliance with US financial sanctions [...]. In close consultation with the US State Department, OFAC controls (compliance with) US financial sanctions and enforces them in collaboration with federal authorities.
>
> (4, author's translation)

16 Author's interview, Washington, 14 February 2014.

17 This was the case despite Yeltsin's official centralisation of foreign policy under the auspice of the President's office. This attempted centralisation in effect sidelined the foreign ministry, thus encouraging Minatom and the defence ministry to engage in their own diplomacy (Belopolsky 2009: 32–33, 35–38).

18 Such as the delivery of a gas centrifuge needed to enrich weapons-grade uranium (see Parker 2009: 116).

19 Russian Prime Minister Viktor Chernomyrdin also sent a confidential letter to US Vice President Al Gore, committing Russia to only complete the first reactor unit of the Bushehr power plant, to supply the fuel, and to train Iranians to operate the plant (Khlopkov and Lutkova 2010: 7).

20 Shifts in framework conditions in connection with the Russian–Western stand-off over the 'Ukraine crisis' demand a caveat here. In the wake of the deterioration of relations between Russia and the West, the Russian government has shown a tendency of economic alienation from US-inspired financial and economic instruments – in addition to the level of political resentment, and in addition and reaction to Western attempts to isolate Russia economically. Examples are the Putin administration's announcement to substitute embargoed manufactured goods from the West by domestic products; indirect taxes and direct product bans; and relevant changes in the customs legislation (Libman 2014).

21 These are official UN trade numbers retrieved from the Statistics Division of the Department of Economics and Social Affairs. Available at http://comtrade.un.org/data/.

22 In 2005, then US Deputy Secretary of State Robert Zoellick has called on China to behave like a 'responsible stakeholder' (Zoellick 2005). See http://2001–2009.state.gov/s/d/former/zoellick/rem/53682.htm. 'A critic may argue that basically he asked China to become like the United States', Amitai Etzioni (2011: 540) remarks, and analyses the 'aspirational standards' that such a normative and essentially simplistic term reveal (542–553). Taking note of what she perceived as a more assertive Chinese foreign policy, former US Secretary of State Hillary Clinton (2014) described China as a 'selective stakeholder' (75).

23 Former US National Security Council official Dr Robert Litwak confirms this in his 2012 book *Outlier States*:

> the Clinton administration diplomatically pressed both countries [China and Russia] to forgo nuclear commerce with Iran, making the cessation a condition for U.S. civil nuclear exports to China and threatening the cutoff of U.S. aid to Russia to get the Kremlin to forgo the sale of fuel-cycle technology.
>
> (163–164)

24 President Erdoğan Addresses the UN General Assembly. Available at www.tccb.gov.tr/news/397/91138/president-erdogan-addrebes-the-un-general-abembly.html (accessed 8 January 2015).

25 Mahbubani (2013: 227) writes of 'democratic global governance', and shows why Western governments oppose such concepts for demographic reasons: Should a representational formula for international organisations based on population be applied, the Western predominance therein would substantially shift in favour of non-Western states.

26 Given Chinese and Russian domestic politics, the choice for such semantics on the part of China and Russia thus does not come without irony. Viatcheslav Morozov (2013) has analysed this co-existence between a democratisation rhetoric on the international level and repressive domestic politics as the expression of what he calls a 'subaltern empire', where Russia portrays itself as a subaltern player in international relations, while its domestic behaviour is colonial.

27 As was also the assessment of a US State Department official. Author's interview, Washington, 30 December 2014.

28 Author's interview with Russian foreign ministry official, Moscow, 18 April 2013; author's interview with Russian diplomat to the EU, Brussels, 7 October 2014; author's interview with Chinese foreign ministry official, Beijing, 18 April 2014; author's interview with Turkish foreign ministry official, Ankara, 20 August 2014.

29 Author's interview with Turkish foreign ministry official, Ankara, 20 August 2014.

30 Author's interview with Russian diplomat to the EU, Brussels, 7 October 2014.

31 Author's interview with Chinese diplomat to the US, Washington, 31 October 2014.

32 Author's interview with Turkish official, London, 30 March 2016; author's interview with Turkish diplomat, Brussels, 15 April 2016.

33 Author's interviews with Dr Mark Fitzpatrick, International Institute for Strategic Studies London, 11 July 2014; with Dr Dina Esfandiary, International Institute for Strategic Studies, London, 11 July 2014; with EEAS official, Brussels, 13 March 2013; with EEAS official, Brussels, 4 June 2013; with former German ambassador to Iran, Berlin, 26 August 2013; with EEAS official, Brussels, 20 March 2014; with European diplomat, Brussels, 18 March 2014; with former Chinese diplomat to Iran, Beijing, 22 April 2014; with European diplomat, Brussels, 6 February 2015; author's conversation with former British Governor at the IAEA, Brussels, 23 June 2015. Deputy Director of the Institute of Middle East Studies of the China Institute for Contemporary International Relations formulates:

> Russia has more leverage over the Iran dossier than China has. China just has economic relations with Iran, which doesn't mean we have influence. We have to take a very pragmatic position. Russia has to stand in the first row.
>
> (Author's interview, Beijing, 24 April 2014)

34 Niall Ferguson (2012) writes of the West and the 'Resterners' (323). Such analyses typically look at material preconditions for the 'rise' of countries and their resulting potential to challenge order. Kupchan (2014) has recently attempted an important step towards taking the focus away from this fixation in presenting 'a synthesis of constructivist and rationalist approaches by demonstrating how ideational, cultural, and material interests combine to shape the social purpose of hegemonic powers' (222).

Three out of the four 'logics of hegemony' that he juxtaposes, however, are themselves material ('geopolitical, socioeconomic, commercial'). Acharya (2014) rejects over-theorisations about the collective rise of 'emerging powers' when he writes of the 'hype of the rest' (5).

35 Author's interview with Russian foreign ministry official, Moscow, 12 November 2013.
36 Author's interview with Turkish official, London, 30 March 2016.
37 Author's interview at the French permanent mission to the IAEA, Vienna, 18 August 2016.
38 This was true at least until the cessation of trade between Russia and Turkey over disagreements in their respective Syria policies in late 2015.
39 Niall Ferguson and Moritz Shularick (2007) have thus coined the term 'Chimerica' to describe the mutual financial dependence between the US and China.
40 Sakwa (2002: 366) has used this formulation to capture Russian foreign policy reorientations following the breakup of the Soviet Union.
41 The debate about the 'decline' of the US has been spurred by Paul Kennedy's seminal book *The Rise and Fall of the Great Powers* (1987).
42 See also Ferguson (2015: 22) about the relativity of statistics in predicting 'rise'.
43 Such an analysis is not reserved exclusively for critical studies. For an analysis of 'cooperative hegemony' from the perspective of 'ideational-institutional realism', see Pedersen (2002).
44 Joshua Cooper Ramo (2010) has used the term 'co-evolution' in this context.

References

Acharya, Amitav. 2014. *The End of American World Order*. Cambridge: Polity Press.

Adib-Moghaddam, Arshin. 2014. *On the Arab Revolts and the Iranian Revolution. Power and Resistance Today*. London: Bloomsbury.

Alcaro, Ricardo and Tabrizi, Aniseh Bassiri. 2014. Europe and Iran's Nuclear Issue: The Labours and Sorrows of a Supporting Actor. *The International Spectator* vol. 49, no. 3: 14–20.

Agamben, Giorgio. 2002. *Homo sacer. Die Souveränität der Macht und das nackte Leben*, Trans. Hubert Thüring. Frankfurt: Suhrkamp.

Antonenko, Oksana. 2001. Russia's Military Involvement in the Middle East. *Middle East Review of International Affairs* vol. 5, no. 1: 1–14.

Arms Control Association. 2014. History of Official Proposals on the Iranian Nuclear Issue. *Fact Sheets & Briefs*, January 2014. Available at www.armscontrol.org/factsheets/Iran_Nuclear_Proposals (accessed 25 September 2016).

Armstrong, David, Lloyd, Lorna and Redmond, John. 2004. *International Organisation in World Politics*. Basingstoke and New York: Palgrave Macmillan.

Balzacq, Thierry. 2010. *Securitization Theory: How Security Problems Emerge and Dissolve*. New York: Routledge.

Barzegar, Kayhan. 2014. Iran–US Relations in the Light of the Nuclear Negotiations. *The International Spectator* vol. 49, no. 3: 1–7.

Belopolsky, Helen. 2009. *Russia and the Challengers. Russian Alignment with China, Iran, and Iraq in the Unipolar Era*. Basingstoke: Palgrave Macmillan.

Bevir, Mark and Gaskarth, Jamie. 2015. Global Governance and the BRICs: Ideas, Actors, and Governance Practices. In: Gaskarth, Jamie (ed.), *Rising Powers, Global Governance and Global Ethics*. Abingdon and New York: Routledge, pp. 74–96.

Biersteker, Thomas and Moret, Erica. 2015. Rising Powers and Reform of the Practices of International Security Institutions. In: Gaskarth, Jamie (ed.), *Rising Powers, Global Governance and Global Ethics*. Abingdon and New York: Routledge, pp. 57–73.

Breslin, Shaun. 2009. Understanding China's Regional Rise: Interpretations, Identities and Implications. *International Affairs* vol. 85, no. 4: 817–835.

Breslin, Shaun. 2013. China and the Global Order: Signaling Threat or Friendship? *International Affairs* vol. 3: 615–634.

Bulychev, Georgy and Vorontsov, Alexander. 2007. North Korea – An Experiment in Nuclear Proliferation. In: Arbatov, Alexei (ed.), *At the Nuclear Threshold. The Lessons of North Korea and Iran for the Nuclear Non-Proliferation Regime*. Moscow: Carnegie Endowment for International Peace, pp. 13–29.

Burns, William J. and Glasser, Susan B. 2015. You Can't Bomb it Away. Interview with William Burns, *Politico*, 15 March 2015. Available at http://carnegieendowment.org/2015/03/15/you-can-t-bomb-it-away/i49d?mkt_tok=3RkMMJWWfF9wsRoiuaTIZKXonjHpfsX64u8rUKCg38431UFwdcjKPmjr1YcAS8p0aPyQAgobGp5I5FEIQ7XYTLB2t60MWA%3D%3D (accessed 19 March 2015).

Bush, George W. 2002. State of the Union Address, 29 January 2002. Available at www.johnstonsarchive.net/policy/bushstun2002.html (accessed 17 March 2011).

Buzan, Barry and Waever, Ole. 2003. *Regions and Powers. The Structure of International Society*. Cambridge: Cambridge University Press.

Chomsky, Noam. 2016. *Who Rules the World?* London: Hamish Hamilton.

Clark, Ian. 2005. *Legitimacy in International Society*. Oxford: Oxford University Press.

Clark, Ian. 2011. China and the United States: a Succession of Hegemonies? *International Affairs* vol. 87, no. 1: 13–28.

Claude, Inis. 1966. Collective Legitimisation as a Political Function of the United Nations. *International Organisation* vol. 20, no. 3: 367–379.

Clinton, Hillary Rodham. 2014. *Hard Choices*. New York: Simon & Schuster.

Cox, Robert W. 1981. Social Forces, States and World Orders: Beyond International Relations Theory. *Millennium: Journal of International Studies* vol. 10, no. 2: 126–155.

Cox, Robert W. 1996. *Approaches to World Order*. Cambridge: Cambridge University Press.

Daly, John. 2013. How Far Will Turkey Go in Supporting Sanctions Against Iran? *The Turkey Analyst* vol. 6, no. 13. Available at www.turkeyanalyst.org/publications/turkey-analyst-articles/item/48-how-far-will-turkey-go-in-supporting-sanctions-against-iran?.html (accessed 8 March 2014).

Defense Industry Daily. 2006. US Ban on Russian Defense Firms Raises the Stakes. *Defense Industry Daily*, 11 August 2006 www.defenseindustrydaily.com/us-ban-on-russian-defense-firms-raises-the-stakes-02522/ (accessed 17 October 2013).

Dupont, Pierre-Emmanuel. 2012. Countermeasures and Collective Security: The Case of the EU Sanctions Against Iran. *Journal of Conflict and Security Law* vol. 17, no. 3: 301–336.

Eichengreen, Barry. 2011. *Exorbitant Privilege: The Rise and Fall of the Dollar and the Future of the International Monetary System*. Oxford: Oxford University Press.

Etzioni, Amitai. 2011. Is China a Responsible Stakeholder? *International Affairs* vol. 87, no. 3: 539–553.

Ferdinand, Peter. 2007. Russia and China: Converging Responses to Globalization. *International Affairs* vol. 83, no. 4: 655–680.

Ferdinand, Peter. 2013. The Positions of Russia and China at the UN Security Council in the Light of Recent Crises. *Briefing Paper*, March 2013. Directorate-General for External Policies of the European Union.

Ferguson, Niall. 2012. *Civilization. The West and the Rest*. London: Penguin Books.

Ferguson, Niall and Schularick, Moritz. 2007. 'Chimerica' and the Global Asset Market Boom. *International Finance* vol. 10, no. 3: 215–239.

Ferguson, Yale H. 2015. Rising Powers and Global Governance: Theoretical Perspectives. In: Gaskarth, Jamie (ed.), *Rising Powers, Global Governance and Global Ethics*. Abingdon and New York: Routledge, pp. 21–40.

Fitzpatrick, Mark. 2006. Iran and North Korea: The Proliferation Nexus. *Survival* vol. 48, no. 1: 61–80.

Fitzpatrick, Mark. 2016. The Iran Deal Shows its Worth. *Politics and Strategy, the Survival Editors' Blog*, 18 January 2016. Available at www.iiss.org/en/politics%20and%20strategy/blogsections/2016-d1f9/january-c129/the-iran-deal-shows-its-worth-ded4 (accessed 2 March 2016).

Gertz, Bill. 2000. *The China Threat. How the People's Republic Targets America*. Washington: Regnery.

Gramsci, Antonio. 1971. *Selections from the Prison Notebooks*, ed. and trans. Hoare, Q. London: Lawrence & Wishart.

Guldimann, Tim. 2007. The Iranian Nuclear Impasse. *Survival*, vol. 49, no. 3: 169–178.

Habertürk. 2012. Turkey to Cut Iran Oil Imports. *Habertürk* March 30. Available at www.haberturk.com/general/haber/729605-turkey-to-cut-iran-oil-imports (accessed 1 April 2013).

Hu, Jintao. 2007. *Documents of the 17th National Congress of the Communist Party of China*. Beijing: Foreign Languages Press.

Hung, Ho-fung. 2016. *The China Boom. Why China Will Not Rule the World*. New York: Columbia University Press.

Hurd, Ian. 1999. Legitimacy and Authority in International Politics. *International Organization* vol. 53, no. 2: 379–408.

International Institute for Strategic Studies. 2016. What the Iran Nuclear Accord Means for Sanctions Today and Tomorrow. *Workshop Report*, 14 January 2016. Available at www.iiss.org/en/events/events/archive/2016-a3c2/january-6318/npdp-london-workshop-545e (accessed 21 April 2016).

Kagan, Frederick W. 2015. Complexities of Iranian Sanctions Relief. *American Enterprise Institute, Article*, 3 April. Available at www.aei.org/publication/complexities-of-iranian-sanctions-relief/ (accessed 26 February 2016).

Katzenstein, Peter. J. (ed.). 1996. *The Culture of National Security. Norms and Identity in World Politics*. New York: Columbia University Press.

Katzman, Kenneth. 2006. The Iran–Libya Sanctions Act (ILSA). *CRS Report for Congress*, 26 April 2006. Available at http://fpc.state.gov/documents/organization/66441.pdf (accessed 27 September 2014).

Kennedy, Paul, 1987. *The Rise and Fall of the Great Powers: Economic Change and Military Conflict from 1500 to 2000*. New York: Random House.

Khajepour, Bijan. 2014. Decoding Iran's 'Resistance Economy'. *Iran Pulse, Al monitor*, 24 February 2014. Available at www.al-monitor.com/pulse/originals/2014/02/decoding-resistance-economy-iran.html# (accessed 20 January 2015).

Khanna, Parag. 2014. Iran Nuclear Sanctions: Why the World Still Does Business with Tehran. *CNN*, 10 June 2014. Available at http://edition.cnn.com/2014/06/10/opinion/iran-nuclear-sanctions-parag-khanna/ (accessed 4 October 2014).

Khlopkov, Anton and Lutkova, Anna. 2010. *The Bushehr NPP: Why Did it Take So Long?* Moscow: Center for Energy and Security Studies. Available at http://a-pln.org/sites/default/files/apln-analysis-docs/TheBushehrNPP-WhyDidItTakeSoLong.pdf (accessed 12 May 2015).

Kissinger, Henry. 2014. *World Order*. London: Penguin Books.

Kupchan, Charles. 2014. The Normative Foundations of Hegemony and the Coming Challenge to Pax Americana. *Security Studies* vol. 23: 219–257.

Lavrov, Sergei. 2016a. Russia's Foreign Policy in a Historical Perspective. *Russia in Global Affairs* vol. 14, no. 2: 8–19.

Lavrov, Sergei. 2016b. *Statement at the 71st Session of the UN General Assembly*, 23 September 2016. Available at https://gadebate.un.org/sites/default/files/gastatements/71/71_RU_en.pdf (accessed 24 September 2016).

Leverett, Flynt. 2013. The Iranian Nuclear Issue, the End of the American Century, and the Future of International Order. *Penn State Journal of Law & International Affairs* vol. 2, no. 2: 240–271.

Leverett, Flynt and Leverett, Hillary Mann. 2013. *Going to Tehran. Why the United States Must Come to Terms with the Islamic Republic of Iran*. New York: Metropolitan Books.

Libman, Alexander. 2014. Aussenwirtschaftlicher Protektionismus in Russland. *SWP-Aktuell* 69, November.

Litwak, Robert. 2008. Living with Ambiguity: Nuclear Deals with Iran and North Korea. *Survival* vol. 50, no. 1: 91–118.

Litwak, Robert. 2012. *Outlier States. American Strategies to Change, Contain, or Engage Regimes*. Washington, D.C.: Woodrow Wilson Center Press.

Litwak, Robert. 2014. *Iran's Nuclear Chess: Calculating America's Moves*. Washington, D.C.: Woodrow Wilson Center Press.

Lohmann, Sascha. 2013. Unilaterale US-Sanktionen gegen Iran. *SWP-Aktuell* vol. 63: 1–8.

Lohmann, Sascha. 2014. Minenfelder der US-Außenwirtschaftspolitik. Unilaterale Finanzsanktionen im Dienst nationaler Sicherheit. *SWP-Aktuell* vol. 71: 1–8.

Lohmann, Sascha. 2015. Zwang zur Zusammenarbeit. *SWP-Aktuell* vol. 54: 1–8.

Lundestad, Geir. 2013. Conclusion: The Future. In: Lundestad, Geir (ed.), *International Relations Since the End of the Cold War*. Oxford: Oxford University Press, pp. 290–306.

Mahbubani, Kishore. 2013. *The Great Convergence. Asia, the West, and the Logic of One World*. New York: Public Affairs.

Meier, Oliver and Pieper, Moritz. 2015. Russland und der Atomkonflikt mit Iran. Kontinuitäten und Brüche bei den russischen Interessen im Zeichen der Ukrainekrise. *SWP-Aktuell* 38, April 2015: 1–4.

Morozov, Viatcheslav. 2013. Subaltern Empire? *Problems of Post-Communism* vol. 60, no. 6: 16–28

Mousavian, Seyed Hossein. 2012. *The Iranian Nuclear Crisis. A Memoir*. Washington: Carnegie Endowment for International Peace.

Mousavian, Seyed Hossein. (with Shahir Shahidsaless). 2014. *Iran and the United States. An Insider's View on the Failed Past and the Road to Peace*. New York, London: Bloomsbury.

Nau, Henry R. 2012. Introduction: Domestic Voices of Aspiring Powers. In: Nau, Henry R. and Ollapally, Deepa M. (eds), *Worldviews of Aspiring Powers. Domestic Foreign Policy Debates in China, India, Iran, Japan, and Russia*. New York: Oxford University Press, pp. 3–35.

Nephew, Richard. 2015. Commentary on the Nuclear Deal Between Iran and the P5+1. Center on Global Energy Policy, Columbia University. *Commentary*, 14 July. Available at http://energypolicy.columbia.edu/sites/default/files/energy/Commentary%20on%20the%20Nuclear%20Deal%20between%20Iran%20and%20the%20P5%2B1.pdf. (accessed 26 February 2016).

Orlov, Vladimir. 2010. *Iranskiy Faktor v Opredelenii Vneshnepoliticheskikh Prioritetov Rossii* (The Iranian Factor in Determining Russia's Foreign Policy Priorities). Moscow: PIR-Center.

Patel, Ketan and Hansmeyer, Christian. 2009. American Power, Patterns of Rise and Decline. In: Parmar, Inderjeet, Miller, Linda B. and Ledwidge, Mark (eds), *Obama and the World. New Directions in US Foreign Policy*. London and New York: Routledge, pp. 275–288.

Parasiliti, Andrew. 2012. Closing the Deal with Iran. *Survival* vol. 54, no. 4: 33–42.

Parsi, Trita. 2007. *Treacherous Alliance. The Secret Dealings of Israel, Iran, and the U.S.* New Haven and London: Yale University Press.

Patrikarakos, David. 2012. *Nuclear Iran. The Birth of an Atomic State*. London, New York: I.B. Tauris.

Pedersen, Thomas. 2002. Cooperative Hegemony: Power, Ideas and Institutions in Regional Integration. *Review of International Studies* vol. 28: 677–696.

Perthes, Volker. 2008. *Iran – Eine politische Herausforderung*. Frankfurt am Main: Suhrkamp.

Pirseyedi, Bobi. 2013. *Arms Control and Iranian Foreign Policy. Diplomacy of Discontent*. Abingdon: Routledge.

Posch, Walter. 2013. The Third World, Global Islam and Pragmatism. The Making of Iranian Foreign Policy. *SWP Research Paper April 2013*. Berlin: Stiftung Wissenschaft und Politik.

Putin, Vladimir. 2014. Speech at the Meeting of the Valdai International Discussion Club. *Kremlin Transcript*. Available at http://eng.kremlin.ru/news/23137 (accessed 26 January 2015).

Ramo, Joshua Cooper. 2004. *The Beijing Consensus*. London: The Foreign Policy Centre.

Ramo, Joshua Cooper. 2010. How To Think About China. *Times*, 19 April 2010.

Rapkin, David P. and Braaten, Dan. 2009. Conceptualising Hegemonic Legitimacy. *Review of International Studies* vol. 35: 113–149.

Reus-Smit, Christian. 2014. Power, Legitimacy, and Order. *The Chinese Journal of International Politics* vol. 7, no. 3: 341–359.

Ria Novosti. 2007. Moscow to Press U.S. for Sanction Cancellation – FM Lavrov. *Ria Novosti* 3 February 2007. Available at http://en.ria.ru/world/20070203/60158997.html (accessed 17 October 2013).

Rouhani, Hassan. 2011. *Amniyat Melli va Diplomasi-ye Hastehi Iran (National Security and Nuclear Diplomacy)*. Tehran: Center for Strategic Research.

Russian foreign ministry. 2013. *Concept of the Foreign Policy of the Russian Federation.* Available at www.mid.ru/bdomp/ns-osndoc.nsf/1e5f0de28fe77fdcc32575d900298676/869c9d2b87ad8014c32575d9002b1c38!OpenDocument (accessed 15 January 2014).

Sakwa, Richard. 2002. *Russian Politics and Society*. 3th edn. London and New York: Routledge.

Sakwa, Richard. 2011. Russia and Europe: Whose Society? *Journal of European Integration* vol. 33, no. 2: 197–214.

Sakwa, Richard. 2012. The Problem of 'the International' in Russian Identity Formation. *International Politics* vol. 49, no. 4: 449–465.

Sakwa, Richard. 2015. *Frontline Ukraine. Crisis in the Borderlands*. London: I.B. Tauris.

Schirm, Stefan A. 2010. Leaders in Need of Followers: Emerging Powers in Global Governance. *European Journal of International Relations* vol. 16, no. 2: 197–221.

Schmitt, Carl. 1993. *Politische Theologie: Vier Kapitel zur Lehre der Souveränität.* Berlin: Duncker & Humblot.

Serfaty, Simon. 2011. Moving into a Post-Western World. *The Washington Quarterly* vol. 34, no. 2: 7–23.

Stokes, Doug. 2014. Achilles' Deal: Dollar Decline and US Grand Strategy after the Crisis. *Review of International Political Economy* vol. 21, no. 5: 1071–1094.

Taylor, Brendan. 2010. *Sanctions as Grand Strategy*. Adelphi series, International Institute for Strategic Studies. Abingdon: Routledge.

Tessman, Brock and Wolfe, Wojtek. 2011. Great Powers and Strategic Hedging: The Case of Chinese Energy Security Strategy. *International Studies Review* vol. 13: 214–240.

Trenin, Dmitri. 2014. Russia: Sealing the New Quality of Its Foreign Policy. *Eurasia Outlook*, 20 January 2014. Available at http://carnegie.ru/eurasiaoutlook/?fa=54244 (accessed 22 February 2014)

Tzanakopoulos, Antonios. 2011. *Disobeying the Security Council. Countermeasures against Wrongful Sanctions*. New York: Oxford University Press.

UN. 2006. Security Council Imposes Sanctions on Iran for Failure to Halt Uranium Enrichment, Unanimously Adopting Resolution 1737. *Resolution 1737*. Available at www.un.org/News/Press/docs//2006/sc8928.doc.htm (accessed 21 December 2012).

UN. 2008. Security Council Tightens Restriction on Iran's Proliferation-sensitive Nuclear Activities, Increases Vigilance over Iranian Banks, Has States Inspect Cargo. *Security Council Resolution 1803*. Available at www.un.org/News/Press/docs/2008/sc9268.doc.htm (accessed 21 December 2012).

UN. 2010. Security Council Imposes Additional Sanctions on Iran, Voting 12 in Favour to 2 Against, with 1 Abstention. *Security Council Resolution 1929*. Available at www.un.org/News/Press/docs/2010/sc9948.doc.htm (accessed 21 December 2012).

US Department of State. 2012. Three Companies Sanctioned Under the Amended Iran Sanctions Act. *Media Note*, 12 January 2012. Available at www.state.gov/r/pa/prs/ps/2012/01/180552.htm (accessed 4 May 2014).

US Department of the Treasury. 2016. *Guidance Relating to the Lifting of Certain U.S. Sanctions Pursuant to the Joint Comprehensive Plan of Action on Implementation Day*. 16 January.

Welch, David E. 1992. The Organizational Process and Bureaucratic Politics Paradigms: Retrospect and Prospect. *International Security* vol. 17, no. 2: 112–146.

White House. 2015. The Iran Nuclear Deal: What You Need to Know about the JCPOA. *Information Sheet*. Available at www.whitehouse.gov/sites/default/files/docs/jcpoa_what_you_need_to_know.pdf (accessed 26 February 2016).

Williams, Michael C. 2003. Words, Images, Enemies: Securitization and International Politics. *International Studies Quarterly* vol. 47: 511–531.

Worth, Owen. 2015. *Rethinking Hegemony*. London: Palgrave Macmillan.

Zarif, Mohamad Javad. 2006. Letter dated 31 July 2006 from the Permanent Representative of the Islamic Republic of Iran to the United Nations addressed to the President of the Security Council. United Nations Security Council, *S/2006/603*. Available at www.iranwatch.org/sites/default/files/unsc-s2006603-irancomm-080206.pdf (accessed 13 April 2015).

Zoellick, Robert B. 2005. Wither China? From Membership to Responsibility? Remarks to National Committee on U.S.–China Relations. New York, 21 September 2005. Available at http://2001-2009.state.gov/s/d/former/zoellick/rem/53682.htm (accessed 6 August 2014).

6 Conclusion

The 'Iran question' and dissent in world order

1 Synthesis and concluding reflections

Powerful narratives can produce their own realities. The battle over interpretations of the Iranian nuclear conflict is a prime example. 'The international community', 'rogue states', 'illegitimacy' and 'fairness' in international relations are no innocent terms, and this book has shown how diverging interpretations over these terms lay at the root of policy contestations between different actors involved in the Iranian nuclear conflict. The latter has been a dispute where international regimes and institutional provisions became subject to political re-interpretations by various actors involved. The nuclear non-proliferation treaty had become a bone of contention not only for its division of the world into nuclear haves and have-nots, but also for a structural imbalance between efforts to combat nuclear non-proliferation and nuclear disarmament ambitions already before the outbreak of the Iran nuclear crisis (ElBaradei 2012: 172). What was introduced with the Iran case was a debate over the tacit recognition of non-explicit 'rights' out of political expediency. Couched in terms of 'justice' and 'fairness' in international politics, the stand-off over Iran's nuclear programme became a disagreement about World Order. Porter (2014) forcefully argues that 'institutional and cognitive distortions' (217) have led to a 'manufactured crisis'. Institutional (UN) structures have become the subject of debates about legality and illegality in international relations, and disagreements about the legality and proportionality of different multilateral and unilateral sanctions adopted against Iran functioned as a proxy debate about the standing of Iran in world politics.

Against this background, this book has analysed the foreign policies of China, Russia and Turkey towards the Iranian nuclear programme. Cutting across the disciplines of International Relations, Security Studies, Area Studies and studies in inter-regionalism, this study has argued that the co-existence between dominant powers and powers that favour revisions to international security governance is characterised by the interacting forces of contestation and accommodation. The purpose of the book has been two-fold. First, much scholarly attention has been paid to foreign policy approaches to the Iranian nuclear crisis. Yet, no comprehensive analyses have been conducted so far that use the Iranian nuclear case as an illustration to conceptualise the interaction between 'established' power structures and those actors resisting them. This book is a first step to fill this gap in the literature. Second, while studies of the interactions between dominant power centres and 'norm challengers' have mainly focused on subversive moments for the global political economy, implications for international security governance have remained understudied. Here, the book's contribution lies in contextualising the Iran nuclear file both as an international security challenge and a crisis in norm dynamics. As such, it emphasises the need to politicise the very ontology of security (Bilgin 2013: 103; Booth 2005). In doing so, the book thus did not proceed from a dichotomous understanding of domination and confrontation in international politics that bedevils so many teleological readings of future interactions between 'the West' and 'emerging' actors from 'the Rest'. Instead, it has attempted to transcend an ontology that borrows from state-centric conceptions of international relations in order to understand how legitimacy, the perpetuation of hegemonic power structures and mechanisms of resistance interact as forces collectively moulding World Order (see also Clark 2005, 2011; Rapkin and Braaten 2009; Reus-Smit 2014).

Rather than constituting a bloc challenge to American dominance in 'the international system', the preceding analysis has shown how China, Russia and Turkey display *norm* contestation and different conceptions of desirable security cultures, but propose an equitable adherence to common *rules* internationally. Situating this research finding in the wider literature on international norm dynamics that conceives of norms as essentially contested narratives (Adler-Nissen 2014; Bob 2012; Epstein 2012; Kersbergen and Verbeek 2007; Towns 2012; Wiener 2008; Wunderlich 2013), it was shown how compliance is a situational and relational term. An adherence to rules agreed on collectively is often accompanied by disagreements about norms. Yet, normatively principled resistance in international relations can also give way to the adoption of policies that fall short of acting upon norm discourses, and to policies that, indeed, run counter to normative positions. This bespeaks political expediency for those actors advocating such policies (and can be referred to as 'horse-trading' in policy parlance), and translates into outright betrayal for those actors regarding themselves as 'revolutionary'. In Iran's nuclear case, the Chinese, Russian, and Turkish reluctance to use sanctions as tools in international diplomacy did not prevent the eventual adoption of international sanctions against Iran. The Chinese and Russian governments, however, regarded chapter VII sanctions as the best means to contain policy options that they anticipated to be even more damaging. Faced with dangerous rhetoric about pre-emptive strikes against Iran especially on the part of the Israeli government, Russia and China began to accept UN sanctions as the lesser evil that would take unilateral military options off the table. It also provided them with a procedural mechanism to influence the shape and outcome of such measures (see also Ferdinand 2013). When the nuclear file was referred to the UNSC in 2006 as a result of failed diplomatic attempts to solve the crisis, Russia worked, together with China, towards slowing down the pressure on Iran and 'watering down' drafts of sanctions resolutions without 'emptying' them. The eventual adoption of sanctions resolutions bore witness to an increasing realisation in Moscow and Beijing of Iranian delaying techniques, but also to a level of receptiveness to US pressure. It also incentivised Russia to propose technical solutions to solve the nuclear crisis. Similarly, the Turkish government seized a momentum in 2009 to engage in pro-active diplomacy that drew on its external mediatory potential. Critical of the use of sanctions yet receptive to US positions, Turkey had been faced with a similar conundrum on how to balance its ideational and geopolitical interests in Iran with expectations about and perceptions of Turkish Iran policies held within the US government.

This observed sensitivity to US positions, the preceding chapters have shown, was due to the dominance of the United States in the perpetuation of 'hegemonic structures'. The leverage power of sanctions is a textbook example. Whilst multilateral Iran sanctions were seen as complying with the 'rules' of the UN system, additional unilateral sanctions were perceived as illegitimate and as an extraterritorialisation of domestic legislation. With 'secondary' sanctions targeting third country entities from engaging with Iran, the US had introduced a powerful mechanism of coercion overseen not by the American diplomatic machinery in the State Department, but by its Treasury Department. The unique position of the US capital market in the global economic and financial architecture had been identified as a useful leverage tool to influence other actors' Iran policies (Taylor 2010: 68). Besides an ideational resistance to unilateral sanctions, the economic impact of these 'secondary sanctions' on third country entities constituted an additional material aspect of Chinese, Russian, and Turkish criticism. At this point, this book has offered a two-level model of foreign policy to understand Chinese, Russian and Turkish Iran policies: Norm advocacy on a discursive level need not be coherently translated into actual policies on a behavioural level. While Chinese, Russian and Turkish officials publicly advocate for an adherence to a security culture that emphasises compliance with the norms of 'sovereignty' and 'non-interference', and thus norm divergence from more intrusive approaches advocated by other actors in the Iran nuclear conflict, their level of perceived material and political dependence on the United States prompts them to follow foreign policies that still comply with parts of US-inspired power structures.

Chinese, Russian, and Turkish instances of compliance with such sanctions lists, then, indicates a level of receptiveness to the economic leverage of US-dominated international financial mechanisms in instances where the respective governments have leeway over commercial decisions. While the precise balance between public and private sector deliberations is not always transparent in the cases analysed, interviewees confirmed that 'compliance' with sanctions that go beyond UN-mandated sanctions can be state-enforced and is often state-induced. Big energy companies in the countries examined here are state-owned, and decisions to, for example, reduce oil imports in response to energy sanctions are often political ones. But also private companies recoil from business in Iran where the Damocles sword of financial sanctions is dangling over their heads. Compliance with Iran sanctions regimes on the part of private companies here forcefully underscores the *structural* dimension of hegemony whose reach extends beyond state and governmental control. The power of labels and of 'reputational costs' works in subtle ways, and often slips the control of central governments. US financial sanctions exert a powerful hold over other governments' decisions to curtail trade with Iran because their companies would otherwise risk losing access to US financial markets. The weight of the US capital market remains a powerful tool to discipline economic actors, and the role of the US dollar as international reserve currency is an important instrument in upholding global financial control. Yet, its enduring credibility relies on a 'multitude of agencies' (Worth 2015: 173). As long as both state and non-state actors self-discipline themselves in order to avoid financial penalties and reputational costs, such structures rely both on coercion and manufactured consent. Hegemony is not territorially bounded.

On a discursive level, then, Chinese, Russian and Turkish foreign policies toward the Iranian nuclear programme have breathed the ambition to partially 'de-Westernise' security cultures and discourses toward Iran. This book has shown how their politics of resistance contests consensual hegemonic arrangements as described above. Public advocacy for 'democratic' international relations works towards the erosion of the monopoly over interpretations of the provisions of international security governance. Chinese, Russian, and Turkish foreign policies display a normative divergence with a US-inspired security culture towards Iran, and make a distinction between legitimate sanctions (those that follow UN rules) and illegitimate ones (those that are adopted unilaterally and thus violate the principle of Sovereign Equality). Such normative divergence from Western policies that go beyond UN provisions has occasionally been met by acts of resistance in foreign policy behaviour. Eventual 'compliance' with US policy preferences on Iran has been situational, selective, and motivated by differing ideational, institutional and material stakes at hand. US accusations of 'sanctions-busting' and 'circumvention' of sanctions against China and Turkey for their recourses to creative payment modalities for Iranian oil deliveries is a case in point.

The JCPOA negotiated in 2015 has introduced a new momentum to the Iran nuclear file. Diplomatic talks have managed to seal a complex deal to peacefully bring the Iranian nuclear conflict to an end. In exchange for Iran honoring its terms of the agreement to ensure the exclusively civilian nature of its nuclear programme, sanctions imposed on Iran over its nuclear programme are lifted. With the IAEA's stamp of approval regarding Iran's compliance with the agreement, the way was paved for the formal implementation of the agreement that saw the lifting of nuclear-related sanctions as well as the unlocking of previously frozen assets (IAEA 2016). While the international sanctions architecture built around Iran is being taken apart, resilient structures of hegemony complicate the proverbial turning of the page on Iran's international standing. The persistence of non-nuclear-related financial sanctions and legal ambiguities about the definition of US 'foreign subsidiaries' continue to complicate renewed business operations with Iran even after the JCPOA's Implementation Day (IISS 2016). Actors wanting to engage with Iran still have to navigate a legal minefield and exhibit resistance to a securitised business environment. They continue to be exposed to powerful financial sanctions leverage, both targeted and extraterritorial.

However, inter-state diplomacy here has created an agreement that generates normative pressure which in turn impacts on national foreign policies. The longevity of the mechanisms created will depend on the joint implementation by state parties in the West and non-West. Under the JCPOA, both Russia and China play active roles in the conversion of Iranian nuclear facilities as part of their implementation commitments. Russia's state-owned nuclear energy agency Rosatom is helping Iran to convert the uranium enrichment facility at Fordow into a production centre for medical isotopes, China took the lead in an international consortium tasked to convert the heavy-water reactor at Arak into a nuclear research centre. In addition, their voices in the Joint Commission will continue to weigh on an equal footing as those of France, the UK, Germany, the US and Iran, and as permanent Security Council members, their veto power will continue to influence any future decisions on the Iran case pursuant to UNSCR 2231 (UN 2015). As such, the international Iran diplomacy has provided a unique window to peek into and make sense of the inner workings of hegemony and resistance in international politics.

The working relationship between former hegemons and 'rising powers' as potential challengers is constantly being re-balanced and re-negotiated. These are no linear processes, and rather than proceeding from hype theorisations about US decline and 'the rise of the rest', this book has offered a more nuanced picture of the workings of norm contestations and their implications for gradual power transitions. In this regard, with different prioritisations and perceptions of legitimacy on the part of the different actors involved, the Iranian nuclear crisis was not only a battlefield for the survival of the NPT regime, but will constitute a debate about differing conceptions of world order and security governance for decades to come.

2 Areas for further research

Observed discrepancies between a state's official rhetoric and its foreign policy behaviour are not limited to the Iranian nuclear case. The previous chapter has pointed to the ambivalence between Chinese and Russian public pledges to the principles of 'sovereignty' and 'non-interference', while their interference in other states may breach the very principles they hold dear in international politics. This ambivalence is neither limited to China, Russia or Turkey, nor to the principles presented in the preceding chapters. The distinction between discourse and behaviour in a two-level model to analyse foreign policy merits further research and can serve as a model to be applied to other issue areas beyond the Iranian nuclear case. It allows for an analytical synthesis between norm dynamics and rule applications in international relations and makes a conceptual contribution to the theoretical minefield of the international norm literature. The analytical as well as policy challenge is to craft models to acknowledge the inherently contested nature of norms while still allowing for a rules-based international order conducive to the furtherance of international peace and security.

Research into policies and international politics is never neutral, as is no social science research (Smith 2002). By necessity, the analysis in this book has been only one way of telling the story, and it is a conscious research decision to choose the conceptual, theoretical and methodological angle that has guided the investigation. However, it has been shown how the combination of a neo-Gramscian conceptual framework with the literature on international norm dynamics can offer theoretically novel interpretations of key questions about the power of norms in security and foreign policies and the co-existence between 'norm-shapers' and 'norm-takers'. On this basis, six potential areas for further research are identified here.

First, it has been suggested how power shifts in international relations take place gradually. The 'power transition paradigm' largely remains a neo-realist hobby horse, and the relationship between 'established powers' and 'emerging powers' will continue to both inspire research projects and obfuscate policy debates at the same time. Labels become 'sticky', develop a dynamic of their own that guide our assumptions about politics and policies and become 'path dependent'. The 'BRICS' label is a prime example of a classification of a group of rather diverse countries into one conceptual unit that has become a political unit in its own right despite fuzziness over status, scope and policy implications. This research has shown how 'resistance' to established power structures on the part of 'emerging powers' is multifaceted. As the analysis of the Iranian nuclear case has demonstrated, the legal contestation over nuclear rights and obligations under the nuclear non-proliferation treaty gave way to a much deeper-seated malaise in international power constellations. Working with these concepts as worked out in this book will continue to enrich our understanding of fault lines between superpowers (US) and self-proclaimed revolutionary actors (Iran), between the modernising and the modernised world (Patrikarakos 2012: 30), and between those actors that substantially shaped world governance in the last century, and those that will likely be 'brics' of new or modified terms of the game. Pronounced economic interests need not automatically lead to a stronger articulation of political divergences. Analytical projections impact on self-perceptions, and more scholarly research is needed to accurately reflect on the nexus between macroeconomics and political clout in international relations at the interstice of the disciplines of economics, IR, security and Area Studies.

Second, the implementation of the nuclear agreement up until 'Termination Day' as defined in the JCPOA will take many years, and the political nature of the Iranian nuclear conflict will be sure to spark further research on policy coordination between those actors bearing responsibility for a successful closure thereof. Domestic dynamics in the seven states tasked with implementing the JCPOA can adversely affect foreign policy coordination, and uncertainty lingers especially over the future of the political and security establishment in Iran and in the United States. The eventual closure of the Iran dossier in a UN Security Council resolution as well as the closure of the case by the IAEA pursuant to its documentation of the exclusively peaceful nature of Iran's nuclear programme is conditioned on robust verification processes by the IAEA and the JCPOA parties as well as compliance with the agreement by all JCPOA parties. The evolution of the complex and fragile arrangements made in the summer of 2015 will need to be closely monitored by researchers and Iran watchers. This also holds true for the implications of the lifting of both multilateral and unilateral Iran sanctions that began in January 2016. The latter aspect has the power to disrupt policy coordination within the P5+1, where disgruntlement and explicit criticism about continued sanctions regimes are becoming noticeable even after 'Implementation Day' of the JCPOA. Likewise, the role of such mechanisms created by the JCPOA such as the Joint Commission and the procurement channel set up to monitor Iranian nuclear purchases needs to be assessed in light of diverging policy preferences examined in this book.

Third, relations between Russia and the West continue to bedevil international politics, and continue to be a topical subject in need of critical and timely research. The 'Ukraine crisis' has political fallout effects that affect cooperation with Russia in the common neighbourhood between the EU and Russia, NATO–Russia relations, and coordination of policies between Russia and the West in containing conflicts elsewhere. The case of the nuclear talks has illustrated a surprisingly 'compartmentalised' foreign policy, where frosty relations in most issue areas did not rule out constructive policy coordination on the Iranian nuclear file. The conceptual tools and various foreign policy motivations presented here can contribute to a better understanding of the complexity of Russian foreign policy, and can be applied to analyse relations between the West and Russia in other issue areas not studied in depth in this book.

Fourth, the analysis of Turkish foreign policy has provided a substantive discussion of the foreign policy of a country that will most likely continue to spark scholarly debates and discussions. Turkey's foreign policy has been shown to be at a crossroads between partially competing regional security conceptions. Turkey watchers will continue to make sense of Turkey's derailed EU integration project, increasingly authoritarian domestic tendencies, Turkish–Kurdish relations, the uneasy Turkish–American relationship, and Turkey's ambiguous stance on the advance of transnational Islamist fundamentalism in the Middle East. The discussion of Turkish Iran policies has underlined how Turkey's position towards Iran's nuclear programme is inseparable from the wider context of Turkish regional policies and their implications for global security governance, and it can be a useful stepping stone in understanding Turkish interests on other foreign policy fronts. The Turkish case study has also shown how an actor that is not formally part of the UN-mandated negotiation format can have an impact on the direction of diplomacy. Similar cases of diplomatic mediation as well as the potential for it can be studied in a comparable framework. Such research projects could build on the analysis provided here on Turkish mediation and expand the existing scholarly literature on mediation and negotiation. Cases need not be limited to Iran diplomacy. Insights into the mechanisms, preconditions and perceptions of third country mediators can shed light onto the structural dimensions of international diplomacy.

Fifth, the analysis of Chinese foreign policy towards Iran has touched upon myriad foreign policy motivations of China and has embedded the discussion of China's position towards the Iranian nuclear programme into the wider historical, regional, and global context of Chinese diplomacy. As China's political weight is growing commensurate with its economic influence in important world regions, China's leadership is adopting its foreign policy planning. Scholarly research is needed to accompany and study this development. The evidence and conclusions presented here can contribute to the vast body of scholarly literature on Chinese foreign policy and provide nuance to the sometimes overheated discussion on 'China's rise'. As this book has argued throughout, the focus on structural forces of World Order can help zoom out from a too narrow actor-centric angle that characterises many studies of contemporary China.

Finally, the previous chapter has referred to changes in US norm discourses that embedded US Iran policy in a more multilateral approach, and indicated how international norm dynamics have the potential to change policy preferences. More research is needed to investigate the effect that dissenting voices can have on US positions, and the extent to which 'resistance to hegemony' can generate 'feedback effects' on hegemonic structures. Rather than conceptualising a one-way 'socialisation' of 'emerging powers' into existing governance structures, the case of Iran's nuclear file can serve as a basis for new research projects to investigate the effects of processes of 'two-way socialisations', whereby resistance to dominant power structures can eventually usher in a shift in these structures themselves in what Gramsci would call *trasformiso*. Such research would nuance the static extremes of 'US decline' versus the rise of something else and would also have to shed light on the nexus between foreign policy learning, internal determinants of foreign policy, and domestic audience costs.

The Iranian nuclear crisis is a proxy for numerous fundamental debates about international relations. The list of possible further research areas presented here is therefore not exhaustive, and should be thought of rather as a sketch board from which to venture out. This study has hopefully managed to demonstrate how the 'Iran question' remains indeed a burning question of our age.

References

Adler-Nissen, Rebecca. 2014. Stigma Management in International Relations: Transgressive Identities, Norms and Order in International Society. *International Organisation* vol. 68, no. 1: 143–176.

Bilgin, Pinar. 2013. Critical Theory. In: Williams, Paul D. (ed.), *Security Studies. An Introduction*. Abingdon and New York: Routledge, pp. 93–106.

Bob, Clifford. 2012. *The Global Right Wing and the Clash of World Politics. Cambridge Studies in Contentious Politics*. New York: Cambridge University Press.

Booth, Ken. 2005. Critical Explorations. In: Booth, Ken (ed.), *Critical Security Studies and World Politics*. Boulder, CO: Lynne Rienner, pp. 1–25.

Clark, Ian. 2005. *Legitimacy in International Society*. Oxford: Oxford University Press.

Clark, Ian. 2011. China and the United States: A Succession of Hegemonies? *International Affairs* vol. 87, no. 1: 13–28.

Cox, Robert W. 1996. *Approaches to World Order*. Cambridge: Cambridge University Press.

ElBaradei, Mohamed. 2012. *The Age of Deception. Nuclear Diplomacy in Treacherous Times*. New York: Bloomsburg.

Epstein, Charlotte. 2012. Stop Telling Us How to Behave: Socialization or Infantilization? *International Studies Perspectives* vol. 13, no. 2: 135–145.

Ferdinand, Peter. 2013. The Positions of Russia and China at the UN Security Council in the Light of Recent Crises. *Briefing Paper*, March 2013. Directorate-General for External Policies of the European Union.

Gramsci, Antonio. 1971. *Selections from the Prison Notebooks*, ed. and trans. Hoare, Q. London: Lawrence & Wishart.

IAEA. 2016. Verification and Monitoring in the Islamic Republic of Iran in Light of United Nations Security Council Resolution 2231 (2015). *IAEA Board Report*, 16 January 2016. Available at www.iaea.org/sites/default/files/gov-2015-53.pdf (accessed 6 September 2016).

International Institute for Strategic Studies. 2016. What the Iran Nuclear Accord Means for Sanctions Today and Tomorrow. *Workshop Report*, 14 January 2016. Available at www.iiss.org/en/events/events/archive/2016-a3c2/january-6318/npdp-london-workshop-545e (accessed 21 April 2016).

Kersbergen, Kees van and Verbeek, Bertjan. 2007. The Politics of International Norms. *European Journal of International Relations* vol. 13, no. 2: 217–38.

Patrikarakos, David. 2012. *Nuclear Iran. The Birth of an Atomic State*. London and New York: I.B. Tauris.

Porter, Gareth. 2014. *Manufactured Crisis. The Untold Story of the Iran Nuclear Scare*. Charlottesville: Just World Books.

Rapkin, David P. and Braaten, Dan. 2009. Conceptualising Hegemonic Legitimacy. *Review of International Studies* vol. 35: 113–149.

Reus-Smit, Christian. 2014. Power, Legitimacy, and Order. *The Chinese Journal of International Politics* vol. 7, no. 3: 341–359.

Schwarzmantel, John. 2009. Introduction. Gramsci in His Time and in Ours. In: McNally, Mark and Schwarzmantel, John (eds), *Gramsci and Global Politics: Hegemony and Resistance*. London and New York: Routledge, pp. 1–16.

Smith, Steve. 2002. The United States and the Discipline of International Relations: 'Hegemonic Country, Hegemonic Discipline'. *International Studies Review* vol. 4, no. 2: 67–85.

Taylor, Brendan. 2010. *Sanctions as Grand Strategy*. Adelphi series, International Institute for Strategic Studies. Abingdon: Routledge.

Towns, Ann E. 2012. Norms and Social Hierarchies: Understanding International Policy Diffusion 'from Below'. *International Organization* vol. 66, no. 2: 179–209.

UN. 2015. *Resolution 2231 (2015), Adopted by the Security Council at its 7488th Meeting, on 20 July 2015*. Available at www.un.org/ga/search/view_doc.asp?symbol=S/RES/2231(2015) (accessed 23 September 2016).

Wiener, Antje. 2008. *The Invisible Constitution of Politics: Contested Norms and International Encounters*. Cambridge and New York: Cambridge University Press.

Worth, Owen. 2015. *Rethinking Hegemony*. London: Palgrave Macmillan.

Wunderlich, Carmen. 2013. Theoretical Approaches in Norm Dynamics. In: Müller, Harald and Wunderlich, Carmen (eds), *Norm Dynamics in Multilateral Arms Control: Interests, Conflicts, and Justice*. Athens: University of Georgia Press, pp. 20–47.

Index

Small Systems and Fundamentals of Thermodynamics

Small Systems and Fundamentals of Thermodynamics

Yu. K. Tovbin

CRC Press is an imprint of the
Taylor & Francis Group, an **informa** business

CRC Press
Taylor & Francis Group
6000 Broken Sound Parkway NW, Suite 300
Boca Raton, FL 33487-2742

Printed on acid-free paper
Printed and bound by CPI Group (UK) Ltd, Croydon CR0 4YY

International Standard Book Number-13: 978-1-138-58724-3 (Hardback)

Visit the Taylor & Francis Web site at
http://www.taylorandfrancis.com

and the CRC Press Web site at
http://www.crcpress.com

Contents

Foreword

Recently, various experimental methods of research have been actively developed, which have made it possible to significantly improve the resolution of measurements of the sizes and properties of small systems. For a long time, only particles of a new emerging phase (liquid drops or microcrystals) and gas bubbles in liquids were considered small systems. For them, for the first time, methods of thermodynamic description were involved, and this state of affairs was long considered acceptable until doubts had been accumulated about their correctness.

The physical basis of these doubt is the well-known result of J.W. Gibbs (1902) that as the size of systems decreases, the role of fluctuations increases, and the thermodynamic description becomes insufficient. This general statement to the present prevails in the qualitative interpretation of numerous measurements, and the work of T.L. Hill (1963) on small systems only emphasizes the absence in the world literature of specialized books in this field. Hill's work drew attention to the need to take into account the effects of fluctuations for the description of small systems and the transition to discrete calculus. The simplest examples of such systems in different statistical ensembles were considered and the influence of fluctuations was demonstrated for them.

Recall that almost one hundred and forty years ago, J.W. Gibbs (1878) practically completed the construction of equilibrium thermodynamics, as a science on the most common thermal properties of macroscopic bodies. Thermodynamics studied the thermal form of the motion of matter – the laws of thermal equilibrium and the conversion of heat into other types of energy. In his work, Gibbs extended thermodynamics to macroscopic non-uniform systems. These provisions to date are the main reference point for many areas of knowledge: physics, chemistry, mechanics, biology, geology, etc., including all related disciplines.

Apparently, Gibbs was also the first to understand the need for a deeper understanding of thermodynamics, as an exclusively model-free field of science, which he did when developing a statistical method for studying the properties of macroscopic bodies, based on model atomic–molecular representations from the very beginning.

The foundations of thermodynamics were based on postulates that are based on experimental measurements of macroscopic systems. They were confirmed by age-old observations, so there are no strict limitations on the scope of their application in them. The shift of the interests of researchers to a small-scale region requires the same rigorous thermodynamic analysis of numerous thermophysical measurements in the submicron and nanometer ranges, as well as for macroscopic systems.

The traditional references in the traditional statistical thermodynamics to the limitations of the use of thermodynamic approaches, connected with the discreteness of the description at the molecular level of the properties of substances and phases, are sufficiently general and allow great arbitrariness in their interpretation. Until now, there was a very wide variation in the interpretation of such restrictions in this matter, since the question of the size and time limitations of the use of thermodynamics for small systems was not so acute before. These limitations exist according to the following features:

1. by the size of the region in which fluctuations are important, and in particular, what sizes of regions appear in the equations of thermodynamics (what is the magnitude of the elementary volume dV);

2. by the degree of homogeneity of the volume within the phases, and what is the minimum phase size, or how it differs from molecular associates;

3. the method of taking into account the curvature factor of curved interfaces, including the question of the applicability of the Kelvin equation;

4. The degree of non-equilibrium deviations of the non-equilibrium thermodynamics described by the equations, and how much these deviations should be small so that the equilibrium state can be considered real. In passing, there are questions about the correctness of the application of the thermodynamic approaches in kinetics, in particular, about the correctness of the use of the activity coefficient for the activated complex in it.

The monograph is devoted to an analysis of the above limitations on the use of macroscopic thermodynamic interpretations for small systems. It is based on the attraction of the methods of statistical thermodynamics which Gibbs founded and, in essence, is a comparison of Gibbs thermodynamic results with the results obtained by modern methods of his own statistical thermodynamics. The main attention is paid to the drops of a vapour–liquid system, for the description of which all existing methods of statistical physics (integral equations and stochastic Monte Carlo and molecular dynamics methods) were used.

Strict statistical approaches based on integral equations turned out to be complex for numerical realization in such a highly non-uniform system as the interface between phases with a density difference of up to two or three orders of magnitude. For the same reason, with additional complication due to strong density fluctuations, stochastic methods proved to be ineffective. The only approach that allowed to solve this problem was a discrete–continuum method based on the many-particle distribution functions introduced by N.N. Bogolyubov (1946), using discrete molecular distributions on the molecular size scale, forming cells, and with a continual description on the scale smaller than the size molecules inside the cells. This allowed us to combine known techniques in the lattice gas model, which is widely used in problems of phase transformations, taking into account all internal motions of molecules within cells, and to find a general approach to solving problems for the equilibrium distribution of molecules in strongly non-uniform systems. This method is applicable to a substance in three aggregate states, so it is the only method that provides an equal description of the three phase separation interfaces. It is an alternative method of molecular dynamics and provides three to five orders of magnitude faster calculations than the molecular dynamics method.

The key issue in solving the problem of small systems was the contradiction between the Kelvin equation (1871) and the Yang-Lee condensation theory (1952). The essence of the contradiction is related to the differences between thermodynamic and statistical constructions. Thermodynamics conditionally consists of four stages: 1) initial statements, 2) the onset of thermodynamics (the First and Second Laws of thermodynamics), 3) consequences from the beginnings of thermodynamics for macroscopic systems (these are all the formulas following from the beginnings given in all textbooks), and 4) additional hypotheses of thermodynamics outside macroscopic systems. Statistical physics differs already in the first point: instead of a small number of macroscopic parameters, the analysis proceeds with a complete ensemble of molecules (their coordinates and impulses). Further, the probability theory with its rules for averaging the distributions of the coordinates and momenta of the molecules, and obtaining the mean values, works. As a rule, the results of statistical physics are approximate, depending on the type of potential interactions and internal approximations. In statistical physics, everything is laid in the first stage, the next steps are automatically performed, and when properly operated, one should not interfere in the fourth stage. This makes it possible to verify the consequences of the thermodynamic constructions introduced for small systems in the fourth stage.

In the construction of the Kelvin equation, the relationship between the pressures in neighbouring phases is used from the condition of mechanical equilibrium, as in the mechanics of continuous media, which has nothing to do with the chemical potential (there was no such concept at the time of Laplace and Kelvin), and this leads to so-called metastable drops. These drops are absent in the Yang–Lee theory as equilibrium objects, which illustrates their contradiction. Historically, the Kelvin equation was the first equation for small systems, and it became decisive for the entire development of the direction of science. Many years of accumulated experience with the use of this equation put him in doubt, without waiting for an explanation of this contradiction. Today, the Kelvin equation is almost completely abandoned in the two most important areas of its practical application: the processes of formation of new phases and adsorption porosimetry.

The monograph deals with all the questions connected with this contradiction, as well as answers to the above questions: the concept of the 'minimum phase size', the size range of systems for which thermodynamic equations are applicable, including the notion of an 'elementary volume' dV, the notion of 'passive forces' introduced by Gibbs, and the field of applicability of the concept of 'local equilibrium' in non-equilibrium processes, and a number of other related questions.

The book consists of seven chapters. The first chapter contains examples of experimental studies of small systems and outlines the main provisions of classical thermodynamics to the extent necessary to consider the limitations of its application on the size factor and to determine the conditions for violation of local equilibrium. This refers to the formulation of the basic principles of equilibrium thermodynamics, including the beginning of thermodynamics, the problem of the Kelvin equation is formulated, and the main provisions of non-equilibrium thermodynamics are given, including the principles of constructing phenomenological equations for the transport of molecules and the rates of chemical reactions on the basis of the law of acting masses.

The multiphase systems and the Gibbs phase rule are discussed as the main object. It is the transition from macrosystems to polydisperse systems, porous bodies, and small systems, which required more detailed consideration of size effects and fluctuations unavoidable for small systems. A joint consideration of classical thermodynamics as a common system of views on equilibrium and non-equilibrium processes inevitably leads to the need to analyze the relaxation times of the thermodynamic parameters of the system.

In the second chapter, the principles of molecular description are presented: the task of this chapter is to show the difference between the molecular description of systems and the thermodynamic one. A continual description of the distribution of molecules in space is introduced (the equilibrium distribution of molecules is described by a chain of Bogolyubov–Born–Green–Kirkwood–Yvon (BBGKY) equations) and a discrete–continual distribution. In it, the description of the coordinates of molecules is divided into two scales: discrete with a characteristic size of the order of the size of the molecules (the initial presentation is constructed for the simplest case of symmetric particles), which leads to the isolation of cells, and to the continual one, which describes the distribution of the centre of mass of molecules inside the cell. The difference of this approach from different versions of previous lattice constructions is explained and its direct connection with the BBGKY chain is shown. The foundations for calculating the non-equilibrium function of molecular distributions in a discrete–continual description are introduced, and its special case is the kinetic equations of chemical reactions in dense phases (Master Equation).

The third chapter is devoted to the presentation of the traditional thermodynamic approach to the account of the interface and the molecular approach for plane and curved boundaries. The principal difference between the molecular and thermodynamic formulation of the problem of the phase boundary is shown. The molecular theory leads to the existence of a system of equations for finding the concentration profile of the density between coexisting phases. The solution of this system of equations makes it possible to calculate the surface tension between vapour and liquid and leads to the existence of equilibrium drops (2010), which can not appear in thermodynamics because of its rough description of the properties of the phase boundaries. The molecular theory makes it possible to find the size dependence of the surface tension $\sigma(R)$ on the drop radius R, which makes it possible to determine the conditions for the appearance of a new phase or the minimum phase size R_0. Traditional metastable drops are discussed in Appendix 1.

The fourth chapter quantitatively characterizes small systems and introduces for them a new method of mathematical description on the basis of a discrete calculus. The necessity of using symmetrized difference derivatives of different orders is shown, which makes it possible to control the traditional differential derivatives used in statistical physics in the search for extrema of statistical sums and the calculation of size fluctuations. Examples of the use of discrete calculus for calculating the equilibrium characteristics of small systems are given. The lower limit of the applicability of a thermodynamic description is found without taking into account the influence of the discrete nature of matter and the contribution of fluctuations. Other examples that limit the size of the regions in which thermodynamics are applicable are discussed.

The fifth chapter is devoted to non-equilibrium processes. For them, kinetic equations are introduced on the basis of the local equilibrium condition. The relaxation times allow one to assess the possibility of realizing an equilibrium process under laboratory conditions. As the degree of non-equilibrium increases, the structure of the system of kinetic equations changes, and in addition to changing the thermodynamic parameters of the state, it is necessary to take into account the relaxation times of the pair functions, which sharply increases the dimensionality of the system. This transition makes it possible to find a criterion for the realization of local equilibrium. The possibility of applying equations for highly non-equilibrium processes is discussed: turbulent flows and frozen states of solids at low temperatures. The type of equations included in the general system of the transport equation is indicated in Appendix 2, and the question of calculating the dissipative coefficients and describing the evolution of the ensemble of small systems in a two-level model is also discussed there. The introduction of relaxation times of various properties allows us to discuss the essence of passive Gibbs forces and the so-called non-equilibrium thermodynamic functions. With their help, a non-equilibrium surface tension is introduced, and the phase boundary relaxation is analyzed for different types of boundaries. In conclusion of Chapter 5 we give examples of the effect of fluctuations in the kinetics of elementary stages of the ideal and give an approach to allowance for fluctuations in non-ideal reaction systems.

In the sixth chapter, the foundations of chemical kinetics in non-ideal reaction systems are described. It is shown that these equations provide conditions for a self-consistent description of the equilibrium and kinetics of chemical reactions and transport processes for any degree of non-equilibrium. This section plays an important role in describing the continuous transition from any non-equilibrium states of systems to their equilibrium for the equations considered in Chapter 5, which allows us to relate the material on the non-equilibrium thermodynamics of Chapter 1 and the kinetic equations of Chapter 2.

The seventh chapter discusses the analysis of thermodynamic interpretations related to the Kelvin equation: an explanation is given of the contradiction between the Yang–Lee theory and the Kelvin equation. The same principle of precedence of the condition of mechanical equilibrium over the chemical equilibrium, as in the Kelvin equation, was used in Gibbs' work and in later works of the twentieth century to refine the Gibbs constructions. Today, there is relatively little that is known from the size properties of small drops, therefore, in Chapter 7, the characteristics of metastable and equilibrium drops are clarified. The discussion of the contradiction between the Yang-Lee theory and the Kelvin equation is concludes and a direct calculation of the relaxation time of the transition from metastable drops to equilibrium ones is carried out. Also, the seventh chapter questions discusses the accuracy of the molecular theories used when considering the thermodynamic characteristics, metastable states of solid-phase systems, and the use of the activity coefficients in equilibrium (Appendix 3) and in kinetics.

The conclusion gives a brief summary of the limitations on the application of thermodynamics and the current state of the molecular-kinetic theory.

The fundamentals of classical and statistical thermodynamics and their achievements for macrosystems are well known in the literature on numerous monographs. The questions of their justification are not discussed. Also, many other more complex systems (mixtures, Coulomb potentials, polymers, etc.) are not discussed. The aim of the work was not to review the numerous modifications of the discussed equations of Kelvin and Gibbs in applications for small systems (this is impossible in the past more than 140 years).

The chapters 1 and 2 present only the material on classical and statistical thermodynamics, which is necessary for the analysis of small systems. The chapters 3, 4, 5 and 7 give examples of calculations of the characteristics of small systems, and how fluctuations affect them, which, as a rule, are poorly discussed in the literature.

The monograph has an interdisciplinary character – the fundamentals of thermodynamics are expounded in many natural sciences: physics, chemistry, mechanics, biology, geology, and materials science. The material is described for specialists in the field of physical chemistry, statistical thermodynamics, physics of surface phenomena and phase transitions, kinetic theory in condensed phases and hydrodynamics, mechanics of solids and technologists engaged in the creation of new materials, as well as for students and graduate students of relevant specialties.

Symbols and Abbreviations

a_f^i –the local Henry constant of the molecule i for the site with the number f (or a_q^i of the site of type q); for a uniform surface a_i is the Henry constant of the molecule i, and a is for one substance;

a_f^{i0} –the pre-exponent of the local Henry constant of the molecule i for the site f.

A similar simplification of the notation for a transition from a non-uniform system to a uniform one and from a mixture to a single substance is retained for all characteristics.

c_i –the concentration of molecules i in the number of molecules N_i per unit volume;

$C_v(f)$ –the specific heat per particle in cell f;

$E(f)$ –the internal energy of molecules in cell f;

$E_{fg}^{iV}(\rho)$ –the activation energy of hopping of a molecule of sort i from site f by distance ρ to a free site g;

$E_{fgh}^{iVj}(\omega_r)$ –the energy contribution of the molecule j at the site h located at a distance of the r-th coordination sphere from the 'central' pair of sites fg with orientation ω_r, to the non-ideal hopping function of molecule i from site f to site g;

$d_{qp}(r)$ –the conditional probability of finding a site of type p at a distance of r-th c.s. from a site of type q;

D –the diffusion coefficient;

$D_{1,2}$ –coefficient of mutual diffusion in a binary mixture

D_{ij} –the diffusion coefficient of component i in a multicomponent mixture under the influence of the gradient of component j;

D_{fg}^* –the self-diffusion coefficient of a pure substance;

f –the site number of the non-uniform system;

F –free energy;

f_q –probability of finding a site of type q in a non-uniform system;

$f_{qp}(r)$ –the probability of finding a pair of sites of the type q and p at a distance r in the non-uniform system;

F_i^0 –the partition function of the molecule i in the gas;
F_q^i –the partition function of a particle i located at a site of type q;
F_q^{i*} –the partition function of the particle i at the site of type q in the transition state;
F_q –the fraction of sites of type q; for the monolayer q of the drop, the fraction of sites of this monolayer;
$F_i(f)$ –the component i of the external force created by the wall potential in cell f;
G –the Gibbs potential;
h –the Planck constant.
h_f –the cluster Hamiltonian;
H –the width of the neck-like pore, the number of monolayers;
i –the sort of the particle in the cell, including the vacancy;
J_i –the thermodynamic flows (Section 9) or the diffusion flux of molecules of sort i;
J_i –the partition function of a molecule of sort i; (Section 30)
k_B –the Boltzmann constant;
$K_{fg}^{iV}(\chi)$ –the rate constant for the hopping of the molecule i from the cell f to the free cell g at a distance χ;
l_i –the mean free path of component i;
m_i –the mass of the molecule i;
$m(\omega_r)$ –the set of neighbouring sites with fixed values of r and ω_r;
M –the number of cells in the system;
N_{den} –the number of densely packed particles in the same volume,
N_q^i –the number of molecules of sort i at sites of type q,
N_{qp}^{ij} –the number of pairs of molecules of sort ij at sites of type qp,
P –pressure;
P_i –the partial pressure of component i;
$P(f)$ –the pressure in cell f;
q –type of site (cells);
Q_N –the partition function of the system states;
Q_{fi} –the binding energy of a molecule of sort i, located at the site f, with an adsorbent;
Q_q^i –the binding energy of a molecule i located in a site of the type (layer) q with the walls of the pore;
$Q_{q,f}$ –the binding energy of a molecule located at a site with number f of a section of type q in a complex porous system;
r –the distance between the particles (in units of λ);

R –is the radius of the drop;
R_{lat} –the radius of the intermolecular interaction potential (in units of λ);
$P(\{p,q\},t)$ –the total distribution function N of the molecules p_1, r_1,..., p_N, r_N, where p_i is the impulse and r_i is the coordinate of the centre of the mass of the molecule i, $1 \leq i \leq N$, at time t;
S –entropy;
s –the number of components of the mixture;
$S_{fg}^{i}(r)$ –factor of the non-ideality function;
$S(i \to k)$ –the vacancy region of rotation of the molecule from orientation i to orientation k,
t –the number of different types of sites in the system;
T –temperature;
T_i^c –critical temperature of component i;
T_f –local temperature,
$t_{fg}^{ij}(r)$ –the conditional probability of finding the molecule j at the site g at a distance r from the molecule i at the site f;
u –flow velocity, m/s;
$\mathbf{u}(f)$ –the velocity vector in cell f with components $u_i(f)$ in the direction $i = x, y, z$;
U –internal energy;
$U(q)$ –the interaction potential of the molecule of the layer q with the wall of the pore;
$U_{fg}^{i}(\chi)$ –the probability that a molecule i jumps from cell f to a free cell g by distance χ;
$U_{\xi fg(j)}^{(i)V}(\chi)$ –the probability of a molecule jumping after collision with a molecule j at a site ξ by a distance χ;
$U(z)$ –the interaction potential of the molecule with the z coordinate with the pore wall;
V –the volume of the system, or the symbol of the vacancy;
V_{fg}^{iV} –the concentration component of the hopping rate of the molecule i from cell f to the free cell g;
$V_{\xi fg}^{(j)iV}(\chi)$ –the concentration component of the hopping rate of molecule i from cell f to the free cell g after collision with molecule j at site ξ by distance χ;
v_0 –the cell volume;
W_α –the probability of realizing the elementary stage α;
$W_i(\rho)$ –the probability of hopping of a molecule of sort i by distance ρ;
w_i –the average thermal velocity of molecules i in the gas phase;
$w_{i(j)}$ –the average relative velocity of the molecule i after its

collision with the molecule j;

w^i_{fg} –the average thermal velocity of the molecules i between the sites f and g;

x^i_f –the mole fraction of component i in site f;

z –the number of nearest neighbours of the lattice structure;

z_q –the number of nearest neighboirs in the q layer;

$z_{fg}(r)$ –the number of neighbouring sites in the layer g at a distance r from the site in the layer f.

$z^*_{fg}(\chi)$ –the number of possible jumps between sites f and g by distance χ;

Greek characters

α –the symbol of the dense phase of the stratified system (Chapter 3);

α –the dimensionless parameter $\varepsilon^*/\varepsilon$; or the stage number;

β –the symbol of the rarefied phase of the stratified system (Chapter 3);

β –inverse value of thermal energy $(k_B T)^{-1}$;

γ^i_f –a variable describing the occupied state i of the cell (site) with the number f;

δ –the width of the monolayer;

$\delta\varepsilon^{ij}_{fh}(r)$ –the difference of the quantities $(\varepsilon^{*\,ij}_{fh}(r) - \varepsilon^{ij}_{fh}(r))$;

ε_i –parameter of interaction potential of adsorbate of sort i - adsorbent;

$\varepsilon^{ij}_{fg}(r)$ –the interaction parameter of the molecule i at the site f with the neighbouring molecule j at the site g at a distance r;

$\varepsilon^{*\,ij}_{fh}(r)$ –the interaction parameter of the activated complex of migration of molecule i from site f to free site g with neighbouring molecule j in the ground state at site h at distance r;

η_{fg} –the local shear viscosity of the mixture flow between the sites f and g;

η^q_f –the function of the correspondence between the number of the site f and its type q;

θ –the total degree of filling of the volume of the system, the dimensionless quantity, $0 \leq \theta \leq 1$;

θ^i_q –the partial degree of filling of sites of type q by molecules of sort i;

$\theta^{ij}_{fg}(r)$ –the probability of finding particles i at site f and j at site g at a distance r;

$\theta^{iV}_{fg}(\chi)$ –the probability of realizing a free trajectory from cell f to cell g of length χ.

$\theta(p_1, r_1, \ldots, p_N, r_N)$ –the total distribution function of N molecules; p_i is the momentum and r_i is the coordinate of the centre of the mass of the molecule i, $1 \leq i \leq N$;

$\kappa(f)$ –the coefficient of thermal conductivity in cell f;

$\kappa_{qp}(\omega_r|\rho)$ –the number of sites with orientation $\omega_r(\rho)$ in the r-th c.s. of the central pair of particles at the sites of type q and p at a distance ρ;

λ –linear cell size;

$\lambda_{qp\xi}(\omega_r|\rho)$ –the number of sites of type ξ with orientation ω_r, located in r-th c.s. around central sites of type qp at distance ρ;

Λ_f^i –the non-ideality function of the system for molecules i at site f;

$\Lambda_{fg}^i(\chi)$ –the non-ideality function of the system for hopping of a molecule i from the site f to a free site g by a distance χ;

μ_i –the chemical potential of component i;

μ_{ij} –the reduced mass of the colliding molecules i and j;

v_f^i –a one-particle contribution to the total energy of the system by a particle of type i located at a site numbered f; v_i – for a uniform lattice structure, v - for one substance;

$\xi(f)$ –the bulk viscosity coefficient in cell f;

π –spreading pressure analog of the equation of state for a lattice system;

$\pi_r(\rho)$ –the number of orientations of the sites in the r-th c.s. around the central pair of sites at a distance ρ;

ρ –the average distance between molecules (in units of λ);

σ_{ij} –distance of the closest approach of two components i and j of the mixture;

$\phi(r) = Ar^n - Br^m$ –Mie pair potential (n, m) where the constants A, B, n, m are positive,

χ –the molecule hopping length (in units of λ);

$\omega(f)$ –the enthalpy (thermal function) of the mass unit related to a given cell f.

$\omega_r(\rho)$ –the orientation of the site h located in the r-th coordination sphere of the 'central' pair of sites fg at distance ρ; $1 \leq \omega_r(\rho) \leq \pi_r(\rho)$.

Brackets

Curly: $\{\eta_f^q\}$ denote a complete list of values, here, for example, for η_f^q, $1 \leq f \leq M$, $1 \leq q \leq t$;

Square: $[m_{qp}]$ fixing the numbers m_{qp} of the sites of type p in the coordination sphere of the central site of type q, $1 \le p \le t$.

Indexes

For a uniform system, the subscripts i, j, k, l are the sorts of particles, including the vacancy V.

For a non-uniform system, subscripts: site numbers, if the indices f, g, h, or the types of sites a non-uniform lattice, if the indices q, p, ξ; superscripts: i, j, k – sorts of molecules or V (vacancy).

Abbreviations used

AC –activated complex
BBGKY –Bogolyubov–Born–Green–Kirkwood–Yvon
c.s. –coordination sphere
DF –distribution function
QCA –quasi-chemical approximation
MD –molecular dynamics
LGM –lattice gas model
TARR –theory of absolute reaction rates
KE –Kelvin's equation

1

Background

1. Small systems

In thermodynamics small systems are systems that have a large surface contribution in comparison with their volume contribution to all thermophysical functions. They have sizes (radii), which can vary over a wide range from nanometers to submicron values. A traditional example of small bodies is liquid drops in a supersaturated vapour and bubbles in liquid phases [1,2].

Today in the literature there is a great deal of activity in discussing various aspects of the study of small bodies that have appeared recently [3–29], especially in connection with the development of measurements in the range of sizes from 1 to 100 nm. This interest is caused by the transition of experimental equipment to a new level of spatial resolution in the last 15–20 years. This affects both traditional fields of knowledge, studying colloidal systems and non-uniform catalysis, and many other areas of physics, chemistry and biology. In the specified size range, many physical and chemical properties change, which opens up new approaches to the study of substances. There was a separation of such concepts as nanoclusters, nanostructures, and related phenomena into a separate area of physico-chemistry. This happened mainly as a result of significant progress in obtaining and researching nanoobjects, the emergence of new nanomaterials, nanotechnologies and nanodevices. New giant nanoclusters of a number of metals, fullerenes and carbon nanotubes, many nanostructures based on them and on the basis of supramolecular hybrid organic and inorganic polymers, etc. have been synthesized. There has been remarkable progress in the methods of observing and studying the properties of nanoclusters and nanostructures associated with the development of tunnel and scanning microscopy, X-ray and optical methods using synchrotron radiation, optical laser spectroscopy, radio frequency spectroscopy, Mössbauer spectroscopy, etc. [3-29].

Small systems also include modifications of 'micro-reactor' systems as their number in a unit volume increases: aerosols, aerogels, porous and non-porous friable bodies of various structures, etc. The classification of nanoclusters and nanostructures is based on the methods of their preparation [17]. This determines their distinction into isolated nanoclusters and nanoclusters, combined in a nanostructure with weak or strong intercluster interactions or cluster–matrix interaction. The group of isolated and weakly interacting nanoclusters includes: molecular clusters, gas ligandless clusters (clusters of alkali metals, aluminum and mercury, clusters of transition metals, carbon clusters and fullerenes, van der Waals clusters), colloidal clusters. Nanoclusters and nanostructures include solid-state nanoclusters and nanostructures, matrix nanoclusters and supramolecular nanostructures, cluster crystals and fullerites, compacted nanosystems and nanocomposites, nanofilms and nanotubes.

The properties of small systems are studied by a thermodynamic approach, with the help of which one attempts to determine the patterns of their formation, growth, properties and their changes in the process of phase transitions. The generality of the thermodynamic approach allows us to consider all the listed systems from a unified position. However, the possibility of using macroscopic thermodynamics to describe the properties of small systems is limited by the very phenomenological nature of the science of the most common thermal properties of macroscopic bodies. Thermodynamics studies the thermal form of the motion of matter – the laws of thermal equilibrium and the conversion of heat into other types of energy. The main content of thermodynamics is the consideration of the laws of thermal motion in systems in thermal equilibrium, when they lack the macroscopic displacements of one part relative to the other, and also the regularities in the transition of systems to the equilibrium state. Thermodynamics considers macroscopic phenomena caused by the combined action of a huge number of continuously moving molecules or other particles that make up the surrounding bodies, without any models for their movement. Thermal motion is characterized by the temperature of the body, which is a measure of the intensity of thermal motion.

The fundamentals of thermodynamics were built on the postulates resulting from experimental measurements of macroscopic systems [30–33]. Speaking of limitations in the use of thermodynamics, we speak about the degree of correctness of application of its approaches for small systems that are not macroscopic objects, or are incorrectly considered if there is a macroscopic ensemble of small bodies. Reducing the size of small bodies increases the role of their surface. Let us recall what new factors appear and influence the properties of systems near the interfaces and on the surfaces themselves in comparison with their bulk properties. These factors increase the heterogeneity of the local properties of small systems at the molecular level.

Near-surface area. The presence of a surface requires the separation of the near-surface region from each side of the interface plane, as well as the specificity of these phases [30, 34–38]. This requires a self-consistent description of both the contacting phases themselves and the nature of the distribution of the components of these phases along the normal to the boundary. The components of volumetric phases can have a wide spectrum of mutual distributions for the given phase structures. In more complex cases, various combinations are possible between the phase transitions of stratification and ordering. For example, a system can go into a disordered or ordered phase or into two different ordered phases. The type of phase transitions and the regions of their realization are determined by the concentrations of the components, the temperature, and the interparticle interaction potentials. Similar transitions are also carried out in multicomponent solutions. The increase in the number of components increases the number of different combinations of phase transitions [39].

The nature of the distribution of particles along the normal to the surface, as a rule, differs from their distribution in the volume [40–44]. This causes local non-uniformities. In thermodynamics there is a phase separation, and all phases are treated in the same way. Although at the microlevel there are qualitative internal differences between the liquid and solid phases. Below we discuss the boundary of a solid body, for which the greatest variety of the surface properties is characteristic in comparison with liquid phases.

In solids, near-surface non-uniformities are associated with the absence or change in the symmetry of the phase in the other half-plane, which disrupts the periodicity of the crystal structure along the normal to the surface and leads to a difference in the forces acting on the atom from the surrounding other atoms of the alloy on the surface and in the volume (the atom in volume has a larger number of neighbours than the atom on the surface). The nature of the distribution of solid-state atoms in the near-surface region has a determining effect on all surface processes: the surface composition determines the properties of the surface on which adsorption of gas-phase molecules occurs and surface reactions take place, and the dissolution process of adsorbed particles and the reorganization of the region itself depend on the distribution of atoms in the near-surface region. A much more complicated situation is realized in the case of a 'frozen' strongly non-equilibrium state of the surface. In these conditions, the methods of forming surfaces are extremely important: by breaking down a part of the crystal in a vacuum, by spraying on a substrate, by precipitation from a solution, by sintering a polycrystalline powder, etc. [45, 46], since the surface preparation technique and subsequent sample processing, preceding the experiment, first of all exert their influence on the state of the surface and near-surface area.

Adsorption. Surface atoms of a solid body have a certain ability to form bonds with molecules of the gas phase, which leads to an increase in the concentration of these molecules near the surface [47–53]. At the atomic level, the surface of a monoatomic crystal is not uniform. It represents a distributed system of atoms. The different positions of the adsorbed molecule over the adsorbent atoms are energetically non-equivalent. The nature of adsorption can also depend on the state of the adsorbed particle. Most solids have on the surface regions of localization of adsorbed particles with different binding energies. The reasons for this are due to the different local chemical composition of the surface and the different geometric arrangement of solid-state atoms. Usually, both these causes are manifested simultaneously. Such surfaces are called non-uniform. The scale of non-uniformity varies widely from point imperfections and impurities to different macro-regions [54, 55].

The physical and chemical prerequisites for the existence of non-uniform adsorption centres include such well known factors as inappropriate periodicity within the bulk crystalline lattice, leading to distortion of the surface structures: Schottky and Frenkel lattice defects, non-stoichiometry of the solid, surface groups of different nature, dislocations and mechanical disturbances of the crystal during growth or under external loads, surface roughness. In addition to these factors, new types of adsorption centres form on the surface which are associated with the method of forming the surface (e.g., the angle at which cleavage takes place in the crystal) and the availability of new types of bonds between atoms of the surface and the adsorbed molecules [56].

In general, the non-uniformity of the flat surface and the crystal surface region may be associated with impaired regularity of the distribution of the surface atoms of the solid (structural non-uniformity) and the difference in the nature of surface atoms (chemical non-uniformity). Any non-uniformity leads to the fact that in the vicinity of each defect there are changes in the properties of the surrounding atoms and this leads to a modification of the properties of neighbouring nodes of the surface structure, which increases the number of types of surface nodes.

An example of chemical non-uniformity is the surface of the A_xB_{1-x} (for example, Pd–Ag), which has unlimited solubility. The structurally uniform surface of the face, let it be (100), of this alloy is non-uniform due to the different adsorption capacities of atoms A and B. The proportion of each component on the surface can vary from zero to unity, depending on the temperature and concentration of particles in the volume. As a rule, chemical and structural non-uniformities are realized simultaneously.

If the binding energy of the adsorbed particle with the surface is commensurable with the binding energy between solid-state atoms, the adsorption of molecules can cause significant changes in the state of the surface. To a large extent, this is due to the fact that the adsorbed particles compensate for the absence of neighbours in surface atoms. As a result, the state of the entire near-surface region of a solid can change. This, in turn, causes a change in the number of different adsorption centres of the surface.

In general, considerable experimental material on the effect of the gas phase on the surface state of adsorbents and catalysts has been accumulated in the literature. This fact is specially used for preliminary creation of the necessary surface composition of catalysts. At the same time, the quantitative characteristics of these changes are mainly present only for binary alloys that are in contact with the H_2, CO, O_2 molecules [57–61]. Studies of recent years show that the effect of adsorbed particles on the state of the surface of a solid body is apparently more significant than is commonly believed now. Structural transformations in surface layers can play a great role (for example, the reorganization of the surface layer of platinum upon adsorption of CO and O_2 molecules [62, 63]) and the processes of the formation of new phases in them, which are analogous to three-dimensional topochemical processes [64-66].

Absorption. Many solids absorb significant amounts of gas phase molecules (H_2, O_2, N_2, etc.) [67, 68]. Dissolved atoms occupy interstices of crystal lattices, forming interstitial alloys [69, 70]. They exert a strong influence on the state of the volume phase of a solid body, changing the short-range and long-range orders in alloys, and for large volume saturations with dissolved atoms the volume phase and the transitions to multiphase systems are possible. Similar changes are possible in the near-surface region of the solid, which change the surface states and affect surface processes.

In the presence of chemical transformations, the state of the system changes due to the migration of particles and their participation in the reactions. The relationship between reaction rates and particle mobilities can vary within very wide limits: cases where the chemical transformation stage (the traditional field of chemical kinetics [71,72]) is limited, and when particle migration (diffusion-controlled reactions [73,74]) is limited, are possible. As a whole, this leads to different regimes of processes on the surface and in the near-surface region and determines the variety of models used to describe them.

Porous bodies. Porous bodies are an example of phase inversion, when small volumes of systems relate to internal regions of solids in which there is no matter. These regions can be filled with gas and liquid molecules. The structure of porous and/or dispersed materials is extremely diverse [75–77], and it significantly affects the equilibrium distributions of molecules, the course of phase transformations and all the dynamic processes occurring inside materials (porous bodies). Here the term 'structure' implies the location and interconnection of the constituent elements of the system in question in space, that is, the term 'structure' implies a set of clearly delineated structural elements with limited autonomy. The structural elements themselves may have a crystalline structure (microcrystals) or be amorphous particles characteristic of such artificial materials as regular web weaving and fabrics, highly ordered polymer systems, etc. In general, dispersed materials can consist of structural elements of both types, for example, in porous carbon of different origin [75].

The internal processes of distribution of the mobile phase are associated with the processes of transport of adsorbed molecules (adsorbate) inside the porous structure, the processes of redistribution of molecules between different sections of the free pore volume, and the processes of phase separation (condensation) of molecules, in the case of the predominance of cooperative behaviour of molecules and the instability of uniform distribution [78, 79]. In many of these systems, it is possible to use external fields (mechanical deformation, acoustic waves, electric and magnetic fields, etc.) that cause various local reconstructions of the matrices and phase states of the systems.

Similar types of non-uniformities can be isolated in various colloidal, including micellar systems, or in other systems of 'soft materials', but the number of non-uniformities on their surfaces is much smaller. Different types of non-uniform systems can be sized both macroscopically and small. The above enumeration gives an idea of the complications in describing the surface properties in comparison with the bulk phase, which inevitably arise when considering small systems. For macroscopic non-uniform systems, thermodynamic approaches are possible. The thermodynamic description negates the differences in the properties of the systems under consideration, which can be quite strongly non-uniform in their local properties. Introducing for them the concepts of surface and/or linear tension, thermodynamics reflects the gross effects caused by the presence of a surface, but it does not allow to give a molecular interpretation of these characteristics, and their experimental determination is extremely laborious. Under these conditions, the only method of theoretical description of non-uniform systems is modelling with the help of atomic-molecular models.

2. Thermodynamic parameters of the state

In sections 2–6 we recall some of the basic theses of thermodynamics [30–33,80–93], which are necessary for analyzing its application to small bodies.

Thermodynamics is a scientific discipline that studies the transitions of energy from one form to another, from one part of the system to another, the energy effects accompanying various physical and chemical processes, their dependence on the conditions of the processes, and the possibility, direction and limits of spontaneous flow of processes in the given conditions. In other words, thermodynamics studies the laws of thermal motion in equilibrium systems and the transition of systems to equilibrium (equilibrium thermodynamics or simply thermodynamics), and also the same regularities in non-equilibrium systems (thermodynamics of irreversible processes or non-equilibrium thermodynamics).

Thermodynamic systems. We will call any material object (and even its parts) a macroscopic system or a definite set of material objects, hypothetically isolated from the environment, consisting of a large number of particles. Such systems can consist of a large number of material particles (for example, molecules, atoms, ions, etc.), or fields, for example, an electromagnetic field. The sizes of macroscopic systems are always much larger than the sizes of atoms and molecules. In thermodynamic systems we are dealing with dynamic systems possessing an extremely large number of degrees of freedom (systems with a small number of degrees of freedom are not considered by thermodynamics).

If a part of the total system is studied, then the rest will be called the surroundings, environment or thermostat, which imposes some conditions on the system under study (for example, the conditions of constancy of temperature, pressure, chemical potential, etc.).

Macroscopic parameters. The totality of all the physical and chemical properties of the system characterizes its state. This term refers to the entire system as a whole in the case of an equilibrium state, or to its locally equilibrium state in a certain region in the case of its non-equilibrium state. The considered state is uniquely determined by a set of independent macroscopic parameters that must be chosen so that they are necessary and sufficient to determine this state. Then the remaining quantities characterizing the state are functions of these variables.

The macroscopic parameters can conveniently be divided into two classes: external and internal. The values determined by the position of external bodies outside our system are called the external parameters a_i, (i = 1, 2, ...), for example, the volume of the system (since their values are determined by the location of external bodies), the intensity of the force field (as it depends on the positions of the field sources – charges and currents not included in our system), etc. Consequently, the external parameters are functions of the coordinates of external bodies.

The values determined by the cumulative motion and distribution in the space of particles entering the system are called internal parameters b_j (j = 1, 2, ...), for example density, pressure, energy, polarization, magnetization, etc. (since their values depend on the motion and position of the particles of the system and the charges entering into them). The spatial arrangement of the particles – atoms and molecules entering the system can depend on the location of external bodies, so the internal parameters are determined by the position and motion of these particles and by the values of the external parameters.

However, depending on the conditions in which the system is located, the same value can be either an external or an internal parameter. For example, for a fixed position of the walls of a vessel, the volume V is an external parameter, and the pressure P is an internal parameter, since it depends on the coordinates and momenta of the particles of the system. If, however, the system is in a vessel with a movable piston at constant pressure, the pressure P will be an external parameter, and the volume V is an internal parameter, since it depends on the position and motion of the particles. Therefore, in general, the difference between the external and internal parameters depends on where the boundary between the system and external bodies is drawn. This circumstance is especially important when considering mechanical contacts.

A set of independent macroscopic parameters that completely determine the state of the system at a given time and are independent of the history of the system are termed *state parameters*. The number of independent parameters describing the thermally equilibrium state is determined empirically [88]. As the state parameters, one can use any of the quantities that serve to characterize the state of the thermodynamic system (or all the physical and chemical properties of the system): temperature, pressure, volume, internal energy, entropy, concentration, polarization, magnetization, etc. The term equilibrium state means that all independent parameters of the system do not change over time (this excludes the possibility of realizing stationary flows due to the action of some external sources). Such a state of the system is called the state of thermodynamic equilibrium, and accordingly, the parameters of such a state are called the *thermodynamic* parameters.

State functions. The physical quantities that have a definite value for each thermally equilibrium state of the system are also called thermodynamic quantities or state functions. These include, for example, temperature, pressure, internal energy, enthalpy and entropy.

The internal parameters of the system are divided into intensive and extensive ones.

Intensive parameters are those parameters whose values do not depend on the mass or size of the system when it is divided into parts that do not violate the equilibrium state. If the system in a thermally equilibrium state is divided into parts by means of impermeable partitions, then each part will remain in an equilibrium state. This means that the equilibrium state of a uniform system is its intrinsic property and is determined by the thermodynamic variables that do not depend on the size of the system. Examples of intensive parameters are all the molar and specific properties, temperature, pressure, chemical potential, etc. The values of these properties are not additive.

The *extensive* parameters are those parameters whose values are proportional to the mass or size of the system when it is divided into parts that does not violate the equilibrium state, i.e. the values of the extensive parameters have the additivity property. The extensive parameters include thermodynamic potentials, entropy, volume, etc. The extensive parameters characterize the system as a whole, while intensive ones can take definite values at each point of the system. A system whose energy depends non-linearly on the number of particles is not thermodynamic, and its study by thermodynamic methods is, generally speaking, approximate or even unjustified [89].

Uniform and non-uniform systems. The whole set of thermodynamic systems is divided into two classes – uniform and non-uniform.

Uniform are such systems within which the properties change continuously when passing from one place to another within a certain region. With the uniform systems, the concept of a phase is defined, which is defined as a uniform part of the system that is uniform in composition and physical state, separated from other parts by the interface on which any properties (and the parameters corresponding to them) change discontinuously. Within the phase, the uniform system has the same physical properties in any local, arbitrarily chosen parts, equal in volume. Examples of such systems are mixtures of various gases and solutions, both liquid and solid. In these systems, reactions can occur between the constituents of the mixture, the dissociation of gas or solute, solvation, polymerization, etc. In equilibrium in such systems the reactions cease in terms of the value of the *macroscopic* parameters.

It is necessary to distinguish between aggregate states and phases. While there are only four aggregate states – solid, liquid, gaseous and plasma, the number of phases is unlimited; even the same chemically pure substance in the solid aggregate state can have several phases (rhombic and monoclinic sulphur, gray and white tin, etc.).

Non-uniform are systems that consist of several physically uniform phases or uniform bodies, so that within the systems there are discontinuities in the continuity of the variation of their properties. These systems are the aggregates or different aggregate states of the same substance (ice–water, water–vapour, etc.), or various crystalline modifications (gray and white tin, etc.), or various products of mutual dissolution (aqueous salt solution – solid salt – vapour), or products of chemical interaction of various substances (a liquid alloy and a solid chemical compound of two metals).

The non-uniform processes include aggregate transformations of individual substances, processes of evaporation, crystallization and stratification of liquid solutions, processes of dissolution and melting of solid solutions, sublimation, etc. A common feature of these processes, which makes them to a certain extent related, is that physically non-uniform substances or solutions of substances that are separated by interfaces take place in these processes. The physically non-uniform systems, delimited by interfaces, are usually called phases, and systems consisting of several phases are non-uniform systems. A feature of the non-uniform systems is the presence of interfaces, through which the interaction between phases occurs. As a result of phase processes, phase masses, their properties and composition change. It is also possible to change the number of phases. Therefore, the doctrine of non-uniform systems can be defined as the science of mutual phase conversions, as a result of which the quantitative ratio and number of phases change, as well as their compositions and properties.

3. Thermodynamic processes

The change of any thermodynamic parameter of the state of the whole system or local area is called a thermodynamic process or briefly, a process. For example, when the volume is changed, the system expands; when the characteristics of the external field change – the process of magnetization or polarization of the system take place, etc. Thermodynamics examines changes in the state of the system, occurring either on their own or under the influence of contacts with other systems.

The processes occurring at a constant temperature are called isothermal, those occurring under constant pressure are isobaric, and those occurring at a constant volume are isochoric. Adiabatic processes are those in which the system does not accept and does not give off heat, although it can be related to the environment by the work received from it and carried out over it.

Quasistatic process. This is the name for ideal processes during which the system and the environment remain in a thermally equilibrium state. Such a process is approximately realized in those cases when the changes occur rather slowly. In the extreme case of very slow changes, both processes occur along the same trajectory in opposite directions. Thus, the quasistatic process is reversible, and the system is at all times in equilibrium states. The most important quasistatic processes are: quasistatic isothermal process (the system is in contact with a thermostat having a constant temperature (thermal reservoir)) and a quasistatic adiabatic process (in which the system has no thermal (and material) contact with the environment, but under the influence of the system on the environment or, conversely, the environment on the system can work be performed).

A process is said to be non-equilibrium or non-static if the change in any parameter a occurs within a time t less than or equal to the relaxation time τ ($t < \tau$), where τ is the relaxation time of the system withdrawn from the equilibrium state and left to itself. Any process of relaxation is a non-equilibrium process.

A *thermodynamic process* is called equilibrium if during the process the system passes only through a continuous series of equilibrium states. In equilibrium thermodynamics, the concepts of the equilibrium process and reversibility coincide: the reversible process is at the same time an equilibrium process, and, conversely, every equilibrium process is reversible. With such a process at any given moment the system is in a state infinitely close to equilibrium, and it is infinitely small to change the conditions so that the process can be reversed, that is, it can flow in the opposite direction. This requirement is general, both for phase transformations and for the realization of chemical transformations. The definition of the *equilibrium* state indicated above is associated with the equilibrium process.

For stable equilibrium it is also characteristic that one can approach it in principle from two opposite directions. Thus, from a molecular point of view, a stable equilibrium is dynamic. The equilibrium is established not because of the absence or termination of the process, but because it flows simultaneously in two opposite directions at the same rate. It is the equality of the rates of the forward and reverse processes that is the cause of the conservation of the system without changing in time (with the external conditions unchanged).

Since the equilibrium process is always a process that is infinitely slow, all practically realizable processes can only approximate the equilibrium ones to some extent. This circumstance imposes certain restrictions on the difference between the concept of 'thermodynamic' characteristics and 'thermophysical characteristics', since the degree of closeness of the states of the system to the intermediate equilibrium states requires control, and, consequently, a criterion for proximity to a given equilibrium. The latter can not be done without analyzing the relaxation stages of approximation to local equilibrium: analyzing the spectrum of the relaxing characteristics (variables) and the characteristic relaxation time for each variable. We recall that in thermodynamics we consider the longest relaxation time during which an equilibrium is established for all the parameters of the given system. In this case, naturally, the question arises of the self-consistency of the description of the relaxation stages and the proper equilibrium state, which is practically never discussed in the framework of the non-equilibrium thermodynamics.

The change in the thermodynamic function in any process depends only on the initial and final states of the system and does not depend on the path of the transition. Any function of the system, the change in which in any process depends on the path of the transition, is *thermophysical*. This connects the current state of the system with the course of its evolution and brings the discussion beyond the framework of classical equilibrium thermodynamics into the region of non-equilibrium thermodynamics. Processes that can not be reversed are called irreversible.

4. Basic conditions of equilibrium thermodynamics

Classical thermodynamics is built on initial *statements* that formalize the conditions for its application, and on the *Laws of Thermodynamics* that give a mathematical notation of the law of conservation of energy for any forms of its existence. The initial statements and the Laws of Thermodynamics reflect the accumulated experience of experimental measurements.

1st statement. With time, an isolated macroscopic system comes to a state of thermodynamic equilibrium and can never leave it spontaneously. This final state is called the state of thermal equilibrium. Material particles continue their complex motion, but from a macroscopic point of view the thermally equilibrium state is a simple state, which is determined by several parameters, such as temperature and pressure. This assumption determines the scope of applicability of thermodynamics: *a*) processes that do not end with the onset of equilibrium, as well as all phenomena associated with large spontaneous deviations of the system from the equilibrium state, are excluded; *b*) systems with a small number of particles are excluded (in them the role of fluctuations is great); and *c*) the size of the systems is restricted from the above, in particular, by the volumes created by the experimenters.

2nd statement. It includes the concept of thermal equilibrium of two systems and the law of transitivity of thermal equilibrium (or the Zeroth Law of Thermodynamics).

Thermal equilibrium of two systems. If two isolated systems A and B are brought into contact with each other, then the complete system A + B eventually goes into a state of thermal equilibrium. In this case it is said that the systems A and B are in a state of thermal equilibrium with each other. Each of the systems A and B individually is also in a state of thermal equilibrium. This equilibrium will not be broken if we remove the contact between the systems, and then after a while restore it. Consequently, if the establishment of contact between the two systems A and B, which previously were isolated, does not lead to any changes, then we can assume that these systems are in thermal equilibrium with each other ($A \sim B$).

The Zeroth Law of Thermodynamics (the law of transitivity of thermal equilibrium). If the systems A and B are in thermal equilibrium and the systems B and C are in thermal equilibrium, then the systems A and C are also in thermal equilibrium with each other: $A \sim B,\ B \sim C \rightarrow A \sim C$.

5. The Laws of Thermodynamics

The First Law of Thermodynamics expresses the quantitative aspect of the law of conservation and transformation of energy in application to thermodynamic systems. The total energy of the system is divided into external and internal. Part of the energy, consisting of the energy of the motion of the system as a whole and the potential energy of the system in the field of external forces, is called external energy. The rest of the energy of the system is called internal energy.

In thermodynamics, the motion of the system as a whole and the change in its potential energy under such motion are not considered, therefore, the energy of the system is its internal energy. The energy of the position of the system in the field of external forces is included in its external energy provided that the thermodynamic state of the system does not change when it moves in the force field. If, however, the thermodynamic state changes when the system moves, then a certain part of the potential energy will be part of the internal energy. The internal energy $U = U(a_1,...,a_n; T)$ is an internal parameter and, therefore, depends in equilibrium on the external parameters a_i and temperature T.

When the thermodynamic system interacts with the surroundings, energy is exchanged. In this case, two different ways of transferring energy from the system to external bodies are possible: with changing external parameters of the system and without changing these parameters. The first way to transfer energy, associated with the change of external parameters, is called work, the second method – without changing the external parameters, but with the change of the new thermodynamic parameter (entropy) – heat, and the process of energy transfer – heat exchange.

The energy transferred by the system with changes in its external parameters is called the work W (and not the amount of work), and the energy transferred to the system without changing its external parameters is the amount of heat Q. As can be seen from the definition of heat and work, these two different methods of energy transfer studied in thermodynamics are not equivalent. Indeed, while the expended work W can directly go on to increase any kind of energy (electric, magnetic, elastic, potential energy of the system in the field, etc.), the amount of heat Q directly, i.e. without prior conversion to work, can only go to increase the internal energy of the system.

The work W and the amount of heat Q represent two different modes of energy transfer, considered in thermodynamics, and, therefore, characterize the energy exchange process between systems. For an infinitesimal equilibrium change in the parameter a, the work done by the system is $\delta W = A\, da$, where A is the generalized force, which is an equilibrium function of the external parameter a, and temperature T. When n external parameters change, the operation of the system

$$\delta W = \sum_i A_i da_i, \tag{5.1}$$

In the case of a non-equilibrium infinitesimal change in the parameter a, the work δW_{ne} performed by the system is also equal to $\delta W_{ne} = A_{ne}\, da_{ne}$, but in this case the generalized force A_{ne} due to the initial statements of thermodynamics is a function of the external parameters a_i, internal parameters b_j, and their derivatives in time.

The First Law of Thermodynamics establishes: the internal energy of the system is a single-valued function of its state and varies only under the effect of external influences. According to the First Law, the change in the internal energy $U_2 - U_1$ of the system upon its transition under the influence of these influences from the first state to the second is equal to the algebraic sum of Q and W, which for the final process is written in the form of equation

$$U_2 - U_1 = Q - W, \tag{5.2}$$

and for the infinitesimal (elementary) process the equation of the First Law has the form

$$\delta Q = dU + \delta W. \tag{5.3}$$

The equation of the First Law in the form (5.2) or (5.3) is valid for both equilibrium and non-equilibrium processes. The equation of the First Law allows us to determine the internal energy $U(a_1,..., a_n, T)$ in the state $(a_1,...,a_n, \mathrm{T})$ only up to the additive constant $U(a_1^\circ, ..., a_n^\circ; T^\circ)$, which depends on the choice of the initial state $(a_1^\circ,..., a_n^\circ, T^\circ)$.

The Second Law of Thermodynamics is a law on *entropy*: entropy, like U, is a function of state, entropy exists for every equilibrium system and it does not decrease for any processes in isolated and adiabatically isolated systems. Mathematically, the Second Law of thermodynamics for equilibrium processes is written as

$$\delta Q = TdS. \tag{5.4}$$

This expression for the heat quantity element has the same form as the expression for the elementary work δW, with temperature T being the intensive heat transfer parameter (thermal generalized force), and entropy S – extensive heat transfer parameter (generalized coordinate). Entropy, by definition, is an additive quantity proportional to the number of particles in the system, i.e. the entropy of the whole system, equal to the sum of the entropy of the individual subsystems.

Basic equation of thermodynamics. For the equilibrium processes Massey [32] introduced a generalized equation for the First (5.3) and Second (5.4) Laws of thermodynamics in the form that was considered to be the basic equation of thermodynamics. Their combination gives the basic equation of thermodynamics for equilibrium processes in the form

$$TdS = dU + \sum_i A_i da_i. \qquad (5.5)$$

For a simple system under the all-round pressure P, equation (5.5) has the form

$$TdS = dU + PdV. \qquad (5.6)$$

Equation (5.5) is the starting point for the analysis of all equilibrium processes in thermodynamic systems with a constant number of particles.

The First Law of thermodynamics is based on the law of conservation of energy in the system under consideration (the law of conservation of mass was known even before the construction of thermodynamics). To close it, the equation of the First Law of thermodynamics must be supplemented by the equations of state (thermal $b_k = f_k(a_1,..., a_n, T)$, where b_k is the equilibrium internal parameter of the system (except for internal energy), which is conjugate to the external parameter a_i, i.e. $b_k = A_i$, and the caloric parameter $U = U(a_1,..., a_n, T)$). Equations of state can not be deduced from the principles of thermodynamics. They are established from experience or are found by methods of statistical physics, since microscopic information on the statistical behaviour of molecules is necessary for their construction. In particular, the absence in thermodynamics of a thermal equation for pressure is due to the lack of consideration of the properties of the impulse of the system.

When calculating many quantities, it is necessary to know both the thermal and the caloric equations of the state of the system. Experimentally, these equations can be obtained independently of each other. The basic equation of thermodynamics (5.5) gives the following differential equation connecting the thermal equations for a variable of type a_i and the caloric equation of state (which makes knowledge of one of them unnecessary):

$$T\left(\frac{\partial A_i}{\partial T}\right)_{a_i} = \left(\frac{\partial U}{\partial a_i}\right)_T + A_i. \qquad (5.7)$$

If the caloric and thermal equations of state are known, then with the help of the principles of thermodynamics one can determine all the thermodynamic properties of the system.

Nernst's theorem (The Third Law of thermodynamics). All the energy characteristics in thermodynamics make sense for relative changes in transitions from a certain initial point (the set of parameters of the system a_i that uniquely determine its state) to any final point. In order to be able to establish a common reference point for all thermodynamic characteristics, *Nernst's heat theorem* was formulated: as the temperature approaches zero, the entropy of any equilibrium system for isothermal processes ceases to depend on any thermodynamic parameters of the state and, in the limit ($T = 0$ K), assumes the same universal system constant for all systems, which can be taken equal to zero by Planck's suggestion. This theorem allows us to define additive constants in expressions for entropy that can not be calculated by any other thermodynamic path.

According to the Third Law, the entropy can be found knowing only the dependence of the heat capacity on temperature and not having the thermal equation of state which is unknown for condensed bodies. The task of calculating entropy is reduced to determining only the temperature dependence of the heat capacity.

At present, the validity of the Third Law is justified for all thermodynamic equilibrium systems, but the main question of its application remains – is equilibrium always being realized during the experiment?

Characteristic functions. Characteristic are the functions of the state of the system in which all the thermodynamic properties can be expressed in the simplest and, at the same time, explicit form. In this case, the thermodynamic properties are understood as physical properties that depend only on temperature, pressure (or volume), and composition. The characteristic of a function is a consequence of the choice of independent variables (state parameters). These variables are called natural independent variables.

All the thermodynamic quantities characterizing a given system can be obtained as partial derivatives of the characteristic functions, and the so-called thermodynamic equations are the relationships between these quantities (the analytical formulation of thermodynamics). Thermodynamics can give only general information on the form of thermodynamic functions, but can not determine their specific form for each particular system. This dependence should be established empirically or by means of statistical mechanics.

We present for the uniform system the expressions of the most frequently used characteristic functions and their natural variables, as well as the total differential of the function:

internal energy U, (S, V, N_i)

$$dU = TdS - PdV + \sum_i \mu_j dN_j; \tag{5.8}$$

enthalpy $H = U + PV$, (S, P, N_i)

$$dH = TdS + VdP + \sum_j \mu_j dN_j; \tag{5.9}$$

Helmholtz free energy $F = U - TS$, (T, V, N_i)

$$dF = -SdT - PdV + \sum_j \mu_j dN_j; \tag{5.10}$$

Gibbs free energy

$$\begin{aligned} G &= F + PV = \sum_j N_j \mu_j, \ (T, P, N_i), \\ dG &= -SdT + VdP + \sum_j \mu_j dN_j. \end{aligned} \tag{5.11}$$

If the characteristic function is given as a function of natural independent variables, then the thermodynamic properties of the system are completely defined. If it is given as a function of another set of independent variables, then it is not enough to determine all the thermodynamic properties. To determine them, additional information is required to find the unknown function obtained in the integration [31–33,87]. Consequently, if the given thermodynamic function is regarded as a function of such variables for which it is not characteristic, then the other conjugate thermodynamic quantities can not be expressed in explicit form with the help of this function.

We also give the so-called Gibbs–Duhem equation

$$SdT - VdP + \sum_j N_j d\mu_j = 0. \tag{5.12}$$

Here, only the work related to the change in pressure is considered in explicit form, as in equation (5.6). Equation (5.12) plays an important role in various applications. It is the same fundamental equation as the equations for the reduced characteristic functions (5.8)–(5.11), which, because they reflect changes that are equal to work done under certain conditions, are also called thermodynamic potentials.

Gibbs called the equations (5.8)–(5.11) fundamental equations to emphasize that they express the connection between the characteristic functions and their variables and, therefore, give an exhaustive thermodynamic characteristic of the n-component mixture. The fundamental equations are equivalent to each other. Therefore, for a complete description of the thermodynamic properties of the mixture,it is necessary to have only one fundamental equation.

6. Interphase equilibrium

Phase. The concept of the phase was introduced into science by Gibbs [30]. Van der Waals [94] clarified this most important concept of the theory of non-uniform systems, in connection with the fundamentals of the derivation of the phase rule. Following these studies, we will determine that *the phase is a uniform region of a non-uniform system consisting of an individual substance or solution/mixture and bounded by an interface whose thermodynamic properties are described by a single fundamental equation and that coexists with other phases of the general non-uniform system.*

This definition bypasses the problem of the dispersity of one phase, if all its regions are macroscopic. Otherwise, this situation can be called as the Gibbs phase approximation for describing a macroscopic non-uniform system. In this definition, the extension of this concept to metastable states, proposed in various papers, in particular, in [87], is excluded. This phase definition is used below for small systems.

The question of how many phases can exist simultaneously in a system under given external conditions and how many state variables can be changed simultaneously is answered by the phase rule [87–90]. To use the phase rule, it is necessary to have an idea of the number of phases and components, as well as the idea of the possibility of chemical reactions in the system. The number of degrees of freedom of the thermodynamic system f consisting of n components participating in k independent reactions within the r phases is $f = n - k - r + b$, where the number $b = 2$ (these numbers correspond to the parameters of the system T and P) in the absence of external fields, and the given number b increases accordingly by the number of external fields affecting the states of the molecules of the system.

Conditions for thermodynamic equilibrium and stability. The theory of thermodynamic equilibrium was developed by Gibbs on the basis of Lagrange's approach to mechanical statics, that is, by generalizing and extending the principle of virtual displacements to thermodynamic systems. It is known from mechanics that a mechanical system with perfect connections is in equilibrium if the sum of the work of all the given forces for any virtual displacement of the system is zero (the principle of virtual displacements). Writing analytically this principle (the general equilibrium condition) in the form of an equation and solving it together with the equations determining virtual displacements, it is possible to find concrete equilibrium conditions for the mechanical system in each given problem.

The equations that are satisfied by virtual displacements and the equation of the principle of virtual displacements are written as follows. Let the state of the mechanical system be determined by the coordinates $q_1,\ldots,q_n$, and the constraints imposed on the system are expressed by the conditions $f_s\ (q_1,..,\ q_n) = 0,\ s = 1,2,\ \ldots,\ k \leq n$. Then the displacements $\delta q_1,\ \ldots,\ \delta q_n$, permitted by these links and called virtual or possible displacements, obviously satisfy the equations

$$\sum_{i=1}^{n} \frac{\partial f_s}{\partial q_i} \delta q_i = 0. \tag{6.1}$$

If Q_i is a generalized force conjugate to the coordinate q_i, then the principle of virtual displacements has the form

$$\sum_{i=1}^{n} Q_i \delta q_i = 0. \tag{6.2}$$

Solving jointly equations (6.1) and (6.2) by the method of uncertain Lagrange multipliers, one can find the equilibrium conditions for a given mechanical system.

This method of determining the equilibrium conditions was extended by Gibbs to thermodynamic systems. The equilibrium state of a thermodynamic system is determined by the temperature T and external parameters $a_1....a_n$, which characterize the ratio of the system to external bodies.

According to the second initial statement of thermodynamics, at equilibrium all internal parameters are functions of external parameters and temperature, and therefore, when a_i and T are given, they are not needed to determine the state of the equilibrium system. If the system is deviated from the equilibrium state, then the internal parameters are no longer functions of only external parameters and temperature. Therefore, the non-equilibrium state must be characterized by additional independent parameters. It is believed that this makes it possible to treat a non-equilibrium system as an equilibrium system, but with a larger number of parameters and corresponding generalized forces that 'hold' the system in equilibrium than in an equilibrium state [84, 89]. In this case, the thermodynamic functions of the system in the non-equilibrium state should be considered equal to the values of these functions for an equilibrium system with additional 'retaining' forces (their role is played by external fields and adiabatic partitions).

For non-equilibrium processes, the basic inequality of thermodynamics that is a sequel of equality (5.5), which expresses the First and Second Laws of thermodynamics, must be rewritten in the form

$$TdS \geq dU + \sum_i A_i da_i \tag{6.3}$$

where the equality sign refers to equilibrium, and inequalities to non-equilibrium elementary processes. On the basis of such a representation, considering the output of a system from an equilibrium state as a result of virtual deviations of internal parameters from their equilibrium values, one can, using the basic inequality of thermodynamics (6.3) for non-static processes, obtain general conditions for thermodynamic equilibrium and stability.

Since now the state of thermodynamic systems is determined not only by mechanical parameters, but also by special thermodynamic (temperature, entropy, etc.) and other parameters, instead of one general equilibrium condition for mechanical systems (6.2) for thermodynamic systems there will be several depending on the ratio of the system to external bodies (adiabatic system, isothermal system, etc.).

The general conditions for the stable equilibrium of thermodynamic systems in various cases are determined by the extremal values of the corresponding thermodynamic potentials. These conditions are not only sufficient but also necessary if all other conditions for establishing equilibrium are ensured (since the conditions we have found are not the only ones for the possibility of processes) – Gibbs associated these constraints on the realization of the equilibrium with 'passive forces' (see below). Thermodynamic potentials can have several extrema (for example, entropy has several maxima). The states corresponding to the largest (entropy) or the smallest

(Helmholtz energy, etc.) of them are called stable (absolutely stable equilibrium states), others are metastable (semi-stable). In the presence of large fluctuations, the system can go from a metastable state to a stable state.

Two-phase system. As the simplest example of determining specific equilibrium conditions, starting from the general equilibrium conditions established above, let us consider an isolated two-phase system of the same substance. This example is given in all textbooks (taking into account the surface contribution, it will be considered in section 45). If S_α and S_β are the entropy of the first and second phases, respectively, then the entropy of the entire system $S = S_\alpha + S_\beta$, its general equilibrium condition $\delta S = \delta S_\alpha + \delta S_\beta = 0$.

Each of the phases is a one-component system with a variable number of particles, and the basic equations of thermodynamics for them will accordingly be:

$$T_\alpha \delta S_\alpha = \delta U_\alpha + P_\alpha \delta V_\alpha - \mu_\alpha \delta N_\alpha, \; T_\beta \delta S_\beta = \delta U_\beta + P_\beta \delta V_\beta - \mu_\beta \delta N_\beta.$$

Determining by these formulas the expressions for δS_α and δS_β, we obtain the general equilibrium condition $\delta S = 0$ in the form

$$\frac{\delta \mathrm{U}_\alpha + \mathrm{P}_\alpha \delta \mathrm{V}_\alpha - \mu_\alpha \delta \mathrm{N}_\alpha}{T_\alpha} + \frac{\delta \mathrm{U}_\beta + \mathrm{P}_\beta \delta \mathrm{V}_\beta - \mu_\beta \delta \mathrm{N}_\beta}{T_\beta} = 0. \tag{6.4}$$

Since the system is isolated, its extensive parameters obey the following coupling equations: $U_\alpha + U_\beta = U =$ const (internal energy of the system), $V_\alpha + V_\beta = V =$ const (volume of the whole system), $N_\alpha + N_\beta = N =$ const (total number of particles). As independent parameters of the system, we choose U_α, V_α, T_α; as dependent – U_β, V_β, T_β. According to these relationships, virtual changes in system parameters are expressed as

$$\delta U_\beta = -\delta U_\alpha, \; \delta V_\beta = -\delta V_\alpha, \; \delta N_\beta = -\delta N_\alpha.$$

Solving jointly the equation of the general equilibrium condition (6.4) with the equations for virtual changes of the indicated extensive parameters of the system, we find that

$$\left(\frac{1}{T_\alpha} - \frac{1}{T_\beta}\right)\delta U_\alpha + \left(\frac{P_\alpha}{T_\alpha} - \frac{P_\beta}{T_\beta}\right)\delta V_\alpha - \left(\frac{\mu_\alpha}{T_\alpha} - \frac{\mu_\beta}{T_\beta}\right)\delta N_\alpha = 0. \tag{6.5}$$

Because of the independence of the variations δU_α, δV_α, δN_α, the following particular phase equilibrium conditions for a two-phase single-component system are obtained, as the equality of three particular phase equilibrium conditions

$$P_\alpha = P_\beta,\ T_\alpha = T_\beta,\ \mu_\alpha = \mu_\beta. \tag{6.6}$$

This entry is a consequence of the analog of the principle of mechanical equilibrium for thermodynamic parameters – the equality of thermodynamic forces in both phases.

Instead of three equalities for particular equilibria, one can write down one condition of complete equilibrium

$$\mu_\alpha(P,T) = \mu_\beta(P,T), \tag{6.7}$$

which shows that when two phases of the same substance are in equilibrium, the pressure is a function of temperature, that is, the parameters T and P cease to be *independent*.

These phase equilibrium conditions (6.6) or (6.7) are valid only for uniform phases, that is, in the absence of a field of external forces. If, on the other hand, the phases are in an external field (for example, in a gravitational field), then in equilibrium only the temperatures are equal in both phases, the pressure and chemical potential in each phase are functions of coordinates – this is a consequence of the fact that the parameter independent of the coordinates is not the chemical potential but it is the total chemical potential, which includes the potential energy of a particle in external fields.

Diagrams of state. The ultimate goal of analyzing the equilibrium of phases in non-uniform systems is to establish strict relationships between the parameters characterizing the state of the system. Knowing the dependence between the state parameters, it is possible not only to determine the equilibrium state of the non-uniform system, but also to predict the nature of the phase transformations with varying temperature, pressure, and concentration in a certain direction.

In the thermodynamics of non-uniform systems, they share analytical and geometric methods [30]. The analytical method is based on partial differential equations. The problem is to establish the regularities of the flow of the equilibrium heterogeneous processes on the basis of differential equations [94].

Graphical interpretation of the change in the characteristic thermodynamic functions as a function of the state parameters and the establishment on this basis of regular graphic relationships between them underlies the theory of the diagrams of the state of non-uniform systems [87,95–97]. A phase equilibrium diagram, or a state diagram, is a graphical representation of the relationships between the state parameters. Each point on the state diagram, called a figurative point, specifies the numerical values of the parameters characterizing the given state of the system. The straight lines connecting the figurative points of two phases in equilibrium are called nodes. All the details of the chemical interaction process, for example, the appearance of new phases and certain compounds, the formation of liquid and solid solutions, find an exact and definite reflection in that geometrical complex of surfaces, lines and points that forms a chemical diagram [97,98]. The state diagram gives an answer to the question of how many and exactly which phases form a system at given values of the state parameters. In accordance with the choice of temperature, pressure and concentration as the main parameters of the state of the non-uniform system, we unambiguously arrive at the choice of the free Gibbs energy as the main thermodynamic function that characterizes it.

To formulate these representations, different physical properties of systems are measured: the temperature of phase transitions, thermal properties (thermal conductivity, heat capacity, thermal expansion), electrical (electrical conductivity, dielectric constant), optical (refractive index, rotation of the polarization plane of light), density, viscosity, hardness and others, and also the dependence of the rates occurring in the system of transformations on its composition [95–99]. 'Self' thermodynamics does not determine the number of phases in the system, their aggregate states, the minimum concentrations of components, etc., which must be taken into account in deciphering the experimental data. Gibbs' introduction of the presence of macroscopic non-uniform phases is the key in practical work on phase transitions, and serves as a tool for the self-consistency of data on the thermodynamic functions in non-uniform systems (as a phase approximation).

When working with nanometer dispersed phases, a more detailed interpretation of the measured characteristics is required due to the description of the properties of the interface and their surface tension, which is currently relatively rare. The listed possibilities on interrelations of various experimental methods are necessary at researches of conditions of real polydisperse systems.

However, the most important questions for small bodies: the minimum phase size and the minimum size at which macroscopic connections of thermodynamics are correctly used, until recently remained open.

7. The problem of the Kelvin equation

The small systems mentioned above include both isolated small bodies and their ensembles (macroscopic and finite) – polydisperse systems, polycrystalline structures, porous bodies, etc. Today, when moving from macrosystems to small systems, the theoretical interpretation of the latter systems is based on the Young–Laplace [100,101] and Kelvin [102] equations.

The Kelvin equation is the first thermodynamic work on phase equilibrium for small systems. It was the basis of all subsequent work. Very often the derivation of the formula for the saturated vapour pressure over the curved drop surface is carried out as follows [34,103]: first, the conditions of mechanical equilibrium on the curved surface of the vapour–liquid interface are sought, and then the chemical potentials in the vapour and liquid are equated.

The Young–Laplace equation. Consider a small element of a curved surface between two phases α (liquid) and β (vapour) having two radii of curvature R_1 and R_2. Let the surface be displaced by a small distance dz under the influence of a tensile pressure equal to $P_\alpha - P_\beta$. The change in its area is $dA = (x + dx)(y + dy) - xy \approx xdy + ydx$. For a system in equilibrium, the total work of a small displacement must be zero. It consists of work on the displacement of the surface by dz (equal to $\sigma dA = \sigma(xdy + ydx)$), where σ is the surface tension, and from the expansion work of the vapour under the action of excess pressure $P_\alpha - P_\beta$ (equal to $(P_\alpha - P_\beta)\, xydz$). That is, $\sigma(xdy + ydx) = (P_\alpha - P_\beta)\, xydz$. Taking into account that in the given geometry $(x + dx) / (R_1 + dz) = x/R_1$, then $dx = xdz/R_1$ and $dy = ydz/R_2$, we obtain the Young–Laplace equation

$$P_\alpha - P_\beta = 2\sigma / R,\ 1/R = (1/R_1 + 1/R_2), \tag{7.1}$$

where $1/R = (1/R_1 + 1/R_2)$ is the mean value of the radius of curvature.

The Kelvin equation is obtained from the condition that the chemical potentials of the liquid and the vapour are equal to $\mu_\alpha = \mu_\beta$ under the condition of mechanical equilibrium according to the Young–Laplace equation. If we go from one equilibrium state to another at a constant temperature, then $dP_\alpha - dP_\beta = d(2\mu/R)$ and $d\mu_\alpha = d\mu_\beta$. For each phase, we can apply the Gibbs–Duhem equation $s_\alpha dT + V_\alpha dP_\alpha + d\mu_\alpha = 0$, where s_α and V_α are the molar entropies and molar volumes of the phase α (similarly for phase β). At T = const, these equations give the relation $V_\alpha dP_\alpha = V_\beta dP_\beta$. Excluding dP_α, this allows us to rewrite the changes in the Young–Laplace equation as $d(2\sigma/R) = [(V_\alpha - V_\beta)/V_\alpha]\, dP_\beta$. The molar volume of liquid V_α is very small in comparison with the molar volume of steam V_β, and if the vapour behaves like an ideal gas ($V_\beta = k_B T/P_\beta$), then the last equation can be transformed as $d(2\sigma/R) = -[k_B T/(P_\beta V_\alpha)]dP_\beta$. Integration from (R, P_β) and (∞, P_s) gives the Kelvin equation (KE)

$$P_\beta(R)/P_s = \exp\{2\sigma V_\alpha/(k_B TR)\}, \tag{7.2}$$

where P_s is the saturated vapour pressure corresponding to $R = \infty$ at a given temperature T (it is assumed that the fluid is not compressible and the quantity V_α is independent of pressure).

It is believed that the KE shows how the vapour pressure around small metastable drops exceeds the analogous vapour pressure around larger metastable drops, the count is taken from the pressure $P_s(T)$ in a system with a planar interface. It follows from the KE that the saturated vapour pressure above the convex (concave) surface of the meniscus is greater (lower) than the saturated vapour pressure above the flat interface. The Kelvin equation and its analogues (Thomson's equation for the phase transition temperature shift and the Ostwald-Frendlich equation for changing the solubility of particles from the curvature of small particles) [35] were actively used for all questions of analysis of the new phase [1] and in problems of adsorption porosimetry [103]. For adsorption in pores, this means that if the meniscus of the adsorbed liquid is concave (convex), then the filling of the pores occurs at pressures lower than (greater than) the saturated vapour pressure in the bulk phase.

The Kelvin equation leads to an unlimited increase in the pressure of the metastable vapour at small droplet radii, which has no physical meaning due to the absence of this concept in molecular associates, and this equation does not allow us to explain the phenomenon of limiting supersaturation for small particles (Oswald ripening).

The application of the Kelvin equation was 'supported' using coarse models for the equations of state, which lead to loops of the van der Waals type in the intermediate region of the parameters of the binodal curve. The van der Waals loop on isotherms exists in many equations of state, and it is well known that its origin is the result of the approximate nature of the equations used. The Maxwell rule [104] corrects this inaccuracy [104], which makes it possible to determine the densities of the coexisting phases of vapour and liquid at a given temperature, which corresponds to the saturated vapour pressure $P_s(T)$. The answer to the question of how to describe metastable states in the framework of molecular theories, statistical physics, speaks of the need to change the type of equations and the transition from equilibrium to non-equilibrium bonds. The use of the concept of metastable states is associated with the convenience of interpreting hysteresis phenomena, when instead of the kinetic theory very simple equations of thermodynamics are applied.

The same attitude is maintained today, when the fundamental results of statistical physics are ignored, such as the Yang–Lee theory [104–107]: all the state parameters inside the binodal curve are non-equilibrium and for them there is no equilibrium equation of state. According to the Yang–Lee theory, the thermodynamic Kelvin equation is erroneous (see section 18).

In order to show the incorrectness of the Kelvin equation, it is necessary to recall the following thermodynamic considerations for macrosystems: 1) in thermodynamics there are concepts of intense and extensive properties that do not depend on the amount of matter in the system; 2) the true state of equilibrium does not depend on the path to the final state; 3) the complete equilibrium of the non-uniform system is the consequence of three kinds of mechanical, thermal and chemical equilibria [30], therefore, in complete equilibrium, it is not important which order of equilibrium is first to be postulated.

Recall that the original statements and the Laws of thermodynamics were not derived from any principles, but were formulated on the basis of experience. Thermodynamics works with statements or hypotheses (and not with proofs like the molecular theory) and their consequences according to the rules for constructing equilibrium conditions and their stability, taken from mechanics and generalized to a greater number of degrees of freedom. Therefore, working with hypotheses applies to the same degree to the conditions for mechanical, thermal and chemical equilibrium of phases in a non-uniform system, as corresponding to the condition of complete equilibrium of the system, and then the consequences of them are looked. This was noted in the Preface as three stages of thermodynamics for macrosystems which are tested by experience.

Despite the explicit presence in the Kelvin equation of the radius of a small system R (a drop or another new phase), it is not indicated anywhere that this equation contradicts thermodynamics. How to understand this obvious mismatch of concepts?

The size of the drop R in the Kelvin equation appeared during the consideration of mechanical equilibrium. For mechanics, the size is a natural characteristic both at the microlevel (when looking at individual atoms/molecules), and for an ensemble of individual particulates – stresses are defined at their boundaries. But in the transition to the mechanics of continuous media, as in thermodynamics, the concept of the particle size disappears, there is only the size of the system that determines the boundary conditions. In the thermodynamics of non-uniform systems, there are concepts of the weight or volume fraction of the phase, depending on the choice of the type of variables, but there is no linear size of the individual particles (for more details, see [33]).

In terms of its meaning, the parameter (drop radius) R in the Kelvin equation can not be an extensive quantity proportional to the number of particles of any subsystem of the size not less than the value of the elementary volume dV (see Section 2). The intensive parameters also formally refer to the subsystem not smaller than the size of the elementary volume dV, i.e. the concepts of the intensive and extensive parameters for macrosystems have nothing to do with the 'size' of the subsystem. In other words, in thermodynamics, the independence of the thermodynamic properties from the sizes of the subsystem is assumed [31–33]. The size of small bodies is transferred to thermodynamics from mechanics, and this transfer refers to the fourth stage of the construction of thermodynamics. The correctness of such a transfer must be verified by the molecular theory.

We should note the difference between the items 2 and 3. Item 2 refers to the general system which before Gibbs was associated with a uniform system. The conditions for the realization of particular equilibria are not discussed in the initial propositions and postulates of thermodynamics (these are its first two stages). They are a consequence of the introduced generalized rules on the equilibrium states in the heterophase system (without allowance for surface contributions). The principle of introducing the description of phase states in the non-uniform systems and the establishment of the phase equilibrium conditions go beyond the framework of a purely technical technique, since the essence of the method of describing the state of the system changes in it – the number of parameters varies (this is the first stage of thermodynamics). Instead of the property of one uniform system, a set of properties of different systems appeared. And these properties had to be differentiated, because just the concept of 'complete equilibrium' of the system became vague. As soon as the properties became different, the very notion of partial equilibria arose (see [82]). After that, they can be established in any order, and this is not the same thing as complete equilibrium. The presence of partial equilibria allows the possibility that these equilibria can be realized or not.

The Kelvin equation is derived on the basis of a consistent application of the postulates of thermodynamics about two equilibria (mechanical and chemical) between a small subsystem and a macrosystem. First, the question of pressures in neighbouring phases in the presence of a curved surface is considered, and then, using the fixed values of these pressures, the condition of equality of chemical potentials is considered (i.e. postulated), or the extreme properties of the Gibbs energy of the complete system are considered, which lead to the same results [33,108] (see Section 44), since an implicit connection between these particular equilibria is used in the derivation.

If we change the order of introduction of equilibrium conditions: first chemical, and then mechanical, the answer will be different. The first condition gives the equality of chemical potentials $\mu_\alpha = \mu_\beta$, and from it for T = const follows the condition that the pressures in both phases are equal $\mu_\alpha (P_\alpha, T) = \mu_\beta (P_\beta, T) \rightarrow P_\alpha = P_\beta$, as for a flat boundary. As a result, the droplet size is excluded from consideration due to the rules of operation of thermodynamics with macroscopic objects. This path contradicted the old knowledge of the Laplace equation for curved surfaces in the case of mechanical equilibrium, although they were introduced *apriori* in the absence of the notion of chemical equilibrium. The way out of this contradiction (with the retention of the drop size) would be the possibility of adjusting the mechanical equilibrium under the chemical equilibrium of the system. But the conditions of interphase equilibrium are determined by the postulates of thermodynamics for macrosystems independently of each other (as in the general interpretation [33,108]), and therefore they can not in principle provide a relationship. It turns out that the complete equilibrium of the system depends on the order of the establishment of equilibria by the types of subsystems, which contradicts the basic postulate of complete equilibrium of the system (item 2).

For equilibrium functions, the influence of the path to the final result implicitly indicates the dependence of the process on the time characteristics. It is shown below that the analysis of this equation leads to the need to take into account temporal characteristics, therefore, for a complete analysis of small systems, we will consider the main points of non-equilibrium thermodynamics.

8. Fundamentals of non-equilibrium thermodynamics

The thermodynamics of non-equilibrium processes is a general theory describing macroscopic thermodynamically non-equilibrium processes [89,109–116]. Classical equilibrium thermodynamics for non-equilibrium processes establishes only inequalities that indicate the direction of the processes. The problem of non-equilibrium thermodynamics is to give a quantitative description of non-equilibrium processes depending on the initial and/or external conditions for states that do not differ much from the equilibrium states. Taking into account the model-free nature of equilibrium thermodynamics, in non-equilibrium thermodynamics all processes are also considered as in the mechanics of continuous media, i.e. in the continuum description, and their state parameters are considered as variables from the continuous coordinates and time.

In thermodynamically equilibrium systems, as is well known, the pressure P, the temperature T, and the chemical potential μ_i are constant along the entire system: grad $P = 0$, grad $T = 0$, grad $\mu_i = 0$. Under non-equilibrium conditions, pressure P, temperature T and chemical potential μ_i (grad $P \neq 0$, grad $T \neq 0$, grad $\mu_i \neq 0$) are not constant, therefore irreversible processes of impulse, energy, mass transfer, etc., appear in the system.

When generalizing the classical thermodynamics to non-equilibrium processes, one starts with the idea of local equilibrium. It is generally accepted that the relaxation time increases with increasing system size, so that individual *macroscopically* small parts of the system come to themselves in an equilibrium state much earlier than the equilibrium between these parts is established. Non-equilibrium thermodynamics assumes that, although in general the state of the system is non-equilibrium, its individual small parts are in equilibrium (more precisely, quasi-equilibrium), but they have thermodynamic parameters that vary slowly with time and from point to point.

The sizes of these physically small equilibrium parts of the non-equilibrium system and the times of the change in the thermodynamic parameters in them are determined experimentally in thermodynamics. It is usually assumed that the physical elementary volume L^3 on the one hand contains a large number of particles ($v_0 \ll L^3$, $v_0 = \gamma_s \lambda^3$ – volume per particle, γ_s – shape factor, λ - average distance between particles), and on the other hand, the non-uniformities of the macroscopic parameters $a_i(r)$ over the length L are small in comparison with the value of these parameters ($|\partial a_i/\partial x|L \ll a_i$), i.e.,

$$v_0^{1/3} \ll L \ll \left| \frac{1}{a_i} \frac{\partial a_i}{\partial x} \right|^{-1}. \tag{8.1}$$

Time τ of the changes in the thermodynamic parameters in physically small equilibrium parts is much longer than the relaxation time τ_l within them and much less than time τ_L, during which equilibrium is established throughout the system:

$$\tau_l \ll \tau \ll \tau_L. \tag{8.2}$$

The properties of a non-equilibrium system are determined by local thermodynamic potentials, which depend on spatial coordinates and time only through the characteristic thermodynamic parameters for which the equations of thermodynamics are valid. Thus, if the local internal energy density $u(\mathbf{r}, t)$, the specific volume $v(\mathbf{r}, t)$ ($v = \rho^{-1}$, ρ is the local mass density of the medium) and the local concentrations $c_i(\mathbf{r}, t)$ are selected as the characteristic variables, the state of a physically elementary volume in a neighbourhood of the point r at time t is described by the local entropy $s = s[u(\mathbf{r}, t), v(\mathbf{r}, t), c_1(\mathbf{r}, t), ..., c_n(\mathbf{r}, t)]$, determined by the basic equation of thermodynamics

$$Tds = du + Pdv - \sum_i \mu_i dc_i. \tag{8.3}$$

This equation (8.3) for specific (by mass) local magnitudes is also the basic equation of non-equilibrium thermodynamics. With its help, all expressions for the mass, impulse and energy fluxes (in a more general case, charges) in the dynamics of non-equilibrium processes are obtained.

The general evolution of entropy is expressed in terms of two contributions as

$$\frac{dS}{dt} = \frac{d_e S}{dt} + \frac{d_i S}{dt}, \quad (8.4)$$

where the first contribution means the flow of entropy, and the second – the production of entropy.

The local entropy s (mass units or ρs – volume units) depends on the thermodynamic parameters $a_i(\mathbf{r}, t)$ like at full equilibrium, so for an irreversible process in the adiabatic system the entropy rate per unit volume (entropy production) is equal to

$$\frac{d_i S}{dt} = \frac{d(\rho s)}{dt} = \sum_i \frac{d(\rho s)}{da_i}\frac{da_i}{dt} = \sum_i J_i X_i. \quad (8.5)$$

where the following notations are introduced: quantities $d(\rho s)/da_i \equiv X_i$ are called thermodynamic forces, and the quantities $da_i/dt \equiv J_i$, determining the rate of change of the parameters $a_i(\mathbf{r}, t)$ are called thermodynamic flows. These names are related to the fact that an increase in entropy is the 'cause' of an irreversible process when the local macroscopic parameters $a_i(\mathbf{r}, t)$ change under adiabatic conditions.

The entropy of the entire non-equilibrium system is composed additively from the entropy of its individual parts

$$S = \int_V \rho s dV. \quad (8.6)$$

9. Equations of non-equilibrium thermodynamics

The so-called equation of the entropy balance (8.4) plays a central role in non-equilibrium thermodynamics. This equation expresses the fact that the entropy of a certain volume element varies with time for two reasons. First, it changes due to the presence of some entropy flow into a given volume element; secondly, due to the presence of some source of entropy, the existence of which is due to irreversible phenomena within the volume element. We always deal with a positive source of entropy, since entropy can only arise, but not be destroyed. With reversible processes, there are no sources of entropy. This is the local formulation of the Second Law of thermodynamics.

To determine the thermodynamics of the non-equilibrium system of entropy production and the change in time of all its other thermodynamic functions with the help of the basic equation (8.3), it is necessary to add to this equation the balance equations for a number of quantities (mass, internal energy, etc.), as well as equations connecting the fluxes J_i with the thermodynamic forces X_i. The conservation laws include such quantities as the diffusion flow and flows of heat ad tensor pressure which characterise accordingly the transfer of mass, energy and an impulse.

Any extensive quantity $B(x, y, z, t)$ of the system is subject to the balance equation

$$\partial B/dt = -\text{div}\mathbf{J}_B + I_B \tag{9.1}$$

where $\mathbf{J}_B$ is the density of the total flux of the quantity $B = \rho b$ (the density of matter, b is the value of B referenced to the mass), I_B is the change in B due to its sources, referred to volume and time. Equation (9.1), in which I_B vanishes, expresses the law of conservation of B. Thus, the law of conservation of mass has the form of the hydrodynamic equation of continuity

$$\partial\rho/\partial t = -\text{div}(\rho\mathbf{u}) \tag{9.2}$$

where $\mathbf{u}$ is the mass velocity at a given point x, y, z at time t.

The density of the total flux $\mathbf{J}_B$, generally speaking, does not reduce to the convective flux $B\mathbf{u}$, i.e., to the transfer of the quantity B to the flow of matter, but also contains the terms J_B of another non-convective portion of the flow (heat flux, diffusion flux, etc.):

$$\mathbf{J}_B = B\mathbf{u} + I_B \tag{9.3}$$

Thus, the balance equation (9.1) of the additive quantity can be written in the form

$$\partial(\rho b)/\partial t = -\text{div}(\rho b\mathbf{u} + J_B) + I_B \tag{9.4}$$

where the partial derivative on the left $\partial(\rho b)$ determines the change in the value $B = \rho b$ at a given fixed point in space. This derivative can be expressed in terms of the total (substantial) derivative of B, which refers to a 'particle' of substance moving in space (as a continuous medium). The change dB in the value B of the substance particle is made up of two parts: from a change in B at a given point in space and from a change in B in transition from a given point to a point distant from it by a distance $d\mathbf{r}$ traversed by the particle under consideration during the time dt. The first of these parts is equal to

$\frac{\partial B}{\partial t}dt$, and the second part is equal to

Consequently

$$\frac{\partial B}{dt} = \frac{\partial B}{\partial t} + (u, \nabla) B. \tag{9.5}$$

Therefore, the law of conservation of mass (9.2) and the balance equation for the quantity B (9.4) can be written accordingly in the form

$$\frac{d\rho}{dt} = -\rho \operatorname{div} \mathbf{u} \text{ and } \rho \frac{db}{dt} = -\operatorname{div} \mathbf{J}_B + I_B. \tag{9.6}$$

In accordance with the general formula (9.6), the equation of the entropy balance will be

$$\rho \frac{ds}{dt} = -\operatorname{div} \mathbf{J}_s + d(\rho s) / dt, \tag{9.7}$$

where $\mathbf{J}_s$ is the entropy flux density, and the local rate of entropy generation (8.5).

To find the explicit form of $\mathbf{J}_s$ and $d(\rho s)/dt$, formula (9.7) is compared with the expression for $\rho ds/dt$, obtained from equation (8.3).

$$\rho \frac{ds}{dt} = \frac{\rho}{T} \frac{du}{dt} + \frac{\rho p}{T} \frac{dv}{dt} - \sum_i \frac{\rho \mu_i}{T} \frac{dc_i}{dt} \tag{9.8}$$

in which the expressions for the derivatives with respect to time and entropy production (8.5) are substituted.

The system of conservation laws, together with the equation of the entropy balance and the equations of state, is non-closed: it must be supplemented by a set of phenomenological equations connecting the reversible fluxes and the thermodynamic forces entering into the expression for the intensity of the source of entropy. In a first approximation, the fluxes are linear functions of the thermodynamic forces. The Fick's law of diffusion, the law of Fourier thermal conductivity and the Ohm electrical conductivity law belong to the class of linear phenomenological laws. These linear laws should, generally speaking, also reflect possible cross effects, since each flux can in principle be a linear function of all the thermodynamic forces that are necessary to characterize the intensity of the source of entropy in the system. In the general case of irreversible processes, the production of entropy is caused both by the phenomena of transport (energy, electric charge, etc.), and by internal transformations in the system (chemical reactions, relaxation phenomena) [112].

Let us give an expression for the production of entropy in chemical reactions in a uniform system. Here, the reactions are not related to the mass transfer processes and the flows to the equilibrium state occur in the coordinates of the composition of the system N_i (the number of particles of type i), but not in spatial coordinates. Let the uniform system consist of n substances i (i = 1, 2,..., n), between which r chemical reactions j (j = 1, 2, ..., r) can occur. If N_i is the number of particles of type i, v_{ij} is the stoichiometric coefficient of substance i in reaction j, then the change in the number d_jN_i of particles of type i over the time interval dt in reaction j is equal to

$$d_jN_i = v_{ij}I_jdt,\ I_j = \frac{1}{v_{ij}}\frac{d_jN_i}{dt} = \frac{d\xi_j}{dt}, \tag{9.9}$$

where I_j is the reaction j rate, and the differential $d\xi_j = d_jN_i/v_{ij}$ determines the 'affinity of the reaction' and has the same meaning and sign for all substances participating in the given reaction. For this reason, ξ_j is taken as the internal parameter of the system and is called the degree of completeness of the reaction j.

The change in the number of particles of the i-th sort for all reactions in a closed system is equal to $dN_i = \sum_{j=1}^{r} d_jN_i = \sum_{j=1}^{r} v_{ij}d\xi_j$. Therefore, the basic equation (8.3) in the absence of transfer processes and for the constant volume of the system, takes the form

$$TdS = -\sum_{i=1}^{n}\sum_{j=1}^{r}\mu_i v_{ij}d\xi_j = \sum_{j=1}^{r}A_jd\xi_j, \tag{9.10}$$

where A_j is the chemical affinity of the reaction j (j = 1, 2,..., r):

$$A_j = -\sum_{i=1}^{n}\mu_i v_{ij}. \tag{9.11}$$

The entropy production in the considered case of chemical reactions in a uniform multicomponent system, according to formulas (9.11) and (9.9), is equal to

$$d(\rho s)/dt = \sum_{j=1}^{r} I_jA_j/T. \tag{9.12}$$

In the case of a separate reaction in a closed system $d(\rho s)/dt = IA/T$.

For further consideration, it is necessary to know the relationship of the fluxes $\mathbf{J}_j$ and/or I_j to the forces X_j which are known functions of concentration. When applying the theory in specific cases, one proceeds as follows: make up the balance of entropy, find from it a definite expression for the dissipative function, then determine the fluxes and forces in such a way that all these quantities are independent and become equal to zero in equilibrium, formulate phenomenological equations, and finally, apply the Onsager reciprocity relations.

Non-equilibrium thermodynamics has found many diverse applications in physics and chemistry. To classify these applications, the different irreversible phenomena were grouped according to their 'tensor properties' [112]. Our task does not include an account of known conditions for non-equilibrium thermodynamics. Below we will consider only on its limitations related to the description of the properties of small systems. To this end, we discuss the question of a self-consistent description of non-equilibrium and equilibrium processes and the concept of Gibbs' 'passive forces'. We also recall that all the results of non-equilibrium thermodynamics are obtained under the condition of local equilibrium realization, and the question of its correspondence with the strongly non-equilibrium states of the system remains open.

10. Self-consistency of equilibrium and dynamics

Irreversible processes take place until either a stationary state or equilibrium is established (excluding the possibility of realization of periodic processes). If several irreversible processes are superimposed and the final state attained corresponds to equilibrium, then in certain cases it is possible to obtain general conditions for the coefficients that describe irreversible processes, without the application of the thermodynamics of irreversible processes.

As an illustrative example, consider the chemical reaction $L \leftrightarrow M$ [109], which can flow in both directions in a uniform volume. If the reaction occurs in an ideal gas phase or in an ideally dilute solution, then it is known from the experiment that the reaction rate w is determined by expression

$$w = \kappa c_L - \kappa' c_M, \tag{10.1}$$

where c_L and c_M are the molar volume concentrations of particles of the form L and M, respectively; κ and κ' are the rate constants for the reactions from left to right and from right to left. Chemical equilibrium occurs when the velocities of both reactions become equal, i.e., when the rate of the entire reaction w becomes zero:

$$w = 0. \tag{10.2}$$

Now, however, the equilibrium conditions of classical thermodynamics in the above case mean:

$$c_M / c_L = K, \tag{10.3}$$

where K is the equilibrium constant. The quantity K, as κ and κ', does not depend on c_L and c_M. It follows immediately from (10.1)–(10.3) that

$$\kappa / \kappa' = K. \tag{10.4}$$

Thus, from equation (10.1), which follows directly from the experiment, the obvious condition (10.2) and relation (10.3), taken from classical thermodynamics, we obtain the dependence between the coefficients κ and κ', which describe irreversible processes, and the equilibrium constant K.

This simplest example indicates that the ratio between the rate constants of the reaction in the forward and backward directions must necessarily give an equilibrium constant, or the equilibrium constant can be determined in different ways: either from equilibrium or from kinetic measurements. In this sense, we can say that the empirical regularities for describing the reaction and equilibrium rates in the system under consideration give a self-consistent description of this process at any time intervals including finite deviations of the equilibrium state, as well as the limiting equilibrium case itself.

The law of mass action. In the general case of more complex elementary stages of chemical reactions in the gas phase, the law of mass action, which was empirically established by Guldberg and Waage (1867), is used to describe reaction rates. For reversible reactions of a general form we can write $\sum_i v_i [A_i] \underset{k_2}{\overset{k_1}{\rightleftarrows}} \sum_j v_j [A_j]$ where the symbols A_i and A_j in parentheses denote different reacting particles, the values of v_j and v_i are equal to the negative and positive values of the stoichiometric coefficient (the sign of the coefficient is determined by their location: on the left or right side of the equation). The constants k_1 and k_2 are the reaction rate constants in the forward and backward directions. Numerically, they are equal to the reaction rate at single values of the concentration of each of their reagents in the forward direction.

The rate of the considered reaction within the framework of the law of mass action [71, 72] will be written as

$$w = k_1 \prod_i c_i^{\nu_i} - k_2 \prod_j c_j^{\nu_j}. \qquad (10.5)$$

In the equilibrium state, the rate is zero $w = 0$, and it follows from (10.5) that the rate constants in the forward and backward directions are related to each other in the form

$$k_1 / k_2 = \prod_j c_j^{\nu_j} / \prod_i c_i^{\nu_i} = K \qquad (10.6)$$

where $K = k_1/k_2$ is the equilibrium constant of this stage.

The law of mass action is justified for an ideal gas mixture and dilute solutions for which the following expression for the chemical potential can be written $\mu_i = \mu_i^0 + k_B T \ln(c_i)$ where μ_i^0 and c_i are the chemical potential of the standard state and the molar volume concentration of component i.

For non-ideal systems, the concept of activity coefficients is used, expressing the chemical potential as

$$\mu_i = \mu_i^0 + k_B T \ln(a_i) = \mu_i^0 + k_B T \ln(\alpha_i c_i) = \mu_i^{id} + k_B T \ln(\alpha_i). \qquad (10.7)$$

The activity coefficient is defined as the ratio $a_i = a_i/c_i$, here a_i is the activity of component i in solution, depending on all concentrations of the solution components (provided the total concentration is determined $c = \sum_i c_i$) and on all their molecular properties, including intermolecular interaction energies.

The intermolecular interactions in non-ideal reaction systems, according to the theory of absolute reaction rates (TARR) [117, 118], are taken into account by using thermodynamic bonds known from the theory of non-ideal solutions [82]. This approach is related to the preservation in the expression for the reaction rates of the concentration factor used in the form of the law of mass action (through the product of reagent concentrations), as in formula (10.5), and with the change in the rate constants of reactions in the form

$$K_i(ef) = K_i^0 \frac{\alpha_i}{\alpha_i^*} \exp(-E_i / k_B T) = K_i \alpha_i / \alpha_i^*,$$
$$K_{ij}(ef) = K_{ij}^0 \frac{\alpha_i \alpha_j}{\alpha_{ij}^*} \exp(-E_{ij} / k_B T) = K_{ij} \alpha_i \alpha_j / \alpha_{ij}^*, \tag{10.8}$$

where α_i is the activity coefficient of molecules of type i, α_i^* and α_{ij}^* is the activity coefficient of the activated complex of the mono- and bimolecular stage; $K_i = K_i^0 \exp(-E_i/k_BT)$ and $K_{ij} = K_{ij}^0 \exp(-E_{ij}/k_BT)$ are the rate constants of mono- and bimolecular stages, K_i^0 and K_{ij}^0 are the pre-exponentials, and E_i, E_{ij} are the activation energies of these stages.

Continuous system. Another example of the connection between the equilibrium and kinetic characteristics relates to the problem of the relationship between diffusion and sedimentation [109]. In a continuous system with two independently moving substances, in the presence of concentration gradients and the action of stationary external force fields (gravitational or centrifugal), the simultaneous occurrence of diffusion processes (the transfer of matter due to the concentration gradient) and sedimentation (the motion of matter under the action of external fields) in the volume element under normal conditions in accordance with experience can be described by the following equation:

$$_w\mathbf{J}_2 = -D \operatorname{grad} c_2 + c_2 s_2 \mathbf{g}, \tag{10.9}$$

where $_w\mathbf{J}_2$ is the appropriately determined diffusion flux (diffusion flux density vector) of substance 2; c_2 is the molar volume concentration of substance 2; $\mathbf{g}$ is the vector of gravitational or centrifugal acceleration; D is the diffusion coefficient; s_2 is the sedimentation coefficient. The last two quantities do not depend on grad c_2 and g. At equilibrium (sedimentation equilibrium), the diffusion and sedimentation in each volume element must obviously be compensated. Consequently, we have

$$_w\mathbf{J}_2 = 0. \tag{10.10}$$

However, now the equilibrium condition for classical thermodynamics for such a case has the form [109]

$$(M_2 - \rho V_2)\mathbf{g} = \left(\frac{\partial \mu_2}{\partial c_2}\right)_{T,P} \operatorname{grad} c_2. \tag{10.11}$$

In this equation, M_2 is the molar mass (molecular weight), V_2 is the partial molar volume, μ_2 is the chemical potential of substance 2, ρ is the density, T is temperature, and P is the pressure in the volume element. Comparison of the equations (10.9) and (10.11) gives

$$\frac{D}{s_2} = \frac{c_2}{M_2 - \rho V_2}\left(\frac{\partial \mu_2}{\partial c_2}\right)_{T,P} \tag{10.12}$$

The equation (10.12) is the relationship between the two transport coefficients D and s_2 and the equilibrium values obtained from the empirical equation (10.9), the obvious condition (10.10) and the relation (10.11), which follows from classical thermodynamics. A similar relationship is extended to the system with any number of substances and in the same way a general relationship between filtration and osmosis for binary systems with a membrane was obtained [109].

Determining the general criterion for obtaining such links in the above way for transport processes, it can be established that when a number of transport processes are superimposed in the system in which the final state of the system corresponds to equilibrium in an non-uniform system, then by means of classical thermodynamics the general relationship between the various transport coefficients is established. This general situation facilitates the derivation of dependences between transport values, since in certain cases the application of more complicated methods of thermodynamics of irreversible processes becomes superfluous. However, this approach is not applicable if the imposition of several transport processes leads to equilibrium in a uniform system.

11. Passive Gibbs forces

Discussing the question of relaxation times, we should note the path that was introduced by Gibbs when considering slow or 'frozen' processes. The criteria he introduced in discussing equilibrium and stability are formulated with respect to only possible changes. In the general case, processes that are 'essentially unrealistic' can be proposed (see pp. 63–64 [30]).

In Appendix 3 in [30] the following comment is given: "Here we are talking about states that, although they are not truly equilibrium, remain unchanged during the observation time. For example, in a mixture of hydrogen and oxygen under normal conditions, no changes are observed, although the state of such a system is not an equilibrium state. In such a system, a weak external action, such as an electric spark, can produce a noticeable change in state, and the degree of this change is in no way related to the intensity of the external action (with the intensity of the spark current or the intensity of the sunlight). In such situations, it is considered that the transition of the system to a state of equilibrium is prevented by some 'passive forces of resistance'."

The mechanical analogue of passive resistances can be the friction force at rest which keeps the body on an inclined plane. Gibbs distinguishes between the equilibrium caused by passive resistances and the equilibrium that is provided by the 'balance of active tendencies acting in the system' (p. 64, [30]). Forces of resistance, similar to viscosity or sliding friction, which only slow the transitions, but do not prevent them, are not enough to neglect the corresponding changes in state.

Thus, the term 'passive forces' means powerful enough force actions, the nature of which is not disclosed, and the introduction of the very notion of passive forces is associated with the time interval that the experimenter has in carrying out own measurements. The main issue of the correctness of the application of this method is the interpretation of the situation, depending on the accuracy and reliability of the measurements. For example, geological rocks can not be considered equilibrium due to their genesis, but the application of thermodynamic methods to them is a common practice and one of the necessary methods for their analysis [33], although the treatment of measurements performed, obviously non-equilibrium systems as thermodynamic rather than thermophysical ones, remains the responsibility of the authors. The same applies to macroscopic samples of metals and alloys consisting of grains of different sizes [10,24]. Another example of a wide range of opinions about relaxation times is the processes of adsorption deformation in which a rigid body is either inert [119] or deformed, but for some reason it is assumed that any initial state of the sample is equilibrium [85,120].

The need to take into account different relaxation times of components in complex systems is explicitly mentioned in the separation of components into mobile and frozen components [30]. The same principle is actively used in considering adsorption and absorption processes, when fixed components (solid-state atoms) create local fields (cells) that are filled with mobile (ad- or absorbed) components.

As the size of the system decreases, the relaxation times should decrease and this aspect remains often controversial due to the limited time of experimental measurements. In real systems, the times of relaxation processes always correlate with the sizes of small bodies and with the nature of the polydispersity of the material.

12. Necessity of taking into account relaxation times

The order of introduction of equilibria (first mechanical, and then chemical) in the derivation of the Kelvin equation is strictly fixed, and they can not be rearranged. The essence of the contradiction lies in the fact that different relaxation times are implicitly compared here: 1st for relaxation of pressure, and 2nd for relaxation of the chemical potential. In physical sense, the 1st relaxation time for pressure is much less than the second relaxation time for the chemical potential. In thermodynamics this remains behind the scenes; it is meant that we are talking about complete equilibrium or about many times longer than the relaxation time for the chemical potential.

Size and time. The concept of 'time' is also excluded from classical equilibrium thermodynamics, like 'size'. The size of the small body was introduced in equilibrium with the help of the Kelvin equation and, as shown above, it also automatically introduces the concept of the relaxation time. Thus, if we take a more general point of view than only the concept of complete equilibrium (without time) and take into account the differences in the relaxation times of different properties of the system, then the incorrectness of the Kelvin equation follows without the results of statistical physics. To do this, it is necessary to include the notion of different relaxation times for mechanical and chemical equilibria.

This is a fairly natural course of describing the properties of real systems, because it is directly connected with the concept of quasistatic transitions between the initial and final states of the system. More precisely, in a real experiment, there should be a consideration of the relaxation times for various properties that characterize the rate of convergence of the current state of the system with the equilibrium one. The specificity of all real processes requires differentiation of the conditions for reaching full equilibrium through three types of equilibria for specific properties of the system: mechanical, thermal and chemical. By virtue of the differences in properties, there is a clear system of relations between the relaxation times of $\tau_{mech} \leq \tau_{therm} \leq \tau_{chem}$ related to the transfer of impulse, energy and mass, which also follows from numerous experiments [82,92].

The appearance of the relaxation time of the properties due to the appearance of 'size' in thermodynamics naturally raises the question of the quasistatic process itself, and on what sizes it is performed. The smaller the body size, the less the relaxation time and the faster the equilibrium is established. Therefore, for the *macroscopic* sizes of thermodynamic objects, the question of the reality of attainability of equilibrium is not abstract. It requires experimental proof (and not just agreement with such an assertion), i.e., the applicability of the postulate of equilibrium must be proved experimentally, only after this it is possible to apply thermodynamic equations.

The question of the time factor plays a particularly important role in solid-phase systems. Real polydisperse materials, which are used in experiments with solids, consist of grains. The smaller the grain size, the faster the redistribution of atoms is achieved in it and the chemical equilibrium is established (but at the same time the role of the size factor and the inapplicability of macroscopic bonds for its recording increase). The larger the grain size, the slower the chemical equilibrium is established in it. As a result, the question of real equilibrium in most polydisperse solid materials often remains open. As examples, one should consider Elliot's data [123–125] and the overwhelming number of other experimental systems with complex structures of numerous phases [95–99]. All in all, this requires the introduction of a number of criteria on the correctness of the application of macroscopic laws of thermodynamics (in micro-non-uniform systems) for small particles and their ensembles.

General view of classical thermodynamics. An analysis of the problem of the Kelvin equation points to the need for a more general view of the conditions for the application of the postulates of thermodynamics, taking into account the inclusion of relaxation times of various properties of the system into consideration, as a necessary element for discussing the progress of any equilibrium establishment processes, which is actually the main task of thermodynamics.

In the transition to non-equilibrium thermodynamics, it is usually assumed [89,109–116] that the formulation of the combined equation of thermodynamics is sufficient for any local region dV. To justify this assumption, it is necessary to indicate the criteria that distinguish the region of locally equilibrium states. This will be done below, but it is prudent to formulate two refinements which are needed in the joint consideration of equilibrium and non-equilibrium processes. In part, these updates have already been introduced earlier, and here they are given for a more precise exposition of the part of the material that has not previously been accented.

For equilibrium Massey systems, a generalized equation was introduced for the First and Second Laws of thermodynamics (5.6) in the form $dU = TdS - PdV$, which became the basic equation of equilibrium thermodynamics. In developing the foundations of non-equilibrium thermodynamics, the condition of local equilibrium for elementary macroscopic volumes of macrosystems was introduced and this equation is written as the inequality $dU \leq TdS - PdV$ [89,109–116]. This automatically transferred all the initial positions of equilibrium thermodynamics to the region of non-equilibrium processes. Today, when discussing the state of classical thermodynamics, it is advisable to consider both types of processes together from a single point of view.

Refinement 1. On the need for a self-consistent description of the dynamics and equilibrium state of the system. Refinement 1 is already implied in the Second Law of thermodynamics when discussing the possibility of an exit and return of the system from an equilibrium state. It is a prerequisite for introducing the concept of 'affinity' in chemical thermodynamics, working both in equilibrium and in non-equilibrium [82,125]. It also serves as a justification for the introduction of the concept of thermodynamic functions in non-equilibrium states [109], since these functions automatically change to their equilibrium functions during the transition of the system to the equilibrium state.

Refinement 1 is necessary to introduce a single elementary volume dV in equilibrium and the dynamics to be found, as well as a criterion for the correctness of the construction of molecular models of kinetic stages and elementary stages. When the rates of reversible reactions are equal in the forward and backward directions, these models should ensure the same expressions for equilibrium constants as the equilibrium constants constructed in the framework of only equilibrium distributions. For ideal systems, this was discussed in section 10.

Refinement 2. On the need to take into account the differences in the characteristic relaxation times τ for the transfer of various properties: impulse (τ_{imp}), energy (τ_{ener}) and mass (τ_{mass}). These mechanical invariants retain their meaning also for macroscopic ensembles, both in equilibrium and in non-equilibrium thermodynamics and in statistical mechanics. Refinement 2 reflects the experience of studying dynamic processes and is necessary as a reflection of real experimental measurements. The relaxation time characterizes the process of establishing thermodynamic equilibrium in the macroscopic system under study. Because of the unequal physical parameters of the thermodynamic system, they tend to their equilibrium at different rates. This fact is reflected as the presence of 'partial equilibrium' when introducing the concept of affinity for non-equilibrium processes with chemical reactions occurring at constant pressures and temperatures (τ_{imp}, $\tau_{ener} \ll \tau_{mass}$) [82,125]. In this case, partial equilibrium with respect to the impulse means that there is no convective flow in the system, and partial equilibrium with respect to temperature means that there is no flow of heat. The complete equilibrium of the system corresponds to the times when the whole system achieves its equilibrium in all its parameters. Or this is achieved at times τ exceeding the relaxation time of the slowest thermodynamic parameter (τ_{imp}, τ_{ener}, $\tau_{mass} \ll \tau$).

Thus, the problem of describing the properties of small bodies leads to the necessity of simultaneously taking into account the basic positions of not only equilibrium but also non-equilibrium thermodynamics.

References

1. Frenkel J., Kinetic theory of liquids, Academy of Sciences of the USSR, 1945.
2. Skripov V.P., Metastable liquid. Moscow, Nauka, 1972.
3. Nanomaterials: Synthesis, Properties and Applications. Edited by A.S. Edelstein, R.S. Commarata. Bristol: Institute of Physical Publishing. Bristol and Philadelphia. 1996..
4. Gusev A.I., Rempel' A.A., Nanocrystalline materials. Moscow, Fizmatlit, 2001.
5. Uvarov N.F., Boldyrev V.V., Usp. khimii. 2001. V. 70. No. 4. P. 307. [Russ. Chem. Rev. 70, 265 (2001)]
6. Petrii O.A., Tsirlina G.A., Ibid. 2001. V. 70. No. 4. P. 330. [Russ. Chem. Rev. 70, 285 (2001)]
7. Haruta M., Date M., Appl Catal. A: General. 2001. V. 222. P. 427.
8. New materials. Team of authors, ed. Yu.S. Karabasov. Moscow, MISIS. 2002.
9. Daniel M.-C., Austric D., Chem. Rev. 2004. V. 104. P. 293.
10. Haruta M., Gold Bull. 2004. V. 37. No. 1–2. P. 27.
11. Chuvil'deev V.N., Non-equilibrium grain boundaries in metals. Theory and applications. Moscow. Fizmatlit. 2004.
12. Smirnov V.V., Lanin S. N., Vasil'kov A. Yu., Nikolaev S. A., Murav'eva G. P., Ty-

urina L. A., Vlasenko E. V., Ross. nakotekhnologii. 2007. V. 2. No. 1–2. P. 47. [Russ. Chem. Bull. 54, 2286 (2006)]
13. Rostovshchikova T. N., Smirnov V. V., Kozhevin V. M., et al., Ross. Nanotekhnol. 2 (1–2), 47 (2007)
14. Ozin G.A., Arsenalt A.C. Nanochemistry. A chemical approach to nanomaterials. Cambridge: The Royal Society of Chemistry. 2005.
15. Rambidi N., Berezkin A., Physical and chemical fundamentals of nanotechnologies. Moscow, Fizmatlit, 2008.
16. Fisher T.S., Lessons from nanoscience: A lecture notes series: Thermal energy at the nanoscale. V. 3, World Scientific, 2014.
17. Suzdalev I.P., Nanotechnology: physical chemistry of nanoclusters, nanostructures and nanomaterials. KomKniga, Moscow. 2006.
18. Cao G., Nanostructures & Nanomaterials. Synthesis, Properties & Applications. University of Washington: Imperial College Press. 2006.
19. Gemming S., Schreiber M., Suck J.B. Materials for Tomorrow: Theory, Experiments @ Modeling. Berlin. Heidelberg: Springer-Verlag. 2007.
20. Suzdalev I.P.. Electric and magnetic transitions in nanoclusters and nanostructures. Moscow. Krasand, 2011.
21. Handbook Springer of Nanotechnology, Bharat Bhushan (Ed.) 2nd revised and extended edition. Berlin - Heidelberg - New York: Springer. Science + Business Media Inc., 2007.
22. Gusev A.I. Nanomaterials, nanostructures, nanotechnologies. Moscow, Fizmatlit, 2007.
23. Ryzhonkov D.I., et al., Nanomaterials. Moscow, Binom, Laboratory of knowledge. 2008.
24. Zhilyaev A.P., Pshenichnyuk A.I., Superplasticity and grain boundaries in ultrafine-grained materials. Moscow. Fizmatlit. 2008.
25. Poole C., Owens F., Nanotechnology. Moscow, Tekhnosfera. 2009.
26. Nanostructured materials. Ed. R. Hannink, F. Hill. Moscow, Tekhnosfera. 2009.
27. Fakhl'man B., Chemistry of New Materials and Nanotechnologies. Dolgoprudnyi, Intellekt, 2011.
28. Eliseev A.A., Lukashin A.V., Functional nanomaterials. Moscow, Fizmatlit. 2010.
29. Vorotyntsev V.M., Nanoparticles in two-phase systems. Moscow, Izvestiya, 2010.
30. Gibbs J.W., Thermodynamics. Statistical mechanics. Moscow, Nauka. 1982.
31. Putilov K.A., Thermodynamics. Moscow, Nauka, 1971.
32. Krichevsky I.R., Concepts and the basics of thermodynamics. Moscow, Khimiya, 1970.
33. Voronin G.F., Fundamentals of thermodynamics. Moscow, Publ. MGU, 1987.
34. Adamson A. W., The Physical Chemistry of Surfaces,Mir, Moscow, 1979. [Wiley, New York, 1976]
35. Rusanov A.I., Phase equilibrium and surface phenomena. Leningrad, Khimiya, 1967.
36. Ono S., Kondo S., Molecular theory of surface tension. Moscow, IL, 1963. [Handbuch der Physik, Vol X (Springer) 1960]
37. Rowlinson, J., Widom B., Molecular theory of capillarity. Moscow, Mir, 1986. p. [Oxford: ClarendonPress, 1982]
38. Tovbin Yu.K., Theory of physico-chemical processes at the gas-solid interface, Moscow, Nauka, 1990. [CRC, Boca Raton, Florida, 1991].
39. Gufan Yu.M., Structural phase transitions. Moscow, Nauka, 1982.
40. Santen R.A. von, Sachtler W.M.H., J. Catal. 1974. V. 37. P. 202.

41. Williams F.L., Nason D., Surface Sci. 1974. V. 45. P. 377–408.
42. Kymar F., et al., Phys. Rev. B. Solid State. 1979. V. 19. P. 1954–1962.
43. Moran-Lopez J.L., Falicov L.M., Ibid. 1978. V. 18. P. 2542–2548.
44. Moran-Lopez J.L., Falicov L.M., Ibid. P. 2549–2554.
45. Farncuort H.E. Interphase gas–solid boundary, Ed. E. Flad, trans. from English. Ed. A.V. Kiselev. Moscow, Mir, 1970, PP. 359–370.
46. Lodiz R., Parker R. Growth of single crystals. Ed.A. A. Chernov, A.N. Lobachev. Moscow, Mir, 1974.
47. De Boer J., Dynamic nature of adsorption. Ed. B. M. Gryaznov. Moscow, IL, 1962. 290 p.
48. Crowell A., Interphase gas–solid boundary, Ed. E. Flad, trans. from English. Ed. A.V. Kiselev. Moscow, Mir, 1970. P. 150–174.
49. Vol'kenshtein, F.F., Physicochemistry of the surface of Semiconductors, Moscow, Nauka, 1973.
50. Kiselev V.F., Krylov O.V., Adsorption processes on the surface of semiconductors and dielectrics. Moscow, Nauka, 1978.
51. Steele W.V. Interphase gas–solid boundary, Ed. E. Flad, trans. from English. Ed. A.V. Kiselev. Moscow, Mir, 1970, P. 260.
52. Morrison S., Chemical physics of the surface of a solid body. Ed. F.F. Vol'kenstein. Moscow, Mir, 1980.
53. Krylov O.V., Kiselev V.F., Adsorption and catalysis on transition metals and their oxides. Moscow, Khimiya, 1981.
54. Roginsky S.Z., Adsorption and catalysis on non-uniform surfaces. Moscow and Leningrad, Publishing House of the USSR Academy of Sciences, 1948.
55. Dunning V.V., Interphase gas–solid boundary, Ed. E. Flad, trans. from English. Ed. A.V. Kiselev. Moscow, Mir, 1970. P. 230.
56. Jaycock M. Parfitt J., Chemistry of interfaces.– Moscow: Mir, 1984. – 270 p. [Wiley, New York, 1981].
57. Leygraf C, Hultquist G., Surface Sci. 1976. V. 61. P. 60.
58. Franken P.E.C., Ponec V., J. Catal. 1974. V. 35. P. 417.
59. Kugler E.L., Boudart M., Ibid. 1979. V. 59. P. 201.
60. Wang T., Schmidt L.D., Ibid. 1981. V. 71. P. 411.
61. Sedlacek J., Hilaire L., Legare P. et al., Surface Sci. 1982. V. 115. P. 541.
62. Ertl G., Norton P.R., Rustig J., Phys. Rev. Lett. 1982. V. 49. P. 177.
63. Griffits K., et al., Surface Sci. 1984. V. 138. P. 113.
64. Delmon B., The kinetics of non-uniform reactions. Ed. V.V. Boldyrev, Moscow, Mir, 1972.
65. Rozovskii A.Ya., Kinetics of topochemical reactions. Moscow, Khimiya, 1974.
66. Barre P., Kinetics of non-uniform processes. Ed. V.V. Boldyrev. Moscow, Mir, 1976.
67. Fast J.D., Interaction of metals with gases. Ed. L. A. Shvartsman. Moscow, Metallurgiya, 1975. V. 2.
68. Gel'd L.V., Ryabov P.A., Mokhracheeva L.P., Hydrogen and physical properties of metals and alloys: Hydrides of transition metals. Moscow, Nauka, 1985.
69. Andrievskii R.A, Umaiskii Ya.S., Interstitial phases. Moscow, Nauka, 1977.
70. Smirnov A.A., The theory of interstitial alloys. Moscow, Nauka, 1979.
71. Emanuel J.M., Knorre D.T., Course of chemical kinetics. 4th ed., Moscow, Vysshaya shkola, 1984.
72. Eremin E.Ya., Fundamentals of chemical kinetics. Moscow, Vysshaya shkola, 1976..
73. Frank-Kamenetsky D.A., Diffusion and heat transfer in chemical kinetics. Moscow, Publishing House of the Academy of Sciences of the USSR, 1967.

74. Ovchinnikov A.A., Timashev S.F., Belyi A.A., Kinetics of diffusion-controlled chemical processes. Moscow, Khimiya. 1986.
75. Fenelonov V.B. Porous carbon. Novosibirsk, Institute of Catalysis, 1995.
76. Modeling of porous materials. Novosibirsk: Siberian Branch of the USSR Academy of Sciences, 1976.
77. Kheifets L.I., Neimark A.V., Multiphase processes in porous bodies. Moscow, Khimiya, 1982.
78. Frevel L.K., Kressey L.J., Annal. Chem. 1963. V. 35. P. 1492.
79. Mayer R.P., Stowe R.A., J. Coll. Sci. 1965. V. 20. P. 893.
80. Plank M., Treatise on thermodynamics. New York, Dover Pub. 1945.
81. Guggenheim E.A., Modern thermodynamics, described by the method of J.W. Gibbs. Leningrad and Moscow, State. Nauch.-Tekh. Izdat. Khim. Lit., 1941.
82. Prigogine I., Defay R., Chemical thermodynamics. Novosibirsk, Nauka, 1966.
83. Sommerfeld A., Thermodynamics and statistical physics. Moscow, IL, 1955.
84. Leontovich M.A., Introduction to thermodynamics. Statistical physics. Moscow, Nauka, 1983.
85. Guggenheim E.A., Thermodynamics. An advanced treatment for chemists and physics, 5th edition. Amsterdam: North-Holland, 1967. PP. 166-169.
86. Semenchenko V.K., Selected chapters of theoretical physics. Moscow, Prosveshchenie, 1966.
87. Storonkin A.V., Thermodynamics of non-uniform systems. Leningrad, LGU, 1967.
88. Kubo R., Thermodynamics. Moscow, Mir, 1970. 304 p.
89. Bazarov I.P., Thermodynamics. Moscow, Vysshaya shkola, 1991.
90. Kireev V.A., Course of physical chemistry. Moscow, Khimiya, 1975. 773 p.
91. Kvasnikov I.A., Thermodynamics and statistical physics. V. 1: Theory of equilibrium systems: Thermodynamics. Moscow, Editorial URSS, 2002.
92. Landau L.D., Livshits E.M., Theoretical physics. V. 5. Statistical physics. Moscow, Nauka, 1964.
93. Van der Waals I.D., Kohnstamm Ph., Thermostatics course. Moscow, ONTI, 1936.
94. Sychev V.V., Differential equations of thermodynamics. Moscow, Vysshaya shkola, 19910.
95. Kurnakov N.S., Introduction to physical and chemical analysis. 4 ed., Moscow and Leningrad, 1940.
96. Anosov V.Ya., Pogodin S.A., The main principles of physical and chemical analysis. Moscow and Leningrad, 1947.
97. Mikheeva V.I. Method of physical and chemical analysis in inorganic synthesis. Moscow, Khimiya, 1975. 272 p.
98. Glazov VM, Pavlova L.M. Chemical thermodynamics and phase equilibria. Moscow, Khimiya. 1981.
99. West A.R., Solid state chemistry and its applications, Chichester, etc, John Wiley & Sons, 1984.
100. Young T., Philos. Trans. R. Soc. London. 1805. V. 95. P. 65.
101. Laplace P.S., Traite de Mecanique Celest; Supplement au dixeme livre, Sur l'Action Capillaire, Courcier, Paris. 1805. V. 4.
102. Thomson W.T., Phil. Mag. 1971, V. 42. P. 448.
103. Greg S., Sing K. Absorption, specific surface, porosity. Moscow, Mir, 1984.
104. Huang K., Statistical mechanics. Moscow, Mir, 1966.
105. Yang C.N., Lee T.D., Phys. Rev. 1952. V. 87. P. 404.
106. Lee T.D., Yang C.N., Phys. Rev. 1952. V. 87. R. 410.
107. Hill T.L., Statistical Mechanics. Principles and Selected Applications. Moscow: Izd.

Inostr. lit., 1960. – 486 p. [N.Y.: McGraw–Hill Book Comp. Inc., 1956]
108. Kvasnikov I.A., ref. [92], pages 113-116.
109. Haase R., Thermodynamics of irreversible processes. Moscow, Mir, 1967. [Dr. Dietrich Steinkopff, Darmstadt, 1963]
110. Onsager L., Phys. Review. 1931. V. 37. P. 405.
111. Prigogine I., Introduction to the thermodynamics of irreversible processes. Izhevsk, Regular and chaotic dynamics, 2001. [Charles C Thomas Sprinfield, Illinois, U.S.A., 1955]
112. de Groot S., Mazur P., Non-equilibrium thermodynamics. Moscow, Mir. 1964. [Amsterdam. North – Holland Publ. Company. 1962]
113. Gyarmati, I., Non-equilibrium thermodynamics. Field theory and variational principles. Moscow, Mir, 1974. [Springer, Berlin, Heidelberg, New York, 1970]
114. Landau, L.D., Lifshitz E.M., Theoretical physics. VI. Hydrodynamics. Moscow, Nauka, 1986.
115. Bird R.B., Stuart W.E., Lightfoot E.T. Transport phenomena. Moscow, Khimiya, 1974. [Wiley, New York, London, 1965]
116. Lykov A.V. Heat and mass transfer. Moscow, Energiya, 1978.
117. Eyring H. J., Chem. Phys. 1935. V. 3. P. 107.
118. Glasston S., Laidler K.J., Eyring H., Theory of absolute reaction rates. Moscow, IL, 1948 [Princeton Univ. Press, New York, London, 1941].
119. Hill T., Catalysis, questions of theory and methods of investigation, Moscow, IL, 1955.
120. Bakaev V.A., Izv. AN SSSR, Ser. khim. 1971. No. 12. P. 2648.
121. Rice O.K., J. Phys. Chem. 1927. V. 31. P.207.
122. Prigogine I., Defay R., J. Chim. Phys. 1949. V. 46. P. 367.
123. Elliott G.R.B., Lemons J.F., J. Phys. Chem. 1960. V. 64. P. 137.
124. Elliott G.R.B., Lemons, J.F., Advanced in chemistry series, No. 39. R.F. Gloud, Ed., Amer. Chem. Soc., Washington, D.C., 1964. P. 144, 153.
125. Roof R.B. Jr., Elliott G.R.B., Inorganic Chemistry. 1965. V. 4. P. 691.

2

Fundamentals of molecular theory

In development to the thermodynamic method, J.W. Gibbs (1902) [1, 2] also proposed a statistical method for studying the properties of macroscopic bodies, from the very beginning based on model atomic–molecular representations. The main task of statistical physics is to establish the laws of behaviour of a macroscopic quantity of matter, knowing the laws of the behaviour of particles from which the system is constructed (atoms, molecules, ions, electrons, photons, etc.). Accordingly, the conclusions of statistical physics are valid only to the extent that the assumptions made about the behaviour of the particles of the system are valid. From the very beginning, this method of solving the problem traces the atomic–molecular mechanism of phenomena. The statistical method makes it possible, on the one hand, to give a rigorous justification for the laws of thermodynamics, and on the other hand, to establish the limits of their applicability, and also to predict the conditions for violation of the laws of classical thermodynamics due to fluctuations and to estimate their scale.

All subsequent works of statistical physics [3–15] continue to use the method developed by Gibbs for the averaging procedures for ensembles of particles with respect to velocities and coordinates [2]. Statistical physics gave statistical interpretations of internal energy and entropy, previously conducted in thermodynamics. It is necessary to single out the general approach in static mechanics proposed by N.N. Bogolyubov [6,16], from a unified position considering equilibrium and non-equilibrium processes in the gas phase. Subsequently, these approaches were transferred to dense phases [17–30]. At present, they make it possible from a unified point of view to consider three aggregate states of a substance and their interface, both for equilibrium and non-equilibrium processes by means of a discrete–continual description of the particle distributions (based on the 'lattice gas model' (LGM)) [31–35].

13. Microscopic states of molecules and their description

The microscopic specification of the N body system includes: a) specifying the thermodynamic parameters of the system that determine its macroscopic state, or the specification of the external conditions into which the system under consideration is placed; b) the actual task of the system based on the atomistic approach: we know all the microscopic characteristics of the system, that is, the masses and structure of its constituent molecules, the charges and spins of the particles, the potentials of their interaction with external fields and with each other, etc.

The task of the system is carried out by specifying the total energy of the system (the Hamiltonian), which includes the kinetic energy T, the potential energy in the external field E_{ex}, and the internal energy of the interparticle interaction E_{in}. If the masses m_i of the particles of the system and their number N are given, then $T = T(p_1 ... p_N) = \sum_{1 \le i \le N} p_i^2 / 2m_i$, where p_i is the impulse of the i-th particle.

The potential energy of interaction with external fields, which are given as external parameters, is defined as the sum of the potential energies E_i of each of the N particles: $E_{ex} = \sum_{1 \le i \le N} E_i$. So for particles in a uniform vertical field of gravity $E_i = m_i g Z_i$, Z_i is the coordinate along the Z axis, etc. The potential E_i includes the potential of the walls of the vessel, fixing the external macroscopic parameter V, the volume of the system. The simplest and most convenient model for E_w is the impermeable wall model (a three-dimensional potential well bounded by an infinitely high vertical barrier): $E_w = 0$ if the point **r** is inside the vessel, or $E_w = \infty$ if the point **r** is outside the vessel [36]. In the case of finding a point **r** on the surface of a vessel, it is required to detail the potential type, or its calculation, for example, based on the atom–atom wall potential, depending on its structure and composition [31].

In determining the energy E_{in}, taking into account the interaction of the particles with each other, we will assume that this quantity is expressed as the total sum of all pair contributions, regardless of whether there are other particles around them or not. Such a potential of pair forces can be determined either experimentally when investigating data on the scattering of two isolated particles of the required sort by each other, or with the help of theoretical calculations.

$$E_{in} = \sum_{1 \le i < j \le N} E_{ij}, \tag{13.1}$$

where in the general case the potential E_{ij} depends on the arrangement of the particles (i.e., on $\mathbf{r}_i$ and $\mathbf{r}_j$), their orientations and the values of their spins, electric moments, if any, etc. Such interaction forces are called non-central, or tensor. When developing a number of specific problems, it is often enough to assume that the interaction of particles is central and isotropic, i.e. what $E_{ij} = E(r_i - r_j) = E(|r_i - r_j|)$. Such an approximation, when the interaction depends only on the modulus of the distance between the particles, from the physical point of view corresponds to the description of systems such as a gas or a liquid. Of course, it is not satisfactory when studying questions related, for example, to crystallization and the appearance of ordered spatial particle configurations.

In the simplest case, for a system N of spinless identical particles with a central interaction, the Hamiltonian is written in the form

$$H = \sum_{1 \le i \le N} \frac{\mathbf{p}_i^2}{2m_i} + \sum_{1 \le i \le N} E(\mathbf{r}_i) + \sum_{1 \le i < j \le N} E(|\mathbf{r}_i - \mathbf{r}_j|). \tag{13.2}$$

If necessary, terms with a more complex dynamic structure involving multiparticle interactions can be included in H.

The task of the statistical theory based on the microscopic task of the system is: (1) to determine all the thermodynamic equilibrium properties and characteristics of the system (the problem of statistical thermodynamics); (2) to evaluate the nature and magnitude of deviations of a given parameter of the system from its average value, that is, to construct a theory of fluctuations; (3) to investigate the simplest types of non-equilibrium processes occurring in statistical systems, their temporal evolution, i.e., to construct a kinetic theory. These tasks are implemented below under the LGM.

14. Continuous functions of molecular distributions

A natural approach to the study of many-particle systems is the use of probabilistic formalism. The microscopic idea of the initial discreteness of the system when using the probabilistic way of describing it is not preserved. The efforts of classical mechanics are aimed at finding the trajectories of the motion of each of the N particles of the system, which always preserves the discrete nature of the system. In statistical theory, the main attention is paid to the consideration of the probability density $w_N (\mathbf{r}_1, ..., \mathbf{r}_N, t)$ continuous in the space of coordinates and momenta of the field, which describes only the probability of detecting a system in a microscopic state corresponding to a set of its arguments. In the equilibrium theory, the time argument t is absent, and in the mechanical formulation of the problem it always remains a dynamic value.

The problem of constructing a statistical theory of equilibrium systems is the establishment of general expressions for statistical distributions, i.e., such distributions when the averages calculated with their help correspond to the observed macroscopic quantities that appear in thermodynamic relationships. Starting with [2], the calculations were constructed by direct averaging of the characteristics using the Gibbs distribution function $w_N (\mathbf{r}_1, ..., \mathbf{r}_N)$. This path is convenient for weakly interacting particles. More general and fruitful was the approach to the theory of non-ideal statistical systems, developed by N.N. Bogolyubov [6,16]. It is based on the idea of investigating not the integral value of the partition function $Q = Q(T, V, N)$, but the correlation properties of the particles of the system expressed in terms of the corresponding correlation functions. This makes it possible not to calculate the infinite–dimensional integral Q in the forehead, but to solve a system of several integro-differential equations for the correlation functions. The idea has acquired such a general significance in statistical mechanics that it embraced not only the theory of non-ideal equilibrium systems, but also the kinetic theory.

Correlation functions. In the equilibrium theory, the initial moment for constructing the correlation functions is the probability density of the particle momentum distributions in $6N$-dimensional phase space of particles and their coordinates $(p, r) = (p_1,..., p_N, r_1, ..., r_N)$ the canonical Gibbs distribution $w(p,r) = \text{const} \exp\{-\beta H(p,r)\}$, where $H(p, r)$ is defined in (13.2), $\beta = (k_B T)^{-1}$, k_B is the Boltzmann constant, and T is the temperature. The value $w_N(r_1,...,r_N)dr_1...dr_N = \dfrac{1}{Q_N}\exp[-\beta H(p,r)]\dfrac{dr_1...dr_N}{V^N}$ is the probability of detecting a state of the system when its particles are located in differentially small volumes $(r_1, r_1 + dr_1), ..., (r_N, r_N + dr_N) = \{r, r + dr\}$

(the curly brackets denote the complete set of quantities), Q_N is the statistical sum of a system of N particles

$$Q_N = \frac{1}{N!h^{3N}} \int_V \cdots \int_V dp_1 dr_1 \cdots dp_N dr_N \exp(-\beta H), \tag{14.1}$$

The function $w_N(r)$ itself is a general symbol, which is convenient for the formulation of connections between correlation functions of different nature and for connection with thermodynamic functions. Using this definition, we can introduce distribution functions that depend on the arguments of a small number of particles. Thus, the single-particle distribution function $F_1(r_1) = V \int_V w_N(r) dr_2 ... dr_N$ determines the probability of detecting a particle of the system in the volume $(r_1, r_1 + dr_1)$. Since all the N particles of the system are the same and the function $w_N(r)$ does not change when the particle indices are rearranged, instead of r_1 in this expression, any of the r_j, where $1 \le j \le N$.

The two-particle distribution function (often called the pair correlation function) $F_2(r_1, r_2) = V^2 \int_V w_N(r) dr_3 ... dr_N$ determines the probability of detecting one particle of the system in the volume $(r_1, r_1 + dr_1)$, and the other in the volume $(r_2, r_2 + dr_2)$. In the general case, the s-partial (s = 1, 2, 3, ...) distribution function

$$F_s(r_1, ..., r_s) = V^s \int_V w_N(r) dr_{s+1} ... dr_N \tag{14.2}$$

determines a similar probability for s selected particles of the system.

From the general properties of the introduced correlation functions, the conditions for their normalization and the relationship between functions of different orders are important, which are written as

$$\int_V \frac{F_s(r_1, ..., r_s)}{V^s} dr_{s+1} ... dr_N = 1, \quad \int_V \frac{F_s(r_1, ..., r_s)}{V} dr_s = F_{s-1}(r_1, ..., r_{s-1}), \tag{14.3}$$

and also the condition for weakening correlations with increasing distances between molecules: so for the simplest pair correlation function for $F_2(r_1, r_2)_{|r_1 - r_2| \to \infty} \to F_1(r_1) F_2(r_2)$ $|r_1 - r_2| \to \infty$, which means that the molecules become statistically independent of each other.

A similar expression for weakening the correlation can also be written not only between individual molecules, but also between their groups, when the distance between them exceeds a certain minimum distance, the so-called correlation radius, less than which the molecules can not be regarded as independent. First of all, it is determined by the radius of the interaction potential R_{lat}, which directly affects the behaviour of a pair of molecules, as well as their indirect influence through neighbouring molecules.

If the system is isotropic, then the pair function $F_2(r_1,r_2) = g(|r_1 - r_2|) \equiv g(r)$ depends only on one argument – the distance between the particles $r = |r_1 - r_2|$. In this case the probability of detecting a particle at a distance from r to $r + dr$ from some other particle $dw(r) = g(r)\frac{4\pi r^2}{V}dr$ is equal, therefore the function $g(r)$is called the radial distribution function of a gas or liquid.

The introduced correlation functions make it possible to express the thermodynamic properties of the system. In particular, the internal energy, the equation of state, etc. are expressed through the function $g(r)$. The total internal energy U, referred to the volume V, is written as follows

$$\frac{U}{V} = \frac{3}{2}\rho kT + \frac{4\pi\rho^2}{2}\int g_2(r)E(r)r^2 dr, \tag{14.4}$$

where ρ is the density of matter, $E(r)$ is the pair interaction potential, and the pressure P is expressed as

$$P = \rho kT - \frac{4\pi\rho^2}{6}\int g_2(r)\frac{\partial E(r)}{\partial r}r^3 dr. \tag{14.5}$$

15. Equations on the continuous functions of distributions

To circumvent the problems of direct calculation of various averages and corresponding statistical sums, it was proposed to construct systems of coupled equations for the distribution functions of different dimensions. This system of equations is called the BBGKY (Bogolyubov–Born–Green–Kirkwood–Yvon) chain of equations. It is obtained by differentiating the expression for the total partition function of the system from the coordinates of one of the particles.

If we denote the total distribution function by $F(\{N\})$, in the form [13]

$$F(\{N\}) = \frac{\lambda^N}{N!}\exp[F - \beta E(\{N\})], \tag{15.1}$$

where $F = -kT\ln(QN)$ is the free Helmholtz energy of a system of N particles, $\lambda = (2\pi mkT)^{3/2}$, then the s-partial correlation function $F(\{s\})$ can be expressed in terms of $F(\{N\})$ $F(\{s\})=\frac{\lambda^N}{(N-s)!}\int...\int \exp[F-\beta E(\{N\})]d\{N-s\}$ as $F(\{s\})$ where is the symmetrized function (14.2), the curly brackets mean the abbreviated entry for integrals whose number is $(N–s)$.

Denote by $D(\{N\})$ the quantity $\exp[F–\beta E(\{N\})]$ and write the identity: $D(\{N\}) = \exp[F–\beta E(\{N\})]$, which is differentiated by the coordinate of the i-th particle r_i, which gives

$$\frac{\partial D(\{N\})}{\partial r_i}+\beta\frac{\partial E(\{N\})}{\partial r_i}D(\{N\})=0. \tag{15.2}$$

Multiply expression (15.2) on the right and left by a factor $\lambda^N/(N–s)!$ and integrate over the coordinates $r_{s+1},...,r_N$. Taking into account the equality for the interparticle interactions (13.1), so that $\frac{\partial E_{in}(\{N\})}{\partial r_i}=\frac{\partial}{\partial r_i}\sum_{1\le i<j\le N}E_{ij}(r_i,r_j)=\frac{\partial E_{in}(\{s\})}{\partial r_i}+\frac{\partial}{\partial r_i}\sum_{s+1\le j\le N}E_{ij}(r_i,r_j)$ we write

$$\begin{aligned}&\frac{\partial F(\{s\})}{\partial r_i}+\beta\frac{\partial E_{in}(\{s\})}{\partial r_i}F(\{s\})+\\&+\beta\sum_{s+1\le j\le N}\int...\int\frac{\partial}{\partial r_i}E_{ij}(r_i,r_j)\frac{\lambda^N}{(N-s)!}D(\{N\})dr_{s+1}...dr_N=0\end{aligned} \tag{15.3}$$

Integrating over all coordinates $r_{s+1},...,r_N$ except one r_k $(k = s+1, ..., N)$ in the last term of the left side of the equation, we obtain the function $F_{s+1}(r_1,...,r_s,r_k)/(N–s)$. In this case, the factor $(N–s)!$ is replaced by $(N–s–1)!$, so that as a result of the summation the factor $(N–s)$ disappears. After this, the function in the integrand is defined by definition, $F(\{s+1\})$which gives the desired expression

$$\begin{aligned}&\frac{\partial F(\{s\})}{\partial r_i}+\beta\frac{\partial E_{in}(\{s\})}{\partial r_i}F(\{s\})+\\&+\beta\int\frac{\partial}{\partial r_i}E_{is+1}(r_i,r_{s+1})F(\{s+1\})dr_{s+1}=0.\end{aligned} \tag{15.4}$$

This structure of equations has the general property of non-closed: the equations for the correlation functions form a chain of successively entangled equations. Each equation for F_s contains in the integral term the function F_{s+1}, so that the solution of these equations must be preceded by the procedure for disengaging the chain, so that the remaining group of equations would be closed. The closure operation is performed in different ways, depending on the type of system under consideration and the physical conditions in which it is located. This question is examined in detail in various specific situations [9,15–29]. In principle, this universal approach allows solving almost any problem.

16. Discrete functions of molecular distributions

For dense phases (liquid, solid in bulks and for adsorbed phases on the surfaces of condensed phases), lattice models in which the states of molecules were characterized by a small number of parameters: the coordinate of the centre of the cell (or the lattice site) and its energy of interaction with neighbouring molecules were widely used. In addition to this set of state parameters, the lattice models differ in the nature of the site's occupancy states.

States of occupation of cells. The experimental data that appeared in the thirties testifying to the proximity of many properties of a liquid and a crystal: comparatively small relative changes in volume and energy during melting, close values of the heat capacity of substances in the liquid and solid states, short-range ordering in a liquid found in X-ray diffraction study, etc. [37], served as a basis for the transfer of the idea of lattice models for describing the liquid state. A model of non-spherical molecules was proposed in [38], and a theory describing all three aggregate states (gas, liquid, and solid) was developed in a series of papers [39, 40].

Lattice models allow us to reflect the most important part of potential functions in the vicinity of the minimum of the potential curve and all the numerous variants of these models are to some extent related to the properties of the intermolecular potential in the vicinity of this minimum. This is most clearly manifested in the fact that all lattice models operate with the number of nearest neighbours z, which is directly related to the local structure of the liquid. If for the adsorbed particles the number z is determined by the potential relief of the substrate, then for a liquid this number is the average statistical characteristic that is associated with the cooperative behaviour of a large ensemble of molecules and the properties of intermolecular interactions.

The central feature of lattice models is that the total volume of the system V is divided into some elementary volumes v_0, which form a periodic lattice. The statistical problem of the spatial distribution of molecules is greatly simplified by introducing the assumption of spatial regularity of the distribution of molecules. This replaces the continual description of the distribution of molecules to a discrete one. In addition, having introduced a number of assumptions, it is possible to simplify so much the problem of calculating the configuration integral of the system with allowance for intermolecular interaction, which is possible for its analytical solution.

Various variants of cell filling with different number of particles were proposed. There have been attempts to bring cellular models closer to real liquids, assuming that the cells in the system have different sizes and are placed irregularly [41,42]. Models with irregular cell structures are used mainly to calculate the radial distribution function; good results are also obtained for thermodynamic functions. It was also assumed that cells can contain one or more molecules, but they do not have the concept of a free cell.

The most important is the situation where each cell is allowed to contain one particle or to be empty – this is the so-called lattice gas model (LGM) that 'came' from the theory of adsorption: the well-known Langmuir model [43] is in fact the first formulation of the LGM, about filled and free sites. It should be clearly divided the use of LGMs from so-called lattice models. The account of vacancies allows to take into account the irregularity of the structure and does not require the artificial introduction of the concept of collective entropy [44]. These models are applicable not only to liquids, but also to gas; they are able to give an equation of state describing both phases simultaneously and transferring a liquid–vapour phase transition. Note that the LGMs and models of irregular cell structures [45] allow us to describe not only the liquid–vapour phase transition (this requires attraction between molecules), but also the liquid–solid phase transition, i.e., the liquid–solid phase transition or a phase transition of the melting type [46,47].

The LGM corresponds to three situations [9,27]: adsorption and absorption of gas-phase molecules, as well as solid solutions (the well-known analogs of the LGM are the binary alloy model without allowance for vacancies, and the Ising model with two directions of spin along and against the field [48]). If we confine ourselves to spherical molecules of the same kind, then the cell size v_0 is directly related to the size of the molecule, which is given by the parameter σ in the potential Lennard-Jones function (LJ). This parameter characterizes a solid incompressible sphere of a molecule whose nature follows their exchange interaction of the electron shells of two molecules as they approach each other, leading to repulsion of the molecules. Such a choice of the quantity $v_0 = \gamma_s\lambda^3$ imposes the condition of a single filling of the cell volume when a molecule is placed in it (γ_s is the cell shape factor, λ is the average distance between the particles). Taking into account the explicit form of the LJ potential, we have $\lambda = 2^{1/2}\sigma \approx 1.12\ \sigma$. The ratio $M = V/v_0$ determines the number of cells in the system under consideration.

Fixation of the molecule in the centre of the cell corresponds to its occupation state. Mathematically, this event is described by the value γ_f^i, where f is the cell number, $1 \le f \le M$, the index i denotes the occupation state of the cell with the number f. If there is a particle A in the site f, then $\gamma_f^A = 1$. If cell f is free, then there is a vacancy in it, therefore $\gamma_f^A = 0$, and $\gamma_f^V = 1$. For two types of occupation states $s = 2$ of any lattice structure site, corresponding to a one-component system for which i = A or V are vacancies, the random variables γ_f^i obey the following relations:

$$\sum_{i=1}^{s} \gamma_f^i = 1 \text{ and } \gamma_f^i \gamma_f^j = \Delta_{ij}\gamma_f^i, \tag{16.1}$$

where Δ_{ij} is the Kronecker symbol, which means that any site is necessarily occupied by some, but only one, particle. These conditions mean the absence of multiple filling of any cell. In particular, the equality $(\gamma_f{}^i)^k = \gamma_f{}^i$ of any integer $k > 1$ holds. In the general case, equations (16.1) hold for any value $s > 2$.

Particle configurations. The number of different states of the system is equal to s^M, where it increases sharply both with increasing number of occupied states of the site s and with increasing number of sites of the system M. In this connection, it is impossible to use the detailed description of real systems with the help of complete sets $\{\gamma_f^i\}$ and have go to their abbreviated description. (For example, detailed description is impossible at sufficiently small values of s and M, for example, for $s = 3$ and $M = 15$.) For a brief description of the state of the system, discrete many-particle distribution functions or correlation functions (simpler, correlators) are defined as follows

$$\theta_{f_1 \ldots f_m}^{i_1 \ldots i_m} = \sum_{k_1=1}^{s} \cdots \sum_{k_M=1}^{s} \prod_{n=1}^{m} \gamma_{f_n}^{k_n} P\left(\{\gamma_f^k\}, \tau\right), \text{ where } \gamma_{f_n}^{k_n} = \begin{bmatrix} 0, & k_n \neq i_n \\ 1, & k_n = i_n \end{bmatrix}. \tag{16.2}$$

where the sums are taken over all sites of the lattice (the types of particles in them are numbered by the symbols k_1, ..., k_M from 1 to s, i_f is the particle sort at site f; m is the order or dimension of the correlator characterizing the probability of the following particle configuration at time τ: in the site f_1 there is a particle i_1, in the site f_2 – a particle of the sort i_2, etc. up to the m-th particle inclusive. The state of occupation of the remaining (M–m) lattice sites is of no interest to us but averaging is carried out over them.

In the LGM, elementary events are described by the quantities γ_f^i (16.1), (16.2), and their various products are complex events. The probabilities of these complex events characterize the probabilities of the corresponding local particle configurations. For $m = M$, the correlator (16.2) is the total M-particle distribution function through which any correlator is expressed as

$$\theta_{f_1 \ldots f_m}^{i_1 \ldots i_m}(\tau) = \sum_{i_{m+1}=1}^{s} \ldots \sum_{k_M=1}^{s} \theta_{f_1 \ldots f_M}^{i_1 \ldots i_M}(\tau). \tag{16.3}$$

This connects correlators of the dimension m and m–1 with each other as

$$\theta_{f_1 \ldots f_{m-1}}^{i_1 \ldots i_{m-1}}(\tau) = \sum_{i_{m+1}=1}^{s} \theta_{f_1 \ldots f_m}^{i_1 \ldots i_m}(\tau). \tag{16.4}$$

Averaging over the states of occupation of cells determines the degree of their filling with particles of type i: $\theta_i = \langle \gamma_f^i \rangle$. This is the simplest single-particle (unary) correlator, which characterizes the probability of finding a particle i at the instant of time τ at the site with the number f – the local density of the component i at the point f at time τ (expressed in mole fractions). The numerical density θ_i is related to the concentration c_i as $c_i = \theta_i / v_0$.

The second-dimensional correlators are pair correlators

$$\theta_{fg}^{i_1 i_2} = < \gamma_f^{i_1} \gamma_g^{i_2} >_\tau \tag{16.5}$$

which characterize the probability of finding a particle i_1 at site f and particle i_2 at site g at time t. By virtue of the projection properties of the quantities γ_f^i (16.1) here $f \neq g$. Similarly, we can consider higher-dimensional correlators: the third, fourth, etc. order [27].

In the case of an equilibrium particle distribution, the normalized distribution function is written as [27]

$$P(\{\gamma_f^i\}) = \frac{1}{Q} \exp[-\beta H(\{\gamma_f^i\})], \tag{16.6}$$

where Q is a normalizing factor, usually called the partition function or the sum over the states of the system; $H(\{\gamma_f^i\})$ is the effective Hamiltonian (or Hamiltonian) function of the lattice system:

$$H(\{\gamma_f^i\}) = \tilde{H}(\{\gamma_f^i\}) - \sum_{f,i} \mu_i \gamma_f^i, \; \tilde{H}(\{\gamma_f^i\}) = H_{kin}(\{\gamma_f^i\}) + H_{pot}(\{\gamma_f^i\}). \quad (16.7)$$

In the formula (16.7) $\tilde{H}(\{\gamma_f^i\})$ is the Hamiltonian function, which characterizes the total energy of the system – the sum of the kinetic $H_{kin}(\{\gamma_f^i\})$ and potential $H_{pot}(\{\gamma_f^i\})$ energies; μ_i – the chemical potential of particles i – the distribution (16.6) is written for a large canonical ensemble. The expression for the partition function of the lattice system has the form

$$Q = \sum_{i_1=1}^{s} \ldots \sum_{i_M=1}^{s} \exp[-\beta H(\{\gamma_f^i\})]. \quad (16.8)$$

The concrete form of writing the Hamiltonian H is uniquely related to the physical model. The change in the physical model requires the construction of a new Hamiltonian. Changing the interaction potential, we will change the form $H_{pot}(\{\gamma_f^i\})$. In old lattice models, the particles are assumed to be fixed at the lattice sites. Their kinetic energy does not contain translational degrees of freedom, but consists of vibrational and, if the particle has a complex structure, from rotational degrees of freedom. Previously, it was believed that: 1) the motion of each particle independently of the motion of other particles is an analogue of the Einstein oscillator model [11]; 2) the expressions for $H_{kin}(\{\gamma_f^i\})$ are determined for the pure component (or for ideal models) and they do not change when interparticle interaction is taken into account.

To calculate the thermodynamic characteristics, averaging over all configurations is required. The possibility of solving exact problems even for discrete distribution functions, as for any other system of many interacting bodies, is quite unique. It is for such discrete productions of problems that few exact solutions have been obtained within the framework of non-one-dimensional structures [49]. One-dimensional structures always admit exact solutions [48,50]. This was first discovered in the work of Ising [48], after which such model systems became known as the Ising model. Nevertheless, in a number of situations, one-dimensional models are also useful in many problems of adsorption of molecules on the edges of faces, at the base of steps of stepped surfaces, on linear polymers, etc.

The exact solution was first obtained for the two-dimensional Ising model only in the absence of an external field. Onsager [51] studied a rectangular lattice; later, solutions were found for other types of two-dimensional lattices [49]. For a two-dimensional Ising model in a nonzero field and a three-dimensional model, the results, which can be called exact, are available only in the form of expansions in density and temperature [9, 52]. The presence of exact solutions for the two-dimensional case and practically exact solutions for three-dimensional systems makes the Ising model (and 'Ising' lattice gas) the most important model basis for studying the regularities of phase transitions and critical phenomena. Significant advances in the theory of phase transitions are associated with the consideration of the generalized Ising model, in which the variable characterizing the state of the site takes not two values, but more (in particular, when there is a continuous series of values).

Approximate methods. In the overwhelming number of situations where there are no exact solutions to the problem, it is necessary to use approximate solutions, which are also important for understanding the behaviour of cooperative systems. The question of the nature and accuracy of approximate calculations occupies the central part of the theory of condensed phases [27].

In the theory of discrete distribution functions, different ways of approximating the probabilities of many-particle configurations through the probabilities of configurations of smaller dimension are used. For simplicity, we consider a one-component system whose sites are occupied by particles A (we will assume that these are adsorbed particles A) or V (vacancies). The complete spectrum of the particle A configuration in the first coordination sphere around the central (not shown) particle on the lattices $z = 4$ and 6; adjacent particles A – blackened circles, the remaining sites of the first coordination sphere are occupied free (or occupied by the particle V). Let us denote by the symbol θ_i $(n\sigma)$ the probability that in the first coordination sphere of the central particle of type i, i = A and V, there are n particles A with an arrangement of σ (Fig. 16.1).

Examples of such approximations are shown in Fig. 16.2 for planar lattices with $z = 4$ (*a*) and 6 (*b*). The first variant of the approximation corresponds to the mean-field approximation when the probability of the multiparticle configuration is expressed in terms of the product of the probabilities of the sites to be occupied by their particles θ_i of type i (or through unary distribution functions), θ_i $(n\sigma) = \theta_i\theta_A^n\,\theta_V^{z-n}$ which are related to each other by the normalization condition

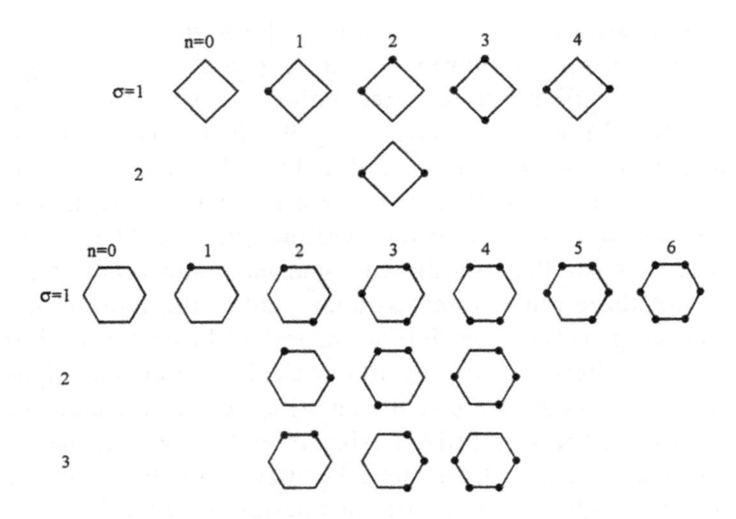

Fig. 16.1. The configurations of particles A in the first coordination sphere around the central particle on the sieves $z = 4$ and $z = 6$; blackened circles – adsorbed particles A; n is the number of particles A, $\sigma(n)$ is the type of configuration for a fixed value of n, $R = 1$ [27].

$\theta_A + \theta_V = 1$. Obviously, the method of arranging the particles, fixed by the symbol σ, does not matter. With this approximation, there are no correlations.

The second variant of approximation is also one-particle – in it the many-particle configuration of neighbours around the central particle is approximated through the product of unary distribution functions (polynomial approximation). The difference from the first way of disengaging is that the correlation itself between the central particle and its nearest neighbour is preserved.

The third variant of the approximation is that the probabilities of the many-particle particle configurations $\theta_i(n\sigma)$ are approximated in terms of the local probabilities θ_i for the detection of particles i (unary distribution functions) and in the pair distribution functions $\theta_{ij} = \left\langle \gamma_f^{\,i} \gamma_g^{\,i} \right\rangle$. This is the so-called Guggenheim's quasi-chemical approximation (QCA) [53] or the Bethe–Peierls approximation [54,55]. The correlation effect is explicitly taken into account via the pair distribution function θ_{ij}, which characterizes the probability of finding a pair of particles ij, ij = A, V at neighbouring sites. In this approximation, we have $\theta_i(n\sigma) = \theta_i t_{iA}^{\,n} t_{iV}^{\,z-n}$, where $t_{ij} = \theta_{ij}/\theta_i$ is the conditional probability of finding the particle j near the particle i; and $\theta_{iA} + \theta_{iV} = \theta_i$ or $t_{iA} + t_{iV} = 1$. Note that in QCA the weights of configurations with different values of σ (for n = const)

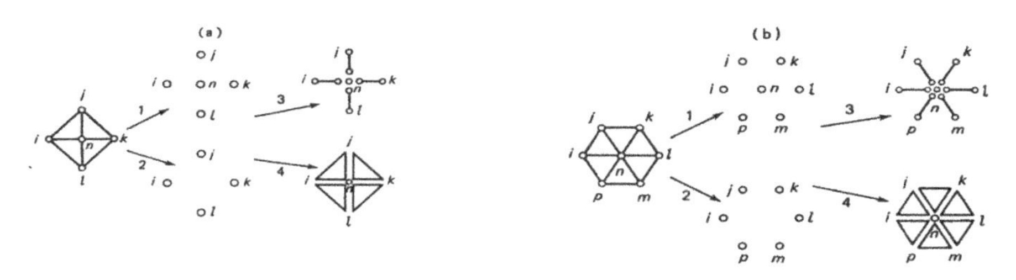

Fig. 16.2. Schemes of approximate calculations of many-particle configurations of neighbouring molecules for two-dimensional lattices $z = 4$ (a) and 6 (b): 1 - mean-field approximations and chaotic (through the concentrations of all molecules); 2 - polynomial approximation (through the concentrations of neighbouring molecules); 3 - quasi-chemical approximation (through the probabilities of pairs of molecules); 4 - allowance for indirect correlations in the quasi-chemical approximation and allowance for triple correlations (through the probabilities of triples of molecules) [27].

are equiprobable, or the symbol σ does not affect the calculation of $\theta_i\,(n\sigma)$.

The fourth variant of approximation is the simplest example of taking into account the correlation effects of three molecules. This approximation is a discrete version of the so-called Kirkwood superposition approximation [56], in which for three particles located at any distances an approximation is introduced of the probabilities of simultaneous realization of triples of particles θ_{ijk} of type *ijk* in the form $\theta_{ijk} = \theta_{ij}\,\theta_{kj}\,\theta_{ki}/(\theta_j\theta_j\theta_k)$ in terms of the probabilities of pairwise and unary distribution functions. Within the framework of such an approximation, the symbol σ determines contributions from pairs of particles at different distances. When using the superposition approximation, the problem is also closed through equations for unary and pairwise distribution functions.

The structure of the fourth approximation allows us to go beyond just the pair approximation, if we assume that additional equations will be built on the probabilities of triple configurations. However, such a path sharply increases the dimensionality of the system of equations, which must be solved in order to find the equilibrium particle distribution [57].

Adsorption systems. The widest application of lattice models is associated with the development of the possibility of simultaneously taking into account intermolecular interactions and heterogeneity of lattice structures. This is a typical case of adsorption systems, which are discussed in detail in monographs [27, 31, 58–62]. Here the lattice models are rigorously justified by the existence of a surface potential, creating regular or irregular regions of localization of adsorbed particles. The nature of the heterogeneity of lattice structure cells is determined by the potentials of the interaction of molecules with the surface of a solid and the chemical properties of atoms that form a surface and their structural organization. Heterogeneity is generated at the supramolecular level by the polydisperse structure of porous bodies, and at the atomic level by non-uniformity of open surfaces and pore wall surfaces: surface potential type, lattice defects near surfaces, the presence of different surface groups on the surface, dislocations and surface roughness, surface amorphization, etc. All these features of adsorption systems are conveniently described by the LGM.

Description of non-uniform systems. We will call the system non-uniform if, under the conditions of the equilibrium distribution of particles, the occupation states of different sites are different. This definition combines two different factors that lead to different occupation states by particles of the same kind of different sites. The first factor is the non-uniformity associated with the difference in interactions with the surrounding particles, which do not change their state at the characteristic relaxation times of the process. Examples are different adsorption surface centres and different absorption centres in the volume of the solid, which differ in binding energies with molecules of the gas phase. The second factor is the non-uniformity, which is related to the nature of the particle distribution due to the interaction between them. Examples are two- and multiphase systems and interfaces of stratifying phases, including the subsurface region of the gas-solid interface, as well as the ordering of particles on an uniform surface or in the volume of a solid. In the second case, the non-uniformity is due to the cooperative properties of the system, leading to the realization of phase transitions of stratification and (or) ordering. We note that the simultaneous 'action' of various factors that cause the non-uniformity of the system also leads to its non-uniformity. As an example, we point out the case of ordering of adsorbed particles (the 2nd factor) on the stepped surface (1st factor) [27].

Non-uniform lattices reflect a wide range of real situations, so in the following we formulate a general approach regardless of the physical nature of the non-uniformity. We introduce the quantities η_f^q characterizing the type q of the lattice site with the number f, $1 \leq f \leq M$, $1 \leq q \leq t$, t is the number of site types [27]. The quantities η_f^q are analogues of the quantities γ_f^i. That is, $\eta_f^q = 1$ if site f is a site of type q, and $\eta_f^q = 0$ otherwise. For them analogous relations

$$\sum_{q=1}^{t} \eta_f^q = 1, \quad \eta_f^q \eta_f^p = \Delta_{qp} \eta_f^q \tag{16.9}$$

The set of values $\{\eta_f^q\} = \eta_1^q, \ldots, \eta_M^q$ uniquely determines the non-uniformity of the lattice sites. However, γ_f^i and η_f^q there is a fundamental difference between the quantities: η_f^q the set is assumed to be fixed and unchanged, whereas over the set γ_f^i averaging is carried out. As in the case of quantities γ_f^i, it is difficult to work with a full set of values η_f^q and a transition to the distribution functions of non-uniform sites is necessary. Their dimension, as will follow from the following, depends on the interaction potential, on the approximation used for taking into account the interaction, on the type of the elementary process, and on the dimensionality of the particle configurations.

The state of occupation of sites of a non-uniform lattice, as above, is characterized by random variables γ_f^i, but the site itself is further characterized by its intrinsic property (type) through a given value η_f^q. In the general form, the correlators on a non-uniform lattice have the form analogous to (16.2):

$$\theta_{f_1 \ldots f_m}^{i_1 \ldots i_m}(q_1 \ldots q_m \mid \tau) = < \gamma_{f_1}^{i_1} \ldots \gamma_{f_m}^{i_m} \mid \eta_{f_1}^{q_1} \ldots \gamma_{f_m}^{q_m} >_\tau \tag{16.10}$$

In the formula (16.10), the types of sites on which the correlation function of dimension m is defined are additionally indicated. The same addition is present in all the above expressions (16.3)–(16.8), so there is no need to rewrite them for non-uniform lattices. The condition of statistical independence for correlators is written in the following form:

$$\theta_{f_1 \ldots f_m}^{i_1 \ldots i_m}(q_1 \ldots q_m \mid \tau) = \prod_{n=1}^{m} \theta_{f_n}^{i_n}(q_n \mid \tau) \tag{16.11}$$

It means that the filling of each site does not depend on the occupation state of other sites, that is, there are no correlation effects. Strictly speaking, such a situation is possible only in the absence of interaction between the particles. A special case of using condition (16.11) is the recording of the law of mass action (surfaces) for calculating the rate of chemical transformations in the kinetic regime.

Knowing the local correlators, it is easy to construct macroscopic characteristics of the particle distribution over the lattice sites. In the equilibrium on non-uniform lattices (surfaces), the nature of the filling of sites of each type is of great importance. To this end, in the expressions (16.10) there is an indication of the types of sites. A more 'active' role is played by the values η_f^q in the expressions for the local energy of the lattice systems, they reflect the influence of the site type on the energy characteristics of the bond and the activation energies.

17. Functions of molecular distributions in a discrete-continual description

The transition to discrete-continual models was connected with the solution of the problem of substantiating lattice models of a liquid [31,63,64] and in connection with the refinement of the description of the thermal motion of molecules inside cells [32]. The question of the validation of lattice models of a liquid was solved using the cluster approach [27, 65].

The molecular–statistical theory of equilibrium systems is based on the Gibbs distribution for microscopic states with a fixed selected set of macroscopic variables (defining the ensemble type) describing the observed state of the system [4–7]. For the canonical ensemble belonging to N particles of one kind that are in the volume V at temperature T, this distribution is indicated above in Section 14. The essence of the cluster approach consists in replacing the calculation of the partition function of the system under study by solving a system of equations for the cluster distribution functions that characterize the probabilities of realization different local configurations of molecules $\theta(1, \ldots, N)$ with coordinates $r_1, \ldots r_N$. This idea is similar to the idea of constructing equations for the correlation functions in Section 15.

To construct this system of equations, we choose m "central" sites in the centre of the region N. The width of the remaining region containing $(N-m)$ sites should be equal to the radius of the interaction potential between the particles R. In this case, particles in the central sites do not interact with particles in the sites outside the considered region. We denote such distribution functions as $\theta(\{m\}|N)$, where the symbol $\{m\}$ stands for the list γ_f^i, $1 \leq f \leq m$ and $1 \leq i \leq s$, the occupation states m of the central sites. If we fix the types of particles in the central sites and consider the relations between the functions $\theta(\{m\}|N)$ with different occupation states of the central sites, then, for example, for $m = 1$, we obtain relations of the type $\theta(i|N)/\theta(j|N)$ with particles of sorts i and j. These relations are very simply expressed in terms of cluster Hamiltonians (or the total energy of the central particles) $h(i|N)$ for a region containing N sites and a particle i at the centre, as

$$\theta(i|N) / \theta(j|N) = \exp\{\beta[h(j|N) - h(i|N)]\}, \qquad (17.1)$$

Using the principle of inclusion–elimination of probabilities and normalization relations, the functions $\theta(i|N)$ can be expressed in terms of a sequence of correlators of lower dimension (16.2)–(16.4).

The discrete distribution functions are in many respects similar to the functions of the continuum distributions (14.2) and (14.3), except for the fact that it is more convenient to introduce their normalization not into the entire volume of the system, but into the volumes of local sites – i.e. each local correlator is defined on the ensemble of copies of identical lattices. This allows us to introduce the concept of a fully distributed model in which each site can be considered as a separate sort of site, and any size group of sites of the same type can be formed from a set of identical sites. Their contributions will be taken into account by the weights of the corresponding functions of the site distributions for the non-uniform lattice (for details, see [27,31,66]).

Accounting for the interaction of nearest neighbours. Below is given a concrete example of constructing cluster equations in the case of interactions between the nearest neighbours. We shall assume that the internal degrees of freedom of the particles are weakly dependent on the change in the energy of the interparticle interaction of neighbouring particles when a particle is replaced one kind by another. Consider a uniform lattice system, any site of which can be occupied by a particle of sort i, $1 \leq i \leq s$, s is the number of components. If we denote the parameters of pair interaction of neighbouring particles of sort i and j, then the potential energy of the system will be composed of all possible pairs of interacting particles and the effective Hamiltonian will be written as [27,65]

$$H = \sum_{f,i} v_i \gamma_f^i - \frac{1}{2} \sum_{f,g} \sum_{i,j} \varepsilon_{ij} \gamma_f^i \gamma_g^j, \qquad (17.2)$$

where the sum over f is taken over all sites of the lattice, and the sum over g with respect to all z neighbours of site f; the sums over i and j are taken over all the states of occupation of the lattice sites (the factor 1/2 takes into account that each pair is calculated twice), v_i is the one-particle contribution of component i to the energy of the system.

In the cluster approach, the original lattice system is presented as a set of clusters. We confine ourselves to clusters with one central site. The initial lattice is presented as clusters consisting of $z + 1$ sites: the central site and its z neighbours. For each cluster, we can introduce the quantities h_f, characterizing the contributions to the total energy of the system of central particles (cluster Hamiltonians). When the interaction of the nearest neighbours is taken into account, the cluster Hamiltonian for a cluster with one central site has the form

$$h_f = \sum_{i=1}^{s} h_f^i, h_f^i = \left(v_i - \sum_g \sum_{j=1}^{s} \varepsilon_{ij} \gamma_g^j \right) \gamma_f^i \tag{17.3}$$

We define two types of cluster correlators. Correlators of the first type can be represented as follows:

$$\begin{gathered} \theta_{fg_1 \dots g_z}^{ij_1 \dots j_z} = \frac{1}{Q} \sum_{i_1=1}^{s} \dots \sum_{i_M=1}^{s} \gamma_f^i \prod_{g=1}^{z} \gamma_g^j \exp(-\beta H) = \\ \frac{1}{Q} \sum_{i_1=1}^{s} \dots \sum_{i_M=1}^{s} \gamma_f^i \prod_{j=1}^{s} (\gamma_g^j)^{n_j} \exp[-\beta(h_f + G_f)] \equiv \theta_{\{[n]\}}^i, \end{gathered} \tag{17.4}$$

where $G_f = H - h_f$. The second equality determines the transition from the indication of the sort of particle located at the site g, where g is any of the sites of the coordination sphere, to the number of energy bonds of the central particle i with neighbouring particles j in the coordination sphere of the cluster. On a uniform lattice, the energy of the central particle does not depend on the number of sites on which the particle and its neighbours are located, therefore, for simplicity, the subscripts of the sites f, g_1 ... g_z are omitted. Direct brackets in the functions $\theta_{\{[n]\}}^i$, where $\{[n]\} = [n_1 \dots n_s]$, mean that there are exactly n_1 particles of sort 1 in the coordination sphere of the cluster, exactly n_2 of particles of sort 2, etc. For unambiguous assignment of the occupation state of the sites of the coordination sphere it is sufficient to set $(s - 1)$ the value of n_j, since

$$n_s = z - \sum_{j=1}^{s-1} n_j, \quad 0 \le n_j \le z, \tag{17.5}$$

Thus, particles of sort s can be considered as an addition to the complete filling of the sites of the coordination sphere of the cluster. The brackets mean the full set of values of n_j. Correlators of the first type (17.4) characterize the probability of finding particle i in the centre of the cluster, and n_j particles of the sort j, $1 \leq j \leq s$, at the sites of its coordination sphere.

The expression for G_f in (17.4) does not depend on the occupation state of site f, but depends on the states of occupation of the sites of its coordination sphere; therefore, for any particular set of particles $\{[n]\}$, formula (17.4) can be rewritten as

$$\theta^i_{\{[n]\}} = \Lambda \exp[\beta(-\nu_i + \sum_{j=1}^{s} \varepsilon_{ij} n_j)], \quad \Lambda = \text{const}(\{[n]\})/Q, \tag{17.6}$$

where the unknown constant const($\{[n]\}$) takes into account the energy contributions from all the lattice sites, except for the central site of the cluster with particle i.

Correlators of the second sort are obtained from the correlators (17.4) when averaging over the occupation states of a part of the coordination sphere sites. Using the numbers of energy bonds of the central particle with its neighbours, these correlators will be written in the form

$$\theta^i_{\{n\}} = \left\langle \gamma^i_f \prod_{j=1}^{s-1} (\gamma^j_g)^{n_j} \right\rangle, \quad 0 \leq n_j \leq z, \tag{17.7}$$

The functions (17.7) characterize the probabilities that there is a particle i in the centre of the cluster, and in the coordination sphere of the cluster there are at least n_1 particles of sort 1, n_2 of particles of sort 2, etc., not less than n_{s-1} particles of the sort s–1. In this case, (17.5) does not hold, and in the notation (17.7), straight brackets are omitted.

All the thermodynamic characteristics of the system are expressed through the correlators (17.7). Formula (17.7) is a special case of formula (16.2), which characterizes the probability of finding an arbitrary configuration of particles on the lattice: there is a particle of sort i_1 at site f_1, a particle of sort i_2 is located at site f_2, and so on, up to the m-th particle inclusive.

So $\theta^i_{\{0\}} = \theta^i_{0\ldots0} = \langle \gamma^i_f \rangle = \theta_i$ represents the probability of finding a particle i at any lattice site; fixation of the value of θ_i in the sorption isotherms (adsorption or absorption) determines the external pressure of the molecules of the gas phase or vice versa.

The functions $\theta^{i}_{0\ldots j\ldots 0} = \langle \gamma^{i}_{f}\gamma_{g}^{\ j}\rangle = \theta_{ij}$ (with j-th index equal to one) are the probabilities of finding two particles i and j at neighbouring sites of the uniform lattice (pair functions). Knowing these averages, one can find the heats of sorption or mixing, heat capacity, etc.

The introduction of correlators of two sorts makes it possible to reduce the entire procedure for constructing systems of equations for correlators only to work with the probability relations between them. Let's consider the steps of this work for a binary system.

We construct equations describing the local distribution of the particles A and B of the binary solution in the equilibrium state [27, 65]. For clarity, we confine ourselves to taking into account the pair interactions between the central particle and its neighbours, according to Eq. (17.2). In the first stage, we use expression (17.6), and consider the ratio of the correlators of the first sort, which differ in the sort of the central particle, but have the same state of occupation of the sites of the coordination sphere:

$$\begin{aligned}&\theta^{B}_{[n]} = M_n\theta^{A}_{[n]},\ M_n = \exp[-\beta(\nu + n\omega)],\\&\nu = \nu_B - \nu_A + z(\varepsilon_{AB} - \varepsilon_{BB}),\ \ \omega = \varepsilon_{AA} + \varepsilon_{BB} - 2\varepsilon_{AB},\end{aligned} \tag{17.8}$$

where n is the number of particles A in the coordination sphere of the cluster. The ratio of the functions (17.6) eliminates the unknowns Q and const([n]}).

In the second stage, we consider the connection of the correlators of the first and second types. By definition of correlators of the second type, containing not less than n particles A in the coordination sphere of the cluster, we represent

$$\theta^{A}_{n} = \sum_{k=0}^{z-n} C^{k}_{z-n}\theta^{A}_{[n+k]},\ 0 \le n \le z, \tag{17.9}$$

The coefficients C^{k}_{z} of (17.9) are easily obtained from the following arguments. Adding one more particle to the n particles, we pass to the configuration containing $n + 1$ particles A and characterized by the correlator $\theta^{A}_{[n+1]}$. Such a transition is possible in (z–n) ways. A similar transition to (n+2) particles is possible with the addition of two particles A (z–n) (z–n–1)/2 ways, etc. The coefficients C^{k}_{m} are the number of combinations of m elements in k: $C^{k}_{m} = m!/(k!(m-k)!)$.

Turning the linear system of equations (17.9) with respect to correlators of the first sort, we find

$$\theta_{[n]}^A = \sum_{k=0}^{z-n} (-1)^k C_{z-n}^k \theta_{n+k}^A . \tag{17.10}$$

A similar expression holds for correlators of the first type with a central particle B.

In the third stage, we substitute the expressions (17.10) for the central particles A and B into relations (17.6):

$$\sum_{k=0}^{z-n} (-1)^k C_{z-n}^k \theta_{n+k}^B = M_n \sum_{k=0}^{z-n} (-1)^k C_{z-n}^k \theta_{n+k}^A \tag{17.11}$$

This linear system of equations connects the unknown functions θ_n^A and θ_n^B with each other. It is solved sequentially, beginning with $n = z$. Its solution has the form

$$\theta_n^B = \sum_{k=0}^{z-n} C_{z-n}^k M_{n+k} \sum_{r=0}^{z-n-k} C_{z-n-k}^r \theta_{n+k+r}^A = \exp[-\beta(\nu + n\omega)] \sum_{k=0}^{z-n} C_{z-n}^k x^k \theta_{n+k}^A \tag{17.12}$$

where $x = \exp(-\beta\omega) - 1$. In the second equality, the terms are rearranged so that the first sum changes the dimension of the correlators, and the energy factors are in the internal sum, and the explicit form M_n (17.10) is taken into account.

The system of equations (17.12) is the initial one for the calculation of cluster correlators. With a suitable sequential search of all lattice sites, the system (17.12) allows in principle to obtain an exact solution. This is done for a one-dimensional lattice [67]. For two- and three-dimensional lattices, the number of equations of the system (17.12) is less than the number of unknowns θ_n^A and θ_n^B (taking into account the normalization condition and the invariance condition for the filling of the central and neighbouring cluster sites), so it must be closed. Depending on the method of closing the system of equations (17.12) by approximating the higher correlators through correlators of lower dimensionality, we obtain different approximations differing in the accuracy of taking into account the correlation effects. The structure of the equations obtained is such that it is necessary to formulate a general rule for approximating all higher correlators through the lower ones, which determines the functional dependence of the closed system for uncoupled correlators. After finding the uncoupled correlators, this, in turn, allows one to self-consistently calculate all higher correlators, both of the first and second sorts in the approximation under consideration.

Continuous distribution of molecules. To describe the distribution of molecules, we use the continuous analog of the cluster approach [63, 64, 68]. For a group with the same number of molecules N, but differing in their coordinates, we consider the ratio $F(\{s\})$ of the functions defined in Section 15 below (15.1) and replaced by the symbol $\theta(\{r_j\})$ (as above, $\{r_j\}$ is the complete set of coordinates particles)

$$\theta(r_1,\ldots,r_N) = \theta(r_1^*,\ldots,r_N^*)\xi \exp\left\{\beta \sum_{1\le i<j\le N} [\varepsilon(|r_i^* - r_j^*| - \varepsilon(|r_i - r_j|)]\right\},$$

$$\xi = \Lambda/\Lambda^*,\ \Lambda = \int_V \cdots \int_V dr_{N+1}\cdots dr_M \exp\left\{-\beta\left[\sum_{1\le i\le N;N+1\le j\le M} \varepsilon(|r_i - r_j|) + \sum_{N+1\le i<j\le M} \varepsilon(|r_i - r_j|)\right]\right\} \tag{17.13}$$

where the coordinates of the molecules in different groups are denoted by r_i and r_i^*, respectively (the coordinates with an asterisk correspond to Λ^*).

Let us select from N molecules a group containing m molecules in a sphere with volume ω, which we will consider to be central. The volume ω is surrounded by a sphere ω_R whose radius is not less than $\omega^{1/3} + R$, where R is the radius of the interaction potential between the molecules. In the volume $(\omega_R - \omega)$ there are the remaining $(N-m)$ molecules. If their coordinates are fixed, then independently of the position m of the central molecules $\xi = 1$ and formula (17.13) is rewritten as

$$\theta(r_1,\ldots,r_m | r_{m+1},\ldots,r_N) = \theta(r_1^*,\ldots,r_m^* | r_{m+1},\ldots,r_N) \exp\{\beta \times$$
$$\times\left(\sum_{1\le i<j\le m} [\varepsilon(|r_i^* - r_j^*|) - \varepsilon(|r_i - r_j|)] + \sum_{1\le i\le m;m+1\le j\le N} [\varepsilon(|r_i^* - r_j|) - \varepsilon(|r_i - r_j|)]\right)\Bigg\} \tag{17.14}$$

Expressions (17.14) are a set of relationships that differ in the different locations of both the central molecules and molecules in the surrounding region $(\omega_R - \omega)$.

In the case of a discrete distribution of molecules, these relationships were used in the cluster approach for lattice structures at $m = 1$ and 2 [23, 25]. The same relationships can be used to describe systems with a continuous distribution of molecules. To prove the latter, we need to consider the system (17.13), in which

N is replaced by m, and the value of M by N, and put $r_i = r_i^*$, $2 \le i \le m$, $r_1^* = r_1 + dr_1$. Expanding the expression in (17.13) in terms of dr_1, we obtain

$$\frac{\partial}{\partial r_1}\ln\theta(r_1,\cdots,r_m)+\beta\frac{\partial}{\partial r_1}E_m+\frac{\partial}{\partial r_1}\xi=0, \qquad (17.15)$$

where E_m is the energy of the group of m molecules, defined in (13.1), and the derivative with respect to ξ (according to (13.1), definition $F(\{s\})$ (14.2) and (17.13)) leads to the following expression

$$\frac{\partial}{\partial r_1}\xi=\frac{\beta(N-m)}{V\theta(r_1,\cdots,r_m)}\int_V dr_{m+1}\theta(r_1,\cdots,r_{m+1})\frac{\partial}{\partial r_1}E(|r_1-r_{m+1}|). \qquad (17.16)$$

Thus, equations (17.15) and (17.16) are a system of integro-differential BBGKY equations [6] (they can be transformed in (15.4)). Equations (17.13) determine the relationship of different configurations of groups consisting of the same number of molecules, which differ in their coordinates. Differential changes of these probabilities with a coordinate change in the positions of *each* of the molecules of the group obey the system of BBGKY equations.

Equations (17.14) are one of the specific methods for locating molecules that obey the integral relations (17.13). The procedure for enumerating molecular arrangements in the regions ω and $(\omega_R - \omega)$ can be organized arbitrarily. This question is discussed in [64, 68] by dividing the volume of the system into a fine grid, smaller than the diameter of the molecules. It is useful to note that for the first time the idea of introducing a shallow 'grid' in the description of continual quantities was introduced by Boltzmann [69] in calculating the entropy for the velocity distribution of molecules in an ideal gas, which led to the well-known results on the increase of entropy under equilibrium conditions (see also [70]).

In the case of a fine grid, the difference system of equations (17.4) for discrete distribution functions gives a description of the structure of the liquid state, as in theories of integro-differential or integral liquid equations. This explains that in [71,72] a procedure was given for constructing the average values of the effective parameters of the interparticle interaction with respect to the continual potential curves in the mean-field approximation. This procedure, which reveals the meaning of effective parameters, retains its value even with the new procedure for constructing successive approximations for different lattice parameters λ_n. Lattice fluid models have a rigorous statistical justification and are directly related to the functions of molecular distributions, which are described by the BBGKY equations. Traditional lattice models are the first step in the sequence of partitioning of space into elementary volumes, which in the limit approach a continual description. This procedure provides an arbitrarily close approach to the continual description.

The first approximation is sufficient to obtain the thermodynamic properties of simple liquids and their mixtures, but in order to describe the structural characteristics, subsequent approximations are required that allow us to take into account not only the average density of the system but also the inhomogeneities of the local density. Approximation to the continual description can be achieved both by increasing the number of places occupied by a molecule in the lattice, which is equivalent to a finer division into cells for a molecule of a fixed size, and also taking into account the displacement of the centre of mass of molecules from the centre of the cell. For lattice systems with molecules that block several cells, the general procedure for their statistical analysis based on the cluster approach is given in [65,73]. The breakdown of the volume into cells in the LGM does not impose any restrictions on the motion of the particles; the latter can move around the entire volume. They take into account not only the 'standard' positions of the molecules in the centre of the cell, but also the displacements of the particles from the indicated positions, which allows obtaining the structural characteristics of the liquid.

The static justification of the LGM is key to the possibility of a unified description of all three phases. The initial setting of the LGM strictly bound to the crystalline lattice of a solid limited its use to solids and their surfaces. The removal of this restriction provides a general approach for liquid and vapour, as well as for all their interfaces.

18. Connection between thermodynamic functions and correlation functions

A system of equations for the pair distribution functions in the QCA, in which the pair functions are related to each other as

$$^*\theta_{fg}^{AA}(r)^*\theta_{fg}^{VV}(r) = ^*\theta_{fg}^{AV}(r)^*\theta_{fg}^{VA}(r), \quad ^*\theta_{fg}^{ij}(r) = \theta_{fg}^{ij}(r)\exp[-\beta\varepsilon_{fg}^{ij}(r)]. \quad (18.1)$$

and this system allows us to express the thermodynamic functions $\sum_{\lambda=1}^{\Phi}\theta_{fg}^{i\lambda}(r)=\theta_f^i$, $\sum_{l=1}^{\Phi}\theta_{fg}^{i\lambda}(r)=\theta_g^{\lambda}$ for a uniform volume with allowance for lateral interactions inside R_{lat} c.s. in the traditional form for a discrete lattice.

The summary energy of a one-component system is written as the sum of the contributions $F = F_{lat} + F_{vib} + F_{tr}$ from the lattice system $F_{lat} = E_{lat} - TS_{lat}$ and the vibrational (F_{vib}) and translational (F_{tr}) motions of the particles. In particular, $E_{lat} = -\langle H\rangle$ – internal energy, and S_{lat} – entropy, are written as

$$E_{lat}=-\sum_f \theta_f^A v_f^A+\frac{1}{2}\sum_r\sum_{f,g}\varepsilon_{fg}^{AA}(r)\theta_{fg}^{AA}(r) \tag{18.2}$$

$$S_{lat}/k_B = \sum_{f,i=A,V}\theta_f^i\ln\theta_f^i+\frac{1}{2}\sum_r\sum_{f,g}\sum_{i,j=A,V}\left[\theta_{fg}^{ij}(r)\ln\theta_{fg}^{ij}(r)-\theta_f^i\theta_g^j\ln\theta_f^i\theta_g^j\right] \tag{18.3}$$

In the expression for E_{lat}, there are only functions θ_f^A and $\theta_{fg}^{AA}(r)$ for particles A, whereas in the expression for S_{lat} there are all sorts of occupation states of sites (particles and vacancies). The expressions for F_{vib} and F_{tr} are concretized in the approximations used [31].

The isothermal relationship between the pressure in the thermostat and the density inside the system (the expression for the chemical potential) has the form

$$\begin{gathered}\theta_f^s=\theta_f^i\exp\beta(v_f^i-v_f^s)\prod_{r=1}^{R_{lat}}\prod_{g\in z_f(r)}S_{fg}^i(r),\ \sum_{j=1}^{s}\theta_f^j=1,\\ S_{fg}^i(r)=1+\sum_{j=1}^{s-1}t_{fg}^{ij}(r)x_{fg}^{ij}(r),\ \ x_{fg}^{ij}(r)=\exp[-\beta\varepsilon_{fg}^{ij}(r)]-1.\end{gathered} \tag{18.4}$$

The equation of state (pressure inside the system)

$$\begin{gathered}P=\frac{kT}{\mathrm{v}_0}\theta-\frac{1}{2d\mathrm{v}_0}\sum_{\chi=1}^{R_{lat}}\sum_{f,g}\sum_{i,j=1}^{s-1}\int_{\mathrm{v}(f)}\int_{\mathrm{v}(g)}\theta_{ij}(r_{fg}\mid\chi)r_{ij}\times\\ \times(\partial\varepsilon_{ij}(r_{fg}\mid\chi)/\partial r_{ij})dr_i dr_j,\end{gathered} \tag{18.5}$$

where d is the dimension of the lattice.

Here it is assumed that the internal degrees of freedom (translational and oscillatory motion) are separated from the configuration states. These expressions are common for three aggregate states of matter. Their generalization in the framework of a discrete–continual description is given in [73, 74].

The essence of the QCA was discussed in Section 16. In the existence of two phases, it leads to the so-called van der Waals loop, and to the necessity of using the Maxwell rule to determine the densities of coexisting phases of gas and liquid [9,11,27]. The meaning of Maxwell's construction and the essence of metastable states that are inside the stratification curve (or the binodal curve) in the statistical theory is provided by the Yang–Lee condensation theory [9,11,105,106].

The Yang–Lee condensation theory and isotherm loops. The Yang–Lee theory [105,106] is based on the study of the number of roots of a large partition function of the system

$$\Xi(z,V)=\sum\nolimits_{N=0}^{\infty} z^N Q_N(V), \text{ where } Q(V)=\frac{1}{N!\lambda^{3N}}\int d^{3N}r \exp[-\beta E(r_1,...,r_N)],$$

$$E(r_1,...,r_N)=\sum_{i<j} u(r_i,r_j), \quad u(r_i,r_j)=u(|r_i-r_j|),$$

for a fairly general sort of potential short-range interactions between molecules. Let the interaction potential be given in the form of a solid sphere ($u(r) = \infty$, $r \le a$) with an attractive potential of radius r_0 and maximum depth $-\varepsilon(u(r) = -\varepsilon$, $a< r <r_0)$, and $u(r) = 0$, $r \ge r_0$. The equation of state is obtained by eliminating the value of z from the two parametric equations $\beta P=\frac{1}{V}\ln\Xi(z,V)$, and $\frac{1}{v}=\frac{1}{V}z\frac{\partial}{\partial z}\ln\Xi(z,V)$, where the quantity v is the specific volume, i.e. a parameter that does not depend on the total volume V; z is activity, defined as $z = \exp(\beta\mu)/\Lambda^3$, μ – chemical potential, $\Lambda^2 = h^2/(2\pi mkT)$, h – Planck's constant.

The idea of the theory is to investigate the singularities of the behaviour of the limiting relations $\beta P=\lim\limits_{V\to\infty}\left[\frac{1}{V}\ln\Xi(z,V)\right]$ and $\frac{1}{v}=\lim\limits_{V\to\infty}\left[\frac{1}{V}z\frac{\partial}{\partial z}\ln\Xi(z,V)\right]$, without making an explicit calculation of $P(v)$, and in the limiting case $V\to\infty$, then the detection of the singularity of the equation of state can be interpreted as a phase transition.

For small particle numbers N, the large partition function $\Xi(z,V)=1+zQ_1(V)+z^2Q_2(V)+...+z^{N_m}Q_{N_m}(V)$ is a polynomial of degree N with positive coefficients (here N_m is the maximum number of spheres that can be placed inside a given volume V), and this polynomial can not have real roots. Such roots can appear only as a result of the limiting transition $N \to \infty$, under the condition $V \to \infty$, so that N/V = const. The passage to the limit $V \to \infty$ must be understood in a strict mathematical sense. In particular, the operations of passing to the limit *lim* with $V \to \infty$ and taking the derivatives $z(d/dz)$ may not be permutable.

The Yang–Lee theory showed that there really are situations when such singularities exist [75]. In the particular case of the two-dimensional Ising model, for which the partition function can be calculated accurately, the correctness of this study [76], independent of a particular type of potential, was shown, and not related to the specific type of closure of higher DFs (distribution functions) through the lower ones. The phase transition is completely characterized by the distribution function of the roots. The Ising model undergoes a first-order phase transition, and mathematical analysis provides answers to many questions of the theory of phase transitions. Omitting the proof of this theory, we formulate its main corollaries for first-order phase transitions.

This theory justified the correct application of Maxwell's rule, to search for coexisting densities and saturated vapour pressure using the secant so-called van der Waals loop, which is obtained in all approximate methods of calculating the equations of state (and / or isotherms). This proof allows completely to refuse any information about the properties of the approximate approach used, since in it there are no intermediate points relating to the two-phase region of the system or points within the stratification curve (or binodal). Note that the Maxwell rule itself uses the isotherms of approximate equations, so it was important to show that the properties of the limit points themselves – the coexisting vapour and liquid densities – do not depend on the type of approximate isotherms (although the type of approximation used determines their concrete position, as well as the saturated vapour pressure P_s).

Figure 18.1 *a* shows the curve (1) of the isothermal relationship between the density and the chemical potential, calculated by the approximate method (according to equation (18.4)), at which a van der Waals type loop appears. The position of the secant Maxwell, determined from the condition of equality of areas above and below the secant, gives the saturated vapour pressure P_s and the density of the coexisting phases of the fluid θ_f and the pair θ_v.

We recall that the Maxwell rule can be applied in the 'pressure-volume' coordinates for a fixed number of molecules, i.e. 'pressure-specific volume', or in the coordinates 'chemical potential–density' (Here, numerical densities are used instead of the usual specific volumes, defined as $v = 1/\theta$ [11]). But Maxwell's rule does not hold in the coordinates 'pressure–density' and 'chemical potential–specific volume' coordinates [9,77]. Figure 18.1 *b* shows shaded areas on an isotherm loop of equal area; v^+ corresponds to the specific volume of the liquid phase, v^- to the gaseous phase (the vertical curve v = 1 corresponds to the limiting value of volume on a rigid lattice in the vapour–liquid system, the value of v for a compressible lattice may be less than unity [31]).

The results of the Yang–Lee theory are obtained on a complete set of molecular configurations, which rejects any treatment of the so-called metastable states as equilibrium states. They indicate that the binodal curve is a set of singular points for which one-sided limits are defined in the direction of the nearest point of the binodal, but there are no limits for any directions from the interior points to the binodal. Or, that only two-phase solutions correspond to strictly equilibrium solutions below the critical point. Only such solutions satisfy the thermodynamic conditions for the phase equilibrium of the phases α and γ: $T_\alpha = T_\gamma$, $P_\alpha = P_\gamma$, $\mu^i_\alpha = \mu^i_\gamma$. All other solutions within the region of two-phase solutions are not equilibrium. Therefore, this theorem uniquely eliminates any metastable state from the number of equilibrium states.

The formal continuation of single-phase solutions to the two-phase region is possible for any phases. If the phases have one symmetry, as in the case of vapour and liquid, then the continuation of the solutions leads to the van der Waals loop, which is the consequence of using the canonical distribution. This distribution follows from the full statistical sum, if only its maximum term is used, instead of completely taking into account all the terms. It is this simplification, which is purely technical in the calculation of isotherms, that is responsible for the appearance of loops. When going over to the full spectrum of particle configurations, this loop disappears even in approximate approaches [9], which completely agrees with the result of Li-Yang's work [76], since all the results in the LGM (in this book) belong to the class of quasi-Ising models for which it was performed, including, if we use the most rough approximation of correlation effects, the mean-field approximation [9]. If crystals with different symmetries participate in the phase transition, as is the case in solid-state transitions of the second kind, then the equations of state and expressions for the thermodynamic functions for each phase are different, and as stated above, the Maxwell rule is not applicable. A similar situation exists for the solid–liquid transition [12, 78].

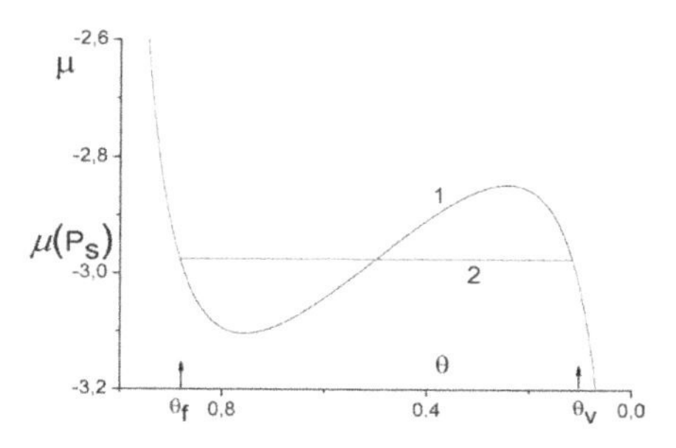

Fig. 18.1a. Isotherm with van der Waals loop (1) and Maxwell cross section (2) in the 'chemical potential–density' coordinates.

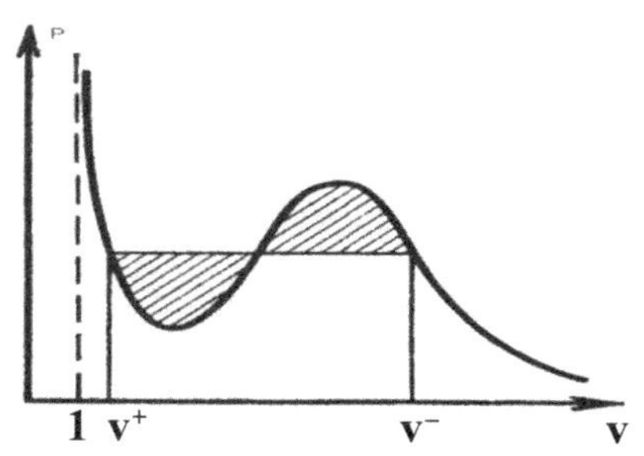

Fig. 18.1b, Isotherm with van der Waals loop and Maxwell cross-section in the 'pressure–specific volume' coordinates.

At the same time, the Yang–Lee theory also answers the question about the nature of metastable states: the existence of any metastable state under experimental conditions is a consequence of natural inhibitions that do not allow the realization of the *complete spectrum of molecular configurations* and, accordingly, the equilibrium of the system.

The exact result of the Yang–Lee theory is important as a result obtained without introducing approximate connections between the distribution functions of all dimensions and preserving exact relations between them. If there are changed relationships between the weights of different configurations, then their change is possible only in time, and this change must be described only by kinetic equations. That is why the exclusion of metastable states from equilibrium concepts opens a real opportunity to apply kinetic approaches and to select the real dynamic variables for which the kinetic equations are constructed. In the absence of dynamic variables, kinetic equations can not be formulated.

In the general case for solid-phase systems in the absence of equilibrium, the number of non-equilibrium states is not limited by anything – it can be anything, because is determined by the mechanical stability and mechanism of the process and the course of evolution of the system starting with some initial conditions. The absence of complete mixing in solids leads to the realization of many intermediate states with different in duration characteristic life times, which are usually called metastable. The evolution of such 'frozen' states should be described only by kinetic equations. This situation is reflected in various kinds of relaxation theories that extend the thermodynamic approach to weakly non-equilibrium states [79–83].

19. Fundamentals of the calculation of non-equilibrium functions of molecular distributions

Continuous distribution. In a macroscopic volume of a gas or liquid, the Liouville equations serve as the starting position for constructing transport equations, which operate with the total distribution function of the molecules of the system. Six variables describing its spatial coordinates and velocities are used to describe the state of each molecule [6,15,21,22,84–86]. In describing the dynamics of the system, we must pass to the generalized chain of coupled Bogolyubov equations [6] for time (non-equilibrium) distribution functions. They are written as follows: the expression for the function $\theta_{(s)}$ can be written as

$$L^0_{x_1,\ldots,x_s}\theta_{(s)} - \sum_{1\le i\le s} B_{ij}\theta_{(s)} = n\sum_{1\le i\le s}\int B_{i,s+1}\theta_{(s+1)}dx_{s+1}, \qquad (19.1)$$

where: t is the time $L^0_{x_1,\ldots,x_s} = \frac{\partial}{\partial t} + \sum_{1\le i\le s}(\mathbf{v}_i\frac{\partial}{\partial r_i} + F_0\frac{\partial}{\partial p_i})$, $F_4 = -\partial u(\mathbf{r}_i)$ / $\partial \mathbf{r}_i$ is the external force, $u(\mathbf{r}_i)$ is the potential energy of the molecule i in the external field (Section 13), $x_i = (\mathbf{r}_i, \mathbf{v}_i)$, $\mathbf{v}_i$ is the velocity of the molecule i, $\mathbf{p}_i$ is its momentum; $B_{ij} = \frac{\partial E_{ij}}{\partial r_i}\frac{\partial}{\partial p_i} + \frac{\partial E_{ij}}{\partial r_j}\frac{\partial}{\partial p_j}$, E_{ij} is the intermolecular interaction potential between molecules i and j (13.1).

Because of the extreme complexity of the chain of equations for the sequence of distribution functions $\theta_{(1)}$, $\theta_{(2)}$, ... it is natural to seek approximate closed equations for the simplest distribution functions, which is achieved by a more crude description of processes in the system under consideration.

Discrete description of distributions. The kinetic equation for the non-equilibrium discrete distribution functions is constructed, as for the continual description, in terms of the reduced functions $\theta_{(s)}(\{i, f, \mathbf{r}^i_f, \mathbf{v}^i_f\}, t)$, here s is the order of the distribution function, $1 \le s \le N$. The state of the system at time t is described by the total distribution function $\theta_{(N)}(\{i, f, \mathbf{r}^i_f, \mathbf{v}^i_f\}, t)$, which characterizes the probability of being in the cell with the number f, $1 \le f \le N$ (the complete list of cells is denoted by curly brackets { }), particles of sort i (the state of occupation of the site f), at the point with the coordinate $\mathbf{r}^i_f$, and having the velocity $\mathbf{v}^i_f$. This designation corresponds to the total distribution function $P(\{\gamma_f^i\},t)$ (16.6), for which the values of the velocities $\mathbf{v}_f^i$ were not determined.

The local distribution functions in the space $\theta_{(N)}(\{i,f,\mathbf{r}_f^i,\mathbf{v}_f^i\},t)$ used here differ from the continual distribution functions by their normalization to the cell volume (rather than the volume of the system). They are not defined outside the cells under consideration: $\theta_{(N)}\left(\{i,f,\mathbf{r}_f^{\,i},\mathbf{v}_f^{\,i}\},t\right)_{r_{f,\alpha}\notin\omega_f}=0$, where $1 \le f \le N$, $\alpha = x, y, z$; $\mathbf{r}_f \in \omega_f$ (here the symbol ω_f is used for the volume of the cell with the number f instead of $\mathbf{v}_f$, in order to distinguish this symbol from the velocity of the particle). The boundary conditions in the velocity space coincide with the usual conditions: these functions vanish when the velocity modulus tends to infinity $\theta_{(N)}\left(\{i,f,\mathbf{r}_f^i,\mathbf{v}_f^i\},t\right)_{v_{f,\alpha}=\pm\infty}=0$.

We will use QCA accounting for intermolecular interactions, which allows the probability of any configuration of molecules, i.e. $\theta_{(1)}(\mathbf{r}_f^i,\mathbf{v}_f^i,t)$ and paired $\theta_{(2)}(\mathbf{r}_f^i,\mathbf{v}_f^i,\mathbf{r}_g^j,\mathbf{v}_g^j,t)$ distribution functions, $1 \le f,\ g \le N$, in the following form [115–117]

$$\theta_{(N)}\left(\{i,f,\mathbf{r}_f^{\,i},\mathbf{v}_f^{\,i}\},t\right)=\prod_{f=1}^{N}\theta_{(1)}(\mathbf{r}_f^{\,i},\mathbf{v}_f^{\,i},t)\times \times\prod_{g\in z_f}\left\{\xi_{fg}^{ij}\left(\mathbf{r}_f^{\,i},\mathbf{v}_f^{\,i},\mathbf{r}_g^{\,j},\mathbf{v}_g^{\,j}t\right)\right\}^{1/2} \tag{19.2}$$

where the exponent 1/2 takes into account that the pairs of cells are listed twice; the index g runs through all z neighbours of the site f; $\xi_{fg}^{ij}(\mathbf{r}_f^i,\mathbf{v}_f^i,\mathbf{r}_g^j,\mathbf{v}_g^j,t)=\theta_{(2)}(\mathbf{r}_f^i,\mathbf{v}_f^i,\mathbf{r}_g^j,\mathbf{v}_g^j,t)/[\theta_{(1)}(\mathbf{r}_f^i,\mathbf{v}_f^i,t)\theta_{(1)}(\mathbf{r}_g^j,\mathbf{v}_g^j,t)]$ is a pair correlation function (we shall omit the argument of time t for simplicity below).

Kinetic equations are constructed for the complete set of these local distribution functions. Taking into account the normalization relations for a single-component substance, it is sufficient to construct the kinetic equations for the local unary $\theta_{(1)}(x_f)$ and the pair $\theta_{(2)}(x_f,\ x_g)$ distribution functions, which have the 'usual' form [6,15,21,22, 84–86].

The kinetic equations for unary functions are written as [87–89]

$$\left(\frac{\partial}{\partial t}+\mathbf{v}_f\frac{\partial}{\partial \mathbf{r}_f}+\frac{F(f)}{m}\frac{\partial}{\partial v_f}-\sum_h\int\frac{\partial \varepsilon_{fh}}{\partial \mathbf{r}_f}\frac{\partial t_{fh}(x_f,x_h)}{m\partial \mathbf{v}_f}dx_h\right)\theta_{(1)}(x_f)= \\ =I=\int\frac{\partial \varepsilon_{fg}}{\partial r_f}\frac{\partial \theta_{(2)}\left(x_f,x_g\right)}{m\partial v_f}dx_g \tag{19.3}$$

where ε_{fg} is the potential interaction function of molecules in cells f and g; m is the mass of the molecule; $\mathbf{F}(f)$ is the vector of the external conservative force in the cell f (in the narrow pores, the main contribution is made by the wall potential, and the gravitational field can be neglected). Here we use the numerical density θ instead of the traditional mass density ρ ($\rho = m\theta/v_0$).

The sum over h is taken over all neighbours of the site f, it describes the terms created by the interactions of neighbouring molecules in the neighbourhood of the site f with which the molecule at the site f does not collide (the so-called Vlasov contributions). The magnitude of this time interval is determined by the time variation in the left-hand side of the kinetic equation in the derivative $\partial/\partial t$. In their structure, these terms are completely similar to the collision integral, which is on the right. The intermolecular interaction potential ε_{fg} corresponds to the molecular arrangements at the sites h at a distance up to R_{lat} c.s. For rarefied gases this term is absent, then the formula (19.3) goes into the Boltzmann equation. The presence of neighbours sharply complicates the form of the kinetic equation and requires knowledge of the pair distribution functions. Usually in the theory of a solid and plasma [19, 86] these terms are considered in the mean-field approximation, in which the closure occurs at the level of unary distribution functions, i.e. Instead of functions $\partial\theta_{(2)}(x_f,x_g)/\partial\mathbf{v}_f$, derivatives $\partial[\theta_{(1)}(x_f)\ \theta_{(1)}(x_g)]/\partial\mathbf{v}_f$ are considered. If we neglect the contribution of the collision integral I, we obtain the Vlasov equation [15, 90]. In this case, the short-range LJ potential is considered for the vapour–liquid system and therefore it is necessary to preserve both types of terms. As the density increases, the role of Vlasov's contributions increases. However, as shown in Chapter 6, unary distribution functions do not provide a self-consistent description of systems with a wide range of density variations, so it is necessary to preserve the pair distribution functions.

The kinetic equations for the pair distribution functions are written out in a similar manner [87–89]

$$\left(\frac{\partial}{\partial t}+\mathbf{v}_f\frac{\partial}{\partial q_f}+\frac{F(f)}{m}\frac{\partial}{\partial \mathbf{v}_f}+\mathbf{v}_g\frac{\partial}{\partial \mathbf{r}_g}+\frac{F(g)}{m}\frac{\partial}{\partial \mathbf{v}_g}--\frac{\partial \varepsilon_{fg}}{\partial r_f}\frac{\partial}{m\partial v_f}-\frac{\partial \varepsilon_{fg}}{\partial r_g}\frac{\partial}{m\partial v_g}\right)\theta_{(2)}(x_f,x_g)-$$
$$-\sum_{\xi}\int\left\{\frac{\partial \varepsilon_{f\xi}}{\partial \mathbf{r}_f}\frac{\partial}{m\partial \mathbf{v}_f}+\frac{\partial \varepsilon_{g\xi}}{\partial \mathbf{r}_g}\frac{\partial}{m\partial \mathbf{v}_g}\right\}\theta_{(3)}\left(x_f,x_g,x_\xi\right)dx_\xi = \tag{19.4}$$
$$=\int\left\{\frac{\partial \varepsilon_{fh}}{\partial \mathbf{r}_f}\frac{\partial}{m\partial \mathbf{v}_f}+\frac{\partial \varepsilon_{gh}}{\partial \mathbf{r}_g}\frac{\partial}{m\partial \mathbf{v}_g}\right\}\theta_{(3)}\left(x_f,x_g,x_h\right)dx_h$$

The potential function ε_{fg} refers to a pair of molecules considered at the sites f and g. At the sites ξ there are neighbouring molecules that interact (simultaneously or separately depending on the distance) with molecules at the sites of f and g, but do not collide with them in the considered time range. As above, as the density of molecules increases, the role of Vlasov's contributions increases. The system of equations (19.3) and (19.4) is closed with the aid of superposition or QCA at the level of paired functions and describes the dynamics of the non-uniform system.

20. Kinetic equations in dense phases

In dense phases, the main role is played by interparticle interactions, and the contribution of translational motion is small. Under these conditions, the kinetic processes of migration and chemical transformations of particles in space are described with the help of the basic kinetic equation (Master Equation) for the total distribution function $P(\{\gamma_f^i\},\tau)$ (for spin systems these are the so-called Glauber type equations [91, 92]. They are obtained under the condition that the contribution of the translational motion of particles is small: migration in a solid, in a liquid at rest, on surfaces of solid and liquid phases, and also in the absence of the effect of translational motion of particles on the rate of their reactions (local equilibrium conditions).

These equations are simpler than the equations of the previous section. Movements of molecules in space are carried out through elementary particle jumps to neighbouring free sites (vacancies) – the so-called migration stage in a multistage description of physicochemical processes. To calculate the thermal migration rate of molecules, a transition state model is used that treats displacements as an activation process to overcome a barrier created by neighbouring particles. This model was proposed for gases by Eyring [93, 94] and later transferred to condensed phases in [27,95–105].

At the present time, the kinetic theory at the atomic–molecular level within the framework of the LGM [27,100–105] can be used practically throughout the entire time range, starting from the characteristic times of atomic vibrations to macroscopic ones, including the times of reaching equilibrium states. The theory considers the complete set of elementary processes of molecular displacements and their chemical transformations taking place in a system of non-equivalent lattice sites. To construct the general structure of the kinetic equations of the lattice model, we shall assume that the lattice sites are non-equivalent. The nature of the heterogeneity of the lattice sites is assumed to be known and unchanged in time. From a physical point of view, the non-uniformities in the distribution of particles are caused both by interactions between the particles of the system itself and by the possible additional influence of external fields or interactions (for example, the field of the substrate potential). This formulation of the problem makes it possible from a unified point of view to cover a wide range of questions (see Section 1) related to the spatial distribution of particles and taking into account chemical transformations in the course of chemical reactions.

According to Section 16, $\{\gamma_f^i\} = \gamma_1^i,\ \gamma_2^i, \ldots, \gamma_N^n$ is the complete collection (or a complete list) of the values of the values γ_f^i of all lattice sites that uniquely determine the complete configuration of particle arrangements on the lattice at time τ. For brevity, we denote this state as $\{\mathrm{I}\} \equiv \{\gamma_f^i\}$. Let the general process under consideration consist of a set of stages and let α denote the step number of the elementary process. The basic kinetic equation (Master Equation) for the evolution of the total distribution function of a system in the state {I}, due to the realization of elementary processes α in condensed phases, has the form

$$\begin{aligned}\frac{d}{d\tau}P(\{\mathrm{I}\},\tau) = \sum_{\alpha,\{\mathrm{II}\}}\big[W_\alpha\big(\{\mathrm{II}\}\to\{\mathrm{I}\}\big)\times \\ \times P\big(\{\mathrm{II}\},\tau\big) - W_\alpha\big(\{\mathrm{I}\}\to\{\mathrm{II}\}\big)P(\{\mathrm{I}\},\tau)\big],\end{aligned} \tag{20.1}$$

where $W_\alpha(\{\mathrm{I}\}\to\{\mathrm{II}\})$ is the probability of the realization of the elementary process α (the probability of a transition through the channel α), as a result of which at the time moment τ the system from the initial state {I} goes to the final state {II}. In formula (20.1), the sum is taken over various types of direct processes (index α) and over all inverse processes {II}, in which the state of occupation of each of the sites of the system changes.

If the elementary process runs on the same site, the lists of occupation states of the sites of the system {I} and {II} differ only for this site. One-site processes are processes associated with a change in the internal degrees of freedom of a particle, with adsorption and desorption of undissociated molecules, with a reaction by a collision mechanism. If the elementary process proceeds at two neighbouring lattice sites, then the lists of states {I} and {II} differ by the occupation states of these two sites. Two-site processes – exchange reactions, adsorption and desorption of dissociating molecules, migration processes by the vacancy and exchange mechanisms, etc. The sum over the states {II} corresponds to a change in the occupation states of all lattice sites. The interrelation of the states {I} and {II} depends on the mechanism of the process, which determines the set of elementary stages α.

Equation (20.1) is written in the Markov approximation, for which it is assumed that the relaxation processes of the internal degrees of freedom of all particles proceed faster than the processes of changing the states of occupation of different sites of the lattice system.

The transition probabilities W_α are subject to the condition of detailed balancing

$$\begin{aligned} &W_\alpha(\{\mathrm{I}\}\to\{\mathrm{II}\})\exp\left(-\beta H(\{\mathrm{I}\}\to\{\mathrm{II}\})\right)= \\ &=W_\alpha(\{\mathrm{II}\}\to\{\mathrm{I}\})\exp\left(-\beta H(\{\mathrm{II}\})\right), \end{aligned} \tag{20.2}$$

where $H(\{\mathrm{I}\})$ is the total energy of the lattice system in the state {I}. In the equilibrium state, $P\left(\{\gamma_f^i\},\tau\to\infty\right)=\exp\left(-\beta H\left(\{\gamma_f^i\}\right)\right)/Q$ where Q is the partition function of the system, the system (20.1) in QCA transforms to equations (18.4).

Expressions for $W_\alpha(\{\mathrm{I}\}\to\{\mathrm{II}\})$ are constructed taking into account all molecular features of the system: 1) each site f is characterized by a definite set of number of particle sort s_f that can be in it; 2) the internal degrees of freedom F_j^i of a particle of some sort i depend on the number of the lattice site; 3) the interaction parameters $\varepsilon_{fg}^{ij}(r)$ of the particles i and j, located at the sites with the numbers f and g at a distance r from each other depend on the site numbers.

The large dimensionality of the system (20.1) does not allow us to use it to study the dynamics of macroscopic systems by direct integration; therefore, kinetic equations are constructed with respect to the functions of lower-order distributions through which higher-order distribution functions are closed. To this end, instead of the total distribution function $P(\{\gamma_f^i\},\tau)$, the abbreviated method of its assignment through time distribution functions (correlators), defined in (16.2), is used to describe the evolution of the system,

The introduced local time functions (16.2) imply that at each instant of time averaging is performed over the complete ensemble of copies of the non-uniform lattice system under consideration over all of its realizable states. This definition is analogous to the determination of the averages in the equilibrium Gibbs statistical theory and the non-equilibrium theory of gases and liquids. The difference from the non-equilibrium theory of gases and liquids is that the local inhomogeneities under consideration are realized on a microscopic atomic scale (instead of small elementary volumes containing a macroscopic amount of matter in the theory of gases and liquids).

The kinetic equations for the correlators are obtained by multiplying expression (20.1) by $\prod_{n=1}^{m}\gamma_{f_n}^{i_n}$ and averaging over all states of the system. The sorts of particles at the sites f_n, leading to nonzero contributions to the change in the correlators (16.2), depend on the direction of the elementary process and are determined by the initial {I} or final {II} states. This leads to the following kinetic equations:

$$\frac{d}{d\tau}\theta_{f_1\ldots f_m}^{i_1\ldots i_m}(\tau)=$$
$$\sum_{\alpha,\{II\}}\left[\left\langle\prod_{n=1}^{m}\gamma_{f_n}^{i_n^*}W_\alpha(\{\mathrm{II}\}\to\{\mathrm{I}\})\right\rangle-\left\langle\prod_{n=1}^{m}\gamma_{f_n}^{i_n}W_\alpha(\{I\}\to\{II\})\right\rangle\right],\tag{20.3}$$

where the sorts of particles i_n^* in the first term on the right correspond to the states of occupation of the lattice sites in the state {II}. Equations (20.3) are the starting points for obtaining the kinetic equations of the processes occurring in the condensed phases.

The construction of the kinetic equation for the local concentration of molecules (m = 1) leads to a large increase in the order of the unknown correlation functions in the right-hand side of the equation. This dimension is equal to the number of all neighbouring molecules that affect the rate of this stage. We recall that in the BBGKY kinetic chain the dimension of the correlators on the right-hand side of the kinetic equations is successively increased by one. This structure of the kinetic equations of the LGM differs from the traditional form of the BBGKY kinetic chains, and accordingly the kinetic theory raises the problem of calculating high-dimensional correlators. In a discrete version of the theory based on the LGM, the system of kinetic equations also has a high dimensionality and, in practice, tends to be limited to a minimal dimension that preserves the effects of correlation between molecules.

The necessity of considering high-order correlation functions is due to the fact that the interaction of activated complexes with surrounding particles not only quantitatively changes the dynamic characteristics of the transient regimes, but can qualitatively change the evolution of the system, for example, the number of stationary states of the system increases [106]. In addition, it was shown [107] that the type of transition probabilities affect the calculated values of the critical dynamic index for the correlation length, which contradicts the dynamic hypothesis of universality [108, 109]. A consistent introduction of the theory of absolute reaction rates into the theory of kinetic equations for an arbitrary character of particle mobility is given in [27,100–105] when considering processes on smooth and rough surfaces.

A closed system of equations for the first $\left(\theta_f^i = \left\langle \gamma_f^i \right\rangle\right)$ and the second $\left(\theta_{fg}^{ij}(r) = \left\langle \gamma_f^i \gamma_g^j \right\rangle\right)$ correlators in general form is written as

$$\frac{d}{dt}\theta_f^i = I_f^i = \sum_\alpha \left[U_f^b(\alpha) - U_f^i(\alpha)\right] + \sum_r \sum_h \sum_j \sum_\alpha \left[U_{fh}^{bd}(r|\alpha) - U_{fh}^{ij}(r|\alpha)\right] \tag{20.4}$$

$$\begin{aligned} &\frac{d}{dt}\theta_{fg}^{ij}(r) = I_{fg}^{ij}(r) = \sum_\alpha \left[U_{fg}^{bd}(r|\alpha) - U_{fg}^{ij}(r|\alpha)\right] + P_{fg}^{ij}(r) + P_{gf}^{ji}(r) \\ &P_{fg}^{ij}(r) = \sum_\alpha \left[U_{fg}^{(b)j}(r|\alpha) - U_{fg}^{(i)j}(r|\alpha)\right] + \\ &\sum_h \sum_m \sum_\alpha \left[U_{hfg}^{(cb)j}(r|\alpha) - U_{hfg}^{(mi)j}(r|\alpha)\right], \end{aligned} \tag{20.5}$$

where $U_f^i(\alpha)$ are the rates of elementary single-site processes $i \leftrightarrow b$ (here $h \in z_f$), $U_{fg}^{ij}(r|\alpha)$ are the elementary two-site processes $i + j_\alpha \leftrightarrow b + d_\alpha$ ($h \in z(r)$) at a distance r; the second term in $P_{fg}^{ij}(r)$

describes the stage $i + m \leftrightarrow b + c$ at neighbouring sites f and h at a distance r. All the velocities of the elementary stages $U_f^i(\alpha)$ and $U_{fg}^{ij}(r|\alpha)$ are calculated within the framework of the theory of absolute reaction rates for non-ideal reaction systems written out in the QCA for accounting for interparticle interaction (see below in Sections 49 and 51).

Equations (20.4) and (20.5) satisfy the normalizing relations (16.3), which are satisfied at any time.

Of great importance is the question of the separation of dynamic variables describing the state of molecules into fast and slow ones. This question is solved depending on the chosen characteristic time scale determined by the values of $W_\alpha(\{I\}\rightarrow\{II\})$. Slow variables are described by kinetic equations, and fast variables are described by algebraic equations, since they usually refer to particles having an equilibrium distribution. In the general case, fast particles (described by fast variables) form a tuning subsystem that exerts its influence on the energy of slow elementary processes in kinetic equations. In turn, slow particles (described by slow variables) determine the character of the distribution of fast particles. They form a spatial region in which fast processes are realized (excluding their common spaces, regions occupied by slow particles), and influence their potential on the character of the distribution of fast particles.

Usually, the spatial coordinates of molecules or quantum numbers related to their electron terms are usually referred to slow variables, and to quantum numbers referring to the rotational and vibrational states of molecules [109]. The latter does not exclude the possibility of considering the population of vibrational levels of molecules in the gaseous phase as slow variables. On the other hand, the stages of migration of molecules can also be fast. For example, at a low activation energy compared to the desorption energy, surface migration is a fast step compared to the desorption stage.

These kinetic equations are used in Chapter 5 to construct a local equilibrium criterion, and in Chapter 6 to discuss the issue of self-consistency of equilibrium and kinetics. Chapter 7 discusses the use of thermodynamic interpretations in kinetics.

References

1. Gibbs J.W., Elementary principles in statistical mechanics, developed with especial references to the rational foundations. New York, 1902.
2. 2. Gibbs J.W., Thermodynamics. Statistical mechanics. Moscow, Nauka, 1982. 584 .
3. 3. Mayer J.E., Goeppert-Mayer, M., Statistical mechanics. Moscow, Mir, 1980. [Wi-

ley, N.Y.-Sydney-Toronto, 1977]

4. Fowler R.H.. Statistical Mechanics. Cambridge. Cambridge Univer. Press, 1936.
5. Fowler R.H., Guggenheim E.A.. Statistical Thermodynamics. Cambridge. Cam¬bridge Univer. Press, 1939.
6. Bogolyubov N.N.. Problems of dynamic theory in statistical physics. Moscow, Gostekhizdat, 1946. [Interscience, New York, 1962]
7. Sommerfeld A.. Thermodynamics and statistical physics. Moscow, IL, 1955. [Academic, New York, 1964].
8. Leontovich M.A., Introduction to thermodynamics. Statistical physics. Moscow, Nauka, 1983.
9. Hill T.L., Statistical Mechanics. Principles and Selected Applications. – Moscow: Izd. Inostr. lit., 1960. – 486 p. [N.Y.: McGraw–Hill Book Comp. Inc., 1956].
10. Hill T.L., Thermodynamics of Small Systems. Part 1. New York Amsterdam: W. A. Benjamin, Inc., Publ., 1963. Part 2. 1964.
11. Huang K., Statistical mechanics. Moscow, Mir, 1966.
12. Landau L.D., Livshits E.M., Theoretical physics. V. 5. Statistical physics. Moscow, Nauka, 1964.
13. Kubo R. Statistical mechanics. Moscow, Mir, 1967.
14. Uhlenbeck J., Ford J. Lectures on static mechanics. Moscow, Mir, 1965.
15. Vlasov A.A., Statistical functions of distributions. Moscow, Nauka, 1966.
16. Gurov K.P., Foundations of the kinetic theory. Moscow, Nauka, 1967.
17. Fischer I.Z., Statistical theory of liquids. Moscow, Fizmatgiz, 1961.
18. Zubarev D.N., Non-equilibrium statistical thermodynamics. Moscow, Nauka, 1971.
19. Bazarov I.P., Statistical theory of the crystalline state. Moscow, Izd-vo MGU, 1972.
20. Statistical physics and quantum field theory. Ed. N.N. Bogolyubov. Moscow, Nauka, 1973.
21. Balescu, R. Equilibrium and non-equilibrium statistical mechanics. V.2. Moscow, Mir, 1978..
22. Croxton K., Physics of the liquid state. Moscow, Mir, 1979.
23. Bazarov I.P., Nikolaev P.N., Correlation theory of a crystal. Moscow, Publishing House of Moscow State University, 1981.
24. Gurov K.P., Kartashkin B.A., Ugaste Yu.E., Mutual theory in multiphase metal systems. Moscow, Nauka, 1981.
25. 25. Bazarov I.P., Gevorkyan E.V., Statistical theory of solid and liquid crystals, Moscow State University Publishing House, Moscow, 1983.
26. Klimontovich Yu.L., Statistical physics. Moscow, Nauka, 1982.
27. Tovbin Yu.K., Theory of physico-chemical processes at the gas–solid interface, Moscow, Nauka, 1990. [CRC, Boca Raton, Florida, 1991].
28. Martunov G.A., Fundamental Theory of Liquids: Method of Distribution Functions. Bristol, A. Hilger, 1992.
29. 29. Martunov G.A., Classical Statistical Mechanics (Fundamental Theories of Physics, V.89). Dordrecht: Kluwer Acad. Publ., 1997.
30. Zubarev D.N., et al., Static mechanics of non-equilibrium processes, V. 1 and 2, Moscow, Fizmatlit, 2002.
31. Tovbin Yu.K., The Molecular Theory of Adsorption in Porous Solids, Moscow, Fizmatlit, 2012. [CRC Press, Taylor@Francis Group, 2017]
32. Tovbin Yu.K., Zh. fiz. khimii. 1995. V. 69. No. 1. P.118. [Russ. J. Phys. Chem. 1995. V. 69. No. 1. P. 105]
33. Tovbin Yu.K., Zh. fiz. khimii. 1998, V.72. №5, P .775 [Russ. J. Phys. Chem. , 1998 V. 72, No. 5, P. 675]

34. Tovbin Yu.K., Senyavin M.M., Zhidkova L.K. Zh. fiz. khimii.1999, V. 73. № 2. P. 304. [Russ. J. Phys. Chem. 1999. V. 73. № 2. P. 245]
35. Tovbin Yu.K., Zh. fiz. khimii. 2017. 91. No. 3. P. 381. [J. Phys. Chem. A 91, 403 (2017)].
36. Kvasnikov I.A., Thermodynamics and statistical physics. V. 2: The Theory of Equi¬librium Systems: Statistical Physics. Moscow, Editorial URSS, 2002.
37. Frenkel Ya.I., Kinetic Theory of Liquids. Moscow, Academy of Sciences of the USSR, 1945.
38. Fowler R.H., Rushbrooke G.S., Trans. Far. Soc. 1937. V. 33. P. 1272.
39. Lennard-Jones J.E., Devonshire A.F., Proc. Roy. Soc., 1937. V. 163A. P. 53.
40. Lennard-Jones J.E., Devonshire A.F., Proc. Roy. Soc., 1939. V. 169A. P. 317.
41. Collins R., Phase transitions and critical phenomena. Ed. C. Domb, M.S. Green. London – New York, Academic Press, 1972. P. 271.
42. Vortler N.L., et al., Physica, A. 1979. V. 99. P. 217.
43. Langmuir I., J. Amer. Chem. Soc. 1916. V. 38. P. 2217.
44. Hirschfelder J. O., Curtiss C. F., Bird R.B., Molecular theory of gases and liquids. Moscow, IL, 1961. [Wiley, New York, 1954].
45. Runnels L.K., Phase transitions and critical phenomena, Ed. C. Domb, MS Green. London–New York, Academic Press, 1972. V. 2. P. 305.
46. O'Reilly D.E., Phys. Rev. A. 1977. V. 15. P. 1198.
47. Shinomoto S.-G., Progr. Theor. Phys. 1983. V. 70. P. 687.
48. Ising E., Zs. Phys. B. 1925. 31. S. 253.
49. Baxter R., Exactly solvable models in statistical mechanics. Moscow, Mir, 1985. [Academic Press, London, 1982].
50. Kac M., Probability and related topics in physical sciences. Moscow, Mir, 1965. [Proceeding of the summer seminar Boulder, Colorado, 1957, Interscience Pub. Ltd. London].
51. Onsager L., Phys. Rev. 1944. V. 65. P. 117.
52. Domb C., Phase transitions and script phenomena, Ed. C. Domb, M.S. Green. London–New York, Academic Press, 1974. V. 3. P. 356-484.
53. Guggenheim E.A., Proc. Roy. Soc. London. A. 1935. V. 148. P. 304
54. Bethe H.A., Proc. Roy. Soc. London. A. 1935. V. 150. P. 552.
55. Peierls R., Proc. Camb. Phil. Soc. 1936. V. 32. P. 471.
56. Kirkwood J.G., Monroe E., J. Chem. Phys. 1942. V. 10. P. 395.
57. Tovbin Yu.K., Progress in Surface Science. 1990. V. 34, No. 1-4, P. 1-236.
58. Dunning W., Interphase boundary gas-solid. Moscow, Mir, 1970. P. 230.
59. Roberts M.W., McKee C.S., Chemistry of the metal-gas interface, Clarendon Press, Oxford, 1978.
60. Somorjai G. A., Chemistry in two-dimension surface, Cornell Univ. Press L., N.Y., Ithaca, 1981.
61. New in the study of the surface of a solid, Ed. T. Jayadevay, R.M. Vanselov, Moscow, Mir, 1977. Issue 2. [CRC Press, Inc., Cleveland, 1974]
62. Jaycock M. Parfitt J., Chemistry of interfaces.– Moscow: Mir, 1984. – 270 p. [Wiley, New York, 1981].
63. Tovbin Yu.K., Doct. Sci. Thesis, Moscow, NIFHI L.Ya. Karpova. 1985.
64. Tovbin Yu.K., Theoretical methods for describing the properties of solutions. Interuniversity collection of scientific works. Ivanovo. 1987. P. 44.
65. Tovbin Yu.K., Zh. fiz. khimii. 1981. V. 55. No. 2. P. 273.
66. Tovbin Yu.K., Zh. fiz. khimii. 1990. P. 64. No. 4. P. 865. [Russ. J. Phys. Chem. 1990. V. 64. № 4. P. 461]

67. Fedyanin V.K., Statistical physics and quantum field theory. Moscow, Mir. 1973.
68. Tovbin Yu.K., Zh. fiz. khimii. 2005. V. 79. No. 12. C. 2140. [Russ. J. Phys. Chem. , 2005 V. 79, No. 12, P. 1903]
69. Boltzman L., Selected Works. Moscow, Nauka, 1984..
70. Rumer Yu.B., Ryvkin M.Sh., Thermodynamics, statistical physics and kinetics. Moscow, Nauka, 1971.
71. Kirkwood J.G., J. Chem. Phys. 1950. V. 18. P. 380.
72. Salsburg Z.W., Kirkwood J.G., J. Chem. Phys. 1952. V. 20. P. 1538.
73. Tovbin Yu.K., Zh. fiz. khimii. 2006. V. 80. No. 10. C. 1753. [Russ. J. Phys. Chem. 2006. V. 80. № 10. P. 1554]
74. Tovbin Yu.K., Zh. fiz. khimii. 2015. V. 89. No. 11. P. 1704. [Russ. J. Phys. Chem. A. 2006. V. 89. № 11. P. 1971]
75. Yang C.N., Lee T.D., Phys. Rev. 1952. V. 87. P. 404.
76. Lee T.D., Yang C.N., Phys. Rev. 1952. V. 87. P. 410.
77. Kvasnikov I.A., Thermodynamics and statistical physics. V. 1: Theory of equilibrium systems: Thermodynamics. Moscow, Editorial URSS, 2002.
78. Landau L.D., Zh. Eksp. Teor. Fiz. 1937. V. 5. P.627.
79. Rostiashvili V. G., Irzhak V. I., Rozenberg B. A., Glass Transitions in Polymers, Leningrad, Khimiya,1987.
80. Leontovich M.A., Zh. Eksp. Teor. Fiz. 1936. V. 6. P. 561.
81. Mandelstam L.I., Leontovich M.A., Zh. Eksp. Teor. Fiz. 1937. V. 7. No. 7. P. 438.
82. Mikhailov I.G., et al., Fundamentals of molecular acoustics. Moscow, Nauka, 1964..
83. Haase R., Thermodynamics of irreversible processes. Moscow, Mir, 1967. [Dr. Dietrich Steinkopff, Darmstadt, 1963]
84. Chapman S., Kauling T., Mathematical theory of non-uniform gases. Moscow, IL, 1960.
85. Ferziger J., Kaper G., Mathematical theory of transport processes in gases. Moscow, Mir, 1976.
86. Klimontovich Yu.L., Kinetic theory of non-ideal gas and non-ideal plasma. Moscow, Nauka, 1975.
87. Tovbin Yu.K., Modern chemical physics, Moscow, Izd-vo MGU, 1998.
88. Tovbin Yu.K., Khim. Fizika. 2002. V. 21. No. 1. P.83.
89. Tovbin Yu.K., Zh. fiz.khimii. 2002. V. 76. No. 1. P.76. [Russ. J. Phys. Chem. 2002. V. 76. № 1. P. 64]
90. Vlasov A.A., Zh. Eksp. Teor. Fiz. 1938. V. 8. P. 291.
91. Glauber J., J. Math. Phys. 1963. V. 46. P.541.
92. Stanley G., Phase transitions and critical phenomena. Moscow, Mir. 1973.
93. Eyring H. J. Chem. Phys. 1935. V. 3. P. 107.
94. Glasston S., Laidler K.J., Eyring H., Theory of absolute reaction rates. Moscow, IL, 1948 [Princeton Univ. Press, New York, London, 1941].
95. Tovbin Yu.K., Fedyanin V.K., Kinetika i kataliz. 1978. Vol. 19. No. 4. P. 989.
96. Tovbin Yu.K., Fedyanin V.K., Kinetika i kataliz. 1978. V. 19. No. 5. P. 1202.
97. Tovbin Yu.K., Fedyanin V.K., Fiz. Tverd. Tela. 1980. V. 22. No. 5. P. 1599.
98. Tovbin Yu.K., Fedyanin V.K., Zh. fiz. khimii. 1980. V. 54. No. 12. P. 3127, 3132.
99. Tovbin Yu.K., Zh. fiz. khimii. 1981. Vol. 55. No. 2. P. 284.
100. Tovbin Yu.K., Dokl. AN SSSR. 1982. V. 267. No. 6. P. 1415.
101. Tovbin Yu.K., Dokl. AN SSSR. 1984. P. 277. No. 4. P. 917.
102. Tovbin Yu.K., Progress in Surface Science. 1990. V. 34, No. 1-4. P. 1-236.
103. Tovbin, Yu.K., Poverkhnost. Fizika. Khimiya. Mekhanika. 1989. No. 5. P. 5.
104. Tovbin Yu.K., Dynamics of Gas Adsorption on heterogeneous Solid Surfaces. Eds.

W. Rudzinski, W.A. Steele, G. Zgrablich. Elsevier: Amsterdam, 1996. P. 240-325.
105. Tovbin Yu.K., Thin Films and Nanostructures. Vol. 34. Physico-Chemical Phenomena in Thin Films and at Solid Surface. Eds. L. I. Trakhtenberg, S. H. H. Lin, and O. J. Ilegbusi, Elsevier, Amsterdam, 2007. P. 347.
106. Tovbin Yu.K., Cherkasov, A. N., Teor. Eksp. Khimiya. 1984. V. 20. No. 4. P. 507.
107. Pandit R., Forgacs G., Rujan P., Phys. Rev. V. 1982. V. 25. P. 1860.
108. Ma Sh., Modern theory of critical phenomena. Moscow, Mir, 1980.
109. Nikitin E.E., Theory of elementary atomic-molecular processes in gases. Moscow, Khimiya. 1970.

3

Phase separation boundary

21. Thermodynamic values of the surface layer

Interfaces between coexisting phases are the simplest example of local non-uniformities in a system. They inevitably arise within the framework of the phase approximation of describing the properties of non-uniform systems by Gibbs as *heterogeneous* systems, when the contributions from the phase boundaries can not be neglected in thermodynamic functions. Our task is to compare the hypotheses of the thermodynamic approach with the results obtained on the basis of the molecular theory, using the same hypotheses. Below, the thermodynamic introduction to the problem is limited to reminding the basic assumptions about the concepts of the dividing surface. Determination of the surface of tension and methods for determining the surface tension based on thermodynamic and hydrostatic approaches in metastable spherical drops, as well as other issues of comparing the continuum description of surface tension with the molecular discrete description in the LGM (lattice gas model), are discussed in Appendix 1 and Chapter 7.

Dividing surface. Consider a two-phase system with a phase boundary that can be flat or curved. Imagine a mathematical surface called the dividing surface, which is introduced in such a way as to accurately divide the two phases, denoted as α or β [1,2]. In general, the position of the dividing surface is chosen in such a way that it is normal to the density gradient in the transition zone and, therefore, in the case of a flat boundary, it must be a plane. The use of a dividing surface has the advantage that it allows us to consider the interphase layer without indicating its thickness.

Any quantity related to one of the phases will be denoted by the lower index of the phase α or β. Let the volume of the system be divided into two volumes V_α and V_β. If the dividing surface is selected lying in the transition layer, then V_α contains the bulk phase α together with a part of the transition zone material, and V_β contains the bulk phase β and the rest of the transition zone material.

Let us imagine a hypothetical system built of two volume phases α and β, which remain strictly uniform up to the above dividing surface and therefore have volumes V_α and V_β, respectively [1,2]. The superscript i will be used to denote the i-th molecular component. Let N_α^i and N_β^i be the numbers of molecules of the i-th sort in the above-mentioned hypothetical phases α and β. Specifically, N_α^i is equal to n_α^i – the number of molecules of the i-th sort per unit volume of the phase α multiplied by the volume occupied by this phase V_α, and N_β^i is, respectively, $n_\beta^i V_\beta$.

In the general case, N_i – the total number of molecules of the i-th sort in the real system – is not necessarily equal to the number $N_\alpha^i + N_\beta^i$ for the above-mentioned hypothetical system; However, it can be represented in the form

$$N^i = N_\alpha^i + N_\beta^i + N_b^i \tag{21.1}$$

where N_b^i can be considered as an additive associated with the existence in the real system of the interface or interphase region. The number N_b^i and any other similar quantity will be called the surface quantity, and mark with the subscript b.

Figure 21.1 shows the schemes of two-phase systems with a flat phase interface located in a vessel, having the shape of a rectangular parallelepiped and arranged so that one of its edges is parallel to the direction of gravity. These schemes sequentially detail the interface region.

The scheme in Fig. 21.1 *a* shows the mathematical surface or the dividing surface between the phases α and β [1,2]. The scheme in 21.1 *b* reflects the finite width of the transition region between the phases α and β, following Guggenheim [3,4]. This region is regarded as a separate layer of matter of finite thickness, bounded by the planes *K'H'* and *K"H"* parallel to the dividing surface *KH*. It is assumed that the parts of the system above the *K'H'* plane and below the *K"H"* plane are completely uniform, so that the spatial variation of the thermodynamic quantities is limited by the surface phase between *K'H'* and *K'H"* and having a thickness of $\Delta = \Delta_\alpha + \Delta_\beta$.

In molecular approaches, the same system will be considered as a system consisting of a set of monomolecular layers parallel to the

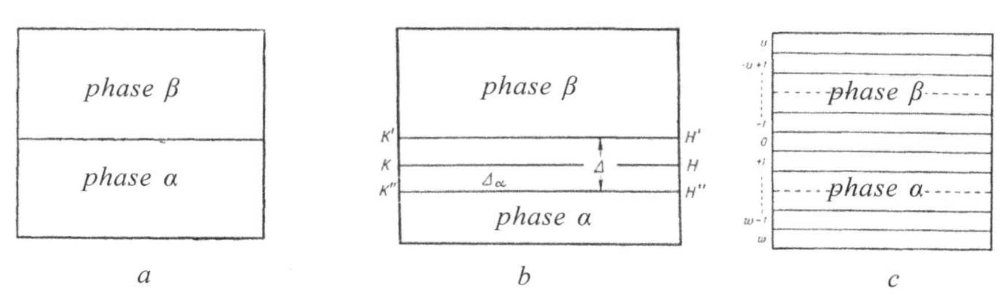

Fig. 21.1. (*a*) Two-phase system with a flat interface. (*b*) The phase separation boundary is regarded as a separate layer of a material of finite thickness bounded by the planes *K'H'* and *K"H"* parallel to the dividing surface *KH*. (*c*) The phase boundary is considered as a set of monomolecular layers parallel to the phase interface.

phase interface (Fig. 21.1 *c*) and perpendicular to the direction of gravity [2]. Plane bounding monolayers will be called equipotential. We denote the width (thickness) of the monolayer by λ; all of them have the same area A. For brevity, we shall assume that one of the equipotential planes is a zero fiber and assign numbers –1, –2,..., $-u$ to successive layers lying above zero, and numbers 1, 2,..., ω to layers lying below zero (Fig. 21.1 *c*). The way in which the zero layer is selected depends on the method for determining the surface tension. One of the most commonly used definitions is an equimolecular surface. The quantities related to the monolayer q will be denoted by the subscript q, for example, the number of molecules in the q-th layer will be denoted by N_q.

The contribution of the interphase layer to thermodynamic quantities. Any extensive thermodynamic property of the whole system, like the number of molecules, can be considered with respect to an arbitrarily chosen dividing surface in the form of a sum of three contributions: the contribution of the phase α, the contribution of the phase β, and the contribution of the interphase region. For example, the free Helmholtz energy for the whole system can be written as

$$F = F_\alpha + F_\beta + F_b. \tag{21.2}$$

where F_α is the free energy of the volume phase α in the case when the latter remains uniform up to the dividing surface, F_β is an analogous value for the volume phase β. In general, $F_\alpha + F_\beta$ is the free energy of a hypothetical system in which intense physical properties jump discontinuously on the dividing surface from the values they have in the bulk phase α to the values they take in the bulk phase β and F_b is the contribution of the transition layer to real free energy. The total free energy F of the two-phase system together with the interphase region is a completely determined quantity and the relation (21.2) should be considered as the determination of the excess free energy F_b.

Similarly, the excess internal energy U_b and entropy S_b are determined respectively as

$$U = U_\alpha + U_\beta + U_b \text{ and } S = S_\alpha + S_\beta + S_b. \tag{21.3}$$

Using the relation $F = U - TS$ and the definitions given above, we obtain

$$F_b = U_b - TS_b, \tag{21.4}$$

The absolute temperature T is assumed constant in all parts of the system.

22. The planar interface of macroscopic phases

Determination of surface tension. Let the area of the boundary A (Fig. 21.1 *a*) increase by a value dA by means of a reversible isothermal displacement of the side walls of the vessel. If the system is in hydrostatic equilibrium at a pressure P, the work done by it in the described process can be divided into two parts: the work PdV spent on increasing the volume dV, and the excess work associated with the increase in the area of the boundary by dA, which we denote $-\sigma dA$. Then the total work done by the system in this process is $PdV-\sigma dA$. Now let's shift the upper and lower covers of the vessel so that we return the system to its original volume. The work done by the system in this process is $-PdV$. At the end of these two processes, the system will have the same pressure, composition and temperature as it originally had.

The only change is that the area of the border has increased by dA. This means that the work done by the system in order to cause an increase in the area of the boundary by dA for constant values of volume, pressure and temperature is σdA. Therefore, σ represents the work done by the system with a single increase in the area of the interface. It is called surface tension. The elementary work dW produced by the system with a change in its volume and boundary area, respectively, on dV and dA should be written as

$$dW = PdV - \sigma dA. \tag{22.1}$$

This expression can be regarded as the definition of the surface tension σ. It can be seen from (22.1) that in the case of a plane boundary the surface tension does not depend on the position of the dividing surface, since in this case the changes of the latter do not affect the area of the boundary in any way.

If the surface tension σ determined by the described method were negative, then the work done by the system with increasing interface area would be positive and, consequently, the interface area could grow spontaneously. This means that a flat interface is stable only for positive σ. Therefore, the surface tension value defined above can not be negative, just as the pressure in the liquid, determined in the usual way, can not be negative.

The internal energy U of the two-phase system also varies with the area of the interphase boundary. If the amount of heat obtained by the system in the process of an infinitely small increase in the area of the boundary dA is denoted by dQ, then the First Law of thermodynamics is written in the form

$$dQ = dU + PdV - \sigma dA. \tag{22.2}$$

Equality (22.2) is also valid for non-equilibrium systems as long as they are in mechanical equilibrium at constant pressure.

Surface free energy. If the system is in thermodynamic equilibrium, then the amount of heat dQ absorbed during the reversible change in state, according to the Second Law, is equal to the increase in the entropy of the system dS, multiplied by the temperature T. Combining this with the first law (22.2), we obtain

$$dU = TdS - PdV + \sigma dA. \tag{22.3}$$

This important relation can be represented in a more convenient form, using the expression for the Helmholtz free energy:

$$dF = -PdV - SdT + \sigma dA. \tag{22.4}$$

In an open system that can exchange with matter and energy with the environment, the number of molecules can vary and therefore the total differential of free energy given by expression (22.4) should have an additional term associated with this change:

$$dF = -PdV - SdT + \sigma dA + \sum_{i=1}^{s-1} \mu_i dN_i, \tag{22.5}$$

where $(s-1)$ is the number of components (the contribution of vacancies is omitted), μ_i is the chemical potential calculated for the molecule of the i-th component, which should be constant throughout the system if it is in chemical equilibrium. The fundamental Gibbs equation (22.5) for a two-phase system can be regarded as an equation that determines the surface tension σ in an open system.

If the free energy F is known as a function of volume, temperature, interface area and number of molecules, then σ can be easily calculated from relation

$$\sigma = \left(\frac{\partial F}{\partial A}\right)_{T,V,\mathbf{N}} \tag{22.6}$$

where the symbol **N** represents the set of numbers N_1, N_2,...,N_s. We will increase the volume of the two-phase system shown in Fig. 1 *a*, while limiting the thermodynamically equilibrium case and keeping the temperature, pressure, composition and height of the vessel unchanged. Then the quantities F, V, **N**, A vary in the same proportion. Thus, F is a uniform function of the first power of the variables V, N, and A, and therefore, by Euler's theorem, we obtain from (22.5)

$$F = \sum_{i=1}^{s-1} \mu_i N_i - PV + \sigma A. \tag{22.7}$$

For the volume phases α and β, the free energy expressions are written as,

$$F_\alpha = \sum_{i=1}^{s-1} \mu_i N_\alpha^i - PV_\alpha, \; F_\beta = \sum_{i=1}^{s-1} \mu_i N_\beta^i - PV_\beta, \tag{22.8}$$

then subtracting (22.8) from (22.7) and using the equalities (21.1), (21.2) and the volume additivity condition, we obtain

$$F_b = \sum_{i=1}^{s-1} \mu_i N_b^i + \sigma A. \tag{22.9}$$

In contrast to the surface tension σ, excess values such as F_b, U_b and N_b^i depend on the position of the dividing surface. If we choose the dividing surface in such a way that the sum $\sum_{i=1}^{s-1} \mu_i N_b^i = 0$, then (22.9) reduces to the relation

$$F_b = \sigma A. \tag{22.10}$$

In the case of a single-component system, this particular dividing surface plays an important role in the theory of surface tension and is called an equimolecular dividing surface. We emphasize that N_b^i

can not always be called the number of molecules of sort i adsorbed in the interphase region, since N_b^i depends on the position of the artificially introduced dividing surface. It is seen from (22.10) that the surface tension becomes equal to the surface density of free energy only with such a special choice of the dividing surface [5,6]. It should be emphasized that surface tension is neither internal energy nor potential energy per unit surface area.

Thermodynamic relationships for a plane boundary. The fundamental Gibbs equations for the volume phases α and β are written as follows:

$$dF_\alpha = -PdV_\alpha - S_\alpha dT + \sum_{i=1}^{s-1} \mu_i N_\alpha^i, \tag{22.11}$$

$$dF_\beta = -PdV_\beta - S_\beta dT + \sum_{i=1}^{s-1} \mu_i N_\beta^i, \tag{22.12}$$

Subtracting (5.1) and (5.2) from (4.3) and using (1.1), (2.1) and (2.3), we obtain the relation

$$dF_b = -S_b dT + \sigma A + \sum_{i=1}^{s-1} \mu_i N_b^i. \tag{22.13}$$

which can be regarded as the fundamental Gibbs equation for the interphase layer. Thus, we see that the interphase thermodynamic quantities are related to each other by the same relation as in the case of the usual volume phase, and, consequently, they can be treated as if they determined a certain third phase. However, this analogy of the resulting expressions for some third phase has a purely formal meaning. Such equations can be used to construct thermodynamic bonds only between surplus surface characteristics, as defined above. But the boundary itself is not an autonomous phase, and under no circumstances can the boundary be considered a phase. This fact has a fundamental limitation on thermodynamic constructions for the total content of molecules in the transition region.

On the basis of equation (22.13), all thermodynamics is constructed for a planar interphase layer [2]. For example, differentiating (22.9) and using (22.13), we arrive at the general form of the Gibbs adsorption equation

$$d\sigma + s_b dT + \sum_{i=1}^{s-1} \Gamma_i d\mu_i = 0, \tag{22.14}$$

where the surface density of the entropy is usually called the surface entropy $s_b = S_b/A$, and the value $\Gamma_i = N_b^i/A$ will be called the surface density of the number of molecules of sort i.

The contribution of σA also appears in all other thermodynamic potentials when it is necessary to take into account the surface properties of the system. Thus, if for a volume phase the Gibbs potential G is written as $G_v = \Sigma_i \mu_i m_i = U - TS + PV$, then for a plane boundary this potential has the form $G_b = \Sigma_i \mu_i m_i = U - TS + PV - \sigma A$, designations for internal energy U, entropy S, temperature T, pressure P, volume V, surface tension σ; μ_i and m_i are the chemical potential and mass of component i. In thermodynamics, the concept of a phase interface is formulated for macroscopic systems, for which the inclusion of surface contributions is only necessary if the surface area A is so developed that the contribution of the term $\sigma A = G_b - G_v$ to all thermodynamic potentials becomes commensurable with the contribution from bulk phases: $\sigma A \sim G_v$ and it can not be neglected. If $G_v \gg \sigma A$, or $G_b - G_v \approx 0$, then the surface contribution to the thermodynamic functions of the systems under consideration is omitted.

Mechanical determination of surface tension for a plane interface. The physical interface of the phases is not a geometric surface of zero thickness, but a transition layer whose thickness is finite (Fig. 21.1 *b*). Since the density in the transition zone varies noticeably in the direction normal to the interface, within this region the pressure tensor, defined in the hydrostatic sense, must also change, referring to the isotropic and constant hydrostatic pressure within each phase.

We introduce a rectangular coordinate system (x, y, z) with the z axis directed along the normal to the flat interface from the phase α to the phase β and the plane xy parallel to the plane of the boundary [2]. In any uniform phase, the pressure is isotropic, i.e., the force acting on the unit area is normal to it and is the same for all unit area orientations. Consequently, within each of the uniform phases α or β, the pressure tensor **P** is reduced to the hydrostatic pressure P multiplied by the unit tensor **I**: **P** = P**I**, or

$$P_{xx} = P_{yy} = P_{zz} = P, P_{xy} = P_{yz} = P_{zx} = P_{yx} = P_{zy} = P_{xz} = 0. \quad (22.15)$$

Inside the interphase region, the force acting along the normal to the unit area is different in different directions. However, from the symmetry requirements it is easy to see that in the planar case, even in the transition region, $P_{xy} = P_{yz} = P_{zx}$ should disappear, and P_{xx} and P_{yy} must be equal to each other and not depend on either x

or y. For the remaining non-zero components (or components) of the pressure tensor, it is convenient to introduce the notation $P_{zz} = P_N(z)$ and $P_{xx} = P_{yy} = P_T(z)$, which we will call the normal and tangential components of the pressure tensor **P** at the point z.

The pressure tensor on a planar vapour–liquid interface is written as

$$\mathbf{P} = P_T(z)(\mathbf{e}_x\mathbf{e}_x + \mathbf{e}_y\mathbf{e}_y) + P_N(z)\mathbf{e}_z\mathbf{e}_z, \tag{22.16}$$

where $\mathbf{e}_x$, $\mathbf{e}_y$, $\mathbf{e}_z$ are unit vectors directed along the Cartesian axes $\alpha = x, y, z$; P is the isotropic pressure in bulk coexisting phases. In general, the tangential component of $P_T(z)$ can vary along z in a complex manner, whereas the values of the normal component $P_N(z)$, according to the hydrostatic equilibrium condition between the coexisting phases, must equal P even in the transition zone: $P_N(z) = P$.

We choose a unit area in the form of a strip having a unit width in the y direction and extending from $-\ell/2$ to $\ell/2$ in the z direction. It is obvious that the total stress $\Delta\Sigma_x$ acting in the direction of the x axis through this unit area can be written as

$$\Delta\Sigma_x = -\int_{-\ell/2}^{\ell/2} P_T(z)dz. \tag{22.17}$$

If the value of $\Delta\Sigma_x$ did not contain a contribution from the transition zone, it would simply be $-P\ell$. For a planar boundary, the excess stress due to the presence of an interphase boundary (here the unit area is chosen as the strip shape as specified above) is [2.7]

$$\sigma = -\int_{-\ell/2}^{\ell/2} P_T(z)dz + P\ell = \int_{-\ell/2}^{\ell/2} [P - P_T(z)]dz \tag{22.18}$$

where the condition that the normal component of the pressure tensor of the plane boundary and pressure in the bulk phase is equal ($P_N(z) = P$).

This equation shows that a real system with a planar boundary between phases can be treated as if it consisted of two uniform phases separated by a plane membrane of zero thickness carrying the tension σ defined by (22.18). Thus, equation (22.18) is a definition of surface tension, and we will take it as the mechanical definition of σ.

It is also obvious that the surface tension determined by (22.18) is independent of ℓ provided that ℓ has a macroscopic length such that for $z = \ell/2$ and $z = -\ell/2$ the value of $P_T(z)$ becomes equal to the hydrostatic pressure P From this consideration, we can assume that ℓ is infinitely large, and rewrite (22.18) in the conventional form:

$$\sigma = \int_{-\infty}^{\infty} (P - P_T)\,dz. \tag{22.19}$$

Using the fact that P_N does not depend on z and is equal to P, it is possible to obtain Becker's equation [7] from (22.19)

$$\sigma = \int_{-\infty}^{\infty} (P_N - P_T)\,dz. \tag{22.20}$$

In ordinary hydrodynamics, the physical quantities at the lengths of the order of the range of intermolecular forces are assumed to be constant, therefore the value of $P_T(z)$ introduced above can not be directly identified with pressure in the hydrodynamic sense. At the microscopic level, the question of the correspondence between $P_T(z)$ and $P_N(z)$ with the local components of the pressure tensor and hydrodynamic concepts was developed much later in the framework of microscopic hydrodynamics [8] (see also Section 37).

Thermodynamics introduced the concept of surface tension, suggesting its definition from experimental data. To calculate its value, it is necessary to use molecular models.

23. Molecular description of a planar interface

The most consistent way in solving this problem for pure substances was associated with the use of molecular models. For this purpose, all the existing methods of statistical thermodynamics [2,9–25] are involved. These include the theory of van der Waals capillarity [9,11,12], the method of integral equations [2,9,10,13,14] and its simplified version without correlation effects – the density functional method [15–19], the method molecular dynamics [20,21] (and Monte Carlo [13]) and the lattice gas model (LGM) [22–25].

Continual description [2]. Consider a system in a vessel that has the shape of a rectangular parallelepiped with edges of length l_1, l_2, l_3, directed along the axes of the rectangular coordinate system x, y, z, respectively; the z axis is directed vertically upwards – opposite to the direction of gravity. The volume of the system is $V = l_1 l_2 l_3$ and the interface is $A = l_1 l_2$ (see Fig. 23.1).

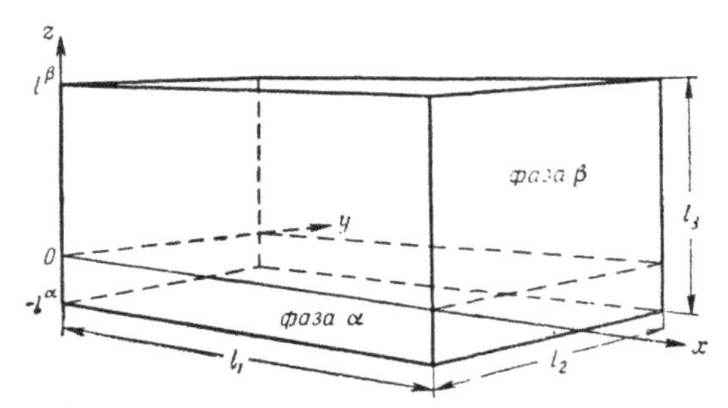

Fig. 23.1. The flat interface of the phases [2].

The free Helmholtz energy of a two-phase system depends on A and V, $\left(\dfrac{\partial F}{\partial l_1}\right)_{l_2,l_3,T,N} = l_2\left(\dfrac{\partial F}{\partial l_1}\right)_{V,T,N} + l_2 l_3\left(\dfrac{\partial F}{\partial l_1}\right)_{A,T,N} = l_2\sigma - l_2 l_3 P$ so from which $\left(\partial F / \partial l_1\right)_{A,T,N} = \left(\partial F / \partial l_3\right)_{l_1,l_2,T,N} / (l_1 l_2)$, using the relationship, we obtain the following relation $\sigma = \dfrac{1}{l_1 l_2}\left[l_1\left(\dfrac{\partial F}{\partial l_1}\right) - l_3\left(\dfrac{\partial F}{\partial l_3}\right)\right]$.

In order to differentiate the partition function Z (because $F = -kT\ln Z$), a device specially developed by Bogolyubov should be used in terms of volume, which allows us to relate the expression for the surface tension for a planar interface through the distribution functions

$$\sigma = \frac{1}{2}\int_{-\infty}^{\infty} dz_1 \int \theta(z_1, \mathbf{r}_{12}) \frac{\partial \varepsilon(r_{12})}{\partial r_{12}} \frac{(x_{12}{}^2 - z_{12}{}^2)}{r_{12}} d\mathbf{r}_{12}$$

where $\theta(z_1,\mathbf{r}_{12})$ is paired distribution function (DF), characterizing the probability of finding two particles at points with coordinates z_1 and $z_1 + \mathbf{r}_{12}$; $\varepsilon(r_{12})$ is the paired interaction potential of the specified particles, $x_{12} = x_1 - x_2$ is the component of the distance vector between two particles $\mathbf{r}_{12} = \mathbf{r}_1 - \mathbf{r}_2$ along the x axis, z_{12} – similarly for the z axis.

The dependence of the surface tension on the distribution functions formally solves the problem. For computations, the values of unary and paired DFs are necessary inside a vapour–liquid interface that is non-uniform in density. The equations for these DFs are given in [2,9,10]. Because of the large computational problems with the use of integral equations for unary and paired DFs, the van der Waals theory of capillarity and the density functional method are widely used. The difficulties in using the molecular dynamics method for calculating the drop characteristics are discussed in [21]. However, these methods give rather different results, so the thermodynamic approach remains the main one in calculating the nucleation energy for the kinetics of first-order phase transitions.

Discrete concentration profile. Consider a macroscopic two-phase system with a liquid–vapour interface (Fig. 21.1 *c*). The simplest version of the LGM, which has the simplest equations for the concentration profile, is discussed. Recall that LGM was the first molecular approach to the theory of surface phenomena [2,9,26–29], which was applied both to simple gases and liquids, as well as to metals and alloys – see the survey papers [29–31]. It makes it possible to calculate the surface tension, considering the parameters of the model either as determined from the data of a direct experiment, or as found from other bulk properties, for example, on the stratification curves. The equations for the volume phase are given in monographs [8, 29–33] (see also Section 18).

The volume of the system is divided into individual elementary cells or volume sites $v_0 = \gamma_s \lambda^3$ (see Section 16). We combine the cells into monolayers of width λ. All sizes in the system (drop radius, width of the transition region, etc.) will be measured in units of λ. The sites located in one monolayer q are assigned the type q. A site of type q is characterized by the numbers of its nearest neighbours z_{qp}, $1 \leq q, p \leq \kappa$, both in the same monolayer $p = q$ and in neighboring monolayers $p = q \pm 1$, for any site in the monolayer q, and $\sum_{p=q-1}^{q+1} z_{qp} = z$. Here κ is the number of monolayers in the flat transition region between the vapour and the liquid, and z is the coordination number of the lattice. For a planar lattice or bulk layer, the number of cells in all monolayers is the same and the numbers $z_{q,p}$ are constant: $z_{q,q-1} = z_{q,q+1} \equiv z_1$, $z_{q,q} = z-2z_1$. For example, for a flat lattice with a coordination number $z = 6$, we have $z_{q,q\pm1} = z_1 = 1$, $z_{q,q} = 4$.

For a given temperature T, according to the Maxwell rule [8, 32, 33], in the bulk phase, we determine the equilibrium pressure P_0 and the numerical densities of the coexisting phases $\theta^{(L)}$ and $\theta^{(V)}$, liquid and vapour, respectively. We will include monolayers from each of the coexisting phases in the number: by the construction of the system of equations, the index $q = 1$ corresponds to the dense phase ($\theta_1 = \theta^{(L)}$), and the index $q = \kappa$ corresponds to the vapour phase ($\theta_\kappa = \theta^{(V)}$).

The equations for the concentration profile $\{\theta_q\}$, $1 \le q \le \kappa$, have the form

$$a_q P = \frac{\theta}{1-\theta}\Lambda_q,\ \Lambda_q = \prod_{p=q-1}^{q+1}\left(S_{qp}\right)^{z_{qp}}, \qquad (23.1)$$

here the equations take into account the energy non-uniformity of the lattice sites along the normal to the surface and the interaction between nearest neighbours ($R_{lat} = 1$); Λ_q is a function of non-ideality, depending on the type of approximation. Henry's constants $a_q = \beta F_q \exp(\beta\varepsilon_q)/F_0$ reflect internal motions of molecules in different regions with variable local density. In the formula (23.1), $s = 2$ are particles A and vacancies. All local partial fillings θ_q are functions of pressure P. The non-ideality functions Λ_q take into account only direct correlations between interacting particles in the QCA. In the case of QCA, the functions $S_{qp} = 1+xt_{qp}$, $t_{qp} = \theta_{qp} / \theta_q \equiv \theta_{qp}^{AA} / \theta_q^A$ where they are defined by the following formulas (here $x = x_{qp}^{AA} = \exp(\beta\varepsilon)-1$ from eq, (18.4))

$$\theta_{qp}^{AA} = \frac{2\theta_q^A\theta_p^A}{\delta_{qp}+b_{qp}}, \delta_{qp} = 1 + x_{qp}^{AA}(1-\theta_q^A-\theta_p^A),$$
$$b_{qp} = \left\{\left[\delta_{qp}\right]^2 + 4x_{qp}^{AA}\theta_q^A\theta_p^A\right\}^{1/2}, \qquad (23.2)$$

$$\sum_{i=1}^{s}\theta_{qp}^{ij} = \theta_p,\ \sum_{j=1}^{s}\theta_{qp}^{ij} = \theta_q,\ \sum_{i=1}^{s}\theta_q^i = 1. \qquad (23.3)$$

The system of equations (23.1) with allowance for the normalizations to the functions θ_f^i and θ_{fg}^{ij} is closed. The concentration profile of the transition region $\{\theta_q\}$ is found by iterating by the Newton method from the solution of the system of equations (23.1), giving $P_q = P_0$ for a given value of the width of the transition region $\{\theta_q\}_{q=1}^{q=\kappa}$ and the initial approximation to the density profile of the transition region that varies monotonically from $\theta_1 = \theta^{(L)}$ to $\theta_\kappa = \theta^{(V)}$ The accuracy of the solution of this system is not less than 0.1%.

Knowing the concentration profile, we can calculate the thermodynamic characteristics of the interface. The expression for the free energy of the system F for the distributed model of the system under consideration with $R_{lat} = 1$, according to (18.2), (18.3), will be written as [34]

$$F = \sum_q \sum_i \theta_f^i M_f^i = \sum_q \sum_i \left\{ \theta_q^i \left(v_q^i + kT \ln \theta_q^i \right) + \frac{kT}{2} \times \right.$$
$$\left. \times \sum_{p=q-1}^{q+1} z_{qp} \sum_{i=1} \left[\theta_{qp}^{ij} \ln{}^* \theta_{qp}^{ij} - \theta_q^i \theta_p^i \ln\left(\theta_q^i \theta_p^1 \right) \right] \right\}, \tag{23.4}$$

where $^*\theta_{fg}^{ik} = \theta_{fg}^{ik} \exp(-\beta\varepsilon_{fg}^{ik})$, and

$$M_f^i = v_f^i + kT \ln \theta_f^i + \frac{kT}{2} \times$$
$$\times \sum_{p=q=1}^{q+1} z_{qp} \ln\left[{}^*\theta_{fg}^{ii} \, {}^*\theta_{fg}^{ik} / \left(\theta_f^i \right)^2 {}^*\theta_{fg}^{ki} \right], \tag{23.5}$$

where the symbol k refers to one of the components of the mixture – a reference sort of particles, $1 \le k \le s$, $k \ne i$. The choice of k is determined by the convenience of calculation.

Calculation of the width of the interphase region. The quantity κ is found by varying the upper limit of the system of equations (23.1)–(23.3) in order to obtain a constant molecular density profile over the layers for given values of the temperature T and the chemical potential μ (or external pressure P). The procedure for finding the width of the transition region and calculating the distributions of molecules at the vapour–liquid interface is demonstrated using the example of Fig. 23.2. The length of the transition region κ is the maximum number of monolayers at which the local densities θ_q change monotonically from the values of $\theta^{(V)}$ corresponding to the vapour phase to the values of $\theta^{(L)}$ corresponding to the liquid phase for all q, $1 \le q \le \kappa$. Figure 23.2 shows the results of calculation of the transition region in the volume and shows the effect of the number of monolayers on the value of the free energy F. The value of the free energy is calculated from the found concentration profile (23.1).

Figure 23.2 *a* shows the change in the numerical density θ in the transition from the vapour phase to the liquid phase. The abscissa is the number of the layer (in the volume it is just the site number) of the transition area. The ordinate is the density θ_q at the site of the layer q, $1 \le q \le \kappa$, $\kappa = 12$ for $\tau = T/T_c = 0.8$. Dotted lines reflect the discrete nature of the transition region, consisting of κ

sites. Figure 23.2 *b* shows the free energy values of the transition region F calculated from the formulas of Ref. [34] for the values of κ obtained during the iterative determination of the length of the transition region. Figure 23.2 *c* shows the nature of the change in the increment $\Delta F = F(\kappa) - F(\kappa - 1)$ with increasing κ. In view of the discreteness of the change in κ, we have $\Delta\kappa = 1$ and the increment ΔF is an analog of the derivative $dF/d\kappa$.

The curve in Fig. 23.2 *c* shows that for the values of $\kappa \geq 11$ the derivative of free energy along the length of the transition region is practically constant. A further increase in the length of the region leads to a certain limiting value of the specific free energy of the system, the concentration profile ceasing to depend on the length of the transition region, which determines the size of the transition region at a given temperature.

The initial section θ_q for $q < 5$ on the profile of Fig. 23.2 *a* remains practically constant. The real density difference between the liquid and the vapour is realized at a length of only $l = 5$–6 monolayers, that is $l \sim \kappa/2$. A further increase in κ can lead to a non-monotonicity of the profiles of local densities in the transition region. This is due to the possibility of formation of fluid structures having a more developed surface between vapour and liquid than the monotonic structure of the transition region at the vapour–liquid interface.

Figure 23.3 shows the density distribution profiles in the transition region between liquid and vapour for a plane boundary for a wide range of reduced temperatures. The ordinate in this figure shows

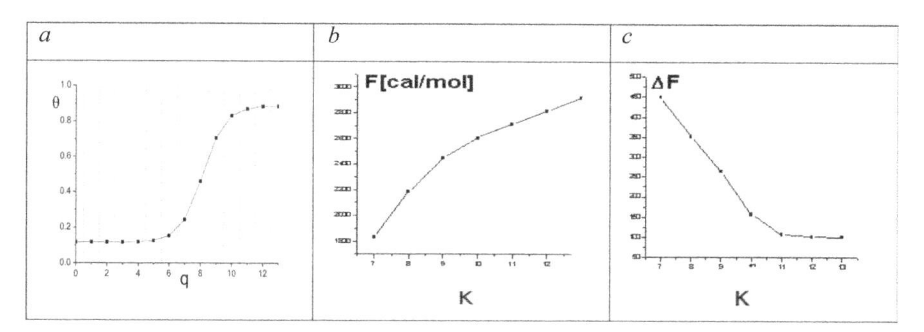

Fig. 23.2. Characteristics of the transition region of argon in the volume at $\tau = 0.8$. (*a*) The density profile θ_q in the transition region of the vapour–liquid interface in the volume for $\kappa = 12$; on the axis of the abscissas, q is the number of the layer in the transition region. (*b*) The free energy of the transition region $F(\kappa)$ as a function of the length of the transition region κ. (*c*) The free energy increment of the transition region ΔF as a function of the length of the transition region κ [8].

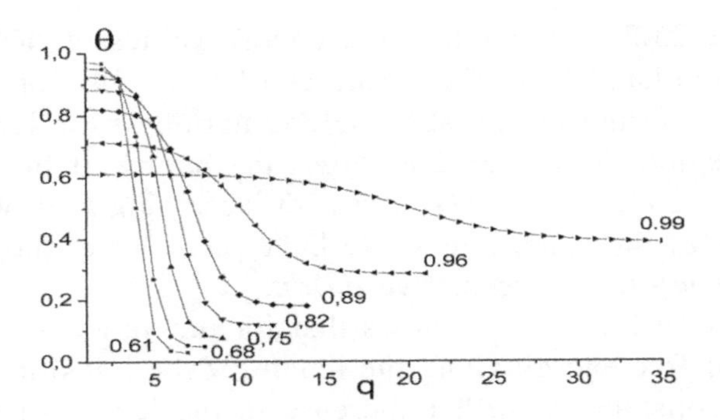

Fig. 23.3. Profiles of the density of the transition region from liquid to vapour at different temperatures for a planar lattice. The values of $\tau = T/T_c$ are indicated on the curves. The abscissa is the number of the monolayer of the transition region q, and the local densities θ_q along the ordinate.

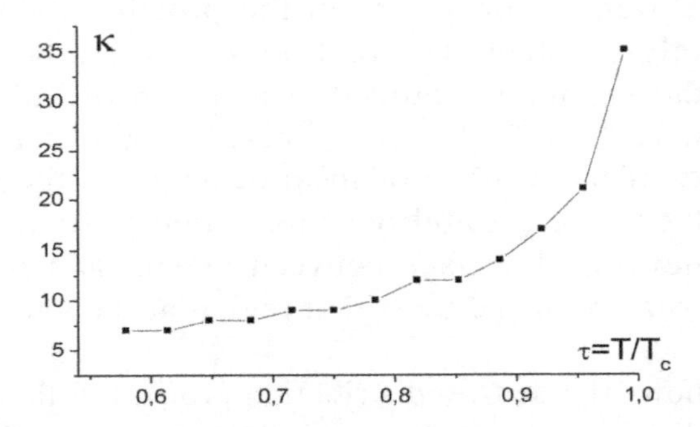

Fig. 23.4. Dependence of the width of the transition region κ on the reduced temperature τ.

the local densities θ_q in the layers of the transition region. The abscissa is the number of the corresponding layer. With increasing temperature, the number of layers in the transition region increases, and the values of $\theta_1 = \theta^{(L)}$ and $\theta_\kappa = \theta^{(V)}$ approach (for $\tau = 1$ they will be equal).

The width of the transition region (the number of its layers) increases with increasing temperature, as shown in Fig. 23.4. For $\tau \to 1$, we have an increase in κ, which corresponds to $\kappa \to \infty$.

Surface tension. The formula for calculating the surface tension for a plane boundary by the thermodynamic definition in terms of excess free energy is written, according to [2,9,28,29,34], as

$$\sigma A = \sum_{q=1}^{\kappa} (M_q^V - \mu_V), \tag{23.6}$$

where the functions M_q^V and have the form

$$M_q^V = kT\ln\theta_q^V + \frac{kT}{2}\sum_{p=q-1}^{q+1} z_{qp}\ln\left[\theta_{qp}^{VV}\theta_{qp}^{Vk}/\left(\theta_q^V\right)^2\theta_{qp}^{kV}\right],$$
$$\mu_V = kT\ln\theta_V + \frac{1}{2}kTz\ln\left[\theta_{VV}/(\theta_V)^2\right] \qquad (23.7)$$

We note that the functions M_q^i defined by formula (23.5) in expression (23.6) can refer to any sort of particle i. The formula (23.6) follows from the direct thermodynamic definition as the excess of free energy in the transition region (22.10) $\sigma A = \sum_{i=1}^{s}\sum_{q=1}^{\kappa}(M_q^i - \mu_i)\theta_q^i$ due to the fulfillment of the local equilibrium equations (23.1).

The internal pressure π in the lattice system corresponds to its connection with the chemical potential of vacancies in the form $\pi v_0 = -\mu_V$. For non-uniform regions this relation has the form $\pi_q v_q^0 = -M_q^V$. Then the local pressure π_q in the non-uniform region q with respect to the layered structure of the interface is written as

$$\beta\pi_q v_q^0 = -\ln\theta_q^V - 1/2\sum_{p=q-1}^{q+1} z_{qp}\ln(\theta_{qp}^V/(\theta_q^V\theta_p^V)) \qquad (23.8)$$

Formula (23.8) corresponds to the average pressure in a non-uniform system. In the uniform phase (23.8), the isotropic pressure P is described. At the interface, the non-uniform distribution of molecules leads to a change in the isotropic properties of the phase, including the violation of the isotropy of the pressure at the interface – the latter differs from the pressure in the bulk phase.

Along with the thermodynamic determination of surface tension, hydrostatic determination of surface tension is actively used [2.9] through the mechanical work of creating the surface. On the basis of expression (23.8), we can select the components of the pressure tensor π_q^α as follows

$$\beta\pi_q^\alpha v_q^0 = -\ln\theta_q^V - 3/2\sum_{p=q-1}^{q+1} z_{qp}\cos^2(qp,\alpha)\ln\left(\theta_{qp}^{VV}/\theta_q^V\theta_p^V\right), \qquad (23.9)$$

where the subscript α denotes the direction of the design axis in space, $\alpha = N$ and T. The symbol (qp, α) denotes the angle between the direction of α and the direction of communication between the sites in the layers q and p. For a planar boundary, N is the direction of the normal to the interface of the vapour–liquid interface; T is one of the two tangential directions, so $\cos^2(qp,N) + 2\cos^2(qp,T) = 1$.

The above structural characteristics z_{qp} allow us to have the identical structure of expression (23.9) for interfaces with different symmetries. Thus, for $z = 6$, the normal components of the pressure tensor are written in the form

$$\beta\pi_q^N v_q^0 = -\ln\theta_q^V - 3/2[z_{qq-1}\ln(\theta_{qq-1}^{VV} / \theta_q^V \theta_{q-1}^V) + z_{qq+1}\ln(\theta_{qq+1}^{VV} / \theta_q^V \theta_{q+1}^V)], \tag{23.10}$$

and the tangential components of the pressure tensor differ in the coefficients of the second term:

$$\beta\pi_q^T v_q^0 = -\ln\theta_q^V - 3z_{qq}\ln(\theta_{qq}^{VV} / \theta_q^V \theta_q^V) / 4. \tag{23.11}$$

According to the hydrostatic definition of surface tension, instead of formula (23.6)

$$\sigma A = \sum_{q=1}^{\kappa} (\pi_{1,\kappa} - \pi_q^T), \tag{23.12}$$

which is identical to the Becker equation (22.20), since because of the equality of pressure in the coexisting phases $\pi=\pi_1=\pi_\kappa=\pi_q^N$.

This model was used to describe various experimental data [28-30,35]. For example, calculations of the temperature dependences of the surface tension for substances whose properties are close to the so-called law of the corresponding states are presented [36]. Details of calculations are indicated in [35], they are performed for 19 substances. Some of them are shown in Fig. 23.5, and the parameters are indicated in Table 23.1. The results of the calculation are plotted with solid lines and are marked with the corresponding sequence number from Table 23.1 (field *a*: 1–6, field *b*: 7–11), the experimental points are marked with symbols.

24. Molecular description of the curved interface

A rigorous statistical theory describing the equilibrium distributions of molecules leads to flat boundaries [2,9,10]. The presence of curved surfaces, from the point of view of thermodynamics [1,40–42], indicates a metastable state of the liquid.

The layer structure of spherical monolayers [24, 25]. The most detailed (distributed) model in the LGM was determined in [33, 34]. With its help, you can form any type of non-uniform systems by combining sites of the same sort into a common group. In particular, by combining sites into layers, the layered model of a flat boundary is formulated above. A similar principle was used in [24] to describe spherical phase interfaces. It considers a layer model of the transition region of a spherical drop, which is divided into monomolecular layers of width λ that are uniform in their properties. These layers are numbered by the index q, where q is the site number pertaining to the monolayer in question, $1 \leq q \leq \kappa$, here κ is the width of the transition region. Each monolayer is characterized by its density θ_q. As above, we restrict ourselves to the interaction of only the nearest neighbours.

The structural characteristics of z_{qp} must be related to the average numbers of bonds between sites in different monolayers q and p, and the number of sites in each monolayer of radius R must be related

Table 23.1. The table data of the substances examined and the model parameters applicable to them

No.		ε_0/k, K [36]	σ_0, Å [36]	T_{cr}, K [37]	τ	T_{melt}/T_{cr} [37]	R_L	$\delta\sigma_0$ %	$\delta\varepsilon$ %	σ_{exp}, mN/m	σ_{theor}, mN/m	δ, %
1	CH_4	148.2	3.817	190.55	0.48	0.47	1	0	−10	18.00 [38]	18.03	0.17
2	N_2	95.05	3.698	126.25	0.51	0.5	1	0	−6.8	11.77 [39]	11.91	1.2
3	O_2	117.5	3.58	154.77	0.42	0.35	1	0	−4.3	19.40 [39]	19.67	1.4
4	C_2H_6	230	4.418	305.3	0.44	0.30	1	0	−4.3	24.62 [38]	24.57	0.20
5	CO	100.2	3.763	132.91	0.53	0.52	1	0	−3.2	12.11 [37]	12.15	0.33
6	n-C_4H_{10}	297	4.971	426.2	0.40	0.32	1	0	−2.2	27.2 [38]	26.91	1.1
7	C_6H_6	440	5.270	562	0.50	0.5	2	0	−10	30.24 [38]	30.25	0.033
8	F_2	112	3.653	144	0.48	0.37	2	0	−4.2	17.90 [38]	17.91	0.056
9	C_{12}	257	4.4	417.15	0.48	0.41	3	0	−8.2	33.00 [37]	33.06	0.18
10	C_2H_4	199.2	4.523	282	0.55	0.37	4	+10	−6.2	19.05 [38]	19.13	0.42
11	C_3H_8	254	5.061	368.8	0.38	0.23	4	+10	−5.1	27.80 [38]	27.62	0.65

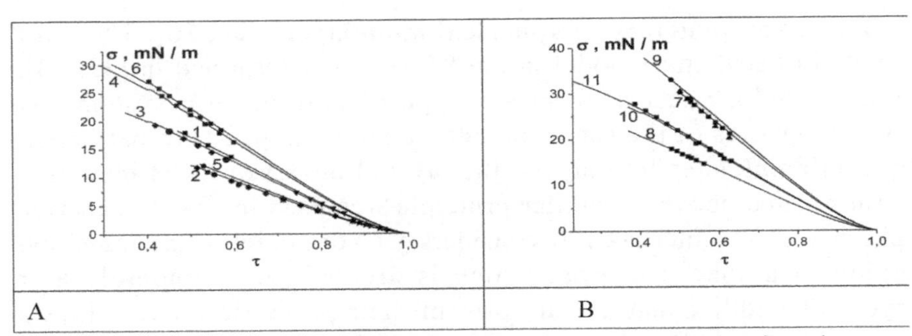

Fig. 23.5. Dependence of surface tension σ on reduced temperature τ for substances from Table 23.1.

to its curvature. As a result, the structure of the entire transition region is given by the average numbers $z_{qp}(R)$, denoting the number of neighboring sites of the layer p around any site of the layer q, where $1 \le q \le \kappa$. Each layer q of the transition region has a radius of curvature $q = R$, $z_{qp}(R) \equiv z_{Rp}$. For a planar interface (for $R \to \infty$), this formulation of the problem automatically goes over to the model of Section 23 described earlier. The construction of equations describing the distribution of molecules in the transition region of the interface of coexisting phases reduces to constructing the numbers $z_{qp}(R)$.

We assume that the drop has a sufficiently large diameter, then the deviations of the lattice structure in the drop from the strictly regular rectangular lattice in the bulk phase will be small. In the 'curved' spherical drop layers one can retain the same layer-by-layer method for describing the number of nearest neighbours $z_{qp}(R)$, which should depend on the number of the monolayer and the radius of the drop R [24, 25]. The count of the number of monolayers comes from the centre of the drop. For simplicity, Fig. 24.1 *a* shows a lattice with $z = 6$ nearest neighbours (see Fig. 24.1 – field *b*). The closest neighbours of the filled cell located in the monolayer q are 2 cells located in the monolayers $p = q-1$ and $p = q+1$ (in the figure – left and right), as well as 4 cells located in the same monolayer $p = q$. Two of them are shown in Fig. 24.1 *b* (top and bottom), and the other two are located in front of the filled cell and behind it along the normal to the plane of the drawing.

Figure 24.1 shows a fragment of a drop of radius R with centre at point O and a transition region consisting of κ monolayers from the radius R to $R+\kappa$. By radius R we mean the radius of the liquid part of the drop (internal, shaded part of the drop (region 1)), it corresponds to the index $q = 1$ of the transition region. The outer region of the 3-vapour, the region of the vapour corresponds to the radius $R + \kappa$.

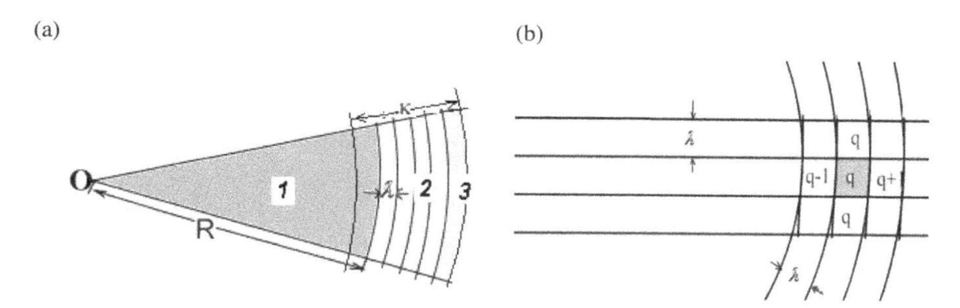

Fig. 24.1. (a) A schematic fragment of a drop of radius R, the inner part of which is filled with liquid, the outer part (3) – with vapour, and the transition region from liquid to vapour consists of κ spherical monolayers of width λ, including one monolayer of liquid and one monolayer of vapour. (b) scheme of the nearest neighboring sites in the 'curved' lattice $z = 6$.

The transition region 2 represents κ spherical monolayers, including one monolayer of liquid and one monolayer of vapour (Fig. 21.1 *a*). In the curved spherical drop layers, the layer-by-layer method for describing the number of nearest neighbours $z_{qp}(R)$ (Fig. 21.1 *b*) is preserved for a plane boundary. Accordingly, all functions become dependent on R.

The overall balance of the sites of bonds between the surrounding molecules is preserved in the form

$$\sum_{p=q-1}^{q+1} z_{qp}(R) = z. \tag{24.1}$$

Each monolayer $R_q=R+q-1$ contains $N(R_q)$ of elementary cells v_0 (sites). The relation between $N(R_q)$ and the radius of the monolayer R_q is determined from a simple geometric relationship:

$$\begin{aligned} N(R_q) &= \left[V(R_q+0.5)-V(R_q-0.5)\right]/v_0, \\ V(R_q+0.5) &= 4\pi/3\left[\lambda(R_q+0.5)\right]^3, \end{aligned} \tag{24.2}$$

where $V(R_q+0.5)$ is the volume of a sphere with a radius $(R_q+0.5)$.

The difference in square brackets determines the volume of the spherical monolayer. The centre of the drop corresponds to the value $R = 0$ in the expression (24.2). The expression (24.2) gives rough estimates of $N(R_q)$ for $R = 1 \div 4$, therefore $N(R_q) = 5$ as the minimum value of $N(R_q) = 5$. The question on the small size of drops should be considered within the framework of a discrete description and here it is not discussed.

It follows from (24.2) that the number of sites of the monolayer R_q is highly accurate in proportion to the area of the sphere S_R passing through the centre of the cell with radius R_q+0.5, where the monolayer radius is measured in the diameter numbers of the molecule λ

$$N(R_q) = 4\pi/3\left[3R_q^{\ 2}+0.25\right] = 4\pi\left(R_q^{\ 2}+1/12\right)\approx 4\pi R_q^{\ 2} = S(R_q). \tag{24.3}$$

Each monolayer q, $1 \le q \le \kappa$ is assigned its weight $F_q(R)$ – the fraction of sites of type q in the interphase layer. The contribution of each monolayer in the weight function of the sites of the transition region is expressed as

$$F_q(R) = N_q(R)\ /\ N(R),\ 1\le q\le\kappa,\ N(R) = \sum_{q=R+1}^{R+\kappa-1} N_q(R), \tag{24.4}$$

For a plane boundary, the values of $F_q = \kappa^{-1}$.

From the condition of the balance of bonds between the nearest sites in adjacent layers q = R and R+1, the equation $N(R_q)$

$$N(R_q)z_{qq+1}(R) = N(R_{q+1})z_{q+1q}(R), \tag{24.5}$$

which connects the number of neighbours between different neighbouring layers in the forward $z_{qq+1}(R)$ and the inverse $z_{q+1q}(R)$ directions from the centre of the drop.

Formulas (24.1) and (24.5) hold for each value of R. Two equations are not sufficient to uniquely determine the three values $z_{qp}(R)$ (p = q, $q \pm 1$) for any R. A strict procedure for determining $z_{qp}(R)$ from the analysis of successive addition of surface atoms and minimization of the surface area of the drop. Such a procedure is an independent problem, so we confine ourselves to determining one of the quantities $z_{qp}(R)$, starting from physical assumptions.

We represent the structure numbers for the curved lattice $z_{qq\pm1}(R)$ in terms of analogous numbers for the planar lattice $z_{qq\pm1}$ in the form

$$z_{qq-1}(R) = z_{qq\pm1}(1 - \alpha / R), \tag{24.6a}$$

then from equations (24.1) and (24.5) we obtain

$$z_{qq+1}(R) = z_{qq\pm1}\left[1 + (2-\alpha)/R + (1-\alpha)/R^2\right], \tag{24.6b}$$

$$z_{qq}(R) = z_{qq} - 2z_{qq\pm1}\ (1-\alpha)\ (1+1/(2R))/R$$

Equations (24.6) can be interpreted as approximation expressions for two cases having a clear physical picture.

In the first case, for $\alpha = 0$, the relation (24.6a) means that for any site of the curved monolayer R_q, the number of bonds to the centre of the drop remains the same as it was in the bulk phase for a plane boundary. This assumption follows from a consideration of the microcrystallite at low temperatures [1.42], for which there are many faces of different orientation of the single crystals. Each face has a flat (three-dimensional) structure, and their docking, forming regions with projected surface atoms between the faces, are regions with an increased number of broken bonds, whereas all the connections deep into the crystal for these surface atoms remain. In this case, for a curved boundary, the number of bonds between neighboring layers oriented from the centre of the drop is larger than for a plane boundary ($z_{qq}+1(R) > z_{qq\pm1}$), and the values of the number of bonds within one layer are smaller ($z_{qq}(R) < z_{qq}$). In the asymptotic limit, the values of $z_{qq+1}(R)$ and $z_{qq}(R)$ tend to their limits $z_{qq\pm1}$ and z_{qq}.

In the second case with $\alpha = 1$, the numbers $z_{qq\pm1}(R)$ are related to the surface area of this monolayer R with the areas of adjacent monolayers $p = q \pm 1$:

$$z_{qq\pm1}\left(R\right) = z_{q\pm1q}S(R_{q\pm1}) \;/\; S(R_q), \tag{24.7}$$

where the plus sign refers to the farther layer from the centre of the drop, and the minus sign to the layer closer to the centre of the drop. This leads to an increase in the number of connections of the drops directed from the centre $z_{qq+1}(R) \approx z_{qq-1}(1+2/R)$ due to a decrease in the fraction of the bonds directed to the centre of the drop $z_{qq-1}(R) \approx z_{qq-1}(1-2/R)$. At the same time, the number of neighbours in the same layer does not change $z_{qq}(R)$ = const = z_{qq}. For $R \rightarrow \infty$, the ratio $S(R_{q\pm1})/S(R_q) \rightarrow 1$, and this relation is consistent with the volume values z_{qp}. This case refers to large drop radii and high temperatures when the spherical surface is slightly curved.

In the general case, the parameter α is a function of the temperature and size of the drop. When it deviates from the limiting cases 0 and 1, formula (24.6) takes into account the change in all bond numbers $z_{qp}(R)$ from the curvature of the monolayer in accordance with the law λ/R, but with different numerical coefficients.

Structural characteristics [43]. The behaviour of the structural characteristics of a drop as a function of its radius R, varying from 2^2 to 2^{10}, is shown in Figs. 24.2–24.4 for a transition region of width

$\kappa = 8$ monolayers and a reference dividing surface located on the average of the 4th monolayer.

The ratio of the number of sites $\eta_\delta = \sum_{q=4-\delta+1}^{4} N(R_q) / \sum_{q=5}^{5+\delta-1} N(R_q)$ in δ monolayers before and after the dividing surface, for different numbers of monolayers $\delta = 1$, 2, 3 and 4 is shown in Fig. 24.2. This characteristic shows the effect of the drop size on the relative contribution of the internal and external monolayers to the energy of the drop, provided that the dividing surface is in the middle. All curves tend to monotone. But for $\delta = 1$ this is achieved at $R \sim 2^8$, and for $\delta = 4$ the ratio close to unity is reached at $R \sim 2^{10}$, purely geometric deviations from unity with respect to the volumes of internal and external groups of monolayers that play an important role in calculating the values of free energy and surface tension are noticeable for sufficiently large drop sizes.

The values of the weight coefficients F_q for different monolayers of the transition region with $\kappa = 8$ are given in Fig. 24.3. All the curves tend to the limit $1/\kappa = 1/8$. For monolayers that are internal with respect to the dividing surface, $F_q \to 1/\kappa$ from below, and for monolayers external to the dividing surface, $F_q \to 1/\kappa$ from above. In the case of $\kappa = 8$, the ratios F_q differ by more than 7 times for small drops, and become commensurable (the differences between them do not exceed 15%) only at $R > 2^6$.

In addition to the influence of the number of sites in different monolayers, structural characteristics play a role in the distribution of molecules through the values of the numbers $z_{qp}(R)$ in the formulas for local isotherms. Figure 24.4 shows the ratios $z_{4p}(R)/z_{qp}$ for the two variants $z_{qp}(R)$ calculated by formulas (24.6) for $\alpha = 0$ and 1. Both variants give the same dependence on the drop size for $z_{45}(R)$,

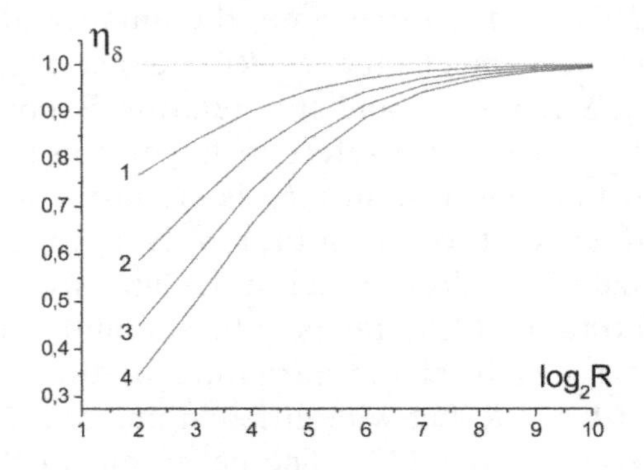

Fig. 24.2. Dependences of structural characteristics $\eta_\delta = \sum_{q=4-\delta+1}^{4} N(R_q) / \sum_{q=5}^{5+\delta-1} N(R_q)$ $\delta = 1, 2, 3{,}4$ on the radius of the drop R.

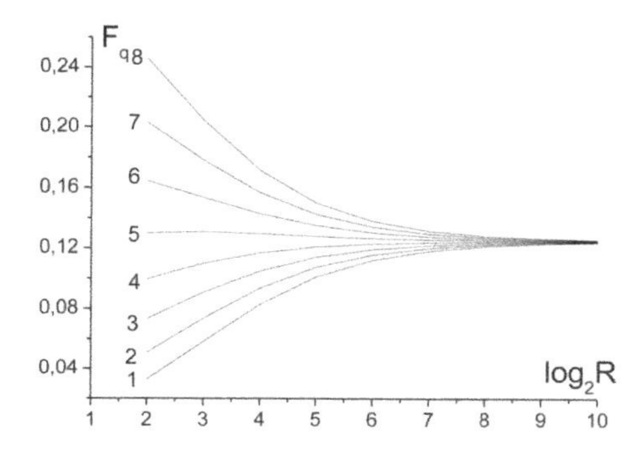

Fig. 24.3. Dependence of the weighting coefficients F_q, $1 \leq q \leq \kappa$ on the radius of the drop R.

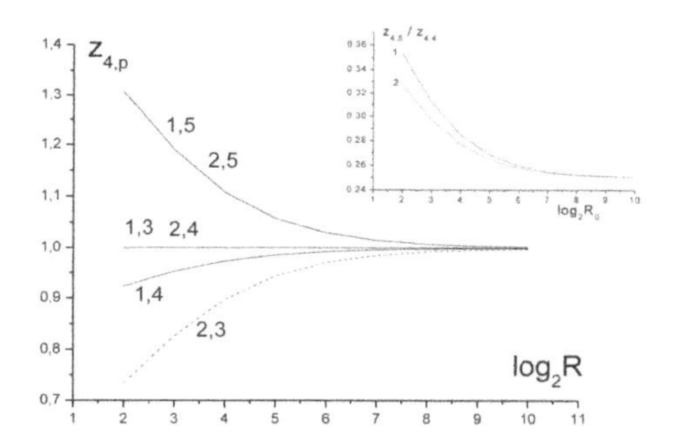

Fig. 24.4. Dependences of the structural characteristics $z_{4p}(R)/z_{4p}$ on the drop radius R for a transition region consisting of $\kappa = 8$ monolayers calculated in two ways. The notation on the curves is b, p, where $b = 1$ for calculations with $\alpha = 0$ and $b = 2 - \alpha = 1$. The insertion gives the ratios $z_{45}(R)/z_{44}(R)$ for the same two variants of constructing the numbers $z_{qp}(R)$.

and the difference between them reduces to redistribution between the numbers of the $z_{44}(R)$ and $z_{43}(R)$ bonds. The insertion gives the relations $z_{45}(R)/z_{44}(R)$ for the same two variants of constructing the numbers z_{qp}. Taking into account that the given numbers $z_{qp}(R)$, like the numbers z_{qp} in (23.1), are exponents in local isotherms, their change with increasing drop radius should cause noticeable differences in the dependences of local densities on the drop size.

Local isotherms. The transition region between phases is a non-uniform system, which exists due to intermolecular interaction. The equation of the local isotherm for the monolayer q relates the probability of filling the cell θ_q and the chemical potential of the molecule μ_q in the layer q, $2 \leq q \leq \kappa-1$. According to [24, 25], we obtain the following system of equations for local isotherms for cells of different types in the transition region of vapour–liquid at a fixed temperature T

$$\beta v_0 a_q P = \exp(\beta\mu_q) = \frac{\theta_q}{1-\theta_q}\prod_p (S_{qp})^{z_{qp}(R)}, \qquad (24.8)$$

$$S_{qp} = 1 + t_{q,p}^{AA} x, \quad x = \exp(-\beta\varepsilon) - 1,$$

where $\beta = (k_B T)^{-1}$, k_B is the Boltzmann constant; ε – parameter of intermolecular interaction of nearest neighbours; t_{qp}^{AA} is defined in (23.2). As for a plane boundary above, the normalization relations (23.3) are satisfied.

The system of equations (24.8) with respect to the local densities θ_q has the dimension equal to the number of layers (κ–2) of the transition region between the vapour and the liquid. It is solved by Newton's iteration method for given values of the vapour density (θ_v) for q = 1 and liquid (θ_f) for $q = \kappa$, as for plane boundaries. In the absence of the Laplace equation, the pressure inside the drop and in the surrounding pair are the same – such drops will be called equilibrium drops. For them, the chemical potential in the drop and vapour is also the same. In this chapter, attention is paid to equilibrium drops, which are absent in thermodynamics and appear only in molecular theories.

The difference from the equations for a plane lattice is associated only with the quantities $z_{qp}(R)$, which depend on the radius of the drop. Accordingly, the formulas for free energy and for local pressures also change. On the basis of the local distributions of molecules $\{\theta_q\}$ found, the expression for the free energy of the transition region is written in the form [25] $F(R) = \sum_{q=2}^{\kappa-1}\sum_{i=A,V} F_q(R) M_q^i(R)\theta_q^i$, where

$$M_q^i(R) = \Delta_{i,A}\left(v_q^i + \sum\nolimits_{p=q-1}^{q+1} z_{qp}(R)\varepsilon_{qp}^{ii}/2\right) +$$

$$+\beta^{-1}\left[\ln\theta_q^i + \frac{1}{2}\sum\nolimits_{p=q-1}^{q+1} z_{qp}(R)\ln\left(\frac{\theta_{qp}^{ii}}{\theta_q^i\theta_p^i}\right)\right], \qquad (24.9)$$

where F_q is defined in (24.4). The contribution of a particle of sort i that fills the site q to free energy is denoted by M_q^i, where Δ_{iA} is the Kronecker symbol, $i = A$, V is the sort of particle filling the site q; this contribution is absent for vacancies. The expression (24.9) for the free energy is normalized to one site of the system. If all p coincide with q, then formula (24.9) determines the chemical potential of particle i in the bulk phase q.

The average value of the local pressure π_q in the layer q with respect to the layer structure of the spherical interface is expressed as

$$\beta\pi_q(R)v_q^0 = -\ln\theta_q^V - 1/2\, z_{qp}(R)\ln(\theta_{qp}^{VV}/(\theta_q^V\theta_p^V)). \qquad (24.10)$$

Formula (24.10) corresponds to the average pressure in the layer q of the non-uniform transition region.

The thermodynamic definition for the surface tension σ can be written in terms of the mean pressures $\pi_q(R)$, which is a direct analog of expression (23.6) for a plane boundary.

$$\begin{aligned}\sigma A &= \frac{1}{F_\rho}\left(\sum_{q\le\rho} F_q(R)(\pi_1 - \pi_q(R)) + \sum_{q>\rho} F_q(R)(\pi_\kappa - \pi_q(R))\right) \\ &= \frac{1}{F_\rho}\sum_{q=1}^{\kappa} F_q(R)(\pi - \pi_q(R))\end{aligned} \qquad (24.11)$$

where the first equation is written in the traditional way for metastable drops, the symbol ρ refers to the radius of the reference dividing surface. (Metastable drops are discussed in Appendix 1 and Chapter 7.) The second equality holds because the equilibrium drop exists on both sides of the interface $\mu_\kappa^V = \mu_1^V = \mu_V = -\pi v_0$. Here the functions $M_q^V(R) = -\pi_q(R)v_0$ have the form (24.9). The rest is like for a flat border.

Macroscopic symmetry of a drop. The drop symmetry requires that, in the absence of an external field, the normal and two tangential components of the pressure tensor P [2.9] are isolated, as in the planar case,

$$P = P_T(r)(\mathbf{e}_\theta\mathbf{e}_\theta + \mathbf{e}_\varphi\mathbf{e}_\varphi) + P_N(r)\mathbf{e}_R\mathbf{e}_R, \qquad (24.12)$$

where $\mathbf{e}_R$, $\mathbf{e}_\theta$, $\mathbf{e}_\varphi$ are orthogonal unit vectors corresponding to the coordinates R, θ, φ; $P_N(R)$ and $P_T(R)$ are the normal and tangential pressures. The values of $P_N(R)$ and $P_T(R)$, according to spherical symmetry, do not depend on the angular variables θ and φ. In the transition to uniform macrophases of vapour and liquid, both quantities coincide with the pressures in both phases P_α and P_β, respectively. In the general case, when the spherical system $P_N(R)$ is mechanically examined, it turns out that it is no longer a constant but a function of R, which can be determined from the hydrostatic equilibrium condition. The functions $P_N(R)$ and $P_T(R)$ depend on the size of the drop. The dependence of P_N on R is determined from the condition of mechanical (hydrostatic) equilibrium. In the absence of external fields, the hydrostatic equilibrium equation has the form

$$\nabla \mathbf{P} = 0. \tag{24.13}$$

It should be noted that Eq. (24.13) is an equation for the mechanics of continuous media – they imply a practically constant density, or the presence of macroscopic discontinuities [44]. Both variants do not correspond to the interface of the vapour-liquid phases with variable density. This question arose long ago in connection with the equation for the plane boundary of Becker [7]. The mechanical equilibrium must be considered at the microlevel of the discrete model. For equilibrium pressure drops inside and outside the drop, the same $P_\alpha = P_\beta$, therefore $P_N(R) = \text{const}$.

Mechanical definition of σ. For the components of the pressure tensor π_q^α, $\alpha = N$ and T, formula (23.9) is rewritten for the curved boundary as follows

$$\begin{aligned} \beta\pi_q^a(R)\, v_q^0 &= -\ln\theta_q^V - 3/2\, z_{qp} \\ &(R)\cos^2(qp,\alpha)\ln(\theta_{qp}^{VV}/\theta_q^V\theta_p^V), \end{aligned} \tag{24.14}$$

As above, the index α denotes the direction of the design axis in space, and the symbol (qp,α) denotes the angle between the direction of α and the direction of communication between the sites in the layers q and p. For a spherical boundary, N is the direction of the normal to the interface of the vapour–liquid boundary; T is one of the two tangential directions, so $\cos^2(qp,N)+2\cos^2(qp,T) = 1$. It follows from (24.13) that the ordinary coupling $\pi_q(R) = (\pi_q^N(R)+2\pi_q^T(R))/3$.

According to the hydrostatic definition of surface tension, we have

$$\sigma = \frac{1}{F_\rho(R)}\left[\sum_{q\le\rho} F_q(R)(\pi_1 - \pi_q^T(R)) + \sum_{q>\rho} F_q(R)(\pi_\kappa - \pi_q^T(R)\right], \tag{24.15}$$

where the functions $\pi_q^T(R)$ are the tangential components of the pressure tensor (24.14).

The boundary value of this function for $q = 1$ in the liquid phase drops is calculated by the formula (24.10) (index of the fluid f is replaced by the index 1). For another boundary value of the function M_κ^V for $q = \kappa$, we have the same formula with the replacement of the subscript 1 by the index κ. Equilibrium drops characterized by the condition that the chemical potential ($\mu_\kappa = \mu_l$) within the drops in the vapour phase at $P_0(T)$ vapour pressure and a given temperature T. This means that the internal pressure in the vapour and liquid in the same and equal π_q^N. These expressions are determined by the formulas indicated in (24.10).

The imposition of an additional condition: the equality of the normal components of local pressures in all layers of the interphase region of equilibrium drops means $M_q^{V(N)}(R) = M_1^V = M_\kappa^V = \mu_V$. It follows that $\pi_q^T(R) = [3\pi_q(R) - \pi_q^N]/2$, therefore, (24.15) follows

$$\sigma = \frac{3}{2F_\rho(R)} \sum_{q=1}^{\kappa} F_q(R)[\pi - \pi_q^T(R)]. \tag{24.16}$$

The additional consideration of the mechanical condition for the equality of the normal components in all layers between the vapour and the liquid leads to an expression that differs from the purely thermodynamic determination of the surface tension (24.11) by a constant coefficient of 3/2 [45]. It is important to note the difference in the two mechanical definitions of surface tension by formulas (24.15) and (24.16), using geometric constraints of the discrete model and the macroscopic symmetry condition at the microlevel.

These differences are due to the fact that in Sections 23 and 24 the simplest way of calculating σ on a rigid lattice is discussed, therefore, the description of the mechanical properties (in particular, of the internal pressure) of the system, which is related to lattice deformability, is approximate [46]. The analysis showed that expression (24.16) gives higher values of σ than formula (24.15) for any R, but this difference has a systematic character and it is of the order of up to 10%. Most importantly, the dimensionless ratio $\sigma(R)/\sigma_{bulk}$, where σ_{bulk} is the value of the surface tension for a planar lattice (Section 23), which is below the goal of our analysis, does not depend on additional accounting for mechanical equilibrium. Thus, the absolute differences in magnitude do not affect any of the size characteristics. The same applies to the ratio $\sigma(R)/\sigma_{bulk}$, calculated by the formula (24.11).

The radius ρ_e of an equimolecular surface for equilibrium drops is determined in the usual way

$$\sum_{q=1}^{\rho_e} F_q(R)(\theta_q - \theta_1) + \sum_{q=\rho_e+1}^{\kappa} F_q(R)(\theta_q - \theta_\kappa) = 0, \tag{24.17}$$

where the index $q = 1$ refers to the drop, the index $q = \kappa$ refers to the pair. For $q \leq \rho$ there are layers with increased density, with $q > \rho$ – layers with reduced density. This surface is a dividing surface that divides the intermediate region into two subregions, related to the liquid and the vapour, respectively.

25. Properties of equilibrium drops [45]

Molecular models make it possible to calculate the concentration density profile in the interphase region. For curved boundaries, the procedure for calculating the concentration profile is identical to the procedure for a flat boundary. The radius of the liquid part of the drop R is an additional parameter of the state of the 'drop–vapour' system. The temperatures below are given in the dimensionless form $\tau = T/T_c$, where T_c is the critical temperature in the bulk phase and is defined as the disappearance temperature of two-phase regions (stratification).

Figures 25.1 and 25.2 show curves showing the influence of the drop radius and temperature on the properties of the interphase region. From the comparison of Figures 25.1 *a* and 25.2 *a*, on which the concentration profiles of the interphase region are given for three drop sizes $R = 30$ (1), 120 (2), 480 (3), it is evident that the width of the interphase region increases with increasing temperature. For the low temperature ($\tau = 0.55$, Fig. 25.1 *a*), the profiles with $R =$ 120 (2) and 480 (3) practically coincide with the volume profile, whereas at high temperatures ($\tau = 0.82$, Fig. 25.2 *a*) the profile even for $R = 480$ (3) differs from the volume profile.

Figures 25.1 *b* and 25.2 *b* show the local values $(\pi_q - \pi_0)$ used in the formulas for calculating the surface tension. The finite values of σ depend on their distribution with respect to q. These values are differently distributed in the transition region for different drop radii R. With increasing temperature and, respectively, the width of the interphase region, their maximum value sharply decreases and the curves become smoother. The curves in these figures, starting from zero ($\pi_1 - \pi_0 = 0$), increase sharply and reach a maximum, almost as sharply decrease to a negative value, after which they slowly increase to zero ($\pi_\kappa - \pi_0 = 0$).

Figures 25.1 *c* and 25.2 *c* show the values of σ/σ_b for different positions of the reference surface $q_r = q$ within the interphase region. The values of $\sigma(q)$ decrease monotonically with increasing number of the layer q, $1 \leq q \leq \kappa$, because the monolayer q determines the

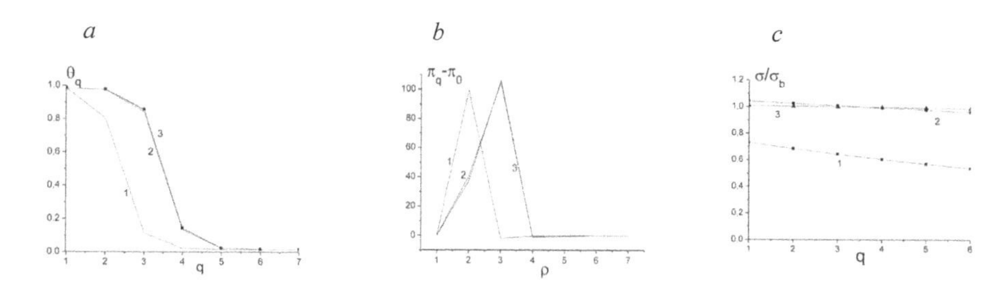

Fig. 25.1. Change of the layer number q in the interphase region to the local concentration value θ_q, (a), the value $(\pi_q-\pi_0)$ (b) and the surface tension value σ (c) for drops of size R = 30 (1), 120 (2), 480 (3) at τ = 0.55.

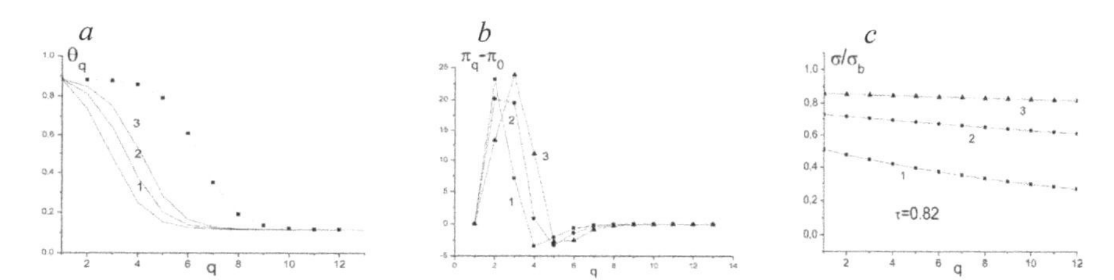

Fig. 25.2. The influence of the layer number q in the interphase region on the value of the local concentration θ_q, (a), the values $(\pi_q-\pi_0)$ (b) and the surface tension σ (c) for drops of size R = 30 (1), 120 (2), 480 (3) at τ = 0.82. Points on 3, and correspond to a flat boundary.

position of the reference surface ρ_r, the denominator in formula (24.15) increases with increasing q, and the numerator remains constant. Consequently, the maximum of $\sigma(q)$ is reached for q = 1. Hence the main conclusion is that for the equilibrium drops, one can not use the traditional thermodynamic concept of the *surface of tension*, as a reference surface located inside the transition region, in which the quantity σ has an extremum. This conclusion agrees with the fact that equilibrium drops in the transition region lack a dividing surface on which a pressure jump occurs according to the Laplace equation. The position of the equimolecular dividing surface $\rho = \rho_r = \rho_e$ is uniquely determined by the concentration profile.

Width of the transition area. Above, a method has been described for determining the width of the interphase region $\kappa(T)$ and the concentration profile of the density of the transition region of a plane boundary. It is completely preserved for the spherical boundary.

The value of κ for a drop depends on the radius R and temperature. As the width of the transition region we select the smallest value of κ for which there exists a monotone solution θ_q, $1 \le q \le \kappa$ of the system under consideration, which we will call the density profile of the transition region. For a given drop radius R, the determined value of κ will be denoted by $\kappa(R)$. We note that the solution of the equilibrium system of equations (24.8) does not depend on the above methods for determining the surface tension σ. The dependences of $\kappa(R)$ for different temperatures are shown in Fig. 25.3.

These curves differ markedly at different temperatures. The lower the temperature τ, the lower the R value, the minimum size of the transition region is established. The stepwise course of the curves is due to a discrete change in the number of monolayers. The main conclusion is that with increasing drop radius, the quantity $\kappa(R)$ increases, and the width of the interphase region of the drop $\kappa(R)$ does not exceed the width κ for a plane boundary at the same temperature: $\kappa(R) \rightarrow \kappa(bulk)$ and $\kappa(R) \leq \kappa(bulk)$.

Figure 25.4 illustrates the effect of the drop size on the width of the interfacial region. For different temperatures, $\gamma = V_\kappa/V_{drop}$ is the fraction of the volume of the interphase layer $V_\kappa = \Sigma_{q=2}^{\kappa-1} N_q(R)$ in the full volume of the drop $V_{drop} = V_{liq} + V_\kappa$. Here $V_{liq} = 4\pi R^3/3$ is the volume of the liquid part of the drop.

With increasing drop size, the share of the interphase region in the total volume of the drop sharply decreases. With increasing temperature, the width of the interphase region κ, and, consequently, the value $\gamma = V_\kappa/V_{drop}$ increase. At $R = 500$, V_κ is 3.5% of the total volume of the drop for $\tau = 0.62$ and 6% of the total volume of the drop for $\tau = 0.82$. For small drop sizes, the fraction of the interphase layer in the total volume of the drop sharply increases. Near $R = 50$, this fraction γ reaches 40–50%, and at $R = 10$, the share of the interphase region in the total volume of the drop prevails. The inset in Fig. 25.4 shows the range of variation of ρ_e for drops of radius

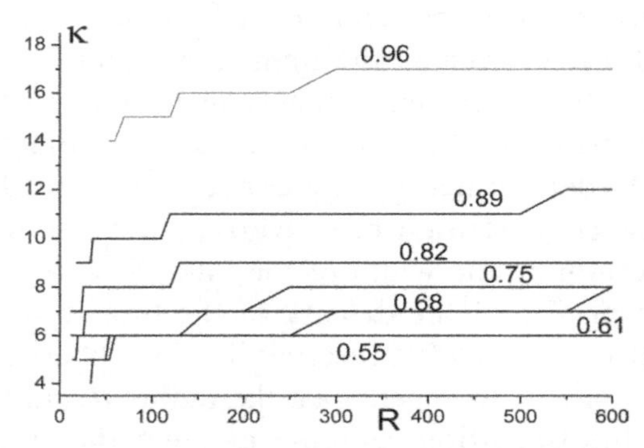

Fig. 25.3. Dependence of the width of the transition region of the spherical drop $\kappa(R)$ on its radius R. The numbers at the curves denote reduced temperature τ.

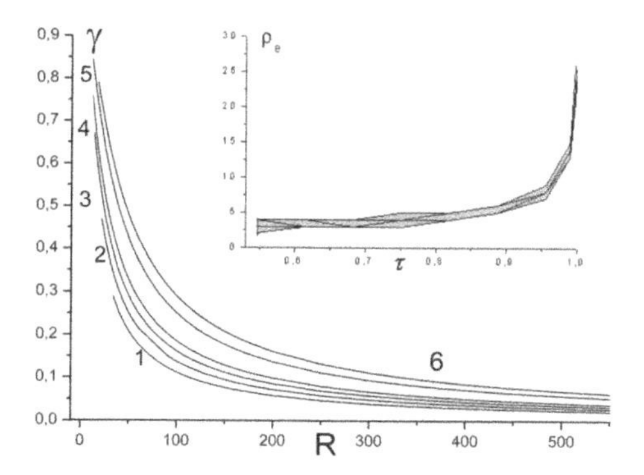

Fig. 25.4. The dependence of the quantity $\gamma = V_\kappa/V_{drop}$ on the drop size R for the temperatures $\tau = 0.55$ (1), 0.61 (2), 0.68 (3), 0.75 (4), 0.82 (5), 0.89 (6), here $\alpha = 0.5$. The inset shows the dependence of the intervals of variation of ρ_e for drops of size from $R = 8$ to $R = 1000$ on the temperature.

from $R = 8$ to 10^3 as a function of temperature. The calculation was carried out for 7 radii (with a practically constant step on a logarithmic scale). It can be seen that the spread of the values of ρ_e is only about 2 monolayers. The behaviour of the curves in Fig. 25.4 coincides with the course of the behaviour of the curve ρ_e in Fig. 23.4, constructed for a plane lattice, for which we can conditionally assume that $\rho_e \approx \kappa/2$.

Size dependence of surface tension. Figure 25.5 shows the calculated temperature dependence of σ on the drop size normalized to the surface tension of the flat lattice σ/σ_b (at the same temperature). The calculations were carried out with two limiting values with the parameter $\alpha = 1$ and 0. It is seen that the surface tension increases monotonically with increasing drop radius, tending to its volumetric value $\sigma(R) \to \sigma_{bulk}$. For relatively low temperatures $\tau < 0.75$, an

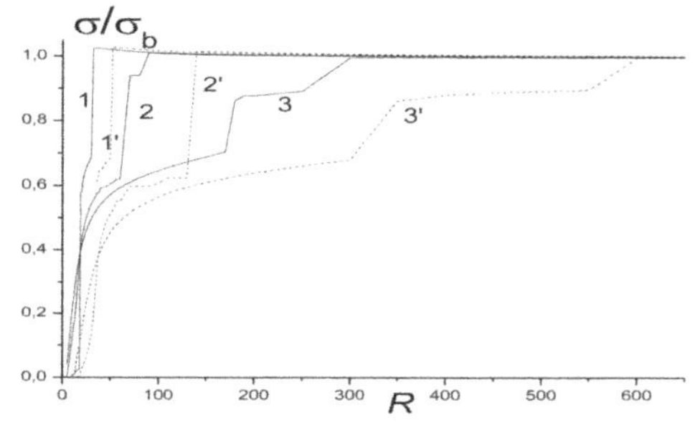

Fig. 25.5. Dependence of σ/σ_b on the drop size; (a): $\tau = 0.55$ for $\alpha = 1$ (1), $\tau = 0.55$ for $\alpha = 0$ (1′) $\tau = 0.61$ for $\alpha = 1$ (2), $\tau = 0.61$ for $\alpha = 0$ (2′) $\tau = 0.68$ for $\alpha = 1$ (3), $\tau = 0.68$ for $\alpha = 0$ (3′).

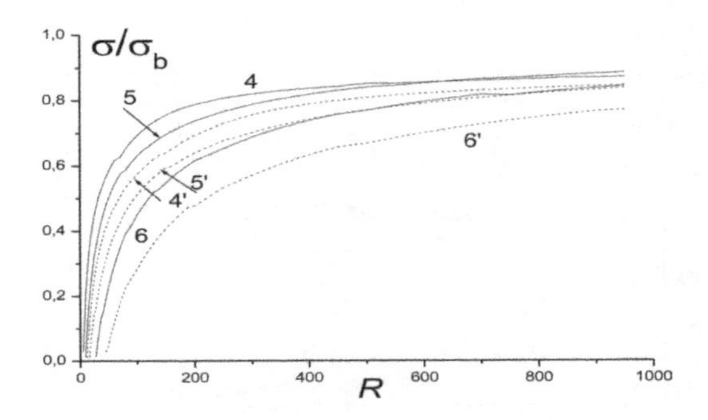

(b): $\tau = 0.75$ for $\alpha = 1$ (4), $\tau = 0.75$ for $\alpha = 0$ (4′) $\tau = 0.89$ for $\alpha = 1$ (5), $\tau = 0.89$ for $\alpha = 0$ (5′) $\tau = 0.96$ for $\alpha = 1$ (6), $\tau = 0.96$ for $\alpha = 0$ (6′).

abrupt change in surface tension is possible. These jumps occur at small values of the radius R in connection, which raises the question of the proportionality of the width of the transition region and the size of the molecules. In traditional thermodynamic constructions it is considered that the size of a molecule can be neglected. With increasing temperature, the number of monolayers κ increases, and the relative change of κ with a growth of R plays a smaller role.

Figure 25.5 shows that the theory allows us to consider a very wide range of drop sizes. This served as a basis for analyzing the characteristic scales of drop sizes: from nanometers to the submicron range (see the next section 26).

Temperature dependence of surface tension. Figure 25.6 gives the temperature dependences of the surface tension of drops $\sigma(T|R$ = const) for a number of fixed values of radii from 16 to 10^3. For each temperature, the normalization was carried out to its volume value σ_b.

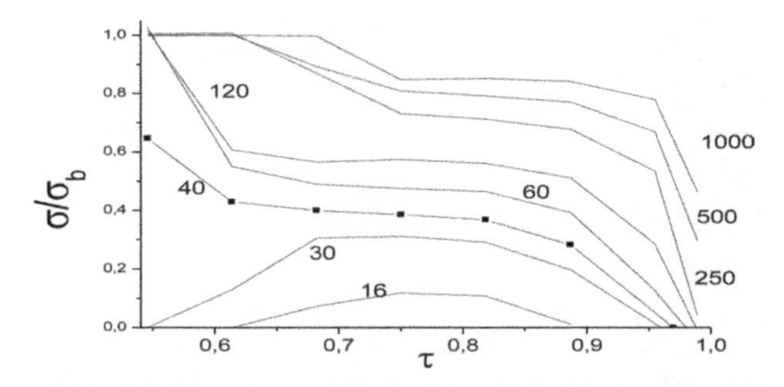

Fig. 25.6. Dependences of σ/σ_b for drops of different sizes R from 16 to 1000 on temperature τ, $\alpha = 1$. The values of the drop radius R are shown on the curves.

There are also two types of curves. The first type of curves refers to sufficiently large radii ($R \geq 40$), for which the surface tension is positive, starting from low temperatures. The general trend of these curves is a decrease in σ with increasing temperature, although a small non-monotonicity of the curves is possible in the intermediate temperature range. Near the critical temperature, all curves have a negative value of $\sigma(R)$. This indicates the instability of this drop size at high temperatures.

For the second type of curves, the dependences $\sigma(T|R = \text{const})$ increase with increasing temperature from negative values of $\sigma(R)$ to a certain maximum and after its passage decrease to negative values, as in the first case. In the case of a positive maximum, there is a range of temperatures for the specified drop sizes, in which the equilibrium drops are stable. These drops can not exist at low and high temperatures because of the negative value of the surface tension. In particular, this refers to $R = 16$ and 30. At the same time, the curve for $R < 10$ has a maximum at negative values of σ which indicates its instability at any temperatures. The limiting value of R separating both types of curves is $R \sim$ from 35 to 40. The reversal of the surface tension value to zero corresponds to the loss of stability of the drop as a liquid phase. Or the condition $\sigma = 0$ means that the molecules of the system under consideration are not split into two phases. Recall that the positive value of surface tension indicates the impossibility of spontaneous unlimited increase in the area of the interface [2,9].

This circumstance indicates a natural (in physical sense) criterion of formation (small in size), since by construction within the drop there is necessarily a region with constant properties, which is necessary by the definition of the phase, which was introduced by Gibbs [1]. For more details, see below in Section 27.

26. Three characteristic drop size scales [46]

A drop is an object of an intermediate size: from molecular associates to macrodrops of a new phase. The physical basis for changing concepts, in fact, is absent. The nature of the intermolecular potential acting in the system of molecular associates implies the absence of saturation and, as a consequence, the absence of a limit for the degree of association [40]. When describing the equilibrium drop there is no Laplace equation, which greatly simplifies mathematical analysis. Let us consider the behaviour of the surface tension on the drop size $\sigma(R)$. Three characteristic sizes can be distinguished which correspond to: 1) the value of the radius R_b, the greater of which the surface tension values are close to the volume value, with small curvatures of the surface of large drops; 2) the value of the radius R_t above which the thermodynamic description of the surface tension of the drop is justified, when both the discrete nature of the matter and the contributions of spontaneous fluctuations can be neglected; and 3) the critical value R_0 of the beginning of the formation of the phase associated with the appearance of the surface tension of the dense phase.

Large radii. For the analysis of large drops, it is convenient to count the drop properties from the equilibrium profile of the concentrations of the plane boundary (we denote its characteristic by index ∞). The deviation of the surface tension of the curved boundary $\sigma(R)$ from σ_∞ can be written as

$$\delta\sigma(R) = \sigma_\infty - \sigma(R) = \frac{1}{F_\rho(R)} \sum_{q=1}^{\kappa} F_q(R)[\pi_q(\infty) - \pi_q(R)]. \tag{26.1}$$

If, as a first approximation, we use the assumption that the density profiles on the curved $\{\theta_q\}_R$ and the flat $\{\theta_q\}_\infty$ separation boundaries are close, then the functions are also close to analogous functions in the volume, and the problem reduces to bulk densities and structural characteristics. Expression (6) gives

$$\delta\sigma(R) = \frac{1}{2N_\rho} \sum_{q=1}^{\kappa} N_q \sum_{p=q-1}^{q+1} \delta z_{qp}(R) \ln[\theta_{qp}^{VV}(R)/(\theta_q^V \theta_p^V)]_\infty, \tag{26.2}$$

here $\delta z_{qp}(R) = z_{qp}(R) - z_{qp}$, where $\Sigma_p \delta z_{qp}(R) = 0$, according to the formulas (24.1). The presence of the logarithmic factor, taking into account expressions (24.6) for the numbers $z_{qp}(R)$, which lead to members of type $1/R$, results in the non-zero value of the second sum over the index p. The ratio under the logarithm sign in (26.2) depends on the temperature; nevertheless, the coefficient before $(1/R)$ in the second sum can be qualitatively estimated as one. Then $\delta\sigma(\mathrm{R}) = \sum_{q=1}^{\kappa} N_q/(2RN_\rho) \approx \kappa/(2R)$ therefore, as the size of the drop decreases, σ decreases.

The main result (26.2) is that the quantity $\delta\sigma(R)$ depends on the radius as $1/R$, i.e. *any* curvature of the surface leads to a change in the surface tension, or the range of the drop size values, within which the surface tension depends on the curvature of the dimension, is large. If we assume that the volumetric value of σ_∞ for a plane boundary can be used as a reference for the quantity $\sigma_b = 0.99\sigma_\infty$, then this condition is reached at $R_b/\lambda > 10^2\kappa/2$. Here, R_b denotes the size of large drops, for which the values of σ differ little from their volume value for the flat interface σ_∞. For all values of $R > R_b$, this difference can be neglected. As the size of the drops decreases, the formula (26.2) ceases to work. In addition to changing the structural characteristics, it is necessary to take into account the change in the $\{\theta_q\}_R$ profile. However, the sign of $\delta\sigma(R)$ is preserved, i.e. the smaller the radius R, the smaller $\sigma(R)$.

The width of the interphase region κ depends strongly on the temperature: from $\sim 3\lambda$ (near the triple point) to 40λ (near the critical point). It follows that even for low temperatures, the size range of drops having surface tension values close to volume values can reach about 10^2 molecular diameters, i.e. a value of the order of 40 nm for argon atoms, and at high temperatures reaches a micron scale. For atoms of inert gases and simple molecules ranging in size from 0.3 to 0.6 nm, this occurs at a drop radius up to the micron size. This fact was not paid attention before, and, as a rule, the bulk value of the surface tension is traditionally attributed to relatively small drops.

Small radii ~R_0. We have shown above that for small drops there exists a radius $R_0 = R(\sigma = 0)$ at which the surface tension vanishes, since a small drop, in fact, becomes a molecular associate or cluster without surface tension [40]. The size of the drops R_0 corresponds to the critical size of the associate, which determines the onset of the formation of the phase and the appearance of the surface tension of the dense phase. For values of $R < R_0$, there is a negative value of the surface tension σ, at the boundary of the drop, which is considered unstable according to thermodynamic concepts. The condition $\sigma = 0$ defines the minimum drop size as a cooperative property of the system, by analogy with the *stable* two-phase equilibrium, in contrast to the condition on the unstable state of the system in thermodynamics [40]. This definition in principle differs from the critical size in thermodynamics, which is introduced through the condition on the maximum of the free Gibbs energy for the *unstable* state of the embryo [40] (and also because thermodynamics does not determine this size, but it is established a priori [40]). The same critical radius R_0^* ($\sigma = 0$) exists for metastable drops [43]. At the point R_0^* the pressure inside the drop also becomes the total pressure

$P_f = P_v = P_s$. The values of R_0^* and R_0 coincide (in more detail in Section 60).

The mathematical analysis of the condition $\sigma = 0$ consists in solving the system of equations for the concentration profile (24.1). In [46], a qualitative estimate was obtained for $R_0 = 7\lambda$ on the basis of the simplest model consisting of two monolayers and a degree of asymmetry of the profile of the order of ~1.3 (from the data of numerical calculations). Thus, it turns out that the critical value of the equilibrium nucleus R_0 must be at least 7 monolayers. This radius corresponds to sufficiently large relative fluctuations in the number of molecules in the whole drop $\eta_V = 2.6\%$ or only on its surface $\eta_s = 4.0\%$ (for more details see Section 34).

Intermediate sizes R_t. The sizes R_0 and R_b cover a wide range of sizes, from the field of existence of molecular associates to the two-phase states of a system with a practically flat boundary. Between them is the intermediate size of the drops R_t, at which it is necessary to take into account, on the one hand, the discreteness of the substance, and, on the other hand, the requirement of absence of spontaneous (thermal) fluctuations in the substance [47]. These two interrelated concepts define the lower limit of the applicability of the thermodynamic approach.

Any thermodynamic potential contains a surface contribution only if it gives a sufficiently appreciable effect on the calculated characteristics of the system. Otherwise, the contribution of the surface can be neglected – it is sufficient to consider the bulk coexisting phases. The presence of a surface contribution means the development of the surface, and the large contribution of the interface between the phases in the thermodynamic characteristics of the system. For the calculation of the latter, the values of the first and second derivatives of the thermodynamic potentials (by temperature for heat capacities and by pressure for the compressibility coefficients) are necessary. Non-planar interfaces are characterized by the curvature of the local surface area, which is also expressed in terms of the first and second derivatives of the equation of the surface $z(x,y)$ along the x and y coordinates. In order for these derivatives to be meaningful, all the thermodynamic functions must be defined at the points of space in question, as well as for plane boundaries [24]. From this it follows that there is some minimal characteristic length that determines the size required for the calculation of spatial derivatives.

From the analysis of the properties of the surface tension of spherical drops, it is possible to distinguish three characteristic sizes of a drop $R_0 < R_t < R_b$. The first characteristic size, equal to $R_0 \sim 10\lambda$, corresponds to the onset of the formation of a dense phase and to the appearance of a surface tension of the drop. For $R < R_0$, we have molecular associates. The second characteristic size, equal to R_t (it is estimated in Section 34 from 40λ to 90λ), separates the range of applicability of the thermodynamic description of the surface tension of a drop. For $R > R_t$, the discreteness of the matter and the contributions of spontaneous fluctuations can be completely neglected. The third characteristic size, equal to $R_b \sim 10^2 \div 10^3\lambda$ for different temperatures, separates the size range of large drops, in which the surface tension values are close to the bulk value.

27. Criterion for the minimum phase size

Figure 25.6 indicates the existence of solutions of the system of equations (24.8) for which the values of σ are negative. This occurs at critical drop sizes. The condition $\sigma = 0$ can be achieved both by raising the temperature near the critical temperature, so by decreasing the size of the drop. In the first case, the system is in the region of large thermal fluctuations and the selected radii are smaller than the average size of the fluctuations at a given temperature. In the second case, the 'phase' inside the drop becomes unstable due to the large fraction of the interphase region in the total volume of the drop (see Fig. 25.4).

Calculation of the surface tension of the drop $\sigma(R)$ of an arbitrary radius R allows one to answer the question posed about the minimum phase size R_0. The condition $\sigma > 0$ means that there is no spontaneous increase in the surface, which is a sign of the thermodynamic stability of the drop as an independent phase. The value of the drop radius $R_0 = R(\sigma = 0)$, at which the equality $\sigma = 0$ is satisfied, is taken as the lower boundary of existence of the liquid phase. With the drop size $R > R_0$, we have a two-phase system; therefore, in fact, the value of R_0 is limited by the possibility of introducing the concept and calculating the single-phase chemical potential on the traditional $\mu(T)$ curves for two coexisting phases. This follows from the fact that direct contact of these phases is an indispensable condition for the two-phase coexistence of molecules. In turn, molecular associates from dimer to drops of size $R < R_0$ can exist for a long time as a result of the dynamic process of formation and decomposition of associates, but they can not be attributed to the properties of the equilibrium phase and equilibrium surface tension. Thus, the concepts of cluster and drop have distinct differences related to the thermodynamic behaviour of these objects – with the value $R_0 = R(\sigma = 0)$.

A simplified analytical estimate gave the smallest value of R_0 equal to 7λ. The calculations for the full model showed slightly larger values of R_0/λ from 8 to 12, depending on the structure of the interphase boundary, i.e. $R_0 \sim 10\lambda$. According to the parameters of the intermolecular interaction, found from the description of the experimental data for plane boundaries, the size dependences of the surface tension for 19 substances were calculated, some of which are given in Section 23. Figure 27.1 shows these dependences of the surface tension $\sigma(R)$ normalized to the surface tension of the flat lattice σ_{bulk} on the radius of the drop R normalized to the width of the monolayer λ, and in Fig. 27.2 values of R_0/λ as a function of reduced temperature τ [35].

These calculations of the values of R_0 should be refined: they refer to the values of the structural parameter $\alpha = 1$, which correspond to macroscopic values of the radii (refinement of α should increase R_0). This question was considered in more detail (the effect of the structural parameter α) in [45]. The calculations also did not take into account the scaling corrections near the critical temperature [8]. Nevertheless, these results show that there is a general universality of the behaviour of 19 substances, and it reflects the characteristic features of the behaviour of the cooperative system at small sizes.

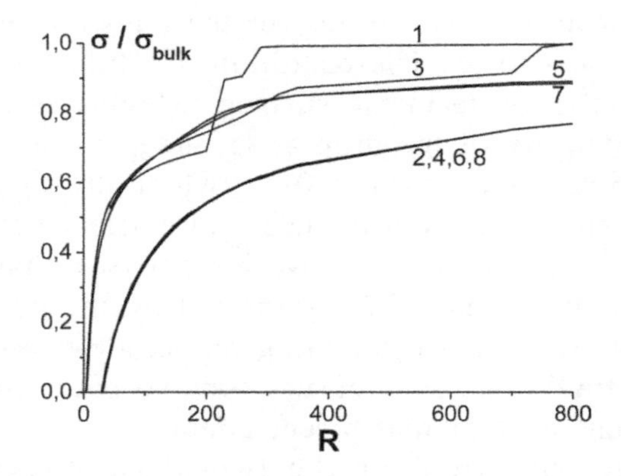

Fig. 27.1. The normalized surface tension $\sigma(R)$ for two reduced temperatures $\tau = 0.6$ (1, 3, 5, 7) and 0.95 (2, 4, 6, 8) with interatomic interaction within the limits of the c.s. $R_{lat} = 1$ (1, 2), 2 (3, 4), 3 (5, 6), 4 (7, 8) (see Table 23.1). [35].

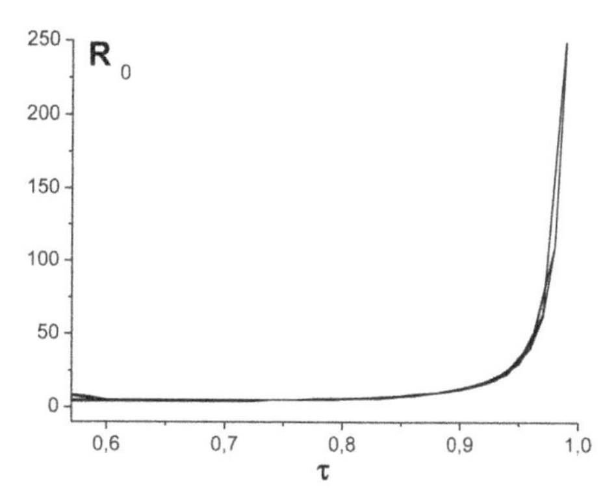

Fig. 27.2. The dependence of the minimum radius of the drop R_0, corresponding to the formation of a new phase, on reduced temperature τ [35].

The universality of estimating the minimum size of the equilibrium phase [48]. The question of the lower boundary of the sizes of the existence of equilibrium phases is important in the description and interpretation of experimental data at the microlevel, since it determines the type of equations that should be used in calculating thermodynamic characteristics (recall that the very concept of 'phase' is thermodynamic [1]). There are three types of experimental data in which the question of the minimum phase size plays a key role: 1) the formation of a new phase in the process of vapour condensation, in which the concept of a critical nucleus is introduced [1,40,49]; 2) capillary condensation in narrow pores, in which adsorption–desorption hysteresis disappears when the width of the pore width decreases [50–52]; and 3) spontaneous magnetization of particles of small size, in which the magnetization hysteresis disappears as the crystallite size decreases [53]. In [48], these systems are discussed from a single point of view of comparing the values of the minimum particle size that have the properties of the equilibrium phase. The systems under consideration have one common property: cooperative behaviour of molecules or spins.

A free drop exists only through intermolecular interaction – this is a purely cooperative property. For isolated drops, the critical size is discussed above.

Fluid in the pores. For porous systems, a systematic analysis of phase transitions in small systems is given in [8, 54]. The exfoliating simple (low-molecular) liquid, which is in a limited volume of the pore, decomposes into two phases of vapour and liquid. A characteristic feature of such systems is that the surface potential rapidly decreases along the normal from the pore wall to its central part. The restriction of the volume to the walls of the pores only affects the wall region in which several dense molecular monolayers can form before the capillary condensation of the molecules in the central part of the pore. Such a potential can not cause condensation in the central part, if there is no cooperative behaviour of the molecules themselves.

The conditions for stratification of fluid in narrow slit-like, cylindrical, globular and other types of pores over a wide temperature range were analyzed in [55]. Of particular importance is the description of the experimental data on the capillary condensation of a number of adsorbed molecules in long cylindrical channels of MCM-41 [56], which have quasi-one-dimensional properties for sizes $R_c^* < 6.5–7\lambda$. Similar results were obtained for channels of other cross sections [54].

The knowledge of R_c^* allows us to find the radius of the spherical pore R_s^* containing the same number of molecules [54], which is in a limited section of the channel with radius R_c^* and length L^* if we assume that the cooperative properties of molecules of an infinite channel are approximated by this section. Approximation is achieved under the condition $2/L < \delta_1^*$, where δ_1^* is a numerical criterion corresponding to the performance of this approximation. In terms of its meaning and size, the quantity δ_1^* corresponds to a value of $1/R_c^*$ [54]. As a result, it was obtained that the minimum critical channel length $L^* = 2R_c^*$. Whence follows the minimum volume of the spherical pore R_s^* equal to $R_0 = R_s^* \sim 10\lambda$. Consequently, the beginning of the formation of an equilibrium liquid phase in a vapour and the stratification of the fluid in bounded volumes of porous systems of different geometry occur at the same smallest size of a new phase: $R_0 \sim 10\lambda$ (or about 4 nm for Ar atoms).

Magnetization of microcrystals. From the point of view of the statistical theory [32,57–59] of the first-order phase transitions, there is a one-to-one correspondence between the phase behaviour of liquid stratifying systems and the spontaneous magnetization of magnetic materials having spin 1/2. In particular, the theory of phase transitions [57–59] ascribes to them the same critical exponents in the vicinity of the critical point. Small solid particles exist due to the interaction between its atoms. Spin–spin interactions are realized for a fixed distribution of atoms inside the particles. They depend on the size and shape of the atomic subsystem, including the dependence on the property of the boundary of a small particle. The number of atoms in a particle must be large enough to ensure the existence of a spontaneous magnetization of the spin subsystem. If the particle is uniform in its central part, then the behaviour of the spin subsystem is equivalent to the behaviour of molecules inside the central part of the pore. The data on the magnetization of small crystals of magnetic substances (Fe, Ni, and Co metals) are given in Ref. [48], indicating the possibility of the appearance of spontaneous magnetization [60-64]. An indication of the spontaneous magnetization is the existence of a hysteresis loop when an external magnetic field is applied. The loop disappears as the temperature rises and/or the size of the microcrystal changes to a paramagnetic state, so the picture is completely analogous to adsorption hysteresis.

In the work [60] nanowires of Fe, Co, and Ni ferromagnetic metals were obtained in porous alumina matrixes with channel diameters from 5 to 40 nm. The course of the hysteresis curves of the magnetization depends on the diameter of the channel. The resulting nanowires were polycrystalline. The smallest crystal sizes were several nanometers. Hysteresis is observed on large crystals. If the wires were uniform in length, then such systems would be analogous to capillary condensation in MCM-41 or for a system of larger pores. It is noted [60] that with a decrease in the pore size, fluctuations in the system are markedly increased.

The study of Fe powders led to an estimate of the particle size in which a magnetization hysteresis appeared exceeding 5 nm [53], which corresponds to the condition $R_0 \sim 10\lambda$, where $\lambda = 0.26$ nm is the crystal lattice size of Fe. For the Co powder used as a catalyst, particle size estimates were obtained for the transition from the paramagnetic state to the ferromagnetic state from 7 to 8 nm [61, 62], which corresponds to values R_0 from 14λ to 16λ at $\lambda = 0.25$ nm. In [63], with clusters of Ni in very narrow carbon channels not exceeding 2.5 nm, only paramagnetic states of the clusters were detected. These examples are in good agreement with the results of capillary condensation in narrow-porous systems.

However, in general, more complex situations are possible in magnetic systems, when magnetism arises due to the joint properties of embedded metal clusters and the original matrix, as is the case in the Pd clusters system in the carbon matrix [64]. Such situations in capillary condensation are not realized.

Thus, by comparing the minimum size of thermodynamically stable liquid drops and the pore sizes in which the fluid stratification into two coexisting phases occurs, as well as data on the hysteresis of the spontaneous magnetization of microcrystals, one can conclude that there exists a universal size of the volume of atoms / molecules that corresponds to the condition for the realization of the equilibrium phase. Estimates [48] indicate a qualitative correspondence between the three types of experimental data and require more detailed analysis by refining the factors affecting the numerical values of the minimum sizes for each of the experiments. The maximum value of R_0 converted to the volume of the sphere does not exceed 15λ; exactly such an increase in R_0 gives the above estimates for magnetic systems for which these deviations are due to the inaccuracy of the polydispersity of the powders, the oxidation state of metals, the non-spherical shape, etc. Of course, the thermodynamic approaches can not be used for this kind of refinement of the properties of small bodies, especially when considering the sizes of small particles smaller than the minimum estimates for the existence of equilibrium phases $R_0 \sim 10\lambda$.

28. Equilibrium drops and phase rule

The molecular theory leads to the existence of a new object of small systems – equilibrium drops. Such drops are impossible in thermodynamics – the problem of the Kelvin equation was discussed in Section 7.

The solutions to the concentration profile of the equilibrium drops completely correspond to the result of the Yang–Lee theory [65], since they do not contain any information about the states of the system inside the binodal curve. The boundary conditions on the concentration profile correspond to the macroscopically stable states of the stratifying phases of vapour and liquid, as for plane interfaces. For them, the conditions for mechanical and chemical equilibrium of the bulk phases are satisfied. As the drop size increases, the concentration profile becomes the profile of the planar interface. Accordingly, the surface tension $\sigma(R,T)$ of the drop is transformed into the surface tension of the plane macroscopic boundary $\sigma(T)$.

Equilibrium drops are an intermediate state of matter in the transition from small associates to macroscopic phases, as their size increases, under the same fixed external conditions T, P. The existence of equilibrium drops indicates the need for a change in the recording of the phase rule for curved boundaries: practically in all publications, and even in the Commentaries on page 506 of [1] that surface tension σ can increase the number of degrees of thermodynamic freedom per unit due to the presence of a curved boundary even in the absence of the external field. Obviously, the drop type should not influence the phase rule for two reasons.

Firstly (in terms of physical meaning), any boundary is a 'non-autonomous phase'. The influence of the boundary extends only to a certain near-surface region. Thus, the surface does not affect the internal properties of the phases, unlike the external fields acting on all molecules within a given volume. In other words, the properties of non-autonomous phases can not be parameters of the state of the system. They necessarily functionally depend on the parameters of the state of the bulk coexisting phases and must obey the same conditions of thermal, mechanical and chemical equilibrium at each point of the transition region, as do the coexisting phases themselves.

Secondly (from a formal point of view), despite the presence of two pressures P_α and P_β on the curved boundary of the metastable drop (i.e., the number of parameters of such a system is increased by one, as it contains more parameters), but for the phase rule this it does not matter – it should take into account all the relationships between the parameters of the system. In the case of a metastable drop, there is an additional connection according to the Laplace equation, which connects these two pressures through $\sigma_{met}(R,T)$. Therefore, the total number of degrees of thermodynamic freedom for a curved surface does not change with respect to such a number for a plane boundary (and equilibrium drops again emphasize this). This conclusion fully corresponds to the physical meaning of the 'phase' as a macroscopic object, for which the state of the surface is secondary, but the surface necessarily exists and limits its size.

The existence of equilibrium drops allowed us to pursue a strict concept of the minimum phase size, and this concept turns out to be the same for both equilibrium and metastable drops, because it satisfies the Laplace equation (Section 60). This fact indicates that the use of rigorous molecular theories makes it possible to extend the thermodynamic concept of 'phase' to small systems, down to values less than 10 nm. This is actually two orders of magnitude smaller than the traditional concept of the phase, as a macroscopic formation of the order of the micron size.

Note that in recent years, compared with the references [11–23], little has changed. As an example, we cite almost randomly selected papers [66–69] from the set of analogous papers. They also almost arbitrarily refer to the concepts of the minimum phase size, the regions of phase transformations and, accordingly, to surface tension. In each of these works, these concepts are interpreted in their own way, depending on the type of system and methods of carrying out calculations without any justification for their ideas and their connection with thermodynamics.

References

1. Gibbs J.W., Thermodynamics. Statistical mechanics. Moscow, Nauka, 1982. 584 c.
2. Ono S., Kondo C., Molecular theory of surface tension. Moscow, IL, 1963. [Handbuch der Physik, Vol X (Springer) 1960]
3. Guggenheim E.A., Trans. Faraday Soc. 1940. V. 37. P. 397.
4. Guggenheim E.A., Thermodynamics, Amsterdam, 1950.
5. Fower R.H., Proc Roy. Soc. A. 1937. V. 159. P. 229.
6. Fower R.H., Physica. 1938. V. 5. P. 39.
7. Bakker G., Kapillaritat und Oberflachenspannung, Handbuch der Experimental physik, Bd. VI, Leipzig, 1928.
8. Tovbin Yu.K., Molecular theory of adsorption in Porous Solids. Moscow: Fizmatlit, 2012. [CRC, Boca Raton, Florida, 2017].
9. Rowlinson, J., Widom B., Molecular theory of capillarity. Moscow, Mir, 1986. [Oxford: ClarendonPress, 1982]
10. Croxton K., Liquid State Physics. A Statistical mechanical Introduction. Moscow, Mir, 1978. 400 p. [Cambridge Univer. Press, 1974]
11. Iwamatsu M., J. Phys.: Condens. Matter. 1994. V. 6. L173.
12. Baidakov V.G., Boltachev G.Sh., Zh. fiz. khim. 1995. V. 69. P. 515.
13. Moody M.P., Attard P., J. Chem. Phys. 2002. V. 117. P. 6705.
14. He S., Attard P., Phys. Chem. Chem. Phys., 2005, V. 7. P. 2928.
15. Oxtoby D.W., Evans R., J. Chem. Phys. 1988. V. 89. P. 7521.
16. Bykov T.V., Shchekin A.K., Neorg. mater.. 1999. V. 35. No. 6. P. 759.
17. Bykov T.V., Shchekin A.K., Kolloid. zh. 1999. V.61. No. 2. P. 164.
18. Bykov T.V., Zeng H.S., J. Chem. Phys. 1999. V. 111. P. 3705.
19. Bykov T.V., Zeng H.S., J. Chem. Phys. 1999. V. 111. P. 10602.
20. Thompson S.M., Gubbins, K.E., Walton, J.P. R., et al., J. Chem. Phys. 1984. V. 81. P. 530.
21. Zhukhovitsky D.I., Kolloid zh. 2003. V. 65. No. 4. P. 480.
22. Appert C., Pot V., Zaleski S., Fields Institute Communications. 1996. V. 6. P. 1.
23. Ebihara K., Watanabe T., Eur. Phys. J. B 2000, V. 18. P. 319.
24. Tovbin Yu.K., Zh. fiz. khim. 2010. V. 84. No. 2. P. 231. [Russ. J. Phys. Chem. A 84, 180 (2010)]
25. Tovbin Yu.K., Zh. fiz. khim. 2010. V. 84. No. 10. P. 1882. [Russ. J. Phys. Chem. A 84, 1717 (2010)]
26. Prigogine I.R., Molecular theory of solutions. Moscow, Metallurgiya.
27. Tovbin Yu.K., Kolloid zh. 1983. V. 45. No. 4. P. 707.
28. Okunev B.N., Kaminsky V.A., Tovbin Yu.K., Kolloid zh. 1985. V. 47. No. 6. P. 1110.
29. Smirnova N.A., Molecular theory of solutions. Leningrad, Khimiya, 1987.
30. Morachevsky A.G., et al., Thermodynamics of liquid–vapour equilibrium. Ed. Morachevsky A.G., Lenigrad, Khimiya, 1989.
31. Prausnitz J.M., Lichtenthaler R.N., de Azevedo E.G., Molecular thermodynamics of fluid-phase equilibria. Second ed., Prentice-Hall Inc., Englewood Cliffs, New Jersey, 1986.
32. Hill T.L., Statistical Mechanics. Principles and Selected Applications. – Moscow:

Izd. Inostr. lit., 1960. – 486 p. [N.Y.: McGraw–Hill Book Comp. Inc., 1956]
33. Tovbin Yu.K., Theory of physico-chemical processes at the gas-solid interface, Moscow, Nauka, 1990. [CRC, Boca Raton, Florida, 1991].
34. Tovbin Yu.K., Zh. fiz. khim. 1992. V. 64. No. 5. P. 1395. [Russ. J. Phys. Chem., 1992 V. 66, No. 5, P. 741]
35. Tovbin Yu.K., Zaitseva E.S., Rabinovich A.B., Zh. fiz. khim. 2017. V. 91. No. 10. P. 1734. Rus. J. Phys. Chem. A, 2017, V. 91, No. 10, P. 1957]
36. Hirschfelder J. O., Curtiss Ch. F., Bird R. B., Molecular theory of gases and liquids. – Moscow: Inostr. Lit., 1961. – 929 p. [Wiley, New York, 1954].
37. Tables of physical quantities. Directory. Ed. I.K. Kikoin. Moscow, Atomizdat, 1976.
38. Vargaftik N.B., Handbook of thermophysical properties of gases and liquids, Nauka, Moscow, 1972.
39. Abramzon A.A., Surface-active substances: properties and applications. 2nd ed. Leningrad, Khimiya, 1981.
40. Frenkel J., Kinetic theory of liquids. Moscow, Academy of Sciences of the USSR, 1945.
41. Kubo R., Thermodynamics. Moscow, Mir, 1970. [Amsterdam: North-Holland Publ. Comp., 1968.]
42. Bazarov, I.P., Thermodynamics. Moscow, Vysshaya shkola. 1991.
43. Tovbin Yu.K., Rabinovich A.B., Izv. AN. ser. khim. 2009. No. 11. P. 2127. [Russ. Chem. Bull. 58, 2193 (2009)]
44. Sedov L.I. Continuum mechanics. V. 1. Moscow, Nauka, 1970.
45. Tovbin, Yu.K., Rabinovich A.B., Izv. AN. ser. khim. 2010. No. 4. P. 663. [Russ. Chem. Bull. 59, 677 (2010).]
46. Tovbin Yu.K., Zh. fiz. khim. 2010. V. 84, No. 4. P. 797. [Russ. J. Phys. Chem. A 84, 705 (2010)]
47. Tovbin Yu.K., Zh. fiz. khim. 2012. V. 86. No. 9. P. 1461. [Russ. J. Phys. Chem. A 86, 1356 (2012)]
48. Tovbin Yu.K., Zh. fiz. khim. 2010. V. 84. No. 9. P. 1795. [Russ. J. Phys. Chem. A 84, 1640 (2010)]
49. Laudise R. A., Parker R. L., The growth of mono-crystals. – Moscow: Mir. 1974. – 540. [Prentice-Hall, Inc., Englewood Cliffs, New Jersy, 1970; Solid State Physics, V. 25, 1970].
50. Dubinin M.M., Usp. khimii. 1955. Vol. 24. P. 3.
51. Gregg, S.J. Sing, K.G.W., Adsorption, Surface Area and Porosity. – Moscow: Mir, 1984. [Academic Press, London, 1982].
52. Karnaukhov A.P., Adsorption. Texture of dispersed porous materials. Novosibirsk, Nauka, Siberian Branch of the Russian Academy of Sciences, 1999.
53. Petrov Yu.I., Physics of small particles. Moscow, Nauka, 1982.
54. Tovbin Yu.K., Petukhov A.G., Izv. AN. ser. khim. 2008. No. 1. P. 18. [Russ. Chem. Bull. 2008. V. 57. No. 1. P. 18]
55. Tovbin Yu.K., Zh. fiz. khim. V. 82. No. 10. P. 1805. [Russ. J. Phys. Chem. A. 2008. V. 82. No. 10. P. 1611]
56. Beck J.S., et al., J. Am. Chem. Soc. 1992. V. 114. P. 10834.
57. Fischer M., Nature of the critical state. Moscow, Mir, 1968..
58. Stanley G., Phase transitions and critical phenomena. Moscow, Mir.
59. Ma Sh.. Modern theory of critical phenomena. Moscow, Mir, 1980.
60. Zeng H., et al., Phys. Rev. B, 2002. V. 65. P. 134426.
61. Chernavsky P.A.. Ross. khim. zh. 2002. V.46. No. 3. P. 19.
62. Chernavskii P.A., Khodakov A.Y., Pankina G.V., Girardon J.-S., Quinet E., Applied

Catalysis A: 2006. V. 306. P. 108.
63. Fedosyuk V.M., et al., Fiz. Tverd. Tela. 2003. V. 45. No. 9. P. 1667.
64. Shanina B.D., et al., Zh. Eksper. Teor. Fiz. 2009. V. 136. No. 4. P. 711.
65. Yang C.N., Lee T.D., Phys. Rev. 1952. V. 87. P. 404.
66. Dolgusheva E.B., Trubitsyn V.Yu., Fiz. Tverd. Tela. 2010. V. 52. No. 6. P. 1163.
67. Factorovich M.H., Molinero V., Scherlis D.A., J. Am. Chem. Soc. 2014. V. 136. P. 4508.
68. Lau G.L.V., et al., J. J. Chem. Phys. 2015. V. 142. P. 114701.
69. Belashchenko D. K., Zh. fiz. khim. 2015. V. 89. No. 3. P. 517. [Russ. J. Phys. Chem. A. 2015. V. 89. No. 3. P. 513]

4

Small systems and size fluctuations

29. Small system fluctuations

Macroscopic thermodynamics, which refers to a large number of small systems, deals with strictly defined thermodynamic functions [1–3]. The current trend, consisting in the transfer of the thermodynamic approach to smaller particles, is connected with the hope of preserving the same thermodynamic functions and their interrelations in a wide range of system sizes. The question of the possibility of such an extension of the domain of the thermodynamic description remains open. To solve it, it is necessary to use the methods of statistical thermodynamics. With a decrease in the volume of the system, equilibrium fluctuations increase for all the thermodynamic characteristics [1,4]. The existence of fluctuations is a direct consequence of the discreteness of the structure of matter at the atomic–molecular level of any system and the thermal motion of molecules [5–7].

As the size of the substance decreases, the proportion of surface particles increases in comparison with their total number, and spontaneous density fluctuations increase. The problem of the role of equilibrium fluctuations was discussed by Hill [8–10], who laid the foundations of the thermodynamics of small systems, and in [11] in which the role of macroscopic fluctuations on non-uniform surfaces was discussed in order to improve the accuracy of the description of adsorption isotherms. However, due to the macroscopic size of the systems [11], the contribution of fluctuations did not allow obtaining fundamentally new information, and this topic has been forgotten for many years.

When formulating the question which systems are considered small, it was assumed that the error from ignoring the surface contribution is 1% [9]. Then for a surface property in a three-dimensional system we have $M^{2/3}/M = 0.01$, or $M = 10^6$. That is, the system is considered 'small' when $M < 10^6$. For the boundary effect in a two-dimensional system: $M^{1/2}/M = 0.01$; or $M < 10^4$. For a logarithmic term, the condition $\ln M/M = 0.01$ implies $M < 600$. For a term of the order of unity (the boundary effect in unit dimension, or the 'net' effect): $1/M = 0.01$; $M < 100$.

If we go over to analyzing the value of the mean-square fluctuation η, then its choice is not arbitrary. Since the problem is posed of comparing the calculated characteristics in classical and statistical thermodynamics, the choice of criterion η refers to the comparison of two different (continuum and discrete–molecular) points of view, and this criterion should not depend on the accuracy of existing experimental techniques. The thermodynamic description of the curved boundary is characterized by the curvature of the mathematical surface to which the magnitude of the surface tension refers. To comply with this situation in molecular theories, it is necessary to abandon all the features of the distribution of molecules at the interface between phases, which depend on the temperature and type of the model.

In these conditions, this choice of one percent is rather crude. In the case of interest to us, Table 29.1 shows the relationship between the number of molecules M and the value of the corresponding radii of spherical drops related to the indicated values $\eta = 100/M^{1/2}$ % with respect to the fluctuations in the number of surface particles (R_s) and the volume (R_v) of the sphere, where the radii are measured in units of λ (λ is the average distance between the molecules in the liquid, equal to $\lambda = 1.12\sigma$, σ is the size of the solid sphere of the spherical molecule, which appears in the Lennard-Jones potential).

From Table 29.1 it follows that the density fluctuations on the surface do not exceed 1%, the particle size R_s should be in the range from 28 to 89 λ.

Volume or surface. The question of what characteristics of a small system should be compared, related to its volume or surface, plays an important role [12]. From Table 29.10 follows an obvious fact: for $M > 2 \cdot 10^2$ we have $R_s > R_v$, i.e. for a given accuracy of describing the properties of the volume of the drop, the radius of the drop R_v is considerably smaller than the radius R_s in the case of describing the surface properties. We note that

Table 29.1. The scale of the particle sizes containing different numbers of drop molecules M related to their surface (R_s) and bulk (R_v) regions at a given level of the relative magnitude of the mean-quadratic fluctuations of the number of molecules $\eta = 100/M^{1/2}$ %

$M =$	10^8	10^6	10^5	10^4	10^3	10^2	49
η, %	0.01	0.10	0.32	1.0	3.2	10	14
R_s	2800	280	90	28	9	2.8	2.0
R_v	275	62	27.5	13.5	6.2	2.9	2.3

the concept of the interface between phases in thermodynamics refers to macrosystems. Allowance for the surface contributions is necessary if the surface area A is so developed that the contribution of the surface contributions to the thermodynamic potentials becomes commensurable with the contribution of the bulk phases. If the Gibbs potential G for a volume phase is written as $G = \Sigma_i \mu_i m_i = U - TS + PV$, then for a plane boundary and an equilibrium drop this potential has the form $G = \Sigma_i \mu_i m_i = U - TS + PV - \sigma A$, and for the curved boundary of the metastable drop $G = \Sigma_i \mu_i m_i = U - TS + P_\alpha V_\alpha + P_\beta V_\beta - \sigma A$, where the usual notations are: internal energy U, entropy S, temperature T, pressure P, volume V, μ_i and m_i is the chemical potential and the mass of component i. The contribution σA, where σ is the surface tension, also appears in all other thermodynamic potentials, when it is necessary to take into account the surface properties of the system. In the case of metastable drops, the phases are in equilibrium at different pressures P_α and P_β.

For an isolated small system, all terms of the thermodynamic potential must be described with equal accuracy, and the equations of thermodynamics treat both the volume and surface terms from the continuum point of view – without taking into account the effects of fluctuations. If the volume of a small system is chosen as a condition for choosing the accuracy of the description, then its surface characteristics will include fluctuation effects, which changes the accuracy of the description of the entire system. Therefore, when comparing the continuum and discrete–molecular points of view for small systems, it is necessary to choose surface characteristics. The most convenient for this purpose is the 'surface tension' itself, which naturally appears in disperse systems. Below we consider the values of the fluctuations η on the surface of tension and not inside the volume of the drop. This is necessary to have not only the volume, but also the surface not experiencing fluctuations, i.e. corresponding to the requirements of thermodynamics. The question of the fluctuations of the internal volume of the phases is considered in Section 34.

30. The discreteness of matter

The discreteness of matter is the only natural limitation on the possibility of using thermodynamic approaches. The effects of the discreteness of matter are most clearly seen in the case of adsorption, which, in the absence of fluctuation effects, fairly well describes the main factors of adsorption systems: surface non-uniformity and lateral interactions between adsorbed molecules [13]. But in the transition to ultradispersed particles, the existing theory should be supplemented by taking into account the equilibrium fluctuations and analyzing their influence on all the molecular characteristics of real surfaces. Traditionally in thermodynamics solid adsorbents are considered as homogenized objects. For this reason, all the results of [8–10] refer to small uniform systems. The actual non-uniformity of solid particles is well known [13] (see Section 1).

We will consider the distribution of molecules in a large canonical ensemble ($\{\mu\}$, V, T) or ($\{\mu\}$, M, T), where the symbol $\{\}$ is the set of values of μ_i chemical potentials of molecules of sort i. The symbol M is used to denote the total number of sites of a system of volume V, since for a rigid lattice structure $V = v_0 M$, where v_0 is the volume of the site. The non-uniform surface consists of sections of sites of different types whose number is equal to M_q, $1 \le q \le t$, t is the number of site types; $M = \sum_{q=1}^{t} M_q$.

The number of occupied states of any lattice site, including vacant sites, is s (the index of sort $i = s \equiv V$ refers to vacancies). Each molecule of sort i (except for vacancies) has an internal statistical sum J_q^i, depending on the type of site q occupied by a given particle. Denote by N_q^i the number of molecules of sort i that are located at sites of type q. The number of free sites of type q is denoted by $N_q^V = M_q - \sum_{i=1}^{s-1} N_q^i$.

The total energy of the non-uniform ideal lattice system in the grand canonical ensemble is written as [11]

$$H = \sum_{f=1}^{M} \sum_{i=1}^{s} v_q^i \gamma_f^i \eta_f^q, \quad v_q^i = -\beta^{-1} \ln J_f^i - \varepsilon_f^i - \mu_f^i \tag{30.1}$$

where J^i_f is the partition function of a particle of sort i located at a site with number f of type q; $1 \le f \le M$, $1 \le i \le s$, ε^i_f is the binding energy of a particle i with a site of type q having the number f on the surface of the adsorbent, μ^i_f is the chemical potential of particle i at site f of type q, which in equilibrium equals the chemical potential of the particle i in the thermostat μ_i outside the lattice system, $\mu_i = \mu^i_f$, $\mu_i = kT\ln\left(P_i / J_i\right)$, J_i is the partition function of a particle of sort i in the gas phase, P_i is its partial pressure. The variable γ^i_f and the quantity η^q_f are defined in Section 16.

δ-shaped distributions. In the absence of lateral interactions, each centre is filled independently, so the probability of its filling is expressed as:

$$\theta_f^{\,i} = N^i_q / M_q = a^i_f P_i / (1 + a^i_f P_i), \quad a^i_f = J^i_f / (kTJ_i)\exp(\beta\varepsilon^i_f). \quad (30.2)$$

where a^i_f is the local Henry coefficient characterizing the degree of retention by the centre f of a particle of sort i.

This result is easily obtained in the grand canonical ensemble for a non-uniform surface in which the large partition function Ξ for an ideal non-uniform systems is expressed in terms of the 'local' partition functions Ξ_q for the sites q as $\Xi = \prod^t_{q=1} \Xi_q$. In turn, each local partition function Ξ_q is connected with the partition function in the canonical ensemble $Q_q(\{N_q^{\,i}\})$, which is expressed in terms of the statistical weight of the given distribution $\Omega_q(\{N^i_q\}) = C^{N_q}_{M_q} = M_q! / \prod^s_{i=1} N^i_q!$, where $N^s_q \equiv N^V_q$. As a result, we have

$$\Xi = \prod^t_{q=1} Q_q(\{N^i_q\})\exp(\beta\mu^i_q) = \prod^t_{q=1} \frac{M_q!}{\prod_{i=1}^{s} N^i_q} \prod_{i=1}^{s-1} J^i_q \exp(\beta\mu^i_q), \quad (30.3)$$

Taking into account the fact that each type of site is filled with molecules of different sort, equation (30.3) yields the exact expression

$$\ln\Xi = \sum^t_{q=1}(1 + \sum^{s=1}_{i=1}\lambda^i_q)^{Mq}, \quad \lambda^i_q = \exp(\beta\mu^{i*}_q),$$
$$\mu^*_i = \mu_i + \beta^{-1}\ln\left(J^i_q\right). \quad (30.4)$$

The spreading pressure in the lattice gas model is given by

$$\pi = \beta^{-1}\ln\Xi / M = \sum_{q=1}^{1}(1 + \sum_{i=1}^{s-1}\lambda_q^i)^{Mq} / M. \qquad (30.5)$$

This quantity is a discrete analog of the ordinary pressure P in three-dimensional systems of volume V; for the rigid lattice in question, the change in volume is due to a change in the number of vacant sites of a fixed size.

The degree of filling of a site of type q by molecules of sort i, expressed by formula (30.2), follows from relation

$$\theta_q^i = \langle N_q^i \rangle / M_q = \partial(kT\ln\Xi) / \partial\mu_q^{i*}, \qquad (30.6)$$

where angular brackets <...> mean averaging over the ensemble, and the symbol <A> means the average value of A. Equation (30.2) is also obtained from the condition that determines the maximum contribution to the sum (30.3), $\partial(\ln\Xi)/\partial N_q^i|_{M,T,\{\mu\}} = 0$.

Equation (30.6) corresponds to the δ-shaped form of the distribution functions $P_q(\{N_q^i\}) = Q_q(\{N_q^i\})\exp\left(\sum_{i=1}^{s-1}\beta\mu_i\right)$ for sites of sort q, which 'cut out' from the total sum (30.3) maximal terms approximating the 'local' sums Ξ_q for sites q.

The equilibrium filling of the entire surface is obtained by weighing the sites of different types by means of the distribution functions F_q characterizing the probability of finding sites of the type q on the surface, $1 \le q \le t$, t is the number of types of different centres, their number is denoted by M_q, so that $F_q = M_q/M$ and $\sum_{q=1}^{t} F_q = 1$. Then the partial filling of the surface by molecules of sort i is written as

$$\theta_i = \sum_{q=1}^{t} F_q\theta_q^i. \qquad (30.7)$$

The equilibrium fluctuations of the local partial fillings η_{qp}^{ij} characterize the mean square deviations of the fillings by the i and j molecules of a pair of sites of the q and p sorts. They are defined as $\eta_{qp}^{ij} = \left\langle \left(N_q^i - \langle N_q^i\rangle\right)\left(N_p^j - \langle N_p^j\rangle\right)\right\rangle$.

For the adsorption of molecules of one sort ($s = 2$) on a non-uniform surface ($t > 1$), the formulas given correspond to $\eta_{qp} \equiv \eta_{qp}^{AA}$ [11]:

$$\eta_{qp} = M\sum_{q=1}^{t} F_q\theta_q(1-\theta_q) = \sum_{q=1}^{t} M_q\theta_q(1-\theta_q). \qquad (30.8)$$

which generalizes the well-known expression for the adsorption of one sort of molecules on a uniform surface (t = 1) $\eta \equiv \eta_{11} = M\theta(1-\theta)$ [11,16].

In the case of adsorption of a mixture of molecules (s > 2) on a uniform surface, the expressions for equilibrium fluctuations $\eta_{11}^{ij} \equiv \eta_{qp}^{ij}$ are constructed by analogy with the known formulas [2,3]. Equilibrium fluctuations η_{qp}^{ij} in the case of adsorption of a mixture of molecules on a non-uniform surface were not discussed prior to the work [14]. The structure of such expressions η_{qp}^{ij} is considered below in view of size effects.

Mathematical apparatus. The mathematical description of the discreteness requires the use of the appropriate apparatus [17,18], based on the use of the difference calculus instead of the traditional continuous calculus. This circumstance, apparently, was first pointed out in [9] – the molecule can not be divided into smaller parts. It was specified in [14, 15] that the ordinary difference (non-symmetrized) derivatives [9] should be replaced by symmetrized derivatives.

For the analysis of small systems, in connection with the increase in the role of fluctuations, it is necessary to take into account the difference between the distribution functions of the δ-shape and the effect of the boundedness of the surface or volume of the system under consideration. Both of these factors must influence the probabilities of the filling of the sites θ_q^i and the equilibrium fluctuations of these fillings.

Derivatives for small systems. The definition of the derivative of the function $F(x)$, where x is the argument of the continuous function F, is written in the differential calculus as [19]

$$\frac{dP(x)}{dx} = \lim_{h \to 0} \left[\frac{P(x+h) - P(x)}{h} \right], \tag{30.9}$$

where h is the increment of the continuous variable x. In numerical calculations, the value of the increment h remains finite, but much less than the region of the characteristic change of the function under consideration, so the original definition of the derivative is rewritten in the difference form $dP(x)/dx = \Delta P(x)/\Delta x = [P(x + h) - P(x)] / h$, provided that the used value of h corresponds to the equality $dP(x)/dx \approx \Delta P(x)/\Delta x$ with the specified accuracy.

LGMs always operate with a discrete set of molecules, and the difference increments are a natural way of counting in this approach. The magnitude of the argument increment for a discrete variable N can not be arbitrary. The minimum increment value ΔN is ± 1, since the molecule can not be broken up. By analogy with the expression (30.9) of the difference derivative $\Delta P(N)/\Delta N$, for any function $P(N)$ of the discrete argument N, the relation $\Delta P(N)/\Delta N = [P(N + \Delta N)-$ $-P(N)]/\Delta N$, where $\Delta N = \pm 1$.

The choice of the sign of $\pm h$ in the increment for the differential calculus does not matter. In mathematics courses [19] it is proved that the limit does not depend on the method of limiting transition inside the domain of definition of a continuous function. For small systems with a discrete change in the value of N, the fixation of the sign (± 1) affects the final result. Usually, a plus sign is conventionally taken as an increment. This choice was adopted in the works on calculus in finite differences [16] and in combinatorics [20,21]. Nevertheless, in practical work on numerical calculations in the case of differential calculus, it is shown [18] that such an asymmetric choice of the method of calculating the difference derivative loses exactly in comparison with the symmetric definition of the derivative when the rule is used: $dP(x)/dx = [P(x + h)-P\ (x-h)]/2h$.

Such symmetrization in the differential calculus is possible by virtue of the continuity of the argument x for any values of the increment h. When using the analog of the symmetric definition for the difference derivative, one should use the expression [14,15]

$$\Delta P(N)\Delta N = \left[P(N+1)-P(N-1)\right]/2. \tag{30.10}$$

To illustrate the differences in these methods for determining derivatives, we consider how the expressions for the chemical potential in the canonical ensemble with discrete values of the number of particles N and the number of sites of the system M on which these molecules are distributed are constructed ($s = 2$). We denote by $F_{N,M}$ the free Helmholtz energy: $F_{N,M} = E_{N,M}-TS_{N,M}$, where $E_{N,M}$ is the energy, and $S_{N,M} = k\ln\Omega_{N,M}$ is the entropy of the system. In accordance with formula (30.3) for an ideal system $E_{N,M} =$ $-\beta^{-1}\ln J$, J is the internal statistical sum of the molecule, and $\Omega_{N,M} =$ $M!/(N!\ N_V!)$, where the number of unoccupied sites $N_V = M - N$.

The chemical potential of a molecule for a given problem can be written in four ways: through the differential derivative (A) and the three difference derivatives: (B) the asymmetric method, $\Delta N =$ $+1$, (B) the asymmetric method, $\Delta N = -1$, and (Γ) the symmetric method, $\Delta N = \pm 1$.

(A) Differential derivative: $\mu_{N,M} = \partial F_{N,M} / \partial N_{|M}$. This approach uses the expansion in the series $\ln N!$ by the Stirling formula

$$\ln N! = \\ = N(\ln N - 1) + \ln(2\pi N)/2 + (12N)^{-1} - (360N^2)^{-1} + o(N), \tag{30.11}$$

where $o(N)$ is the remainder of the series.

Restricting ourselves to the account in the Stirling formula of the first smallness correction, we have

$$F_{N,M} = -N\beta^{-1}\ln J - \beta^{-1}\{M\ln M - N\ln N - N_V \ln N_V + \ln[M/(2\pi N N_V)]/2\},$$

what gives

$$\mu_{N,M} = -\beta^{-1}\ln J - \beta^{-1} \times \\ \times\{\ln(N_V/N) - (1/N - 1/N_V)/2 - (1/N^2 - 1/N_V^2)/12 + \ldots\} \tag{30.12}$$

(B) The traditional asymmetric difference derivative with sign (+1) [9] is written as $\mu_{N,M} = F_{N+1,M} - F_{N,M}$. This gives

$$\mu_{N,M} = -(N+1)\beta^{-1}\ln J - \beta^{-1}\ln\Omega_{N+1,M} + \\ +N\beta^{-1}\ln J + \beta^{-1}\ln\Omega_{N,M} = -\ln J - \beta^{-1}\{\ln(N_V/N) - \\ 1/N + 1/(2N^2) - 1/(3N^3) + 1/(4N^4) + \ldots\} \tag{30.13}$$

(C) Asymmetric difference derivative with sign (−1):

$$\mu_{N,M} = F_{N,M} - F_{N-1,M}. = -N\beta^{-1}\ln J - \\ -\beta^{-1}\ln\Omega_{N,M} + (N-1)\beta^{-1}\ln J + \beta^{-1}\ln\Omega_{N-1,M} = -\ln J - \\ -\beta^{-1}\{\ln(N_V/N) + 1/N_V - 1/(2N_V{}^2) + \\ +1/(3N_V^3) - 1/(4N_V^4) + \ldots\} \tag{30.14}$$

(D) Symmetric difference expression for the chemical potential [14]:

$$\mu_{N,M} = [F_{N+1,M} - F_{N-1,M}]/2 = [-(N+1)\beta^{-1}\ln J - \beta^{-1}\ln\Omega_{N+1,M} + \\ +(N-1)\beta^{-1}\ln J + \beta^{-1}\ln\Omega_{N-1,M}]/2 = -\beta^{-1}\ln J - \beta^{-1}\{\ln(N_V/N) - \\ -(1/N - 1/N_V)/2 - (1/N^2 - 1/N_V^2)/4 + \\ +(1/N^3 - 1/N_V^3)/6 + \ldots\}. \tag{30.15}$$

A comparison of formulas (30.12)–(30.15) shows that the first two terms relating to macroscopic contributions in all variants of chemical potential determinations are the same, and the contributions from the size of the system depend on the method of determining the derivative. This difference begins with the third term relating to the first correction for the size of a system of type M^{-1} or N^{-1}. The asymmetric definitions of the difference derivatives (30.13) and (30.14) differ from the differential derivative (30.12), beginning with the first correction from the size of the system. Between themselves, both difference definitions (30.13) and (30.14) differ drastically. Thus, for small systems, the increment sign $\Delta N = \pm 1$ plays an important role. Simultaneously it is shown that the transition to a more accurate symmetric difference derivative leads to a different result (30.15): it differs from expressions (30.13) and (30.14) already in the third (i.e. first correction) term, which indicates the incorrectness of the asymmetric definitions. The differences (30.15) from the differential derivative (30.12) begin only with the fourth summand (ie in the second order of smallness), while the third terms in both expansions are the same. An increase in the accuracy of calculating the chemical potential is associated with an increase in the number of correction terms. The contributions from the subsequent correction terms of the expansion in powers of M^{-k} or N^{-k} in more precise expressions (30.15) have lower coefficients than in formula (30.12).

Thus, the way of constructing derivatives strongly affects the values of the chemical potential $\mu_{N,M}$. Similar constructions for the spreading pressure also lead to conclusions about the need to use only symmetric difference derivatives.

Higher derivatives. These symmetric definitions of difference derivatives should be used consistently and when calculating higher orders of derivatives [15]. In particular, the second difference derivative, which is necessary for finding the fluctuations, must be constructed as the first symmetric difference from the symmetrized first difference derivatives. This gives

$$\frac{\Delta^2 P(N)}{\Delta N^2} = \frac{\Delta}{\Delta N}\left(\frac{\Delta P(N)}{\Delta N}\right) = \frac{\Delta}{\Delta N}\left[\frac{1}{2}\big(P(N+1) - P(N-1)\big)\right], \quad (30.16)$$

where increments $\Delta N = \pm 1$ refer to each of the arguments in the terms inside the bracket. The consistent application of rule (30.10) in formula (30.16) leads to expression

$$\frac{\Delta^2 P(N)}{\Delta N^2} = \frac{1}{4}\left[\left(P(N+2) - P(N)\right) - \left(P(N) - P(N-2)\right)\right]$$
$$= \frac{1}{4}\left[P(N+2) - 2P(N) + P(N-2)\right]. \tag{30.17}$$

Recall that in analogy with the recording of the first derivative, in the differential calculus the second derivative can be formally written in the asymmetric form $d^2P(x) / dx^2 = [P(x + 2h) - 2P(x + h) + P(x)]/h^2$ or in the symmetric form $d^2P(x)/dx^2 = [P(x + h) - 2P(x) + P(x-h)]/h^2$. It is obvious that the accuracy of the second (symmetrized) expression for the second derivative in point x is much higher than the accuracy of the first asymmetrized expression. For the second symmetric derivative constructed through the asymmetric first difference derivatives for $\Delta N = \pm 1$ we have

$$\frac{\Delta^2 P(N)}{\Delta N^2} = \frac{1}{2}\left[P(N+1) - 2P(N) + P(N-1)\right]. \tag{30.18}$$

The difference between the definitions (30.17) and (30.18) is in the different length of the interval on which the second derivative is constructed. The longer the interval length, the higher the accuracy of the expression $\Delta^2 P(N)/\Delta N^2$. The formula (30.18) has an interval length equal to 2 between three values of N and $N \pm 1$, on which the second derivative is determined, whereas in the completely symmetric discrete definition (30.17) the second derivative has an interval length between five values of N equal to 4.

In the general form, the difference derivative of the n-th order necessary for calculating the higher contributions of the fluctuations will be written in the form

$$\frac{\Delta^n P(N)}{\Delta N^n} = \frac{1}{2^n}\sum_{k=0}^{n}(-1)^{n-k} C_n^k P(N + n - 2k), \tag{30.19}$$

where the length of the interval for calculating the derivatives is $2n$. In the particular case $n = 3$, formula (30.19) has the form

$$\frac{\Delta^3 P(N)}{\Delta N^3} = \frac{1}{8}\left[P(N+3) - 3P(N+1) + 3P(N-1) - P(N-3)\right] \cdot$$

31. An ideal system, one component [15]

Uniform surface. We begin our discussion of the thermodynamic properties of small systems with the simplest case of adsorption of molecules of one kind (N is their number) distributed over sites of a uniform lattice with size M size in the absence of lateral interactions between molecules ($s = 2$). In this example, we will consider the basic problems of the analysis of small systems. To obtain expressions for the mean-square fluctuations, it is necessary to find the second derivative in the vicinity of the maximum of the distribution function $\ln P(N, M)$. Further, knowing the fluctuations of the distribution function, one can obtain the characteristics of interest taking into account the contribution of the fluctuations and determine their effect on the calculated characteristics (and on the observed experimental data).

For the problem under discussion, formula (30.3) is rewritten as

$$\begin{aligned} \Xi &= \textstyle\sum_{N=1}^{M} P(N,M),\ P(N,M) = Q(N,M,T)\exp(\beta\mu N), \\ Q(N,M,T) &= J^N M!/(N!N_V!), \end{aligned} \qquad (31.1)$$

where J is the internal statistical sum of the adsorbed particle, $N_V = M-N$, here μ is the chemical potential of the molecule, fixed by the state of the thermostat and let us introduce $\mu^* = \mu + \beta^{-1}\ln J$.

The traditional procedure for constructing macroscopic equations for the equilibrium distribution of molecules consists in replacing the sum over N in the expression (31.1) by the maximum term of the sum $\Xi^* = P(N^*)$, where $P(N^*) = Q(N^*, M, T)\exp(\beta\mu N^*)$, which approximates the value of the sum Ξ.

$$\begin{aligned} \ln P(N,M) &= \ln\Omega(N,M) + \beta\mu^* N = \\ &= \ln M! - \ln N! - \ln N_V! + \beta\mu^* N. \end{aligned} \qquad (31.2)$$

The desired expression for $P(N^*)$ is determined from the condition $\partial \ln P(N)/\partial N = 0$. The use of differential derivatives is justified for $M, N \to \infty$ and $N/M = \text{const}$, when the number of molecules is described within the framework of continuous calculus. The search for $P(N^*)$ in (31.2) is based on the use of the Stirling formula for a large number of molecules in the asymptotic form $\ln N! = N(\ln N-1)$.

For small systems, the calculation of the maximum term Ξ^* is based on the use of a more accurate expansion (30.11), in which $u(N) = \ln(2\pi N)/2 + (12N)^{-1} - (360N^2)^{-1} + o(N)$, $\Delta \ln P(N, M) / \Delta N = 0$. This gives the following equation for the most probable value of N^*, which is a function of temperature and a fixed value of the number of sites M and chemical potential μ:

$$\beta\mu^* = \ln\left[\frac{N}{N_V}\right] - \frac{1}{2}\ln\left[\frac{N(N_V+1)}{(N+1)N_V}\right]. \tag{31.3}$$

The first term corresponds to the well-known Langmuir equation for a macroscopic uniform surface. The second term takes into account the limited size of the surface. It more fully reflects the dimensional contribution to expression (30.15) for $\beta\mu_{N,\,M}$ (without expansion in a series). As the values of M and N increase in comparison with unity (M and $N \gg 1$), the second contribution decreases and, in the macroscopic size limit, it is zero.

The second derivative is calculated at constant values of M and μ^*

$$\frac{\Delta^2 \ln P(N,M)}{\Delta N^2} = \frac{1}{4}\ln\left[\frac{N(N-1)N_V(N_V-1)}{(N+1)(N+2)(N_V+1)(N_V+2)}\right]_{N^*}. \tag{31.4}$$

Under the sign of the logarithm, all the factors in the numerator are smaller than the corresponding factors in the denominator, so a negative value is on the right. Formally, for N, $N_V >> 2$, this expression (31.4.) becomes zero, which means a sharp narrowing of the distribution function – it goes into the δ-function.

For analytical estimates, in order to take into account the size factors, it is necessary to leave the corresponding contributions in powers of N^{-k} (similar to N_V^{-k}), where $k = 1, 2, 3$, etc., in the series expansion, as in (30.15). In particular, the first correction with terms $k = 1$ for the dimensional contribution is written as

$$\frac{\Delta^2 \ln P(N,M)}{\Delta N^2} = -\frac{M+2}{NN_V}\bigg|_{N^*}, \tag{31.5}$$

that $M \gg 2$ remains $(-M/NN_V)$ on the right-hand side of the formula (31.5), which coincides with the expression of [9] obtained in continuous calculus. This demonstrates the coincidence of the results in the first order in M^{-1} for discrete and continuous descriptions in the second order of expansion for $P(N)$ in the number of molecules, as in the first order of expansion for the first derivatives. Those. the allowance for the fluctuation contributions in the first order in M^{-1} preserves the coincidence of the two calculation methods.

In the expansion of the expressions (31.2)–(31.4) in a series, one must take into account the same accuracy of the expansion as in the logarithmic factors: $\ln(1+x) = \sum_{m=1}^{k}(-1)^{m+1}x^m/m$, where $-1 < x < 1$, with a given value k, and in the expression for u (...) by the Stirling formula (30.11). After agreeing the two calculation methods on the basis of discrete and continuous calculus in small systems, one can use the usual integral relations and take into account the change in the form of the distribution function due to the contribution of Gaussian fluctuations, which is expressed as [8]

$$\ln P(N) = \ln\left[P(N^*,M)\right] - \eta(N-N^*)^2,$$
$$\eta(N^*,M) = -\frac{1}{2}\frac{\Delta^2 \ln P(N,M)}{\Delta N^2}\Big|_{N=N^*}. \tag{31.6}$$

In the first order of smallness of the size contributions, we obtain

$$\eta = M\theta(1-\theta)/(1-D),$$
$$D = \frac{1}{2M}\left(\frac{M+N^2+(M-N)^2}{(N+1)(M-N+1)}\right). \tag{31.7}$$

For a macroscopic uniform surface, (31.7) is transformed into $\eta = M\theta(1-\theta)$ [16,21].

It follows from (31.6) that

$$\Xi = \int_{-\infty}^{+\infty} P(N^*,M)\exp[-\beta_1(N-N^*)^2]d(N-N^*) =$$
$$= P(N^*,M)\pi^{1/2}\eta^{-1/2}, \tag{31.8}$$

or

$$\ln\Xi = \ln[P(N^*,M)] + 1/2\ln\pi - 1/2\ln[\eta(N^*,M)]. \tag{31.8a}$$

For the calculations of $\ln[P(N^*, M)]$, using formulas (30.11) and (31.6), we obtain

$$\ln \Xi = M\ln M - N\ln N - N_V\ln N_V + \beta\mu^* N + u(M) - u(N) - \\ -u(N_V) + 1/2\ln\pi - \frac{1}{4}\ln\left[\frac{N(N-1)N_V(N_V-1)}{(N+1)(N+2)(N_V+1)(N_V+2)}\right] \quad (31.9)$$

Spreading pressure. The last expression (31.9) can be used to calculate the so-called spreading pressure, which is traditionally applied in the LGM and serves as the equation of state $\pi = \beta^{-1}\ln \Xi/M$. It follows the equation for the pressure of the macrosystem with allowance for the fluctuation contributions and all the dimensional corrections

$$\beta\pi = \ln\left[M/N_V\right] - \ln\left[(N_V+1)N\eta(N^*,M)/(N+1)N_V\right] /2M + \\ + \left[1/2\ln\pi + u(M) - u(N) - u(N_V)\right]/M. \quad (31.10)$$

For small systems, substitution of formula (31.3) into the expression for $P(N, M)$ (31.2) give

$$\ln P(N,M)|_{N^*} = \ln\Omega(N^*,M) + \beta\mu^* N^* = M\ln(M/N_V) + \\ + u(M) - u(N) - u(N_V) - \frac{N}{2}\ln\left[\frac{N(N_V+1)}{(N+1)N_V}\right].$$

where all the terms are calculated for $N = N^*$, which allows us to obtain an expression for the average spreading pressure $\beta P_a = \ln P(N, M)|_{N^*}/M$

$$\beta\pi_a = \ln(M/N_V) + \left[u(M) - u(N) - u(N_V) - \frac{N}{2M}\ln\left[\frac{N(N_V+1)}{(N+1)N_V}\right]\right]. \quad (31.11)$$

In the macroscopic limit, both values of P and P_a go over into the well-known relationship between the number of adsorbed molecules and the spreading pressure [22,23]: $\pi = \pi_a = -kT\ln(1-N^*/M)$. We note that the formula (31.11), in contrast to (31.10), refers to the δ-shaped form of the distribution function $P(N, M)$ in a small system without taking into account the influence of fluctuations.

Degree of filling the surface. The average values of the number of adsorbed molecules are found from the known value of Ξ by the Gibbs formula [4,8]

$$N = \partial(kT\ln\Xi)/\partial\mu_{|T,M} = M\lambda^*/(1+\lambda^*), \tag{31.12}$$

since the magnitude of the chemical potential is a continuous quantity, whence we obtain the well-known Langmuir equations (30.2).

Calculation of the average number of adsorbed particles with allowance for fluctuations is carried out according to an analogous formula via the constraints (31.3) and (31.9), reflecting the contributions of the dispersion of the distribution function Ξ

$$\begin{aligned} <N> &= \sum_{N=0}^{M} NP(N,M)/\sum_{N=0}^{M} P(N,M) = \partial\{kT\ln\Xi\}/ \\ /\partial\ln\lambda &= \partial\{\ln[P(N^*,M)] + 1/2\ln\pi - 1/2\ln[\eta(N^*,M)]\}/ \\ /\partial\ln\lambda &= N^* - \Delta N, \text{ where } \Delta V = \partial\ln\left[\eta\left(N^*,M\right)\right]/2\partial\ln\lambda. \end{aligned} \tag{31.13}$$

These formulas readily allow us to answer the question of what molecular densities are most noticeable in approximation with terms of the first order of smallness M^{-1}. They give

$$\theta_M = \langle N\rangle/M = \theta_\infty + (M-2N)/(2M^2). \tag{31.14}$$

From the last formula it follows that for $N = M/2$ we have $\theta_M = \theta_\infty$, i.e. when the surface is half filled, the dimensional corrections are completely absent for any of the largest areas. The maximum value of the size correction is realized for small ($\theta_M = \theta_\infty + (2M)^{-1}$) and large ($\theta_M = \theta_\infty - (2M)^{-1}$) fillings. However, taking into account the eigenvalue of the degree of filling of the surface (that is, in comparison with θ_∞), the contribution of the fluctuations is most noticeable for small $\theta \to 0$.

On small faces of microcrystals, the influence of fluctuations can be commensurable with the degree of filling θ. For example, for $M = 100$ and the degree of filling $\theta_\infty = 10^{-2}$ (that is, of the order of ~1%) we have that the filling of the surface can differ up to one and a half times ($\theta M = 3\theta_\infty/2$) or the value $\Delta\theta = \theta_M - \theta_\infty = 0.5\theta_\infty$ is 50% of the degree of filling of the macroscopic section. A similar influence of fluctuations ($\Delta\theta = 0.5\theta_\infty$) is realized with a simultaneous increase in the area of the section of the surface M and a decrease in the degree of its filling (θ_∞) ~ $1/M$. This factor can be important when considering experimental data for small microcrystals of sensors and catalysts. With increasing surface area, the value of the fluctuation contribution $\Delta\theta$ for a fixed value of θ_∞ decreases as $1/M$. (In the general case, for small values of M, it is necessary to use more precise expressions than formula (31.14).)

Numerical examples. Below are illustrations of the obtained equations for regions of different sizes [24]. The chemical potential is related to the number of adsorbed molecules by different adsorption isotherms in Fig. 31.1. These six curves are shown in Fig. 31.1 in the relative coordinates $\Delta\theta = (\theta-\theta_0)/\theta_0$ for $M = 100$. Here $\ln(aP) = \ln(N/N_V) + A_k$, $k = 1-6$, $a = J\beta/J_0$, where the index k = 1−4 corresponds, respectively, to the expressions for the corrections (30.12)–(30.15), and $k = 5$ is the first size correction into these expressions $A_5 = (1/N-1/N_V)/2$ corresponding to the first term and $k = 6$ is the exact expression (31.3) $A_6 = \ln\{[(N+1)N_V]/[N\,(N_V+1)]\}/2$. The values of θ_0 on the abscissa in the range from zero to unity refer to macroscopic values. Hill's recommendations on the use of non-symmetric difference derivatives $\Delta N = +1$ (30.13) and $\Delta N = -1$ (30.14) give the maximal differences from the exact solution (31.3). The calculation of the first correction turned out to be the most approximate to the exact solution, while taking into account the third corrections both in the discrete (30.15) and in the continuum description (A) has larger deviations than curve 5 for the variant ($k = 5$). Thus, a simple increase in the number of the terms in the series in M^{-k}, $k > 1$, does not automatically lead to an increase in the accuracy of the calculation.

The inset in Fig. 31.1 shows the curves for the variants ($k = 5$) and ($k = 6$) in the cases $M = 10^m$, where the exponents $m = 1$, 3 and 4 are indicated on the curves. The solid lines correspond to the exact contributions of A_6, and the dashed lines correspond to the first corrections from the size effects A_5. With decreasing M, the range in which the values of the dimensionless density θ reflecting allowance for the boundedness of the size of the surface area $1/M \leq \theta \leq (M-1)/M$ is sharply reduced.

To determine the effect of density fluctuations, one must know the variance $\eta = -D/2$ of the distribution function $P(N, M)$, which is expressed in terms of the second difference derivatives as $D = \Delta^2P(N, M)/\Delta N^2{}_{|T,M}$. The following expressions for the exact value of D^E (31.4) were obtained in [15] and taking into account the contribution of only the first correction D^1. The formula for D^E corresponds to the exact expression A_6 in the isotherm equation (31.3). In the value of D^1, the effects of the bounded size of the faces of microcrystals are

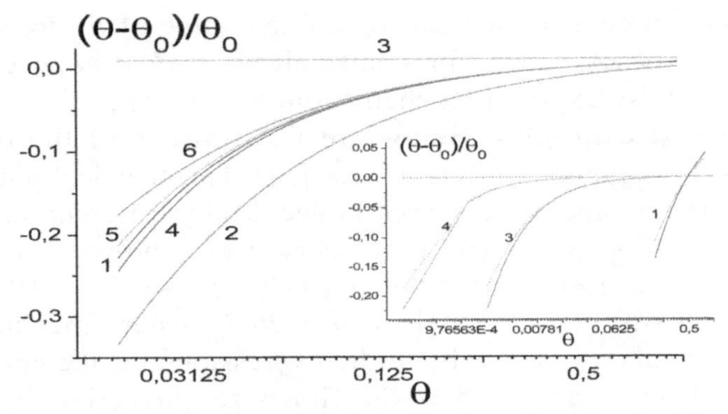

Fig. 31.1. Dependence $(\theta-\theta_0)/\theta_0$ on θ in logarithmic scale: (a) – for $M = 100$ using A_k, $k = 1, .., 6$, the values of k are shown on the curves. On the inset – similar curves are shown by solid lines for A_5, contour – for A_6 at $M = 10^m$, $m = 1, 3, 4$, the values of m are shown on the curves

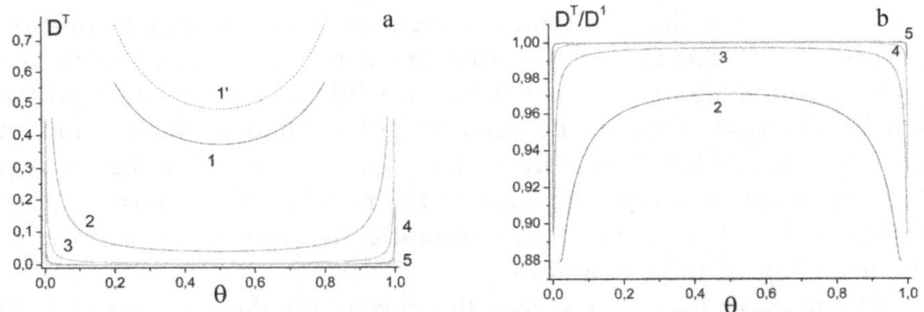

Fig. 31.2. The dependence of D^E (a) and the ratio D^E/D^1 (b) on θ for a uniform surface at $M = 10^m$ using A_6, the values of m are shown on the curves; curve 1′ in Fig. 31.2 *a* corresponds to D^1 using A_5 for $M = 10$.

taken into account in the first order in M^{-1} (calculation is performed using expression A_5).

Figure 31.2 *a* shows the concentration dependences of the coefficients D^E and D^1. With increasing M, the dispersion decreases sharply. Thus, for $M = 10^5$, practically remains zero throughout the entire density range, both for macroscopic dimensions. For $M = 10^4$, in the region of small and large fillings, there are differences from zero values for 2–3% density. Further, with a decrease in M to 10^3, the domain of difference of D^E from the zero value extends to 20%. At even smaller values of M, the value of D^E differs from the macroscopic size in the entire density range, starting at $M =$ 500–600. The curves for $M = 10$ show the maximum dispersion differences for small sections. The dashed curve 1′ refers to a similar dependence obtained with the first dimensional correction. Those. approximate solutions overestimate the variance. The D^E / D^1 ratios are shown in Fig. 31.2 *b* for $M = 10^2 - 10^5$.

The curves given for the adsorption of particles on a uniform surface indicate the importance of taking into account the fluctuation effects in the case of small faces of microcrystals.

Non-uniform surfaces. In real materials, the non-uniformity of the surface due to the presence of different faces and their edges is well known [13, 25, 26]. Therefore, for small particles, the contributions of different faces can not be neglected. The description of non-uniform ideal systems is reduced to the summation of the contributions of any characteristics calculated for individual faces. All formulas obtained above can easily be generalized by summation with weights F_q and replacement of the complete filling θ by local filling $\theta_q = N_q/M_q$, relating to the sites of type q, $1 \leq q \leq t$, t – the number of types of sites (here, they are not written). Complete adsorption isotherm written as $\theta(P) = \sum_{q=1}^{t} F_q q_q(P)$, wherein $F_q = M_q/M$ is the fraction of sites of type q on the whole surface, $\sum_{q=1}^{t} F_q = 1$.

Differences in the binding energies Q_q of the particle with surface portions of type q must be distinguished explicitly via local Henry constant $a_q = a_0 \exp(\beta Q_q)$, here a_0 is the pre-exponent having the back pressure dimension. Local fillings are related to the chemical potential as [14,15]

$$ln\left(a_q P\right) = ln\left(N_q / N_q^V\right) - \frac{1}{2}\ln\left[\frac{N_q(N_q^V+1)}{(N_q+1)N_q^V}\right] \qquad (31.15)$$

where $N_q^V = M_q - N_q$, the second summand is a generalization of expression (31.3) with exact accounting of all size contributions (expressions A_6 for different faces).

The filling of different faces of the crystal takes place in accordance with the values of Henry's constants, so the effect of fluctuations for small and large fillings on different faces leads to a qualitatively new situation. Its meaning is illustrated in Fig. 31.3, which shows the number of adsorbed molecules on two small faces of the crystal ($\Delta Q = Q_1 - Q_2$). The abscissa is the degree of filling of the surface θ of the small crystal. The dotted curve refers to a macroscopic lattice. One can see the influence of the size factor on the local filling of different faces of a non-uniform surface. The results of the calculation show that 1) the differences in the degree

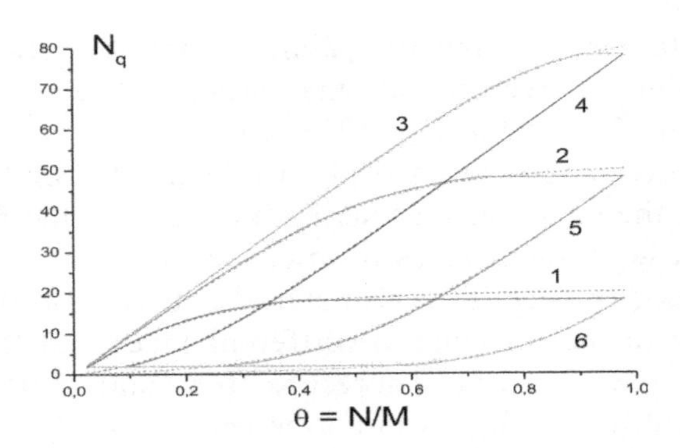

Fig. 31.3. Dependences of N_q, $q = 1{,}2$ on $\theta = N / M$ for $t = 2$ for the values of the molecular parameters $\beta\varepsilon = 1$, $Q_1 = 3\varepsilon$, $\Delta Q = 3\varepsilon$. Curves 1, 2, 3 correspond to N_1, curves 4, 5, 6 correspond to N_2 for $F_1 = 0.2, 0.5, 0.8$, respectively. The solid lines denote the curves for $M = 100$, the dotted lines the curves for $M = \infty$.

of filling of small and macroscopic systems are quite appreciable, and 2) the extent of the region θ in which the deviation data are observed is large enough, and one can speak about the possibility of their experimental detection.

The equation for the density fluctuations of adsorbed molecules on the non-uniform surface of small particles in the first order in N^{-1} will be written as

$$\eta = \sum_{q=1}^{t} M_q \eta_q, \ \eta_q = \theta_q(1-\theta_q)/(1-D_q),$$
$$D_q = \frac{1}{2M_q}\left(\frac{M_q + N_q^{\,2} + (M_q - N_q)^2}{(N_q+1)(M_q - N_q + 1)}\right), \tag{31.16}$$

where η_q is the contribution of local density fluctuations in the section of sites of type q. With increasing M_q, the second term of the denominator D_q in η_q turns to zero, and the known expression [8] is obtained.

Analysis of this expression shows: 1) The maximum local dimensional density fluctuations exist for $\theta_q \to 0$ and $\theta_q \to 1$. (For $\theta_q \sim 1/2$, maximum density fluctuations not related to dimensional effects, including, for macroscopic systems.) 2) An important role in the value of η is played by the relations M_q/M_p. 3) The density fluctuations on a non-uniform surface oscillate – the maximum number of oscillations is equal to the number of types of centres

t, which makes it possible to estimate them from the experiment if there is a complete separation of the contributions of η_q from different parts of the surface.

Figure 31.4 shows concentration dependences of the density fluctuations of adsorbed molecules on a non-uniform surface consisting of two types of centres. The calculation is made for three surface compositions. The difference between the curves for small ones with $M = 100$ (solid lines) and macroscopic (dotted lines) of systems is observed at any densities. In the field of medium fillings the differences are small, but for small and large fillings they increase. For clarity, curves *5*, relating to local fluctuations corresponding to the sites of the first type of curve 1, are presented. The variation of the surface composition sharply affects the form of the concentration dependence of the mean-square fluctuation η. The presence of peaks in the value of η is associated with a transition from preferential filling of one site to another and from the ratio in the binding energies ΔQ. If the differences are small (curve 4), then both sections are filled simultaneously and there are no maxima. This fact was noted earlier for macrosystems in [11].

The accuracy of calculations. Let us discuss two questions on increasing the accuracy of calculating adsorption characteristics on non-uniform surfaces. The first question is related to the fact that the increase in fluctuations with decreasing size of a small system leads to the influence of fluctuations on the average values of the filling of the faces. Such a fact is absent in macroscopic systems, but for small systems it can be shown that the discrete nature of statistics

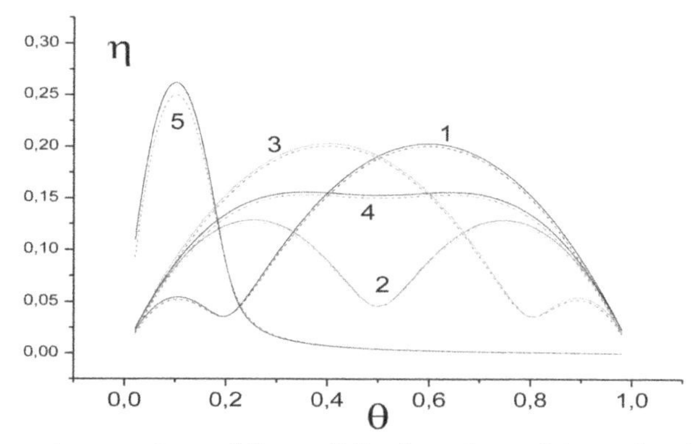

Fig. 31.4. Dependences of η on θ for $t = 2$ for the values of the molecular parameters $\beta\varepsilon = 1$, $Q_1 = 3$; curves 1, 2, 3 correspond to $\Delta Q = 6$ and $F_1 = 0.2, 0.5, 0.8$, respectively, curve 4 corresponds to $\Delta Q = 3$ and $F_1 = 0.5$. Curves 5 refer to local fluctuations η_1 corresponding to the sites of the first type of curve 1 for $F_1 = 0.2$ and $\Delta Q = 6$. Solid lines denote curves for $M = 100$, dotted lines for curves for $M = \infty$.

leads to violation of well-known expressions for the moments of Gaussian distribution functions [16], which approximate the discrete distribution functions in small systems [15]. The nature of such a violation is related to the asymptotic ($M \to \infty$) character of the Gaussian distribution functions and the complexity of determining the value of M from which this approximation becomes quite accurate.

The average number of adsorbed particles with an allowance for fluctuations is obtained from the well-known Gibbs formula [1,4]: $\langle N \rangle = \partial\{kT\ln\Xi\}/\partial\ln(P)$, which takes into account the variance of the distribution function $P(N, M)$. This Gibbs equation follows directly from the distribution functions and does not depend on the type of variables (discrete or continuous) for the number of molecules.

$$\begin{aligned}\langle N_q \rangle &= \partial\left\{\ln[P_q(N_q,M_q)] + 1/2\ln\pi - 1/2\ln(\eta_q)\right\} / \partial\ln(P) = \\ &= N_q - \Delta N_q,\end{aligned} \tag{31.17}$$

where $\langle N_q \rangle$ is the average number of particles on the surface for a given chemical potential μ, and $\Delta N_q = \partial\ln(\eta_q)/[2\partial\ln(P)]$ is the change in this number due to fluctuations. Figure 31.5 *a* shows the correction curves for local adsorption isotherms for two types of centres with allowance for the fluctuation density contributions related to the surface of the number of sites $M = 100$. Vertical and horizontal sections of the curves demonstrate the course of the changes in ΔN_q. For small densities, the corrections are negative, which leads to an increase in the degree of filling of each section of the surface, and for large degrees of filling, the fluctuation corrections reduce the overall filling. The value $q = 0$ corresponds to the mean corrections that refer to the whole surface: $\Delta N_0 = \sum_{q=1}^{t} F_q \Delta N_q$.

The analysis showed that with increasing surface size, the influence of fluctuations decreases rapidly. For values exceeding $M = 10^5$, we can speak of negligibly small deviations from the macroscopic behaviour of the system. In the particular case of a uniform surface and restricting to taking into account only the first size correction, we obtain the expressions for ΔN in the work [8].

The second question is related to the specifics of allowance for fluctuations on non-uniform surfaces. The structure of equations (31.9) and extension of equation (31.4) at any $t > 1$ shows [15,27] that the main problem in calculating the effects of fluctuations on non-uniform surfaces is the so-called limiting or 'forbidden' values of $N_q^b = 0, 1, M_q-1, M_q$ for which the right-hand sides of the mentioned expressions are not defined. Figure 31.1 shows how, with a decrease in M, the range of the value of the dimensionless density that corresponds to a given size of the surface area $1/M \leq \theta \leq (M-1)/M$ decreases. In Figs. 31.3 and 31.4 it is shown that when the local density approaches small and large fillings, the role of fluctuations always increases, and it is with these values of N_q that local bonds are absent. In the general case, an increase in the number of types of centres t inevitably leads to an increase in the total number of surface sites, which include the 'forbidden' numbers of $N_q^b = 0, 1, M_q-1, M_q$ molecules for large and small fillings of the faces of each type of centre. Their number is $N^b = 4t$, so with increasing t the total number of boundary values N^b increases.

The more non-uniform the surface with respect to bond energies, the more significant this factor is, since during the filling of one face the other faces will remain practically free or vice versa, almost completely filled. In both situations, the role of fluctuations increases. This situation is always realized when there are differences in local fillings related to the difference in the values $\beta\Delta Q = \beta\,(Q_1 - Q_2) > 3$, for example, for $t = 2$ (see Fig. 31.4).

The presence of 'forbidden' N_q^b values should be monitored in all calculations and this represents one of the problems in calculating adsorption on non-uniform surfaces. It is obvious that the direct summation of contributions in a large partition function for non-uniform surfaces [15] will take into account all configurations without exception, but this is a very laborious procedure. The transition to the use of information about maximum contributions and their variances significantly simplifies the calculation. But even in this case the problem for non-uniform surfaces remains complicated because of the large contribution of the 'forbidden' values of N_q^b. Therefore, it seems possible to introduce approximations of the boundary values of the occupation numbers on each of the faces in order to carry out the calculation in the entire density range with an allowance for fluctuations. We recall that, as a limiting value for small N (i.e., for an ideal gas or vacancies), we have $\langle \Delta N^2 \rangle = 1$ [1–3]. As the simplest method of approximating the density description in the region of 'forbidden' values of N_q^b, it is possible to define the limiting values of ΔN_q so that they are also equal to one. This will correspond to a doubled value of the current density in the region of limiting small and large fillings. Calculations of the fluctuation corrections ΔN_q with this approximation are shown in Fig. 31.5 *b*.

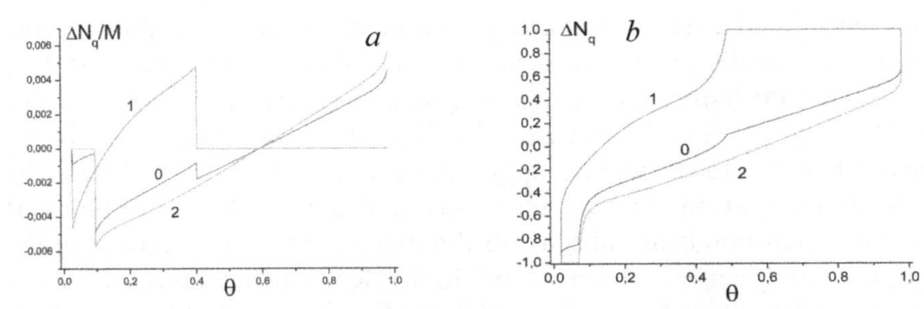

Fig. 31.5. Dependences of the initial values $\Delta N_q/M$ (a) and the corrected values ΔN_q (b), $q = 0, 1, 2$, on θ for $t = 2$ for the values of the molecular parameters $\beta\varepsilon = 1$, $Q_1 = 3\varepsilon$, $\Delta Q = 3\varepsilon$, $F_1 = 0.2$ on the surface $M = 100$. The values of q are shown on the curves.

Thus, it is obtained that the greatest relative influence of density fluctuations is manifested with small fillings of each face of the particle. The magnitude of the fluctuating contribution for large filling is the same as for small filling. As the total filling of the non-uniform surface increases, the root-mean-square (rms) value of density fluctuations have an oscillating character at any face dimensions, taking into account oscillations due to differences in the binding energies of molecules at different sites of the surface. An estimate is obtained – the surface size containing not less than 10^5 adsorption centres can be conditionally considered as the size larger than which the fluctuation corrections are small.

32. An ideal system, two components [15]

Uniform surface. Let us illustrate the specifics of constructing equations for the adsorption of mixtures on a uniform lattice ($t = 1$) containing two sorts of molecules with allowance for vacant sites, $s = 3$. Let N_i, $i = 1$ and 2 denote the number of particles of the first and second sort on the lattice of M sites. We denote the number of free lattice sites by $N_V = M - N_1 - N_2$ (the index V refers to the last of the number of occupied states of the sites s). The search for the maximum term in the large partition function Ξ is carried out by varying the numbers N_1 and N_2 for a fixed value of M and given chemical potentials μ_1, μ_2 and the temperature of the system T. $\ln P(N_1, N_2, M)$ will be considered:

$$\ln P(N_1,N_2,M) = \ln\Omega(N_1,N_2,M) + \beta\textstyle\sum_{i=1}^{s-1} N_i\mu_i^* =$$
$$= \ln M! - \ln N_1! - \ln N_2! - \ln N_V! + \beta\textstyle\sum_{i=1}^{s-1} N_i\mu_i^*,$$

and find the difference symmetric derivatives with respect to the numbers N_1 and N_2. When calculating the partial derivatives, only one variable changes, therefore, according to formula (30.10), we write

$$\frac{\Delta \ln P(N_1,N_2,M)}{\Delta N_1} =$$
$$= \frac{1}{2}[\ln P(N_1+1,\, N_2,M) - P(N_1-1,N_2,M)] =$$
$$= \beta\mu_1^* + \frac{1}{2}\ln\left[\frac{N_V(N_V+1)}{N_1(N_1+1)}\right] = D_1 + \delta P_1; \quad (32.1)$$
$$P_1 = \beta\mu_1^* + \ln(N_V / N_1),\ \delta P_1 = \frac{1}{2}\ln\left[\frac{N_1(N_V+1)}{N_V(N_1+1)}\right].$$

In the last equation, the macroscopic contribution P_1 and the correction for the limited volume of the system (size contribution) ΔP_1 are distinguished. For macrosystems with N_1, $N_V \to \infty$, the size contribution is zero. An analogous equation for the second partial difference symmetric derivative with respect to molecules of the second kind can be written in the form

$$\frac{\Delta \ln P(N_1,N_2,M)}{\Delta N_2} = \frac{1}{2}[\ln P(N_1,N_2+1,M) - P(N_1,N_2-1,M)] = P_2 + \delta P_2,$$

$$P_{(i)} = \beta\mu_i^* + \ln\left(N_V / N_i\right), \qquad \delta P_{(i)} = \frac{1}{2}\ln\left[\frac{N_i(N_V+1)}{N_V(N_i+1)}\right] \quad (32.2)$$

Equation (32.2) is written in a general form – it refers to any ideal system containing an arbitrary number of components (s–1). If the set of numbers of molecules of such a mixture is denoted by $\{N_i\}$, where $1 \le i \le s-1$, and the number of free sites $N_V = M - \sum_{i=1}^{s-1} N_i$, then we can express in general form the conditions on the maximum summand in the sum for Ξ: $\frac{\Delta \ln P(\{N_i\},M)}{\Delta N_i} = P_i + dP_i = 0$, whence follows

$$\beta\mu_i^* = \ln(N_i / N_V) - \frac{1}{2}\ln\left[\frac{N_i(N_V+1)}{N_V(N_i+1)}\right]. \quad (32.3)$$

The system of equations (32.3) describes the partial isotherms of the adsorption of a multicomponent mixture on a uniform surface that is limited in area. In the particular case of adsorption of a single-component substance, we pass to the equation (31.3). The solution of the system (32.3) determines the set of the most probable values of the numbers of the adsorbed molecules of the mixture $\{N_i^*\}$. This system is nonlinear with respect to the relationship between the given values of the chemical potentials of the mixture components $\{\mu_i\}$ and the numbers of adsorbed $\{N_i^*\}$ molecules. With an increase in the surface area, the value of ΔP_i decreases, and for macroscopic systems we have the well-known expressions $\beta\mu_i^* = \ln(N_i/N_V)$ (Langmuir partial isotherms).

The second partial derivatives are taken at constant values of the chemical potentials $\{\mu_i^*\}$ and the size of the system M

$$\frac{\Delta^2 \ln P(N_1,N_2,M)}{\Delta N_1^{\,2}} = \frac{\Delta}{\Delta N_1}\frac{\Delta \ln P(N_1,N_2,M)}{\Delta N_1} = \\ = \frac{1}{4}\ln\left[\frac{P(N_1+2,N_2,M)P(N_1-2,N_2,M)}{P(N_1,N_2,M)^2}\right] = \\ = \frac{1}{4}\ln\left[\frac{N_1(N_1-1)N_V(N_V-1)}{(N_1+1)(N_1+2)(N_V+1)(N_V+2)}\right]. \tag{32.4}$$

This equation is identical in its form to equation (31.4) for the adsorption of one substance. The differences are in different ways of calculating N_V. For one substance, the value of N_V is uniquely determined by N_1 and M, whereas for a multicomponent mixture the value of N_V depends additionally on the values of N_k of the other components of the mixture.

Equation (32.4) shows that for any number of components of the mixture, the structure of the expression for the second derivative with respect to any component i is preserved.

It is convenient to represent equation (32.4) in the form of two contributions from the component i ($P_{2(i)}$) and the number of vacant sites ($P_{2(V)}$)

$$\frac{\Delta^2 \ln P(\{N_i\},M)}{\Delta N_i^{\,2}} = P_{2(i)} + P_{2(V)}, \\ \text{where } P_{2(i)} = \frac{1}{4}\ln\left[\frac{N_i(N_i-1)}{(N_i+1)(N_i+2)}\right]. \tag{32.4a}$$

The formula for $P_{2(V)}$ is the formula (32.4a) for $P_{2(i)}$, in which the index i is replaced by the index V. This expression is fundamental for ideal systems, since all mixed derivatives are expressed through it, which are defined as

$$\frac{\Delta^2 \ln P(N_1,N_2,M)}{\Delta N_2 \Delta N_1} = \frac{\Delta}{\Delta N_2}\frac{\Delta \ln P(N_1,N_2,M)}{\Delta N_1} = \\ = \frac{1}{4}\ln\left[\frac{P(N_1+1,N_2+1,M)P(N_1-1,N_2-1,M)}{P(N_1-1,N_2+1,M)P(N_1+1,N_2-1,M)}\right] = \\ = \frac{1}{4}\ln\left[\frac{N_V(N_V-1)}{(N_V+1)(N_V+2)}\right] \equiv P_{2(V)} \tag{32.5}$$

Thus, for any number of mixture components, all the mixed second difference derivatives have the same form. This is due to the fact that the degrees of filling of any surface sites with different molecules of the mixture are related to each other by normalization conditions. Their 'engagement' is due to competition for filling free surface sites.

In the transition to macroscopic systems, all the second derivatives vanish because of the narrowing of the width of the distribution function $P(N_1, N_2, M)$ and its transition to the δ-shaped form. For bounded systems, it is necessary to use the normal distribution (s–1) of order. For the case s = 3, we will represent the distribution function in the form (since the number M here is fixed and to simplify the record we omit it)

$$P(N_1,N_2) = P(N_1^*,N_2^*)\exp[-\Sigma_{i=1}^{s-1}\Sigma_{j=1}^{s-1}\eta_{ij}(N_i-N_i^*)(N_j-N_j^*)],$$

where the parameters of the normal two-dimensional distribution η_{ij} are determined by the equations (32.4) and (32.5) with $\{N_i\}=\{N_i^*\}$

$$\eta_{ij} = -1/2\frac{\Delta^2 \ln P(N_1,N_2,M)}{\Delta N_j \Delta N_i}. \tag{32.6}$$

In this case, a large statistical sum is written as

$$\Xi^* = \int_{-\infty}^{+\infty}\int_{-\infty}^{+\infty} P(N_1^*,N_2^*)\exp\left[-\sum_{i,j=1}^{s-1}\beta_{ij}(N_i-N_i^*)(N_j-N_j^*)\right] = \\ = d(N_i-N_i^*)d(N_j-N_j^*) = P(N_1^*,N_2^*)\pi\det(\beta_{ij})^{-1}, \tag{32.7}$$

or the logarithm, the maximum term of the partition function, taking into account the fluctuations from which all the thermodynamic expressions are obtained, has the form

$$\ln \Xi^* = \ln\left[P(N_1^*, N_2^*)\right] + \ln \pi - \ln \det(\eta_{ij}). \qquad (32.71a)$$

The first term refers to the macroscopic characteristics of the system. All size properties of small systems are reflected through the elements of the matrix $\det(\eta_{ij})$.

The case of an ideal mixture of adsorbed molecules on a uniform surface corresponds formally to a well-developed theory of fluctuations in the bulk phase for small filling [10]. The differences associated with high densities in the LGM change the specific expressions for the equations for the maximum of the distribution function $(s-1)$ of order and their variance. But the general methodology is preserved, although the calculations become much more complicated, so we do not dwell on these questions.

As the simplest examples demonstrating the role of the size of a surface, we give analytical expressions in the first order of smallness of the contributions (for which the difference and continuous derivatives coincide) for 1) the determinant $\det(\eta_{ij})$, and 2) the corrections for the partial filling of the surface of a two-component mixture more accurate results should be obtained numerically.)

1) It can be proved that the expression for the determinant $\det(\eta_{ij})_{s-1}$ of dimension $(s-1)$, referring to the $(s-1)$ number of components of the mixture, is expressed as

$$\det(\eta_{ij})_{s-1} = M \Big/ \left(\prod\nolimits_{i=1}^{s} N_i\right). \qquad (32.8)$$

This formula determines the denominator to calculate all the mean-square fluctuations, which are expressed in terms of the corresponding contributions of the inverse and attached matrices from the original matrix η_{ij} by the known technique [2,3,10].

2) The expressions for the fluctuation corrections for the partial fillings of a uniform surface by a two-component mixture are written as

$$\Delta\theta_i = \left(N_V - N_i\right) / 2M\left(N_V + N_i\right), \qquad (32.9)$$

where $\Delta\theta_i = \theta_i - \theta_i^\infty$ and $\theta_i = \langle N_i \rangle / M$, and θ_i^∞ is the partial filling of the surface by molecules of sort i for the macrosystem. Differences in the partial contributions of components are determined by the

Henry constants. The fluctuation correction for complete filling of the surface has the form $\Delta\theta = \Delta\theta_1 + \Delta\theta_2$. For $s = 2$, formula (32.9) becomes the formula (31.14).

Calculation of the mixture. Everywhere below, the calculations are performed at a temperature corresponding to $\beta\varepsilon = 1/2$, where ε is the interaction parameter between the molecules of the first sort. For an ideal system, the parameter $Q_1 = 4\varepsilon$ serves as a measure of the binding of the first sort molecule to the surface. It is assumed in the calculations that the binding energy of molecules of the first sort with the surface, which corresponds to a sufficiently strong coupling of the adsorbed particle to the surface. The binding energy of molecules of the second sort with the surface is assumed to be: $Q_2 = \gamma Q_1$, where $\gamma = 1.4$ [28].

The influence of the size effects on the adsorption of a binary gas mixture on a uniform surface has been investigated numerically by varying the value of M in the range from 10^5 to 10. The argument is given by the total filling of the surface θ and the molar fraction of the first component $x_1 = \theta_1/(\theta_1 + \theta_2)$. Using partial differential equations (32.3), the partial pressures of the P_i components in the vapour phase and the total pressure in the system $P = P_1 + P_2$ are calculated.

Figure 32.1 *a* shows the relative deviations at $(\theta_i - \theta_i^\infty)/\theta_i^\infty$ different sites from the value θ_i^∞ for the macroscopic system (1) with $M = \infty$. These deviations depend on both the M value and the mole fraction of the first component in the mixture. The calculations were carried out for three values $x_1 = 0.15$ (1), 0.5 (2) and 0.85 (3) at $M = 10^4$. Relative deviations sharply increase for small degrees of filling (for $P \to 0$). In order to demonstrate the dependence of the relative difference on the value of M, the inset data in an enlarged form shows the results of calculations for $x_1 = 0.15$ for $M = 10^4$ (*4*), 10^3 (*5*), and 10^2 (*6*) for small values of the pressure P.

Figure 32.1 *b* considers partial isotherms for $M = 10^4$. Their form corresponds to the well-known curves of Langmuir isotherms. The isotherms for other values of M slightly deviate from them. At x_1 = 0.15 and $x_1 = 0.85$, the partial isotherms 1 and 2 change in places. At $x_1 = 0.5$, both partial isotherms coincide with each other (therefore there is one curve 3).

Figure 32.2 shows that all the η_{ij} values are positive, and the diagonal elements exceed off-diagonal ones. Thus, the determinant is always positive.

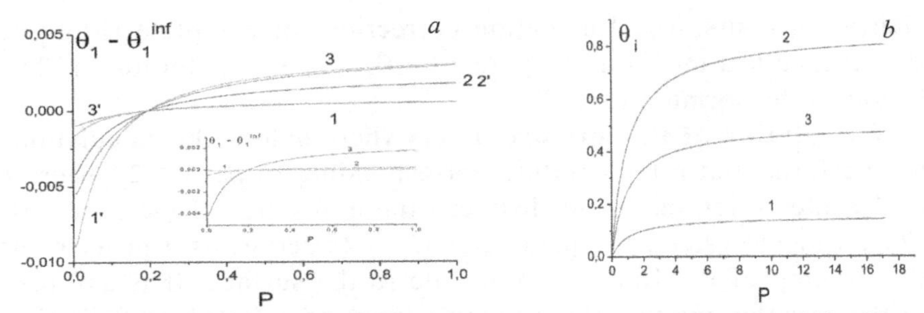

Fig. 32.1a. Deviations $\theta_i - \theta_i^\infty$ as a function of the pressure P at $M = 10^2$ with a change in the molar composition of the mixture $x_1 = 0.15$, 0.5 and 0.85 in the region of low pressures. The notation on the curves, respectively, is 1, 2, 3 for $i = 1$ and 1′, 2′, 3′ for $i = 2$. On the inset: the same deviations at $x_1 = 0.5$ for $M = 10^4$ (1), 10^3 (2) and 10^2 (3).

Fig. 32.1b (right). The dependencies of local densities θ_i on total pressure P. Symbols on the curves: 1 – θ_1 for $x_1 = 0.15$ and θ_2 for $x_1 = 0.85$, 2 – θ_2 for $x_1 = 0.15$ and θ_1 for $x_1 = 0.85$, 3 – $\theta_1 = \theta_2$ for $x_1 = 0.5$.

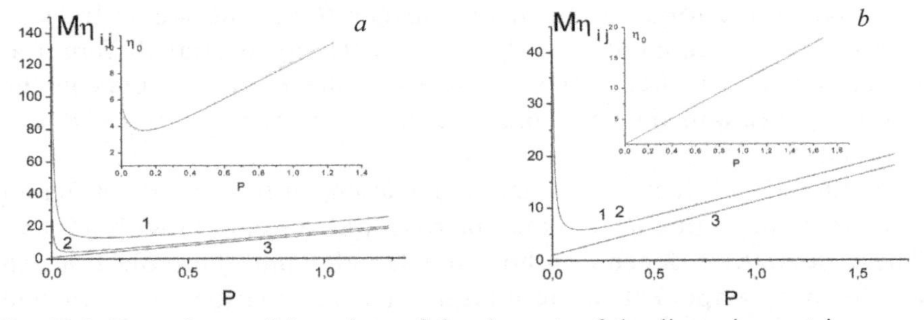

Fig. 32.2. Dependence of the values of the elements of the dispersion matrix η_{ij} on pressure P, at $x_1 = 0.15$ (a) and $x_1 = 0.5$ (b). The notation on the curves: 1 – η_{11}, 2 – η_{22}, 3 – $\eta_{12} = \eta_{21}$. On the insets: the number of conditionality η_0 of the matrix η_{ij}.

The size dependences of the fluctuations can be considered on the graphs connecting the values of the elements $M\eta_{ij}$ of the matrix of root-mean-square deviations with the total degree of filling of the surface θ, shown in Fig. 32.2 $x_1 = 0.15$ (a) and 0.5 (b). The curves are presented in the above form, so that the values of $M\eta_{ij}$ are commensurable. Thus, the magnitudes of the elements of the matrix decrease with increasing size of the section as M^{-1}.

The matrix η_{ij} is symmetric and positive definite. The number of conditionality of the matrix $\eta_0 = \max |\lambda_i| / \min |\lambda_i| \geq 1$, where λ_i are the eigenvalues of this matrix. In numerical methods, the number of conditionality of the matrix determines the sensitivity of the solution of the system of linear equations to the errors of the initial data.

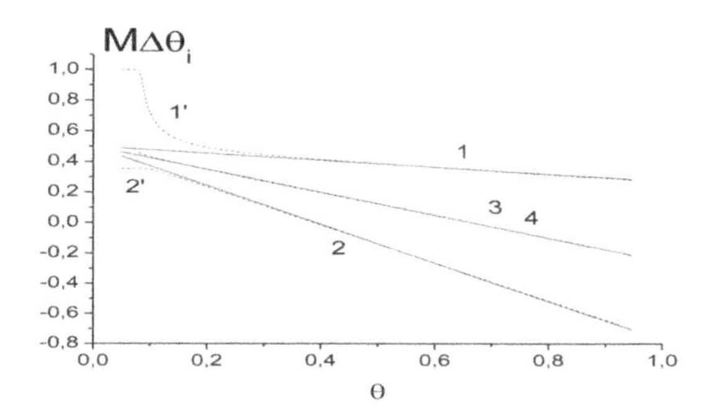

Fig. 32.3. Fluctuation corrections for local densities $\Delta\theta^i$ as a function of the degree of filling θ for $M = 10^4$ are continuous lines and $M = 10^2$ are dotted lines. The notation on the curves 1 – $\Delta\theta^1$, 2 – $\Delta\theta^2$ at $x_1 = 0.15$, 3 – $\Delta\theta^1$, 4 – $\Delta\theta 2$ at $x_1 = 0.5$ for $M = 10^4$. The same notation, but with a prime for $M = 10^2$.

The higher η_0, the more sensitive is the system to the errors of the original data (i.e. to the values of the elements of the matrix).

The effect of the density fluctuations $\Delta\theta_i$ on the values of partial degrees of filling is demonstrated by Fig. 32.3. The reduced values of the fluctuation corrections $M\Delta\theta_i$ are shown. These contributions are maximal in the region of small and large degrees of filling. The eigenvalues of the fluctuation corrections $\Delta\theta_i$ depend on the size of the region as M^{-1}.

The values of the corrections also depend on the molar fractions of the components of the mixture. The larger the differences in molar fractions, the more they differ from each other. By reducing the molar fraction of the first component in the mixture and the total amount of adsorbed material, one can reach N_1 values that correspond to the forbidden values of 0 and 1 in the formulas (32.4) and (32.5). Similarly, for a small fraction of the second component and for small fractions of free sites at large fillings $\theta \to 1$. In this case, the conditions for using the continual description are violated, even with symmetric difference derivatives, and the limiting values of the corresponding filling degrees correspond to a maximum value of $\Delta N_i = 1$ [24], which corresponds to the fixation of all molar fractions upon reaching one of N_i, $i = 1, 2$, and V, values $\Delta N_i = 1$. This factor is important for small systems and its role increases during the transition to non-uniform surfaces.

Non-uniform surfaces. This general case of ideal systems retains the properties indicated above for the adsorption of one substance on the non-uniform surfaces and for the adsorption of mixtures on a uniform surface. Because of the independence of the adsorption process at the sites of different types, all characteristics are summed up by contributions from each type of sites q, and by virtue of the 'meshing' of the molecules of the mixture with each other when competing for free sites of a surface of the same type, we have that the preceding formulas must be rewritten with additional lower indices of sites of type q. The description of non-uniform ideal systems is reduced to the summation of the contributions of any characteristics on individual faces. The formulas obtained above can be easily generalized by summing with weights F_q for the contributions of different types of centres q ($M_q = MF_q$) and replacing the partial fillings θ_i by the local partial fillings $\theta_q^i = N_q^i/M_q$ related to sites of type q, $1 \le q \le t$, t is the number of types of sites.

The generalization of formulas (32.3) for local partial isotherms to the case of non-uniform surfaces is written as

$$a_q^i P_i = \frac{\theta_q^i}{\theta_q^V}\left[\frac{\theta_q^V(\theta_q^i + 1/M_q)}{\theta_q^i(\theta_q^V + 1/M_q)}\right]^{1/2}, \tag{32.10}$$

The elements of the dispersion matrix η_{qp}^{ij} in which the mean values θ_q^i are found from the solution of the system of equations (32.10), are determined in the form

$$\eta_{qp}^{ij} = -1/2\frac{\Delta^2 \ln P(N_q^i, N_p^j, M)}{\Delta N_p^j \Delta N_q^i} = \begin{cases} D_q(i) + D_q(V), & i = j, q = p \\ D_q(V), & i \ne j, q = p \\ 0, q \ne p, \text{any } i,\ u,\ j \end{cases} , \tag{32.11}$$

where $D_q(i) = \frac{1}{4}\ln\left[\frac{\theta_q^i(\theta_q^i - 1/M_q)}{(\theta_q^i + 1/M_q)(\theta_q^i + 2/M_q)}\right]$, including $= V$.

As for a uniform surface, the size effects appear in small areas of individual faces due to their limited size and due to fluctuations of these local partial degrees of density filling. In general, the degree of filling of type q sites by molecules of type i can be represented in the form $\theta_q^i(\text{fl}) = \theta_q^i + \Delta\theta_q^i$, where the degree of filling θ_q^i is found from the solution of the system of equations (32.10), and the fluctuation corrections of the local partial fillings $\Delta\theta_q^i$ are found from the following expression

$$\Delta\theta_q^i = -\frac{1}{2M_q}\frac{\partial \ln \det}{\partial \ln(p_i)} = -\frac{1}{2M_q}\sum_{q=1,k=1}^{t,(s-1)}\frac{\partial \ln \det}{\partial \theta_q^k}\frac{\partial \theta_q^k}{\partial \ln(a_q^i p_i)} \qquad (32.12)$$

This formula is a generalization to the non-uniform surfaces of the fluctuation correction for a pure substance, in which the double sum explicitly takes into account the number of independent components of the mixture $1 \leq i \leq (s-1)$ located on sites of the type q, $1 \leq q \leq t$.

The noted peculiarities of the influence of the boundedness of segments of uniform faces are mainly preserved for a non-uniform surface. To illustrate this fact, we give, for example, Fig. 32.4, as a generalization of the calculations in Fig. 32.3, for local partial fluctuation corrections to the degree of filling of $\Delta\theta_q^i$ on a non-uniform surface consisting of two types of centres: $t = 2$, $F_1 = F_2 = 0.5$, $Q_2^i = Q_1^i/2$.

The local partial and average partial isotherms of adsorption are shown in Fig. 32.5. These isotherms have a simple Langmuirian form of saturation curves. By varying the proportion of sites of different types and the binding energies, a wide spectrum of the isothermal binary mixture can be obtained (curves 3 and 6 refer to the average partial isotherms).

A characteristic feature of the curves in Fig. 32.4 is the presence of large sections with a fixed limiting value of the deviation of the fluctuation increment of the local partial density. This feature is enhanced with a decrease in the mole fraction of any of the components of the mixture. This tendency is preserved with increasing number of types of different centres t.

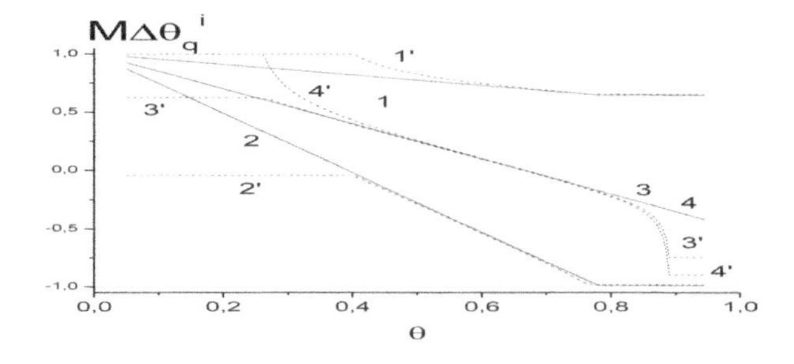

Fig. 32.4. Fluctuation corrections for local densities $\Delta\theta_q^i$ as a function of the degree of filling θ for $M = 10^4$ are continuous lines and $M = 10^2$ are dotted lines. The notation on the curves 1 – $\Delta\theta_q^1$, 2 – $\Delta\theta_q^2$ at $x_1 = 0.15$, 3 – $\Delta\theta_q^1$, 4 – $\Delta\theta_q^2$ at $x_1 = 0.5$ for $M = 10^4$. The same notation, but with the prime for $M = 10^2$.

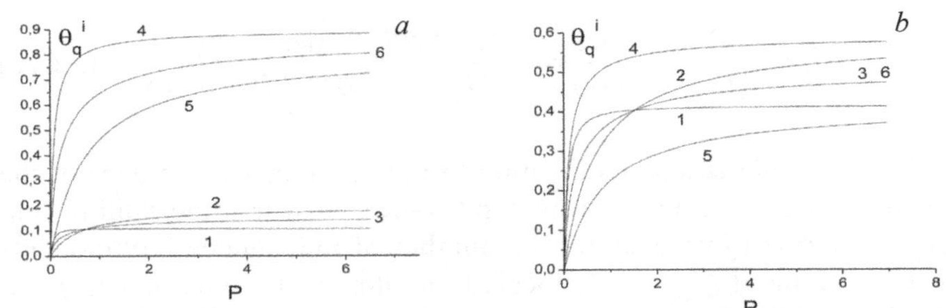

Fig. 32.5. Dependences of the local densities θ^i_q on the total pressure P at $x_1 = 0.15$ (a) and $x_1 = 0.5$ (b). The notation on the curves: $1 - \theta^1_1$, $2 - \theta^1_2$, $3 - \theta^1$, $4 - 1 - \theta^2_1$, $5 - \theta^2_2$, $6 - \theta^2$.

Equations (32.11) show that the so-called limiting or 'forbidden' values $N^b_q = 0, 1, M_q - 1, M_q$ play an important role in calculating the effects of fluctuations on non-uniform surfaces, for which the right-hand sides of expressions (32.11) are not defined . An increase in the number of types of centres t inevitably leads to an increase in the total number of surface sites, which include the 'forbidden' numbers of molecules $N^b_q = 0, 1, M_q-1, M_q$, for large and small fillings of the faces of each type of centre. Their number is $N^b = 4t$, so with increasing t the total number of boundary values N^b increases.

The more non-uniform the surface with respect to bond energies, the more significant this factor is, since during the filling of one face the other faces will remain practically free or vice versa, almost completely filled. In both situations, the role of fluctuations increases. In this paper, as the simplest method for approximating the density description in the region of 'forbidden' values of N^b_q, the limit values of ΔN^i_q are used so that they are equal to one – the limiting value for the minimum value of $N^i_q = 1$ (that is, for an ideal gas of particles or vacancies).

33. Non-ideal systems

Mean-field approximation. Let us consider the fluctuation effects in the simplest case of adsorption of one substance ($s = 2$) on different faces of a single crystal, taking into account lateral interactions in the mean-field approximation. This leads to the fact that the parameters of the lateral interaction ε_{qq} can depend on the type of the face q, $1 \le q \le t$, t is the number of site types [28], but each face is considered isolated. The symbol M is used to denote the total number of surface sites consisting of sections of size M_q, $\sum_{q=1}^{t} M_q = M$. In this version of the theory, the Hamiltonian (30.1) can be rewritten in the form [13]

$$H = \sum_{f=1}^{M}\sum_{i=1}^{2}\left(v_q^i \gamma_f^i \eta_f^q - \frac{1}{2}\sum_{g,j} \varepsilon_{qq} \gamma_f^i \gamma_g^j \eta_f^q \eta_g^q \right). \tag{33.1}$$

where the index g refers to z_{qq} the number of nearest neighbours of the site f on the face q.

We will consider the distribution of molecules in the grand canonical ensemble (μ, M, T), where the symbol μ is the chemical potential of the molecules in the bulk phase. Each site can be occupied or free. We denote by N_q the number of adsorbed molecules at sites of type q. The number of free sites of type q is denoted by $N_q^V = M_q - N_q$.

The expression for the local functions Q_q (N_q, M_q, T) for faces of type q is written as

$$Q_q\left(N_q, M_q, T\right) \; = \; M_q! / \left(N_q! N_q^V!\right) J_q^N \exp(z_{qq}\beta\varepsilon_{qq} N_q^2 / 2M_q)). \tag{33.2}$$

This implies the change of equation (31.2) to the following expression

$$\begin{aligned} \ln P_q(N_q) \; &= \; \Sigma_{q=1}^{t}\{\ln M_q! \; - \ln N_q! \; - \\ &-\ln N_q^V! \; + \beta\mu_q^{i*} N_q + z_{qq}\beta\varepsilon_{qq} N_q^2 / (2M_q)\}. \end{aligned} \tag{33.3}$$

Omitting the intermediate calculations, we write out the equation for the local isotherm, taking into account both lateral interactions and size effects [15]

$$\beta\mu_q^* = \ln\left[\frac{N_q}{N_q^V}\right] - \frac{1}{2}\ln\left[\frac{N(N_V+1)}{(N+1)N_V}\right] - z_{qq}\beta\varepsilon_{qq} N_q / M_q. \tag{33.4}$$

The second derivative is defined as

$$\frac{\Delta^2 \ln P_q(N_q, M_q)}{\Delta N_q^2} = \frac{1}{4}\ln\left[\frac{N_q(N_q-1)N_q^V(N_q^V-1)}{(N_q+1)(N_q+2)(N_q^V+1)(N_q^V+2)}\right] + z_{qq}\beta\varepsilon_{qq}. \tag{33.5}$$

It follows from Eqs. (33.4) and (33.5) that in the absence of correlation effects between adsorbed molecules, the contributions constructed for the size effects retain their form. However, the presence in these equations of the contribution from lateral interactions can substantially change the solutions of the equations obtained in comparison with the solution of the equations of Section 31.

As an example, let us indicate how the equation (31.14) changes by the fluctuation correction $\Delta\theta_q = \theta_q - \theta_q^{\infty}$ in the first order in M_q^{-1}:

$$\Delta\theta_q = \frac{M_q^2(M_q - 2N_q)}{2[M_q^{\ 2} - \beta\varepsilon_{qq}z_{qq}N_q(M_q - N_q)]^2} = \frac{(1-2\theta_q)}{2M_q[1-\beta\varepsilon_{qq}z_{qq}\theta_q(1-\theta_q)]^2}. \tag{33.6}$$

In the second equation, the size component is explicitly distinguished. As the size of the face increases, the differences in the fillings decrease. The second equation also allows us to rewrite this correction by selecting the equation $\beta\varepsilon_{qq}z_{qq}\theta_q(1-\theta_q) = 1$ for the spinodal curve in the denominator, which separates the 'metastable' region and the region of thermodynamic instability of the stratifying molecules. The magnitude of the correction on a section of type q depends on the nature of the intermolecular interaction.

If $\varepsilon_{qq} < 0$, which corresponds to the case of ordered chemisorbed molecules, then with the contribution of the lateral interaction $|\beta\varepsilon_{qq}z_{qq}|$ the value of the denominator increases with respect to the denominator for an ideal system and the effect of lateral interaction coincides with the effect of increasing the size of the face. If $\varepsilon_{qq} = 0$, then we return to equation (31.14).

In the case of attraction between $\varepsilon_{qq} > 0$ molecules, the denominator decreases with increasing $\beta\varepsilon_{qq}z_{qq}$, so the influence of the lateral interaction and the increase in the size of the face are directed in opposite directions. As the current value of the local density θ_q approaches the given face q to the spinodal curve, the behaviour of the correction depends on the ratio of the quantities M_q and the square bracket. It can be seen that the decrease in the denominator of the density occurs very rapidly (according to a quadratic law), therefore, for any fixed value of M_q there exists a density such that the correction (33.6) increases sharply. Obviously, this cooperative behaviour of the system depends on the size of the region according to equation (33.4). A more precise description of the phase behaviour of molecules for small systems requires the use of more accurate approximations than the approximation of the molecular field.

Quasi-chemical approximation. The formulation of the problem remains, but now we need to take into account the correlation effects [29]. The mean value of the number of pairs of sites of different types qp is denoted by M_{qp}. The connection between the number of pairs of sites of different types has the form $\sum_{p=1}^{t\prime} M_{qp} + 2M_{qq} = z_q M_q$ where z_q is the number of neighbours of a site of type q, the sign of the prime of the sum means the absence of a term with $p = q$, $1 \le p \le t$. If we introduce the numbers z_{qp} of sites of type p near the site of type q that characterize the local structure of the non-uniform surface, then their relation to the numbers of pairs of M_{qp} sites of type qp is given as $z_{qp} = (1+\Delta_{qp})M_{qp}/M_q$ or $M_{qp} = z_{qp}M_q/(1+\Delta_{qp})$, where Δ_{qp} is the Kronecker symbol. The total balance of pairs of sites is written in the form of two sums $\sum_{q=1}^{t} M_{qq} + \sum_{qp=1}^{t^*} M_{qp}$.

As above, J_q is the partition function of a particle in a site of type q; $1 \le q \le t$, J is the partition function of the molecule in the gas phase, $\mu = \beta^{-1}\ln(\beta P/J)$, $\beta = (kT)^{-1}$, P is its pressure; ε_q is the binding energy of a particle with a site of type q on the surface of the adsorbent. The parameters of the lateral interaction ε_{qp} depend on the type of the pair qp, $1 \le q, p \le t$, on which there are two adjacent adsorbed molecules.

The expression for the partition function of the non-uniform system in the QCA, taking into account interactions only between nearest neighbors, is written as $Q = Q_1 Q_2$, where

$$Q_1 = \prod_{q=1}^{t} \sum_{N_q, N_{qq}=0}^{M_q, M_{qq}} Q_{qq}, \quad Q_2 = \prod_{qp=1}^{t^*} \sum_{N_{qp}=0}^{M_{qp}} Q_{qp}, \tag{33.7}$$

The first factor Q_1 refers to the distribution of adsorbed molecules at centres of types q: the sum over particles A is taken over all possible values of N_q from zero to the full filling of all sites of M_q of a given type q, and their pairs ij = AA, AV, VA and VV on pairs of sites of one type qq. The second factor Q_2 refers to the cross pairs of sites of different types qp ($q \ne p$), on which the same particles ij = AA, AV, VA and VV are found. The number of such pairs of sites for $q \ne p$ is renumbered (the order of the indices does not play the role of $M_{qp} = M_{pq}$) and their total number is denoted by t^*. The summation is carried out over the numbers of pairs N_{qp} of neighbouring particles AA from zero to M_{qp}.

These two types of factors in (33.7) consist of the following contributions

$$Q_{qq} = C_{M_q}^{N_q} J_q^{N_q} \left(C_{M_q}^{N_q}\right)^{-z_{qq}} \frac{M_{qq}!}{N_{qq}^{AA}!N_{qq}^{VV}![(N_{qq}^{AV}/2)!]^2} \exp(\beta\varepsilon_{qq}N_{qq}^{AA}), \qquad (33.8)$$

$$Q_{qp} = \left(C_{M_q}^{N_q}\right)^{-z_{qp}} \left(C_{M_p}^{N_p}\right)^{-z_{pq}} \frac{M_{qp}!}{N_{qp}^{AA}!N_{qp}^{AV}!N_{qp}^{VA}!N_{qp}^{VV}!} \exp(\beta\varepsilon_{qp}N_{qp}^{AA}) \qquad (33.9)$$

where $N_q^V = M_q - N_q$ is the number of free sites of type q, and $C_{M_q}^{N_q} = \frac{M_q!}{N_q!N_q^V!}$ is the number of combinations of M_q sites of type q by the number of adsorbed N_q particles at these sites.

Correcting factors $\left(C_{M_q}^{N_q}\right)^{-z_{qq}}$ and $\left(C_{M_q}^{N_q}\right)^{-z_{qp}} \left(C_{M_p}^{N_p}\right)^{-z_{pq}}$ are necessary for the refinement of the entropy factor, since the number of independent pairs is overestimated in the QCA [3, 13]. In the case of $\beta\varepsilon_{qp} \to 0$, they lead to an exact solution corresponding to a chaotic distribution of molecules over sites of different types $\tilde{N}_{qp}^{ij} = M_{qp}N_q^iN_p^j/(M_qM_p) = z_{qp}N_q^iN_p^j/[(1+\Delta_{qp})M_p]$, where i, j = A, V [13, 30].

The balance of the number of pairs N_{qp}^{ij} of different types entering into expressions (2) and (3) for statistical weights in the quasi-chemical approximation is expressed as

$$N_{qq}^{AV} = N_{qq}^{VA} = z_{qq}N_q^A - 2N_{qq}^{AA}, \quad N_{qq}^{VV} = z_{qq}M_q/2 - z_{qq}N_q^A + N_{qq}^{AA} \qquad (33.10)$$

for pairs of sites consisting of sites of the same type qq, and

$$\begin{aligned} N_{qp}^{AV} &= z_{qp}N_q^A - N_{qp}^{AA}, N_{qp}^{VA} = z_{pq}N_p^A - N_{qp}^{AA}, \\ N_{qp}^{VV} &= z_{pq}M_p - z_{pq}N_p^A - z_{qp}N_q^A + N_{qp}^{AA} \end{aligned} \qquad (33.11)$$

for pairs of sites consisting of sites of different types ($q \neq p$). The sum of all pairs of particles of different sorts satisfies the normalization condition $N_{qp}^{AA} + N_{qp}^{AV} + N_{qp}^{VA} + N_{qp}^{VV} = z_{qp}M_q = z_{pq}M_p = M_{qp} = M_{pq}$.

To analyze the fluctuational contributions, we consider the probability of the system $P(\{N_q, N_{qp}\}) = Q\exp[\beta\mu N]$ to be in concrete states in the grand canonical ensemble, where $N = \sum_{q=1}^{t} N_q$:

$$\ln P(\{N_q, N_{qp}\}) = \Sigma'_{q=1} \ln P_q(\{N_q, N_{qq}\}) + \\ + \Sigma^{t^*}_{qp=1} \ln P_{qp}(\{N_q, N_p, N_{qp}\}), \tag{33.12}$$

where two types of summands correspond to the contributions of Q_1 and Q_2.

Let us find the minimum conditions for ln$P$ ($\{N_q, Nq_p\}$), depending on the independent variables $X(= N_q, N_{qp}, N_{qp})$ and for its second derivatives characterizing the dispersion matrix of the distribution function $P(\{N_q, N_{qp}\})$. For small particles it is necessary to use symmetric difference derivatives instead of the usual differential derivatives [14,15]. The extremum condition $\Delta \ln P(\{N_q,N_{qq}\})/\Delta N_q = 0$ condition gives the equation for the local adsorption isotherm, which connects the chemical potential of the system ($\mu_q = \mu + \beta^{-1}\ln(J_q)$) specified by the thermostat, with the number of molecules on each face N_q (or the degree of local filling $\theta_q = N_q/M_q$)

$$\beta\mu_q = \frac{1}{2}\times \\ \times\left[\left(1-\sum_p z_{qp}\right)\ln\frac{N_q^A(N_q^A+1)}{N_q^V(N_q^V+1)} + \ln\frac{(N_{qq}^{VV}-z_{qq})![(N_{qq}^{AV}+z_{qq})/2!]^2}{(N_{qq}^{VV}+z_{qq})![(N_{qq}^{AV}-z_{qq})/2!]^2} + \right. \\ \left. + \sum_{p\neq q}{}'\ln\frac{(N_{qp}^{VV}-z_{qp})![(N_{qp}^{AV}+z_{qp})!]}{(N_{qp}^{VV}+z_{qp})![(N_{qp}^{AV}-z_{qp})!]}\right]. \tag{33.13}$$

The extremum conditions for each type of N_{qq}^{AA} $1 \leq q \leq t$, and N_{qq}^{AA}, $1 \leq (qp) \leq t^*$ $(\Delta \ln P(\{N_q, N_{qp}\})/\Delta N_{qp}^{AA} = 0)$ pairs are followed by equations on the relationship between pair functions:

$$\beta\varepsilon_{qq} = \frac{1}{2}\ln\frac{N_{qq}^{AA}(N_{qq}^{AA}+1)N_{qq}^{VV}(N_{qq}^{VV}+1)}{[N_{qq}^{AV}/2(N_{qq}^{AV}/2+1)]^2}, \\ \beta\varepsilon_{qp} = \frac{1}{2}\ln\frac{N_{qp}^{AA}(N_{qp}^{AA}+1)N_{qp}^{VV}(N_{qp}^{VV}+1)}{N_{qp}^{AV}(N_{qp}^{AV}+1)N_{qp}^{VA}(N_{qp}^{VA}+1)}. \tag{33.14}$$

These equations transform into known macroscopic expressions for $N_{qq}^{AA} \gg 1$. Equations (33.13) and (33.14) define a system of equations taking into account the limited size of the faces of microcrystals and the lateral interaction of molecules.

The description of the adsorption of molecules in small regions of a non-uniform surface with an allowance for lateral interactions and taking into account quadratic equilibrium fluctuations is constructed by analogy with the known formulas [2,3]. We construct formulas for the elements of the dispersion matrix $\eta_{km} = -1/2 \dfrac{\Delta^2 \ln P(\{N_q, N_{qq}\})}{\Delta X \Delta Y}$ where the symbols X and Y refer to all independent variables N_q, N_{qq} and N_{qp}, (the index k is determined by a list of θ_q, θ_{qq}, θ_{qp}) whose mean values are found from solutions of the system of equations (33.13) and (33.14).

Knowing η_{km} one can obtain the corrections ΔX_k (i.e. $\Delta\theta_q$, $\Delta\theta_{qq}$ and $\Delta\theta_{qp}$) to the degrees of filling of sites of different types and their pair probabilities due to fluctuations as

$$\Delta X_k = -\frac{1}{2}\frac{\partial \ln \det}{\partial \ln(\lambda_k)} = -\frac{1}{2}\sum_{b=1}^{T_D} \frac{\partial \ln \det}{\partial X_b}\frac{\partial X_b}{\partial \ln(\lambda_k)} \tag{33.15}$$

where det is the determinant of the dispersion matrix made up of the elements η_{km}, its dimension $T_D = 2t + t^*$ is equal to the number of independent variables N_q, N_{qq} and N_{qp}, and express the degree of filling of sites with an allowance for the fluctuations X_k (fl) = $X_k + \Delta X_k$. In the formula (33.15) we use $\lambda_k = \exp(\beta\mu)$ for N_q^A; $\lambda_k = \exp(\beta\mu_{qq})$ for N_{qq}^{AA}; $\lambda_k = \exp(\beta\mu_{qp})$ for N_{qp}^{AA}. To calculate the derivatives, we use the system $\partial X_b/\partial \ln \lambda_k$ of equations (33.13) and (33.14).The derivatives $\partial \ln \det/\partial X_b$ are calculated numerically.

A small drop. The theory was applied to calculate the influence of fluctuations on the characteristics of drops, including the value of the minimum radius R_0 of the drop formed [31]. Taking into account the reality of the quantities $z_{qp}(R)$ (see Section 24), the gamma functions [32] were used instead of the factorials in the calculations using formulas (33.12)–(33.15). The calculation was carried out for both an equilibrium and a metastable drop.

It is found that in the vicinity of the appearance of thermodynamically stable drops for different temperatures, while the volume of individual monolayers and their fluctuations are limited, the size of the drop is larger than when using macroscopic averages for local densities and mean pairs of particles [33]. The nature of this change in the dimensions of R_0 is reflected in Table 33.1 (the value of T_{crit} refers to the critical temperature in a macroscopic volume). The data of the table demonstrate a rather strong influence of the limited size of the system on the values of the drop radii corresponding to the condition that the surface tension σ of the drop is zero, i.e. when

Table 33.1. The minimum dimensions of the liquid phase of the drop (R_0) in the metastable vapour at different temperatures $\tau = T / T_{crit}$ [31]

τ	0.60	0.66	0.73	0.79	0.86
R_0	8	5	5	6	9
R_0(fluct)	16	15	13	12	13

the volume of spherical monolayers is limited, the drop remains thermodynamically unstable for a larger size.

The analysis showed that dimensional effects (smallness of the system and its fluctuations) appear at values of the radii of drops R of the order up to 40λ. This value is in full agreement with the lower limit of drop sizes, for which it is incorrect to apply the equations of thermodynamics at radii smaller than $R_t \sim 41\lambda$ [12] (see next Section).

34. The lower limit of the applicability of thermodynamics [12]

The mathematical apparatus used in thermodynamic equations to describe the curvature of any local area of the interface [19, 34] is an apparatus of differential geometry that operates second-order continuum derivatives. In all the equations of thermodynamics and the mechanics of continuous media for volume phases and interfaces, it is assumed that there are no fluctuations. A comparison of the continual and discrete descriptions should lead to conditions under which the contributions of fluctuations disappear. For a more rigorous estimate of R_t, we compare the thermodynamic properties of a certain surface region from the point of view of (A) thermodynamics and (B) molecular theory.

(A) From the point of view of the theory of a continuous medium, we consider the section representing an elementary minimal area, by means of which the 'total surface' is covered (by Borel's lemma). Recall that, according to this lemma [19], if the complete closed interval $[a, b]$ is covered by an *infinite* system of open intervals (that is, without inclusion of limit points of a given interval), then it is always possible to extract a *finite* subsystem of open intervals from it, which also covers the entire interval $[a, b]$. This means that for a continuum of interior points of the finite interval under consideration one can construct a covering consisting of a finite number of interior intervals. This formulation of Borel's lemma is written out for simplicity in the case of one dimension. Its generalization to a two-dimensional surface is verbatim, replacing the term 'interval' by the term 'region', and has the same meaning.

(B) From the point of view of a discrete medium, the same matching region is the minimum region on which the thermodynamic function, its first and second derivatives must be determined (the second derivative is also needed to describe the curvature of the surface). According to the approach to discrete systems described above, such a region is a local domain containing in one dimension a minimal number of sites $L = 4\lambda$. For a cylindrical surface, we have one dimension and two dimensions for a spherical surface.

Comparison of the two methods of describing the minimum portion of the surface (A) and (B) gives the answer to the desired radius of the drop.

Cylindrical surface. In any cross section of a cylindrical drop, we have a circle that is approximated by a correctly inscribed polygon containing N_s – the number of sides, with the base length $L = 4\lambda$ (Fig. 34.1 *a*). We denote by the symbol ξ the accuracy of the coincidence of the length of the polygon and the circumference of the circle, then it is easy to see that $R_t/\lambda = N_s L$, where $N_s = \pi/(6\xi)^{1/2}$ is determined by the accuracy ξ of the description of the length of circumference by the broken line. From the physical point of view, the accuracy ξ determines the number of particles on the dividing surface, more precisely, the line in the cross section of the cylinder.

The procedure for writing a polygon into a circle quickly converges. When $N_s = 4$, we have the accuracy $\xi \sim 10\%$, and for $N_s = 6$ – the accuracy is $\xi \sim 4.5\%$, which corresponds to $R_t = 16$ and 24λ, for which $\eta \sim 1.8\%$ and 1.2%. Obviously, these values refer to fairly crude approximations. The accuracy of calculating the concentration profiles of drops $\{\theta_q\}$ is, as a rule, no lower than $\sim 0.1\%$, therefore, increasing the accuracy of the number of particles on the dividing line, we have for $N_s = 12$ (accuracy $\xi \sim 1\%$, $R_t = 48\lambda$ and $\eta = 0.6\%$) and $N_s = 24$ (accuracy $\xi \sim 0.5\%$, $R_t = 96\lambda$ and $\eta = 0.3\%$). We note that in the latter case the rms fluctuation η is even greater than the accuracy of calculation of the concentration profile.

Thus, with an increase in the accuracy of matching the number of particles on the dividing line, the size of the drop corresponding to the lower limit of the applicability of thermodynamics increases quite rapidly.

Spherical surface. For a spherical drop, the second dimension must be taken into account. In this case, the area $S_0 \sim L^2$ corresponds to an elementary area. We approximate the surface of a spherical drop

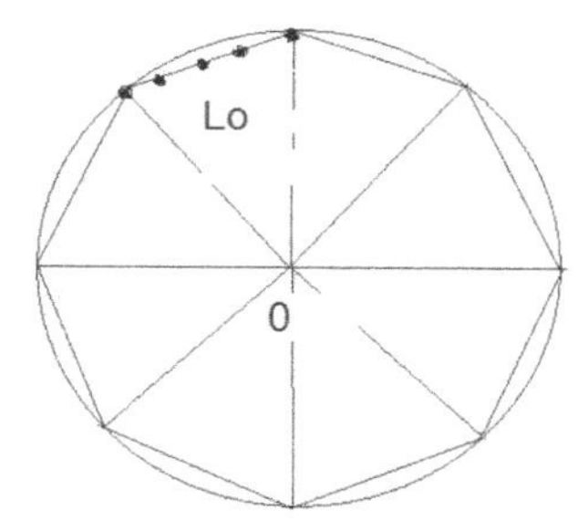

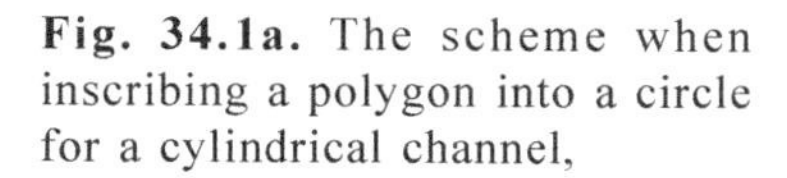

Fig. 34.1a. The scheme when inscribing a polygon into a circle for a cylindrical channel,

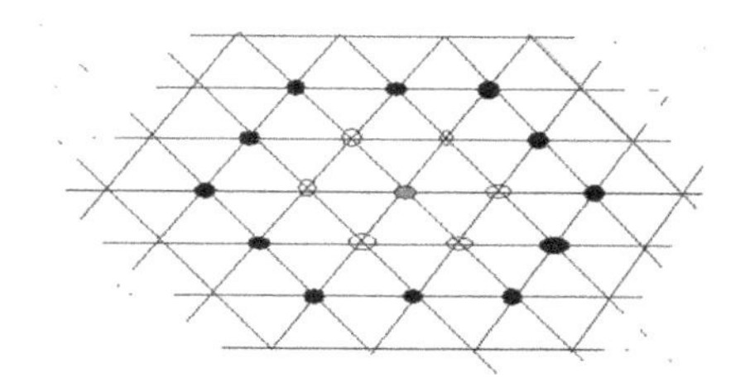

Fig. 34.1b. Elementary platform for the surface of the sphere. The gray circle is the centre, the light circles are the first neighbours, the dark circles are the second neighbours on the lattice z = 6.

of radius R by the set of such elementary areas (as 'mathematical points – analogues of coverings in Borel's lemma). The most dense two-dimensional packing of molecules corresponds to a structure of $z = 6$ with a communication length between the cells λ. This structure has an anisotropic area formed by cells in the redistribution of two coordination spheres around the central cell. The maximum diameter of an elementary area with such a structure is equal to a segment of length L (Fig. 34.1 *b*). The minimum size of the site is $2\times3^{1/2}L$. This leads to a value $S_0 = 6\times3^{1/2}\lambda^2$.

Each flat elementary area can serve as the basis for a cone, which is drawn from the centre of the sphere. To simplify the estimates, we assume that the elementary area has the form of a circle, then the surface of the spherical part of the cone resting on this area is a figure of rotation and can be easily determined [19]. We denote it by S_c. The total number of cones on the surface of a sphere of radius R is equal to the number of molecules $N_c = 4\pi R^2/S_c$, coinciding with the number of sites approximating the sphere. You can enter the effective radius of the circle, equal in area of the considered area, defining it as $R_{ef} = (S_0/\pi)^{1/2}$. Then the quantity $S_c = 2\pi R^2$ $[1-\{1-(R_{ef}/R)^2\}^{1/2}]$. Expanding the expression for S_c under the radical to the second term of smallness, we obtain the relationship between (S_0-S_c) and the ratio (R_{ef}/R).

Taking into account the proportionality of the number of molecules and the area of the flat elementary area, we require that the difference in the areas of all flat areas and all the spherical parts of inscribed cones satisfy condition

$$(S_0 - S_c) / S_0 < \xi_s, \tag{34.1}$$

where ξ_s is the relative accuracy of the surface area of a drop of a given radius R_t (where $\xi_s \sim \xi^2$). This condition is easily achieved by increasing the radius of the sphere. The condition for the smallness of the density fluctuations at the boundary of the minimum sizes of the applicability of thermodynamics is verified indirectly in comparison with the value of η – the relative mean square fluctuation [2] of the number of molecules N_c located on the surface of a sphere of radius R_t. The quantity η must certainly be less than the possibility of an accurate experimental determination of quantity σ.

As a result, the expression connecting the size of the diameter of the sphere R_t and the accuracy of the description of the surface area of the sphere ξ_s: $R_t/\lambda = [6\times3^{1/2}/(2\pi\xi_s)]^{1/2} = 1.29/\xi_s^{1/2}$ follows from the condition (34.1) measured in units of cell length λ.

This expression corresponds to the following sets of values: $\xi_s \sim 0.1\%$, $R_t = 41\lambda$ and $\eta \sim 0.7\%$; $\xi_s \sim 0.05\%$, $R_t = 82\lambda$ and $\eta = 0.34\%$; $\xi_s \sim 0.01\%$, $R_t = 129\lambda$ and $\eta \sim 0.2\%$.

Taking into account that the accuracy of calculations of the system of equations in the molecular theory [35–37] is not lower than 10^{-3} or 0.1%, we get $R_{t1}/\lambda = 41$, which, apparently, satisfies the existing experimental methods [22,23,38,39]. In the case of a significant increase in the experimental possibilities for a more accurate determination of the surface of tension, it is necessary to reduce the criterion value to $\xi = 10^{-4}$, which leads to an increase in the radius to $R_{t2}/\lambda = 129$ and corresponds to a value of $\eta \sim 0.2\%$. .

The obtained estimate is not related to the details of the molecular distribution and does not depend on either the choice of the surface of tension or the temperature. Near the critical point, large spontaneous fluctuations of the molecules are realized, and the probability of the existence of an isolated spherical drop in this situation is negligibly small. Nevertheless, this estimate correlates well with the results of numerical calculations [40]: up to temperatures $\tau = 0.99$, the width of the transition layer, both for a plane boundary and for drops, does not reach this value. Therefore, it should be assumed that for $R_{t1} < 41\lambda$ the thermodynamic description is not justified. For argon atoms, this corresponds to $R_{t1} \sim 16$ nm. Assuming that at $\xi_s \sim 0.05\%$, both the discrete nature of the matter and the contributions of the fluctuations can be neglected, this gives grounds for using the thermodynamic description for $R_{t2} > 80$–100λ (or $R_{t2} \sim 40$ nm for argon atoms). Thus, the minimum size of the drop radius R_t, from which the thermodynamic description can be used, is 16 to 40 nm. These estimates are in good agreement with the initial postulates of thermodynamics about the need for the presence of sufficiently large amounts of matter in the system.

From surface to volume. Traditionally, a qualitative discussion of the role of fluctuations uses the estimate of the relative mean-square fluctuation for the Poisson distribution $\eta_p = \langle N^2 \rangle^{1/2}/N = N^{-1/2}$ [1–4]. The main condition in this distribution is the absence of correlations between molecules. The same estimate corresponds to an ideal gas of molecules. It is constructed for a macroscopically small subsystem in which any number of molecules can exist, provided that it is small in comparison with the total number of molecules in the complete system from which this subsystem is allocated.

Above in Table 29.1 and in the analysis of ξ and ξ_s, when M was varied, η was used, which is the analogue of the numbers η_p, which was calculated for the number of molecules N in the system in the form of a small compact body (rather than isolated molecules). Let us discuss the correspondence between the values of η and η_p for small systems. To do this, let us consider how the magnitude of the mean-square fluctuation η of a uniform system containing M centres changes in the problem of equilibrium filling of a monolayer on a dividing surface. These calculations refer to a small system of non-interacting molecules [24]. (Similar results are obtained by the mean-field approximation, which takes molecular interactions into account, but in a crude manner [15].)

Figure 31.2 *a* shows the concentration dependences of the function D directly related to the value of η as $D = -2\eta$ [15,24], which depends on the size of the region M and on the number of molecules N in the given region ($N_V = M$–N). The relation with the quantities η_p follows from the equality $N = \theta M$. The maximum values of D refer to the region of small and large degrees of filling θ. In the region of average fillings, the value of η is much lower. Fluctuations for a small number of molecules in small systems behave in a similar way, as in macroscopic systems, increase rapidly with decreasing N. As the density increases, for small systems, the same increase in fluctuations is observed as for a small number of molecules. In this case, $N \sim M$, which allows us to use the quantity η instead of η_p for the usual Poisson distribution as in Table 29.1. This fact is related to the presence of vacancies, whereas in real liquids there should be a fluctuation of the discharge regions identical to the fluctuations in the number of low-density molecules–phase inversions correspond to processes of bubble formation in a liquid. (For macrosystems, fluctuations in the region of large densities are usually not discussed.)

For $M = 10^4$, in the region of small and large fillings, there are differences from zero values for 2–3% density. Further, with a decrease in M to 10^3, the difference region D extends from the zero value to 20%. At even smaller values of M, the value of D differs from the macroscopic size in the entire density range, starting at $M = 500–600$. The curves for $M = 10$ show the maximum dispersion differences for small sections. The dotted curve 1′ in Fig. 31.2 refers to a similar relationship obtained with the first dimensional correction $D_1 = [-M/(NN_V)]$. Thus, approximate estimates overestimate the variance.

In order that the density fluctuations can be completely ignored, it is necessary to decrease D in the entire range of θ. With increasing M, the dispersion decreases sharply. So, for $M = 10^5$, the dispersion remains close to zero in the entire density range, then this size can be considered analogous to the value of R_{t1}. For a dividing drop surface, the given M corresponds to a radius $R/\lambda \sim 46$. This value is close to the earlier estimate $R/\lambda = 41$ for $\xi_s = 0.1\%$, which corresponds to $\eta \sim 0.7\%$ (note that $\eta \sim 1\%$ corresponds to $R/\lambda = 28$, see Table 29.1).

The above calculation confirms the universality of the estimates obtained. They are based on the discrete nature of the substance, are not related to the details of the molecular distribution, and do not depend on either the choice of the surface of tension or the temperature. However, a special consideration is required near the critical point.

Internal regions of phases. To go to macrophases, it is necessary to increase the drop radius in order to neglect the contribution of σA to the Gibbs potential (Section 29). Such a transition is consistent with the condition, for example, that the contribution of σA is 0.3 or 0.1%. According to Table 29.1 this refers to the number of molecules of the system $M_{mac} = 4\times10^7–10^9$, which exceeds the number of molecules in the surface regions of drops with radii R_{t1} and R_{t2} from 4×10^2 to 5×10^4 times. It was shown above that the use of the simplest estimate for determining the density fluctuations η in the dense phase (instead of the number of molecules in the gas phase η_P) makes it possible to control the magnitude of the fluctuations depending on the size of the system. For macroscopic volume phases, one can correlate the continual and discrete descriptions of a small region of a three-dimensional lattice around the selected site, and repeating the procedure used above for surface properties by word-by-word, obtain estimates for the quantities R^v_{t1} and R^v_{t2} for the volume phase, which in their sense are completely analogous to the values of R_{t1} and R_{t2}, but without taking into account the state of the surface. Restricting ourselves to the continuum constraints $R^v_{t1} = (3R^2_{t2})^{1/3}$ (similarly for R^v_{t2}) we have the radii of the inner regions $R^v_{t1} = 17\lambda$ and $R^v_{t2} = 29\lambda$.

For macrophases, the question of the size of the internal local region is related to the number of molecules in the region $dV = V_m(R^v_t)$, which is implied in all methods of a continuous medium:

1) In the theory of the formation of new phases (nucleation) and the theory of chemical processes in condensed phases, expressions are constructed for the probability of realizing these processes in a fluctuation manner. It is important to distinguish between the concept of fluctuations as the realization of a multiparticle event consisting of a multitude of phase molecules (for example, the appearance of a phase nucleus, or a collective reorganization of a medium in the process of electron transfer), and as a factor requiring corrections to allow for the limited size of the region in which the elementary process proceeds, describe these processes.

In the first case, it is sufficient to know the probability of multiparticle configurations, whereas in the second case it is additionally required to take into account the size effects of the fluctuations with respect to the indicated multiparticle configurations.

2) For problems of non-equilibrium thermodynamics, it is necessary to have an estimate of the local region in which the concept of local equilibrium is defined. All non-equilibrium flows are constructed in the form of an expansion of the chemical potential gradient, which is the basic thermodynamic characteristic, and it is necessary to determine the conditions for its correct calculation. At present, the concept of a local area does not have a specific connection to the size V_m, which does not allow linking molecular models to the real conditions of most processes.

3) Intermediate systems comprising 10^5 to 10^9 molecules constitute a wide class of transition systems with a characteristic linear size from $10^2\lambda$ to $10^3\lambda$ that can have spatial non-uniformity in each direction, and the efficiency of modelling such non-uniform systems largely depends on the possibility of using continual or discrete models.

The constructed estimates give answers to the questions posed.

1. The estimates of the radii R^v_{t1} and R^v_{t2} refer to the isolated region within the phase. To analyze the properties of local volumes with a radius less than R^v_{t1}, fluctuations must be taken into account. If the radius of the region is larger than R^v_{t2}, then when analyzing the properties of isolated local volumes, one can neglect the allowance for fluctuations. When there is a large number of such local areas within the system, this is not an isolated small system. Local areas are not distinguished in any way, therefore contributions from them are all equivalent. As a result, in a large ensemble of identical regions with a radius less than R^v_{t1}, small-scale fluctuations can be ignored – it suffices to confine oneself to the average values characterizing the most probable distribution.

2. If we consider local internal regions near macrophase interfaces (for example, in membranes), then the boundaries themselves must in any case be separated from the properties of the internal volume. The contribution of fluctuations depends on the set of local regions in the near-surface regions, i.e. of the cross-sectional area. If the surface is macroscopic, then density fluctuations are also not taken into account – averaging over the cross section removes the effects of fluctuations. However, for small cross-sectional areas, when the areas are commensurable or exceed R^v_{t1} or R^v_{t2} by at least an order of magnitude, fluctuations effects are necessary for $R < R^v_{t1}$ and they can be neglected for $R > R^v_{t2}$.

3. It follows that for a drop of radius $R_{t1} \leq 41$, the effect of density fluctuations on the surface can not be neglected, although fluctuations can be ignored at the centre of the drop with the characteristic size of the sphere $R^v_{t2} = 29$. But in order to obtain the thermodynamic characteristics of a drop, it is necessary to take into account its total volume, and for regions with $R > R^v_{t2}$ fluctuations must already be taken into account because these regions include a surface. In the more general case, even the proximity of the boundaries to the local region (without its inclusion) requires the inclusion of fluctuations.

Thus, in macrophases, the allowance for fluctuations depends on the degree of uniformity of the internal volume of the system: in uniform regions the role of fluctuations is negligible, and in non-uniform macro-regions, the necessary condition for neglecting fluctuations is that the size of the non-uniform region is large, i.e. the total volume exceeds M_{mac}. Otherwise, an independent analysis of the role of fluctuations is required.

We note that the non-uniformity of the system is a natural state of the phase in the case of deformation of solids and in porous bodies, as well as with the combined influence of pores and deformations during the transport of molecules through solid membranes. The estimation of R^v_{t2} actually determines the concept of V_m for the volume phase introduced in the mechanics of continuous media. Accordingly, this situation is generalized to the situation of chemical reactions and nucleation processes, when the role of fluctuations can be neglected, if the total volume of the system is macroscopic, and not for non-equilibrium states: for $R > R^v_{t2}$, local fluctuations can be neglected, and for $R < R^v_{t1}$ – it is impossible. The obtained sizes indicate that for characteristic linear sizes of the interior regions up to $10^2\lambda$, a continual description for transport processes can not be used. While for internal regions of the order of $10^3\lambda$ and more, a continual description of such processes will be justified.

The restriction on the minimum number of macrophase molecules $M_{mac} = 4\times10^7$–10^9 is important for the molecular dynamics and Monte Carlo methods, in which periodic boundary conditions are used, with respect to the 'central' design cell in the modeling of the volume phase. The size of the cells must be larger than R^v_{t2} in order to completely neglect the fluctuations. For smaller cells, theoretical methods for small systems are, in fact, a tool for monitoring numerical methods. Note that, traditionally, in modelling by Monte Carlo and molecular dynamics methods, the bulk properties of small systems are mainly considered, since these methods are technically difficult to perform an analysis of surface properties due to a strong fluctuation.

Dilute gases. Figure 31.2 shows that the dependence $\eta(\theta)$ with decreasing θ in the region of small θ and with increasing θ in the region of large θ has a similar form. In order to proceed to the traditional treatment of a rarefied gas, we recalculate the number of particles M by the volume occupied by the rarefied gas, i.e. is $\sim10^3$ times greater than the volume of the dense phase. Then the substance in the drop with the radius R_{t1} corresponds to the number of 2.1×10^4 molecules, and the drop of R_{t2} corresponds to 10^5 molecules, which corresponds to $(2.1-10)\times10^7$ cells. The latter value corresponds to the lower value of the estimate $M_{mac} = 4\times10^7$.

In the gas phase, the main dimensions are the mean distance between molecules ρ and the mean free path ℓ. The first value is used in equilibrium characteristics, and the second is used in kinetic characteristics. The connection with the degree of filling of the volume θ of these quantities is expressed as $\rho = \lambda/\theta^{1/3}$ and $\ell = \lambda/(2^{1/2}\theta)$ [41,42].

Hence for the rarefied gas range $\theta = 10^{-4}$–10^{-3} it follows that the value of ρ varies from 10 to 22λ. Thus, the region with a radius R^v_{t1} is commensurable with the volume of the gas in which the molecules are at an average distance ρ, but this region is smaller than R^v_{t2}. If we compare the total number of molecules in a cube with a side equal to the mean free path, then $N = \theta L^3 = \theta/(2^{1/2}\theta)^3 = (2^{1/2}\theta^2)^{-1}$, or $N = 7.1\times10^5$–7.1×10^7. The indicated sizes go beyond the permissible values of R^v_{t2}, and this amount corresponds to the meaning of the lower macrophase size.

The meaning of this comparison follows from the traditional condition in the theory of a rarefied gas that the relaxation of the gas to equilibrium occurs at characteristic times of the order of the time of a single collision of molecules. It should be recalled that the very concept of time and mean free path is statistical, and they refer to macroscopic regions in the size and in the number of molecules. The fact that within the volume of a cube with a size equal to the mean free path is $N = 7.1\times10^7$ at $\theta = 10^{-4}$, is in complete correspondence with the kinetic theory of gases.

35. Micro-non-uniform systems

In addition to the fluctuation estimates that affect the characteristics of the system, an important role is played by energy non-uniformities that change the relationship between the number of particles and the values of the thermodynamic functions. In this section, two such examples are considered: the presence of edge atoms between the faces of small microcrystals that affect the values of the surface tension of the entire system and the limited size of the volume of the cavity in which the included or adsorbed phase can be located.

Estimation of the conditions for the application of Wulff's theorem. The equilibrium form of a single crystal, which is in contact with its saturated vapour or melt, is determined by Wulff's theorem, which is written as σ_i/h_i = const [43–49]. Thus, the equilibrium shape of a single crystal is characterized by the fact that its faces are removed from a certain point (the Wulff point) by distances proportional to the surface tension of the faces. This point is unique for a single crystal. The theorem expresses the concrete equilibrium condition of the crystal under isothermal conditions for the constant volume of the system (T = const and V = const). An important basis of this theorem is the assumption of the existence of the most equilibrium concept of 'surface tension' of the faces of an anisotropic solid σ_i, where different faces i have different surface tensions. In Wulff's theorem, only the contributions of individual faces are considered, and the contributions of the edge faces are neglected (they are taken into account in Ref. [49] – they change the position of the Wulff point). In our analysis, the question is posed differently: for what size of a single crystal of a given form can we assume that the contributions from the edges can be neglected practically (the contribution of a finite number of vertices is omitted). That is, we consider the size effect in an non-uniform system in which the edges, characterized by linear tension, change the surface tension of the entire system.

This question is considered on the example of the simplest forms of single crystals: tetrahedral, cubic and octahedral [50]. The cubic form consists of faces (100) and 12 edges, and the remaining two consist of faces (111) and have 6 and 8 edges, respectively, which greatly simplifies the analysis [51]. Any interface is a non-uniform region. To describe the distribution of molecules in the lattice gas model (LGM), structural functions of cell distributions in the transition macroscopic region are constructed, which are characterized by a set of nearest neighbor numbers z_{qp}, denoting the number of neighbouring sites of the layer p around the sites of the layer q; $\sum_{p=q=1}^{q+1} z_{qp} = z$. Thus, for a cubic structure with $z = 6$, the numbers z_{qp} are equal to 4 for $p = q$ and 1 for $p = q\pm1$, $\gamma = 1$; and for the FCC structure with $z = 12$, the numbers z_{qp} are equal to 6 for $p = q$ and 3 for $p = q\pm1, \gamma = 1/\sqrt{2}$.

Surface tension characterizes the excess free energy of the transition region. The surfaces of small microcrystals are non-uniform systems: on their surfaces there are different faces and edges between the faces. We represent $\sigma_{av}S_{av} = \Sigma_i\sigma_iS_i + \Sigma_m\xi_mL_m$, where $S_{av} = \Sigma_iS_i + \Sigma_mL_m$, S is the total surface of the microcrystal, σ_{av} is the average value of the surface tension, ξ_m is the linear tension of the edge of type m. For our simplest structures there are only single types of faces, and the signs of the sums are removed. The average value of the surface tension σ_{av} will be written as $\sigma_{av} = \sigma S\,(1+ \xi L/\sigma S)/S_{av}$. The condition for Wulff's theorem is satisfied if the second term ($\xi L/\sigma S \ll 1$) can be neglected – this will be the energy estimate for the condition of applicability of Wulff's theorem. The simplest geometric estimate of the fulfillment of this inequality can be the ratio $L/S \ll 1$, since the quantities S and ξ are of the same order. To do this, we need to consider how the length of the edge and the area change with the growth of the crystal.

The unit of length should be the lattice parameter λ, related to the potential parameter as $\lambda = 2^{1/6}\sigma_{AA}$, which characterizes the average distance between atoms. We shall measure the crystal size in the numbers of the N atoms forming the single crystal under consideration. In this case, length measurements using the lattice parameter λ and the number of N atoms give the smallest values of the number of atoms located on the edges (N_1) and on the faces (N_2) above the indicated figures: $L/S = N_1/N_2$.

In the field of Fig. 35.1, the N_1/N_2 ratio characterizes the geometric estimate of the contribution of all the edge atoms to all surface atoms of the faces of the lattices under consideration. It can be seen that the value of the L/S ratio can be neglected only in the case of N exceeding $N \sim 150000 \sim (53)^3$ atoms in the crystal – the ratio should be compared with the accuracy of the experimental measurements, and at least should be less than 1%.

For the energy estimate, we take into account that between the atoms of the edge and the atoms of the face there exists a transition region κ as between the surface and the volume. When passing to small crystals by analogy with drops [52, 53], we assume that the width of the transition region decreases approximately by a factor of two. Thus, κ is a quantity of the order of two monolayers. Then the region of edge atoms increases from two sides $(1 + 2\kappa)$ times. At the same change in the assignment of sites, the number of atoms on the faces decreases. In this case, the ratio N_1/N_2 increases approximately 5.1 times. At the same time, the ratio of the energy contributions ξ/σ per site from the free energy of the sites on the edge and on the faces

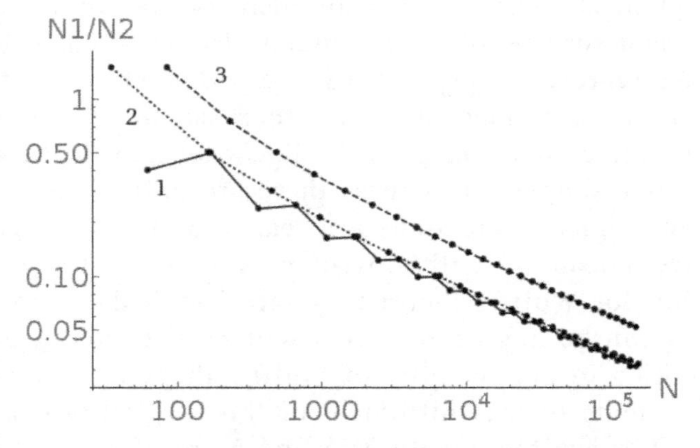

Fig. 35.1. The ratio of the number of edge atoms N_1 to the number of surface atoms N_2 on the faces, depending on the number of atoms of the entire crystal, N (for geometric estimation) [51].

decreases primarily due to a change in the number of bonds for the edge atoms with respect to the average number of sites of the face (which is approximately 1/3 to 1/2). Then the total energy change $\xi L/\sigma S$ can be estimated as a value of the order of 2. The energy estimate gives an overestimate for a given level of deviation accuracy by a factor of two in comparison with the geometric estimate, which is N approximately 3×10^5 or $(67)^3$.

The obtained estimates are in agreement with the earlier obtained results on the analysis of density fluctuations on the drop surface, which made it possible to obtain the lower bound for the applicability of the thermodynamic relationships [12]: for the range of the region radius from 41λ to 90λ (which corresponds to the number of particles from 3.8×10^5 to 29×10^5, accuracy of the description of surface tension is about 0.1%). For a bulk phase, fluctuations in a sphere with a radius of 29λ can be neglected, which is $N \sim 10^5$, and the calculation of van der Waals type clusters has shown [54] a transition to a stable FCC lattice structure at $N \sim 2.5\times10^5$ for $T = 0$ K without allowance for vibrational contributions, which can somewhat reduce this value of N, if we take into account the oscillations.

The estimates obtained for the number of atoms in single crystals are approximately several times higher than estimates based on the obtained value of R_{t2}, which indicates the qualitative agreement of different estimates. It follows from such estimates that as the Miller index increases, the areas of faces decrease, and the contributions of their edge atoms increase. This aspect is ignored in the continual consideration of Wulff's rule [46–48]. This aspect is very important in the transition to more complex forms of surfaces. Thus, in the rhombic dodecahedron, the contribution of the edges decreases noticeably only with an increase in the total number of atoms by approximately an order of magnitude, compared with the numbers obtained above for the simplest geometries considered: tetrahedral, cubic, and octahedral.

A drop in a spherical pore [53]. The limited volume factor plays a role in: 1) adsorption of molecules on small open faces of microcrystals, 2) adsorption of molecules in bound and isolated pores of polydisperse material, and 3) spinodal decomposition of the fluid at short times for large supersaturations of the volume phase, when each local region operates on the average the same. A general approach to the phase state of matter in limited volumes was formed in [55]. Below we consider an example of the effect of volume in the simplest case of a 'drop in a spherical pore' on the phase state of a substance and its surface tension within a pore in the absence of influence of pore walls. In this system there is a single dimensional parameter – the radius of the system.

The parameters of the system are variables: V – volume, N – number of particles and T – temperature. Under isothermal conditions, a decrease in the volume of the system is associated with a change in the quantity V. In this case, the quantity N, or the ratio N/V, can remain constant. In the first case, as the volume decreases, the density of the system changes, and in the second case the number of molecules decreases. With a decrease in the volume of the system, the contribution of the vapour–liquid transition region begins to manifest itself, which in the macroscopic limit is much smaller than the contributions from each of the phases, so the magnitude of the volume of the vapour–liquid transition region should affect the characteristics of the phase state. This, in turn, should influence the surface tension values σ.

The description of equilibrium drops in a macrovolume of a vapour is discussed in detail in Section 24. Let us denote the volume of the system by R_{sys}. If the volumes of the transition region and the coexisting phases are commensurate, we introduce the weights f_q that fix the volume of a region of type q: $1 \le q \le \kappa$. Here, in addition, there are two values of q related to the phase of the liquid $q = 1$ and the phase of the vapour $q = \kappa$, in comparison with the system (24.8). Now we can determine the average density of the system $\theta_{av} = \sum_{q=1}^{\kappa} f_q\theta_q$, where f_q is the fraction of sites in the phase region or the spherical monolayer q; by the definition $f = N_q/N$, N_q is the number of sites in the region q of radius R_q, $N = \sum_{q=1}^{k} N_q$. Quantity θ_{av} is retained when the volume V of a system is simultaneously varied and the amount of substance N is changed, or changes when the volume V is changed and the amount of substance N is maintained.

We confine ourselves to an analysis of the first case. As stated above, fixing the size of the drop R means fixing the size of the region with a constant fluid density. The value of the transition region (κ–2) refers to the region with variable density $V(\kappa-2) = \sum_{q=2}^{\kappa-1} f_q$ and the remaining part of the system $V_\kappa = V(R_{sys}))-V(\kappa-2)$ is the vapour volume. Thus, the volume of the vapour not only depends on the volume of the drop, but also on the volume of the transition region, which in turn depends on the temperature.

Equations (24.8) can be written for the fluid phase sites $q = 1$ and the vapour $q = \kappa$, and the equality condition of pressures must also be satisfied $\pi_{q=1} = \pi_{q=\kappa}$. These two equations complement the system (24.8) when the complete system is closed, if the average value of the density θ_{av} is fixed. Setting this value is necessary to eliminate uncertainty by varying the width of the transition region κ. This allows us to use the functional connection between the densities of the non-uniform system, generated by the condition of constancy of the chemical potential in the complete limited system, in order to trace the effect of a change in the volume of the entire system on its characteristics without violating the biphase condition. This formulation eliminates the need to have a complete view of the isotherm and determine the position of the Maxwell secant, and provides analysis of any system radii from macroscopic (on the order of 1 cm) to several nanometers.

To calculate the surface tension, we used the definition (24.15). As a criterion for the position of the reference line, we select the condition that there is no adsorption of molecules inside the transition layer, then the fixation of this position corresponds to the equimolecular separating line ρ_e. Calculations were made for the simplest argon liquid [56]. We consider a drop of radius R in a bounded system of radius R_{sys} at the reduced temperatures $\tau = T/T_{melt} = 0.68$ and 0.89. Molecules interact within the first coordination sphere on the FCC lattice ($z = 12$). The calculation of pairs of sites of different sorts *zqp* on a curved lattice is performed with a lattice parameter $\alpha = 1$. For concreteness, we confine ourselves to states of the system with a matter density $\theta_{av} = \theta^* \pm \delta$, where $\theta^* = 0.5$, $\delta = 0.01$. Dependences of the properties of the system on its size were studied.

Figure 35.2 shows the mean density of the system θ_{av} (curve 1) for the average density of the transition region (curve 2) and the average phase density (curve *3*) for varying R_{sys}.

In the course of the solution, the average density of the system (curve 1), according to the thermodynamic condition, remains constant up to the second significant number. At the same time, as the size of the system decreases, the average density over phases (curve 3) increases, and the average density along the transition region (curve 2) decreases. The average density in the transition region decreases with a decrease in the size of the system, since molecular reallocations occur in such a way that the reference surface ρ_e is displaced closer to the drop (see below).

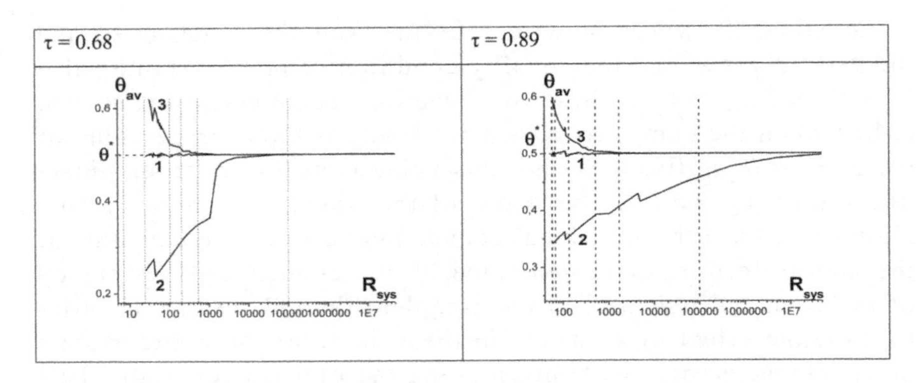

Fig. 32.2. Dependence of the densities averaged over the entire system (1), in phases (3) and in the transition region (2) on the radius of the system R_{sys}.

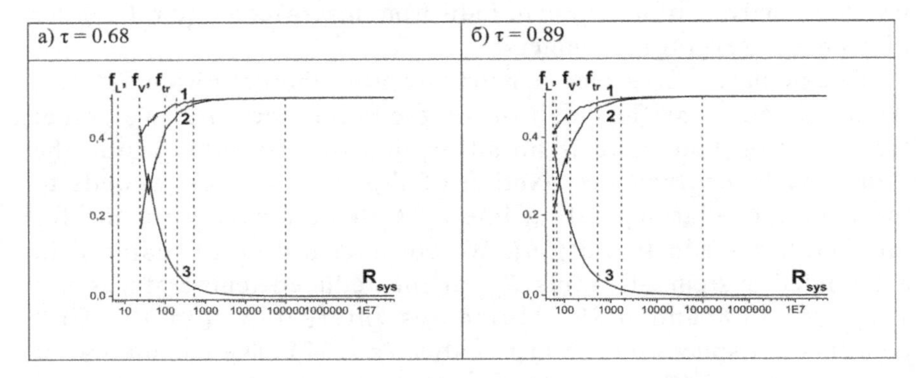

Fig. 35.3. Dependence of the fractions of the sites occupied by the liquid f_L or the vapour f_V phase or the transition region f_{tr} on the radius of the system R_{sys}.

The average density in the phases increases, primarily due to the increasing excess of the fraction of sites occupied by the drop over the fraction of vapour sites with a decrease in the size of the system, as shown in Fig. 35.3, and to a lesser extent due to the fact that the vapour density increases faster than the drop density decreases (see below). Figure 35.3 characterizes the fractions of the drop sites (curve 1), the fractions of the vapour sites (curve 2) and the transition region (curve 3).

With a decrease in the size of the system, the fraction of sites occupied by the drop and vapour decreases, and the fraction of the transition region increases. The vertical lines in Figs. 35.2 and 35.3 indicate the values at which, with a decrease in the size of the system, the deviation from the set value of the average density first becomes greater than 1) 0.0001 (R^1), 2) 0.001 (R^2), 3) 0.005 (R^3),

4) 0.01 (R^4), and also the radii below which there is no solution for the discrepancy for the mean density with an error less than 0.01 (R^*) and below which the width of the transition region (R^{**}) is artificially lowered, because it occupies almost the entire volume.

These values are given in Table 35.1, which also gives the corresponding radii of the drop R, the values of the width of the transition region κ, and the normalized surface tension $\sigma(R_{sys})/\sigma_{bulk}$ for all the indicated radii except for R^* and R^{**}, since for them there is no solution with a given accuracy with respect to the supplied thermodynamic condition.

Analysis showed that with a decrease in the size of the system: 1) the density of the liquid decreases, and the faster the higher the temperature; 2) the vapour density increases and the faster the lower the temperature; 3) the width of the transition region and the position of the reference surface are reduced, and at higher temperatures it is somewhat faster, but at low temperatures it shows greater discreteness; 4) the surface tension decreases and the faster, the higher the temperature; 5) the chemical potential and internal pressure increase, and with the higher speed, the lower the temperature.

Figure 35.4 gives the dependences of the normalized surface tension of the drop in the bounded (curves 1 and 2) and unrestricted systems (curves *3* and 4) on its radius R at τ = 0.68 (curves 1, 3) and 0.89 (curves 2, 4) .

Table 35.1.

$\tau = 0.68$						
	R^1	R^2	R^3	R^4	R^*	R^{**}
R_{sys}	10^5	523	200	103	24	7
R	8×10^4	412	162	65	–	–
κ	10	8	8	8	–	–
$\sigma(R_{sys})/\sigma_{bulk}$	1.00	0.86	0.80	0.74	–	–
$\tau = 0.89$						
	R^1	R^2	R^3	R^4	R^*	R^{**}
R_{sys}	105	523	137	69	59	58
R	8×10^4	411	105	51	–	–
κ	19	14	14	13	–	–
$\sigma(R_{sys})/\sigma_{bulk}$(unlim)	0.98	0,81	0,67	0,58	–	–

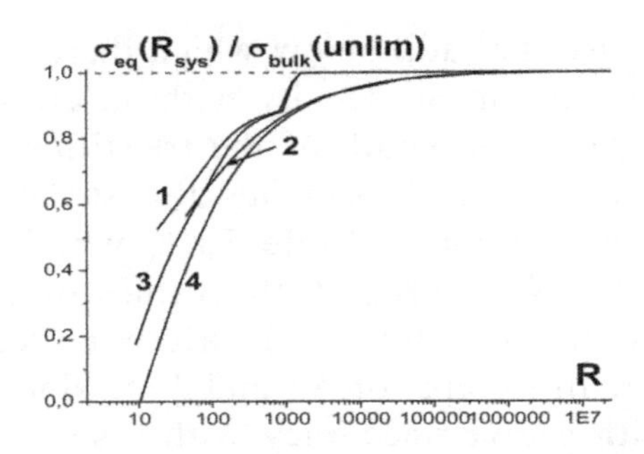

Fig. 35.4. Dependence of the normalized surface tension on the radius of the drop.

According to Fig. 35.4, the surface tension of a drop in a bounded system (curves 1 and 2) is greater than in an unbounded system (curves 3 and 4), which is apparently associated with high pressures in a limited system. The curves 1 and 2 break due to loss of solutions, and curve 3 due to the roughness of the parameter α = 1 in the region of small dimensions. Curve 4 shows that the surface tension turns to zero – the two-phase state of the phase breaks up, but the value of this two-phase existence limit should be higher for a more correct parameter value than $\alpha = 1$.

Thus, a direct molecular calculation of the characteristics of equilibrium drops shows that when the size of the system decreases (at condition $\theta_{av} = 0.5$), the critical temperature decreases and the surface tension decrease and the internal pressure and the chemical potential increase.

The phase rule and micro-non-uniform systems. In concluding this Section, we will point to an example of the influence of the micro-non-uniformity of the system on the phase states of matter. In [57], the correctness of using the macroscopic rule of Gibbs phases in micro-non-uniform systems, in particular, in polydisperse adsorption systems, was considered. Adsorption systems are characterized by a strong short-range surface potential. The joint influence of the surface potential and intermolecular interaction forms subregions of the general system in which phase transitions of the first sort are realized. The surface potential of the walls tends to stratify the adsorbate molecules along two-dimensional subdomains of the general space in order to form phases of approximately the same density in these two-dimensional subregions with the same potential. This trend is analogous to the behaviour of bulk phases with respect to the phase rule in external fields.

The diagrams of stratification in micro-non-uniform systems are characterized by a multiplicity of the vapour–liquid phase transitions of the first sort [34,58]. Accordingly, each curve of stratification has its own critical point. Such systems include adsorption, both on open non-uniform surfaces and inside porous bodies.

It is shown that the type and number of phase diagrams of adsorption systems are determined by the surface potential, which depends on the composition and structure of the non-uniform surfaces of the adsorbent. The number and parameters of the critical points depend on the composition and structure of the non-uniform systems. The maximum possible number of critical points is due to the presence of macroscopic regions of different types. In the case of a locally ordered structure of regions consisting of adsorption centres of different types, one critical point with modified values of critical parameters is realized, as in uniform systems. If the influence of the surface potential predominates over the intermolecular interaction, then there are no critical points in the system.

Those. in micro-non-uniform systems many phase states can be realized, similar to the set of phase states created by extended external fields (gravitational, electric, magnetic) in macroscopic volumes. At the same time, the presence of interfaces between phases in non-uniform systems and their curvature does not introduce any additional influence as that of the 'external' fields on the character of stratification.

References

1. Gibbs J.W. Thermodynamics. Statistical mechanics. Moscow: Nauka, 1982.
2. Landau L.D., Livshits E.M., Theoretical physics. V. 5. Statistical physics. Moscow: Nauka, 1964.
3. Hill T.L., Statistical Mechanics. Principles and Selected Applications. – Moscow: Izd. Inostr. lit., 1960. – 486 p. [N.Y.: McGraw–Hill Book Comp. Inc., 1956].
4. Gibbs J.W.. Elementary principles in statistical mechanics, developed with especial references to the rational foundations, NewYork, 1902.
5. Fowler R.H. Statistical Mechanics. Cambridge: Cambridge Univer. Press, 1936.
6. Klimontovich Yu.L., Statistical physics. Moscow: Nauka, 1982.
7. Heer C.V., Statistical mechanics, kinetic theory and stochastic processes. Moscow: Mir, 1976. [Academic Press New York, London, 1972].
8. Hill T.L., J. Chem. Phys. 1962. V. 36. P. 3182.
9. Hill T.L., Thermodynamics of Small Systems. Part 1. New York Amsterdam: W. A. Benjamin, Inc., Publ., 1963.
10. Hill T.L. Thermodynamics of Small Systems. Part 2. New York Amsterdam: W. A. Benjamin, Inc., Publ., 1964.
11. Oh B.K., Kim S.K. J. Chem. Phys. 1977. V. 67. P. 3427.

12. Tovbin Yu.K., Zh. fiz. khimii. 2012. V. 86. No. 9. P. 1461. [Russ. J. Phys. Chem. A 86, 1356 (2012)]
13. Tovbin Yu.K., Theory of physico-chemical processes on the gas-solid interface, Moscow: Nauka, 1990. [CRC, Boca Raton, Florida, 1991].
14. Tovbin Yu.K., Zh. fiz. khimii. 2010, Vol. 84. No. 11. P. 2182. [Russ. J. Phys. Chem.A. 86, 1356 (2012)].
15. Tovbin Yu.K., Khim. fizika. 2010, V. 29. No. 12. P. 74.[Russ. J. Phys. Chem. B. V. 4. № 6. P. 1033 (2010)].
16. Feller W., Introduction to Probability Theory and its Applications. Volume 1. Moscow: Mir, 1984.
17. Gelfond A.O., Calculus of finite differences. Moscow: Nauka, 1967.
18. Godunov S.K., Ryaben'kii V.S.. Difference schemes (introduction to theory). Moscow: Nauka, 1973.
19. Fikhtengol'ts G.M. Course of differential and integral calculus. Moscow: Fizmatgiz, 1963. V. 1. and V. 3.
20. Kofman A., Introduction to applied combinatorics. Moscow: Nauka, 1975.
21. Kolchin V.F., Sevastyanov B. A., Chistyakov V. P., Random Allocations (Nauka, Moscow, 1978; Winstonand Sons, Washington, DC, 1978).
22. Adamson A. W., The Physical Chemistry of Surfaces,Mir, Moscow, 1979. [Wiley, New York, 1976].
23. Summ B.D., Goryunov Yu.V., Physico-chemical basis of wetting and spreading. Moscow: Khimiya, 1976.
24. Tovbin Yu.K., Rabinovich A.B., Zh. fiz. khimii. 2010. V. 84. No. 12. P. 2366. [Rus. J. Phys. Chem. A, 2010, Vol. 84, No. 12, P. 2166].
25. Roberts M.W., McKee C.S., Chemistry of the metal-gas interface. Oxford: Clarendon Press, 1978.
26. Somorjai, G.A., Chemistry in two-dimension surface. N.Y., Ithaca: Cornell Univ. Press L., 1981.
27. Tovbin Yu.K., Rabinovich A.B., Zh. fiz. khimii. 2011. V. 85. No. 11. P. 2118. [Rus. J. Phys. Chem. A, 2011, V. 85, No. 11, P. 1977].
28. Tovbin Yu.K., DAN SSSR. 1977. V. 235. P. 641.
29. Tovbin Yu.K., Zh. fiz. khimii. 2012. V. 86. No. 7. P. 1301. [Rus. J. Phys. Chem. A, 2012, V. 86, No. 7, P. 1180] .
30. Tovbin Yu.K., Fizikokhimiya poverkhnosti i zashchita materialov. 2011. V. 47. No. 2. P. 115. [Protection of Metals and Physical Chemistry of Surfaces, 2011, V. 47. No. 2. P. 141].
31. Tovbin Yu.K., Rabinovich A.B., Zh. fiz. khimii. 2013. V. 87. No. 2. P. 337. [Russ. J. Phys. Chem. A, 2013, Vol. 87, No. 2, pp. 329].
32. Lebedev N.N., Special functions and their applications. Moscow: Fizmatlit, 1963.
33. Tovbin Yu.K., Rabinovich A.B. Zh. fiz. khimii. 2011. V. 85. No. 8. P. 1514. [Russ. J. Phys. Chem. A, 2011, Vol. 85, No. 8. P.1398].
34. Sychev V.V., Differential equations of thermodynamics. Moscow: Vysshaya shkola, 1991.
35. Tovbin Yu.K., Zh. fiz. khimii. 2010. V. 84. No. 2. P. 231. [Rus. J. Phys. Chem. A, 2010, V. 84. № 2. P. 180].
36. Tovbin Yu.K., Ibid. 2010. V. 84. No. 4. P. 797. [Rus. J. Phys. Chem. A, 2010, V. 84. № 4. C. 705].
37. Tovbin Yu.K., Ibid. 2010. V. 84. No. 10. P. 1882. [Rus. J. Phys. Chem. A, 2010, V. 84. № 10. C. 1717].

38. Jaycock M. Parfitt J., Chemistry of interfaces.– Moscow: Mir, 1984. – 270 p. [Wiley, New York, 1981].
39. Zimon A.D. Fluid adhesion and wetting. Moscow: Khimiya, 1974.
40. Tovbin Yu.K., Rabinovich A.B., Izv. AN. Ser. khim. 2009. No. 11. P. 2127. [Russ. Chem. Bull. 58, 2193 (2009)].
41. Tovbin Yu.K., Teoret. osnovy khim. tekhnologii. 2005. V. 39. No. 5. P. 523. [Theor. Found. Chem. Engin. 2005. V. 39. No 5. P. 493].
42. Tovbin Yu.K., The Molecular Theory of Adsorption in Porous Solids, Moscow, Fizmatlit, 2012. [CRC Press, Taylor & Francis Group, 2017].
43. Bazarov I.P., Thermodynamics. Moscow: Vysshaya shkola,1991.
44. Semenchenko K. V., Selected Chapters of Theoretical Physics. Moscow: Prosveshchenie, 1966.
45. Hilton H., Zentralblatt fur Mineralogie,Goelogie und Palaeontologie, 1901. S. 753.
46. Liebmann H., Z. Kristallogr., 1914. B. 53. S. 171.
47. Laue M., Z. Kristallogr., 1943. B. 105. S. 124.
48. Chernov A.A., Givargizov E.I., Bagdasarov X.C., et al., Advanced crystallography. V. 3. Crystal formation. Moscow: Nauka, 1980.
49. Rusanov A.I., Phase equilibria and surface phenomena. Leningrad: Khimiya, 1967.
50. Sirotin, Yu.I., Shaskol'skaya M.P., Fundamentals of crystal physics. Moscow: Nauka, 1979.
51. Titov S.V., Zaitseva E. S., Tovbin Yu. K., Zh. fiz. khimii. 2017. V. 91. No. 12. P. 2155. [Rus. J. Phys. Chem. A, 2017, Vol. 91, No. 12, P. 2481].
52. Tovbin Yu.K., Rabinovich A.B., Izv. AN. Ser. khim. 2010. No. 4. P. 663. [Rus. Chem. Bull. 2010. T. 59. No. 4. P. 677].
53. Tovbin Yu.K., Zaitseva E.S., Fiziko-khimiya poverkhnosti i zashchita materialov. 2017. V. 53. No. 5. P. 451. [Protection of Metals and Physical Chemistry of Surfaces, 2017, Vol. 53, No. 5, P. 765].
54. Doye J.P.K., Calvo F., J. Chem. Phys. 2002. V. 116. P. 8307.
55. Tovbin Yu.K., Zh. fiz. khimii. 2018. V. 92. No. 1. P. 36. [Russ. J. Phys. Chem. A 92, 29 (2018)].
56. Hirschfelder J. O., Curtiss Ch. F., Bird R. B., Molecular theory of gases and liquids. – Moscow: Inostr. Lit., 1961. – 929 p. [Wiley, New York, 1954].
57. Tovbin Yu.K., Zh. fiz. khimii. 2013. V. 87. No. 6. P. 928. [Rus. J. Phys. Chem. A, 2013, V. 87, No. 6, P. 906].
58. Tovbin Yu.K., Zh. fiz. khimii. 2009. V. 83. No. 10. P. 1829. [Rus. J. Phys. Chem. A. 2009. V. 83. No. 10. P. 1647.

5

Non-equilibrium processes

This chapter discusses issues related to the non-equilibrium state of the system. Chapter 1 presents the main provisions of the thermodynamics of non-equilibrium processes, which are equally related to different aggregate states and phases. The attraction of concepts about relaxation times in the non-equilibrium thermodynamics turned out to be necessary for understanding the connection between the Kelvin equation and the equilibrium state of small drops of a liquid in a vapour (Section 7). Moreover, the fundamentals of non-equilibrium thermodynamics should be analyzed to understand the fundamentals of thermodynamics associated with the very concept of equilibrium and an implicit postulate regarding time constraints in the notion of Gibbs 'passive forces'.

This chapter also discusses two other issues that go beyond traditional non-equilibrium thermodynamics: under what conditions does the local equilibrium condition used in all thermodynamic constructions is broken and how to describe in a self-consistent way the rate of stages in three aggregate states (at any densities, including transient region of the interface), so that the molecular description gives a complete interpretation of the assumptions of non-equilibrium thermodynamics in all aggregate states.

We begin with a discussion of the characteristic relaxation times for mass, impulse, and energy transfer for the gas phase and for mass transfer in solid phases to show the range of relaxation times in three aggregate systems.

36. Relaxation times

The processes of relaxation characterize the processes of the transition of the system to the equilibrium state and are described by kinetic equations. The transfer of mass, impulse, and energy is described by the equations of continuum mechanics, which for vapour–liquid systems are traditionally called hydrodynamic equations [1,2], and in solids – elasticity equations [2,3]. They often give a formal answer that a mathematically correct solution of these equations corresponds to an infinite time of full equalization of all characteristics. However, any changes are considered with a certain degree of accuracy and, naturally, there are no infinite relaxation times. These discussed processes in the gas have their own specific ranges of values, and their estimates for the characteristic length L of the order of 1 cm are considered below [4–6].

Transfer properties. Impulse transfer is responsible for the process of establishing equilibrium pressure. The basic physical mechanism of pressure equalization is the wave of gas density generated by the initial perturbation $\partial^2 P/\partial t^2 = c^2\partial^2 P/\partial x^2$, i.e., the equation describing this process is the wave equation of the hyperbolic type, here for simplicity it is one-dimensional along the x axis, and c is the velocity of the sound wave. Its solution in an infinite medium has a solution of the type of propagating waves $P(x,t) = f_1(x + ct) + f_2(x-ct)$. Therefore, the scale of the pressure relaxation time is the time during which this wave passes the path L (here from one wall of the vessel to the other wall), i.e. $\tau_p \sim L/c$. In air, the velocity of sound is $c \approx 300$ m/s or 3×10^4 cm/s, and the time for establishing the pressure is of the order of $\tau_p \sim 3\times10^{-5}$ s.

Equalization of the concentrations of gas mixture components is a process of diffusion type, described in the simplest case by an equation of parabolic type (as above, we look at one-dimensional motion) $\partial n/\partial t = D\partial^2 n/\partial x^2$. In an unbounded medium, this equation has a simple solution of the type of a spreading Gaussian distribution $n(t,x) = \exp\{-x^2/4Dt\}/(4\pi Dt)^{1/2}$ (if the perturbation at the initial time $t = 0$ had the form $n(0,x) = \delta(x)$). Therefore, the average square of the size of the cloud of particles creating a perturbation of a uniform system is equal to $\overline{x^2} = \int x^2 n(t,x)dx = 2Dt$. If the system is bounded by walls, then the size of the degradation region of increased density $\left(\overline{x^2}\right)^{1/2}$ reaches the value L after the time $\tau_n \sim L^2/2D$ elapses. In the three-dimensional case $\left(\overline{r^2}\right)^{1/2} = \left(\overline{x^2}\right)^{1/2} + \left(\overline{y^2}\right)^{1/2} + \left(\overline{z^2}\right)^{1/2} = 3\left(\overline{x^2}\right)^{1/2}$ we shall have a consequence $\tau_n \sim L^2/6D$. According to tabular data, the diffusion coefficients D for gases such as O_2, N_2, CO_2, etc. are $D = \sim 0.14–0.2$ cm^2/s, therefore at $L \sim 1$ cm, we obtain that $\tau_n \sim 1/6 \cdot 0.2 \approx 1$ s.

The equalization of temperature T is also described by an equation of the parabolic type, but instead of the coefficient D there is a coefficient of thermal diffusivity K, which is equal to the coefficient of thermal conductivity divided by the product of the gas density by its specific heat. For an air-type gas, this quantity is of the order of $K \sim 0.2$ cm^2/s, so that $\tau_T \sim L^2/6K \approx 1$ s (for air, the coincidence of the order of this quantity with coefficient D is purely random).

These simple estimates of the relaxation times demonstrate: 1) the time estimate τ is produced by methods that go beyond quasi-static thermodynamics (as always, the criteria for any approximation should be determined within the framework of a more general consideration); 2) the relaxation processes of various thermodynamic characteristics are different in the mechanism of relaxation itself, therefore, the times τ can differ significantly in magnitude since they depend in different ways on macroscopic parameters (for example, $\tau_P \sim L$, $\tau_n \sim L^2$, etc.); 3) only the 'gas' version of the system is considered above, the relaxation processes in which are determined by the macroscopic equations of motion of a continuous medium. In the study of relaxation processes involving the consideration of external fields (electromagnetic, surface, etc.), the estimation requires a transition to a microscopic level, since it is necessary to involve certain parameters already at the atomic–molecular level (time of molecule rotation in the field, absorption time and emission of a photon, etc.). An analogous situation arises also in the consideration of relaxation processes in the local regions of thermodynamic systems associated with taking into account the interaction of the particles with one another during the transition to condensed phases. In particular, for a solid body, the ratios of the relaxation times (Section 12) are transformed into strongly pronounced inequalities, the differences in which between impulse and mass relaxation times reach up to 10–15 orders of magnitude.

Diffusion of vacancies. Let us consider the relaxation time t^* of the concentration equalization process in solid spherical samples of radius R_s due to the transfer of atoms by the vacancy mechanism in the course of the process, which is as close to equilibrium as the temperature decreases from the melting point to the current value of T. Let us denote by t_0 the time necessary for establishment in the crystal equilibrium distribution of vacancies $C_0(T_t)$ from the moment of its formation at $T = T_t$. We denote by δT the step in temperature with its decrease. The cooling process starts from the surface $R = R_s$ (on which a new concentration of vacancies

$C_0(T_t-\delta T)$ is established) and extends deep into the sphere. The thermal equilibrium is established much faster than the concentration equilibrium, so the process proceeds at a fixed temperature. We denote the concentration of vacancies in the centre of the sphere $R = 0$ at the initial instant of time (equal to the equilibrium vacancy concentration at temperature T_t) by C_1. During the establishment of the vacancy equilibrium after a certain time t at the centre of the sphere, the concentration should become sufficiently close to the value of $C_0(T_t-\delta T)$ in order to consider the entire sample having one vacancy concentration equal to $C_0(T_t-\delta T)$ corresponding to the temperature $T_t-\delta T$. Further, the process of cooling the sample proceeds in the same stepwise manner.

For a spherical sample, this formulation of the problem corresponds to a one-dimensional problem for the radial diffusion of vacancies with the boundary condition $C_0(T)$) for $R = R_s$, which is described by equation

$$\frac{\partial C}{\partial t} = D\left(\frac{\partial^2 C}{\partial R^2} + \frac{2}{R}\frac{\partial C}{\partial R}\right), \tag{36.1}$$

where D is the diffusion coefficient, C is the concentration of vacancies at time t at a point with radius R, $0 \le R \le R_s$. If at the initial instant of time $t = 0$ the concentration of vacancies on a spherical surface is $C = C_0$, and the concentration $C = C_1$ for any R, then according to [7,8], the solution at any time t will have the form:

$$\frac{C - C_1}{C_0 - C_1} = 1 + \frac{2R_s}{\pi R}\sum_{n=1}^{\infty}\frac{(-1)^n}{n}\sin\left(\frac{n\pi R}{R_s}\right)\exp\left(-\frac{Dn^2\pi^2 t}{R_s^2}\right).$$

If we denote by C_1^* the concentration of vacancies in the centre of the sphere at any time, then it is defined as the limit at $R \to 0$:

$$\frac{C_1^* - C_1}{C_0 - C_1} = 1 + 2\sum_{n=1}^{\infty}\frac{(-1)^n}{n}\exp\left(-\frac{Dn^2\pi^2 t}{R_s^2}\right).$$

For large times (in this case, all the terms of the sum multiplied by exponents with exponents proportional to n^2t, are much less than the first term corresponding to $n = 1$) close to the time of exhaustion of non-equilibrium vacancies at a given temperature T, we give a finite expression for the time necessary for the difference between the quantities C_0 and C_1^* was sufficiently small – i.e. the whole sample was in a new equilibrium state, in the following form

$$D\pi^2 t = R_s^2 U, \quad U = -\ln\left((C_0 - C_1^*)/[2(C_0 - C_1)]\right). \qquad (36.2)$$

Under the assumption of the differential nature of cooling $C_1 - C_0 \sim 10^{-2} C_1$ and the considerable degree of exhaustion of the density of non-equilibrium vacancies at the centre of the sphere, where δ_c is a given small value $C_1^* - C_0 = \delta_c C_0$, this solution relates the time t to the temperature dependence of the diffusion coefficient. It follows from formula (36.2) that with decreasing R_s by an order of magnitude the value of the time t decreases by two orders of magnitude, and simultaneously, taking into account the dependence $D = D_0 \exp(-\beta E)$ [8–13], which decreases the temperature for which can be obtained the equilibrium state of vacancies in the entire sample for a given time interval.

Diffusion of the label. In the complete equilibrium of the system, the flow of matter is characterized by a self-diffusion coefficient, which really represents a flux of 'labelled' isotopes. The described process can be used to analyze the conditional diffusion process of a label if it is assumed that the diffusion coefficient refers to a label and the equation describes its transport from the surface to the centre at an equilibrium state of matter density and vacancies, respectively. (The process is modelled in the fact that the time of changing the vacancy density with a change in temperature does not enter into it). In this case, Eq. (36.1) is related to the analysis of the thermal motion of atoms as a function of temperature. The diffusion coefficient is written, as before, in the form $D^* = D_0 \exp(-\beta E^*)$ [8-13], where $\beta = (RT)^{-1}$, R is the gas constant, T is the temperature, E^* is the activation energy of self-diffusion, D_0^* is the pre-exponent of this coefficient. As the temperature decreases, the value of D^* decreases. The process of label mixing is associated with the concentration of vacancies, which also decreases with decreasing temperature throughout the body and as the temperature decreases the vacancy density θ_V decreases as $\theta_V = 1 - \theta_A = [1 + K_a \exp(\Delta H_c / RT)]^{-1} \sim K_c \exp(-\Delta H_c / RT)$ where $K_a \approx K_c^{-1}$, $K_c = J_A^0 / J_A$ is the pre-exponent associated with the partition functions of the atom in the solid J_A and in the vapour J_A^0, ΔH_c is the energy of removal of the atom into vacuum (during sublimation) θ_A is the number of sites of the crystal lattice occupied by the particle A, $\Delta H_c = z\varepsilon_{AA}$, z is the number of the nearest neighbours, and ε_{AA} is the energy of their interaction. When T is reduced to zero, the equilibrium defect crystal must be free of vacancies: $\theta_V \to 0$. This conditional statement of the problem allows us to use Eq. (36.2) for a model-free analysis of the relaxation times using experimental data on self-diffusion of the label [8–13]. Diffusion of the label makes the process of thermal motion 'observable'. With a decrease in the share of vacancies thermal motion is inhibited. All other methods for calculating the process of diffusion and relaxation times require the mandatory introduction of model parameters and the provision of a self-consistent description of the transport process with a description of the phase state of the element. The latter condition is realized quite rarely [10, 13].

The question of the possibility of realizing thermal motion and achieving a conventional value for mixing the label for a 'reasonable' experiment time at different temperatures is a natural question. The times of establishing a uniform label distribution in the crystal, starting from the time of its solidification at the temperature of the triple point T_t (or the melting point) were estimated in [14–16], with a step-by-step decrease in temperature to a certain temperature for a number of elements. The maximum time during which an equilibrium state can be reached in a real experiment is defined as the process time from T_t to the lowest achievable temperature in the experiment. We denote this temperature by T^* and the corresponding time by $t^* = t(T^*)$. The time t^* answers the question of the possibility of realizing the equilibrium state of the label for the 'reasonable' time of the experiment. The time to establish the uniformity of the label within the entire sample depends on its size. The reading starts from a radius of 1 cm. The smaller the R_s, the lower temperatures can cool the sample, observing the thermal motion (1 year corresponds to the order of 3.15×10^7 s). Below we will characterize the temperature through the ratio $\delta = T/T_t$.

Figure 36.1 shows the dimensionless time curves for reaching the label equilibrium for ten single crystals as a function of the dimensionless temperature at $R_s = 1$ cm [15]. The calculation was carried out at $\delta T = 1$ K and $\delta_c = 10^{-4}$. The time to equilibrium at the melting point is denoted by t_0. In the calculations it was estimated by the formulas (36.2) as the time of establishment of the equilibrium concentration of the label with decreasing temperature from T_t to $T_t - \delta T$. With decreasing temperature the curves increase almost exponentially.

The values of the parameters E and D_0, found by the authors of [8–20] from measurements in the indicated temperature ranges, are given in Table 36.1. These data were used for finding t_0 and

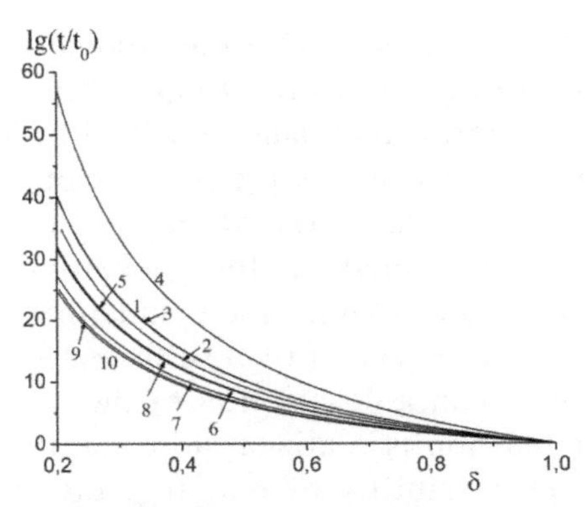

Fig. 36.1. The generalized dependence of the reduced time for the establishment of vacancy equilibrium t/t_0 in a spherical sample of radius $R_s = 1$ cm on the reduced melting point $\delta = T/T_t$ for single crystals: 1 – Ar, 2 – Kr, 3 – Xe, 4 – Si, 5 – Ag, 6 – Al, 7– Au, 8 – Cu, 9– Na, 10 – K.

Table 36.1. Literature data on measurements of the diffusion coefficient in temperature ranges (the upper limit is maximally close to melting point T_t) and the parameters of diffusion coefficient D_0 and E for a number of crystals

Crystal	T_t, K	$T_{min} - T_{max}$, K	D_0, cm²/s	E, kcal/mol	Reference
Ar	83.8	72–83	4.0	3.9	8,18,19,20
Kr	116	90–115	5.0	4.8	8,18,19,20
Xe	133	130–Tt	9.7	7.4	8,18,19,20
Na	370.7	274–367	0.242	104.5	8,17,21
K	336.1	273–333	0.033	9.4	8,17,22
Al	931.5	729–916	1.71	33.9	8,17,23
Ag	1235	914–1228	0.67	45.9	8,17,24
Au	1337	1031–1333	0.084	41.6	8,17,25
Cu	1356	1163–1334	0.78	48.9	8,17,26
Si	1699	1498–1673	0.8	109.8	8

t^*. The values of t_0 for samples of radius 1 cm are given in Table 36.2, as well as the times of emergence of the equilibrium state t_0 for samples with sizes from 1 cm to 1 nm (the first number in each cell) t^*.The second number is the time to reach equilibrium when the temperature is lowered from the minimum temperature in the experiment under T_{max} consideration T_{min}. Values t_0 (with the exception of Na and K) are commensurate with the duration of one year ~3.5×10^7 s or exceed it. Thus, even at the melting point, in order to reach the equilibrium state of the label, times that exceed the time of ordinary laboratory measurements are needed. Moreover, this refers to the minimum temperature of the experiment – the time t^* exceeds t_0 by approximately two or three orders of magnitude. Therefore, for any further lowering of the temperature, the curves in Fig. 36.1 (the decimal logarithms on the ordinate axis!) correspond to sufficiently large times $t > t^*$, which can not always be realized in the laboratory.

Table 36.2. The calculated characteristics of the process of establishing the label equilibrium t_0, t^* and δ^* for samples of crystals of different sizes R_s by the vacancy mechanism

Crystal	Para-meter	R_s							
		1 cm	1 mm	0.1 mm	10 μm	1 μm	100 nm	10 nm	1 nm
Ar	t_0, s	$2.59\cdot10^{9}$	$2.59\cdot10^{7}$	$2.59\cdot10^{5}$	$2.59\cdot10^{3}$	$2.59\cdot10^{1}$	$2.59\cdot10^{-1}$	$2.59\cdot10^{-3}$	$2.59\cdot10^{-5}$
	t^*, s	$1.32\cdot10^{11}$	$1.32\cdot10^{9}$	$1.32\cdot10^{7}$	$1.32\cdot10^{5}$	$1.32\cdot10^{3}$	$1.32\cdot10^{1}$	$1.32\cdot10^{-1}$	$1.32\cdot10^{-3}$
	δ^*	0.86	0.73	0.64	0.57	0.52	0.46	0.43	0.39
Kr	t_0, s	$1.85\cdot10^{8}$	$1.85\cdot10^{6}$	$1.85\cdot10^{4}$	$1.85\cdot10^{2}$	$1.85\cdot10^{0}$	$1.85\cdot10^{-2}$	$1.85\cdot10^{-4}$	$1.85\cdot10^{-6}$
	t^*, s	$8.27\cdot10^{10}$	$8.27\cdot10^{8}$	$8.27\cdot10^{6}$	$8.27\cdot10^{4}$	$8.27\cdot10^{2}$	$8.27\cdot10^{0}$	$8.27\cdot10^{-2}$	$8.27\cdot10^{-4}$
	δ^*	0.78	0.66	0.58	0.52	0.46	0.42	0.40	0.35
Xe	t_0, s	$8.51\cdot10^{8}$	$8.51\cdot10^{6}$	$8.51\cdot10^{4}$	$8.51\cdot10^{2}$	$8.51\cdot10^{0}$	$8.51\cdot10^{-2}$	$8.51\cdot10^{-4}$	$8.51\cdot10^{-6}$
	t^*, s	$2.19\cdot10^{11}$	$2.19\cdot10^{9}$	$2.19\cdot10^{7}$	$2.19\cdot10^{5}$	$2.19\cdot10^{3}$	$2.19\cdot10^{1}$	$2.19\cdot10^{-1}$	$2.19\cdot10^{-3}$
	δ^*	0.81	0.70	0.62	0.55	0.52	0.45	0.41	0.38
Na	t_0, s	$4.81\cdot10^{6}$	$4.81\cdot10^{4}$	$4.81\cdot10^{2}$	$4.81\cdot10^{0}$	$4.81\cdot10^{-2}$	$4.81\cdot10^{-4}$	$4.81\cdot10^{-6}$	$4.81\cdot10^{-8}$
	t^*, s	$7.51\cdot10^{8}$	$7.51\cdot10^{6}$	$7.51\cdot10^{4}$	$7.51\cdot10^{2}$	$7.51\cdot10^{0}$	$7.51\cdot10^{-2}$	$7.51\cdot10^{-4}$	$7.51\cdot10^{-6}$
	δ^*	0.76	0.60	0.50	0.43	0.38	0.34	0.30	0.28
K	t_0, s	$5.56\cdot10^{6}$	$5.56\cdot10^{4}$	$5.56\cdot10^{2}$	$5.56\cdot10^{0}$	$5.56\cdot10^{-2}$	$5.56\cdot10^{-4}$	$5.56\cdot10^{-6}$	$5.56\cdot10^{-8}$
	t^*, s	$1.83\cdot10^{8}$	$1.83\cdot10^{6}$	$1.83\cdot10^{4}$	$1.83\cdot10^{2}$	$1.83\cdot10^{0}$	$1.83\cdot10^{-2}$	$1.83\cdot10^{-4}$	$1.83\cdot10^{-6}$
	δ^*	0.81	0.65	0.54	0.46	0.40	0.36	0.32	0.29
Al	t_0, s	$4.04\cdot10^{7}$	$4.04\cdot10^{5}$	$4.04\cdot10^{3}$	$4.04\cdot10^{1}$	$4.04\cdot10^{-1}$	$4.04\cdot10^{-3}$	$4.04\cdot10^{-5}$	$4.04\cdot10^{-7}$
	t^*, s	$7.85\cdot10^{9}$	$7.85\cdot10^{7}$	$7.85\cdot10^{5}$	$7.85\cdot10^{3}$	$7.85\cdot10^{1}$	$7.85\cdot10^{-1}$	$7.85\cdot10^{-3}$	$7.85\cdot10^{-5}$
	δ^*	0.78	0.65	0.56	0.49	0.44	0.40	0.36	0.33
Ag	t_0, s	$9.65\cdot10^{7}$	$9.65\cdot10^{5}$	$9.65\cdot10^{3}$	$9.65\cdot10^{1}$	$9.65\cdot10^{-1}$	$9.65\cdot10^{-3}$	$9.65\cdot10^{-5}$	$9.65\cdot10^{-7}$
	t^*, s	$6.94\cdot10^{10}$	$6.94\cdot10^{8}$	$6.94\cdot10^{6}$	$6.94\cdot10^{4}$	$6.94\cdot10^{2}$	$6.94\cdot10^{0}$	$6.94\cdot10^{-2}$	$6.94\cdot10^{-4}$
	δ^*	0.74	0.63	0.54	0.48	0.43	0.39	0.36	0.32
Au	t_0, s	$4.95\cdot10^{7}$	$4.95\cdot10^{5}$	$4.95\cdot10^{3}$	$4.95\cdot10^{1}$	$4.95\cdot10^{-1}$	$4.95\cdot10^{-3}$	$4.95\cdot10^{-5}$	$4.95\cdot10^{-7}$
	t^*, s	$5.89\cdot10^{9}$	$5.89\cdot10^{7}$	$5.89\cdot10^{5}$	$5.89\cdot10^{3}$	$5.89\cdot10^{1}$	$5.89\cdot10^{-1}$	$5.89\cdot10^{-3}$	$5.89\cdot10^{-5}$
	δ^*	0.77	0.63	0.53	0.46	0.41	0.36	0.33	0.30
Cu	t_0, s	$6.42\cdot10^{7}$	$6.42\cdot10^{5}$	$6.42\cdot10^{3}$	$6.42\cdot10^{1}$	$6.42\cdot10^{-1}$	$6.42\cdot10^{-3}$	$6.42\cdot10^{-5}$	$6.42\cdot10^{-7}$
	t^*, s	$1.31\cdot10^{9}$	$1.31\cdot10^{7}$	$1.31\cdot10^{5}$	$1.31\cdot10^{3}$	$1.31\cdot10^{1}$	$1.31\cdot10^{-1}$	$1.31\cdot10^{-3}$	$1.31\cdot10^{-5}$
	δ^*	0.86	0.71	0.60	0.52	0.46	0.41	0.38	0.34
Si	t_0, s	$6.16\cdot10^{14}$	$6.16\cdot10^{12}$	$6.16\cdot10^{10}$	$6.16\cdot10^{8}$	$6.16\cdot10^{6}$	$6.16\cdot10^{4}$	$6.16\cdot10^{2}$	$6.16\cdot10^{0}$
	t^*, s	$3.65\cdot10^{16}$	$3.65\cdot10^{14}$	$3.65\cdot10^{12}$	$3.65\cdot10^{10}$	$3.65\cdot10^{8}$	$3.65\cdot10^{6}$	$3.65\cdot10^{4}$	$3.65\cdot10^{2}$
	δ^*	0.89	0.79	0.71	0.65	0.60	0.55	0.51	0.48

Size effects. This approach allows you to follow the size effects. The decrease in the size of the sample sharply reduces the time for the establishment of vacancy equilibrium. It is possible to speak about the practically rapid establishment of equilibrium only for samples with a radius of order and less than 1 μm (with the exception of Si).

The third number in each cell of the second column of Table 36.2 – the value $\delta^* = T^*/T_t$. It is obtained by simulating the cooling of a sample of radius R_s (for given values of the parameters C_0, C_1, δT, δ_c) for the same time for which the minimum temperature was reached in the actual experiment at $R_s = 1$ cm (see Table 36.1). The lower R_s, the lower temperatures to which it was possible to cool the sample during the actual experiment (real experiments were conducted with a poorly defined dispersion composition). This means that these values δ^* characterize the maximum reduction in temperature that can be reached in the experiment, with the achievement of a label equilibrium. At lower temperatures, under the experimental conditions, it would be impossible to achieve equilibrium. All examples indicate the existence of a temperature range from zero to T^* K, in which complete equilibrium of the sample is not achieved.

It should be noted that with the reduction of R_s to 10 nm for small systems it is no longer possible to apply the laws of classical thermodynamics [or the continuous medium model] [27] and from the continual description of migration it is necessary to go over to discrete models for describing the equilibrium state of the crystal and the migration of atoms. The value $R_s = 1$ nm is given for a methodical purpose – it is devoid of physical meaning, not only with respect to the diffusion equation, but also with respect to the spherical shape of the sample. Finally, as the size of the microcrystals decreases, it should be borne in mind that the contribution of surface diffusion along the grain boundaries increases sharply, and the material becomes finely dispersed, i.e. its properties differ from the bulk properties of single crystals. Even for $R_s = 10$ nm, the temperature range from which the equilibrium state is achievable in principle starts from $\delta^* = 0.3{-}0.4$ (we are not talking about quantum crystals).

Mass transfer coefficient of the substance. The relationship between the mass transfer coefficient and the label coefficient is usually established on the basis of the postulates of non-equilibrium thermodynamics, which leads to the expression for the binary mixture connecting the two factors as [8,10–13] $D_i = D_i^* g_i$, i.e. for each component, the mass transfer coefficient D_i is expressed as the product of the self-diffusion coefficient D_i^* and the thermodynamic factor g_i. We note that this expression can formally be applied to a pure substance – the diffusion of vacancies also takes place in it.

When the label is transferred, there is no time factor related to the process of equalization of the vacancies themselves over the sample – their diffusion accelerates the transfer of atoms. Therefore, an estimate using the experimental values of the self-diffusion coefficient D^* gives an overestimated range of relaxation times. In order to eliminate it, we must pass to the account of the influence of the thermodynamic factor g. Then equation (36.1) will correspond to the real mass transfer of the density or vacancies of the pure element. Here the question arises as to the method of creating a concentration gradient. The postulate of non-equilibrium thermodynamics [28-30] on the presence of local equilibrium, also fixes the condition for the equilibrium distribution of vacancies between local regions ($g = g_e$). In one-component systems, any non-equilibrium distribution of atoms, uniquely due to the material balance, is associated with vacancy disequilibrium $\theta_A+\theta_V= 1$ (at any site there must be a vacancy or atom). Therefore, the hypothesis of the equilibrium distribution of vacancies in the presence of a density gradient does not correspond to a pure substance.

The kinetic approach [31–33] to the transport of matter shows that the local equilibrium can be realized in the presence of an equilibrium relationship between the pair function θ_{AA} and the concentration θ within the local region, but under the condition that there is no equilibrium vacancy distribution between neighboring regions. The density gradient θ leads to a change in the values of the pair function θ_{AA} in different concentration regions due to gradθ. As a result, different types of concentration factors g_e and g_k are obtained depending on the accuracy of the transfer dynamics description. From the kinetic point of view, it is more rigorous to take into account the influence of the density gradient on the change in the pair function in neighboring regions through the thermodynamic factor g_k. In this case, two ways of calculating g_k can be formulated: a direct analysis of equations (20.4) and (20.5), taking into account the effect of the substance flux on the change in the connection between the functions θ_{AA} and θ [31], or if the equilibrium coupling between the pair function θ_{AA} and concentration θ [32,33]. Below we restrict ourselves to the second method of calculating g_k: the first method is very cumbersome and it is difficult to extend it to relaxation to equilibrium in non-uniform

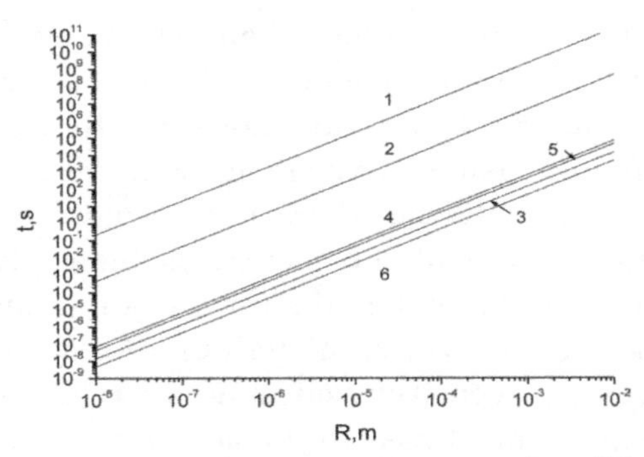

Fig. 36.2. Dependence of the time of establishment of equilibrium on the radius R, m, for samples of Ni and Cu in the process of mass transfer when using different values of g_i in calculations: 1 – g_e(Ni), 2 – g_e(Cu), 3 – g_k(Ni), 4 – g_kCu), 5 – g_k(Ni), 6 – g_v(Cu) at T = 1273K.

systems. For strongly non-equilibrium processes, both methods for calculating g_k are equally violated and direct use of the equations for the kinetics of paired distribution functions is required (20.5).

Figure 36.2 compares the following options for the calculation of the thermodynamic factor for copper and nickel: 1) the traditional assumption about the equilibrium distribution of components in substitutional alloys g_e, when the fraction of vacancies in the values of the activity coefficients of the alloy components can be neglected; 2) the kinetic version of the theory under the condition of local equilibrium between pair and unary distribution functions in the non-equilibrium flow g_k, and 3) the model expression for the local vacancy density g_V, in which it is assumed that the vacancy formation energy is approximately equal to half the activation energy of the label transfer [8,10]. The calculation is made for T = 1273 K, which corresponds to vacancy concentrations $\theta_V^{\mathrm{Ni}} = 6.8 \cdot 10^{-8}$ and $\theta_V^{\mathrm{Ni}} = 1.57 \cdot 10^{-4}$. This gives three types of g_i for nickel when calculating the mass transfer coefficient: $g_e = 0.99998$, $D_e = 0.31 \cdot 10^{-15}$ m^2/s, $g_k = 0.147 \cdot 10^8$, $D_k = 0.457 \cdot 10^{-8}$ m^2/s, $g_V = 0.47 \cdot 10^7$, $D_V = 0.146 \cdot 10^{-8}$ m^2/s, and, respectively, for copper: $g_e = 0.979$, $D_e = 0.14 \cdot 10^{-12}$ m^2/s, $g_k = 0.624 \cdot 10^4$, $D_k = 0.9 \cdot 10^{-9}$ m^2/s, $g_V = 0.92 \cdot 10^5$, $D_V = 0.133 \cdot 10^{-7}$ m^2/s. Figure 36.2 shows that taking into account the thermodynamic factor sharply reduces the value of the mass transfer coefficient. Here this difference reaches up to seven orders.

Despite such a sharp acceleration of the transfer process due to taking into account mobility of vacancies compared with the mobility of the label at an equilibrium number of vacancies, the overall effect of lowering the temperature can lead to a wide range of values δ^* exceeding zero temperatures. This issue has not been studied practically today. Here, various situations are possible, connected with the need to jointly take into account the effect of temperature on the equilibrium vacancy density and the diffusion coefficient. The vacancies have different manifestations themselves in diffusion processes for pure components and alloys. It is no accident that the experimental data in [13] were approximated by the empirical dependences of the mass transfer coefficients on temperature and density. In many ways, this depends on the mismatch between the activation energies and the heats of sublimation.

The result obtained on the impossibility of achieving zero temperatures in real conditions does not contradict the Nernst heat theorem. As T tends to zero θ_V also tends to zero, and in the low-temperature limit the equilibrium defect crystal must be freed of vacancies. This expression is obtained from the analysis of statistical sums, in which the number of configurations realized for all possible arrangements of atoms and vacancies along lattice sites is counted [8–13]. Each of the configurations is realized by permuting atoms and vacancies, but the real trajectory of their displacements is not concretized, just as the time necessary for such a rearrangement is not concretized. Since there are no restrictions on the way of enumeration of all configurations, the final result agrees with the Nernst theorem. In the actual process of lowering the temperature in a single crystal, during its relaxation, there is a single diffusion mechanism for the redistribution of atoms and vacancies. It can be shown [33] that the limiting result of such a process at infinite times leads to the same expression for the vacancy density, as well as the direct calculation of all configurations. However, the relaxation times become so large (see Figure 36.1 and the data in Table 36.2) that during the experiment for about ten years such states at low temperatures are not realized.

37. Motions in three aggregate states

The molecular theory based on the LGM allows one to describe, from a single point of view, the characteristics of matter in three aggregate states. This property of the molecular theory applies to both equilibrium and non-equilibrium processes. This brings it to the level of generality of thermodynamics, which does not differentiate its methods between any phases and aggregate states. The unified point of view not only unifies the structure of the kinetic equations of physical and chemical processes within each of the phases, but also allows us to consider from the same positions all processes at the interfaces of different phases.

A characteristic feature of the use of kinetic equations is the different scale of the characteristic times of the transport stages and chemical reactions depending on the properties of the particles, as well as the strong influence of the environment on the rates of these elementary stages in the condensed phases. In order not to encumber the presentation, as in Sections 19 and 20, we divide the elementary stages of transport and chemical reactions in vapour-liquid systems and in solids (including their boundaries). In solid bodies, the migration of atoms occurs by the vacancy or interstitial mechanisms, whereas the vapour–liquid systems are interpreted as a diffusion of molecules in diffuse regions in dense phases, as a free passage in rarefied gas phases.

As a basis for the development of the theory, the concept of the thermal velocity of molecules is used, which is used in all aggregate states, beginning with a solid body in which there are vacancies (low concentrations), up to the gas where vacancies constitute the bulk of the volume. For the intermediate liquid phase, the concepts of vacancy regions are introduced, through which the molecule can move at a thermal velocity. The physical prerequisite for the concept of vacancies in a liquid is a reduced density in comparison with a solid. This rarefaction has a fluctuation character on the average in the volume of the system, and its magnitude can be comparable with the volume of the molecule, which is sufficient for the shift of the molecule. This point of view is suitable for any density and molecules of any size, as well as for their rotational motion.

The concept of the thermal velocity of molecules is a traditional 'gas' point of view in the kinetic theory. For dense phases, this concept is preserved with decreasing length of the displacement of the molecule, which allows us to speak about the concreteness of each trajectory (its extent and direction). The calculation of the thermal velocity itself uses a probabilistic description, which is natural from the kinetic theory of the fluid. The thermal velocity is introduced in terms of the hopping probabilities $W_a = W_{fg}^{AV}(\chi)$ by the considered distance χ. The meaning of the hopping probability $W_{fg}^{AV}(\chi)$ is the average of the continuous process of motion of the molecule A (located at the initial instant t_0 at site f) into the free cell g, which the particle reaches in time τ. The average is taken along all its trajectories, taking into account lateral interaction with all neighbors and collision with them. The length of the jump by the vacancy mechanism depends on the density: $\chi \gg 1$ is characteristic for vapour, $\chi \sim 1$ for liquid and solid. The calculation of the probability of hopping W_a is traditional in the LGM model discussed in Section 20 (see Section 56).

We recall briefly the main provisions of 'microscopic' hydrodynamics, which in principle solved the main problem of constructing balance fluxes of impulse, energy, and mass in two (vapour–liquid) from the three discussed aggregate states [34]. The connection of the solid phase is associated with a greater localization of molecules than in the liquid. This question was considered in [34] in the discussion of flows near solids, and in [32] when considering the processes of dissolution in solids or their internal rearrangement. The generality of the construction procedure $W_{fg}^{AV}(\chi)$ is most obvious when $\chi = 1$. The exponential form of writing expressions for W_a (20.2) is a consequence of the thermalization of the molecule with a thermostat. Solid wall atoms and neighbouring molecules act in the role of a thermostat for any molecule. The thermostat is considered immobile in the absence of a flow of molecules and mobile if a convective flow of molecules is realized in the system (the walls of the pores are always considered immovable). In the initial positions of the LGM it is always assumed that the thermostat is stationary, and only the processes of redistribution of molecules inside the thermostat are realized in it.

Using the example of a surface monolayer, one can explain the difference between these two concepts if one considers the lattice structure of surface adsorption centres with variable height of the activation barrier for hopping from one centre to another [34]. When the barrier is high, the hopping of each particle is determined by the temperature and the probability that the molecule will gain enough energy to overcome the barrier. This energy comes from the substrate, which is a given thermostat. All other molecules of the adsorbed monolayer are located in a similar situation. Their mutual influence is manifested through lateral interactions that change the height of the barrier, but the molecules themselves move individually all the time, staying in different centres. After the molecule jumps, its excess energy is given back to the thermostat (this is the case of the diffusion mechanism of any flow).

We shall decrease the height of the hopping barrier. In the limit, it can be reduced to such an extent that it formally determines only the position of the molecule. In this case, the state of the molecule is mainly determined by its lateral interactions with neighbors, and their mutual arrangement plays a major role in the state of the adsorbed film. The thermostat for an arbitrarily chosen molecule is the substrate (the potential of the substrate remains common to all molecules) and surrounding molecules. The molecule receives energy for movement both from the substrate and from neighboring molecules, just as the molecule discharges excess energy to the substrate and to the neighbours. In this case, any organization of the flow will be accompanied by a general redistribution of molecules, in which the isolated molecule and its neighbours participate equally in collective motion (therefore the thermostat is considered mobile).

Averaging the thermal velocities of the molecules in the cell volume (i.e., in the direction of the forward and backward directions along the axes $\alpha = x,y,z$), which is the production of the resulting flux, allows us to introduce a new micro-level characteristic $\boldsymbol{u}_f$, which is an analog macroscopic flow velocity. The average value of microhydrodynamic velocity is defined as

$$\mathbf{u}_f \equiv \mathbf{u}_A(f) = \int \mathbf{v}_f^A \theta_f^A \left(\mathbf{v}_f^A\right) d\mathbf{v}_f^A / \theta_f^A. \tag{37.1}$$

The components $u_{f\alpha}$ describe the average velocity of molecules in cell f in the direction α at a characteristic length scale equal to the diameter of the molecule. This makes it possible to take into account the smallest flows (average displacements of molecules) in the convective flow of matter, which appears in the continuity equation. Accordingly, for its description it is necessary to introduce all the traditional transport coefficients.

If we formally consider the forward and backward jumps of molecules in the fixed thermostat in the form of an algebraic sum

$$I_f = \lambda \sum_{g \in z(f)} \left[U_{fg}^{VA} - U_{fg}^{AV} \right], \tag{37.2}$$

then they form diffusion [31, 35, 36] rather than convective flows with respect to the thermostat. Therefore, by setting a pressure gradient (or molecular densities) in a stationary thermostat, the diffusion coefficient can be determined from an analysis of such a non-equilibrium flow.

The kinetic theory defines for a single-component substance the only type of flow of moving molecules – a convective flow described by the continuity equation. To consider the convective flow, it is necessary to match the movement of the thermostat with the thermal motion of the molecules. This is achieved by using an equation for the distribution of pairs, describing their relative position in the coordinate space (as in the theory of a liquid) and depending on the state of the thermostat. The presence of a convective flow at a microscopic level changes the process of redistribution of molecules in comparison with a stationary thermostat, therefore at the microlevel there are two channels for the transport of molecules: micro-convective (described by microhydrodynamic velocity u_f) and thermal.

To determine the relationships of these flows, we should consider the local stationary state in the course of the evolution of the pair function. Then one can find a correction to the equilibrium pair distribution function, through which all real distributions of the molecules are recalculated. Considering the shifts of neighbouring regions of matter, an analysis of the non-equilibrium impulse flux can be used to find the shear viscosity (similarly, the bulk viscosity at deformation/compression of the volume under consideration). This allows us to calculate the convective flow rate using a Navier–Stokes-type equation in which the viscosity coefficients are calculated within the framework of the molecular theory based on the LGM, but taking into account the motion of the thermostat. A similar principle is used to construct microscopic equations for the energy transfer process.

In solid bodies, atoms/molecules do not have degrees of freedom translationally and the kinetic equations discussed below pass into the equations of Section 20. However, this does not exclude the separation of types of atomic motions in solids into diffusion jumps at the microlevel (37.2) and macroscopic motions of planes (37.1) that are well known in the Darken interpretation of the Kirkendall effect [10, 13, 31] or in the displacements of the pore boundaries (the Frenkel effect) [31, 37].

In general, this approach leads to a system of difference equations for ordinary dynamic variables in hydrodynamics (densities, three components of the vector $\mathbf{u}_f$ and temperature) defined at lattice sites. Taking into account the possibility of a detailed description of the molecular distribution of molecules over a wide range of time from picosecond to microseconds without the loss of all molecular information and calculation of microhydrodynamic information on the indicated dynamic variables and transport coefficients, this method is alternative with respect to the molecular dynamics method.

Below we will consider the questions of the way out for the condition of local equilibrium, traditionally used in hydrodynamic equations, including in microscopic hydrodynamics, and self-consistent description of processes in different modes and different aggregate states for any degrees of deviations from equilibrium.

38. Equations of conservation of molecular properties

Transition to transport equations. The kinetic equations for unary and paired distribution functions are given in Section 19. These equations are constructed by closing the Bogolyubov chain [38, 39], which takes into account collisions (Boltzmann terms), intermolecular interactions (Vlasov contributions), and exchange properties (mass, impulse, energy) between different cells due to the thermal motion of molecules in space [34]. Exchange flows are composed of direct movements of molecules into neighbouring free cells and from the transfer of properties (impulse and energy) through collisions with their neighbours. Such a system of equations represents the most detailed description of molecular systems, and with its help, when the scale is larger or rough, one can proceed to transport equations on any space-time scales depending on the problem under consideration. These equations can be applied to three aggregate states. As a result, a complete system of transport equations will describe all time intervals (from picoseconds at the microlevel to seconds at the macrolevel), spatial scales and the concentration range from gas to liquid and solid. With their help, one can justify the equations of non-equilibrium thermodynamics from Section 9, which are also used for any aggregate states, and in addition obtain expressions for the dissipative coefficients.

The transport equations are obtained from the system of kinetic equations [34,40,41] by the unary (19.2) and pair (19.3) DFs in the complete phase space $x_f = (r_f, v_f)$, multiplying these equations on the left and right by the remaining properties (mass, impulse and energy): $S_f^{(r)}(S_f^{(r)} = m_f (r=1), m_f v_{fj}, j=1,2,3 (r=2-4); mv_f^2/2 (r=5))$ and $S_{f\psi}^{(r,n)} = S_f^{(r)} S_g^{(n)} (S_g^{(n)} = m_{g,} m_g v_{gi}, mv_g^2/2)$ and subsequent averaging over the thermal velocities of the molecules v_{fj} and v_{gi}. To illustrate the idea, the evolution of only one component is considered. The resulting system is the equation of conservation of the mean unary and pair moments at local velocities, describing the transfer of different properties. This is the standard way of describing the flows for unary DFs [42–45]. To analyze the applicability of the local equilibrium condition, we need the means for paired DFs.

The averages for any local values of A_f and A_{fg} with respect to local velocities are defined as

$$\begin{aligned}\langle A_f\theta_f\rangle &= \int A_f\theta_f(\mathbf{r}_f,\mathbf{v}_f)d\mathbf{v}_f = \\ &= \theta_f(\mathbf{r}_f)\int A_f\Psi(\mathbf{v}_f)d\mathbf{v}_f = \theta_f(\mathbf{r}_f)\langle A_f\rangle.\end{aligned} \tag{38.1}$$

$$\begin{aligned}\langle A_{fg}\theta_{fg}\rangle &= \iint A_{fg}\theta_{fg}(\mathbf{r}_f,\mathbf{r}_g,\mathbf{v}_f,\mathbf{v}_g)d\mathbf{v}_f d\mathbf{v}_g = \\ &= \theta_{fg}(\mathbf{r}_f,\mathbf{r}_g)\iint A_{fg}\Psi(\mathbf{v}_f,\mathbf{v}_g)d\mathbf{v}_f d\mathbf{v}_g = \theta_{fg}(\mathbf{r}_f,\mathbf{r}_g)\langle A_{fg}\rangle,\end{aligned} \tag{38.2}$$

or $\langle A_f\rangle = \int A_f\Psi(\mathbf{v}_f)d\mathbf{v}_f$ and $\langle A_{fg}\rangle = \iint A_{fg}\Psi(\mathbf{v}_f,\mathbf{v}_g)d\mathbf{v}_f d\mathbf{v}_g$ – the mean is taken over the velocity part of the unary and paired DFs with weight functions $\psi(\mathbf{v}_f)$ and $\psi(\mathbf{v}_f,\mathbf{v}_g)$.

The unary distribution function in (38.1) can be represented as $\theta_f(\mathbf{r}_f,\mathbf{v}_f)=\theta_f(\mathbf{r}_f)f_f^0\xi_f(\mathbf{v}_f)$ where the function $\xi_f(\mathbf{v}_f)$ characterizes the non-equilibrium correction to the Maxwellian distribution function. Dynamic variables $\mathbf{x}_f$ are variables of the phase space for each particle of the system. When averaging over the coordinates inside the cells and the velocities $\mathbf{x}_f$, equations with a spatial resolution of the particle positions in the cells of the order of their molecular size are obtained [34, 40]. Such a 'coarsening' of the coordinates often proves to be sufficient for a uniform volume phase, but for strong external fields, for example, in the field of the surface potential, one should keep the dependence on the coordinates $\theta_f(\mathbf{r}_f)$.

For paired DFs, the presence of correlated thermal velocities

$$\theta_{fg}(\mathbf{r}_f,\mathbf{r}_g,\mathbf{v}_f,\mathbf{v}_g)\equiv\theta_{fg}(\mathbf{r}_f,\mathbf{r}_g)\Psi(\mathbf{v}_f,\mathbf{v}_g)=\theta_{fg}(\mathbf{r}_f,\mathbf{r}_g)f_f^0f_g^0\xi_{fg}(\mathbf{v}_f,\mathbf{v}_g)$$

is taken into account in (38.2), where $\Psi(\mathbf{v}_f,\mathbf{v}_g)=f_f^0f_g^0\xi_{fg}(\mathbf{v}_f,\mathbf{v}_g)$ is the dependence of the pair DF on the thermal velocities, f_g^0 is the Maxwellian equilibrium distribution function with respect to velocities at the site g, $\xi_{fg}(\mathbf{v}_f,\mathbf{v}_g)$ is the correlation function of the velocity distribution of the molecules at the sites f and g with respect to the Maxwellian function velocities.

We will take into account that the averaging over spatial and velocity variables is carried out independently. In order to simplify the DF recoding, we additionally include in the averaging symbol $\langle ...\rangle$ the procedure averaging over the coordinates $\mathbf{r}_f^i$ and $\mathbf{r}_g^j$ inside the cells f and g, then

$$\left\langle A_f\theta_f^i\right\rangle = \theta_f^i\left\langle A_f\right\rangle,\ \left\langle A_{fg}\theta_{fg}^{ij}\left(\mathbf{r}_f^i,\mathbf{r}_g^j\right)\right\rangle = \\ = \left\langle \theta_{fg}^{ij}\left(\mathbf{r}_f^i,\mathbf{r}_g^j\right)\right\rangle\left\langle A_{fg}\right\rangle = \theta_{fg}^{ij}\left\langle A_{fg}\right\rangle. \tag{38.3}$$

where θ_f^i and θ_{fg}^{ij} are the probabilities of finding a particle i at the site f and two particles i, j at the sites f and g (the usual lattice pair DF [32–34]).

We outline the general idea, omitting intermediate transformations. The transition to the transport equations is related to the separation of contributions to the kinetic equations (19.2) and (19.3) into contributions of different nature [40,41]: Boltzmann (associated with particle collision), Vlasov (unconnected with particle collision) and 'migration' corresponding to the escape of particles from cells through free adjacent cells. Thus, for unary DFs (19.2), it is written as

$$\left(\frac{\partial}{\partial t}+\mathbf{v}_f\frac{\partial}{\partial \mathbf{r}_f}\right)\theta_{(1)}\left(\mathbf{x}_f\right)=I_f=I_{f\zeta}+I_{fg}\left(M\right)-I_{fg}(M), \\ I_{f\zeta}=I_{f\zeta}\left(B\right)+I_{f\zeta}(V) \tag{38.4}$$

where on the left side are the usual terms in the absence of external fields for the evolution of the density of molecules in the Boltzmann equation. The right-hand side of equation (38.4) is called the collision integral I_f, which is divided according to the indicated contributions: Boltzmann ($I_{f\zeta}(B)$) and Vlasov ($I_{f\zeta}(V)$) and $I_{fg}(M)$ describing the migration flow. The Vlasov term describes the interaction of a particle at a site f with all its neighbors $h \in z_f$, but $h \neq \zeta$ [40], with particles that do not collide. That is, the directions of particle velocities at the sites f and ζ are oriented towards each other, providing collision, and the particle velocity directions at the sites f and h are oriented in different directions, except for collisions in the considered time interval. The magnitude of this time interval is determined by the time variation in the left-hand side of the kinetic equation in the derivative $\partial/\partial t$. By its structure, the terms in the term $I_{f\zeta}(V)$ are completely similar to the collision integral $I_{f\zeta}(B)$. The term $I_{fg}(M)$ describes the thermal displacements of particles between cells: from the occupied cell f the particle goes to the free cell g. Sites $g \in z_f^*$ are located around site f, excluding sites ζ occupied by particles A, with which there is a collision (B) and excluding sites h that are near to f and with them there is Vlasov interaction (V). The transfer of properties in the exchange term through particle collisions is taken into account by the molecules at the site ζ with which the particle collides at the site f (i.e., through the analog of the Boltzmann contribution at high densities). The term $I_{gf}(M)$ describes the process associated with the thermal motion of the particles in the opposite direction. These two last terms in (38.4) will be called exchange contributions.

From the kinetic equation (19.3), contributions for the collision integral I_{fg}, which is on the right in the kinetic equation for the pair DFs [41], are similarly distinguished. This collision integral I_{fg}, consists of the following contributions:

$$I_{fg} = I_{(f)g} + I_{(g)f} + I_{(f\psi)g}(M) - \\ -I_{(\psi f)g}(M) + I_{f(g\psi)}(M) - I_{f(\psi g)}(M), \tag{38.5}$$

where $I_{(f)g} = I_{(f\zeta)g}(B) + I_{(f\zeta)g}(V)$. Here, the $I_{(f\zeta)g}(B)$ term describes the Boltzmann collisions of the particle f from the pair fg with particles at the sites ζ that do not coincide with the site g. The particle at site g is near and can influence by its potential the collision of particles f and ζ. The same effect through the interaction potential is exerted by particles that are near at sites h that do not coincide with the sites g and ζ. This situation is described by Vlasov's contribution $I_{(f\zeta)g}(V)$. The contribution $I_{(g)f}$ for another particle at site g from the given pair fg has a similar meaning. The remaining terms (38.5) describe the exchange terms. The term reflects $I_{(f\psi)g}(M)$ two exchange mechanisms by moving particles from site f to site ψ or passing a property from site f to site ψ, as described above for the unary function. The parentheses indicate participants in the process – the particle at site g is nearby and can affect the course of the transfer of properties, but there is no direct collision with it.

The term $I_{f(g\psi)}(M)$ has the same meaning as the term $I_{(f\psi)g}(M)$. The brackets indicate the transfer mechanisms of the properties transfer between the sites g and ψ, and the particle f is close by. Finally, the term $I_{(f\psi)g}(M)$ describes the process inverse to the process given by the term $I_{f(g\psi)}(M)$.

After separating the contributions and multiplying the kinetic equations by local properties, we obtain the corresponding transport equations. The separation of contributions by different types is necessary for constructing expressions for dissipative coefficients, which have different forms on different spatial scales.

Complete system of transport equations. If we confine ourselves to the traditional assumption for hydrodynamics [1,36,46] that the transferred properties $S_f^{(r)}$ and $S_{fg}^{(r,n)}$ themselves remain practically constant on the characteristic scale of the change in the coordinates: $\partial S_f^{(r)} / \partial r_{fj} = 0$ (This eliminates the strong non-uniformity of the density of the system inside the region *f*), similarly $\partial S_{fg}^{(r,n)} / \partial r_{fj} = 0$ and $\partial S_{fg}^{(r,n)} / \partial r_{gj} = 0$, then the following system of equations of conservation of unary and pair properties is constructed

$$\frac{\partial\langle \theta_f S_f^{(r)}\rangle}{\partial t} + J_f^{(r)} = I_f^{(r)}, \quad J_f^{(r)} =$$
$$= \sum_{j=1}^{3} \frac{\partial\langle \theta_f v_{fj} S_f^{(r)}\rangle}{\partial r_{fj}} - \theta_f \left\langle \frac{F(f)_j}{m} \frac{\partial S_f^{(r)}}{\partial v_{fj}} \right\rangle, \tag{38.6}$$
$$I_f^{(r)} = \left\langle \left[\sum_{g \in z_f} \left(S_g^{(r)} U_{fg}^{VA} - S_f^{(r)} U_{fg}^{AV} \right) + \sum_{g \in z_f} \left(S_g^{(r)} U_{fg}^{AA^*} - S_f^{(r)} U_{fg}^{A^*A} \right) \right] \right\rangle$$

$$\frac{\partial\langle \theta_{fg} S_{fg}^{(r,n)}\rangle}{\partial t} + J_{fg}^{(r,n)} = I_{fg}^{(r,n)},$$
$$J_{fg}^{(r,n)} = \sum_{j=1}^{3} \left[\frac{\partial\langle \theta_{fg} v_{fj} S_{fg}^{(r,n)}\rangle}{\partial r_{fj}} + \frac{\partial\langle \theta_{fg} v_{gj} S_{fg}^{(r,n)}\rangle}{\partial r_{gj}} - \theta_{fg} \left\langle \frac{F(f)_j}{m} \frac{\partial S_{fg}^{(r,n)}}{\partial v_{fj}} \right\rangle - \theta_{fg} \left\langle \frac{F(g)_j}{m} \frac{\partial S_{fg}^{(r,n)}}{\partial v_{fj}} \right\rangle \right],$$
$$I_{fg}^{(r,n)} = \left\langle \sum_{h \in z_f^*} \left(S_{hg}^{(r,n)} U_{(hf)g}^{(AV)A} - S_{fg}^{(r,n)} U_{(hf)g}^{(VA)A} \right) + \sum_{h \in z_f^*} \left(S_{hg}^{(r,n)} U_{(hf)g}^{(A^*A)A} - S_{fg}^{(r,n)} U_{(hf)g}^{(AA^*)A} \right) + \right. \tag{38.7}$$
$$\left. + \sum_{h \in z_g^*} \left(S_{fh}^{(r,n)} U_{f(gh)}^{A(VA)} - S_{fg}^{(r,n)} U_{f(gh)}^{A(AV)} \right) + \sum_{h \in z_g^*} \left(S_{fh}^{(r,n)} U_{f(gh)}^{A(AA^*)} - S_{fg}^{(r,n)} U_{f(gh)}^{A(A^*A)} \right) \right\rangle$$

where t is time, the sum over j is taken along the axes x, y, z (the index j = 1–3) – it refers to the components j of contributions from the cell variables f or g, $F(f)_j$ is the component j of the external force in the cell f; instead of the traditional mass density ρ ($\rho = m\theta / v_0$), a numerical density θ [40,41] is used. The left-hand sides of $J_f^{(r)}$ and $J_{fg}^{(r,n)}$ equations determine the hydrodynamic flows. Equations (38.6) and (38.7) differ from the expressions in [40,41] by the presence of contributions to the left of the external forces, and also by the explicit form of the collision integrals, expressed in terms of the thermal velocities of migration of the molecules to the free neighboring cells U_{fg}^{AV} (see Section 56) and the collision rate in case of occupation of these cells U_{fg}^{A*A} [34]:

$$U_{fg}^{ij} = K_{fg}^{ij}\theta_{fg}^{ij}\exp(-\beta\varepsilon_{fg}^{ij})\Lambda_{fg}^{lj},\Lambda_{fg}^{ij}\Lambda\left(S_f^i S_g^j\right)^{z-1}, \tag{38.8}$$

where the expressions for the non-ideality functions S_f^i are defined on the basis of TARR (theory of absolute reaction rates) in Sections 53 and 55. To calculate the terms in Eq. (20.5), the expressions $U_{hfg}^{(ij)A} = U_{hf}^{ij}\Psi_{fg}^{jA}, \Psi_{fg}^{jA} = t_{hg}^{jA}\exp\left(\beta\delta\varepsilon_{fg}^{jA}\right)/S_{fg}^{j}$ are used in QCA. The structure of the equations is discussed in more detail in Appendix 2.

The dimension of the complete system of transport equations b (38.6) and (38.7): $b = b_1 + b_2$, consists of the number of equations for the transfer of unary properties $S_f^{(r)}$, $b_l = 5$, and the number of equations for the transfer of paired properties $S_{fg}^{(r,n)}$, $b_2 = 15$. The remaining properties of pairs (write these variables without permuting the indices f and g) are the following functions: 5 functions of the type $\left\langle m_f S_g^{(n)}\right\rangle$ (mass – property $S_g^{(n)}$); 6 functions of type $\langle m_f v_{fi} m_g v_{gk}\rangle$ (impulse – impulse); 3 functions of the type $\langle m_f v_{fi}/m_g v_g^2/2\rangle$ (impulse – energy); and one function $\langle m_f v_f^2/2m_g v_g^2/2\rangle$ (energy–energy). We denote by y_j the dynamic variable for unary and pairwise moments, $j \in b$. To indicate that the dynamic variable y_j refers to a particular group of these variables, we use the notation $j \in b_l$ or $j \in b_2$.

39. The hierarchy of Bogolyubov times

Practically in all theoretical approaches in the kinetics of physicochemical processes in various phases based on the equations of chemical kinetics [47–50], hydrodynamics [1, 2, 35, 36], non-equilibrium thermodynamics [28–30], and kinetic theory [42–35], for any differences in the non-equilibrium with respect to concentrations, the assumption of local equilibrium for paired DFs and the smallness of deviations from it are possible when proceeding to the description of processes at the macroscopic level. In this case, the number of parameters of the state of the system in dynamics coincides with their number in equilibrium.

From the molecular point of view, the local execution of the combined equation for the First and Second Laws of thermodynamics is due to the unique dependence of the pair distribution function of the molecules θ_2 on the density (or on the unary DF) at a fixed temperature at any time t. Usually such a dependence is expressed in the form of algebraic or integral equations connecting the First and Second Laws DFs without the time argument $\theta_2(\theta)$. This leads to the fact that at a fixed temperature and numerical density θ of the particles the pair DF is time-independent and $d\theta_2/dt = 0$.

An analysis of the condition for the realization of this condition was considered in [51] for a one-component system, which allowed us to introduce a criterion fixing the absence of the condition of local equilibrium; when the critical value of this criterion is reached, the description of the transport process within the framework of classical non-equilibrium thermodynamics becomes incorrect.

The organization of the flow of pairs between the regions *f* and *g*. Consider the flow of molecules between regions with the coordinates of the planes *f* and *g* (Fig. 39.1). Suppose that the linear size of the region in which the local equilibrium is realized is L, so that $dV \sim L^3$. In Fig. 39.1 *a* this area is selected as a rectangle; therefore, for the coordinates *f* and *g*: $d\theta_2(f)/dt = 0$ and $d\theta_2(g)/dt = 0$, and a flow of molecules (concentrations) and their pairs occurs between these regions with local equilibrium. If the distance between the coordinates *f* and *g* is less than the size L, then there is a common region for local equilibrium of the pairs with respect to the local density. In this case, there is no transfer of pairs of molecules between the planes *f* and *g*, and, consequently, of the density transfer: the local equilibrium region is incorrectly chosen. If the distance between the coordinates *f* and *g* exceeds the size L, then formally the flow of pairs is realized only between two regions of local equilibrium.

A more accurate way of describing the flow of pairs between the coordinates *f* and *g* is obtained if it is functionally related to the value of the density gradient, without specifying the size of the local

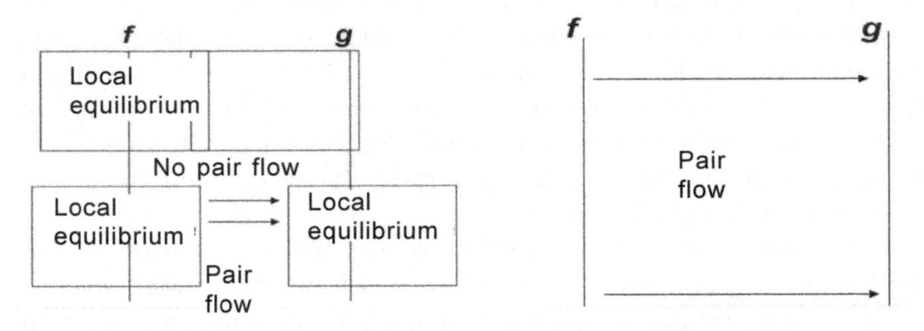

Fig. 39.1. Formal representation of the flows of molecules and their pairs between two regions of local equilibrium with coordinates *f* and *g*.
(right) Flows of molecules and their pairs between the two coordinates *f* and *g* in the case of a functional connection of the flow of pairs to the molecular density gradient.

equilibrium region L (Fig. 39.1 b). This possibility gives an idea of the hierarchy of characteristic relaxation times between the first and second DFs in the local equilibrium condition, proposed by N.N. Bogolyubov [38]: the time dependence of the pair DF is expressed in terms of the time dependences of unary DFs

$$\theta_2(x_f, x_g, t) = \theta_2(\theta(x_f, t),\ \theta(x_g, t)), \tag{39.1}$$

here $x_f = (r_f, v_f)$ and we consider the total phase space of the coordinates r_f and velocities v_f in the neighborhood of the coordinate f; In this case, local quasi-equilibrium of the pairs

$$d\theta_2(\theta(x_f, t),\ \theta(x_g, t)) / \partial t = A = 0, \tag{39.2}$$

where A is some function that must be constructed from the kinetic equation for the pair DF with the condition of a small deviation of the pair DF θ_2 from its equilibrium θ_2^e value for a given density θ. Therefore, $\theta_2(t) = \theta_2^e + \delta\theta_2(\theta(t))$, where $\delta\theta_2(\theta(t))$ is the correction to the value of the equilibrium paired DF due to the influence of the flow of molecules changing the density as a function of time θ(t). Or, the main difference between the notions of a stationary (quasi-equilibrium) and equilibrium is the appearance of $\delta\theta_{(2)}(\theta(t))$ a correction depending on the properties of the flow, with their general property – the independence of the pair DF on time $d\theta_2/dt = 0$.

40. The criterion for local equilibrium

Consider the reduction of the constructed system of equations due to the transition to the local equilibrium condition from the positions of the hierarchy of times of N.N. Bogolyubov [38]. According to the expression (39.2) for a pair DF in the total phase space, this transforms the kinetic system of equations (38.7) for paired moments in the equations that are not explicitly dependent on time.

Local equilibrium for a complete system of equations. This condition means $d\left\langle S_{fg}^{(r,n)}\theta_{fg}\right\rangle / dt = 0$, moreover, as, then it allows to $\left\langle S_{fg}^{(r,n)}\theta_{fg}\right\rangle = \left\langle S_f^{(r)}\theta_f S_g^{(n)}\theta_g\right\rangle$ say that the time dependence of the moment related to site f is only on average and $\left\langle S_f^{(r)}\theta_f\right\rangle$ to site g on average $\left\langle S_g^{(n)}\theta_g\right\rangle$ (but this does not mean the loss of spatial correlations between molecules at the sites f and g). Then for the summands $J_{fg}^{(r,n)}$ and $I_{fg}^{(r,n)}$ in the system (38.7) we have

$$d\left\langle S_{fg}^{(r,n)}\theta_{fg}\right\rangle / dt = A_{fg}^{(r,n)} = J_{fg}^{(r,n)} - I_{fg}^{(r,n)} = 0 \qquad (40.1)$$

where the functions $A_{fg}^{(r,n)}$ must be constructed at the following positions, which determine small perturbations of dynamic variables in the local equilibrium region.

When describing the flows of unary properties in the volume of the system, the usual gradient expressions are used from Eq. (38.6):, where $y_j(1) = y_j(0) + \text{grad}(y_j(0))\lambda/2$ where λ is the cell size, the symbols 0 and 1 indicate the numbering of the sites in a certain direction. Such expansions determine the main contributions of the flows for $j \in b_1$ (through $\text{grad}\left\langle \theta_g S_g^{(n)}\right\rangle$), taking into account the dependence of the thermal velocities (38.8) on unary and paired DFs.

When describing the flows of paired properties, the usual gradient expressions are also used: $y_j(-1,0) = y_j(0,1) - \text{grad}(y_j(0,1))\lambda/2$ and $y_j(1,2) = y_j(0,1) + \text{grad}(y_j(0,1))\lambda/2$, where $j \in b_2$. As explained above, the paired dynamic variables $y_j = \left\langle S_{fg}^{(r,n)}\theta_{fg}\right\rangle$, $j \in b_2$, can be represented at local equilibrium in the form $y_j = y_j^e + \delta y_j$, where y_j^e is the value of the given variable in the state of strict equilibrium (i.e., in the absence of flow), and the correction δy_j to equilibrium values y_j^e is due to the influence of the flow and we confine ourselves to the linear approximation $\text{grad}(y_j) = \text{grad}(y_j^e)$; the perturbations of the values of the variables due to the flows are small. The smallness of these quantities coincides in order with the smallness of the gradients of the properties that cause this flow.

Using the superposition approximation for third-order hydrodynamic moments $J_{fg}^{(r,n)}$ in the left-hand sides and substituting the gradient expansions of all the properties $J_{fg}^{(r,n)}$ on the right-hand sides for each of the equations of the system (38.7) allows us to construct expressions $A_{fg}^{(r,n)} = 0$. Such expansions, also taking into account the dependence of the thermal velocities (38.8) on the unary and paired DFs, isolate additional contributions from the flows (through $\delta\left\langle S_{fg}^{(r,n)}\theta_{fg}\right\rangle$) from them. Whence, dropping the cumbersome technical details of taking into account the specifics of the transfer of properties along two mechanisms (through displacements and collisions) and solving the algebraic system of equations $A_{fg}^{(r,n)} = 0$, we write out the main result – the structure of the connection between the gradients of dynamic variables $\left\langle \theta_f S_f^{(n)}\right\rangle$ and non-equilibrium corrections to the moments of paired DFs

$$\delta\left\langle S_{fg}^{(r,n)}\theta_{fg}\right\rangle = \sum_{n=1}^{b_1} B_{fg}^{(r,n)} \operatorname{grad}\left\langle \theta_g S_g^{(n)}\right\rangle. \tag{40.2}$$

The coefficients $B_{fg}^{(r,n)}$of this expansion represent additional contributions to the dissipative coefficients of the transfer of properties. In the particular case $r = n$ they are analogs of the well know diffusion coefficients at $r = n = 1$ [31,52] (by analogy with diffusion in multicomponent systems – although here we are looking at a one-component system). For $r \neq n$, such coefficients are called crosses: they characterize the flows of the property r under the influence of the gradient of the property n.

In [38], the concept of a hierarchy of characteristic relaxation times between the first and second DFs under the condition of local equilibrium was introduced for a gas with a Maxwellian velocity distribution function, i.e. without taking into account the correlations between the molecular velocities. This same assumption has always been used for any laminar flow in dense phases [45-45]. The presence of correlations of the hydrodynamic velocities appeared in the Karman–Howarth equation at the mesolevel [36,46], whereas here they arise at the microlevel [40]. In [31], the concept of a hierarchy of characteristic times under the condition of local equilibrium is formulated for the pair moment of masses with respect to the processes of mutual diffusion of the components of the alloys.

The obtained structure of expressions (40.2) for the pair moments ($j \in b_2$) shows that the correlation values of the velocity properties will be of the order of the magnitude of the gradient causing this flow. In fact, it is through such small fluctuations that nucleation and the development of macroscopic correlations are realized, both in size and intensity. The very existence of large correlations is possible only in the case of a local equilibrium. The introduced criterion for the non-equilibrium of the partial paired moments separates the regions of thermodynamic parameters related to local equilibrium. If there were no velocity correlations for the equilibrium state, then in the presence of a flux these correlations are non-zero. Their role depends strongly on the density and on the hierarchy of the characteristic relaxation times (see below).

For dense phases, equilibrium density correlations play an important role in the thermodynamic characteristics of the system and, with respect to them, density fluctuations can play a decisive role in the course of diffusion processes. The proposed approach generalizes the equations [34] used to study the dynamics of low-intensive flows in strongly non-uniform porous systems. It is of fundamental importance that the dimension of the reduced system remains the same as in ordinary hydrodynamics or non-equilibrium thermodynamics equal to $b_1 = 5$. We note that the cross dissipative coefficients for the transfer of unary properties appear from the kinetic equation to paired DFs even in the absence of accounting for collisions with neighbouring particles [53] (that is, $U_{hfg}^{(A^*A)A} = U_{hfg}^{(VA)A} = 0$).

Local disequilibrium. In intense fluxes $d < S_{fg}^{(r,n)} \theta_{fg} > / dt \neq 0$, the correlations can be much larger than in accordance with formula (40.2), and therefore it is necessary to have a criterion indicating the inability to use the reduced equations in place of the complete system of equations (38.6), (38.7), i.e. go to the full system with dimension b. If we can single out unary DFs and the explicit (relaxation) dependence of the pair DF on the time as independent arguments, then instead of (39.1) we have:

$$\theta_{(2)}(x_f, x_g, t) = \theta_{(2)}\left(x_f, x_g, t, \theta_{(1)}(x_f, t),\ \theta_{(1)}(x_g, t)\right) \tag{40.3}$$

Here the choice of independent variables in the full-pair DF corresponds to the structure of the time hierarchy between the distribution functions of different dimensions: the independent variables for the time dependences are the non-equilibrium functions $\theta_f(x_f, t)$, $\theta_g(x_g, t)$ and the paired DFs.

The evolution of a pair DF in the total phase space describes how

$$\frac{d\theta_{(2)}(t, x_f, x_g)}{dt} = \frac{\partial\theta_{(2)}(t, x_f, x_g)}{\partial t} + \frac{\partial\theta_{(2)}(t, x_f, x_g)}{\partial\theta_{(1)}(x_f, t)} \frac{d\theta_{(1)}(x_f, t)}{dt} + \\ + \frac{\partial\theta_{(2)}(t, x_f, x_g)}{\partial\theta_{(1)}(x_g, t)} \frac{d\theta_{(1)}(x_g, t)}{dt}. \tag{40.4}$$

Hence it follows that the condition of non-equilibrium of the states of the system that go beyond the condition of local equilibrium is the inequality

$$\partial\theta_{(2)}(x_f, x_g, t) / \partial t \neq 0, \tag{40.5}$$

connected with the explicit time dependence of the pair DF.

Denoting the three terms on the right-hand side of (40.4), respectively $d\theta_{(2)}(t, x_f, x_g)/dt = d_{fg} + D_{fg} + D_{gf}$, we introduce the minimum value $D_{\min} = \min\left(\left|D_{fg}\right|, \left|D_{gf}\right|\right)$ use it to write the expression for the ratio of the relaxation and synchronous variation of the pair DF: $\omega = |d_{fg}| / D_{\min}$. This ratio characterizes the relaxation contribution of the direct variation of the second DF with time in comparison with the change due to a synchronous change in the pair DF from the unary DF, which varies with time (as in eq, (39.1)). It is natural to take it for the degree of disequilibrium. The value ω = 0 corresponds to the case of local equilibrium.

Similarly, when passing to other moments in equations (38.7), we can introduce the same time arguments for the moments of the pair DF

$$\left\langle S_{fg}^{(r,n)}\theta_{(2)}(x_f, x_g, t)\right\rangle = \left\langle S_{fg}^{(r,n)}\theta_{(2)}(x_f, x_g, t, \theta_{(1)}(x_f, t),\ \theta_{(1)}(x_g, t))\right\rangle \quad (40.6)$$

Then the derivative for the pair moment is written as follows

$$\frac{d\left\langle S_{fg}^{(r,n)}\theta_{fg}\right\rangle}{dt} = \frac{\partial\left\langle S_{fg}^{(r,n)}\theta_{fg}\right\rangle}{\partial t} + \frac{\partial\left\langle S_{fg}^{(r,n)}\theta_{fg}\right\rangle}{\partial\left\langle S_{f}^{(r)}\theta_{f}\right\rangle}\frac{d\left\langle S_{f}^{(r)}\theta_{f}\right\rangle}{dt} + \\ + \frac{\partial\left\langle S_{fg}^{(r,n)}\theta_{fg}\right\rangle}{\partial\left\langle S_{g}^{(n)}\theta_{g}\right\rangle}\frac{d\left\langle S_{g}^{(n)}\theta_{g}\right\rangle}{dt}. \quad (40.7)$$

Introducing in (40.7) $d\left\langle S_{fg}^{(r,n)}\theta_{fg}\right\rangle / dt = d_{fg}^{(r,n)} + D_{fg}^{(r,n)} + D_{gf}^{(n,r)}$, then with their help one can obtain a system of relations responsible for $\left\langle S_{fg}^{(r,n)}\theta_{fg}\right\rangle$ the direct relaxation effect on the change in the time of the moment in comparison with its synchronous variation $\omega_{fg}^{(r,n)} = |d_{fg}^{(r,n)}| / D_{fg,\min}^{(r,n)}$, where $D_{fg,\min}^{(r,n)} = \min(|D_{fg}^{(r,n)}|, |D_{gf}^{(n,r)}|)$.

Such partial criteria, which fix the absence of the condition of partial local equilibrium, when certain critical values are reached, indicate an incorrect description of the transport process in the framework of classical non-equilibrium thermodynamics.

More complicated cases of constructing analogues of criteria for the non-equilibrium state of the system in the case of a multicomponent mixture were considered in [54]. This allows us to consider processes with a strong deviation from local equilibrium and to discuss the concept of 'passive forces' introduced by Gibbs [55].

41. Strongly non-equilibrium states and the structure of transport equations

The formulated structure of the complete system of equations (38.6) and (38.7) allows, as a solution, to obtain, as a function of time, the average values of the five dynamic variables $\langle S_f \rangle$ and their 15 pair combinations $\langle S_{fg} \rangle$ (see Appendix 2). These equations represent a microscopically modified hydrodynamic theory of mass, impulse, and energy transfer due to allowance for fluctuations arising when the flow velocity increases, and the kinetic theory of mean square fluctuations of dynamic variables describing the processes of mass, impulse, and energy transfer. The theory relies on closed expressions for the decoupling of multiparticle probabilities through unary and paired DFs and does not have a small thermodynamic parameter (in contrast to the approaches discussed in Ref. [56]), and can therefore be used for various strongly non-equilibrium processes. Introducing in the usual way [57], deviations of one dynamic variable $\Delta S_f = S_f - \langle S_f \rangle$, and their standard deviations as $\Delta S_{fg} = \langle S_f - \langle S_f \rangle \rangle (S_g - \langle S_g \rangle) = \langle S_{fg} \rangle - \langle S_f \rangle \langle S_{fg} \rangle$, as a result of the solution of the constructed system of equations, the time dependences of the root-mean-square fluctuations of the dynamic variables S_f will be obtained. In contrast to the correlation functions that can be introduced for one or two different times [1,36,46], all the new 15 dynamic variables $\langle S_{fg} \rangle$ refer to one time point that corresponds to the time scale of the evolution of the variables $\langle S_f \rangle$.

The resulting system of transport equations for unary properties (38.6) is supplemented by equations for the pair properties (38.6). The general property of the resulting new transport equations for pair properties is that for each unknown function new unknown functions appear with increasing dimension in hydrodynamic velocities, or for both sites of the pair due to the displacement of this property by the hydrodynamic flow, or for one of the sites. In the constructed system of equations, this effect is due to the non-equilibrium nature of the pair DF.

This increase in the dimension of the hydrodynamic variables is a known property of the coupled equations at the hydrodynamic level [46,58], which is completely analogous in its nature to the linkage for the DF at the microscopic level [38] (but known prior to the establishment of an analogous fact for molecular DFs [38]). In both cases, the problem arises of decoupling the higher DF from the lower ones, because otherwise the dimension of the system increases indefinitely. To close it, we need a single approximation of all the equations constructed to preserve the description of the evolution of the transfer of all properties by a self-consistent method (see Appendix 2).

Apparently, for the first time the attraction of additional invariants S_{fg} (except for the five known S_f) in the kinetic theory of ideal gases was proposed in [59, 60]. In this way, the kinetic equations for the unary and paired DFs were constructed. However, it was not possible to really advance in this direction, because the basis was the modification of the unary non-equilibrium distribution function (in the manner of the Maxwellian function with additional invariants in the exponent). Moreover, this way had no prospects for dense systems taking into account intermolecular interactions. In the approach [40,41], qualitatively different from [59, 60], there is not only the transition to dense systems, but also to the concept of correlations, both between thermal velocities and between microhydrodynamic velocities, which are implicitly dependent on each other. Recall, once again, that a strictly equilibrium distribution is realized only in the absence of flows. The traditional assumption of the independence of thermal velocities from local hydrodynamic velocities is associated with the consideration of sufficiently large sizes of regions, as discussed in Section 34. In this case, an almost equilibrium distribution of pairs is established inside the region R_v.

The appearance of correlated hydrodynamic variables can be associated with a violation of the local equilibrium condition when mechanical, energy and/or chemical equilibrium is disrupted. In terms of physical meaning, the condition of local equilibrium uniquely relates the probabilities of unary and paired distribution functions (DF) at a constant temperature. Any violation of local equilibrium should lead to a disruption of the equilibrium coupling between pair and unary DFs and for the description of fluxes it is necessary to use kinetic equations for both unary and pairwise DFs.

The constructed system of transport equations for strongly non-equilibrium processes can be treated as at the initial instant of time under mechanical, temperature and concentration perturbations with respect to the initial state, and when they arise during the evolution of the process. For a single-component substance, the local equilibrium is primarily disturbed by an increase in the linear flow velocity.

Such closure procedures should be solved for each specific statement of the problem. For these purposes, the constructed system of equations represents a rigorous basis for a correct analysis of the nature of disengagement. In concrete formulations, this system of equations is drastically reduced by the allocation of space-time scales, but it remains common for any matter densities. In the general case (with the exception of rarefied gases), the following characteristic scales of the relaxation times for impulses, temperature and density [57] can be considered fulfilled: $\tau_{\text{impulse}} \ll \tau_{\text{temperature}} \ll \tau_{\text{density}}$.

In the presence of local equilibrium, not only the equations (38.6) go over into the traditional system of equations of hydrodynamics, but under these conditions all effects of the correlation of the hydrodynamic variables in equations (38.7) disappear and all the introduced means for the pair functions vanish. This makes it possible to use the conditions for realizing local equilibrium as a necessary and sufficient criterion for neglecting the inclusion of correlation effects between the considered combinations of local mean characteristics, which justifies the existence of a local generalized equation (8.3) for the First and Second laws of thermodynamics [30].

Simplifications of the dynamic part of the system (38.6) and (38.7) can also be realized if the flows are not very intense, but are not in equilibrium with a strong non-uniformity of the system, for example, for the wall regions of solids – they have a strong gradient of velocities and densities (although the flow velocity along the surface is not necessarily zero) [34]. In this case, the first transfer equation in (38.7) becomes the equation of [34,61,62] $\dfrac{\partial\left\langle <\theta_{fg}>\right\rangle}{\partial t}+\sum_{j=1}^{3}\left\{\dfrac{\Delta<\theta_{fg}v_{fj}>}{\Delta r_{fj}}+\dfrac{\Delta<\theta_{fg}v_{gj}>}{\Delta r_{gj}}\right\}=I_{fg}^{(1,1)}$ describing the dynamics of low-intensity flows in porous systems.

Analogous simplifications can also be realized in subsequent equations of the system for paired DFs, if the fluxes are not very intense, but non-equilibrium due to the strong non-uniformity of the system. Conversely, in the situation with non-equilibrium flows of sufficient intensity, an analogue of the Karman–Howarth equation is obtained from equations (38.6) (see Appendix 2), which is well known in the theory of turbulence [36]. The equations of Appendix 2 (A2.9) and (A2.10) agree with the results of fluctuation hydrodynamics [64] in the sense that the fluctuations of the velocities and the heat flux do not depend on each other and can be considered independently.

Kinetic equations of similar structure are used in problems of plastic deformation in solids [65–67]. In these papers, we propose the analogy of turbulence in solids and liquids. In solid bodies deformation is regarded as a non-equilibrium transition in an ensemble of defects such as microscopic shifts. This allowed us to consider a number of structural–scaling transitions in the description of thermodynamics and kinetic effects in materials in the bulk submicro- (nano-) crystalline state, including the processes of plastic deformation of solids.

The question of the kinetic equations for the evolution of solids, apparently, is one of the most complex, since in comparison with gas and liquid systems, deformation states play an additional important role in them, which determine spatio-temporal correlations in the local mechanical and density characteristics [65–72]. Taking into account the rapid thermal relaxation in solids and the establishment of the total temperature, the system of equations written out is sharply reduced when describing 'frozen' or metastable states. In this case, the so-called flicker noise is often observed, which is associated with the rearrangement of the solid matrix over long times [73, 74]. The physical nature of this phenomenon can be different [75–77], and requires analysis in specific processes. The constructed equations make it possible to pass from phenomenological constructions [74] to atomic–molecular models of such processes.

In a general case, the equations for constructed for the pair correlations between the masses, velocities, and energies of molecules describe a broad class of cooperative processes at intermediate (supramolecular) levels in three aggregate states. The possibilities of the new system of equations are oriented to the study of physicochemical processes in turbulent flows and in frozen matrices of solids, in which the state of the system can differ very much from the equilibrium states.

42. Relaxation times and passive forces

Different state parameters differ by their relaxation times. In real conditions, different relationships between them are possible [57,64]: from $\tau_{imp} \leq \tau_{ener} \leq \tau_{mass}$ for rarefied gases up to

$$\tau_{imp} \ll \tau_{ener} \ll \tau_{mass}, \tag{42.1}$$

which allows partial reduction of the complete system of equations (38.6), (38.7) for any degree of deviation from equilibrium, since part of the fastest dynamic variables will correspond to the stationarity condition $d\left\langle S_{fg}^{(r,n)}\theta_{fg}\right\rangle / dt = 0$ with zero time derivative.

The constructed equations allow us to discuss the concept of so-called 'passive forces' well known in thermodynamics, introduced by Gibbs [55]. In the complete system of equations (38.6) and (38.7) all forces present in real systems are explicitly taken into account: intermolecular interactions and external potentials. The kinetic theory of the atomic–molecular level does not work with fictitious forces. As the strongly non-equilibrium system approaches the equilibrium state, new forces can not appear in it. In the equilibrium state, dynamic equilibrium is realized, and all reversible processes have equal in magnitude but opposite directed velocities of elementary stages, which allows the system to maintain its unchanged state in time. Therefore, equations (38.6) and (38.7) exclude the appearance of any unknown forces in thermodynamic equilibrium, and the 'appearance' of passive ones can be due only to the lack of information and/or the inability (unwillingness) to establish the cause of the observed unchanged states of the system under study. By the time the 'passive forces' [55] were introduced (~1875), a number of factors that were established later were unknown. Gibbs pointed to the existence of two classes of situations for which he introduced 'passive forces'. These are 1) chemical reactions under conditions that do not allow the realization of chemical transformations, and 2) mechanical effects on solids, for example, in the example of sliding friction.

In the first case, many situations are known where the process is impeded by high activation barriers, which can not be overcome at low temperatures or in the absence of a catalyst. The example with the process of oxidation of hydrogen, chosen by Gibbs, is additionally significantly complicated by the multistage chain mechanism of the process [47,78]. However, today this is the usual 'working' non-equilibrium situation, connected with a large difference in the internal states of reagents, described by the modern theory of chemical transformations [47,49,50,78–80]. Naturally, at the time intervals (42.1), when the role of the chemical reaction is negligible, the state of the reagents can be described by thermodynamic constraints (without a time argument).

When transferring the mass, it is necessary to distinguish between methods for changing concentrations – reactions or diffusion. In the kinetic regime, the diffusion relaxation times are shorter than the relaxation times of the chemical transformations. In the diffusion regime, on the contrary, the relaxation times of the chemical transformations are less than the diffusion relaxation times. All this requires detailing the relaxation times of τ_{mas} both by process type and by each of the components i of the mixture, $1 \leq i \leq s$, where s is the number of components in the mixture: $\tau_{den}^{dif}(i)$ and $\tau_{den}^{reac}(i)$. Differences between $\tau_{den}^{dif}(i)$ and $\tau_{den}^{reac}(i)$ for each component are associated with changes in its energy in the course of redistributions in space or in internal states. The former usually requires less energy change due to lateral interactions, which are usually less than the energy of internal bonds. In the second case, the changes are associated with internal rearrangements and with disconnection.

In equilibrium thermodynamics, the quantity $\tau_{den}^{reac}(i)_{\max}$ refers to the times for the maximum value of the relaxation time related to the relaxation time of the component i, having the maximum value: $\tau \gg \tau_{mas}^{reac}(i)_{\max}$. If $\tau < \tau_{mas}^{reac}(i)_{\max}$, then this refers to the situation for which Gibbs introduced passive forces of the first kind associated with inhibition of chemical transformations. The range of relaxation times extends to $\tau < \tau_{mas}^{dif}(i)_{\min} << \tau_{mas}^{reac}(i)_{\max}$, where $\tau_{mas}^{dif}(i)_{\min}$ is the minimum relaxation time of the diffusion redistribution from all components of the mixture. This set of characteristic times allows us to give a strict interpretation of the separation of components into mobile and immovable and 'decipher' all versions of Gibbs' 'passive forces' [22].

When differentiating the relaxation times of masses in solids, we take into account that $\tau_{mas}^{dif}(i)$ can decay into two different stages: the relaxation of the density of component i with time $\tau_{mas}^{dif}(i)$ for mass transfer and the relaxation of the pair DF for components ij with time $\tau_{mas}^{dif}(ij)$. They are related to each other by the conditions of the hierarchy of time as $\tau_{mas}^{dif}(ij) < \tau_{mas}^{dif}(i)$ (the relaxation of the impulse proceeds much faster than the relaxation of temperature, and the latter is much faster than the relaxation processes of mass transfer). This specificity of solids leads to an increase in the number of dynamic variables, since paired DFs do not enter into the number of ordinary thermodynamic variables, therefore, in real conditions, it is necessary to prove experimentally the existence of a mass transfer equilibrium and equalization of the chemical potential throughout the system. Only then can the conditions for the application of postulates and equations of classical equilibrium thermodynamics be satisfied. In reality, especially for dense phases, such a proof is very rare, and in an overwhelming number of situations the equations of thermodynamics are applied postulatively.

In the second case, we are dealing with a non-equilibrium in the displacements of molecules in space, or with their deformation states in a solid, described by Eqs. (38.6) and (38.7). The question of the generality of methods for describing a rigid body and hydrodynamics within the framework of the theory of a continuous medium is discussed in [2]. The absence of the continuity equation in Lamé's equations for displacements in solids is due to the condition of invariance of the density, which contradicts all mechanochemical processes discovered later, in which the density changes due to diffusion and reactions. For the relative displacement of two solid macrobodies of the general system, discussed by Gibbs, it is necessary to spend efforts on overcoming frictional forces. Tribology is actively engaged in this process today, and the same molecular models [81] are used for it as for other catalysis, the physical chemistry of the surface and the chemical physics of solids and mechanochemistry [70,82,83]. The situation depends on the time ranges of mechanical influences: whether diffusion mixing of atoms of contacting bodies begins or not. We should note (see numerous references in [82,84,85]) the work on the development of a theory for the diffusion–viscous flow of solids, the evolution of structural imperfections due to diffusion processes in polycrystals, the formation of discontinuities/ruptures, as well as the formation of porous bodies and the dynamics of pores. In these papers, the theory included the continuity equation for a rigid body, as in hydrodynamics, and replaced the coefficient of sliding friction by the coefficient of viscosity [3] (which, as Gibbs noted, is not related to the concept of passive forces.) Therefore, today the level of solid state theory processes is significantly different from the ideas of the Gibbs times. For our time, the concept of 'passive force' is not only archaic, but also incorrect – without an explicit indication of the driving force, it is impossible to build kinetic equations.

From the point of view of the kinetic theory (38.6) and (38.7), all the problems of mechanochemistry satisfy the condition (42.1) [2,3,81–85]. To apply thermodynamic concepts to the description of solid-phase processes, a preliminary analysis is needed on the possibility of using thermodynamics (or on the absence of diffusion inhibitions). Estimates of section 36 of the characteristic relaxation times for the density of solid elements in spherical samples of radius from 1 cm to 1 nm have shown that even the simplest vacancy density equalization mechanism can not always be realized for real time of experiments [14–16].

From the differences in the characteristic relaxation times of the impulse, energy, and mass of the particles (42.1), it follows that reversible deformation changes in finite times can be realized only for fixed compositions and structures of solids. If, during prolonged mechanical disturbances of solids, changes occur in the distribution of atoms, then during the removal of mechanical disturbances, the process of relaxation of the spatial distribution of atoms of a solid, due to a change in the short-range order described by equations (38.7) for paired moments, is realized along other trajectories of the variation of dynamic system parameters. Therefore, always, if the equilibrium state was not reached in the forward direction, the movement in the forward direction under load, and the motion in the reverse directions without load occur along different trajectories. This is the natural hysteresis of any dynamic variables in non-equilibrium processes. Only the achievement at the final point of a strictly equilibrium state allows the process not to depend on the previous trajectory of the process. Refusal to bind to a strictly equilibrium state as a reference [86] means an arbitrariness in the choice of the reference state, and a return to it is practically impossible, because the state of a rigid body is described by a much larger number of variables in the system (38.7) than in a strictly equilibrium state (in fact, the entire list of deformation links of the entire body must be taken into account). Only in a state of strict equilibrium is the thermodynamic state of a solid body described by the usual number of thermodynamic variables. A formal sign of the disequilibrium of a mechanical system is the presence of non-diagonal components of the stress tensor – only the diagonal elements of the stress tensor must be in equilibrium [87].

In the case of variations in the pairwise DF (or paired moments) beyond the domain of definition of parameters satisfying the local equilibrium condition specified in Section 40, the process can never be reversible. For its description it is necessary to know the dynamics of a pair DF, which does not include the number of independent thermodynamic parameters of macroscopic systems, and without it the equations of continuum mechanics become non-closed.

Thus, the concept of 'passive forces' does not agree with the atomic-molecular theory that operates with real forces due to intermolecular interactions and external potentials. Instead, it is necessary to operate with the concepts of sets of relaxation times of elementary stages describing the dynamics of the process under consideration.

Formally, by analogy with situations for which Gibbs introduced passive forces, the same forces can be associated with metastable states in vapour–liquid systems, as well as for many situations with deformation effects of active phases in solid-phase systems (such as films, membranes, catalysts, nanocomposites, etc. of a wide variety of materials, including polymeric ones).

43. Non-equilibrium thermodynamic functions

We confine ourselves to a discussion of the structure of equations (20.4) and (20.5) for solid-phase systems supplemented by equations from system (38.6) and (38.7) describing their deformation in the course of physicochemical processes (these equations are analogous to the elasticity theory equations [2,3], but contain mechanical modules, depending on the evolution of the concentrations of components). Knowing the solution of these kinetic equations with respect to θ^i_f and $\theta^{in}_{fg}(r)$, it is possible to calculate all the thermodynamic functions depending on them as arguments, including non-equilibrium thermodynamic potentials at each instant of time.

We discuss this question using the example of the Helmholtz energy $F = E-TS$, where E is the energy and S is the entropy of the system. The energy of the system is expressed in terms of these non-equilibrium functions (for simplicity we look at a uniform system) as

$$E = N\left\{\sum_{i=1}^{s-1}\left[\theta_i\beta^{-1}\ln(a_i) + \tfrac{1}{2}\sum_{r=1}^{R} z(r)\sum_{j=1}^{s-1}\varepsilon_{ij}(r)\theta_{ij}(r)\right]\right\}. \tag{43.1}$$

According to the definition of entropy in the classical case, the quantity S is determined by the expression ln $\Delta\Gamma$, where $\Delta\Gamma$ is called the statistical weight of the macroscopic state of the subsystem. The quantity S is a dimensionless quantity. Because the number of Δ states can not be less than unity, then the entropy can not be negative. The entropy of the whole system as a whole can also be written

as the average value of the logarithm of the distribution function $S = -\sum_{\{\gamma_f^i\}} P(\{\gamma_f^i\},t)\ln P(\{\gamma_f^i\},t) + S_0$, where the list of all configurations $\{\gamma_f^i\}$ is equivalent to the phase space, and the constant S_0 is chosen so that in the absence of correlations it corresponds to the distribution $P(\{\gamma_f^i\},t)$ expressed by the product of the distributions of individual particles [64,73]. This definition refers to an arbitrary (equilibrium and non-equilibrium) state of the system. In the QCA this leads to the following formula:

$$S = Nk\sum_{i=A}^{s}\left\{\theta_i \ln(\theta_i) + \tfrac{1}{2}\sum_{r=1}^{R} z(r)\sum_{j=A}^{s}\left[\theta_{ij}(r)\ln\theta_{ij}(r) - \theta_i\theta_j \ln(\theta_i\theta_j)\right]\right\}. \quad (43.2)$$

Thus, expressions (43.1) and (43.2) give a notation for the non-equilibrium Helmholtz energy at any time, including the equilibrium state of the system. The difference between the non-equilibrium state and the equilibrium state lies in the method for calculating the functions θ_f^i and $\theta_{fg}^{in}(r)$, as a function of time, and in the limiting case for large times their distribution is described by the equations (18.4) and (18.1). It is this that determines the fundamental importance that the kinetic equations and equilibrium distribution equations are self-consistent (see Chapter 6).

Local pressure. For Gibbs energy we need to add contributions for local mechanical work – the local equation of state. This question is considered in detail in [88]. For particles with translational motion of components, the expression for local pressure will be written as

$$P_{kn}(f) = n_f kT_f - \frac{1}{v_f^0}\left\langle \sum_{h\in z(f)}\sum_{i,j=1}^{s-1}\theta_{f_k h}^{ij} r_{f_n h}^{ij}\frac{\partial\varepsilon_{fh}^{ij}(r)}{\partial r_{f_k h}^{ij}}\right\rangle \quad (43.3)$$

The calculation of expression (43.3) requires knowledge of the time dependences of unary and paired DFs. Differences in the components of the pressure tensor $P_{kn}(f)$ are determined by the nature of the distribution of neighbouring particles in space and in time, which in non-equilibrium is non-uniform along different directions. They depend on the current quantities $\{\theta_{fh}^{ij}\}$, which obey the transport equations (38.6) and (38.7). Their solution connects all non-equilibrium unary and paired DFs, which makes it possible to calculate the $P_{kn}(f)$ values of local pressure as a function of the non-equilibrium states of the system in the region f. In the course of solving the equations of the system (38.6) and (38.7), it is necessary to take into account the linkage of all local quantities for analogous local values in neighbouring sites and regions, which leads to the necessity of formulating the boundary conditions for the component concentrations, the components of the pressure tensor and the energy flux vector on the entire surface, which limits the system.

In contrast to the mechanics of a continuous medium, symmetrization is absent on the hydrodynamic scale of dimensions (see [88] and Appendix 2, for which the symmetry of the components of pressure tensors $P_{kn}(f) = P(f)\,\delta_{kn}$ [1–3]) occurs in the bulk phase and for micro-non-uniform systems in non-equilibrium states there is no symmetrization $P_{kn}(f) \neq P_{nk}(f)$. The lack of symmetrization is also manifested in the construction of dissipative coefficients in locally non-uniform porous systems even under the condition of local equilibrium [34, 61, 62].

In the case under discussion, expression (43.3) can vary depending on the properties of the system. The simplest example of a non-uniform system is the interface of coexisting phases. Let us denote by q the number of the monolayer within the transition region, $p = q \pm 1$ – the numbers of neighbouring layers, $z_{qp}(R_d)$ – the number of neighbours of the central site q with neighbouring layers p, $\sum_{p=q-1}^{q+1} z_{qp}(R_d) = z$ the definition of numbers $z_{qp}(R_d)$ is given in Section 24 ($R_d \equiv R$) [90]. For $R_d \to \infty$ we obtain an expression for the pressures inside a plane interface. The indices kn are related to the macroscopic symmetry of the drop boundary with the numbers of the layers q and p (here they are left for clarity).

$$P_{kn}(q \mid R_d) = n_q kT_q - \frac{1}{v_f^0}\left\langle \sum_{p=q+1}^{q-1} z_{qp}(R_d) \sum_{i,j=1}^{s-1} \theta_{q_k p}^{ij} r_{q_n p}^{ij} \frac{\partial \varepsilon_{qp}^{ij}(r)}{\partial r_{q_k p}^{ij}} \right\rangle \qquad (43.4)$$

If in this expression we go from the local (non-symmetrized) DF to the symmetrized DF with respect to the local volume on the second size scale, as is used in the formulas [90, 91] with a continual description of the forces, then in the second term the coefficient 1/2 should appear. The difference between the symmetrized and asymmetrized characteristics can be easily explained by the example of the pair contribution to the total energy of the system: if the total system is considered, each bond fg enters it twice (so that it is not duplicated by the coefficient 1/2) if the linkage fg counts from the site f or g, then you do not need to enter a factor of 1/2.

Taking into account the coefficient 1/2, formula (43.4) coincides with the well-known Irving–Kirkwood expression [92], obtained by direct calculation of the forces between molecules of the mixture located on different sides of the selected plane of the fluid, or by the virial theorem [91,79,80]. For planar and spherically curved vapour –liquid interface, expression (43.4) can be rewritten by isolating the normal and tangential components of the pressure tensor ($\alpha = N, T$) for each layer q of the transition region

$$P_q^{\alpha} = n_q kT_q - \frac{1}{2\mathrm{v}_f^0}\left\langle \sum\nolimits_{p=q+1}^{q-1} z_{qp}(R_d)\cos^2(qp,\alpha)\sum_{i,j=1}^{s-1}\theta_{qp}^{ij} r_{qp}^{ij}\frac{\partial\varepsilon_{qp}^{ij}(r)}{\partial r_{qp}^{ij}}\right\rangle \quad (43.5)$$

the symbol (qp, α) denotes the angle between the direction of α and the direction of communication between the sites in the layers q and p (for simplicity, only the nearest neighbors are indicated).

If the system is completely uniform, then Eq. (43.3) constructed through asymmetrized DFs with respect to the distinguished site f goes into the equation for the bulk pressure of a multicomponent mixture, analogous to equation (18.5)

$$P = nkT - \frac{1}{6\mathrm{v}_0}\left\langle z\sum_{i,j=1}^{s-1}\theta_{ij} r_{ij}\frac{\partial\varepsilon_{ij}(r)}{\partial r_{ij}}\right\rangle. \quad (43.6)$$

Here, the indices of the coordinates of the particles in the isotropic volume are omitted, and the coefficients 1/2 and 1/3 are introduced in connection with the symmetrization of the pair DFs in the bulk phase and in the transition to an arbitrary orientation of the pair in space (or 1/2 in the two-dimensional system). Equation (43.6) is usually derived in two equivalent ways: from the virial theorem [95] and from the differentiation with respect to the volume of the partition function of the system Z ($F = -kTln\ Z$ is the free energy) [38], which corresponds to the thermodynamic definition of pressure as $\partial F/\partial V|_{T,N} = -P$ [1–3,55].

Equations (43.3) are obtained under the assumption that all neighbors of the central component i are mobile components and the terms from external $F_j^i(k)$ forces created by the fixed components of the system were not taken into account. If these forces are related to surface forces, equations (43.3) are modified by the fact that part of the surrounding volume is occupied by particles of fixed components that do not change the state of occupation of those sites that they fill (denoted by the symbol d). Then the pressure inside the mobile components that are in the field of the wall is written as

$$P_{kn}(f) = n_f kT_f - \frac{1}{v_f^0}\left\langle \sum_{h\in z(f)-w} \sum_{i,j=1}^{s-1} \theta_{f_k h}^{ij} r_{fh_n}^{ij} \frac{\partial \varepsilon_{fh}^{ij}(r)}{\partial r_{f_k h}^{ij}} \right\rangle - \frac{1}{v_f^0}\left\langle \sum_{d\in w} \sum_{i,j=1}^{s-1} \theta_{f_k w}^{id} r_{fw_n}^{id} \frac{\partial \varepsilon_{fw}^{id}(r)}{\partial r_{f_k w}^{id}} \right\rangle \quad (43.7)$$

Here there are two contributions from the intermolecular interaction of mobile components and from their interaction with the wall (the symbol w). In the last term, the atoms of the wall d belong to the near-surface region of the solid ($d \in w$), whereas the area occupied by the wall is subtracted from the total set of sites $z(f)$ surrounding the mobile particle($h \in z(f)-w$). In this way, it is possible to describe different types of non-uniform surfaces [32, 34], repeating word-for-word the conditions for the size of regions and the time scales described after formula (43.3).

In a state of non-equilibrium, the thermodynamic potentials lose their priority properties inherent in the equilibrium state, when their minima determine the most probable values of the distribution functions of all dimensions. Non-equilibrium thermodynamic potentials play the role of the characteristics of the total energy in the absence of equilibrium. In this case, they do not include the kinetic energy of the centre of mass of the whole system and its potential energy of position as a whole (this definition coincides with the usual definition of the total energy of the system in equilibrium). The kinetic approach lacks the priority of entropy. In the absence of the kinetic theory, the direction of the process was determined through the values of S. In the presence of kinetic theory, the course and direction of the process in full volume is determined from the solution of the equations by the current values of the dynamic parameters. *Entropy is an accompanying information* on the course of the dynamic process at any times and spatial scales. (Although its local production retains its meaning in any non-equilibrium state).

In particular, in order to operate with the value of entropy and / or its production, it is necessary to know the local temperature T, which is determined from the solution of the complete system of equations, including the transfer of energy (or temperature). The determination of the non-equilibrium thermodynamic potentials includes the contributions due to the product $T_{non\text{-}eq} S_{non\text{-}eq}$, provided that the quantity $T_{non\text{-}eq}$ is known at every point of the system in the non-equilibrium state from the solution of the kinetic equations for the energy/temperature transfer. The specificity of solid-phase processes due to large differences in the characteristic times of the processes of establishing thermal equilibrium (by relaxation of vibrational motions) and concentration changes (due to diffusion redistribution of components) is that in $T_{non\text{-}eq} \cong T$, where T is the average temperature of the system on the scale of laboratory samples (the case where $T_{non\text{-}eq} \neq T$ refers to intense perturbations and requires separate consideration). But the value of $S_{non\text{-}eq}$ remains unbalanced due to the 'frozen' diffusion process even in long-term observation.

44. Non-equilibrium surface tension

The formulas for the relationship between surface tension and unary and paired DFs are indicated in Sections 23 and 24. The knowledge of non-equilibrium thermodynamic functions and local components of the pressure tensor indicated in Section 43 allows one to calculate the non-equilibrium surface tension for any type of phase interface. The function $F(t)$ (43.1) and (43.2) is a non-equilibrium analog of the equilibrium free energy for a given time t [88, 90]. The same analogs are introduced for other Gibbs thermodynamic potentials G, entropy S, internal energy $U = E_{lat} + E_{vib} + E_{tr}$ ($E_{lat} = \sum_q E_q^{pot}$, and the last two contributions denote the internal energy of the vibrational and translational motions).

For a solid–vapour or liquid interface, the mobile subsystem is always in an equilibrium distribution. It is 'tuned' to a given non-equilibrium state of a solid (including an adsorbent). Formulas for the function $F(t)$, $G(t)$, $S(t)$, $U(t)$ can be expressed in the same way in terms of the local concentrations $\theta_q^i(t)$ and the pair functions $\theta_{qp}^{ij}(t)$, regardless of whether they are equilibrium or non-equilibrium [90]. The functions $\theta_q^i(t)$ and $\theta_{qp}^{ij}(t)$ must be determined by the kinetic equations of the processes that formed the boundary of the phases to a given instant of time. Therefore, the corresponding non-equilibrium analogs of surface tension can be calculated through them as an excess of the surface free energy $\sigma(t)$, which depends on time. These characteristics, being a function of time, pass to the limiting values of the equilibrium surface tension over long times (for this, it is important to fulfill the self-consistency condition – Chapter 6).

The concept of dynamic surface tension for a vapour–liquid σ_{dyn} system was introduced much later than the creation of thermodynamics for the particular case of creating a freshly formed system (without relaxation of the surface of a liquid [91,97,98]).

As indicated above, the concentration in the interphase layer of the solution in thermodynamic equilibrium differs from the concentration in the bulk phase. However, immediately after the formation of a fresh surface, it can have almost the same concentration as inside the bulk phase. The surface tension of such a surface is called the dynamic surface tension σ_{dyn} [97]. Its calculation is simple enough, although the calculation of the surface tension of the system at subsequent intermediate stages is impossible in the framework of traditional equilibrium concepts. For this, non-equilibrium approaches were used [98,99], since the mentioned interphase region is not in thermodynamic equilibrium, and the evolution of the dynamic surface tension should be calculated using the equations of thermodynamics of irreversible processes.

The concept of non-equilibrium surface tension is much broader than the particular case mentioned above, and it can refer to any kind of the non-equilibrium state of the system. Its magnitude varies depending on the evolution of the state of the system. The phase boundary attracts constant attention [100–114], but the overwhelming number of papers is based on the macroscopic equations used to find the effective parameters of time dependences for dynamic surface tension (see for example [114]). This is an important question for practice, but it leaves aside microscopic ideas about the essence of the relaxation process. The attraction of molecular modelling methods [115] meets the usual problems of their application for the interface boundaries, when the density difference at the boundary to two or three orders does not allow a reliable study of the given process. Therefore, below we briefly discuss the possibility of involving the molecular theory.

The non-equilibrium thermodynamic functions introduced above for bulk phases can be transferred to surface characteristics by the same principle. The surface tension is determined through the excess free energy of a real and a hypothetical system extended to the same separation surface (its principle must be consistent with the calculation of the equilibrium distribution of components (formulas of the Sections 21–24 and Appendix 1). The dynamic problem allows finding all current distributions and obtaining a non-equilibrium surface tension as functions of time.

Gibbs [55] introduced mechanical (γ) and thermodynamic (σ) determinations of the surface tension for solids. He defined γ and σ, respectively, as the work of forming a unit of surface area by mechanical action with an unchanged composition and surface structure (for example, when cutting a body), or in the process of dissolution–precipitation, in which no mechanical stress occurs. In fact, Gibbs allowed the existence of two ways of forming the same value of the new area of the formed surface δA in different ways, or it is a question of the effect of the evolution of the system on its identical final state (with a finite value of δA). This topic was least developed in thermodynamics, and its refinements are still being continued. It should be noted that work was performed both on thermodynamic refinements [116–119] and on modelling at the microscopic level [120–124]. The original meaning of the term 'surface tension' was formulated in the mechanics of continuous media, as a purely mechanical stress on the boundary of the body in question.

The difference between the value of γ and the dynamic surface tension σ_{dyn} introduced later is due to the absence of the effect of external loads in the σ_{dyn} value, whereas this aspect is not specified in relation to the value of γ. For a liquid, this state is instantaneous, whereas for a solid body, due to long relaxation times, its state can be 'frozen' for a long time. Both characteristics introduced by Gibbs, γ and σ, relate to the relaxation times of solid state states at large times. If we are talking only about purely mechanical states, then relaxation takes place quickly. But the possibility of changing the composition and structure, by analogy with the relaxation of σ_{dyn}, is also not discussed. All this leads to different interpretations of Gibbs' concepts. For a long time there was a problem even with the sign of the value of γ at the solid–gas interface. Only recently [119], the quantity γ was assigned to positive quantities, like σ, which should be positive under the stability conditions.

At that time, there was no other alternative way of changing the properties of a solid and its surface due to the diffusion of atoms, which was discovered much later [8–13,82]. This channel for changing the state of a solid body is the main channel in many situations and it is absolutely necessary for it to introduce relaxation times for the limiting subsystems. As in the discussion of the concept of passive forces, it is easier to formulate the questions that a microscopic theory should answer than to analyze its correspondence with existing thermodynamic interpretations.

First of all, this refers to the accounting for the defectiveness of a solid body. Most often in calculations of surface tension [120], taking into account the temperature influence of vibrational motions, we are talking about ideal structures in which defects are absent (including vacancies). The technique for calculating the vibrational states of highly defect structures remains to this day complex [16,125], although the theoretical problems of its solution are known [126–131]. Non-equilibrium analogs of thermodynamic potentials, reflecting the effect of material defectiveness, are more stringent than Gibbs' proposal to use the excess surface tension $U^s(t)$ as a definition of γ, and they not only reflect the essence of the non-equilibrium states of solids, but also give a rigorous way of calculating these characteristics through kinetic equations in the course of their temporal evolution.

Accounting for external mechanical loading complicates the situation, since an additional factor appears that affects the distribution of components both inside the solid body and on its surface, and, consequently, affects the distribution of the mobile phase with respect to the mechanically disturbed adsorbent. The introduction of non-equilibrium analogs of thermodynamic potentials requires a refinement of the very concept of the process of creating a new surface under the action of mechanical loads: is it about the value of $\sigma(t)$ relating to the created surface after removal of the applied load or in the course of the external load itself. Further, there should be a concretization whether the surface is created from the volume of the solid phase (for example, by cleavage), or it is created by applying a load to an already existing surface. In the second case, if, after removing the load, the surface does not relax to the initial value of the surface area, then we are dealing with an analog of plastic deformation, and it is necessary to indicate what changes occurred in the solid body to describe them. For this, during the mechanical load, it is necessary to detail the type and method of creating the contact interaction. Different variants are possible here, ranging from direct perturbations of neighbouring solids (or neighbouring phases) to indirect perturbations at the 'far' ends of the crystal, when the deformation interaction is transmitted along the crystal lattice to the entire volume and to the surface. Each non-equilibrium process is characterized by its kinetic scheme and dynamics specificity, which must reflect microscopic models – otherwise the process under the influence of mechanical disturbances is not determined [81].

These questions are naturally important for the analysis of ensembles of small bodies, which are described in the framework of two-level models (see Appendix 2) [54,88,90]. A microscopic theory based on two-level models is only formulated, although for macroscopic systems the problems of contact interactions are well developed [132–137].

In addition, the problem of the size of the created surface is also important and has a simple meaning only in the case of ideal (planar, spherical, etc.) geometries. Any more or less realistic mechanical disturbance causes rough surfaces of various kinds to arise, and for them the estimation of the surface area is an independent problem. This detailing should clarify the microscopic interpretation of real solid-phase objects, depending on the method of their formation [138]. The transition to a strictly statistical description of the non-equilibrium states of solids makes it possible to relate thermodynamic constructions to other measurements: structural, kinetic, mechanochemical, etc.

45. Relaxation of the interface

The equilibrium conditions of the two-phase system are considered in Section 6. We discuss them in more detail, taking into account the contribution of the phase interface, and in addition take into account the process of relaxation of the intense parameters. Let the equilibrium state be established in the isolated system ($d_eS/dt = 0$), and the progress of the process in the vicinity of the equilibrium point is considered $dS/dt = d_iS/dt > 0$, and let the relaxation stage belong to the state of the interface (within each phase, a fast alignment of the properties is assumed). The main attention is paid to the very process of establishing a local equilibrium, in order to analyse the relaxation of the intensive parameters from it. (Such problems are not considered in traditional non-equilibrium thermodynamics [30], using the conditions of local equilibrium).

Taking into account the properties of non-equilibrium functions, which are analogs of equilibrium potentials (Section 43), we write, by analogy with Section 6, the expression for the time evolution of entropy. All functions of this expression (extensive and intensive) are functions that depend on the time argument. For example, $U(t)$ is a function of time in terms of the time dependence of unary and paired DFs, and in the long-time $\lim U(t)_{|t\to\infty} = U$, where U denotes the equilibrium value of the internal energy discussed in Section 6. Then, near the full-equilibrium neighbourhood, the expression for entropy evolution will be written as

$$\delta S(t) = \delta S_\alpha(t) + \delta S_\beta(t) + \delta S_\beta(t), \qquad (45.1)$$

where the first two terms refer to the two phases α and β, and the term $\delta S_b(t)$ refers to the surface contribution (Section 21). Although all functions are functions of time, the conditions for fixing the internal energy, volume and number of particles do not change the form of expressions related to the choice of independent variables. As above, the virtual changes of the system parameters are expressed as $\delta U_\beta = -\delta U_\alpha$, $\delta V_\beta = -\delta V_\alpha$, $\delta N_\beta = -\delta N_\alpha$. Fixing the internal energy, volume and number of particles inside the complete system, and choosing as the independent variables the extensive characteristics of phase α, we have an expression for the time dependence of entropy

$$\delta S(t) = \left(\frac{1}{T_\alpha(t)} - \frac{1}{T_\beta(t)}\right)\delta U_\alpha(t) + \left(\frac{P_\alpha(t)}{T_\alpha(t)} - \frac{P_\beta(t)}{T_\beta(t)}\right)\times$$
$$\times \delta V_\alpha(t) - \left(\frac{\mu_\alpha(t)}{T_\alpha(t)} - \frac{\mu_\beta(t)}{T_\beta(t)}\right)\delta N_\alpha(t) + \frac{\sigma(t)}{T_b(t)}\delta A(t) \tag{45.2}$$

where $\sigma(t)$ is the non-equilibrium surface tension, $A(t)$ is the interface area, and $T_b(t)$ is the effective temperature of the transition region of the interface. This expression as a function of time t tends to its limiting state, which we denote by $\lim \delta S(t)_{|t\to\infty} = \delta S = 0$.

Our task is to consider the properties of the $\delta S(t)$ function as a function of the conditions on the ratio of the relaxation times for the pressure and the chemical potential at the limiting transition $t \to \infty$. The analysis using thermodynamic hypotheses in Section 7 showed that formally here it is possible to obtain two different limiting values for the intensive variables.

To simplify the analysis of the expression $\delta S(t)$, we consider the isothermal process of relaxation of the system. This allows us to write down $T_\alpha = T_\beta = T_b = \text{const} = T$, and formula (45.2) is rewritten as

$$\delta S(t) =$$
$$= \left[\left(P_\alpha(t) - P_\beta(t)\right)\delta V_\alpha(t) - \left(\mu_\alpha(t) - \mu_\beta(t)\right)\delta N_\alpha(t) + \sigma(t)\delta A(t)\right]/T \tag{45.3}$$

We simplify the notation by introducing the notation for differences between pressures and chemical potentials in two phases $\Delta P(t) = P_\alpha(t) - P_\beta(t)$ and $\Delta\mu(t) = \mu_\alpha(t) - \mu_\beta(t)$, then the time derivative for entropy evolution will be written in the form (45.4)

$$\frac{d\delta S(t)}{dt} = \frac{1}{T}\left[\frac{d\Delta P(t)}{dt}\delta \mathrm{V}_\alpha(t) + \Delta P(t)\frac{d\delta V_\alpha(t)}{dt} - \frac{d\Delta\mu(t)}{dt}\delta N_\alpha(t) - \right.$$
$$\left. -\Delta\mu(t)\frac{d\delta N_\alpha(t)}{dt} + \frac{d\sigma(t)}{dt}\delta A(t) + \sigma(t)\frac{d\delta A(t)}{dt}\right] \tag{45.4}$$

To analyze the effect of the temporary relaxation of partial equilibria on the state of complete equilibrium of the system, it is necessary to separate the temporal evolution of extensive and intensive parameters: $\frac{d\delta S(t)}{dt} = \frac{d\delta S_{\mathrm{int}}(t)}{dt} + \frac{d\delta S_{ext}(t)}{dt}$. Temporal changes of all extensive variables $d\delta S_{ext}(t)/dt$, i.e. $d\delta V_\alpha(t)/dt$, $d\delta N_\alpha(t)/dt$ and $d\delta A(t)/dt$ provided that their current values remain within specified limits within the system under consideration, they do not play any role in the evolution of the intensive parameters. These contributions are analogs of the external influence on the entropy fluxes in the isothermal process during the establishment of complete equilibrium with respect to particular (mechanical and chemical) equilibria with respect to the intensive parameters. They need to be subtracted from expression (45.4) in order to analyze only the time dependences of the intensive parameters, which in the limit $t \to \infty$ determine the state of complete two-phase equilibrium

$$\frac{d\delta S_{\mathrm{int}}(t)}{dt} = \frac{d\delta S(t)}{dt} - \frac{d\delta S_{ext}(t)}{dt} = \frac{d\delta S(t)}{dt} -$$
$$-\left[\Delta P(t)\frac{d\delta V_\alpha(t)}{dt} - \Delta\mu(t)\frac{d\delta N_\alpha(t)}{dt} + \sigma(t)\frac{d\delta A(t)}{dt}\right]/T, \tag{45.5}$$

$$T\frac{d\delta S_{int}(t)}{dt} = \frac{d\Delta P(t)}{dt}\delta V_\alpha(t) - \frac{d\Delta\mu(t)}{dt}\delta \mathrm{N}_\alpha(t) + \frac{d\sigma(t)}{dt}\delta A(t). \tag{45.6}$$

We rewrite expression (45.6), taking into account that, by the definition of extensive variables $\delta V_\alpha(t)$, $\delta N_\alpha(t)$, and $\delta A(t)$ between them there are proportionality coefficients with respect to the contributions from these quantities to thermodynamic functions that are not related to the concrete state of the system: $V = K_{NV}N$, $K_{NV} = dN/dV = \mathrm{const}_{NV}$ and $V = K_{AV}A$, $K_{AV} = dA/dV$, where $K_{AV} \ll 1$. Then

$$\frac{T}{\delta V_\alpha(t)} \frac{d\delta S_{int}(t)}{dt} = \left[\frac{d\Delta P(t)}{dt} - K_{NV} \frac{d\Delta\mu(t)}{dt} + K_{AV} \frac{d\sigma(t)}{dt} \right] \tag{45.7}$$

This form of recording reduced all the relationships to the relations between the relaxation times of the pressure τ_p, the chemical potential τ_μ, and the non-equilibrium surface tension τ_σ, respectively, for the three terms (45.7). By mechanical determination of σ the relaxation time of surface tension is related to the pressure relaxation time [88]. All quantities in (45.7) are described by different kinetic partial differential equations (hyperbolic for impulse transfer (pressure) and parabolic for mass transfer (chemical potential)). For qualitative analysis, let us recall the example of the simplest monomolecular chemical reaction of the transformation of the component $A \rightarrow$ *products*, the relationship between the relaxation time and the rate of change of the corresponding parameters.

The aspiration of concentration $\theta(t)$ to its equilibrium value θ_e at $t \rightarrow \infty$, can be characterized by an exponential dependence of the type $\theta(t) = \theta_e + C \exp(-t/\tau)$, where C is a constant related to the magnitude of the parameter $\theta_0 = \theta(t = 0)$ at the initial time $t = 0$ ($C = \theta_0 - \theta_e$), τ is the relaxation time (which in its physical sense corresponds to the average lifetime of the original reagent A). Within the framework of the mass action law, the evolution of the parameter θ is described by the kinetic equation of the first order $\frac{d\theta}{dt} = K_A(\theta_0 - \theta)$, where K_A is the rate constant of the reaction of component A. The solution of this equation has the form $\theta_0 - \theta = \theta_0 \exp(-K_A t) = \theta_0 \exp(-t/\tau)$, or $\tau = 1/K_A$. The larger the magnitude of K_A and the faster the parameter $\theta(t)$ changes with time, the shorter is its relaxation time. This principle is also preserved for other variables described by equations of a more complex type (see Section 36).

Let us consider the relationships between the relaxation times of pressure τ_P associated with the time dependence $d\Delta P(t)/dt$ and the chemical potential τ_μ (related to $d\Delta\mu(t)/dt$) with respect to the relaxation time of the surface contribution τ_σ (related to $d\sigma(t)/dt$), and discuss the limiting cases of establishing the equilibrium of a non-uniform system.

1. The case of coexisting equilibrium contacting phases. In Section 28 it is noted that any phase boundary does not affect the internal properties of the phases. The boundary properties can not be parameters of the state of the system, since they necessarily functionally depend on the parameters of the state of the bulk

coexisting phases. Therefore, the relaxation time of the surface tension τ_σ will be less than the relaxation time of the pressure τ_P in the coexisting phases ($\tau_P \gg \tau_\sigma$). Therefore, the third term in (45.7) can be neglected, and the experimental data (42.1) [57] from the ratio between the relaxation times of the pressure (impulse) and the chemical potential (mass) leads to the inequality $\tau_P < \tau_\mu$ (with the exception of the rarefied gas) or $\tau_P \ll \tau_\mu$ in dense phases. It follows that the time of the relaxation process of the system is completely determined by the mass transfer relaxation time and the evolution of the pressure is written as $\Delta P(t) = \Delta P(\Delta\mu(t))$ – in the form of a functional dependence on the evolution of the chemical potential. By construction, in the limit $\Delta\mu(t \to \infty) = 0$, which leads to the limit $\Delta P(t \to \infty) = 0$ for any sizes of coexisting phases, or $P_\alpha = P_\beta$. This is the case of the equilibrium drop described in Section 24 (phase inversion in Section 24 refers to vapour bubbles in the liquid phase).

2. The case of a foreign film separating two phases. If the foreign film excludes the exchange of molecules between phases that are in the internal equilibrium states, then the film relaxation time is much longer than the relaxation time of the mobile phases. This is a typical case of a film of dense material (latex, rubber, etc.) [117], which limits the mobile phase of vapour or liquid. Moreover, the relaxation time of the film itself, determined by its internal properties, is in no way connected with the state of matter in both phases. Here σ_m is the mechanical surface tension of the material. In the absence of an exchange of a substance there is no chemical equilibrium, i.e. $d\Delta\mu(t)/dt = 0$, and only mechanical equilibrium is considered when the right-hand side of equation (45.7) is written as

$$\frac{T}{\delta V_\alpha(t)}\frac{d\delta S_{int}(t)}{dt} = \frac{d}{dt}\left(\Delta P(t) + K_{AV}\sigma(t)\right) \tag{45.8}$$

The process of establishing equilibrium is determined by the relationship between the relaxation times of the pressure and the surface tension of the film. As $t \to \infty$, we have $\Delta P(t \to \infty) = 0$ as above, while $\sigma(t \to \infty) = \sigma_m$, which obviously does not coincide with σ_e discussed in Chapter 3.

In the limit, another state of the system is obtained, which differs from the strictly equilibrium state for a plane vapour–liquid boundary $P_\alpha = P_\beta$ by an amount $K_{AV}\sigma_m = \sigma_m dA/dV = 2\sigma_m/R$, i.e. $P_\alpha = P_\beta + 2\sigma_m/R$ (the Laplace equation). To analyze the size dependences of $\sigma_m(R)$, there must be a molecular model for taking into account the mechanical characteristics in the actual film between phases, determined by the properties of the film material.

3. If the foreign film does not prevent the transfer of molecules, but the relaxation time of the exchange of molecules (chemical potential τ_μ) is less than the relaxation time of pressure τ_P, then we consider again the case $\tau_P \gg \tau_\mu$ in which equation (45.8) is again obtained. The relation $\tau_P \gg \tau_\mu$ contradicts the experimental data (42.1) [57], and this case is considered formally. But it corresponds to the Kelvin equation in the accepted sequence of establishing first a mechanical and then a chemical equilibrium. The functional relationship $\Delta\mu(t) = \Delta\mu\ (\Delta P(t))$ is realized, when the chemical potential is adjusted to the value of the pressure established by the moment. As a result, again $P_\alpha = P_\beta + 2\sigma_m/R$, where the value σ_m can not be determined experimentally due to $\tau_P \gg \tau_\mu$.

This situation corresponds to the so-called case of metastable drops (see Appendix 1 and Chapter 7) when, in practice, the experimentally measured value of the surface vapour–liquid tension for a plane boundary is substituted for the unknown value of σ_m. In practice, this substitutes the properties of a foreign film by an amount σ for the interface of a real vapour–liquid system.

46. Influence of fluctuations on the rate of stages

The effect of equilibrium fluctuations on microcrystalline particles on the adsorption isotherms was shown in Chapter 4. Here, their effect on the adsorption–desorption rate and surface reaction rate on small microcrystalline particles is discussed. As above, for simplicity, the main attention is paid to the absence of lateral interactions between adsorbed molecules. Equations of the equilibrium distribution of molecules and their mean square displacements on a uniform surface are described by the equations of Section 31 and 32 (for a binary mixture). Equilibrium fluctuations affect the rates of elementary stages if fast surface diffusion takes place, for a fixed number of adsorbed particles their distribution can be regarded as practically equilibrium.

In the absence of interaction between adsorbed particles on a uniform surface, the rates of elementary stages are described within the framework of the law of mass action as

$$U_m = K_m W_m,\ W_m = \theta_i\,(m=1) \text{ and } \theta_i\theta_j\ (m = 2) \tag{46.1}$$

where m is the molecular order of the reaction, K_m is the rate constant, W_m is the concentration component of the rate of the stage, in the particular case of ideal reaction systems.

The knowledge of the average number of adsorbed molecules $N_i = M\theta_i$ makes it possible to calculate the rates of elementary reactions. According to the theory of fluctuations [57], in the expressions for the fluctuations of the rates of elementary reactions W_m is the concentration component of the reaction rate, depending on the number s of variables, taking into account the definition of $\Delta\theta_i = (N_i - \langle N_i \rangle)/M$, will be written as [139,140]

$$\Psi(U_m) = \sum_{i,k=1}^{s} \frac{\partial W_m}{\partial \langle N_i \rangle} \frac{\partial W_m}{\partial \langle N_k \rangle} \langle \Delta N_i \Delta N_k \rangle, \qquad (46.2)$$

where the calculation of the mean-square means $\langle \Delta N_i \Delta N_k \rangle$ is carried out taking into account the limited size of the system M.

Uniform surface. In [139] the value of the temperature corresponding to $\beta\varepsilon = 1/2$ is assumed, where ε is the interaction parameter between molecules of the first sort. For an ideal system, the parameter ε serves as a measure of the binding of the molecule of the first sort to the surface. It is assumed in the calculations that the binding energy of molecules of the first sort with the surface is $Q_1 = 4\varepsilon$, which corresponds to a sufficiently strong binding of the adsorbed particle to the surface. The binding energy of molecules of the second sort with the surface is assumed to be $Q_2 = \gamma Q_1$, where $\gamma = 1.4$.

Figure 46.1 shows the rates of monomolecular desorption processes of the first and second components U_1 and U_2 (a) and the fluctuations of the desorption rates Ψ_1 and Ψ_2 (b) ($m = 1$). Since the densities θ_i are entered linearly in the expressions (46.1), the role of the rate constants reduces to the coefficients before the concentration dependence W_m (for simplicity, their values are put to unity). The change in M has little effect on the behaviour of the rates, and therefore their behaviour is shown for small fillings, where these changes are more noticeable. At the same time, the dependence on x_1 is very significant. With increasing M, the range of values of θ in which the rate fluctuations are determined increases. In general, varying the size of the region M weakly shifts the curves relative to each other at a given scale.

These changes are somewhat more noticeable on fluctuations. The results are given for $x_1 = 0.15$ and $x_1 = 0.5$. For ease of comparison,

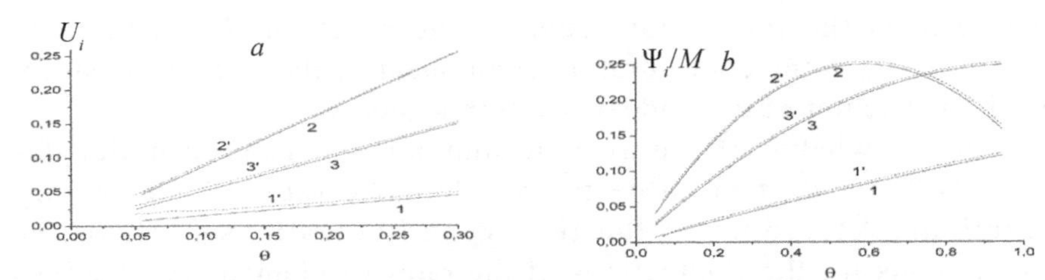

Fig. 46.1. The rates of the desorption processes U_i (a) and their fluctuations Ψ_i (b) $i = 1, 2$ ($m = 1$) as a function of density θ for $M = 10^4$ (solid lines) and $M = 10^2$ with density fluctuations (dashed lines). The notation on the curves: 1, 2 corresponds to $i = 1, 2$ for $x_1 = 0.15$, 3 – corresponds to $x_1 = 0.5$. The same notation, but with a prime for $M = 10^2$.

the desorption rate fluctuations are given in scale Ψ_i/M. If the molar fractions of the components are the same ($x_1 = 0.5$), then both the rates and the fluctuations are monotonous. But if the rates U_1 and U_2 are monotonic at any ratio of components, then for the molar fraction $x_1 = 0.15$ for component 1 (with a smaller fraction), the rate fluctuation curve has a monotonous form, whereas for component 2 (with a larger fraction) it passes through an extremum. For $M = 10^4$, the contributions of the density fluctuations to the values of the rate fluctuations are weakly noticeable at a given scale, whereas for $M = 10^2$ they are quite noticeable.

The behaviour of the relative deviations for elementary reaction rates calculated with allowance for only the size contribution and with an additional allowance for the density fluctuations from the size contribution is more evident than in the case of $M = \infty$. For monomolecular reactions, this comparison is shown in Fig. 46.2. The inset in Fig. 46.2 shows their absolute deviations. These changes agree with the course of the curves in Fig. 32.3, when the deviations change sign during the transition from small values of pressure to large.

Non-uniform surface. The description of non-uniform ideal systems is reduced to the summation of the contributions of any characteristics on individual faces. The formulas obtained above can be easily generalized by summing with weights F_q for the contributions of different types of centres q ($M_q = MF_q$) and replacing the partial fillings θ_i by the local partial fillings $\Delta\theta^i_q = N^i_q/M_q$ related to sites of type q, $1 \le q \le t$, t is the number of types of sites.

Figure 46.3 shows the corresponding desorption rates and their fluctuations on the same non-uniform surface (here $t = 2$, $F_1 =$

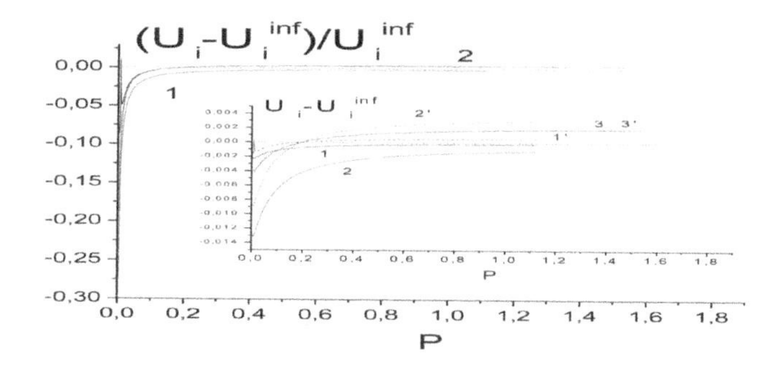

Fig. 46.2. The relative deviations of the reaction rates U_i at $M = 10^2$, the notation on the curves: 1 – corresponds to $x_1 = 0.15$, 2 – $x_1 = 0.5$. On the inset – their absolute deviations, without taking into account the contributions of $\Delta\theta^i_q$ – solid curves, taking into account these contributions – dotted line; the notation on the curves: 1, 2 corresponds to $i = 1, 2$ for $x_1 = 0.15$, 3 – corresponds to $x_1 = 0.5$. The same notation, but with a prime, for curves with allowance for the contributions of $\Delta\theta^i_q$.

$F_2 = 0.5$, $Q^i_2 = Q^i_1/2$). Figure 46.3 is similar to Fig. 46.1, however, the differences between the curves related to different regions and to the account of the effects of fluctuations are more clearly shown (Fig. 46.1 is given on a smaller scale). This is due to the fact that the surfaces with the same total area M are compared, which additionally differ in the size of the portions M_q having different binding energies.

Thus, an increase in the degree of non-uniformity of the surface leads to an increase in the role of the factor of boundedness of its individual regions, and an increase in the contributions of local fluctuations. The obtained results indicate the need to take into account fore the nanosized crystalline particles the size limitations of individual faces and density fluctuations of adsorbed particles, which affect both the equilibrium adsorption isotherms and the surface reaction rates.

Reactions of the Langmuir–Hinshelwood type. Fluctuations for the Langmuir–Hinshelwood type reaction $Z_qA + Z_pB \rightarrow$ *products in the gas phase* flowing on a non-uniform surface depend not only the surface composition, but also the structure of the non-uniform surface [140]. For a Langmuir–Hinshelwood type reaction, the rate of the surface reaction between the components of the first and second sort on a non-uniform surface will be written as [140]

$$U_{AB} = \sum_{q,p=1}^{t} F_{qp} K_{qp}^{AB} \theta_q^{\ A} \theta_p^{\ B}, \qquad \sum_{p=1}^{t} F_{qp} = F_q \ \text{ and } \sum_{q=1}^{t} F_q = 1, \tag{46.3}$$

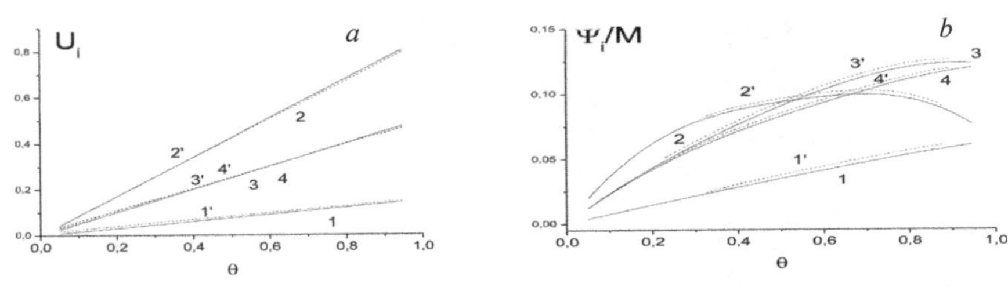

Fig. 46.3. The rates of the processes U_i (a) and their fluctuations Ψ_i (b) $i = 1,2$, as a function of the density θ for $M = 10^4$ – solid line and $M = 10^2$ – dotted line . The notation on the curves: 1, 2 corresponds to $i = 1, 2$ for $x_l = 0.15$, 3, 4 corresponds to $i = 1, 2$ for $x_1 = 0.5$. The same notation, but with a prime, for $M = 10^2$.

where K_{qp}^{AB} is the rate constant of the reaction between the first and second components located at the sites of the q-th and the p-th type in the case of ideal reaction systems, F_{qp} is the probability of finding a pair of centres of the q and p type on the surface, which are related to each other and the surface composition by the normalization relations. The equations (46.3) give normalized expressions for the rate, which do not reflect the real size of the system. In order to take into account the number of adsorption centres, it is necessary to take into account the connection of F_{qp} with the number of pairs of sites of different types $F_{qp} = (1+\Delta_{qp})M_{qp}/z_qM$, where Δ_{qp} is the Kronecker symbol and z_q is the number of the nearest neighbours for a site of type q. For simplicity of analysis, we will assume that there are three types of sites on the surface ($t = 3$), all sites have the same number of neighbours $z_q = z$, and only particles that are at sites of the first and second types (sites of the third type play the role of the carrier of active centres) enter the reaction, then in the expression (46.3) the summation over the types of the site extends to the pairs 11, 12, 21 and 22. Let the model reaction $A + B$ on the non-uniform surface be characterized by a specific matrix of local rate constants: $K_{11}^{AB}:K_{12}^{AB} = K_{21}^{AB}:K_{22}^{AB} = 1:5:1$.

To demonstrate the role of the surface structure in bounded systems, the following surface structures are examined:

1) $F_1 = F_2 = 1/2$, $F_3 = 0$; $F_{11} = F_{22} = 1/2$. The remaining $F_{qp} = 0$.

2) $F_1 = F_2 = 1/2$, $F_3 = 0$; $F_{11} = F_{22} = F_{12} = F_{21} = 1/4$. The remaining $F_{qp} = 0$. This lattice refers to the random lattice ($F_{qp} = F_q * F_p$).

3) $F_1 = F_2 = 1/2$, $F_3 = 0$; $F_{12} = F_{21} = ½$; The remaining $F_{qp} = 0$. The 'chessboard' type lattice

4) $F_1 = F_2 = F_3 = 1/3$; all $F_{qp} = 1/9$. The random lattice.

Figure 46.4 shows the concentration dependence of the Langmuir-Hinshelwood reaction rate on the pressure of the first component p_1, at a fixed value of the pressure of the second component, $p_2 = 1$. The field in Fig. 46.4 *a* shows the $A+B$ reaction rates on a non-uniform surface of macroscopic size U_{AB} (∞). With an increase in the degree of filling of the first component the reaction rate passes through a maximum, and with the predominant filling of the surface with the first component, the rate begins to decrease. This is the typical course of the reaction rate of the Langmuir–Hinshelwood type [25]. The variant of the considered surface of the catalyst is marked with the corresponding number.

The highest values of the rate are observed in the case of the maximum number of pairs of different types 1 and 2, which is realized on the chessboard type structure. With a chaotic arrangement of active centres (curve 2) of the same surface composition, the rate is noticeably lower since the probability of finding active pairs of sites is smaller. With a decrease in the surface composition of the active centres with the same random arrangement of the active centres, the reaction rate decreases even more (curve 4).

a

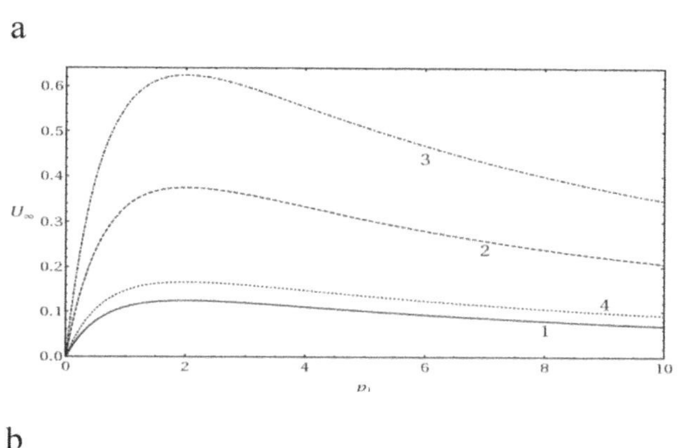

b

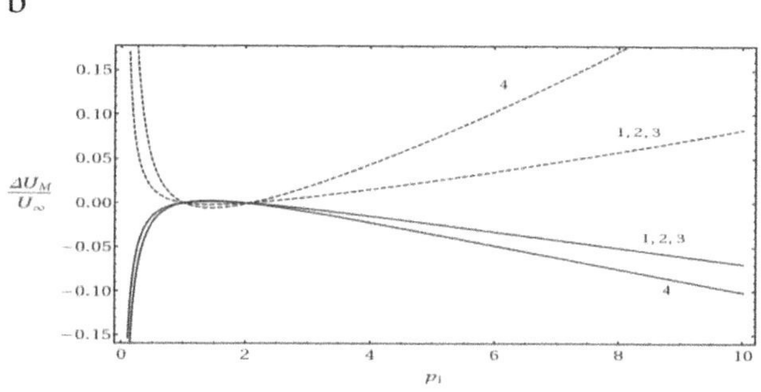

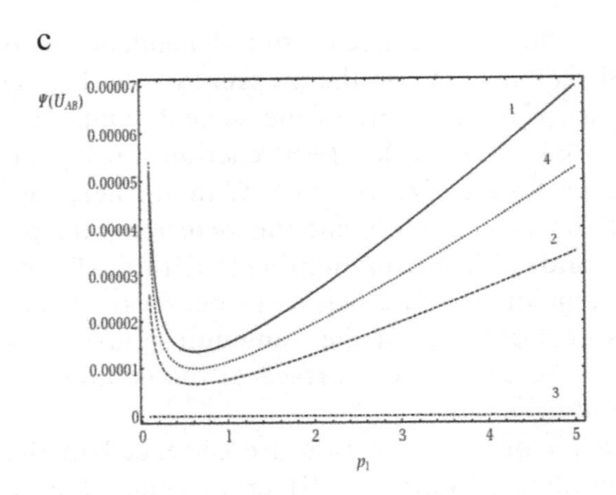

Fig. 46.4. (*a*) The dependence of the reaction rate U_{AB} (∞) in an unrestricted system on the pressure of the first component. (*b*) Dependence of the relative change in the reaction rate in the final system as compared to the infinite one $[U_{AB}(M)-U_{AB}(\infty)]/U_{AB}(\infty)$ without taking into account the fluctuations (solid lines) and $[U_{AB}(\mathrm{fl})-U_{AB}(\infty)]/U_{AB}(\infty)$, taking into account the fluctuations (dashed lines), on the pressure of the first component. (*c*) Dependence of the reaction rate fluctuation $\Psi(U_{12})$ on the pressure of the first component.

Curve 1 corresponds to the case of a 'spotted' arrangement of centres of the same type, in which there are no most active centres of type 1 and type 2 close to each other. We note the similarity of the values of the rates on the curves 1 and 4, although they refer to different compositions and structures of the surface of the catalyst. Relative changes in the surface reaction rates under the same conditions as for macroscopic systems (field in Fig. 46.4 *a*), in the case of bounded systems $[U_{AB}(M)-U_{AB}(\infty)]/U_{AB}(\infty)$ and with additional consideration of the effect of density fluctuations of reagents $[U_{AB}(\mathrm{fl})-U_{AB}(\infty)]/U_{AB}(\infty)$ are shown in Fig. 46.4 *b*. Here, the symbols $U_{AB}(M)$ and $U_{AB}(\mathrm{fl})$ denote the rates obtained in calculations in which local partial fillings respectively relate to bounded areas of the surface and to an additional account of the effects of fluctuations. In field 3*b*, the solid lines refer to the limited system $M = 100$ without fluctuations, and the dashed lines to the case in which the fluctuations are taken into account.

With a decrease in the size of the system the reaction rate decreases in those filling regions in which the partial fillings of the reagents are reduced – the region of small and large surface occupancies. Conversely, an additional consideration of the effects of fluctuations in a bounded system increases the reaction rate in full accordance with those changes in the partial fillings discussed in Section 32. Note that for given simplified model parameters, in the case of an ideal reaction system at $F_3 = 0$ the values of the relative deviations of the reaction rate for the first three types of surfaces coincide.

Figure 46.4 *c* shows the fluctuations of the $A+B$ reaction rate itself, which according to the formulas [139, 141] can be represented in the form

$$\Psi(U_{AB}) = \sum_{q,p=1}^{t} F_{qp} K_{qp}^{AB} \Psi(\theta_q^A \theta_p^B) = \sum_{q=1}^{t} F_{qq} K_{qq}^{AB} \Psi(\theta_q^A \theta_q^B). \qquad (46.4)$$

In the second equality, it is taken into account that for ideal reaction systems there is no mutual influence of reagents located at different sites. The expression for the fluctuation of the function is $\Psi(\theta_q^A \theta_q^B)$ constructed according to the usual rules for calculating fluctuations in individual areas of the surface [57, 142]: $\Psi(U_{AB}) = \sum_{q=1}^{t} K_{qq} F_{qq} \left(\theta_q, \eta_{qq} \theta_q\right) / (M^2 F_q^2)$ where the matrix product of the vector $\theta_q = \left(\theta_q^B, \theta_q^A\right)$ and the fluctuation matrix (the average of the matrix on the vector θ_q) are given in the parentheses.

The values of the rate fluctuations calculated in the normalized form (through partial degrees of filling) for a bounded system $M = 100$ with an allowance for density fluctuations of partial $\theta_q^i(M)$ are shown in Fig. 46.4 *c*. The numbers of the curves correspond to the four types of surfaces considered. In the case of a completely ordered arrangement of centres of different types, there are no pairs of the same type, so curve 3 has a zero value. As above, the displacements of the degrees of surface filling in the region of small and large fillings increase the fluctuations of all quantities – in this case the fluctuations of the rate of the reaction itself.

We note that the transition to a greater number of types of surfaces increases the degree of surface non-uniformity, which leads to an increase in the role of the factor of boundedness of its individual regions and an increase in the contributions of local fluctuations.

47. Fluctuations of rates in small non-ideal reaction systems

In the general case, the molecular–statistical theory for arbitrary non-equilibrium states with an allowance for size fluctuations requires detailed analysis for different characteristic relaxation times of the realized stages α of the general process. Below we shall limit ourselves to an outline of the basis of this approach. It has been pointed out above that the very concept of local equilibrium refers to the relationship between local concentrations and paired DFs (Sections 39 and 40). When transferring the kinetic equations to small bodies, one should take into account the relationships of local volumes, the value of *M*, as indicated above for equilibrium fluctuations, and the number of molecules N_q^i within a given volume.

In the LGM the function $P(\{\gamma_f^i\},\tau)$ is the concentration part of the total non-equilibrium distribution function with respect to coordinates and impulses, which remains after averaging over the impulses (Sections 20 and 43). The form of this function turns out to be universal for any aggregate state of the system; therefore, there exists an explicit expression (43.2) for the non-equilibrium entropy in QCA (for simplicity, the formula (43.2) is written out for a uniform region). This expression allows us to construct a one-to-one correspondence between formula (43.2) and statistical sum Q (16.8) (or a specific form in the QCA (33.7)), which reflects the statistical weight of any local equilibrium configurations. The universality of the expression for entropy means that one can directly use this kind of connection between S (and the text before (43.2)) and the statistical weight in Q for non-equilibrium states. This allows us to generalize the concept of a partition function to non-equilibrium states if we confine ourselves to a small scale of the time interval (this is the value of dt in the left-hand sides of equations (20.4), (20.5)) and restrict ourselves to those changes in the system that can occur in time dt. Then, instead of calculating the total partition function Q, one can select that part of the reaction subsystem from the general non-equilibrium system that participates in a particular equilibrium for a given instant of time. Such time-local equilibria fully correspond to the idea of the activated complex of the stage, which is in equilibrium with the reaction subsystem, in the theory of absolute reaction rates (TARR) [32–34]. This interpretation uniquely relates the stage-by-stage description of the dynamics of the spatial redistributions in the LGM described by equations (20.4) and (20.5), and the change in the non-equilibrium function $P(\{\gamma_f^i\},\tau)$.

The reformulation of the equilibrium type changes the expression for the function $\ln P(\{N_q^i, N_{qp}^{ij}\})$ that was introduced for ideal systems in the sections 30–32, by simply replacing the chemical potential in them with the original reactants at $P(\{N_q^i, N_{qp}^{ij}(r)\}) \equiv Q\exp(\beta\sum_{i=1}^{s-1}\mu_q^{i*}(\alpha)N_q^i)$ where $\mu_q^{i*}(\alpha)$ is the chemical potential of the activated complex: it has its internal degrees of freedom $J_q^{i*}(\alpha)$ and the energy of lateral interaction with surrounding neighbours ε_{ij}^* [32]. In this formulation, the search for difference derivatives is carried out at fixed ratios N_q^i/M_q and $N_{qp}^{ij}(r)/M_{qp}(r)$, since the appearance of the activated complex occurs in the elementary stage when one reagent is converted into another.

As a result, repeating verbatim the procedure for deriving expressions for elementary process rates in the TARR framework for non-ideal reaction systems [32], we obtain [143], taking into account the procedure of the sections 30, 32 for the mixture and 33, that

$$\beta\mu_q^{i*} = \frac{1}{2}\left[\left(1 - \sum_{r=1}^{R}\sum_{p=1}^{t(r)} z_{qp}(r)\ln\frac{N_q^i(N_q^i+1)}{M_q(M_q+1)}\right) + \sum_{r=1}^{R}\ln\frac{(M_{qq}(r)-z_{qq}(r))![(N_{qq}^{iV}(r)+z_{qq}(r))/2!]^2}{(M_{qq}(r)+z_{qq}(r))![(N_{qq}^{iV}(r)-z_{qq}(r))/2!]^2} + \right.$$
$$\left. + \sum_{r=1}^{R}\sum_{p=1,p\neq q}^{t^*(r)}{}'\ln\frac{(M_{qp}(r)-z_{qp}(r))![(N_{qp}^{iV}(r)+z_{qp}(r))!]}{(M_{qp}(r)+z_{qp}(r))![(N_{qp}^{iV}(r)-z_{qp}(r))!]}\right] \quad (47.1)$$

where μ_q^{i*} is the chemical potential of the activated complex of the single-site stage $i \rightarrow$ *product*. Note that in square brackets the first term is normalized to the number of sites of type q; while the second and third terms are normalized to the corresponding numbers of pairs of sites qq and qp. This expression is in complete agreement with expression (33.13) for the equilibrium distribution of molecules in volume-limited systems: equating the elementary process rates in the forward and backward directions, we obtain expression (33.13). For example, for the adsorption stage $\beta\mu_q^i = \beta(\mu_q^{i*} - \mu_q^{V*})$. Substitution of the formula (47.1) in equation (20.1) leads to the appearance of the terms $U_f^i(\alpha)$ in (20.4), (20.5). Similarly, the appearance of local rates of two-site stages is considered.

The generalization of the equations for the pair functions (33.14) leads to the following relationships written out in general form for all pairs in terms of a coefficient $\Delta_{qp}^{ij} = 1 - \Delta_{qp}(1-\Delta_{ij})/2$ that depends on two Kronecker symbols for the types of sites and sorts of particles in the pair (Δ_{qp}^{ij} = 1/2 for $q = p$ and $i \neq j$; and Δ_{qp}^{ij} = 1 for other cases),

$$\beta\delta\varepsilon_{qp}^{ij}(r) = \frac{1}{2}\ln\frac{\Delta_{qp}^{ij}N_{qp}^{ij}(r)(\Delta_{qp}^{ij}N_{qp}^{ij}(r)+1)}{\Delta_{qp}^{iV}N_{qp}^{iV}(r)(\Delta_{qp}^{iV}N_{qp}^{iV}(r)+1)} \quad (47.2)$$

As above, the symbol $\delta\varepsilon_{qp}^{ij}(r) = \varepsilon_{qp}^{*ij}(r) - \varepsilon_{qp}^{ij}(r)$ reflects the difference in energy between $\varepsilon_{qp}^{*ij}(r)$ the activated complex and $\varepsilon_{qp}^{ij}(r)$ the ground states of the central molecule i interacting with its neighbours.

Equation (47.2) connects all pairs (ij) with pair (iV) through the energy parameter.

For dynamic variables in small volumes, the same $X_k(\mathrm{fl}) = X_k + \Delta X_k$ bonds should be performed between the average values of the dynamic variables describing the most probable state of the molecules and the fluctuation corrections (33.15) reflecting the boundedness of the system that were in the equilibrium state, i.e. in dynamics it is a question of considering the connection $\partial X_k(\mathrm{fl})/\partial t = \partial X_k/\partial t + \partial \Delta X_k/\partial t$ throughout the process in the form where the kinetic equations (20.4) and (20.5) are used to calculate the most probable states of the molecules X_k.

The calculation of the fluctuation corrections requires a consideration of the dynamic analogue of expression (33.15), reflecting only those changes in the system states during the time dt which are compatible with the stages in the equations (20.4) and (20.5). In this case, it is required to calculate the time evolution of the dynamic corrections ΔX_k by means of functional derivatives with respect to the variables ΔX_k on the basis of the expressions (47.1) and (47.2)

$$\partial \Delta X_k/\partial t = -\frac{1}{2}\sum_{b=1}^{T_D} \frac{\partial \partial \ln \det{}^*}{\partial t \partial X_b} \frac{\partial X_b}{\partial \ln(\lambda_k{}^*)} \tag{47.3}$$

where the det* = det*($\{X_k\}$) determinant is constructed from the second derivatives with respect to the functions (47.1), (47.2) (in place of (33.13) and (33.14) in (33.15)); Here the time derivative of $\partial \ln \det^* = \partial X_b$ at each step is taken from the right-hand sides of the system of kinetic equations (20.4) and (20.5); and the values λ_k^* have the following meaning: $\lambda_k^* = \exp(\beta \mu_q^{i*})$ for N_q^i; $\lambda_k^* = \exp(\beta \delta \varepsilon_{qq}^{ij}(r))$ for $N_{qp}^{ij}(r)$; $\lambda_k^* = \exp(\beta \delta \varepsilon_{qp}^{ij}(r))$ for $N_{qp}^{ij}(r)$.

From the well-known fact that the entropy in a non-equilibrium state is less than the entropy in the equilibrium state, it follows naturally that fluctuations in the equilibrium state exceed fluctuations in the non-equilibrium states X, so that the above expressions for equilibrium fluctuations can serve as upper estimates for the size kinetic fluctuations. All size estimates on the range of application of thermodynamics and on the minimum phase size obtained from equilibrium distributions remain in force. In this case, the flow of any property is related to the size of the cross-section area S_{ar}, through which the flow passes, and the length of the region L along the flow determines the gradient of the property (density) under consideration. The smaller the length L, the smaller the volume $M = S_{ar}L$, in which the fluctuations are of interest.

Lattice and vibrational contributions of subsystems in dense phases differ sharply with their characteristic relaxation times (by more than 10 orders of magnitude). The lattice subsystem is non-equilibrium at low temperatures, and the vibrational contributions are practically always in equilibrium. Taking into account these regularities, it is possible to construct the size dependences for unary and paired DFs, and through them to calculate the thermophysical and thermodynamic characteristics of local solid bodies (mechanical modules, heat capacities, thermal conductivity and friction coefficients) and average for the entire small body. Under intense perturbations, one can also go on to describe the transfer of impulse under deformations and fractures [144].

The availability of kinetic equations opens the possibility of introducing dynamic exit criteria for local and complete equilibria with the goal of a transition from the thermophysical characteristics to thermodynamic ones. The equations (20.4) and (20.5) can be easily concretized when considering particular problems of dynamics and the transition to fluctuating equilibrium characteristics. We list the most general formulations of the problem: the relaxation of internal uniform drop regions to the equilibrium state; the allowance for fluctuations in the process of relaxation and in the equilibrium state; relaxation in microcrystals for a rigid lattice, taking into account the connection with the gas phase at the solid–vapour or liquid interface; adsorption on solid non-re-arrangeable surfaces of solids; adsorption on solid-tunable surfaces of solids, absorption in interstitials and other issues.

References

1. Landau L.D., Lifshitz E.M. Theoretical physics.VI. Hydrodynamics. Moscow, Nauka, 1986.
2. Sedov L.I., Continuum mechanics. Moscow, Nauka, 1970. V. 1.
3. Landau L.D., Lifshitz E.M. Theoretical physics. VII. Theory of Elasticity. Moscow, Nauka, 1987.
4. Reif F., Statistical physics. Moscow, Fizmatlit, 1977. [McGraw-Hill Book Comp. 1964]
5. Gurov K.P., Phenomenological thermodynamics of irreversible processes (Physical basis). Moscow, Fizmatlit, 1978.
6. Kvasnikov I.A., Thermodynamics and statistical physics. V. 1: Theory of equilibrium systems: Thermodynamics. Moscow, Editorial URSS, 2002.

7. Crack J., The Mathematics of Diffuion. Oxford: Oxford Univer. Press. 1975.
8. Mehrer H., Diffusion in solids. Dolgoprudny. Ed. House. Intellekt, 2011. [Springer – Verlag Berlin Heidelber, 2007].
9. Frenkel J., Introduction to the theory of metals. Moscow and Leningrad, GITTL, 1950.
10. Bokshtein B,S,, Bokshtein S,Z,, Zhukhovitsky A,A,m Thermodynamics and kinetics of diffusion with solids. Moscow, Metallurgiya, 1974.
11. Kitel Ch., Introduction to Solid State Physics. Moscow, Nauka, 1978.
12. Girifalco, L.A., Statistical Mechanics of Solids. Moscow, Mir, 1975. [Oxford, New York: Oxford Univ. Press, 1973].
13. Borovskii, I.B., Gurov, K.P., Machukova, I.D., and Ugaste, Yu. E., Processes of Inter-Diffusion in Alloys, Moscow: Nauka, 1973.
14. Tovbin Yu.K., Komarov V.N., Rus. Chem. Bull., 2013, V. 62, No. 12 P. 2620.
15. Tovbin Yu. K., Komarov V.N., Fiz. Tverd. Tela. 2014. V. 56. No. 2. P. 337. [Phys. Solid State 56, 341 (2014)].
16. Tovbin Yu.K., Titov S.V., Komarov V.N., Fiz. Tverd. Tela, 2015. Vol. 57. No. 2. P. 342. [Phys.Solid State 57, 360 (2015)].
17. Neumann G., Tuijn C., Self-Diffusion and Impurity Diffusion in Pure Metals: Handbook of Experimental Data. Pergamon Materials Series 14, 2009.
18. Tishchenko N.P.. Physica Status Solidi (a). 1982. 73, 279 (1982).
19. Berne A., Boato G., De Paz M., Nuovo Cimento 24, 1179 (1962).
20. 20. Berne A., Boato G., De Paz M., Nuovo Cimento 46, 182 (1966).
21. Nachtrieb N.A., E. Catalano, J.A. Weil. J. Chem. Phys. 20, 1185 (1952).
22. Mundy J.N., Barr L.W., F.A. Smith., Phil. Mag. 15, 411 (1967).
23. Lundy, T.S., Murdock, J.F., J. Appl. Phys. 33, 1671 (1962).
24. Rothman N.L. et al., Phys. Stat. Sol. 39, 635 (1970).
25. Herzig Ch., Eckseler H., Bussmann W., Cardis D., J. J. Nucl. Mater. 69/70 61 (1978).
26. Rothman S.J., Peterson N. L., Physica status solidi (b) 1969 V. 35. No. 1. P. 305.
27. Tovbin Yu.K., Zh. fiz. khimii. 2012. V. 87. No. 9. P. 1461. [Russ. J. Phys. Chem. A 86, 1356 (2012)].
28. Prigogine I., Introduction to the thermodynamics of irreversible processes. Izhevsk: 'Regular and chaotic dynamics', 2001.
29. de Groot S., Mazur P., Non-equilibrium thermodynamics. Moscow, Mir. 1964.
30. Haase R., Thermodynamics of irreversible processes. Moscow, Mir, 1967. [Dr. Dietrich Steinkopff, Darmstadt, 1963].
31. Gurov K.P., Kartashkin B.A., Ugaste Yu.E., Mutual diffusion in multiphase metal systems. Moscow, Nauka, 1981.
32. Tovbin Yu.K., The theory of physical and chemical processes at the gas-solid interface. Moscow, Nauka, 1990. 1990. [CRC, Boca Raton, Florida, 1991].
33. Tovbin Yu.K., Progress in Surface Science. 1990. V. 34, No. 1-4. P. 1-236.
34. Tovbin Yu.K., The Molecular Theory of Adsorption in Porous Solids, Moscow, Fizmatlit, 2012. [CRC Press, Taylor & Francis Group, 2017].
35. Mason E., Malinauskas A., Transport in porous media: a model of a dusty gas. Moscow, Mir, 1986.
36. Bird R. B., Stewart W., Lightfoot E. N., Transport Phenomena. – Moscow: Khimiya, 1974. – 687 p. [John Wiley and Sons, New York, 1965].
37. Geguzin Ya.E. Diffusion zone. Moscow, Science. 1979, 344 p.
38. Bogolyubov N.N. Problems of dynamic theory in statistical physics. Moscow, Gostekhizdat, 1946.
39. Gurov K.P. Foundations of the kinetic theory. Moscow, Nauka, 1967. 460 p.

40. Tovbin Yu.K., Zh. fiz. chemistry. 2014. T. 88. No. 2. P. 261. [Rus. J. Phys. Chem. A, 2014, Vol. 88, No. 2, pp. 213].
41. Tovbin Yu.K., Teoret. Fundamentals of chemical. technol. 2013. T. 47. No. 6. P. 672.
42. Chapman S., Kauling T. Mathematical theory of non-uniform gases. Moscow, Izd-vo inostr. lit., 1960. 510 p.
43. Huang, K., Statistical Mechanics. Moscow, Mir, 1966. 520 p.
44. Ferziger, J., Kaper, G., Mathematical theory of transport processes in gases. Moscow, Mir. 1976. p. [North-Holland, Amsterdam,1972]
45. Hirschfelder J. O., Curtiss Ch. F., Bird R. B., Molecular theory of gases and liquids. – Moscow: Inostr. Lit., 1961. – 929 p. [Wiley, New York, 1954].
46. Monin A.S., Yaglom A.M., Statistical hydrodynamics. Moscow, Nauka, 1965. Part 1; 1967. Part 2.
47. Eremin E.N., Fundamentals of chemical kinetics. Moscow, Vysshaya shkola, 1976. 374 p.
48. Glasston S., Laidler K.J., Eyring H., Theory of absolute reaction rates. Moscow, IL, 1948 [Princeton Univ. Press, New York, London, 1941].
49. Nikitin E.E., Theory of elementary atomic-molecular processes in gases. Moscow, Khimiya, 1970.
50. Eyring H., Lin S.G., Lin S.M., Fundamentals of chemical kinetics. Moscow, Mir, 1983.
51. Tovbin Yu.K., Zh. fiz. khimii. 2015. V. 89. No. 9. P. 1347. [Rus. J. Phys. Chem. A, 2015, Vol. 89, No. 9, P. 1507].
52. Manning J. Kinetics of the diffusion of atoms in crystals. Moscow, Mir, 1971.
53. Lebed' I.V., Khim. fizika. 1996. V. 15. No. 6. P. 64.
54. 54. Tovbin Yu.K., Zh. fiz. khimii. 2017. 91. No. 3. P. 381. [J. Phys. Chem. A 91, 403 (2017)]
55. Gibbs J.W., Thermodynamics. Statistical mechanics. Moscow, Nauka, 1982.
56. Zubarev D.N., Non-equilibrium statistical thermodynamics. Moscow, Nauka, 1971.
57. Landau L.D., Livshits E.M., Theoretical physics. V. 5. Statistical physics. Moscow, Nauka, 1964.
58. Keller L., Fridman A.D., Proc. 1st Intern. Congr. Appl. Mech., Delft., 1924.
59. Zhigulev V.N., Tr. TsAGI. 1969. V. 1135. P. 45.
60. Zhigulev V.N., Teor. Matem. Fiz., 1971. V. 7. No. 1. P. 101.
61. Tovbin Yu.K., Khim. Fizika. 2002. V. 21. No. 1. P.83.
62. Tovbin Yu.K., Zh. fiz.khimii. 2002. V. 76. No. 1. P.76. [Russ. J. Phys. Chem. 2002. V. 76. № 1. P. 64].
63. Tovbin Yu.K., Khim. Fizika.. 2004. V. 23. №12, C. 82.
64. Landau, L.D., Lifshitz E.M. Theoretical physics. IX. Statistical physics. Part 2. Moscow, Nauka, 1966.
65. Naimark O.B., Pis'ma Zh. Eksper. Teor. Fiz. 1998. V. 67. No. 9. P. 714.
66. Naimark O.B., Pis'ma Zh. Eksper. Teor. Fiz. 1997. V. 23. No. 13. P. 81.
67. Naimark O.B., et al., Fiz. mezomekhanika. 2009. V. 12. No. 4. P. 47.
68. Panin V.E., Egorushkin V.E., Ibid. 2008. V. 11. No. 2. P. 9.
69. Butyagin P.Yu., Usp. khimii. 1984. V. 53. No. 11. P. 1769.
70. Butyagin P.Yu., Chemical physics of solids. Moscow, Publishing House of Moscow State University, 2006.
71. Tomashev N.D., Chernova G.P., Theory of corrosion and corrosion-resistant structural alloys. Moscow, Metallurgiya, 1986.
72. Zhilyaev A.P., Pshenichnyuk A.I., Superplasticity and grain boundaries in ultrafine-grained materials. Moscow, Fizmatlit, 2008.

73. Klimontovich Yu.L., Turbulent motion and the structure of chaos. Moscow, Nauka, 1990.
74. Timashev S.F., Flicker-noise spectroscopy. Information in chaotic signals. Moscow, Fizmatlit, 2007.
75. Kogan Sh.M., Usp. fiz. nauk. 1985. V. 145. No. 2. P. 285.
76. Weissman M.B., Rev. Mod. Phys. 1988. V. 60. P. 537.
77. Zarkhin L.S., et al., Usp. khimii. 1989. V. 58. P. 644.
78. Semenov N.N., Chain reactions. Leningrad. ONTI, 1934.
79. Zel'dovich Ya.B., et al., Mathematical theory of combustion and explosion. Moscow, Nauka, 1980.
80. Frank-Kamenetsky D.A., Diffusion and heat transfer in chemical kinetics. Moscow, Nauka, 1987.
81. Dedkov G.V., Usp. fiz. nauk, 2000. V. 176. No. 6. P. 585.
82. Lifshitz I.M., Selected works. Physics of Real Crystals and Disordered Systems. Moscow, Nauka, 1986.
83. Suzdalev I.P., Nanotechnology: physical chemistry of nanoclusters, nanostructures and nanomaterials. KomKniga, Moscow, 2006.
84. Cheremskoi P.G., Slezov V.V., Betekhtin V.I., Pores in a solid body. Moscow, Energoatomizdat, 1990.
85. Zhilyaev A.P., Pshenichnyuk A.I., Superplasticity and grain boundaries in ultrafine-grained materials. Moscow, Fizmatlit. 2008.
86. Rusanov A.I., Thermodynamic basis of mechanochemistry. St. Petersburg, Nauka, 2000.
87. Sirotin Yu.I., Shaskol'skaya M.P., Fundamentals of crystallophysics. Moscow, Nauka, 1979.
88. Tovbin Yu.K., Zh. fiz. khimii. 2017. V. 91. No. 8. P. 1243. [Rus. J. Phys. Chem. A, 91, No. 8, P. 1357 (2017)].
89. Tovbin Yu.K., Zh. fiz. khimii. 2010. P. 84. No. 2. P. 231. [Rus. J. Phys. Chem. A, 84, No. 8, P. 1788 (2017)].
90. Tovbin Yu.K., Zh. fiz. khimii. 2014. T. 88. No. 11. P. 1788. [Rus. J. Phys. Chem. A, 88, No. 11, P. 1965 (2014)].
91. Ono S., Kondo S., Molecular theory of surface tension. Moscow, IL, 1963. [Handbuch der Physik, Vol X (Springer) 1960].
92. Rowlinson J., Widom B., Molecular theory of capillarity. Moscow, Mir, 1986. [Oxford: Clarendon Press, 1982].
93. Irving J.H., Kirkwood J.G., J. Chem. Phys. 1950. V. 18. P. 1950.
94. Schofield P., Chem. Phys. Lett., 1966. V. 62. P. 413.
95. Grant M., Desia R.C., Molec. Phys. 1981. V. 41. P. 1035.
96. Hirschfelder J., Curtis C., Byrd R. Molecular theory of gases and liquids. Moscow, IL, 1961.
97. Rice O.K., J. Phys. Chem. 1927. V. 31. P.207.
98. Prigogine I., Defay R., J. Chim. Phys. 1949. V. 46. P. 367.
99. Defay R., J. Chim. Phys. 1954. V. 51. P. 299.
100. Summ B.D., Fundamentals of Colloid Chemistry. 2007. Moscow, Akademiya.
101. Shikhmurzaev Y.D., Capillary Flows with Forming Interfaces. 2007, Taylor & Francis.
102. Shikhmurzaev Y.D., J. Fluid Mech. 1997. V. 334. P. 211.
103. Blake T.D., J. Colloid Interface Sci. 2006. V. 299. P. 1.
104. Dussan E.B., J. Fluid. Mech. 1976, V. 77, 665.
105. Rasmessen D.H.J., J. Chem. Phys. 1986. V. 85. P. 2272.

106. Hua X.Y., Rosen M.J., Journal of Colloid and Interface Science. 1988. V. 124. No.2. P. 652.
107. Dynamics of surfactant self-assemblies: micelles, microemulsions, vesicles, and lyotropic phases. ed. R. Zana, Taylor & Francis, 2005.
108. Ross S., Morrison I.D., Colloidal systems and interfaces. Wiley-Interscience publication, New York, Toronto, 1988.
109. Adamczyk Z., Journal of Colloid and Interface Science. 2000. V. 229. No. 2. P. 477.
110. de Gennes P.G., Rev. Mod. Phys. 1985, 57, 827–863.
111. Blake T.D., Shikhmurzaev Y.D., J. Colloid Interface Sci. 2002. V. 253. P. 196–202.
112. Khaidarov G.G., et al., Vestnik Sankt-Peterb. Univ., Series 4. 2011. Issue 1. P. 3–8.
113. Khaidarov G.G., et al., Vestnik Sankt-Peterb. Univ. Series 4. 2011. Issue. 1. P. 24–2.
114. Filippov L.K., Filippova N.L., Journal of Colloid and Interface Science. 1997. V. 187. P. 352.
115. Lukyanov A.V. Likhtman A.E., J. Chem. Phys. 2013, 138, 034,712.
116. Eriksson J.C., Surface Sci. 1969. V. 14. P. 221.
117. Adamson A. W., The Physical Chemistry of Surfaces,Mir, Moscow, 1979. [Wiley, New York, 1976].
118. Rusanov A.I., Surface Science Reports. 1996 Vol. 23, Nos. 6–8.
119. Rusanov A.I., Surface Science Reports. 2005. Vol. 58. P. 111.
120. Benson G.G., Yun K.S.S., The solid-gas interface. Ed. E. Alison Flood, Marcel Dekker, Inc., New York, 1967.
121. Dunning W., Interphase boundary gas-solid. Ed. E. Flad. Publishing house Mir, Moscow 1970. P. 230.
122. Sparnaay M.J., Surface Sci. Reports. 1984. V. 4. N 3/4. P. 103–269.
123. Cammarata R.C., Progress in Surface Sci. 1994. V. 46. No. 1. P. 1.
124. Ibach H., Surface Science Reports. 1997. V. 29. P. 193.
125. 125. Tovbin Yu.K., Zh. Fiz. Khimii. 2014. V. 88. No. 7–8. P. 1266. [Rus. J. Phys. Chem. A, 88, No. 7-8, P. 1438 (2014)].
126. Born M., Huang K., Dynamic theory of crystal structures. Moscow, IL, 1958.
127. Leibfried G., Microscopic theory of the mechanical and thermal properties of crystals. Moscow and Leningrad, GIFML, 1963. [Handbuch der Physik, Band 7, Springer, Berlin, 1955, Vol. 2].
128. Maradudin A.A., Theoretical and Experimental Aspects of the Effects of Point Defects and Disorder on the Vibrations of Crystals (Academic Press, New York, London, 1966; Mir, Moscow, 1968).
129. Kosevich A,M,, Fundamentals of the Mechanics of the Crystal Lattice. Moscow, Nauka, 1972.
130. Khachaturyan A.G., The theory of phase transitions and the structure of solids. Moscow, Nauka, 1974.
131. Katznelson A.A., Olemskoy A.I., Microscopic theory of non-uniform structures. Moscow, Izd-vo MGU, 1987.
132. Coussy O., Poromechanics. John Wiley & Sons, Ltd. Chichester, 2004.
133. Fischer-Cripps A.C., Introduction to contact mechanics. 2ed., Springer Science + Business Media, LLC. 2007.
134. Popov V.L., Mechanics of contact interaction and physics of friction. Moscow Fizmatlit, 2013.
135. Aizikovich S.M., et al., Contact problems of the theory of elasticity for non-uniform media. Moscow, Fizmatlit, 2006.
136. Aleksandrov V.M., Chebakov M.I., Introduction to the mechanics of contact interactions. – Rostov-on-Don: Publishing House of the Central Research and Develop-

ment Centre, 2007.

137. Johnson K.L., Contact mechanics. Cambridge Univer. Press, Cambridge London, New York, New Rochelle, 1985.
138. Suzdalev I.P., Nanotechnology: physical chemistry of nanoclusters, nanostructures and nanomaterials. KomKniga, Moscow. 2006.
139. Tovbin Yu.K., Rabinovich A.B., Zh. fiz. khimii. 2011. V. 84. No. 12. P. 2366. [Rus. J. Phys. Chem. A, 84, No. 12, P. 2166 (2011)].
140. Tovbin Yu.K., Titov S.V., Zh. fiz. khimii. 2013. V. 87. No. 1. P. 77. [Rus. J. Phys. Chem. A, 87, No. 1, pp. 93 (2013)].
141. Tovbin Yu.K., Zh. fiz. khimii. 2010. V. 84. No. 11. P. 2182. [Rus. J. Phys. Chem. A, 84, No. 11, P.1993 (2013)].
142. Hill T.L., Statistical Mechanics, Principles and Selected Applications, NewYork, McGraw-Hill Book Comp., Inc., 1956.
143. Tovbin Yu.K., Zh. fiz. khimii. 2015. V. 89. No. 3. P. 551. [Rus. J. Phys. Chem. A, 89, No. 3, P. 547 (2017)].
144. Ionov V.N., Selivanov V.V., Dynamics of destruction of a deformed body. Moscow,Mashinostroenie, 1987.

6

Elementary stages of the evolution of the system

Fundamentally important for constructing probabilities $W_\alpha(\{I\}\rightarrow\{II\})$ is the use of models describing the elementary process within the framework of the theory of absolute reaction rates (TARR) or the theory of the transition state [1-3]. For the first time, TARR was used for the W_α functions in Eq. (20.1) for non-ideal systems in adsorption studies [4–8] and label transfer in a liquid [9]. Later, this theory was extended to a wide variety of processes [10–17]. The velocities of all elementary motions will be described in the framework of this theory.

In the kinetic theory, for any phases, the key is the problem of self-consistency of the expressions for the rates of elementary reactions (stages) and the equilibrium state of the reaction system. The essence of this statement is that, equating the expressions for the reaction rates of any of the stages occurring in the forward and backward directions, equations describing the equilibrium distribution of the molecules of the given system must be obtained. Within the framework of the LGM (lattice gas model), it is possible to find the conditions under which these self-consistency conditions for the description of reaction rates and system equilibrium are fulfilled [10, 14–17]. This question is discussed successively for elementary processes occurring on one (section 50) and two (section 52) sites of the lattice system, taking into account the interaction in quasi-chemical approximation (QCA) for nearest neighbours, and then (section 54) for any interaction radius between the neighbours R.

For an entire non-uniform lattice structure, the self-consistency conditions will be satisfied if they are satisfied for each site and for each pair of sites. Therefore, when we consider the self-consistency of the expressions for the rates of stages, we will operate on specific individual sites, i.e. before using the averaging procedure with different distribution functions for the composition and structure of the non-uniform structure [14–17]. Section 49 shows how the rates for single-site processes are obtained – these are the terms $U_f^i(\alpha)$ in equation (20.4). These terms are not closed – in their right-hand parts there are unknown correlators of a higher (second) dimension. To close Eq. (20.4), it is necessary to construct new kinetic equations for unknown correlators (20.5) or to express them through correlators of lower dimensionality – the latter can be done if the pair functions belong to the equilibrium distribution (Section 16). This principle is common to a system of interacting particles.

48. The rate of elementary stages

The basic idea of TARR is the relationship between the concentration θ^* (the number of activated complexes (AC) per unit volume) with the rate of the elementary process $U = \theta^* \nu$, where ν is the frequency of crossing the activation barrier (s^{-1}). For translational motion [1–3], $\nu = u_t/b$, where u_t is the average velocity of motion of an AC of mass m^* equal to $u_t = (kT/2\pi m^*)^{1/2}$, b is the length of the activation barrier.

The concentration of AC at the top of the barrier can be written as follows: $\theta^* = \theta F_t$, where θ is the concentration of AC after the replacement of one vibrational degree of freedom of movement by translational motion, $F_t = (2\pi m^* kT)^{1/2} b/h$; h is the Planck constant. Hence we obtain that the translational velocity $U = \theta kT/h$. Or, introducing in the usual manner the specific rate of the elementary process $K_i = U/\theta_i$ for the motion of the activated complex formed from particle i, we have that $K_i = \theta\, kT/(h\theta_i)$.

The ratio $\theta/\theta_i = F^*/F_i$ is expressed [1–3] in terms of sums over the states AC (F^*) and molecules in the ground state (F_i), which leads to the standard form for the rate constant of the elementary motion $K_i = kTF^*/(F_i h)$ for the gas phase.

As a result, the type of motion is concretized only in the expressions for the sums over the states of the AC and the molecule in the ground state. For a rarefied phase $F^*/F_i = F_t$. For non-ideal reaction systems this ratio varies, and the influence of neighbouring molecules is additionally required.

The reaction rate in the theory of the transition state is defined as [1–3]:

$$U_i = \chi w \theta_i^* / b, \tag{48.1}$$

where χ is the transmission coefficient, w is the average velocity of passage of the barrier of length b, $w = (2\pi M^* \beta)^{-1/2}$, M^* is the mass of the complex, U_i is the rate referenced to one site of the structure. The problem reduces to the calculation of the concentration of AC θ_i^* for the monomolecular process and θ_{ij}^* for the bimolecular process.

For ideal reaction systems (in the absence of the influence of lateral interactions) the rates of elementary stages, mono- and bimolecular reactions are described within the framework of the law of mass action

$$U_i = K_i \theta_i , \qquad U_{ij} = K_{ij} \theta_i \theta_j . \tag{48.2}$$

where K_i and K_{ij} are the rate constants of elementary processes (stages) that characterize the specific rates of elementary processes:

$$K_i = K_{ij}^0 \exp\left(-E_i / k_{\mathrm{B}} T\right) = \kappa \frac{kT}{h} \frac{F_i^*}{F_i} \exp\left(-E_i / kT\right), \tag{48.3}$$

$$K_{ij} = K_{ij}{}^0 exp\left(-E_{ij} / k_{\mathrm{B}} T\right) = \kappa \frac{kT}{h} \frac{F_{ij}^*}{F_i F_j} \exp\left(-E_{ij} / kT\right),$$

where K_i and K_{ij}^0 are the pre-exponentials of the rate constants, E_i and E_{ij} are the activation energies of the reaction $i \rightarrow$ *product* and $i + j \rightarrow$ *products*; F_i and F_j are the statistical sums (sums over the internal states) of the original molecules, F_i^* and F_{ij}^* are the statistical sums of AC calculated for all degrees of freedom except for the 'reaction path'.

It is assumed here that the number of sites of the structure does not change during the reaction, and the concentration of particles can be characterized as the 'degree of filling' of the surface θ_i. In the derivation of the equations (48.1)–(48.3), it is assumed that the equilibrium distribution of molecules in the reaction system is realized, and that the chemical transformation stage is limited. It also assumes: 1) the absence of diffusion transport at the macroscopic level (uniform distribution over the macrovolume), (2) the absence of the influence of external fields, (3) the absence of a diffusion-controlled regime of the reaction at the molecular level, (4) the absence of influence of intermolecular interactions, and (5) the fraction of particles reacting per unit time is so small that it does not distort the equilibrium distribution of molecules on the surface.

However, the use of thermodynamic approaches in problems of kinetics for non-ideal systems (Section 10), which expresses the use of the concepts of activity coefficients for reagents and activated complexes, leads to contradictions (Section 64). Below we discuss the expressions for the rates of elementary stages in molecular theory based on the LGM (lattice gas model) [14–17] for non-ideal systems, which is applicable to all aggregate states of substances and interfaces, and will ensure their self-consistency with a description of the equilibrium state of the reaction system in the QCA.

In the molecular theory of non-ideal reaction systems, it is necessary to consider successively the entire spectrum of configurations of neighbouring molecules that can influence the course of the reaction for the reagents at the central sites under consideration and weigh the probability of realizing the elementary stage for each of the neighbours' configurations on the surface. Lateral contributions affect the probability of formation of ACs through a change in the activation energy, so the number of neighbours and the way they are located are important for the rate of the process.

49. One-site processes

The multiparticle (cooperative) nature of elementary processes is clearly manifested in the calculation of their rates. The reason for this is illustrated by the examples of mono- ($iZ \rightarrow product$) and bimolecular ($iZ+jZ \rightarrow products$) reactions that are realized in a two-component system ($s = 2$, i = particle A and vacancy V, here Z is the symbol of the reagent-containing site). The generalization of the expressions (48.2) and (48.3) in the case when the interaction of neighbouring particles is taken into account leads to a change in the concentration dependences of the rates of elementary processes.

In the absence of interaction between particles, the rate of the monomolecular reaction ($iZ \rightarrow product$) is expressed in terms of the equations (48.2) and (48.3), where θ_i is the concentration of component i (its mole fraction) which characterizes the probability that any lattice site is occupied by a particle of sort i. Let us discuss the process by the example of adsorption. If $i = A$, then U_A denotes the desorption rate; if $i = V$, then U_V is the adsorption rate. In this case, K_A is the desorption rate constant and K_V is the adsorption rate constant. To unify the form of the recording U_i, we will include the partial pressure of the molecules of the gas phase P_A in the effective adsorption rate constant K_V.

Consider a uniform lattice each site of which has z neighbours. For flat lattices, the sites correspond to adsorption centres, for which $z = 3$, 4, or 6; in the bulk lattices the sites correspond to the absorption centres and for them $z = 4$, 6, 8 or 12. In the simplest case, the influence of the interaction of neighbouring particles manifests itself primarily in the fact that, depending on the local composition of the neighbours around the central particle i, the height of the activation barrier of the elementary one-site reaction varies. The number of the nearest neighbours around the central particle i will be denoted by z. The arrangement of neighbouring particles A around the reagent i is given by the symbols n and σ, where n is the number of particles A in the first coordination sphere z around particles i; $\sigma(n)$ is the index that determines the way the particles are arranged for a fixed n (Fig. 16.1).

Denote by $E_i(n\sigma)$ the interaction energy of the particle i with its neighbours in the ground state, and by $E_i^*(n\sigma)$ the interaction energy of the activated reaction complex with its neighbours in the transition state. The difference $E_i^*(n\sigma) - E_i(n\sigma)$ is the change in the height of the activation barrier for a given local configuration $n\sigma$. At a fixed position of all neighbouring particles, the rate of the elementary process at one central site is written as follows:

$$U_i(n\sigma) = K_i\theta_i(n\sigma)\, F_i(n\sigma), \quad F_i(n\sigma) = \exp\left\{\beta[E_i^*(n\sigma) - E_i(n\sigma)]\right\}, \quad (49.1)$$

where $\theta_i(n\sigma)$ is the probability of existence of the configuration of neighbours $n\sigma$ around the reagent i. This quantity is the probability of a complex event consisting in the simultaneous arrangement of n particles A and $(z-n)$ vacancies in the configuration $n\sigma$ around the central particle i. To calculate the function $\theta_{ij}(n\sigma)$, it must be expressed in terms of reagent concentrations θ_i. The method of calculating $\theta_{ij}(n\sigma)$ through the concentration of reagents depends on the intermolecular interaction accounting approximation used (see Section 16).

The expression for the average velocity of the considered reaction U_i on a uniform surface (lattice) and pertaining to the unit surface is obtained by averaging (49.1) over all ways of arrangement of the particles. If we denote by $\alpha_i(n\sigma)$ the statistical weight of the configuration $n\sigma$ and assume that the rate constant K_i depends on the influence of neighbours only through the value of the activation energy determined by the function $F_i(n\sigma)$, then we get

$$U_i = \Sigma_n \Sigma_\sigma \alpha_i(n\sigma) U_i(n\sigma) = K_i \Sigma_n \Sigma_\sigma \alpha_i(n\sigma) \theta_i(n\sigma)\, F_i(n\sigma), \quad (49.2)$$

We assume that all the energy characteristics of the system can be expressed through pair interactions of the nearest neighbours [14]. We denote by ε_{ij} the interaction parameter of particles i and j in the ground state. Obviously, $\varepsilon_{ij} = \varepsilon_{ji}$. If one of the particles is a vacancy, then $\varepsilon_{iV} = 0$. The formation of the activated complex is associated with the transition of the particle from the ground state to the activated one to which the change in the position of the reagent relative to the positions of its neighbours corresponds. In this case the interaction energy of the particles changes. We denote by ε^*_{ij} the interaction parameter of the particle i, which is in the activated state with the neighbouring particle j in the ground state. For the activated complex $\varepsilon^*_{ij} \neq \varepsilon^*_{ji}$, since these parameters refer to different reagents and their neighbouring particles.

Taking into account only pair interactions, we write out the expression for the energy of the ground state for a two-dimensional lattice $z = 4$ (analogous expressions are satisfied for a two-dimensional lattice with $z = 3$ and three-dimensional lattices with $z = 4, 6, 8$):

$$E_i(n\sigma) = n\varepsilon_{iA} + (z - n)\varepsilon_{iV}, \quad (49.3)$$

i.e. $E_i(n\sigma)$ does not depend on the location of neighbouring particles. Both terms correspond to the interaction of the particle i with its neighbours (particles A and vacancies), $0 \leq n \leq z$.

For the activated complex, the analogue of expression (49.3) will be written as

$$E^*_{ij}(n\sigma) = n\varepsilon^*_{iA} + (z-n)\varepsilon^*_{iV},$$

This leads to the expression for the function $F_{ij}(n\sigma)$:

$$F_i(n\sigma) = \exp\left\{\beta\left[n\delta\varepsilon_{iA} + (z-n)\delta\varepsilon_{iV}\right]\right\}, \ \delta\varepsilon_{ij} = \varepsilon^*_{ij} - \varepsilon_{ij}. \quad (49.4)$$

Formula (49.4) characterizes a linear change in the energy in the exponent of the function $F_i(n\sigma)$ with a change in the number of neighbouring particles. Then in this approximation

$$U_i = K_i \Sigma_n \Sigma_s \alpha_i(n\sigma) \theta_i(n\sigma) \exp\left\{\beta\left[n\delta\varepsilon_{iA} + (z-n)\delta\varepsilon_{iV}\right]\right\}, \quad (49.5)$$

The calculation of the single-site reaction rate was reduced to determining the weight contributions $\alpha_i(n\sigma)$ and calculating the many-particle functions $\theta_i(n\sigma)$. In the general case of a non-equilibrium particle distribution, it is necessary to solve the corresponding kinetic equations. In the case of an equilibrium distribution of reagents, which is realized for the limiting stage under consideration in the kinetic regime (with a large activation energy of the reaction), these many-particle functions can be calculated in concrete equilibrium approximations for the interaction of particles. In order to take into account, as closely as possible, the real properties of the cooperative behaviour of interacting particles, it is necessary to take into account the correlation effects in their spatial distribution. This requirement is achieved if in calculating the functions $\theta_i(n\sigma)$ we use the pair distribution functions θ_{ij} characterizing the probability of finding particles i and j at neighbouring sites (the so-called quasi-chemical approximation (QCA)). As a result, the probabilities of multiparticle particle configurations $\theta_i(n\sigma)$ are approximated through the local probabilities θ_i for detecting particles i (unary distribution functions or the probability of detection of particles of a particular sort) and through the pair distribution functions. We express the equilibrium function of a particular distribution $\theta_i(n\sigma)$ in the following form:

$$\theta_i(n\sigma) = \theta_i t_{iA}^n t_{iV}^{z-n}, \tag{49.6}$$

where $t_{ij} = \theta_{ij}/\theta_i$ is the conditional probability of finding the particle j near the particle i; $\theta_{iA} + \theta_{iV} = \theta_i$ or $t_{iA} + t_{iV} = 1$. For the case under discussion, the transition to pair probabilities θ_{ij} means that the weights of configurations with different σ for n = const are equiprobable. As a consequence, the statistical weight of the distributions of particles A at neighbouring sites relative to the central particle i will be written as the number of combinations of z sites with n particles A: $\Sigma_\sigma \alpha_i(n\sigma) = C_z^n$. Then the formula (49.5) can be rewritten as

$$\begin{gathered} U_i = K_i\theta_i \sum_{n=0}^{z} C_z^n \Phi_{iA}^n \Phi_{iV}^{z-n} = K_i\theta_i S_i^{\,z}, \\ S_i = \Phi_{iA} + \Phi_{iV},\ \Phi_{ij} = t_{ij}\exp(\beta\delta\varepsilon_{ij}). \end{gathered} \tag{49.7}$$

In the function S_i the summation is carried out over all sorts of neighbouring particles. The S_i functions are the factors of the non-ideality function of the reaction system $\Lambda_i^* = S_i^z$, which reflect the influence of each neighbouring particle on the height of the activation barrier of the reaction. In the absence of interaction, the factors $S_i = 1$ and formula (49.7) goes over into the reaction rate for an ideal reaction system (Section 10).

50. Self-consistency of the rates of single-site stages with an equilibrium distribution of molecules

We confine ourselves to the case of fast mobility of reagents in a mixture consisting of s components, when they are equilibrated over the sites of a uniform surface [10]. The stages of adsorption, desorption, or chemical transformations involving adsorbed particles can be limiting. Let the elementary stage of the chemical transformation $A \leftrightarrow B$ be realized, which corresponds to the equality of rates in the forward and backward directions $U_A(\alpha) = U_B(\alpha)$ – the symbol α corresponds to a certain stage number in the general multi-stage process. In conditions of complete equilibrium all stages must be in dynamic equilibrium. With the example of expressions in QCA, we show that the expressions obtained for the reaction rates are self-consistent with a description of the equilibrium state of the reaction systems. That is, equating the rates of direct and reverse directions of each stage of the general process, we obtain the equations of the equilibrium distribution of particles.

The structure of the equations for the reaction rates $U_A(\alpha) = K_A(\alpha)V_A(\alpha)$, where $K_A(\alpha)$ is the reaction rate constant, and $V_A(\alpha)$ is the concentration component of the reaction rate. For an ideal reaction system $V_A(\alpha) = \theta_A$ and for a non-ideal reaction system it is described by the formulas (49.7).

The ratio of the rate constants determined by formulas (48.3) gives the equilibrium constant K_e and excludes the partition function of the activated complex which appears in the pre-exponents of the rate constants of the direct and reverse reactions. But still there is the characteristic of the activated complex $\varepsilon^*_{ij} \equiv \varepsilon_{ij}(\alpha)$, which is included in the non-ideality function $S_i(\alpha)$

$$K_1 = \frac{K_A(\alpha)}{K_B(\alpha)} = \frac{V_B(\alpha)}{V_A(\alpha)} = \frac{\theta_B[S_B(\alpha)]^z}{\theta_A[S_A(\alpha)]^z}. \quad (50.1)$$

To prove the self-consistency of the expressions, it is necessary that the ratio of the non-ideality functions $S_B(\alpha)/S_A(\alpha)$ does not depend on the parameters of the interaction of the activated complex with the neighbours ε^*_{ij}. To this end, we consider the equations in QCA [18]

$$\theta_A = \theta_B \exp[\beta(v_B - v_A)] S_B(A)^z,$$
$$S_B(A) = \sum_{j=1}^{V} t_{Bj} \exp[\beta(\varepsilon_{Aj} - \varepsilon_{Bj})] \tag{50.2}$$

$$\theta_{AC} = \theta_{BC} \exp[\beta(v_B - v_A + \varepsilon_{AC} - \varepsilon_{BC})] S_B(A)^{z-1}. \tag{50.3}$$

here $S_B(A)$ in the lower index is the sort of the considered central particle, and the symbol in brackets is the 'reference' sort of the mixture – often this symbol refers to the vacancy, and $t_{ij} = \theta_{ij}/\theta_i$. Equation (50.2) is the equation (18.4), and equation (50.3) is a different form of the equation (18.1) [14,18].

Eliminating the exponent $\exp[\beta(v_i - v_s)]$ from the equations (50.2) and (50.3), we obtain a connection between the pair and unary DFs

$$t_{AC} = t_{BC} \exp[\beta(\varepsilon_{AC} - \varepsilon_{BC})] / S_B(A). \tag{50.4}$$

This expression shows that the denominator does not depend on the type of the neighbouring molecule C, and the normalization condition is satisfied

$$\sum_{j=1}^{s} t_{Aj} = \sum_{j=1}^{s} t_{Bj} \exp[\beta(\varepsilon_{Aj} - \varepsilon_{Bj})] / S_B(A) = 1. \tag{50.5}$$

Using (50.3), we prove the following identity

$$\sum_{j=1}^{s} t_{nj} \exp[\beta(\varepsilon_{pj} - \varepsilon_{nj})] = \sum_{j=1}^{s} t_{ij} \exp[\beta(\varepsilon_{pj} - \varepsilon_{ij})] / \sum_{j=1}^{s} t_{ij} \exp[\beta(\varepsilon_{nj} - \varepsilon_{ij})], \tag{50.6}$$

which allows one to move from one sort of central molecule to another. To prove this identity, it suffices to put $A = n$, $B = i$, $C = j$ in (50.4) and substitute it in the left-hand side of equation (50.6). This identity will be abbreviated as $S_n(p)=S_i(p)/S_i(n)$. (In the particular case, as a consequence of (50.6), we get that $S_n(p) = S_p(n)^{-1}$ or $S_n(p)S_p(n) = 1$).

In formula (50.1), from the equality of the rates of elementary reactions, equating the rates of the direct and inverse one-site stages $ZA \leftrightarrow ZB$, the ratio

$$S_B(\alpha) / S_A(\alpha) = \sum_{j=1}^{s} t_{Bj} \exp[\beta(\varepsilon^*_{Bj} - \varepsilon_{Bj})] / \sum_{j=1}^{s} t_{Aj} \exp[\beta(\varepsilon^*_{Aj} - \varepsilon_{Aj})].$$

It involves one activated complex between reaction products A and B, so its properties are the same, or $\varepsilon^*_{Aj} = \varepsilon^*_{Bj} \equiv \varepsilon^*_{(AB)j}$. This allows us to obtain from the identity (50.6) that

$$\left(\frac{\sum_{j=1}^{s} t_{Bj} \exp[\beta(\varepsilon^*_{Bj} - \varepsilon_{Bj})]}{\sum_{j=1}^{s} t_{Aj} \exp[\beta(\varepsilon^*_{Aj} - \varepsilon_{Aj})]} \right)^z \equiv \left(\frac{S_B(AB^*)}{S_A(AB^*)} \right)^z = \tag{50.7}$$
$$= \left(\sum_{j=1}^{s} t_{Bj} \exp[\beta(\varepsilon_{Aj} - \varepsilon_{Bj})] \right)^z = (S_B(A))^z$$

which is exactly equal to the concentration component of the isotherm (50.2). This shows that regardless of the path of the process through a chemical reaction or through the exchange of molecules with a thermostat, the equilibrium state is the same.

In the chaotic approximation, the analog of equation (50.6) has the form

$$\sum_{j=1}^{s} \theta_j \exp\left[\beta(\varepsilon_{pj} - \varepsilon_{nj})\right] = \sum_{j=1}^{s} \theta_j \exp\left[\beta(\varepsilon_{pj} - \varepsilon_{ij})\right] / \sum_{j=1}^{s} \theta_j \exp\left[\beta(\varepsilon_{nj} - \varepsilon_{ij})\right] \tag{50.8}$$

and for the mean field approximation, which is obtained from the random approximation by the limiting transition $\beta\varepsilon_{ij} \to 0$, but without the limiting values of such a transition, the analog of equation (50.6) will be expressed as

$$\exp\left[\sum_{j=1}^{s} \beta(\varepsilon_{pj} - \varepsilon_{nj})\theta_j\right] = \exp\left[\sum_{j=1}^{s} \beta(\varepsilon_{pj} - \varepsilon_{ij})\theta_j\right] / \exp\left[\sum_{j=1}^{s} \beta(\varepsilon_{nj} - \varepsilon_{ij})\theta_j\right]. \tag{50.9}$$

These expressions also lead to the conclusion of self-consistency for single-site reactions due to the identity of the interactions of the activated complex with neighbours in the forward and reverse directions of the reaction, or $\varepsilon^*_{Aj} = \varepsilon^*_{Bj} \equiv \varepsilon^*_{(AB)j}$.

51. Two-site processes

Similarly, we can consider the rate of the bimolecular reaction $iZ+jZ$, which is expressed in terms of the equations (48.2) and (48.3), where θ_i and θ_j are the concentrations of the components i and j, which characterize the probabilities that any lattice site is occupied by a particle of sort i and j. If $i = j =$ A, then U_{AA} is the desorption rate; at $i = A$, $j = V$ (or vice versa) U_{AV} is the migration or hopping rate of particle A; if $i = j =$ v, then U_{VV} is the adsorption rate. In this case, K_{AA} is the desorption rate constant, K_{AV} is the migration rate constant and K_{vv} is the rate constant of adsorption. As above, in order to unify the recording form, the values of U_{ij} will include the partial pressure of molecules of the gas phase P_{A2} in the effective adsorption rate constant.

The influence of the interaction of neighbouring particles is manifested primarily in the fact that, depending on the local composition of neighbours around the central pair ij, the height of the activation barrier of the elementary reaction between reagents i and j varies. The number of the nearest neighbours around the central pair ij will be denoted by z^*. We will characterize the arrangement of the neighbouring particles A around the reagents ij by the symbols n and σ, where n is the number of particles A in the first coordination sphere z^* around the pair of particles ij (respectively, (z^*-n) of sites is occupied by vacancies); $\sigma(n)$ is the index that determines the way the particles are arranged for a fixed n (Fig. 51.1).

We denote by $E_{ij}(n\sigma)$ the interaction energy of the reactants ij with all its neighbours in the ground state, and by $E^*_{ij}(n\sigma)$ the interaction energy of the activated reaction complex $i + j$ with its neighbours in the transition state. The difference $E^*_{ij}(n\sigma) - E_{ij}(n\sigma)$ is the change in the height of the activation barrier for a given local configuration $n\sigma$. For a fixed position of all neighbouring particles, the rate of the elementary process at two central sites will be written as follows:

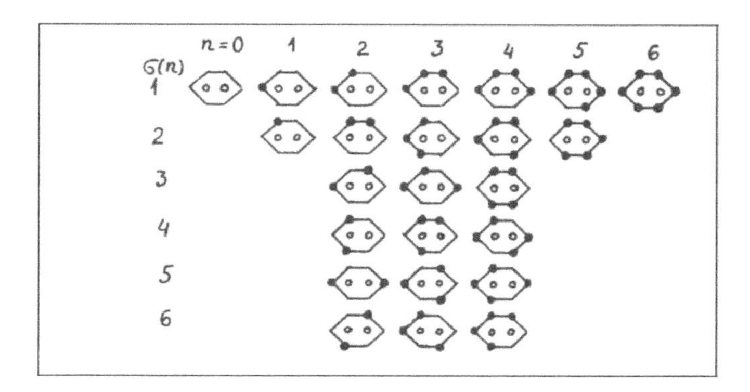

Fig. 51.1. The configurations of particles A in the first coordination sphere around two central particles on a square lattice; light circles — central particles, black circles — adsorbed particles A.

$$U_{ij}(n\sigma) = K_{ij}\theta_{ij}(n\sigma)\, F_{ij}(n\sigma),$$
$$F_{ij}(n\sigma) = \exp\left\{\beta\left[E^*_{ij}(n\sigma) - E_{ij}(n\sigma)\right]\right\}, \tag{51.1}$$

where $\theta_{ij}(n\sigma)$ is the probability of the neighbours configuration $(n\sigma)$ around the reagents ij. This quantity is the probability of a complex event consisting in the simultaneous arrangement of n particles A and (z^*-n) vacancies in the configuration $n\sigma$ around the central pair of particles ij. To calculate the function $\theta_{ij}(n\sigma)$, it must be expressed in terms of reagent concentrations θ_i and θ_j. The method of calculating $\theta_{ij}(n\sigma)$ through the concentration of reagents depends on the approximation of the intermolecular interactions used.

The expression for the average velocity of the considered reaction U_{ij} on a uniform surface (lattice) and pertaining to the unit surface is obtained by averaging (51.1) over all the arrangements of the particles. If we denote by $\alpha_{ij}(n\sigma)$ the statistical weight of the configuration $n\sigma$, then we obtain

$$U_{ij} = \Sigma_n\Sigma_\sigma\alpha_{ij}(n\sigma)U_{ij}(n\sigma) = K_{ij}\Sigma_n\Sigma_\sigma\alpha_{ij}(n\sigma)\theta_{ij}(n\sigma)\, F_{ij}(n\sigma), \tag{51.2}$$

For a planar lattice $z = 4$, the complete set of configurations is shown in Fig. 51.1. Here it is assumed that the rate constant K_{ij} depends on the influence of neighbours only through the value of the activation energy determined by the function $F_{ij}(n\sigma)$.

As above, taking into account only pair interactions, we write out the expression for the energy of the ground state for a two-dimensional lattice $z = 4$:

$$E_{ij}(n\sigma) = \varepsilon_{ij} + n_1\varepsilon_{iA} + (z-1-n_1)\,\varepsilon_{iV} + n_2\varepsilon_{jA} + (z-1-n_2)\varepsilon_{jV}, \tag{51.3}$$

i.e. $E_{ij}(n\sigma)$ does not depend on the method of location of neighbouring particles. Here n_1 and n_2 are the numbers of neighbouring particles A around the initial reagents i and j, which correspond to the configuration number of neighbouring particles σ, $n_1 + n_2 = n$. The first term ε_{ij} reflects the interaction of the initial reagents in the initial state. The two following terms correspond to the interaction of the particle i with its neighbours (particles A and vacancies). The last two terms correspond to the interaction of the particle j with its neighbours. The value $(z-1)$ appears due to the condition that for each reagent i or j one of the z adjacent sites is occupied by another reagent of the given bimolecular reaction j or i; and the remaining

$(z-1)$ neighbouring sites are occupied by neighbouring particles, $0 \le n_{1(2)}$ or $n_1, n_2 \le z-1$.

For the activated complex, the analogue of (51.3) is written as

$$E_{ij}^*(n\sigma) = n_1\varepsilon_{iA}^* + (z-1-n_1)\varepsilon_{iV}^* + n_2\varepsilon_{jA}^* + (z-1-n_2)\varepsilon_{jV}^*, \quad (51.4)$$

where there is no interaction between the 'reagents' i and j, since for the activated complex the interaction between the parts of the complex enters its internal energy. The other terms have the same structure. This leads to the expression for the function $F_{ij}(n\sigma)$:

$$F_{ij}(n\sigma) = \exp\{\beta[-\varepsilon_{ij} + n_1\delta\varepsilon_{iA} + (z-1-n_1)\delta\varepsilon_{iV} + n_2\delta\varepsilon_{jA} + (z-1-n_2)\delta\varepsilon_{jV}]\},$$

where $\delta\varepsilon_{ij} = \varepsilon_{ij}^* - \varepsilon_{ij}$, and in accordance with the configuration number σ, the connection between $n_1 + n_2 = n$ is fulfilled. Then in this approximation

$$U_{ij} = K_{ij}\exp(-\beta\varepsilon_{ij})\Sigma_n\Sigma_\sigma\alpha_{ij}(n\sigma)\theta_{ij}(n\sigma)\exp\{\beta[n_1\delta\varepsilon_{iA} + (z-1-n_1)\delta\varepsilon_{iV} + n_2\delta\varepsilon_{jA} + (z-1-n_2)\delta\varepsilon_{jV}]\}, \quad (51.5)$$

In the quasi-chemical approximation of the interaction between particles [10, 14, 15], the equilibrium distribution function $\theta_{ij}(n\sigma)$ is expressed as

$$\theta_{ij}(n\sigma) = \theta_{ij}t_{iA}^{n_1}t_{jA}^{n_2}t_{iV}^{z-1-n_1}t_{jV}^{z-1-n_2}, \quad (51.6)$$

where $t_{ij} = \theta_{ij}/\theta_i$ is the conditional probability described after formula (49.6). As above, the weights of configurations with different values of σ for n = const are equiprobable (for example, $\theta_{ij}(31) = \theta_{ij}(34)$ in Fig. 51.1). As a consequence, for the statistical weights of the particle distributions of A at neighbouring sites relative to both central reagents i and j, we can write $\Sigma_\sigma\alpha_{ij}(n\sigma) = C_{z-1}^{n_1}C_{z-1}^{n_2}$, where each of the factors is the number of combinations of $z-1$ sites in n_1 (or n_2) by the particles A.

Taking into account the above, the final formula can be written as

$$U_{ij} = K_{ij}\exp(-\beta\varepsilon_{ij})\theta_{ij}\sum_{n_1=0}^{z-1} C_{z-1}^{n_1}\Phi_{iA}^{n_1}\Phi_{iV}^{z-1-n_1}\times$$
$$\times\sum_{n_2=0}^{z-1} C_{z-1}^{n_{21}}\Phi_{jA}^{n_2}\Phi_{iV}^{z-1-n_2} = K_{ij}\exp(-\beta\varepsilon_{ij})\theta_{ij}\left(S_iS_j\right)^{z-1}. \quad (51.7)$$

The influence of intermolecular interactions is manifested through a change in the probability of encounter of reagents described by the function θ_{ij} ($\theta_{ij} \neq \theta_i\theta_j$), and through the non-ideality functions of the reaction system.

In the case of a two-dimensional lattice $z = 6$, expression (51.3) must be rewritten, since there are two sites simultaneously located at a distance of the first neighbours from both reagents. We denote the number of such neighbouring particles by n_3, and their energy contribution through $E_{ij}(n_3\sigma)$. Taking these contributions into account, formula (51.3) can be rewritten in the form

$$\begin{aligned} E_{ij}(n\sigma) &= \varepsilon_{ij} + n_1\varepsilon_{iA} + \left(z-3-n_1\right)\varepsilon_{iV} + \\ &+ n_2\varepsilon_{jA} + \left(z-3-n_2\right)\varepsilon_{jV} + E_{ij}(n_3\sigma), \\ E_{ij}(n_3\sigma) &= n_3(\varepsilon_{iA}+\varepsilon_{jA}) + \left(2-n_3\right)(\varepsilon_{iV}+\varepsilon_{jV}), \end{aligned} \quad (51.8)$$

where $0 \le n_{1(2)} \le z-3$, $0 \le n_3 \le 2$, and $n_1 + n_2 + n_3 = n$. Similarly, for a three-dimensional lattice $z = 12$.

For an activated complex, the analogue of expression (51.8) is written as

$$E^*_{ij}(n\sigma) = n_1\varepsilon^*_{iA} + (z-3-n_1)\varepsilon^*_{iV} + n_2\varepsilon^*_{jA} + (z-3-n_2)\varepsilon^*_{jV} + E^*_{ij}(n_3\sigma),$$
$$E^*_{ij}(n\sigma) = n_3(\varepsilon^*_{iA} + \varepsilon^*_{jA}) + (2-n_3)(\varepsilon^*_{iV}+\varepsilon^*_{jV}),$$

As a result, we obtain an expression for the rate of the bimolecular reaction

$$U_{ij} = K_{ij}\exp(-\beta\varepsilon_{ij})\Sigma_n\Sigma_\sigma\alpha_{ij}(n\sigma)\theta_{ij}(n\sigma)\exp\{\beta[n_1\delta\varepsilon_{iA}+(z-3-n_1)\times$$
$$\times\delta\varepsilon_{iV}+n_2\delta\varepsilon_{jA}+(z-3-n_2)\delta\varepsilon_{jV}+n_3(\delta\varepsilon_{iA}+\delta\varepsilon_{jA})+(2-n_3)(\delta\varepsilon_{iV}+\delta\varepsilon_{jV})]\}.$$

In this case, the contributions from the sites, which are simultaneously at the distance of the first neighbours from both reagents, are not expressed in terms of the functions S_i. The modified expressions for the factors in the concentration component of the reaction rate have a more complex form – see below in Section 54.

As a result of a similar averaging, without taking into account the mutual influence of the neighbours among themselves in the QCA for the mixture with any number of components, the following expressions are obtained for the velocities of mono and bimolecular elementary processes

$$U_i = K_i \theta_i S_i^z = K_i \theta_i \Lambda_i^*, \;\; S_i = \sum_{k=1}^{s} t_{ik} \exp(\beta \delta \varepsilon_{ik}) \quad (51.9)$$

$$U_{ij} = K_{ij} \exp(-\beta \varepsilon_{ij}) \theta_{ij} (S_i S_j)^{z-1}, \quad (51.10)$$

where $\delta\varepsilon_{ik} = \varepsilon_{ik}^* - \varepsilon_{ik}$ is ε_{ik}^* the interaction parameter between the activated complex (the particle i located at the site f in the transition state) of the elementary process α with the neighbouring particle k at the neighbouring site g; where $t_{ij} = \theta_{ij}/\theta_i$ is the conditional probability of finding the particle j near the particle i, calculated in the quasi-chemical approximation. In the derivation of expressions (51.9) and (51.10), it was assumed that the rate constants K_i and K_{ij} weakly depend on the density.

In the function, S_i summation is carried out over all sorts of neighbouring particles. The functions S_i are the factors of the non-ideality function of the reaction system $\Lambda_i^* = S_i^z$, which reflect the influence of each neighbouring particle on the height of the activation barrier of the reaction in (51.9).

A similar situation occurs in the case of a bimolecular reaction (51.10). The structure of the functions S_i is not related to the type of potential lateral interaction functions and the radius of the potential, it is determined by using the quasi-chemical approximation of the account of interactions. The number of functions S_i involved in the reaction rate in the form of factors is determined by the size of the coordination sphere z. For bimolecular reactions, this number of neighbours around two reagents equal to $2(z-1)$ practically doubles. The difference between the expression (51.10) and the formula (51.9) is that the influence of intermolecular interactions is manifested not only through the non-ideality functions, but also through the change in the probability of the encounter of reagents described by the function θ_{ij}. If the potential of interparticle interaction exceeds the nearest coordination sphere, contributions from the subsequent coordination spheres enter into expressions for U_i and U_{ij} in the form of additional factors of functions S_i belonging to different distances [12–17].

52. Self-consistency of the rates of two-site stages with an equilibrium distribution of molecules

The proof of self-consistency of the rates of the two-site stages with the equilibrium distribution of molecules for the two-site reaction is somewhat more complicated than for single-site reactions. This is due to the need to consider the influence of neighbouring molecules (via the non-ideality function) and to connect the probability of encountering reagents with the law of mass action through concentrations.

Let the products $C + D$ be obtained in the direct reaction $A + B$. Using the example of stage $A + C \leftrightarrow B + D$, we verify the self-consistency of the description of the kinetics and equilibrium of non-ideal reaction systems. Here it is necessary to follow the progress of the elementary process and indicate which of the products is obtained from the initial reagent, i.e. indicate that from A we receive C and B produces D. This is due to the structure of the activated complex and its properties should not depend on the reaction path. As above we denote $\varepsilon^*_{Aj} = \varepsilon^*_{Cj} \equiv \varepsilon^*_{(AC)j}$ and $\varepsilon^*_{Bj} = \varepsilon^*_{Dj} \equiv \varepsilon^*_{(BD)j}$, the sort of neighbours can be any $1 \le j \le s$. For a two-site stage, the rate is expressed as $U_{AB}(\alpha) = K_{AB}(\alpha)\exp(-\beta\varepsilon_{AB})V_{AB}(\alpha)$, where $K_{AB}(\alpha)$ is the pre-exponent of the rate constant and $V_{AB}(\alpha)$ is the concentration component of the two-site reaction rate, ε_{AB} is the energy of the lateral interaction of the reagents. It follows from the equality $U_{AC}(\alpha) = U_{BD}(\alpha)$ that the equilibrium constant of the reaction does not depend on the parameters of the activated complex.

$$K_2 = \frac{K_{AC}(\alpha)}{K_{BD}(\alpha)} = \frac{V_{BD}(\alpha)}{V_{AC}(\alpha)} = \frac{\theta_{BD}\exp(-\beta\varepsilon_{BD})[S_B(\alpha)S_D(\alpha)]^{z-1}}{\theta_{AC}\exp(-\beta\varepsilon_{AC})[S_A(\alpha)S_C(\alpha)]^{z-1}}$$
$$= \frac{\theta_{BD}\exp(-\beta\varepsilon_{BD})}{\theta_{AC}\exp(-\beta\varepsilon_{AC})}[S_B(A)S_D(C)]^{z-1} \qquad (52.1)$$

where the identity (50.6) or (50.7) is used in the last equality, as above for the relations of non-ideality functions in single-site stages. This follows from the properties of the activated complex, $\varepsilon^*_{Aj} = \varepsilon^*_{Cj} \ne \varepsilon^*_{Bj} = \varepsilon^*_{Dj}$, $1 \le j \le s$. Consider in the detailed record a single factor for one bond from (z–1) neighbours around each reagent inside the central dimeric particle – the ratio of the concentration factors gives

$$\frac{\sum_{j=1}^{s} t_{Aj} \exp\left[\beta(\varepsilon^{*}_{(AC)j} - \varepsilon_{Aj})\right] \sum_{j=1}^{s} t_{Bj} \exp\left[\beta(\varepsilon^{*}_{(BD)j} - \varepsilon_{Bj})\right]}{\sum_{j=1}^{s} t_{Cj} \exp\left[\beta(\varepsilon^{*}_{(AC)j} - \varepsilon_{Cj})\right] \sum_{j=1}^{s} t_{Dj} \exp\left[\beta(\varepsilon^{*}_{(BD)j} - \varepsilon_{Dj})\right]} = \tag{52.2}$$
$$= \sum_{j=1}^{s} t_{Aj} \exp\left[\beta(\varepsilon_{Cj} - \varepsilon_{Aj})\right] \sum_{j=1}^{s} t_{Bj} \exp\left[\beta(\varepsilon_{Dj} - \varepsilon_{Bj})\right],$$

which excludes the presence of the properties of the activated complex in this respect, as well as for single-site reactions.

To transform the pair functions in (52.1), we multiply equation (50.3) on the right and left by $\theta_A\theta_B$, we write it in the form

$$\theta_{AC} = \theta_{BC}\theta_A \exp\left[\beta(\varepsilon_{AC} - \varepsilon_{BC})\right] / (\theta_B S_B(A)) = \theta_{BC}\varphi(A,B,C) \tag{52.3}$$

Taking into account that $\theta_{nm} = \theta_{mn}$, we represent the required relation with the form (we consider the general case of the function θ_{nm} and θ_{kd})

$$\theta_{kd} = \theta_{mn}\varphi(k,B,d) / \left[\varphi(n,d,B)\varphi(m,B,n)\right], \tag{52.4}$$

and since $\varepsilon_{nm} = \varepsilon_{mn}$, once again using the identity (50.6) to exclude the sort B, we obtain

$$\theta_{kd} = \theta_{mn} \frac{\theta_k\theta_d}{\theta_m\theta_n} \exp\left[\beta(\varepsilon_{kd} - \varepsilon_{mn})\right] \sum_{j=1}^{s} t_{dj} \times \tag{52.5}$$
$$\times \exp\left[\beta(\varepsilon_{nj} - \varepsilon_{dj})\right] \sum_{j=1}^{s} t_{kj} \exp\left[\beta(\varepsilon_{mj} - \varepsilon_{kj})\right].$$

As a result, the ratio of the rate constants of the forward and reverse reactions is expressed in terms of (52.2) and (52.5)

$$K = \frac{\theta_{AB}\exp(-\beta\varepsilon_{AB})}{\theta_{CD}\exp(-\beta\varepsilon_{CD})} \times$$
$$\times \left[\frac{\sum_{j=1}^{s} t_{Aj} \exp\left[\beta(\varepsilon^{*}_{(AC)j} - \varepsilon_{Aj})\right] \sum_{j=1}^{s} t_{Bj} \exp\left[\beta(\varepsilon^{*}_{(BD)j} - \varepsilon_{Bj})\right]}{\sum_{j=1}^{s} t_{Cj} \exp\left[\beta(\varepsilon^{*}_{(AC)j} - \varepsilon_{Cj})\right] \sum_{j=1}^{s} t_{Dj} \exp\left[\beta(\varepsilon^{*}_{(BD)j} - \varepsilon_{Dj})\right]}\right]^{z-1} = \tag{52.6}$$
$$= \frac{\theta_A\theta_B}{\theta_C\theta_D} \left[\sum_{j=1}^{s} t_{Aj} \exp\left[\beta(\varepsilon_{Cj} - \varepsilon_{Aj})\right] \sum_{j=1}^{s} t_{Bj} \exp\left[\beta(\varepsilon_{Dj} - \varepsilon_{Bj})\right]\right]^{z}$$

Formally, taking into account the redefinition of the meaning of the equilibrium constant K, one can consider the flow of a two-site reaction as two independent single-site processes at different sites: $A \leftrightarrow C$ and $B \leftrightarrow D$. The expression (52.6) has the form of a product of two independent processes, each of which is described by its equilibrium (50.1). Therefore, the equilibrium condition of the reaction coincides with the condition for equilibrium of the medium as a whole.

Single-particle approximations. This type of approximation closes the system of equations for equilibrium and kinetic processes through unary DFs (i.e., only through the concentration of components). In this case, all pair functions $\theta_{ij} = \theta_i\theta_j$ are expressed as they do not need to have additional equations (equilibrium and kinetic). The same is true for higher-dimensional DFs. However, by construction, the rate of bimolecular reaction rates is expressed as before: $U_{AB}(\alpha) = K_{AB}(\alpha)\exp(-\beta\varepsilon_{AB})V_{AB}(\alpha)$, which must be substituted $\theta_{ij} = \theta_i\theta_j$. This leads to the following relations (see (51.1))

$$K_2 = \frac{K_{AC}(\alpha)}{K_{BD}(\alpha)} = \frac{V_{BD}(\alpha)}{V_{AC}(\alpha)} = \frac{\theta_B\theta_D\exp(-\beta\varepsilon_{BD})}{\theta_A\theta_c\exp(-\beta\varepsilon_{AC})}[S_B(A)S_D(C)]^{z-1} \qquad (52.7)$$

where the non-ideality functions $S_B(A)S_D(C)$ do not contain the interaction parameters of the activated complex, but the right-hand side of expression (52.7): 1) does not break up into two independent processes with concentration factors $\frac{\theta_B}{\theta_A}S_B(A)^z$ and $\frac{\theta_D}{\theta_c}S_D(C)^z$, as follows from (50.1) and as would be the case for single-site independent stages, and 2) contains energy factors that change the activation energies of the elementary stages E_{AC} and E_{BD}, which are included in the pre-exponentials of the rate constants K_{AC} and K_{BD}.

Thus, from the analysis of similar expressions for chaotic and mean-field approximation, it follows that they do not provide a self-consistent description of equilibrium and kinetics, since as a result of equating the reaction rates, analogous equations for isotherms in the same approximation are not obtained. Therefore, these approximations can not be used to describe kinetic processes. Even in the special case of migrations by the vacancy mechanism the exponential factor with the parameters $\varepsilon_{iV} = 0$ turn to zero, but this does not compensate for the absence of the exponent $(z-1)$ instead of z. (In addition, if the properties of the system are taken into account more accurately, even with the initial values of the parameters

$\varepsilon_{iV} = 0$, due to the internal motions of the molecules, these parameters become non-zero.)

53. Correlation effects in stage velocities

As a result of averaging over local environments around the reagents, the following expressions are obtained for the velocities of mono- and bimolecular elementary processes

$$U_i = K_i\theta_i S_i^z,\ U_{ij} = K_{ij}\exp(-\beta\varepsilon_{ij})\theta_{ij}(S_i S_j)^{z-1} \qquad (53.1)$$

where $\delta\varepsilon_{ik} = \varepsilon_{ik}^* - \varepsilon_{ik}$, and ε_{ik}^* is the interaction parameter between the activated complex (the particle i located at the site f in the transition state) of the elementary process α with the neighbouring particle k at the neighbouring site g, where $t_{ij} = \theta_{ij}/\theta_i$ is the conditional probability of finding the particle j near the particle i, calculated in the quasi-chemical approximation. In the derivation of expressions (53.1), it was assumed that the rate constants K_i and K_{ij} weakly depend on the density. In the function S_i summation is carried out over all types of neighbouring particles. The functions S_i are the factors of the non-ideality function of the reaction system, which reflect the influence of each neighbouring particle on the height of the activation barrier of the reaction: $S_i = \sum_{k=1}^{s} t_{ik}\exp(\beta\delta\varepsilon_{ik})$ – QCA, $S_i = \sum_{k=1}^{s} \theta_k\exp(\beta\delta\varepsilon_{ik})$ – chaotic approximation, $S_i = \exp\left(\beta\sum_{k=1}^{s}\delta\varepsilon_{ik}\theta_k\right)$ – mean-field approximation. In the absence of intermolecular interactions, the multipliers $S_i = 1$ and the formulas go into the reaction rate for the ideal reaction system (48.2).

Equations (53.1) show that the mutual influence of reagents in the case of non-ideal systems complicates their cooperative behaviour (in contrast to traditional chemical kinetics), and this behaviour changes the 'sensitivity' of the change in velocity from concentration and temperature. Both factors simultaneously influence each other, so the concentration dependences become as determinative as the temperature ones. A similar situation occurs in the case of a bimolecular reaction.

Important in the construction of expressions for functions of non-identity is the inclusion of correlation effects between interacting molecules. It has been shown above that taking into account the correlations between the nearest neighbours ensures the self-consistency of the description of reaction rates and the equilibrium distribution of the components of the reaction mixture. If there are no correlation effects, then there is no such self-consistency, and model parameters, determined from equilibrium and kinetic measurements, do not coincide with each other.

The presence of correlation effects also leads to significant differences in the concentration dependences of the velocities at a fixed temperature. This is illustrated by the concentration dependence of the desorption rate in Fig. 53.1 *a*. The calculation was carried out for fixed parameters of monomolecular desorption. The differences are related only to the accuracy of describing the correlation effects in different approximations. In Fig. 53.1 *b* these same approximations are compared with the experimental data on desorption by a potassium atom from the tungsten surface (the shaded regions reflect the experimental error in the estimation of the parameters).

An even greater difference in the curves due to correlation effects is observed in the course of non-isothermal processes. Figure 53.2 lists the so-called thermodesorption curves, which are obtained by linear heating of the surface according to the law $T = T_0 + bt$, where b is the heating rate (deg/sec), T_0 is the initial temperature for which, at the initial time, θ_0. Those. a differential equation is solved $d\theta/dt = -U_A$, where U_A is the desorption rate, for a given initial filling $\theta(t = 0) = \theta_0$. Heating the surface accelerates the desorption process, and as the substance is depleted, the desorption rate decreases. The calculation is carried out at the same value of all the parameters of the model. The vertical axis represents the amount of matter

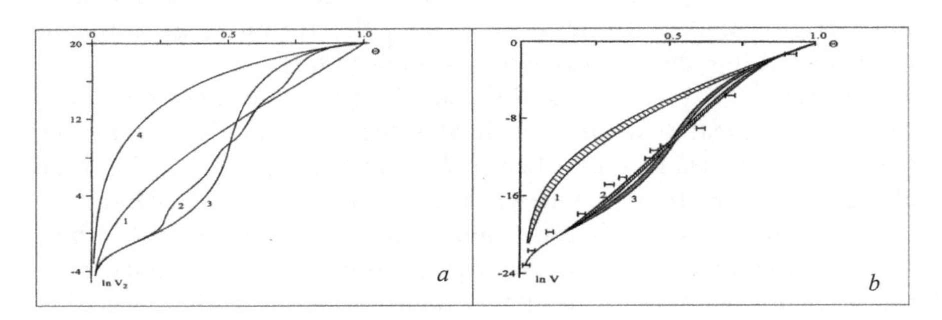

Fig. 53.1. (*a*) Influence of the approximation of accounting for intermolecular interactions on the form of the desorption rate [7]: correlation effects are taken into account in the polynomial (*2*) and quasi-chemical approximations (*3*), and are not taken into account in the molecular field (*1*) and chaotic (*4*) approximation. (*b*) Description of the desorption rate *K* from the surface of W [4,5], calculated in the chaotic (*1*), polynomial (*2*), and quasi-chemical (*3*) approximations. The symbols indicate the experimental desorption rates taking into account the experimental error.

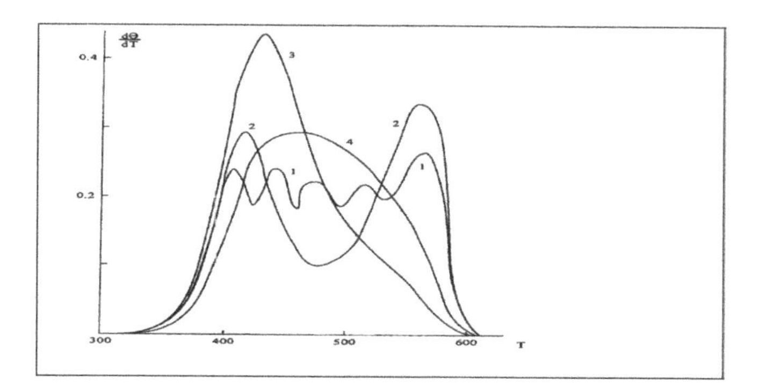

Fig. 53.2. The effect of taking into account the effects of correlation on the form of TDS, calculated in polynomial (*1*), quasi-chemical (*2*), chaotic (*3*) and mean field (*4*) approximations on a uniform surface [19].

that is desorbed at each time point. The chaotic and mean-field approximations do not take into account the correlations, while the QCA and the polynomial approximation preserve correlation effects. In the absence of this accounting, both curves have a single maximum, while taking into account correlations leads to splitting of the curve into several peaks. Thus, the differences are related only to the methods of calculating the non-ideality function. Correlation effects between repulsive particles lead to the splitting of thermal desorption curves on a uniform surface [14–17,19]. It is obtained that the splitting is due to the correlation effects (curves *1* and *2*). Neglecting the correlations for curves *3* and *4* for the same values of the interaction parameters does not split the thermal desorption curves.

For single-particle approximation, it is seen that when self-consistency is absent, then the course of the concentration dependences of the stage rates varies sufficiently with respect to correlations in which self-consistency is conserved. A similar situation in elementary processes for the model of an non-uniform structure – after averaging the velocity in the forward and backward directions can be mismatched. This depends on the ratio of the distribution functions of different types of centres and on the rate of redistribution of reagents between them between them. If surface diffusion is affected, self-consistency will not be ensured. To describe the general multi-stage processes, the relaxation times for each dynamic variable of the reaction system are important.

54. Accounting for the second and next neighbours (uniform systems)

Single-site stages. Let the single-site elementary process $A \to C$ [12-17,20] be realized, where A and C are the components of the lattice solution containing s sorts of particles. The kinetic equations for such a reaction were first constructed in [20]. They use the same approximations that were considered above when closing the system of equations for equilibrium correlators [18].

In QCA, the rates of single-site reactions are expressed as $U_A(\alpha) = K_A(\alpha)V_A(\alpha)$, all values are defined above; only the concentration component of the reaction rate $V_{AB}(\alpha)$ changes, which in this case is written in the form (here and up to the end of Chapter 6 $R = R_{lat} > 1$)

$$V_i(\alpha) = \theta_i \prod\nolimits_{r=1}^{R} S_i\left(r \mid \alpha\right)^{z_r}, S_i(r \mid \alpha) = \sum_{j=1}^{s} t_{Bj}(r) \exp\left[\beta(\varepsilon^*_{Bj}(r) - \varepsilon_{Bj}(r))\right], \quad (54.1)$$

Hence the ratio of the factors in the non-ideality functions gives

$$S_B\left(r \mid \alpha\right) / S_A(r \mid \alpha) = \sum_{j=1}^{s} t_{Bj}(r) \exp\left[\beta(\varepsilon^*_{Bj}(r) - \varepsilon_{Bj}(r))\right] / $$
$$/ \sum_{j=1}^{s} t_{Aj}(r) \exp\left[\beta(\varepsilon^*_{Aj}(r) - \varepsilon_{Aj}(r))\right] = S_B(A \mid r), \quad (54.2)$$

that, as above, excludes the properties of the activated complex and leads to isotherms in the QCA approximation when taking into account the interparticle interaction within R c.s.:

$$K_1 = \frac{K_A(\alpha)}{K_B(\alpha)} = \frac{V_B(\alpha)}{V_A(\alpha)} = \theta_B \prod\nolimits_{r=1}^{R} S_B(A \mid r)^{z_r} / \theta_A,$$
$$S_B(A \mid r) = \sum_{j=1}^{s} t_{Bj} \exp\left[\beta(\varepsilon_{Aj} - \varepsilon_{Bj})\right]. \quad (54.3)$$

Thus, we obtain an analogue of the expression (50.2) in the form (for more details, see [14, 20])

$$\theta_A = \theta_B \exp[\beta(\nu_B - \nu_A)] \prod\nolimits_{r=1}^{R} S_B(A \mid r)^{z_r}. \quad (54.4)$$

Two-site reaction A + B → C + D. The activated reaction complex occupies two adjacent lattice sites (denoted by n and k) and interacts with neighbouring particles in the r-th coordination sphere, $1 \leq r \leq R$, both with respect to the site n with particle A, so and with respect to site k with particle B.

We introduce the concept of a single r-th coordination sphere of a dimer molecule AB as a set of sites located at the same distance r from the closest of sites A or B. The coordination spheres of isolated n and k sites overlap, so the single r-th coordination sphere contains a part of the sites related only to r coordinate sphere of one of the sites n or k, and a part of the sites related to the $(r + \lambda)$ coordination sphere of the neighbouring site k or n. The quantity λ depends on the size of the lattice and the number of nearest neighbours.

We divide the set of sites of the uniform r-th coordination sphere into equivalent groups of sites, which are given by their orientation. The orientation of the sites is determined by the angle formed by the straight line connecting the central sites and the straight line between the site in the r-th coordination sphere of the dimeric molecule and the midpoint between the central sites.

We denote the number of different orientations in the unique r-th coordination sphere by π_r, and the number of sites with a given orientation ω_r in terms of $\kappa_{\omega r}$ $(1 \leq \omega_r \leq \kappa_{\omega r})$. Figure 54.1 shows the distribution of the sites of the plane lattices for $R = 2$.

The interaction potential of the activated complex with the particle j located at the site of the r-th unified coordination sphere of a dimer molecule AB with orientation ω_r, will be denoted by $\varepsilon^*_{ABj}(\omega_r)$. Values $\varepsilon^*_{ABj}(\omega_r)$ are expressed in terms of paired $\varepsilon^*_{ij}(r)$ potentials as follows: $\varepsilon^*_{ABj}(\omega_r) = \varepsilon^*_{Aj}(r_{nm}) + \varepsilon^*_{Bj}(r_{km})$, where m is the number of the site containing the particle j. For example, for a flat lattice $z = 4$ and $R = 2$, we have

$$\varepsilon^*_{ABj}(\omega_1 = 1) = \varepsilon^*_{Aj}(1), \varepsilon^*_{ABj}(\omega_1 = 2) = \varepsilon^*_{Aj}(1) + \varepsilon^*_{Bj}(2), \quad \varepsilon^*_{ABj}(\omega_1 = 3) = \varepsilon^*_{Aj}(2) + \varepsilon^*_{Bj}(1)$$

$$\varepsilon^*_{ABj}(\omega_1 = 4) = \varepsilon^*_{Bj}(1), \ \varepsilon^*_{ABj}(\omega_2 = 1) = \varepsilon^*_{Aj}(2), \ \ \varepsilon^*_{ABj}(\omega_2 = 2) = \varepsilon^*_{Bj}(2)$$

Similarly, the values $\varepsilon^*_{ABj}(\omega_r)$ are constructed for other z and R. If $A = B$, then the symmetric positions of the sites of the single r-th coordination sphere with respect to the plane passing through the middle of the straight line connecting the central sites and perpendicular to this line correspond to the same values $\varepsilon^*_{AAj}(\omega_r)$, $1 \leq \omega_r \leq \pi_r/2$, which corresponds to twice the values of $\kappa_{\omega r}$.

The expressions for the velocities of the two-site stages and the proof of their self-consistency when taking into account the

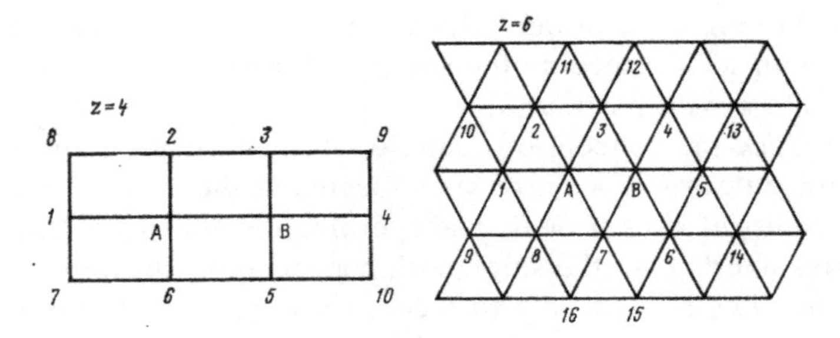

Fig. 54.1 Distributions of sites of planar lattices for $z = 4$ and 6 by equivalent groups of the first two coordination spheres. For $z = 4$, the first c.s. includes sites 1–6 ($\pi_1 = 4$), the second c.s. 7–10 ($\pi_r = 4$): $\omega_1 = 1$ corresponds to site 1, $\omega_1 = 2$ to sites 2 and 6, $\omega_1 = 3$ – sites 3 and 5, $\omega_4 = 4$ – site 4; $\omega_2 = 1$ correspond to sites 7 and 8, $\omega_2 = 2$ – sites 9 and 10. For $z = 6$ the first c.s. ($\pi_1 = 5$), the second – the sites 9 – 16 ($\pi_r = 4$): $\omega_l = 1$ corresponds to site 1, $\omega_l = 2$ corresponds to sites 2 and 8, $\omega_1 = 3$ – sites 3 and 7, $\omega_4 = 4$ to sites 4 and 6, $\omega_l = 5$ corresponds to site 5; $\omega_2 = 1$ correspond to sites 9 and 10, $\omega_2 = 2$ – sites 11 and 16, $\omega_2 = 3$ – sites 12 and 15, $\omega_2 = 4$ – sites 13 and 14.

interaction of neighbours at distances greater than nearest neighbours are given below immediately for an non-uniform lattice.

55. Non-ideal non-uniform systems

In a 'distributed' model for a non-uniform lattice, each site is considered a separate type of the region of localization of the molecule – this is the most common type of non-uniform lattice [14,21–23]. Let us consider the question of the self-consistency of the expressions for the rates of elementary reactions (stages) occurring at one and two sites and for the equilibrium state of the reaction system in the QCA, taking into account any radius of the interaction potential between the neighbours R on the distributed model of the non-uniform lattice system. To this end, we will operate with specific individual sites, i.e. before the averaging procedure with different distribution functions for the composition and structure of the non-uniform surface. The averaged models are obtained from the expressions for the distributed model by weighting using the functions discussed in [14,21–23].

Single-site reaction. First, consider a single-site reaction between the adsorbed particle A and the gas particle B: $ZA + B \leftrightarrow ZC + D$, flowing at a site of type q with the number f. The reaction proceeds in the forward and reverse directions. The equation for the velocity of a single-site stage has the form

$$U_{AB} = \frac{1}{M}\sum_{f=1}^{M} U_f^{AB},\ U_f^{AB} = \hat{K}_f^{AB}\theta_f^A S_f^A,$$
$$S_f^i = \prod_{r=1}^{R}\prod_{h\in z_f(r)} S_{fh}^i(r),\ i = A;$$
$$S_{fh}^i(r) = \sum_{j=1}^{s}\theta_{fh}^{ij}(r)\exp\left[\beta\delta\varepsilon_{fh}^{ij}(r)\right]/\theta_f^i = 1 + \sum_{j=1}^{s-1}\theta_{fh}^{ij}(r)x_{fh}^{ij}(r)/\theta_f^i \quad (55.1)$$
$$x_{fh}^{ij}(r) = \exp\left[\beta\delta\varepsilon_{fh}^{ij}(r)\right] - 1,\ \delta\varepsilon_{fh}^{ij}(r) = \varepsilon_{fh}^{ij*}(r) - \varepsilon_{fh}^{ij}(r)$$

Here M is the number of sites of the distributed system, $\hat{K}_f^{AB}$ is the rate constant of the elementary reaction between the adsorbed particle A and the gas particle B, including the product with the partial pressure of component B in the gas phase of the P_B; the upper symbol '^' means that a pressure factor is applied to the velocity expression for a particle from the gas phase (which is not reflected in the symbol of site f); $\varepsilon_{fh}^{ij*}(r)$ is the interaction parameter of the neighbouring particle j, in the ground state at a distance r, with the activated complex of the reacting molecule i; $\varepsilon_{fh}^{ij}(r)$ is the interaction parameter of two neighbouring particles at a distance r in the ground state.

In the equilibrium condition, the equations for the reaction rates in the forward and backward directions $U_f^{AB} = U_f^{CD}$ give the following relation

$$\hat{K}_f^1 = \hat{K}_f^{AB}/\hat{K}_f^{CD} = \theta_f^C\Psi_f^{CA}/\theta_f^A,\ \Psi_f^{CA} = \prod_{r=1}^{R}\prod_{h\in z_f(r)} S_{fh}^C(r)/S_{fh}^A(r) \quad (55.2)$$

In the absence of interaction, the function $\Psi_f^{CA} = 1$ and $\hat{K}_f^1$ represents the effective equilibrium constant for the site f of the ideal adsorption system; $\hat{K}_f^1 = K_f^1 P_{\hat{A}}/P_D$, where K_f^1 is the true equilibrium constant of the single-site stage $A \leftrightarrow C$.

For an non-uniform lattice in a QCA, instead of (50.4) we have the following relations for the pair distribution functions

$$\theta_f^i\theta_{fh}^{jk}(r) = \theta_f^j\theta_{fh}^{ik}(r)\exp[\beta\{\varepsilon_{fh}^{jk}(r) - \varepsilon_{fh}^{ik}(r)\}]/S_{fh}^i(j|r)$$
$$S_{fh}^i(j|r) = \sum_{k=1}^{s}\theta_{fh}^{ik}(r)\exp[\beta(\varepsilon_{fh}^{jk}(r) - \varepsilon_{fh}^{ik}(r))] \quad (55.3)$$

Where $S_{fh}^i(j|r) = \sum_{k=1}^{s}\theta_{fh}^{ik}(r)\exp[\beta(\varepsilon_{fh}^{jk}(r) - \varepsilon_{fh}^{ik}(r))]$

$$\text{Then } \theta_{fh}^{kl}(r) = \theta_{fh}^{mn}(r)\exp[\beta\{\varepsilon_{fh}^{kl}(r) - \varepsilon_{fh}^{mn}(r)\}]\times \\ \times S_{fg}^{k}(m|r)S_{fg}^{l}(n|r)\theta_f^k\theta_h^l/(\theta_f^m\theta_h^n) \quad (55.4)$$

Using the fact that the properties of the activated complex do not depend on the direction of the reaction ($\varepsilon_{fh}^{Aj*}(r) = \varepsilon_{fh}^{Cj*}(r)$ for any index j), it follows from (55.1) and the equations for the QCA (55.3) of the constraints that

$$S_{fh}^{C}(r)/S_{fh}^{A}(r) = \left[S_{fh}^{A}(C|r)\right]^{-1}, \quad (55.5)$$

and therefore, $\Psi_f^{CA}(r) = [S_{fh}^{A}(C|r)]^{-1}$ that is, the right side of expression (55.2) does not depend on the interaction of AC with its neighbours, and the effective equilibrium constant is expressed only through the parameters of the interaction of particles in the ground state (and not in the transient state) and the equilibrium particle concentrations. This result is completely consistent with the concept of the equilibrium particle distribution, and the resulting expression (55.2) is an equilibrium constant on an non-uniform system. In other words, the equations for the equilibrium distribution are obtained independently of the method of their construction: from the kinetic analysis or directly for the equilibrium state of the molecules of the mixture.

Two-site reaction. A similar conclusion can be obtained from the relations (55.3)–(55.5) if we equate the rates of the forward and backward directions of the two-site reaction stage ($ZA + ZB + C \leftrightarrow ZE + ZD + F$) occurring at two neighbouring sites f and g [1–3]. The reaction rate data are recorded as

$$U_{fg}^{ABC}(1) = \hat{K}_{fg}^{ABC}(fg)\theta_{fg}^{AB}(1)\exp[-\beta\varepsilon_{fg}^{AB}(1)]\prod_{r=1}^{R}\prod_{\omega_r=1}^{\pi(r|qp)}\prod_{h\in z(\omega_r|qp)} S_{fgh}^{AB}(\omega_r) \quad (55.6)$$

$$S_{fgh}^{AB}(\omega_r) = \sum_{j=1}^{s} t_{fgh}^{ABj}(\omega_r)\exp[\beta\delta\varepsilon_{fgh}^{ABj}(\omega_r)],\ \ \delta\varepsilon_{fgh}^{ABj}(\omega_r) = \varepsilon_{fgh}^{ABj*}(\omega_r) - \varepsilon_{fgh}^{ABj}(\omega_r)$$

$$t_{fgh}^{ABj}(\omega_r) = \theta_{fh}^{Aj}(r_1)\theta_{gh}^{Bj}(r_2)/\theta_f^A\theta_g^B\theta_h^j,\ \ \varepsilon_{fgh}^{ABj}(\omega_r) = \varepsilon_{fh}^{Aj}(r_1) + \varepsilon_{gh}^{Bj}(r_2)$$

And for large distances $\theta_{fh}^{ij}(r > R) = \theta_f^i\theta_h^j$, which consequently gives $S_{fgh}^{AB}(\omega_r)|_{r_1>R} = S_{gh}^{B}(r_2)$.

To prove the self-consistency condition, it is necessary to prove the following relation

$$S_{fgh}^{ED}(\omega_r)/S_{fgh}^{AB}(\omega_r)=S_{fh}^{E}(r_1)S_{gh}^{D}(r_2)/[S_{fh}^{A}(r_1)S_{gh}^{B}(r_2)], \qquad (55.7)$$

where the values of ω_r are uniquely related to r_1 and r_2. Then the equilibrium distribution of the molecules will not depend on the method of achieving it through the kinetics of processes in the forward and backward directions or in direct consideration of only the equilibrium configurations. In order to prove equality (55.7), we must remember that $\varepsilon_{fh}^{Aj*}(r_1)=\varepsilon_{fh}^{Ej*}(r_1)$ and $\varepsilon_{gh}^{Bj*}(r_2)=\varepsilon_{gh}^{Dj*}(r_2)$ due to the independence of the properties of the activated complex from the direction of the process. We introduce the notation $\tilde{\theta}_{fh}^{ij}(r)=\theta_{fh}^{ij}(r)\exp[-\beta\varepsilon_{fh}^{ij}(r)]$ and rewrite the left-hand side of (55.7) as

$$\theta_f^A\theta_g^B\sum_{j=1}^{s}F_j\tilde{\theta}_{fh}^{Ej}(r_1)\tilde{\theta}_{gh}^{Dj}(r_2)/\{\theta_f^A\theta_g^B\sum_{j=1}^{s}F_j\tilde{\theta}_{fh}^{Aj}(r_1)\tilde{\theta}_{gh}^{Bj}(r_2)\}$$

Where $F_j=\exp[\beta\{\varepsilon_{fh}^{Aj*}(r_1)-\varepsilon_{gh}^{Bj*}(r_2)\}]/\theta_h^j$, and it must be shown that the ratio $N/L=\tilde{\theta}_{fh}^{Ej}(r_1)\tilde{\theta}_{gh}^{Dj}(r_2)/\left[\tilde{\theta}_{fh}^{Aj}(r_1)\tilde{\theta}_{gh}^{Bj}(r_2)\right]$ does not depend on the index j.

To do this, we express the functions $\tilde{\theta}_{fh}^{Ej}(r_1)$ and $\tilde{\theta}_{gh}^{Dj}(r_2)$ through the functions $\tilde{\theta}_{fh}^{Aj}(r_1)$ and $\tilde{\theta}_{gh}^{Bj}(r_2)$, respectively, using the general QCA for the pair functions (for example, $\tilde{\theta}_{fh}^{Ej}(r_1)=\tilde{\theta}_{fh}^{Aj}(r_1)\tilde{\theta}_{fh}^{EA}(r_1)/\tilde{\theta}_{fh}^{AA}(r_1)$ and so on). This allows you to write $N/L=\Lambda_{fh}^{EA}(r_1)\Lambda_{gh}^{DB}(r_2)$, where $\Lambda_{fh}^{ik}(r)=\{\tilde{\theta}_{fh}^{ii}(r)\tilde{\theta}_{fh}^{ik}(r)/[\tilde{\theta}_{fh}^{ki}(r)\tilde{\theta}_{fh}^{kk}(r)]\}^{1/2}$.

Then, considering the obvious equality: $\sum_{j=1}^{s}F_h^jN/\sum_{j=1}^{s}F_h^jL=N/L$ we can see that the left side of the relation (55.7) is equal to $\theta_f^A\theta_g^B\Lambda_{fh}^{EA}(r_1)\Lambda_{gh}^{DB}(r_2)/\{\theta_f^E\theta_g^D\}$.

On the other hand, the formulas $S_{fh}^{E}(r_1)/S_{fh}^{A}(r_1)=\Lambda_{fh}^{EA}(r_1)\theta_f^A/\theta_f^E$ and $S_{gh}^{D}(r_2)/S_{gh}^{B}(r_2)=\Lambda_{gh}^{DB}(r_2)\theta_g^B/\theta_g^D$ can be proved in the same way, therefore, summing, we have the proof of equality (55.7).

The use of approximations that do not take into account the effects of spatial correlation, such as in the mean-field approximation or the random approximation, does not meet the self-consistency conditions, as explained in Section 54, in particular, this leads to renormalization of the activation energy of the process by a value $\varepsilon_{ij}^{fg}(1)$.

56. The velocity of thermal motion of molecules

The need for simultaneous description of the transport of molecules in multiphase systems, including within adsorbents, in rarefied gas and dense liquid phases, makes it inconvenient to use the concept of the mean free path of molecules varying from $10^4\lambda$ for a gas to λ for a liquid. In the lattice model, as in the kinetic theory of liquids, the basic concept is the probability of hopping (or displacement) of the molecule $W(\chi)$ by the considered distance χ, rather than the mean free path. This concept is also the main one in Chapter 5 when discussing the question of the averaging of various properties in obtaining a single description of transport processes in three aggregate states.

The key to calculating the hopping probability of a molecule $W(\chi)$ is the concept of a vacancy region through which a molecule moves from one cell to another. As an illustration, Fig. 56.1 shows the scheme for hopping a molecule from site f to the nearest free site g, as well as all neighbouring sites in the first two coordination spheres of the lattice structure with $z = 6$, which simultaneously influence the probability of such displacement. All neighbours located at sites 1 through 25 and in site ξ simultaneously interact with a moving molecule from site f to neighbour site g (hopping length $\chi = 1$). In the general case $\chi > 1$, a vacancy trajectory is necessary to move inside the condensed phase, through which the molecule moves with the thermal velocity $W(\chi)$. Such a trajectory is created in a fluctuation manner. The longer the trajectory χ, the more neighbours participate around it in the formation of the environment which affects the probability of a jump. The task of the theory is to calculate the probability of its formation, so that the molecule can move, and take into account the influence of neighbours on this displacement. For this, it is required to calculate in a self-consistent manner the probabilities of the many-particle configurations forming the vacancy trajectory.

The formulation of the transition state model for non-ideal reaction systems, when the barrier to be overcome is created by the potentials of neighbouring particles and the surface of a solid, is given in [24]. As follows from the ratio of the dimensions, the average thermal velocity of the molecules is expressed in terms of the hopping velocity of the molecule $U_{fg}(\chi)$ between the sites f and g by the distance χ in the form

$$W_{fg} = \chi U_{fg}(\chi) / \theta_f, \qquad (56.1)$$

In the state of thermal equilibrium, the probability of molecular hopping between sites is described by formulas

$$U_{fg}^{AV}(\chi) = K_{fg}^{AV}(\chi) V_{fg}^{AV}(\chi), \tag{56.2}$$

where $K_{fg}^{AV}(\chi)$ is the rate constant of the molecule hopping from the site f to the free site g at a distance χ and $V_{fg}(\chi)$ is the concentration dependence of the hopping rate of the molecule;

$$K_{fg}^{AV}(\chi) = K_{fg}^{*AV}(\chi)\exp\left[-\beta E_{fg}^{AV}(\chi)\right], \tag{56.3}$$

$K_{fg}^{*AV}(\chi)$ is the pre-exponent of the rate constant $E_{fg}^{AV}(\chi)$ is the activation energy of the hop determined by the potential of the wall surfaces (the influence of interparticle interactions on the value of the activation jump is determined by the non-ideality function $\Lambda_{fg}^{AV}(\chi)$);

$$V_{fg}^{AV}(\chi) = \theta_{fg}^{AV}(\chi)\Lambda_{fg}^{AV}(\chi), \theta_{fg}^{AV}(\chi) = \theta_{fg_1}^{AV}(1)\prod_{k=1}^{\chi-1} t_{g_k g_{k+1}}^{VV}(1). \tag{56.4}$$

The concentration dependence of the migration rate of the molecule $V_{fg}(\chi)$ is expressed in terms of the product of two factors: 1) $\theta_{fg}^{AV}(\chi)$ is the probability of the vacancy trajectory from χ free sites from the cell f of length χ through the sequence of cells $g(1)$, $g(2)$ and so further to the cell $g \equiv g(\chi)$, for $\chi = 1$ the cell $g(1)$ is finite; 2) $\Lambda_{fg}^{AV}(\chi)$ is a non-ideality function that takes into account the influence of the interactions of molecules around the given trajectory on the probability of a jump inside it:

$$\Lambda_{\xi fg}^{AV}(\chi) = \prod_{r=1}^{R}\prod_{\omega_r=1}^{\pi_r}\prod_{h\in m(\omega_r)}\sum_{k=1}^{V} t_{fgh}^{AVk}(\omega_r \mid \chi) E_{fgh}^{AVk}(\omega_r \mid \chi),$$
$$t_{fgh}^{AVk}(\omega_r \mid \chi) = \frac{\theta_{fh}^{Ak}(r_1)\theta_{gh}^{Vk}(r_2)}{\theta_f^A \theta_g^V \theta_h^k}. \tag{56.5}$$

The $t_{fgh}^{AVk}(\omega_r|\chi)$ multiplier describes the probability of finding a neighbouring particle j at a site h at a distance r_1 from reagent A and r_2 from vacancy V. The particles A and V themselves are at a distance χ. The environment of the particles around the activated complex is conveniently numbered with the help of the number of sites $m(\omega_r/\chi)$ in the orientations $\omega_r(1 \le \omega_r \le \pi_r)$, π_r is the number of orientations in the r-th unified coordination sphere around the dimer AV on sites of central f and g at a distance χ (a single coordination sphere of radius r, $1 \le r \le R$, is defined as the set of sites located at a distance r from either site f or site g); R is the radius of the

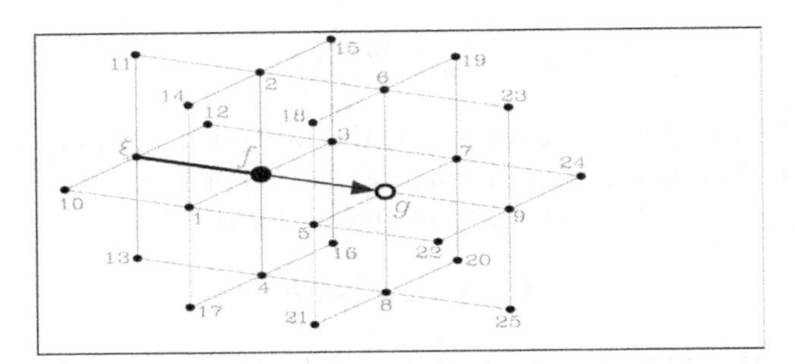

Fig. 56.1. The scheme for hopping a molecule from site f to a free site g and neighbouring sites in the first two coordination spheres of a lattice structure with $z = 6$.

interaction potential. Orientations are measured from the centre of the dimer AV: the point of intersection of the line connecting the central sites fg, and the line connecting the site h to the centre of the dimer AV.

In the process of hopping (in the middle of the fg coupling, see Fig. 56.1), the moving molecule A experiences the influence of neighbouring particles – the energy of this interaction is denoted by the parameters $\varepsilon^{*Aj}_{fh}(r)$ which differ from the analogous energy parameters for particles in the ground state $\varepsilon^{AVi}_{fgh}(\omega_r)$ (i.e., at the lattice sites). The multiplier $E^{AVi}_{fgh}(\omega_r)$ reflects the difference in the interaction of the activated complex with its neighbours through the difference of pair interactions $(\varepsilon^{*ij}_{fg}(r) - \varepsilon^{ij}_{fg}(r))$.

$$
\begin{aligned}
&E^{AVi}_{fgh}(\omega_r \mid \chi) = \exp\beta\left[\delta\varepsilon^{Ai}_{fh}(r_1) + \delta\varepsilon^{Vi}_{gh}(r_2)\right],\\
&\delta\varepsilon^{ij}_{fg}(r) = \varepsilon^{*ij}_{fg}(r) - \delta\varepsilon^{ij}_{fg}(r).
\end{aligned}
\tag{56.6}
$$

Traditionally, the pre-exponent of the rate constant $K^{*AV}_{fg}(\chi)$ is expressed in terms of the ratios of the statistical sums of the reactants and the activated complex as [1–3] $K^{*AV}_{fg}(\chi) = \dfrac{kT}{h}\dfrac{F^{*AV}_{fg}}{F^{A}_{f}F^{V}_{g}}$ where F^{*AV}_{fg} and F^{A}_{f} are the statistical sums of the molecule in the transition and ground states for the vacancy $F_V = 1$, h is the Planck constant. In the absence of lateral interactions, formula (56.2) can be written in the form $U_{fg}(\chi) = K_{fg}\theta_f(1-\theta_g)^{\chi}$.

Far from the surfaces of the pore walls, the ratio F_A/F^{*}_{AV} is the translational degree of freedom in the direction of particle motion and is equal to $(2\pi m\beta^{-1})^{1/2}\chi/h$ then $K_\chi = [(2\pi m\beta)^{1/2}\,\chi]^{-1}$ or $K_\chi = w/4\chi$ where $w = (8/\pi m\beta)^{1/2}$ [25].

Knowing the hopping rates $U_{fg}(\chi)$, we can calculate self-diffusion coefficients and other transport coefficients [24]. In sections 48–54, the elementary process occurred only between reagents at the nearest sites. In this example, it is shown that this approach can be extended to any lengths of distances between reagents (in this case it is related to the length of trajectories) and taking into account the interaction of any radius, the interaction potential between the components of the system. Similarly, it can be used for any type of collision between particles and types of chemical transformations.

References

1. Eyring, H.J., Chem. Phys. 1935. V. 3. P. 107.
2. Temkin M.I., Zh. fiz. khimii. 1938. V. 11. P. 169.
3. Glasston S., Laidler K.J., Eyring H., Theory of absolute reaction rates. Moscow, IL, 1948 [Princeton Univ. Press, New York, London, 1941]
4. Tovbin Yu.K., Thesis. Moscow, Karpov Institute of Physical Chemistry, 1974.
5. Tovbin Yu.K., Fedyanin V.K., Fiz. Tverd. Tela. 1975. V. 17. P. 1511.
6. Tovbin Yu.K., Fedyanin V.K., Kinetika i kataliz. 1978. V. 19. No. 4. P. 989.
7. Tovbin Yu.K., Fedyanin V.K., Kinetika i kataliz. 1978. T. 19. No. 5. With. 1202.
8. Tovbin Yu.K., Fedyanin V.K., Fiz. Tverd. Tela. 1980. V. 22. No. 5. P. 1599.
9. Tovbin Yu.K., Fedyanin V.K., Zh. fiz. khimii. 1980. T. 54. No. 12. P. 3127, 3132.
10. Tovbin Yu.K., Zh. fiz. khimii. 1981. V. 55. No. 2. P. 284.
11. Tovbin Yu.K., Dokl. AN SSSR. 1982. V. 267. No. 6. P. 1415.
12. Tovbin Yu.K., Dokl. AN SSSR. 1984. P. 277. No. 4. P. 917.
13. Tovbin Yu.K., Poverkhnost'. Fizika. Khimiya. Mekanika. 1989. No. 5. P. 5.
14. Tovbin Yu. K., Theory of physical and chemical processes at the gas-solid interface. – Moscow: Nauka, 1990. – 288 p. [CRC, Boca Raton, Florida, 1991].
15. Tovbin Yu.K., Progress in Surface Sci. 1990. V. 34, No. 1-4, P. 1-236.
16. Tovbin Yu.K., Theories of adsorption–desorption kinetics on flat non-uniform surfaces., Dynamics of Gas Adsorption on Non-uniform Solid Surfaces / Eds. by W.Rudzinski, W.A. Steele, G. Zgrablich. Elsevier, Amsterdam, 1996. P. 240–325.
17. Tovbin Yu. K., Thin Films and Nanostructures. Vol. 34. Physico-Chemical Phenomena in Thin Films and at Solid Surface. Eds. L.I. Trakhtenberg, S. H. H. Lin, and O. J. Ilegbusi, Elsevier, Amsterdam, 2007. P. 347.
18. Tovbin Yu.K., Zh. fiz. khimii. 1981. V. 55. No. 2. P. 273.
19. Tovbin Yu.K. Kinetika i kataliz. 1979. V. 20. No. 5. P. 1226.
20. Tovbin Yu.K., Kinetika i kataliz. 1982. V. 23. No. 4. P. 813, 821; No. 5. P. 1231.
21. Tovbin Yu.K., Kinetika i kataliz. 1983. V. 24. No. 2. P. 308, 317.
22. Tovbin Yu.K., Zh. fiz. khimii. 1990. V. 64. No. 4. P. 865. [Russ. J. Phys. Chem. 1990. V. 64. No. 4. P. 461.
23. Tovbin Yu.K., Zh. fiz. khimii. 1992. V. 66. No. 5. P. 1395. [Russ. J. Phys. Chem., 1992 V. 66, No. 5, P. 741.
24. Tovbin Yu.K., The Molecular Theory of Adsorption in Porous Solids, Moscow, Fizmatlit, 2012. [CRC Press, Taylor & Francis Group, 2017.
25. Hirschfelder J. O., Curtiss Ch. F., Bird R. B., Molecular theory of gases and liquids. – Moscow: Inostr. Lit., 1961. – 929 p. [Wiley, New York, 1954].

7

Analysis of thermodynamic interpretations

The material of this chapter combines one common property: all the interpretations discussed are related to the incorrect implicit use of relaxation times. This refers to the priority of establishing a mechanical equilibrium over the chemical equilibrium for the Kelvin equation and the so-called metastable drops, to the equilibrium interpretations of metastable states when changing the thermodynamic parameters of the system and to introducing the concept of the activity coefficient for activated complexes in the theory of absolute reaction rates in describing the rates of elementary stages.

57. The Yang–Lee theory and the Kelvin equation

The Kelvin equation (KE) $P/P_{\infty} = \exp\{2\sigma\, v_{liq}\beta/R\}$ [1] was discussed in detail in Section 7: two different ways of establishing particular equilibria give two different solutions for complete equilibrium. Chapter 3 shows the existence of equilibrium drops, which are absent in thermodynamics. Analysis of the relaxation times (Section 43) not only gives two states of complete equilibrium, but also explains the reason for this difference, which is related to different properties of the boundary. In this section, these different states of two-phase systems are discussed for their correspondence to the Yang–Lee theory [2–5] together with the states of the metastable drop from the Kelvin equation.

Figure 57.1 shows the isotherm curve (T = const) with a stratification loop in the two-phase region for the volume phase (curve *1*) and the secant curve (curve *2*), constructed according to Maxwell's rule, at the saturated vapour pressure $P_s(T)$ for a given temperature T [6,7]. It relates the coexisting densities of the fluid θ_f and the vapour θ_v. In addition to Fig. 18.1 the densities of the transition zone layers are plotted on the secant *2* in the chemical potential–density coordinates: dots indicate the values of the concentration profile at the vapour–liquid interface $\{\theta_q\}$ in different monolayers, as in Fig. 23.1 *a* (these densities refer to the intermediate region between θ_f and θ_v). Figure 57.1 also shows curve *3* that relates the states of the liquid inside the drop and within the part of its boundary that have the same chemical potential values. This 'node' 3 is shifted with respect to the secant 2 by the magnitude of the chemical potential associated with the jump in the Laplace pressure (P_x) with respect to the pressure P_s for a drop of some radius R. Because of the unique relationship between the pressure P_x and the chemical potential $\mu(P_x)$ the pressure jump automatically leads to a jump in chemical potential!

The point (x) of the intersection of the node 3 at the pressure P_x inside the drop with the isotherm curve refers to the figurative points of the equilibrium states of the system with density θ_x. Continuation of node 3 (absent in this figure) towards the vapour spinodal branch of the isotherm before the intersection with curve 1 is indicated by the symbol (o). The point (o) does not belong to the figurative points of the phase diagram. The density θ_o corresponding to the pressure P_x corresponds to the unstable state of the metastable system and refers to the *macroscopic planar* interface, i.e. quantity θ_0 is not related to any drops.

The equilibrium drop corresponds to the density of the coexisting liquid and vapour phases θ_f and θ_v at a pressure P_s for a given temperature T; here $\mu_{liq} = \mu_{vap}$ and $P_{liq} = P_{vap}$, which satisfies the Yang–Lee theory [2–5]. Equilibrium drops are, in fact, the limiting

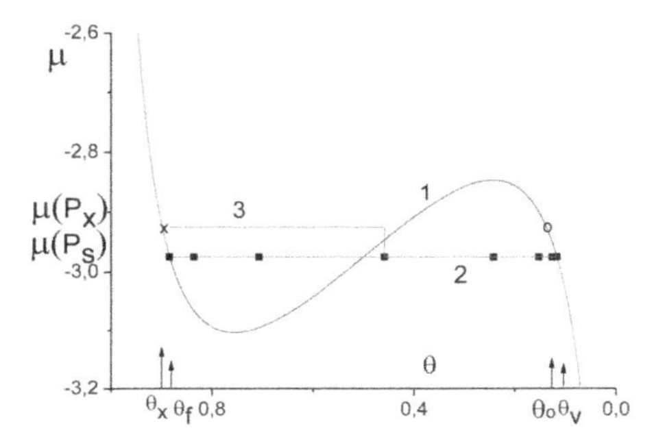

Fig. 57.1. An isotherm with a van der Waals loop (1), a Maxwell secant (2) and a 'node' (3), corresponding to the radius of the drop R.

case of usual stratifying of macroscopical coexisting phases, when the fraction of the new phase is extremely small, and it is necessary to take into account the effect of the interface on the state of the complete system. The metastable drop in the theory corresponds to the densities of the liquid and vapour phases θ_x and θ_v at pressures P_x for the drop and P_s for the vapour. Here $\mu_{liq} \neq \mu_{vap}$ and $P_{liq} \neq P_{vap}$, because the chemical potential and pressure jump *simultaneously* on the dividing surface. These states also satisfy the Yang–Lee theory, since both states of vapour and liquid do not enter the binodal.

For a metastable drop in the Kelvin equation it is assumed that $P_{liq} \neq P_{vap}$, but $\mu_{liq} = \mu_{vap}$. The Yang–Lee theory excludes such states: there are equations of state, both in a drop and in a vapour, connecting their internal densities and pressures. Change of the pressure in any phase, without changing the chemical potential, it is impossible. The density of the point θ_o is on the continuation of single-phase isotherms and falls inside the curve of the binodal. This means, according to the Yang–Lee theory, that the equilibrium state of the system does not correspond to it.

Thus, the molecular theory, even admitting the existence of metastable drops, shows their principal difference from metastable drops in the Kelvin equation. Metastable drops in theory reflect the introduction of a perturbation associated with the curvature of the surface into the very conditions of stratification of coexisting phases due to taking into account the surface tension. This is done through the use of the Laplace equation under the condition that the surface tension is calculated from the concentration profile of the molecules in the transition region.

A metastable drop in the Kelvin equation is constructed purely thermodynamically: first, a condition of mechanical balance between the vapour and the drop is found, and then the equality of their chemical potentials is *declared or allowed*. We note that mechanics, like thermodynamics, does not determine the equations of state themselves in different phases, but the condition of the mechanical equilibrium of the interface allows one to relate the stress at the interface between the phases without the connection with the state of chemical equilibrium of the neighbouring phases. In Gibbs' approach, the mechanical equilibrium is linked by analogy with a film or membrane (see page 229 in [8]), which is another material, and with which the molecules of vapour and drops can not be in chemical equilibrium. The subsequent use of the chemical equilibrium between the vapour and drop molecules is realized in the presence of a film that is constant in time and can not be adjusted by the chemical equilibrium condition (in particular, the exchange of molecules between the vapour and the drop can be excluded). This is confirmed by the analysis in section 45.

In any experiments with flexible membranes, mechanical equilibrium is determined by their properties, and not only by the properties of the phases on both sides of it. For mechanical equilibrium, the surface tension value is introduced without the chemical equilibrium condition on both sides of the surface. Therefore, the surface tension for mechanical equilibrium in the presence of a film (as a foreign body) between phases on both sides of the surface and surface tension for coexisting phases in direct contact are different concepts, and they can not be confused! Thus, both ways of the establishment of complete equilibrium in Section 7 operate with different concepts of surface tension.

It is in the *analogy*, and not in *identity*, of the state of the surface under mechanical equilibrium and the subsequent introduction of the postulate of chemical equilibrium with the determination of the magnitude of the surface tension in the coexisting phases that is the fundamental logical inaccuracy in obtaining the Kelvin equation. This refers to the implicit use of the concept of non-equilibrium or dynamic surface tension. The latter was introduced much later for a particular case of the creation of a freshly formed (without relaxation) surface of a liquid [9–11]. The non-equilibrium surface tension is specified in the derivation of the Kelvin equation by an incorrect relationship between the relaxation times between mechanical and chemical equilibrium (Sections 12 and 45). This conclusion fully applies to the metastable drop in the theory into which the Laplace equation is introduced reflecting the influence of mechanical equilibrium on the chemical equilibrium (Appendix 1 and Section 59). The incorrectness of the Kelvin equation is due to the confusion of the concepts of surface tension in the non-equilibrium and equilibrium states. Moreover, the non-equilibrium state of the system itself is *artificially* introduced in the derivation of the Kelvin equation through the establishment of mechanical equilibrium with an undefined degree of chemical equilibrium, whereas the known relationships between the relaxation times of pressure and the chemical potential require adjustment of the pressure to the evolution of the chemical potential.

As a result, only at the cost of distorting the meaning of the molecular property of the system (which is σ) in thermodynamics can the size of the drop R be introduced. The answer to the question of Section 7 on the essence of the linear size of the drop radius is that it is the result of confusing concepts from thermodynamics and mechanics: parameter R is not an intensive parameter of thermodynamics – it is taken from mechanics. This example shows that for a 'model-free' classical thermodynamics even a curved boundary is a complex object, and the initial statements of thermodynamics are more general concepts than a particular case of curved surfaces. Therefore, any deviation from the three stages of thermodynamics should be controlled by the methods of statistical physics. For more than a century, different correlation ratios have been built on the basis of the erroneous Kelvin equation. Only recently have we started to get rid of the use of the Kelvin equation in the problems of adsorption porosimetry [12] and condensation processes [13], where it possessed the maximum prevalence.

58. Small systems by J.W. Gibbs

In the works of J.W. Gibbs there are two types of additions to the postulates of thermodynamics for bulk phases, which are used in obtaining widely known results on surface phenomena.

The first type of additions is the use by J.W. Gibbs of the same sequence of establishment of equilibria (first mechanical equilibrium, then chemical equilibrium) when considering the thermodynamics of a curved boundary, as in the Kelvin equation. In particular, this sequence is emphasized in the analysis of the stability of curved boundaries (pp. 220–230 in [8]) and in the treatment of other problems in the thermodynamics of small bodies. (The original texts of Gibbs are not repeated: they refer to the relevant pages.) This led to the fact that the thermodynamic criterion of stability $dP/dV < 0$, referring to macroscopic systems, significantly overstates the range of real parameters corresponding to complete equilibrium for small bodies under the condition of the equality of all three quantities in different phases independent of each other: P, T and μ (corresponding, respectively, to mechanical, thermal and chemical equilibria). During the following time this served as a justification for introducing the concepts of metastable states, and the possibility of their interpretation using thermodynamic equations.

The second type of assumption is specially introduced for small bodies, for which Gibbs rejects the same own definitions introduced for phases and phase equilibria (pp. 252–253 in [8]). In particular, he believes that for small bodies it is possible to preserve previously introduced concepts and mathematical relationships, although it emphasizes – in its small bodies there are no inner uniform regions and the properties of a small body are variable inside it. This assumption immediately excludes the concept of 'phase' and 'surface tension' introduced by Gibbs for heterogeneous systems (i.e., the phase approximation does not occur – Section 6). Nevertheless, to this object all the thermodynamic equations are applied, including the concept of the work of formation of a small body W, and the Laplace equation for the connection of surface tension and the pressure difference inside the small body and outside it. As a result, the famous formula for the work of the formation of a new phase $W = \sigma A/3$, associated with the formation of the phase interface and the change in the internal state of the new phase, is obtained. Gibbs used the same rejection of his own definitions for two-phase states and surface tension when considering two neighbouring small drops from immiscible phases (p. 261 and 295 in [8]).

If we use the concept of two-phase systems with internally uniform properties (according to Gibbs), then the work of the formation of a new phase in equilibrium rather than artificially introduced metastable (more precisely, non-equilibrium) conditions will be $W = \sigma A$. This difference causes three times the large numerical divergence when using this expression in the equations of condensation kinetics or the like. since it is included in the exponent of the rate of the nucleation process [14].

When going over to analogous constructions for solid microcrystals, one should once again pay attention to the convention of applying thermodynamic functions to a small system because of the absence of internal uniform regions associated with the concept of phase. From the standpoint of the mechanics of a continuous medium, it is well known that under such conditions a small body is in a state of internal stresses, variable in magnitude along the radius of the body [15]. These include Lamé's work [16] on the distribution of stresses in a medium with internal and external pressures. Recall that the work of Lamé became classic, and after his work the basic mechanical moduli of compression and shear in the theory of elasticity became established. These same questions are also dealt with in modern mechanics courses (see, for example, [17]).

Gibbs, as the founder of the tensor calculus (see the section on page 263 in [8]), was aware of these results. Therefore, when he constructed the thermodynamics of small crystals, he could not use the same arguments as for liquid drops. For a solid, Gibbs introduced two concepts of surface tension: one for a deformed state when a new surface is created in the course of mechanical stresses γ, and a second concept for the equilibrium process of creating a new surface by crystallization and dissolution σ. Therefore, it was no longer possible to use the notion of deformations equally, and for small solid bodies Gibbs gave another justification for the formula $W = \sigma A/3$. Instead of resorting to the presence of a deformed small body, he uses opposing arguments. Gibbs introduces two non-thermodynamic concepts (pp. 255–257–258 in [8]): the assumption of a linear relationship between the size and surface tension of the crystal, and the condition for preserving the similarity of the crystal to itself when its size changes.

Recall that the fundamentals of the thermodynamics of the Gibbs' macrosystems do not allow the introduction of any additional ideas about the structure or body states that are not related to the energy characteristics of the First and Second Laws.

All subsequent work on the thermodynamics of solids was conducted in terms of calculating each of these two contributions. However, the statistical theory leads to the conclusion about their simultaneous participation in all processes that change the state of a solid body, since the surface always changes the state of deformation of the near-surface region with respect to the volume state (i.e., these concepts can not in principle be divided, although for macroscopic bodies this effect is negligible) [17].

At the same time, it should be noted that Gibbs allowed for small microcrystals the possibility of mutual influence of the state of different faces on each other. In subsequent works this aspect of Gibbs' work was omitted and was no longer discussed. In particular, in the formulation of Wulff's rule, such macroscopic systems are considered in which such mutual influence of different faces is absent.

After Gibbs' work, the order of first mechanical and then chemical equilibrium (according to Kelvin), when considering the curved boundaries of phases and small bodies, became a thermodynamic axiom. It can not be ruled out that precisely because of the need to introduce these additional assumptions for small bodies, and also because of the question of the inconsistency of the mechanical and chemical equilibrium conditions for the curved interfaces (there was no concept of 'chemical potential' before Gibbs), the need for more a detailed description of the thermodynamic characteristics of systems, which ultimately led Gibbs in 1902 to the development of a statistical approach [8].

59. Molecular theory of metastable spherical drops

The molecular theory of equilibrium drops is presented in Chapter 3. To pass to metastable drops, it is necessary to add to it the Laplace equation, which relates the internal pressure in the drop to the pressure of the surrounding phase (Appendix 1):

$$\pi_1 - \pi_\kappa = 2\sigma / (R + \rho_r - 1), \tag{59.1}$$

where R is the radius of the drop without the contribution of the monolayers of the transition region, ρ_r is the number of the reference layer or the surface of tension on which the pressure jump in the metastable system between the vapour π_κ and the liquid π_1 in the drop is realized, $\pi_1 \neq \pi_\kappa$. On the same layer ρ_r, a jump in the chemical potential (fixed through the external pressure of the thermostat aP) is realized: up to $q = \rho_r$ we have $\pi_q = \pi_1$, and from $q = \rho_r+1$ to $q = \kappa$ we have $\pi_q = \pi_\kappa$. For equilibrium bulk coexisting phases, $\pi_1 = \pi_\kappa$. As a result, we obtain a closed system of equations (24.8) and (59.1), in which a dividing surface with a pressure rupture on the so-called surface of tension is introduced. The equation of the local isotherm for the monolayer q relates the probability of filling the cell θ_q and the chemical potential of the molecule μ_q in the layer q, $2 \leq q \leq \kappa-1$. Equation (59.1) assumes knowledge of the surface tension σ, which is obtained only after determining the concentration profile and fixing the position of the dividing surface.

The expressions for surface tension σ through local pressures are indicated in Appendix 1 (see also [11, 19]). For the metastable drops existing at metastable pressure P_{met} during the construction of the concentration profile, it is necessary to determine the position of the reference phase interface ρ_r on which the jump between the internal vapour pressure π_κ and the liquid π_1, described by the Laplace equation (19), is realized. These provisions are defined as: a) the equimolecular reference surface ρ_e [11,19,20] – the fixation of the coordinate of the dividing surface through the material balance at the equimolecular interface

$$\sigma_e = \frac{1}{F_{\rho_e}} \left[\sum_{q \leq \rho_e} F_q(\pi_1 - \pi_q^T) + \sum_{q > \rho_e} F_q(\pi_\kappa - \pi_q^T) \right] \tag{59.2}$$

The quantity $\rho_r = \rho_e$ is found from condition

$$\sum_{q=1}^{\rho_e} F_q(\theta_q - \theta_1) + \sum_{q=\rho_e+1}^{\kappa} F_q(\theta_q - \theta_\kappa) = 0 \tag{59.3}$$

The quantities F_q are defined in (24.4) and, for the sake of simplicity of notation, they omit the symbol R.

b) the reference surface ρ_s through the equilibration of forces and moments of forces on the dividing surface [11, 19, 21]

$$\sigma_s = \frac{1}{R_{\rho_e}}\left[\sum_{q\le\rho_e} R_q(\pi_1 - \pi_q^T) + \sum_{q>\rho_e} R_q(\pi_\kappa - \pi_q^T)\right]. \tag{59.4}$$

The quantity $\rho_r = \rho_s$ is found from the condition of equilibration of the moments of forces

$$\sum_{q\le\rho_s} R_q(R_q - R_{\rho_s})(\pi_1 - \pi_q^T) + \sum_{q>\rho_s} R_q(R_q - R_{\rho_s})(\pi_\kappa - \pi_q^T) = 0; \tag{59.5}$$

c) the reference surface ρ_G by Gibbs (this definition of surface tension is purely thermodynamic [8])

$$\sigma_G = \frac{1}{F_{\rho_G}}\left[\sum_{q\le\rho_G} F_q(\pi_1 - \pi_q^T) + \sum_{q>\rho_G} F_q(\pi_\kappa - \pi_q^T)\right]; \tag{59.6}$$

The quantity $\rho_r = \rho_G$ is found from the condition $\min(\sigma_G)$ of the surface tension when the position ρ_G is varied inside the transition region and when the concentration profile $\{\theta_q^i\}$ is recalculated for each value of the reference surface, $1 < \rho_G < \kappa$. The states of the coexisting phases (but not the transition region) are considered fixed. The quantities F_q are defined (24.4).

d) the reference surface ρ_K by Kondo [11,19,22,23] is also determined purely by the thermodynamic method, but its calculation occurs with a fixed concentration profile

$$\sigma_K = \frac{1}{F_{\rho_K}}\left[\sum_{q\le\rho_K} F_q(\pi_1 - \pi_q^T) + \sum_{q>\rho_K} F_q(\pi_\kappa - \pi_q^T)\right]. \tag{59.7}$$

The quantity $\rho_r = \rho_K$ is found from the condition $\min(\sigma_K)$ of the surface tension when the position ρ_K, $1 < \rho_K < \kappa$, for a fixed concentration profile $\{\theta_q^i\}$ is varied. Here the displacement of the position ρ_K reflects a purely mathematical procedure.

Properties of metastable drops. As a first example, let us consider the influence of the position of the surface of tension according to Gibbs (59.6). The pressure jump occurs on the ρ_r-th spherical monolayer within the transition region, $1 \le \rho_G \le \kappa$. The influence of the position of the dividing surface on the profiles of the transition region is demonstrated by the curves in Fig. 59.1 [24]. Shown are the changes in the profiles at the extreme positions of the reference surface: in the second ($\rho_r = 2$) and the penultimate ($\rho_r = \kappa-1$) monolayers for the three values of the drop radii $R = 40$ (1), 100 (2), 250 (5) for $\tau = 0.7$, $\alpha = 1$. The change in the number of the layer q at which the pressure jump occurs necessarily changes the nature of the distribution of molecules within the transition region. The degree of this change in the values of σ can be different, but the state of the transition layer will change with any of the molecular parameters. The curves show the most important difference between molecular models [6.25] from thermodynamic postulates: a change in the value of ρ_r changes the nature of the distribution of molecules within the transition region (or different concentration profiles are obtained). According to the molecular theory, changing any of the molecular parameters changes the properties of the interphase layer.

The curves in Fig. 59.1 are typical examples of concentration profiles, and they show that the width of the transition region κ in the drop is always less than κ for a plane boundary. In this case, the differences in the values of κ in the calculation of metastable and equilibrium drops rarely exceed one monolayer, with the exception of the critical temperature region.

The dependence of the width of the transition region $\kappa(R)$ on the drop radius R over a wide temperature range is shown in Fig. 59.2. A stepwise change in the curves is a consequence of the discrete nature of $\kappa(R)$ as the number of monolayers in the transition region.

Figure 59.3 shows the dependences of the normalized surface tension σ/σ_{bulk} on the radius of the drop R for $\tau = 0.69$ (*a*) and 0.82 (*b*). The variants of the calculations of $z_{q,p}(R)$ are given by the formulas of Section 24 for $\alpha = 0$ and $\alpha = 1$ and for two ways of choosing the number of the monolayer ρ_r that defines the reference surface: $\rho_r = \rho_e$ defines an equimolecular surface (59.2), and $\rho_r = \rho_m$ corresponds to the extremum σ. All curves start from small values of surface tension and $\sigma \to \sigma_{bulk}$ at $R \to \infty$. Naturally, the curves corresponding to $\rho_r = \rho_m$ are located above the curves corresponding to $\rho_r = \rho_e$.

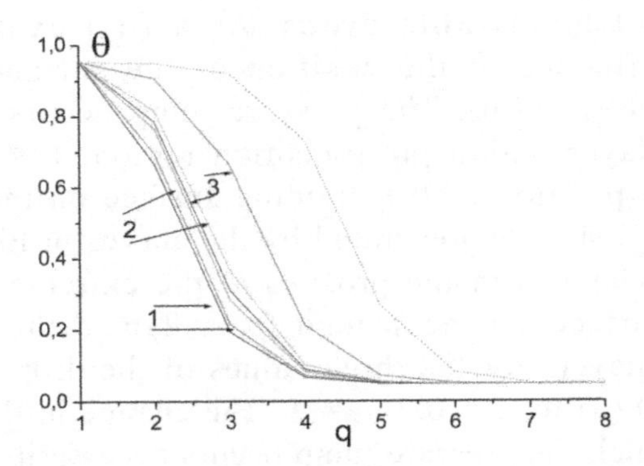

Fig. 59.1. The range of profile changes at the positions of the reference surface in the second ($\rho = 2$) and penultimate ($\rho = \kappa - 1$) monolayers for the three values of the drop radii $R = 40$ (*1*), 100 (*2*), 250 (*3*) is indicated by arrows related to one size drop. Calculation for $\tau = 0.7$, $\alpha = 1$. As above, the points are the profile in the bulk phase.

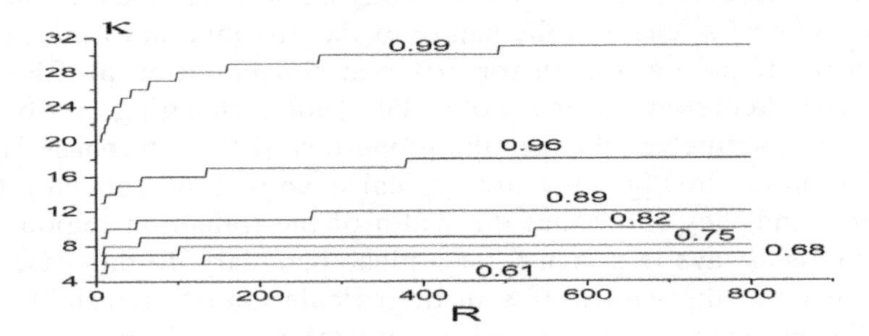

Fig. 59.2. Dependence of the width of the transition region of the liquid–vapour boundary κ on the radius of the spherical drop R for different temperatures $\tau = T/T_c$. The values of τ are shown on the curves.

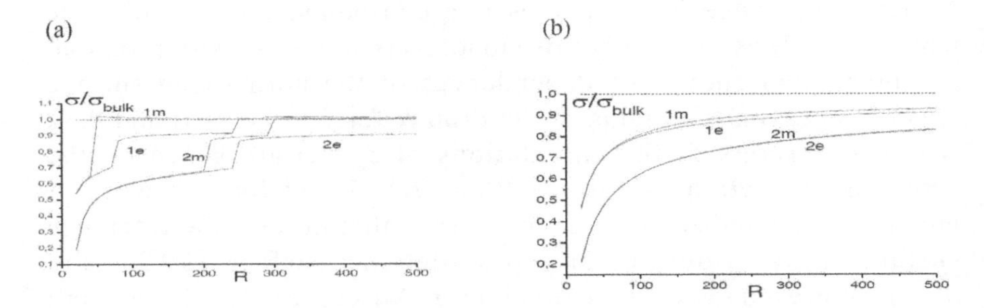

Fig. 59.3. Dependence of the normalized values of the surface tension $\sigma/\sigma_{\text{bulk}}$ on the radius of the drop R. Results are given for two methods of calculating the structural characteristics: 1 – by formulas with $\alpha = 0$, 2 – by formulas with $\alpha = 1$, and for two methods of choosing ρ_r: e – $\rho_r = \rho_e$ and m – $\rho_r = \rho_m$. The results are given for two values $\tau = T/T_c$: $\tau = 0.68$ (a), 0.82 (b).

The curves in Fig. 59.3 can be broken down into three types. The first type includes increasing curves without jumps (Fig. 59.3 *b*). These curves are realized at relatively elevated temperatures ($\tau > 0.75$). The second type includes curves that contain a jump in the surface tension, as a result of which they reach the value $\sigma > \sigma_{\text{bulk}}$ and then $\sigma \rightarrow \sigma_{\text{bulk}}$ at $R \rightarrow \infty$, remaining $\sigma > \sigma_{\text{bulk}}$. Curves of this type are observed only at low temperatures, near the temperature of the triple point $\tau = 0.55 \div 0.62$ (curves 1m, 2m and 1e in Fig. 59.3 *a*). The third, intermediate type includes curves that, like curves of the second jump type of σ, but, like curves of the first type, do not reach the value $\sigma = \sigma_{\text{bulk}}$. They are realized in the intermediate temperature range.

Kondo drops. The thermodynamic definition of σ by Gibbs [8] was refined in Kondo's paper [22]. To calculate the surface tension: instead of real displacement of the position of the surface of tension in the transition region, it was assumed that the density profile in the transition region was fixed, and only the position of the surface of tension was displaced. Molecular theory allowed us to analyze this refinement and compare it with the equimolecular definition [26].

The essence of the analysis consists in investigating the fulfillment of the thermodynamic conditions determining $\sigma(q)$ at a given temperature: 1) fixing the supersaturation of the vapour $\Delta\mu$; 2) the fulfillment of the Laplace equation $\Delta_L = 0$, where $\Delta_L = \pi_1 - \pi_\kappa - 2\sigma/R$, for $\sigma > 0$; and 3) determination of the position of the dividing surface q_r, $2 \leq q_r \leq \kappa-1$. The layer number q_r is fixed to the minimum of $\sigma(q)$ in the inner part of the transition region between the vapour and the liquid. The value of q corresponding to the minimum of $\sigma(q)$ determines the position of the surface of tension q_m, and the value of $\sigma(q_m)$ itself is the required surface tension. This minimum $\sigma(q_m)$ and the drop size $\rho = R + q_m-1$ are used in the Laplace equation.

For comparison, an equimolecular surface is also used. In this case, the profile θ_q, in contrast to $\sigma(q)$, does not depend on q_m, therefore the position of q_e is uniquely determined from the supersaturation of the vapour $\Delta\mu$ and R (T = const).

For Kondo's definition (59.7), it was found that for each fixed value of R, there are two values corresponding to the maximum P_{max} and the minimum $P_{\min}$ supersaturation that are realized in metastable systems. Figure 59.4 shows the regions of admissible values in the coordinates (R, P_{met}/P_s) corresponding to the equilibrium state between metastable vapour and liquid, for medium and high temperatures τ = 0.86 (*1*), 0.79 (*2*), 0.73 (*3*), 0.66 (*4*). The lower the temperature, the

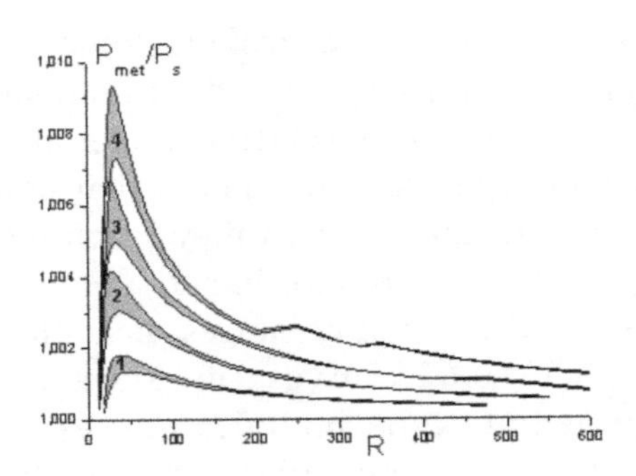

Fig. 59.4. The regions of existence of solutions with respect to R and P_{met}/P_s, which correspond to the metastable states of the vapour–liquid system, for $\tau = 0.86$ (1), 0.79 (2), 0.73 (3), 0.66 (4).

greater the difference between P_{met}/P_s and higher is the corresponding two-valued region. The presence of this area indicates the multiplicity of solutions connecting the parameters of the state of the system. The value $P_{\max}$ corresponds to the corresponding value of R_m and the maximum value of the internal pressure in the drop $\pi_1(R_m)$ at a given temperature.

The essence of the polysemy is explained as follows. Since density profile θ_q is uniquely determined for each R and P_{met} in the range of admissible values, then changing the position of the reference surface $\rho = R + q_r - 1$, we determine for this profile the minimum value of $\sigma(q_r)$ and the corresponding value of q_m. In addition, we determine the position of the reference surface $\rho = R + q_L - 1$, on which the Laplace equation is satisfied. The values of q_L as a function of $(P_{\text{met}} - P_{\text{min}})/P_s$ at $\tau = 0.79$ for $R = 50$ (1), 100 (2), 200 (3) are shown in Fig. 59.5 *a*. On the same field, the symbols show the values of q_m. It can be seen from the figure that for the same pressure the values of q_L differ from q_m by not more than one. In Fig. 59.5 *b* the values of $\sigma(q)$ and $1 < q < \kappa$ are given for integer values of q_L, that is, for $q_L = q_m$. The symbols show the values of $\sigma(q)$ for $q_L = q_m = q_e$, i.e. on an equimolecular surface.

Figure 59.5 shows the fundamental difference between the two methods of selecting a dividing surface. For each radius there is a set of values of P_{met} and for each of them a minimum surface tension from the layer number of the transition region is realized (Fig. 59.5 *b*). The highest value of vapour supersaturation at $R = 50$ on the field (59.5 *b*) corresponds to $P_{\text{met}} = P_{\max}$ – the upper curve corresponds to it. At a minimum, the value of σ is shifted from the right edge to the left when moving from P_{min} to P_{max}. All the curves refer to the fixed value of the liquid part of the drop R. This figure demonstrates that at R = const the quantities P_{met} (or $\Delta\mu$) and q_m are interrelated, thanks to the Laplace equation, and the entire set of solutions satisfies the three thermodynamic conditions listed above. At the same time, the condition for choosing an equimolecular dividing surface leads to a single value of q_e.

Figure 59.6 *a* shows at a temperature of $\tau = 0.79$ the intersection of the line $P_{met}/P_s = 1.003$ with the region of existence of solutions defining the region of admissible values of R at a given pressure. This range of values of R varies from 20 to 65. In Fig. 59.6 *b* the q_m values corresponding to the surface of tension for the specified range of R are given. First, the q_m values increase until they reach the value corresponding to R_m, then remain constant for some time, and then decrease to $q_m = 2$.

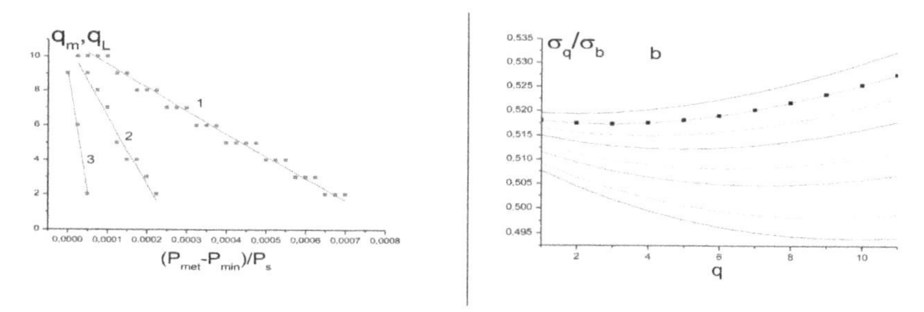

Fig. 59.5. (a) Dependences of q_m and q_L on the pressure P_{met} in a metastable drop at $\tau = 0.79$ for $R = 50$ (1), 100 (2), 200 (3); lines refer to q_L, symbols to integer values q_m. (b) Dependences of the surface tension value σ on the layer number q at pressure values corresponding to integer values of q_L, i.e. at $q_L = q_m$ for $R = 50$. The symbols show the σ-profile at a pressure for which $q_L = q_m = q_e$.

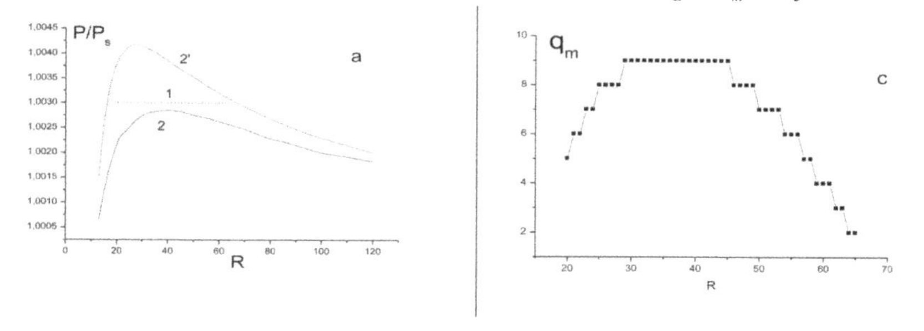

Fig. 59.6. (a) The intersection of the line $P_{met}/P_s = 1.003$ with the region of existence of solutions at a temperature $\tau = 0.79$, which determines the region of permissible values of R at a given pressure. (b) the values of ρ_m as a function of R in the range of acceptable values of R.

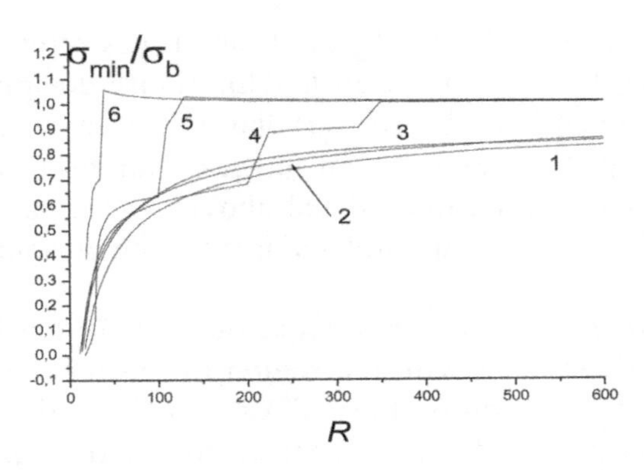

Fig. 59.7. The dependence of surface tension $\sigma_{\min}$ on the radius of the liquid part of the drop R at τ = 0.86 (1), 0.79 (2), 0.73 (3), 0.66 (4), 0.60 (5), 0.53 (6).

The examples in Figs. 59.4–59.6 demonstrate the inadequacy of the thermodynamic conditions [22] for a unique relationship between the degree of supersaturation and the drop radius.

To select a single value of q_m, an additional condition is necessary, which is absent in thermodynamic conditions [22]. For example, if R = const is chosen to be the minimum of all the values of $\sigma(q_m)$ for the admissible values of P_{met}/P_s, then such a dependence of $\sigma_{\min}(R)$ is shown in Fig. 59.7 for a wide range of drop sizes. This dependence largely coincides with the dependence obtained in [24, 27]: at low temperatures the value of the surface tension abruptly exceeds the volume value of σ_b and gradually decreases to it, and at high temperatures, σ_b gradually approaches. Another way to select a single value from the range of admissible P_{met} values is to choose a pressure for which $q_L = q_m = q_e$, where q_e is the equimolecular position of the reference surface. In this case the values of $\sigma(q_e)$ differ somewhat from the values of $\sigma(q_m)$, but the discrepancies are small and commensurate with other methods of calculating σ [24,27].

The analysis of Kondo drops [22] showed that their characteristics are qualitatively similar to those of other metastable drops. In particular, they give the largest values of drop radii, which correspond to the applicability of thermodynamics: the minimum values of the radii of R_{t2} are in the range from 80 to 150λ, which fully agrees with the estimates in Chapter 4.

60. Comparison of the properties of equilibrium and metastable drops

The systems of equations for the concentration profile for equilibrium and metastable drops differ by the presence of an additional Laplace equation for metastable drops. Equations (24.8) for the concentration profile remain unchanged, so it is natural to expect that the properties of metastable and equilibrium drops will not differ very much. Because $P_{met} > P_{eq}$, then, by virtue of the equation of state for the interior of the drop, $\theta_1^* > \theta_1$ and therefore $\sigma_{met}(R|T) > \sigma_{eq}(R|T)$. For both types of drops, the theory gives the dependence of the surface tension on the radius. For $R \to \infty$, both the quantities $\sigma_{met}(R|T)$, $\sigma_{eq}(R|T) \to \sigma_{bulk}(T)$, where $\sigma_{bulk}(T)$ is the volume value of the surface tension on a flat boundary at a given temperature T. With decreasing radii $\sigma_{met}(R|T)$ and $\sigma_{eq}(R|T)$ decrease to zero. The values of the radii R_0 corresponding to the equality $\sigma(R_0) = 0$ represent the lower boundary of existence of the phase. The minimum values of the metastable drops R_0^* according to the molecular theory correspond to the minimum phase formation size for the equilibrium drops $R_0^* = R_0$, because $P_\alpha = P_\beta$, which follows directly from the Laplace equation for $\sigma_{met} = 0$ and $R > 0$.

Below are given examples of comparisons of the properties of equilibrium and metastable drops [24]. The following properties are compared: local densities, pressure, surface tension, total free energy and total mass of drops of the same size, the difference between chemical potentials inside metastable and equilibrium drops of the same size at a given temperature. A comparison of the properties of different drops was carried out for a wide range of temperatures $\tau = T/T_c$, where T_c is the critical point in the bulk phase, from the triple point ($\tau = 0.55$) to the near-critical region $\tau < 0.9$. When comparing the traditional definitions of the surface tension of metastable drops with respect to three types of a dividing surface are used: an equimolecular surface, a surface chosen from the balance of moments of forces, and a dividing surface selected by the surface tension extremum. For an equilibrium drop, the surface tension is determined on an equimolecular surface. Let's start by comparing the concentration profiles, through which all of the listed properties are calculated.

Figure 60.1 shows the results of calculating concentration profiles in the transition region for a drop of radius $R = 100$ over a wide temperature range from a relatively low temperature of $\tau = 0.68$ to a near-critical temperature ($\tau = 0.96$). The following profiles are compared: equilibrium (N) and metastable (L) for the equimolecular dividing surface $\rho_r = \rho_e$. For comparison, the corresponding profiles for a planar lattice are indicated by a prime. The profiles for

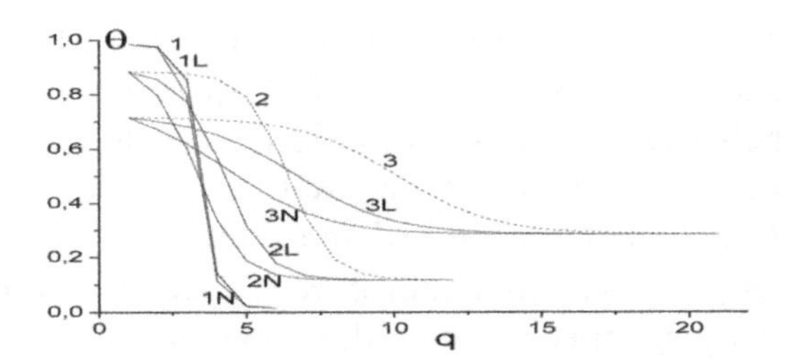

Fig. 60.1. Profiles of the density of the transition region from liquid to vapour at temperatures τ = 0.68 (1), 0.82 (2), 0.96 (3). For a drop of radius R = 100. The notation on the curves: ((*N*) – without taking into account the Laplace equation, (*L*) – taking into account the Laplace equation (8) and $\rho_r = \rho_e$. For comparison, the corresponding profiles for the planar lattice are given by curves 2 and 3.

metastable states are closer to the profiles for a planar lattice than in the case of complete equilibrium of the system [24].

The general results can be formulated as follows: firstly, the profiles for the metastable states (*L*) are closer to the profiles for the planar lattice than in the case of complete equilibrium of the system (*N*). For the lowest temperature (τ = 0.54), all variants of calculations practically coincide, and the width of the transition region is only about 2–4 monolayers. The profiles for different drop radii differ slightly from each other. These calculations are qualitatively correlated with calculations of the profiles obtained by the density functional method [28, 29], although the numerical values differ. In many respects this depends on the differences in the densities of the coexisting vapour and liquid phases obtained by different methods. At the same time, the width of the transition region obtained in this work and in the calculations [28–31] agrees satisfactorily. For high temperatures, the differences between the profiles of the two types of drops and a flat lattice increase.

Figure 60.2 shows the relative properties of metastable drops as a function of their radius R (the radius of the liquid part of the drop is understood as the radius of R) [32]. The region of the vapour corresponds to the radius $R + \kappa$. The value of $R + \rho$ is between R and $R + \kappa$. Figure 60.2 *a* shows the ratio of the densities of the liquid part of the metastable and equilibrium drops, Fig. 60.2 *b* shows the ratio of their internal pressures in the liquid, and Fig. 60.2 *c* gives the difference of the chemical potential $\Delta\mu = \mu_1 - \mu_\kappa$ on the same dividing surface ρ of the metastable drop, to which the surface tension relates. Figure 60.2 *d* shows the dependence of the relative surface tension σ/σ_b on the radius of the drop R. Here

$\sigma_b = \sigma_\infty$ is the value of the surface tension at $R \to \infty$, or, which is the same, for a planar lattice.

The curves in Fig. 60.2 *a*–60.2 *c* for the equimolecular surface and the balance-of-forces surface have the same form: first, the decrease in radius increases the considered value to a certain maximum, after which it decreases to zero. This behaviour corresponds to the presence of maximum supersaturation. This general trend of the presence of a single maximum for the properties under discussion may have differences (local jumps at low temperatures and small radii are possible), because of the discreteness of the change in the number of layers of the transition region $\kappa(T)$ with decreasing temperature [24]. But the general nature of the decrease in the characteristics under discussion with decreasing drop radius remains, and it demonstrates the presence of maximum supersaturation.

The free Helmholtz energy pertaining to the transition region in the QCA and normalized to one site is calculated by the formulas (24.9).

Figure 60.3 shows the logarithmic dependences of the total energy ratios of a metastable and an equilibrium drop of the same size (a)

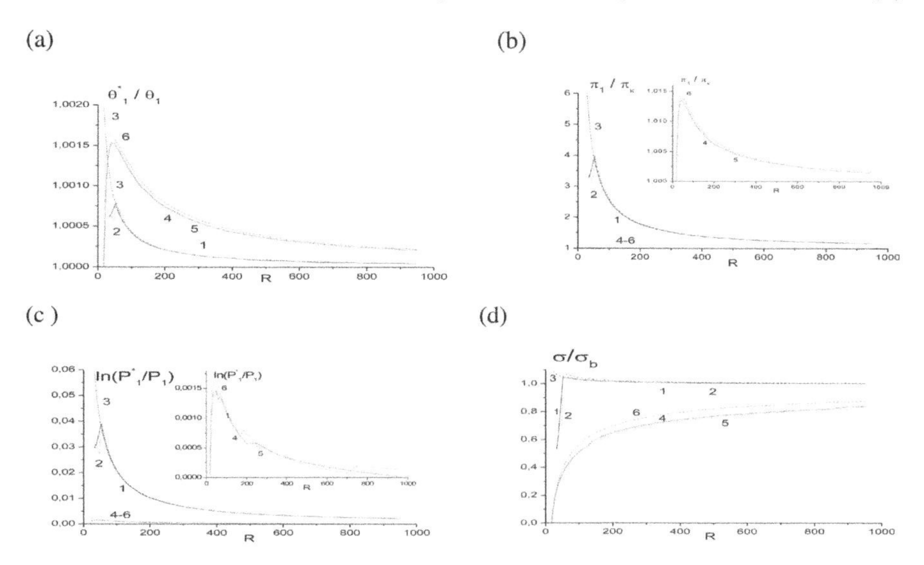

Fig. 60.2. The properties of metastable drops as a function of their radius R: the ratio of the densities of the liquid part of the metastable and equilibrium drops θ_1^*/θ_1(a), the ratio of their internal pressures in the liquid π_1^*/π_1 (b), the difference of these pressures (c), the relative surface tension σ/σ_b (d). The results are given for two values of the reduced temperature $\tau = 0.55$ (1, 2, 3) and 0.89 (4, 5, 6) for three methods of selecting the dividing surface: equimolecular (1.4), balance of moments of forces (2.5) and surface tension (3,6), $\alpha = 1$.

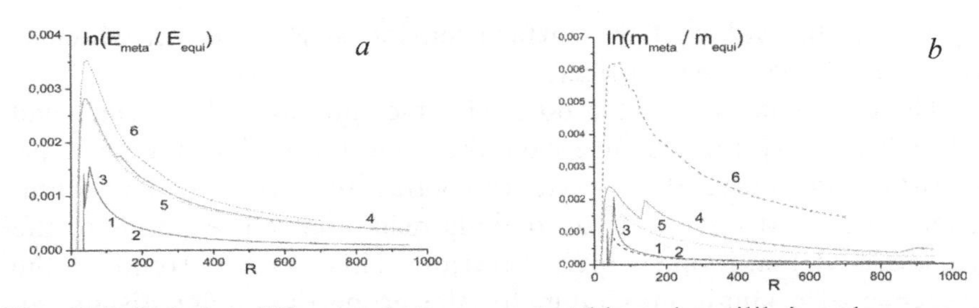

Fig. 60.3. Relative difference between metastable and equilibrium drops as a function of their radius R for temperatures $\tau = 0.55$ and 0.89 for the total free energy E_{meta}/E_{equi} (a) and the drop mass m_{meta}/m_{equi} (b). The results are given for three methods of selecting the dividing surface – the designations of the curves are the same as in Fig. 60.2.

and the ratio of the mass of the same drops (b) to their size R. The subscript indicates the type of drop.

In view of the non-uniformity of the drop, its mass m is defined as the product of local densities by the local volumes of different regions of the drop $m = V_{liq}\theta_1 + \Sigma_{q=2}^{\kappa-1} 4\pi(R+\rho)^2 \theta_q$ where the drop volume $V = V_{liq} + V_\kappa$ consists of the liquid volume $V_{liq} = 4\pi R_{liq}^3/3$ and the volume of the transition layer from liquid to vapour of width κ ($V_\kappa = \Sigma_{q=2}^{\kappa-1} 4\pi(R+\rho)^2$), q is the layer number of the transition region $2 \leq q \leq \kappa-1$. The value $q = 1$ refers to the liquid part of the drop, $q = \kappa$ refers to the vapour phase.

Accordingly, the total free energy of the drop E is expressed as $E = V_{liq}E_1 + \Sigma_{q=2}^{\kappa-1} 4\pi(R+\rho)^2 E_q$, where E_1 is the free energy inside the drop per one fluid site, and E_q is the analogous value for the site in the layer q of the transition region from liquid to vapour.

Calculations based on the molecular model of a spherical drop located in the vapour phase showed fairly close values of the internal properties of metastable drops of different sizes for all three traditional methods for calculating the properties of a metastable drop, differing in the position of the dividing surface (equimolecular, by the balance of the torques and on the surface of tension).

Methodically, the most important value is the magnitude of the discrepancy between the total free energy of the metastable and equilibrium drops, and also between the total mass of the substance contained in these drops. It was found in a wide temperature range from the triple point to the near-critical region that the total free energy and the total mass of the substance in such drops differ by not more than 0.3% by using an equimolecular dividing surface and a surface chosen by the balance of the moments of forces.

A similar difference between an equilibrium drop bounded by an equimolecular surface and a metastable drop bounded by a surface of tension showed a divergence of 0.6% at high temperatures. At low temperatures ($\tau < 0.7$), such a comparison is impossible because of the absence of a surface of tension in the metastable drop, which automatically leads also to the absence of maximum supersaturation [24, 27]. The absence at the low temperatures of the concept of the surface of tension and maximum supersaturation indicates that the conditions for thermodynamic constructions are not fulfilled for all possible parameters of the state of the system.

The small difference between the total free energy and the total mass of the substance between the metastable and equilibrium drops of equal size indicates that: 1) under modern experimental methods of investigation, it is difficult to distinguish the differences in the contributions for the drops of these two types in the thermodynamic characteristics of the system, and 2) the presence in the system drops that can be in different thermodynamic states, requires the use of more perfect kinetic models to describe the condensation process than traditional approaches, taking into account only the metastable drops. The very fact found in the works [24, 27] of the existence of equilibrium drops indicates a fundamental inadequacy of the use of thermodynamic description in the process of vapour condensation.

Comparison of the properties of metastable drops. The presence of metastable drops in the molecular theory makes it possible to compare their properties with metastable drops in thermodynamics. Here the differences between the properties of the metastable drops are more significant than the differences between the equilibrium and metastable drops in the molecular theory. The qualitative difference between the two types of metastable drops (associated with the presence of state equations in the theory and its absence in thermodynamics) is differently related to the states of these drops with the Yang–Lee theory. Further:

1) The metastable drop in thermodynamics corresponds to a single value of the surface tension σ_{bulk}, measured in an experiment for macrosystems. The theory, in contrast to thermodynamics, gives the size dependences of the surface tension $\sigma_{met}(R)$.

2) The size dependence $\sigma_{met}(R)$ significantly reduces the difference in pressure in the liquid and vapour (supersaturation) and leads to the existence of a maximum supersaturation for small drop radii in comparison with the unlimited growth of the internal pressure in

a metastable drop from thermodynamics according to the Kelvin equation.

3) From the mathematical point of view, the solution for metastable drops in the theory is less stable than for equilibrium drops, and at low temperatures finite supersaturations are not always realized for them, as in the metastable drops of thermodynamics.

4) It is natural to raise the question of whether it is possible to pass to the Kelvin equation from the metastable drops in theory. The molecular theory operates a multilayer model, regardless of the type of boundary: flat or curved. An analysis of the system of coupled equations on the molecular densities (24.8) showed that the Kelvin equation can be obtained [7] if we confine ourselves to taking into account only one intermediate monolayer between two phases and exclude the equation of state for it. In a multilayer model, the equations for the concentration density profile have a sufficient number of degrees of freedom to describe a gradual transition from one phase to another, whereas in the thermodynamic approach there is only one variable (pressure). As a result, the effect of the change in the density profile on it with respect to the profile of the plane boundary is attributed to a change in this thermodynamic variable. This change in the real profile can be attributed with the same success to a change in the other phase parameter. For example, the liquid density θ_1, as discussed in [7, 33], was exactly the same. Therefore, any one-layer models contradict the molecular pattern of the multilayer interphase region.

5) In thermodynamics there are no expressions for either the size of the metastable drop, except for the intense parameter R introduced because of the confusion of the concepts of mechanics and thermodynamics (Section 57), nor on the minimum size of the embryo. The well-known Frenkel's interpretation [14] about the maximum of the Gibbs potential corresponding to the unstable state of the embryo is a reformulation of the well-known Gibbs result $W = \sigma A/3$ [8] corresponding to the work of the formation of a new phase (the method of obtaining which was indicated in Section 58). Interpretation of Fig. 50 in [14] has a purely methodological meaning to show the relationship between the internal and surface contributions for a new phase under the conditions of a metastable system for a *predetermined* drop size. Thus, any value can be used as a given size of the embryo.

From the Gibbs' two non-thermodynamic concepts (pp. 255-258 [8]) on the assumption of a linear relationship between the size and surface tension of the crystal and the condition for the similarity of the crystal to itself when its dimensions change, it follows unambiguously that for $R \to 0$ the contribution from the surface tension should vanish. In this limit, the value of the surface contribution σA vanishes not only because of the decrease in the surface area $A = 4\pi/3R^2$ for $\sigma(R \to 0) > 0$, but also because of the decrease in the surface tension $\sigma \to 0|_{R\to 0}$ (for $A > 0$). Gibbs' requirement can only be met by the value of the minimum size of phase R_0 (Sections 25–28). It should be understood as the term of the thermodynamic limit $R \to 0$, because it is uniquely related to the very concept of the phase. The quantity R_0 also follows from the theory for metastable drops, and it is the only value at a given temperature which coincides with the value of R_0 for the equilibrium drops.

61. Quasi-thermodynamics

In this section, we summarize the use of the scheme for calculating the thermodynamic characteristics of Kelvin and Gibbs surface tension of curved surfaces in small systems. These topics include: 1) modifications of thermodynamic equations for refinements of Gibbs constructions, 2) constructions that are intermediate between classical and statistical thermodynamics – approaches to which molecular parameters are introduced; 3) statistical theories that use concepts of the surface of tension by analogy with thermodynamics [11, 19].

In Appendix 1 and Sections 59 and 60, examples of the first direction are given. They regard the interface of phases not as a mathematical plane, but following the ideas of the van der Waals molecular theory [34, 35] (see also [19]) as a region of finite width. In addition, the term 'quasi-thermodynamics' is the theory of surface layers based on the local formulation of thermodynamics. This version of the quasi-thermodynamic theory is presented in [22], it is close to the version originally developed by Tolman [36], and serves as a bridge between the macroscopic thermodynamics and the molecular theory based on statistical mechanics.

The standard position of quasi-thermodynamics is the preservation of the kind of functional relationships established in thermodynamics for the thermodynamic characteristics and in the mechanics of continuous media for mechanical deformations, for each local region with the coordinate r, where r is located inside the transition region. When developing a quasi-thermodynamic theory, ideas of the width of the transition region were not yet known, and this assumption was assumed to be natural. However, later the molecular statistical calculations showed [11,19] that the surface layer is only a few molecular sizes in thickness, and the main postulate of the quasi-thermodynamic theory has been violated. Briefly recall the main provisions of the quasi-thermodynamic theory assuming that, at least near the critical point, when the width of the transition region is much larger than the size of the molecules, these bases are correct. On the other hand, this example clearly shows the qualitative shortcomings of thermodynamics in comparison with molecular approaches.

Quasi-thermodynamics itself only fixes the necessity for the existence of a non-uniform density profile $\theta(z)$ in the transition region (z is the coordinate along the normal inside the transition region), but does not determine the equation for the concentration profile and width of the transition region. The basic postulate of quasi-thermodynamics modifies the concepts of macroscopic thermodynamics in that any intense thermodynamic quantity is uniquely determined at each point as a function of the temperature and density of the number of molecules at the same point z. We denote by $c(z)$, $v(z)$ and $f(z)$, respectively, the number of molecules per unit volume, the volume per molecule, and the free energy per molecule (by definition, we have $v(z) = c(z)^{-1}$), which corresponds to the condition (22.16). From an increase in the area A by an amount δA by reversible isothermal displacement of the vessel sidewall, we obtain a relationship between the local specific volume and the tangential pressure component inside the plane transition region in the form

$$\frac{\partial f(z)}{\partial v(z)} = -P_T(z) \tag{61.1}$$

where $f(z,T) = f(c(z,T);\ T)$, i.e., at a given temperature, the free energy F of the system as a whole is a functional of $c(z)$. Therefore, the equilibrium form $c(z)$ must be determined from the condition of a minimum of F for a given temperature and volume.

The quasi-thermodynamic condition of local equilibrium (see [22, 36]) is written as

$$f(z) + P_T(z)/c(z) = \mu. \tag{61.2}$$

where μ is the total chemical potential for the bulk phases α and β. Inside the bulk phases $P_T(z)$ is reduced to the usual hydrostatic pressure $P_{\alpha,\beta} = P$. Equation (61.2) is one of the fundamental relations of quasi-thermodynamics. It can be shown from this that $s(z) = -\partial f(z)/\partial T$ is the entropy per molecule at point z. Consequently, the internal energy per molecule $u(z)$ at the point z is given by the equality $u(z) = f(z) + Ts(z)$. Equation (61.2) also implies (comparing with (22.7)) the expression for the surface tension

$$\sigma = \int_{-\infty}^{\infty} (P - P_T(z))dz, \tag{61.3}$$

identical to (22.20), derived from the mechanical definition of surface tension for a plane boundary. It follows that the expression (22.20) can be derived from the quasi-thermodynamic equilibrium condition (61.2).

We will not dwell on the case of spherical drops described in the quasi-thermodynamic approach, since their concrete interpretation in a discrete description is given in Appendix 1. Any real expression of quasi-thermodynamic equations for a profile is given only by statistical mechanics, the solution of which allows us to find the equilibrium density profile $\theta(r)$ inside the transition region. A rather detailed critique of quasi-thermodynamics is given in Section 2.5 [19].

The presence of an interface also imposes limitations on the molecular models involved in describing the transition region. An important role is played by the equations of state used – they must satisfy the actual experiment. Otherwise, formal thermodynamic constructions for finding the surface tension do not allow them to calculate. It is well known that the van der Waals equation is extremely important from a methodological point of view. But it has not found its practical application for describing vapour–liquid systems; instead, the virial equation of state was actively used [37, 38]. Similar problems arise when using the van der Waals equation as the basic equation of state for the bulk phase when considering the interface of phases. In [39, 40], according to the quasi-thermodynamic calculations of the transition region, Hill replaced this equation because of its roughness by the Tonks equation [41]. Such a replacement made it possible to obtain a reasonable form of the concentration profile of the vapour–liquid interface.

We note that quasi–thermodynamic constructions began long ago: this includes the theory of the thermodynamic melting point of Pavlov [42, 43] on the account of size effects in a first-order phase transition; all the thermodynamic work described above for different positions for dividing surfaces; many molecular approaches use the thermodynamic position of metastable drops [11, 19], and even use the properties of metastable nuclei [28–30].

Even the very presence of the crystal lattice in crystals is a definite compromise between the absence of the structure of matter in 'model-free' thermodynamics and the constancy of the crystal's properties in terms of volume, because of its isotropy at the level of elementary cells (although the difference from this isotropy inside the cells is well known from the optical branches of the vibrational spectra) [44]. It is for this reason that by introducing the concept of phase, in thermodynamics all phase transitions are treated identically regardless of the aggregate state of the substance, whereas in molecular approaches a specific consideration of the symmetry of the substance is necessary for the correct description of phase transitions [45].

The Tolman model. For quasi-thermodynamic constructions it is necessary to classify different works on taking into account the dependence of the surface tension on the drop radius, which leads to a modification of the Kelvin equation. To do this, a derivative $d\sigma/dR$ (or correction in another form of record) is introduced, which formally reflects the change in the magnitude of the surface tension with the size of the drop. The term 'formally' means the absence of any means of determining this dependence within the framework of thermodynamics itself, with the exception of experimental measurements, which is practically impossible. Various attempts have been made to thermodynamically construct the dependence $\sigma(R)$ [23].

All problems of small bodies rest on the way to determine 'surface tension'. The introduction into thermodynamic approaches of any molecular specificity distorts their essence and leads to hybrids that can not give satisfactory results. This can be seen from the example of the well-known Tolman equation [11, 46].

From the consideration of two types of dividing surfaces (equimolecular and surfaces of tension), the author succeeded in expressing the dependence of the surface tension of the drop on the curvature of the dividing surface in the form $\ln(\sigma(R)/\sigma_\infty) = \int_\infty^R A(R)dR/(1+A(R))$ where $A(R)=2\xi(1+\xi+\xi^2/3)$, $\xi = \delta/R$, δ = const is a quantity of the order of the molecule size. Calculation of this formula leads to the fact that at the minimum drop size $\delta = R$ and $\xi = 1$, corresponding to the molecular associate, consisting of an atom with its nearest neighbours, the ratio $\sigma(R)/$

σ_∞=0.28 is obtained. Thus, the 'associate' has a noticeable surface tension, which contradicts the physical meaning of such a concept.

Approximate calculations based on statistical mechanics for liquid argon showed that δ should be of the order of 3 A, and in the case of a liquid drop it should be positive. Thus, taking the value $\delta = 3$ A, we can expect that the surface tension on a spherical boundary with a radius of 100 A will be 6% less for a drop (larger for a bubble) than for a flat boundary. Thus, the formulas obtained give a purely qualitative estimate indicating a decrease in the surface tension with decreasing drop radius.

Asymptotic molecular theory. Problems with the use of molecular models arise not only for rough state equations, but also for models with an incomplete interaction potential: the asymptotic theory [47,48] operates with the long-range part of the potential, but excludes the short-range (repulsive) part. In [49], expressions for the normal and tangential components of the pressure tensor were obtained using the asymptotic theory of accounting for the interaction between bodies in the liquid phase. This truncated potential yielded the result: the tangential component of the pressure tensor is observed in the transition region at a distance of about 80λ, while the width of the transition region itself is only $\sim 4\lambda$ [23,49]. The microscopic theory shows a unique relationship between the width of the transition region and the difference in the components of the pressure tensor [6,25,27] and (Appendix 1), so in the transition region of the pressure tensor component can not differ from the pressure in the bulk phase. The authors [23, 49] obtained that between the phases there is a transition region with a variable average pressure different from the pressure in these phases, which contradicts the mechanical stability of the system.

The use of estimates for the width of the transition region and its separation into the liquid and vapour parts in this theory together with the equations of thermodynamics led in the limit $R \to 0$ for small drop sizes to the existence of two different pressures in the vapour and inside the drop [23]. This fact can be understood by comparing the value of the thermodynamic limit $R \to 0$ with the value of R_t (Section 26), when the limitation on the applicability of the equations of thermodynamics comes. Indeed, in a metastable drop there is a difference between the internal pressure and the pressure in the vapour, and at a radius R_t these values will differ. At the same time, we note that the separation of the transition region into two subregions provided the limitation of the internal pressure in the drop instead of its unlimited increase in the Kelvin equation.

We also note that asymptotic molecular theory can not be used to analyze the deformation states of any phase due to the lack of taking into account the repulsive potential. Consequently, the molecular models involved in describing the properties of interfaces must be sufficiently correct.

Statistical theories and thermodynamics. The equations of statistical physics for curved interfaces are considered in monographs [11, 19]. All theories took as their basis the assumption of thermodynamics about the presence of a surface of tension, despite the ambiguity of the existing definitions of surface tension. They *a priori* laid the existence of a pressure jump between the vapour and the liquid phases, and further work was done on the construction of surface tension bonds with equations for the distribution of unary and pair functions. Doubts in Section 4.8 of [19] concerning the correctness of quasi-thermodynamic constructions do not, however, affect the very fact of the *a priori* introduction of mechanical equilibrium, as in the Kelvin equation.

The resulting complex system of equations for unary and pair distribution functions was not solved in full integral form, and it was used in the form of a density functional theory that does not take into account correlation effects (the latter excludes its use in problems of kinetics). The mathematical complexity of the problem did not allow earlier: neither to check the correspondence of metastable drops with the Yang–Lee theory [2–5] nor to detect equilibrium drops. In discrete form, this continuum material from the statistical physics of curved interfaces is set forth in Appendix 1 and in Section 24.

Concluding the discussion of metastable drops (Sections 57–60), we note that the molecular theory [6.25] allowed us to investigate all possible ways of determining σ and verify the correctness of thermodynamic hypotheses. The difference between the molecular theory and thermodynamics is that, in fact, there are two parameters in the theory that characterize the state of the drop for a given supersaturation degree P_{met}/P_s: this is the radius of the drop R and the width of the transition region κ. However, in fact, we are talking about a much larger number of degrees of freedom related to the profile within the transition region $\theta(r)$, through which the value of κ is determined. Such a difference in the degrees of freedom leads to more diverse variants of molecular distributions. As a consequence, it was found that different definitions lead to different concentration profiles and values of σ. The analysis showed a significant difference in molecular distributions from the assumptions on the thermodynamic hypotheses that were mentioned above.

We note the quasi-thermodynamic determination of the surface tension through the equilibration of forces and moments of forces on the dividing surface [21], which qualitatively differs from the methods of introducing σ previously considered on the basis of purely thermodynamic hypotheses [11,19,22]. This difference is well known [11]: the conditions for the equality of forces and moments of forces with respect to the dividing surface are sufficient to introduce local pressures inside the transition region, and to obtain equations that ensure the internal mechanical stability of the interface. In this case, the local stability conditions imply the macroscopic Laplace equation, and it should not be used *a priori* as a postulate, with respect to the molecular theory. Here it emphasizes that microscopic molecular equations independently describe all properties without attracting additional connections from thermodynamics or mechanics of continuous media.

It was found in the calculations that at all temperatures the criterion for the applicability of thermodynamics exceeds the width of the transition region $R_t > \kappa$. This indicates the inappropriateness of the application of thermodynamic methods in describing the properties of the transition region along the normal to the interface of any phase. An exception is the critical temperature range. Thus, the molecular theory not only defined the lower bound of the applicability of the thermodynamics R_t in problems with small systems, but also showed that the finding of σ can not also be associated with the use of thermodynamic constructions.

62. Relaxation times of metastable drops to equilibrium states

Let us consider the results of a direct molecular calculation of the relaxation time of metastable drops to their equilibrium state at saturated vapour pressures, depending on the size of the drops. In terms of physical meaning, both types of drops – equilibrium and metastable – are determined for asymptotically large times, which, in principle, can exceed the characteristic times of existence of real local states in a non-equilibrium system during nucleation.

The existence of a metastable drop follows from the condition of mechanical equilibrium in the absence of chemical equilibrium. Such a state can not exist for a long time, since the condition for the minimum free energy of the local subregion necessary for true equilibrium is not satisfied for it [14, 50]. When analyzing various macroprocesses, an important role is played by the characteristic times of processes in local volumes. One of such possible local processes is the transition between the metastable and the equilibrium states of the drops [51].

We use the equations papers [6, 25] (Sections 24 and 59) describing the state of both types of drops at the molecular level: the boundary is a region non-uniform in density separating uniform regions of vapour and liquid. During the relaxation process of the metastable drop, its density, internal pressure, free energy and surface tension for drops of a fixed radius R change. There is no inhibition on the drop surface – the equilibrium between the layers is established more rapidly than inside the volume of the drop. (This follows directly from the notion of 'non-autonomy' of the liquid–vapour boundary, and also from the analysis of Section 44). The density variation inside the drop is due to the diffusion of molecules to the interface and the outflow of molecules from the surface. The total relaxation time of a drop with the establishment of an equilibrium density inside it depends on the size of the drop R. We will consider this process without being connected with the mass transfer processes in the vapour phase – i.e. we estimate the minimum relaxation time of density from an increased value to a smaller value when the size of the drop changes during its partial desorption, which is not complicated by the outflow of molecules around the drop.

The process of mass transfer inside the drop is described by the diffusion equation, as indicated in Section 36. The calculation was carried out for argon atoms with the Lennard-Jones potential [38] for five temperatures $\tau = T/T_c$ (T_c = 151 K). Figure 62.1 shows the dependences of the density inside metastable drops calculated from the equimolecular dividing surface, $z = 12$, taking into account only the interactions of the nearest neighbors. All of them have the same form: as the radius of the drop R increases, first the density $\theta_1(R)$ sharply increases for small radii, and then decreases monotonically. The equilibrium densities $\theta_1^{eq}(\tau)$ at the same temperatures are 0.9760 (τ = 0.55), 0.94852 (τ = 0.65), 0.9051 (τ = 0.75), 0.83566 (τ = 0.85), and 0.70551 (τ = 0.95). These quantities in equilibrium drops are determined by the densities of the binodal curve and remain constant for all radii.

To determine the time to establish the equilibrium value of the density within the entire drop one must know the mass transfer coefficients D. Most literature data on the coefficients refer to self-diffusion coefficients (or label transfer) D^*. For argon drops, experimental data were used for the bulk liquid phase [52]. The values of the mass transfer coefficients in order to obtain the necessary estimates of the relaxation times from Eq. (36.2) depend on the specific conditions of the organization of the process, so different approaches are possible for its evaluation (see Section 36).

We specify the methods for calculating the thermodynamic factor $D_e = g_e D^*$ [53–55], where the thermodynamic factor g_e is related to the chemical potential of the component μ_A as $d\mu_A/d\theta_A$. Or $g_e = 1 + \partial \ln\gamma_i/\partial \ln\theta_i$, where the activity coefficient in the quasi-chemical approximation used for taking into account the lateral interactions [56,57] is equal $\gamma_A = (t_{AA}/\theta_A)^{z/2}$, here $t_{AA}(\theta_A)$ is the probability of finding the particle A next to another particle A: $t_{AA} = 2\theta_A/(\delta + b)$, $\delta = 1 + x(1 - 2\theta_A)$, $b = (\delta^2 + 4x\theta_A^2)^{1/2}$, $x = \exp(-\beta\varepsilon_{AA}) - 1$. When calculating g_e in binary alloys [53–55], the contribution of vacancies (due to their smallness) is neglected in the thermodynamic characteristics. In dense one-component substances, the fraction of vacancies is also small, but the vacancies themselves can not be considered equilibrium in the conditions of mass flow. In this case, we should use another expression for the function g_k, which relates the self-diffusion coefficient and the mass transfer coefficient, which follows from the kinetic analysis (formula (32.6) in [56]): $D_k = g_k D^*$, where $g_k = 1 + \frac{\partial t_{AA}}{\partial \theta_A}\left\{\frac{1}{1 - t_{AA}} + \frac{(z-1)x}{1 + x t_{AA}}\right\}$ is the «kinetic» factor in the same quasi-chemical approximation, connecting D_k and $d(\mu_A - \mu_V)/d\theta_A$. We also consider model estimates of the value of D [54] if one considers that the differences in the activation energies of the self-diffusion coefficients E^* and the mass transfer E_m are due to the formation energy of the vacancy H_V: $E^* = E_m + H_V$. Then $D_V = g_V D^*$, where $g_V = (1 - \theta)^{-1}$.

For the calculation of the functions g_e and g_k, independent experimental data on the heats of sublimation, $H_{\text{subl}} = 1840$ cal/mole [58], were used to determine the value $\varepsilon_{AA} = H_{\text{subl}}/z$ at $z = 12$. The mass transfer coefficients in the listed approaches with functions g_e, g_k and g_V are given in Table 62.1. The calculations yield qualitative estimates, so the experimental data for D^* and H_{subl} and the phase diagram used to calculate the density of the coexisting vapour–liquid phases were taken from different sources and not self-consistent

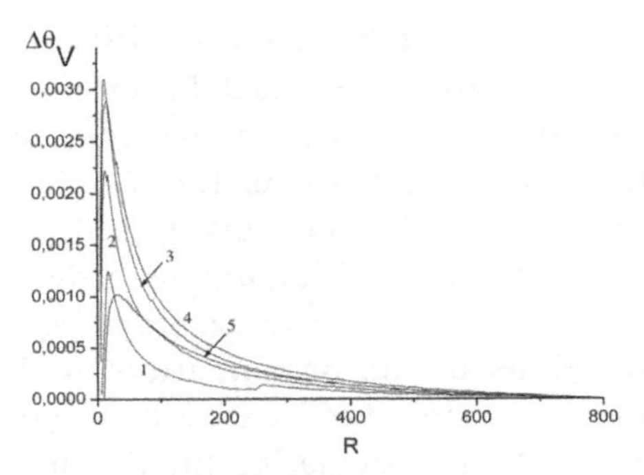

Fig. 62.1. The deviation of the values of the vacancy density (θ_V=1–θ) of a metastable drop as a function of its radius R (calculation with an equimolecular surface) from the equilibrium value $\Delta\theta_V = \theta_V(R) - \theta_V^{eq}$ of the vacancy densities for different reduced temperatures τ: τ = 0.555 (1), 0.65 (2), 0.75 (3), 0.85 (4), 0.95 (5).

Table 62.1. The self-diffusion coefficients (D^*) [52] and the mass transfer in liquid argon at different temperatures on the stratification curve calculated from the equilibrium (g_e) and kinetic (g_k) transition functions to the mass flux and from the model estimate (g_V).

τ	D^*, m²/s	g_e	D_e	g_k	D_k	g_V	D_V
0.55	$1.68\cdot10^{-9}$	0.62	$1.042\cdot10^{-9}$	26.22	$4.40\cdot10^{-8}$	41.67	$7.00\cdot10^{-8}$
0.65	$3.60\cdot10^{-9}$	0.41	$1.476\cdot10^{-9}$	8.45	$3.04\cdot10^{-8}$	19.61	$7.06\cdot10^{-8}$
0.75	$5.30\cdot10^{-9}$	0.20	$1.06\cdot10^{-9}$	2.37	$1.26\cdot10^{-8}$	10.53	$5.58\cdot10^{-8}$
0.85	$7.35\cdot10^{-9}$	–0.02	$-1.47\cdot10^{-10}$	–0.16	$-1.18\cdot10^{-9}$	6.10	$4.48\cdot10^{-8}$
0.95	$9.10\cdot10^{-9}$	–0.25	$-2.275\cdot10^{-9}$	–1.31	$-1.19\cdot10^{-8}$	3.39	$3.08\cdot10^{-8}$

with one another. This led to negative values of the mass transfer coefficients D_e and D_k as they decrease for high temperatures. The decrease in the mass transfer coefficients in the near-critical region is a well-known fact [59], as well as the appearance of negative values of this coefficient describing the so-called 'upward' diffusion associated with the process of phase formation [56], and not with the spreading of an ensemble of non-interacting particles [53, 54]. The model estimate does not work near g_V in the near-critical region, because it reflects only the tendency to change the vacancy density.

Typical results of calculating the relaxation times of metastable drops in a wide range of radii are shown in Fig. 62.2. Here, relaxation times are given only for positive values of the mass transfer coefficients. The solid curves correspond to the calculations with the mass transfer coefficient from the functions g_V, the dashed curves correspond to the mass transfer coefficient from the functions

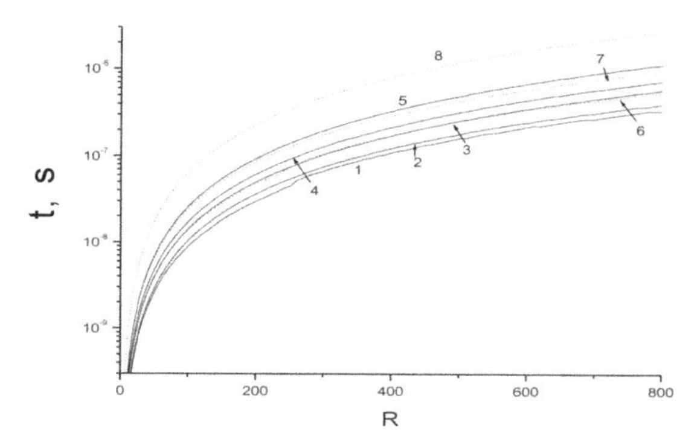

Fig. 62.2. Dependence of the equilibrium establishment time on the radius of the metastable drop at the saturation line at various reduced temperatures τ: 1 and 6 – $\tau = 0.55$, 2 and 7 – 0.65, 3 and 8 – 0.75, 4 – 0.85, 5 – 0.95. The solid lines correspond to the calculations of the mass transfer coefficient D from the model estimate with g_V, the dashed lines correspond to the kinetic factor g_k.

g_k. The unit of measurement of length is the number of monolayers (monolayer size δ).

All curves increase with the size of the drops. For small R, the curves start from nanosecond range fractions and reach a range of up to several microseconds. In both cases, the relaxation time is greatest for the lowest temperature and with increasing temperature the relaxation time decreases, but this change fits within one order of magnitude in time. (If we look at the near-critical temperature region with the thermodynamic factor g_e at $\tau = 0.85$, then the relaxation times increase by about two orders of magnitude.)

For small drops with a radius $R = 6$ of the molecule diameters (or of the order of 2 nm), the relaxation times are ~1 nanosecond. This estimate is obtained without taking into account the fluctuation corrections to the drop size, which increase the lower values of their stable sizes, but this can not increase substantially the range of the characteristic relaxation times of small drops. Therefore, we can conclude that the lower range of values of the relaxation times does not exceed the nanosecond range, and the process of internal relaxation of the states of the metastable drop during the times of several collisions of molecules in the gas phase takes it to its equilibrium state.

For large drops with a radius of the order of $R = 800\delta$ (~0.3 μm), the relaxation times increase by 3–4 orders of magnitude. However, this range is much smaller than the often allowed range of millisecond time resolution in the modeling of nucleation processes

[60]. At a fixed drop radius ($6 \leq R \leq 800$), the range of relaxation time variation from the melting point to the critical temperature does not exceed one order: the lower the temperature, the slower the relaxation process.

Thus, the relaxation times of metastable drops show that with a decrease in their size, a rapid change in the internal state of the drop occurs. This process is much faster than previously thought. The presence of equilibrium drops broadens the spectrum of physically realizable drop states during nucleation. In the general case, in order to calculate the dynamics of the nucleation process, it is necessary to take into account all types of drops that can form within a certain local region of the vapour during diffusion inhibition of its mixing.

63. Metastable states

One of the most problematic questions of physical and chemical analysis is the question of the nature of metastable states and the methods of its description. Classical thermodynamics included metastable states in the number of concepts related to small perturbations to equilibrium [8]. The basis for such inclusion was experimental data on chemical processes in which there were no reactions without a catalyst and mechanical processes associated with friction of sliding of solids (Section 42). At the same time, metastable systems included systems with the formation of a new phase (supersaturated or supercooled vapour, supercooled liquid, etc.), which were described by the equations of state of single-phase systems.

As shown above, in vapour–liquid systems the concept of metastable states is associated with the use of the Kelvin equation, which artificially introduces non-equilibrium through the mixing of the concepts of dynamic and equilibrium surface tension. Therefore, for simple vapour–liquid systems, such an interpretation has a conventional meaning. In more complex situations: amorphous systems, liquid crystals, polymer solutions, etc. Representations of metastable states are often introduced as frozen states at low temperatures [14, 61, 62]. An even greater variety of states is realized in solid-phase systems – for them there is a very wide spectrum of relaxation times of internal rearrangements, especially for different deformation states (Section 43), thus the metastable states, like frozen non-equilibrium states, are very widespread.

In all textbooks, when talking about the possible states of physico-chemical systems in a metastable state (see for example [63, 64]), Fig. 62.1 *a*. "The curve represents the dependence of the Gibbs energy G on the parameter x under the conditions of the existence of the system. The equilibrium condition $dG = 0$ is satisfied at all points of the maximum and minimum. The difference between them is determined by the value of the second derivative, which must be positive at the minimum points ($d^2G > 0$) and negative at the maximum points ($d^2G < 0$). The minimum points (A', A'' and A''') correspond to *stable equilibrium*, and the maximum points to *unstable equilibrium*. The different levels of stability of the equilibrium state correspond to different levels of the minima position on the curve of Fig. 62.1 *a*. (A''') will be thermodynamically more stable in comparison with the states to which higher positions of the minimum (A' and A'') correspond, and the transition from the first position to the other requires work. The state A''' is the most stable for the given conditions of existence of the system. The states (A' and A'') corresponding to small relative stability are called metastable". This text explains the difference between metastable states and the most stable (equilibrium) state.

The field in Fig. 62.1 *b* shows the graph of the dependence of the Gibbs thermodynamic potential on the molar volume v [65]. "At the critical values of the molar volume v_i ($i = 1, 2, ..., 6$), the thermodynamic potential $G(v_i,\alpha)$ undergoes local minima reflecting local non-equilibrium potentials in zones of hydrostatic stretching of various scales. The critical values of v_i correspond to the following states in a deformable solid: v_0 is an equilibrium crystal; v_1 - zones of stress microconcentrators, in which nuclei of dislocations are nucleated; v_2, v_3 – zones of meso- and macroconcentrators of stresses, in which local structural-phase transitions occur with formation of meso- and macrobands of localized plastic deformation, respectively; v_4 corresponds to the intersection of the curve $G(v,\alpha)$ with the abscissa; with a further increase in the local molar volume, the change in the Gibbs thermodynamic potential occurs under the conditions $G(v) > 0$ and the system becomes unstable: various types of material destruction develop in it; at $v > v_4$, two phases can exist: the atom–vacancy phase (for $v = v_5$) and the local vacuum (for $v \sim v_6$) in the form of micropores, cracks and discontinuities". The curve in Fig. 62.1 *b* from [65] is similar in shape to the curve on the field in Fig. 62.1 *a*, but here the situation refers to very strongly non-equilibrium states of a solid, and, nevertheless, it is characterized by the free Gibbs energy, as in the equilibrium state. The reason for including a strongly deformed state in the metastable state was a reference to Leontovich [66], who allowed the representation of non-equilibrium states as equilibrium states in some external field. Such an interpretation is permissible in general thermodynamic constructions [67] when discussing conditions for the minimality of the non-equilibrium free energy as a function of the external field.

It is obvious that the non-equilibrium distribution of particles corresponds to their non-uniform distribution in space even in the bulk phase. However, the converse statement about the non-equilibrium of the system in the case of inhomogeneity is incorrect. At the interface of coexisting phases in the equilibrium state, a non-uniform distribution of particles within the transition region is always realized. Therefore, the construction of A.M. Leontovich [66], who introduced non-equilibrium states with the help of an external field, is not correct. It is not accidental, he was forced in the molecular-statistical substantiation of his statement for thermodynamics to introduce an additional restriction on infinitesimal deviations from equilibrium [68]. Section 40 shows that for small deviations from equilibrium the number of dynamic variables does not increase in comparison with the number of thermodynamic parameters in the equilibrium state. Additional dynamic variables appear in strongly non-equilibrium states, but they can not be represented as an effect of external fields [66].

Recall that the strictly equilibrium state of a solid corresponds to a state in which the internal stresses are completely absent [16]. This condition is certainly not satisfied in examples in Fig. 62.1 *a* and *b*, with the exception of points related to a strictly equilibrium state – the point with the deepest minima (point A‴ in the field in Fig. 62.1 *a*). In the case of internal stresses, the system is not considered to be in equilibrium – these are deformed states. The degree of deviation of the deformed state from the equilibrium state can vary within very wide limits. Thus, in the fields in Fig. 62.1 *a* and *b*, different variants of the energy curves representing the properties of solids are represented along the ordinate axis. They characterize the average value of the energy characteristic of a system relating to an arbitrary large time interval. For all curves of the type shown in Fig. 62.1 there is usually no information on the time range under consideration. The central question is the meaning of the energy curves presented, on which a certain parameter of the

a

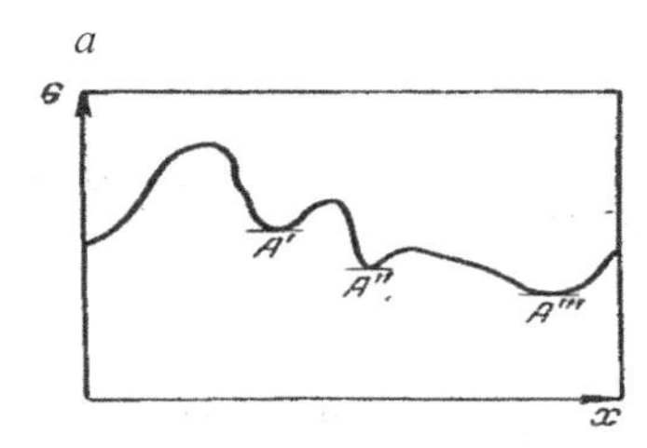

b

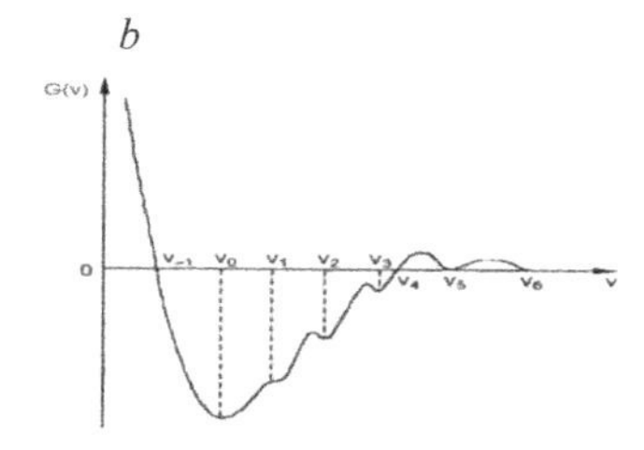

Fig. 62.1. a – Change in *Gibbs energy* $G(x)$ in an arbitrary physicochemical process, x – state parameter [63]. b – dependence of the Gibbs thermodynamic potential $G(v)$ on the molar volume v, taking into account the local zones of stress concentrators of different scale [65].

system (coordinates, concentrations, specific volume, etc.) is present on the abscissa axis, but not time.

The answer to this main question of the interpretation of the experiments is given by the formulated method for calculating non-equilibrium thermodynamic potentials.

In solid bodies, two types of motion are basic: oscillatory and translational in diffusion, which are separated by characteristic times by about ten or more orders of magnitude. (In the general case of intense mechanical perturbations of a solid, it is necessary to additionally include equations for the dynamics of the deformation of the sample [69].) On the ordinate axis, for any system under discussion, the energy characteristic should be a non-equilibrium analog of the Gibbs free energy, and not a potential or internal energy. Averaging over a large time interval inevitably involves vibrational motions that, due to their rapid relaxation, are in locally equilibrium states, depending on the particle configurations in the coordinate space, and their free energy consists, as usual, of the potential energy and entropy of the vibrational motion.

According to the rules of statistical mechanics [4,5,8,67], in order to calculate any equilibrium function $G(x)$, it is necessary that the free energy for a given state parameter x has a minimum: $\partial G/\partial x = 0$. This condition allows us to find x^*, which forms the required dependence $G(x^*)$. Therefore, the true minimum value of free energies, as a function of x, is not directly related to the total curves in Fig. 62.1.

An equilibrium state is a state that differs by an infinitesimal amount, from a state of strict equilibrium that arises from natural mean square fluctuations. The answer to the question of the extent to which the equilibrium potentials G, F, U can be extended to non-equilibrium states depends on the accuracy of the description of the state of the system by means of non-equilibrium and equilibrium distribution functions. If all elementary processes manage to reach their limit values on a given time scale, then the time dependences are insignificant in their contributions, and they can be neglected. From the mathematical point of view, the answer is: $G_{\text{real}}(\theta_i,\theta_{ij},t) = G_{\text{equi}}(\theta_i) + \delta G(\theta_i,\theta_{ij},t)$, where $\delta G(\theta_i,\theta_{ij},t) < \varepsilon$, here ε is the preassigned accuracy of the description of functions; The deviation $\delta G(\theta_i,\theta_{ij},t)$ of the non-equilibrium G_{real} function (θ_i,θ_{ij},t) from the equilibrium value $G_{eqil}(\theta_i)$ depends on the concentration θ_i and on the average number of neighbor pairs θ_{ij} between the different components ij. In practical situations, taking into account the transition from δG to $\delta\theta_i$, the error in local concentrations should not exceed one or a fraction of percent of the characteristic values of equilibrium values. For a solid, these values are less than 10^{-3}–10^{-6} (depending on the temperature) [70]. All other states of the system are non-equilibrium, and their evolution should be described by kinetic equations.

If the kinetic theory and the equilibrium equations are self-consistent (Chapter 6), then for a long time, when an equilibrium state can be reached at elevated temperatures in solids, this non-equilibrium free energy becomes the equilibrium function of the 'ordinary' Gibbs thermodynamic potential. Only this state corresponds to the only real point of the minimum value of the potential indicated in the fields of Fig. 62.1 *a, b* (this point is described by equations for the equilibrium distribution). All other points on the curves in Fig. 62.1 *a, b* are not equilibrium, and they should be described by the kinetic equations (38.6), (38.7). Or the discussed drawings are schemes that are in no way connected with equilibrium thermodynamics, and sufficiently small deviations of the state parameters require a transition to dynamic models.

Usually it is considered that as an argument (on the abscissa axis) any values can be chosen, as listed above: parameters of state, concentration, coordinates, time, etc. But, this is an inaccurate wording. For relaxation processes, only the 'time' is the natural coordinate, and as a functional connection between the thermodynamic potential and, for example, the concentration, the kinetic equation describing the change in concentration over time is used. In the general case, as an argument along the abscissa axis, a vector of dynamic variables describing the evolution of the system should be deposited. It is important to emphasize that for non-equilibrium processes the fixation of only the state parameters used for the equilibrium state is insufficient to describe the dynamic state of a solid. When calculating the dynamics, it is necessary, as a minimum, to take into account the changes in all pair distribution functions that are directly related to the structural characteristics. As a result, the whole set of conditions leads to the need to explicitly take into account the time factor of the process. Thus, we are talking about the connections $G(t)$ and $v(t)$, but these connections are usually absent.

To the metastable states one can not apply equations of strictly equilibrium systems. The fact that the rigorous Yang–Lee mathematical theory separates metastable states from the equilibrium state of matter means that for any physically realizable metastable state, the non-equilibrium parameter should be clearly formulated. As a rule, this parameter is associated with the existence in the system of diffusion restrictions on the mixing of molecules in its volume or with the existence of high activation barriers for the realization of the chemical transformation.

A correct description of the dynamics of the vapour–liquid equilibrium has long been based on kinetic equations for the distribution function of a new phase in terms of sizes [71–73], and porous systems based on molecular models [12]. The experiment indicates the need to use kinetic equations for the distribution functions of phases in terms of size and for all alloys [74], which have different degrees of dispersion. In describing them, analogous kinetic equations [75, 76] are being introduced (see also [77] – the diffusion processes inevitably require the consideration of the time factor). For the physicochemical analysis of solids, kinetic approaches (usually diffusion type) have not found wide application. The resulting equations are based only on intermolecular interaction potentials of particles and do not contain additional thermodynamic representations.

64. Incorrect use of the coefficient of activity in kinetics

One of the main problems in calculating the rates (liquid-phase, solid-phase, surface) reactions in condensed phases is to take into account the influence of the medium. In the condensed phase, the molecules of the reagents are constantly in the field of action of neighbouring molecules. The change in the internal states of the reagents during the formation of the activated complex (AC) of the elementary process can cause a response in the environment, which in turn affects the course of this process.

The equilibrium properties of non-ideal systems in thermodynamics are actively described by the concept of the activity coefficient (10.7) . The relationship between this coefficient and the molecular theory is discussed in Appendix 3 using the example of associated solutions. The importance of taking into account the real properties of potential functions and the need to take into account correlation effects between interacting molecules in solution is demonstrated. We consider how the equations for the non-ideal mixture change in the presence of associates, when the volume of associates exceeds the volume of the initial components of the mixture, and the accuracy of the traditional methods for accounting for residual contributions in the 'theory of associated solutions'.

Equation (10.8) demonstrates the way in which the reaction system, which was formed at the end of the thirties, was taken into account in the framework of Eyring's TARR [78]. For the activated complex, a similar activity coefficient was introduced, as for the ordinary component of the solution. It was believed that this allowed for taking into account the transition from the gas phase process to the process in the condensed phase. Alternative was the view of M.I. Temkin [79.80] who believed that it is necessary to directly take into account interparticle interactions and their effect on the rate of the elementary stage without using the concept of an activated stage complex (for more details on the history of accounting for the non-ideality of reaction systems, see the review [81]). Temkin demonstrated this with the example of non-uniform reactions [79] at ordinary pressures, and was particularly vividly demonstrated in the description of the process of ammonia synthesis at elevated pressures (more than 300 atm) [80]. For high pressures of the gas phase, the pressures themselves must be replaced by volatility, and the rate constants present in the rates of elementary reactions must be modified by taking into account the effect of the displacement of the adsorption equilibrium on the catalyst surfaces. In [80] this was done through an indirect account of the change in the equation of state of the adsorbed layer, but without introducing the concept of an activity coefficient.

The interpretation of the Eyring school has become widespread in the calculations of the rates of liquid-phase reactions based on the assumption of an equilibrium distribution of the components of the medium around the reagents [82–84]. Equilibrium adjustment of the medium during the formation of AC is possible only with a slow process of activation of the initial reagents, i.e. slow motion of the reaction subsystem from its energy states during the thermofluctuation activation process.

Both points of view existed in different processes. In the theory of surface processes at the gas–solid interface, practically all the models are based on the assumption that the reagents surrounding the reagent in the course of the elementary reaction [56, 81] are in the invariance states; the elementary act is instantaneous and the surrounding molecules, without changing their state, create only an 'external field' in which this process proceeds (although the electronic subsystem of neighboring molecules can respond to the reaction process [85, 86]). Both situations can arise during the kinetic course of the reaction, when there is no diffusion inhibition and the components of the solution are distributed in equilibrium.

A comparison of the two types of non-ideality accounting was first carried out in Ref. [88]. Expressions of the rates of elementary stages for fast reactions with representations of the activated complex as an intermediate state of the reaction subsystem, but without changing the configurations of neighboring molecules are given in Chapter 6, and it is shown that they are self-consistent with a description of the equilibrium state of the entire system.

The reaction rate at an equilibrium state of the environment. In the case of a slow reaction, the characteristic time of the elementary process is much greater than the time for the reorganization of the surrounding molecules. The rapid mobility of the environment leads to an equilibrium distribution of neighbouring molecules around the initial reagents, which corresponds to Eyring's hypothesis [78] about the existence of an activity coefficient. Introducing the notion of an activity coefficient for AC α_i^* (or α_{ij}^*), we mean averaging over all possible equilibrium states of the environment. However, the change of the nearest neighbours of the reagents always changes the potential relief of the reaction and, in principle, another channel for realizing the reaction between the initial reagents becomes possible. Together, this makes the use of thermodynamic concepts contradictory to the dynamics of reaction systems.

In the case of a slow reaction, one can construct an expression for the rate of the reaction stage, in which AC is regarded as an analog of a real molecule (according to [78]) and around which an equilibrium distribution of neighbouring molecules is realized. This was done in [88]: $U_i = K_i^* \theta_i$, $U_i = K_i^* \theta_i$, where $K_i^* = K_i \exp(-\beta \delta F_i)$, the exponential factor K_i^* in the rate constant takes into account the change in free energy during the elementary reaction δF_i of the molecule i (the bimolecular stage was also considered there)

Using the QCA of accounting between the components of the solution, the following equation was obtained for the rate of the monomolecular reaction $A \to B$

$$U_i = K_i\theta_i(S_i^*)^z,\ S_i^* = \exp\sum_{j=1}^{s}\left\{t_{ij}\ln\left[t_{ij}\exp(-\beta\varepsilon_{ij})\right]-t_{ij}^*\ln\left[t_{ij}^*\exp(-\beta\varepsilon_{ij}^*)\right]\right\} \quad (64.1)$$

where $t_{ij} = \theta_{ij}/\theta_i$ and $t_{ij}^* = t_{ij}\exp(\beta\delta\varepsilon_{ij})/\sum_{k=1}^{s} t_{ik}\exp(\beta\delta\varepsilon_{ik})$, and the energy difference $\delta\varepsilon_{ij}$ is determined above in the formulas (49.4), (51.10). As a result, the structure of formula (49.7), (51.9) is preserved, but in it the function S_i, which takes into account the imperfection of the reaction system, changes.

The formulas show [87] that the rates of elementary processes $\delta\varepsilon_{fg}^{A\lambda}$ depend on the difference between the intermolecular interaction of AC and the initial reagents with the environment, which form the values of the activation barriers for each specific local composition. If $\varepsilon_{fg}^{*ij} = \varepsilon_{fg}^{ij}$, then $S_{hg}^{*A} = 1$, and formula (64.1) for U_A has the same form as for ideal reaction systems $U_A^{id} = k_A\theta_A$ (the law of mass action).

The role of the relaxation of the medium can be naturally traced by considering the ratio of velocities in the case of slow (sl) and rapid (ra) reactions: $\eta_A = I_A(\mathrm{sl})/U_A(\mathrm{ra})$. The structure of the relations η can be analytically studied only for an isotropic particle distribution for **s** = 2. In this case, $t_{AA} = 1-2(1-\theta)/(1+b)$, $b = \{1 + 4\theta(1-\theta)[\exp(\beta\varepsilon_{AA})-1]\}^{1/2}$ (here $\varepsilon_{AR} = \varepsilon_{RR} = 0$, where the lower symbol R refers to the solvent).

For small densities, we obtain $\eta_A = \{\exp[-c\theta\ln\theta]/(1+c\theta)\}^z$, where $c = \exp(\beta\varepsilon_{AA}^*)-\exp(\beta\varepsilon_{AA})$, and for large densities $\eta_A = \exp[-z\beta\varepsilon_{AA}(1-\theta)]$. Hence it follows that in the region of small fillings, if $\varepsilon_{AA}^* > \varepsilon_{AA}$ ($c > 0$), then the relaxation of the medium accelerates the reaction rate and this acceleration increases with increasing component A concentration. If $\varepsilon_{AA}^* < \varepsilon_{AA}$ ($c < 0$), then the relaxation of the medium reduces the reaction rate compared to its absence, and with an increase in the concentration of component A this slowing increases. At high densities, the situation is reversed: if $\varepsilon_{AA}^* > \varepsilon_{AA}$, the relaxation of the medium slows down the reaction rate, and if $\varepsilon_{AA}^* < \varepsilon_{AA}$, then – accelerates the reaction rate. Thus, the relaxation of the medium can have a different effect on the reaction rate for small and large reagent densities.

Figure 64.1 shows the calculated rates of fast and slow monomolecular reaction and the value of $-\ln(\eta_A)$ for the whole range of the change in the density of reagent A (for $s = 2$, the medium is represented by other molecules of A and the relaxation consists in the redistribution of molecules around the AC during their migration). These calculations correspond to the rate of non-dissociative desorption on a square lattice ($z = 4$).

Note that $-\ln(\eta_A) = -\ln\left(U_A(\mathrm{sl})/U_A(\mathrm{ra})\right) = \beta\left[\Delta E_A^{\mathrm{ef}}(\mathrm{sl}) - \Delta E_A^{\mathrm{ef}}(\mathrm{ra})\right]$, where $\beta\Delta E_A^{\mathrm{ef}} = -\ln(U_A/U_A^{\mathrm{id}})$. Here, the effective activation energies characterize the degree of deviation of the reaction rate in the non-ideal reaction system compared to the ideal one. Therefore, instead of the dependences $\eta(\theta)$, it is convenient to consider the dependences $\ln(\eta(\theta))$. The inset in Fig. 64.1 shows the concentration dependences of the ratios $T_{AA} = t^*_{AA}/t_{AA}$ that characterize local changes in the distribution of components A due to their migration under the influence of AC for a slow reaction. This ratio tends to unity at $\theta \Rightarrow 1$, for small θ the maximum effect of AC: $T_{AA} = \theta\,\exp(\beta\delta\varepsilon_{AA})$. For a quick reaction $T_{AA}=1$ for all θ.

Figure 64.1 also illustrates the qualitative difference between the concentration dependences of the reaction rates for various relaxation of the medium: in the absence of relaxation, $\ln U_A(\mathrm{ra})$ changes practically linearly with increasing θ, and for equilibrium

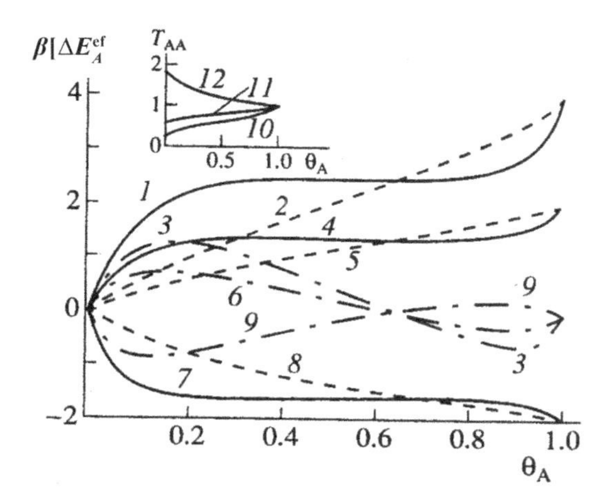

Fig. 64.1. Effective activation energies for slow (1.4.7) and rapid (2.5.8) monomolecular reactions, their difference $\beta[\Delta E_A^{\mathrm{ef}}(\mathrm{sl})–\Delta E_A^{\mathrm{ef}}(\mathrm{ra})]$ (3.6.9) and function $T_{AA}=t^*_{AA}/t_{AA}$ (10–12) (insert); $s = 2$, $\beta\varepsilon_{AA} = 1$, $\varepsilon^*_{AA}/\varepsilon_{AA} = 0$ (1–3,10); 0.5 (4–6, 11); 1.5 (7–9,12) [87].

relaxation of the medium $\ln U_A(\text{sl})$ changes sharply in the regions $\theta < 0.2$ and $\theta > 0.8$ and remains practically constant at $0.2 < \theta < 0.8$.

This general property of the influence of the character of the relaxation of the medium is also preserved for other situations: mono- and bimolecular stages in solutions, which were investigated in [89, 90] (see also below). It can be used as a basis for analyzing the experimental concentration dependences of the logarithms of the reaction rates over a wide range of concentrations.

Dipole system. Both ways of taking into account the effect of the non-ideality of the reaction system in the dipole solvent on the rate of the stages were considered in [90] within the framework of the molecular theory (lattice gas model) [12, 56]. Potential functions of the solvent and reagents take into account short-range Lennard-Jones contributions, which stabilize the system with a dipole interaction, and dipole–dipole contributions, which depend on the orientation of the molecules. As a lattice structure for water simulation (for brevity, we will call so the dipole solvent), we often use the tetrahedral structure of diamond [91, 92], in which the mean-field approximation was used. In [93] this approach was generalized to the case of taking into account spatial correlations between interacting molecules in the quasi-chemical approximation. The latter made it possible to proceed to the calculation of the rates of elementary reactions, which was impossible because of the discrepancy between the description of the equilibrium distribution of molecules and reaction rates with their participation in the mean-field approximation, as in the original model for the dipole solvent [91, 92]. Note that lattice models are relatively often used to calculate the vapour–liquid equilibrium of non-aqueous systems [94,95].

The crystal structure of diamond with the number of neighbors z = 4 was chosen as the lattice. Such a lattice corresponds well to short-range order in the water system and the tetrahedral form of hydrogen bonds. The force field of all molecules (reagents and solvent) was modeled by a superposition of the Lennard-Jones potential for non-electrostatic and dipole–dipole interactions for electrostatic interactions. The dipoles are oriented in space in one of eight discrete directions [93]; i.e. here $\sigma_j = 8$. The energy parameters of the model were found from a comparison with the experimental curve of water stratification [96,97]. Interactions were taken into account up to 4 c.s., details of the calculation are given in [90].

The potential parameters for reagents *A, B, C, D* were chosen so that the rate of interaction of molecule *A* with water would be 2 times less than the water–water interaction at the same molecular arrangement, *B* with water 3 times less, *C* with water 4 times less, *D* with water 5 times less. To ensure that the presence of reagents in the solution had little effect through lateral interactions on the course of the reaction, the concentrations of the components $i = A, B, C, D$ were chosen to be $\theta_i = 0.001$. The calculations were carried out for the case of a low concentration of reagents in order to study the role of dipole solvent molecules. The concentration of water varies from 0 to 0.998 in the case of monomolecular and up to 0.996 in the case of a bimolecular reaction. Thus, the main contribution of the influence of the environment on the reaction rate is exerted by solvent molecules.

The main parameter of the model is the energy of the transition state ε^*_{ij}, which is assumed to be proportional to the binding energy of the molecule with the ground state ε_{ij}: $\varepsilon^*_{ij} = \alpha\varepsilon_{ij}$. The dependences of the reaction rate ratio U on the solvent concentration for different values of the energy parameter $\alpha_{ij} = \varepsilon^*_{ij}/\varepsilon_{ij}$ were calculated, where ε^*_{ij} is the energy parameter of the interaction of the AC formed from the reagent i with the neighbouring particle of the type j. For simplicity, the same α was adopted for all components of the solution. If $\alpha = 1$, the formation of AC does not significantly affect the course of the reaction and the process proceeds as in the case of weakly interacting reaction components (an ideal reaction system). A value of $\alpha = 0$ means that in the transition state, the AC does not interact with the environment, which is an almost unrealistic omission for liquid-phase reactions.

Monomolecular reaction. Figure 64.2 shows how the rate of the monomolecular reaction depends on the concentration of the solvent. The rate U_A of the monomolecular reaction is plotted in relative units as a function of the solvent concentration θ_{H_2O} for fast (solid line) and slow (dots) reactions at 300 K and the values of the parameter $\alpha = 0.25$ (*1*), 0.5 (*2*), 0.75 (*3*). For a fast reaction, the change in the rate as a function of the concentration of the solvent is monotonic and reaches its lowest value when the water concentration reaches its maximum value in the liquid phase. With a slow reaction, an increase in the water fraction leads to a rapid decrease in the reaction rate in the region of low densities and when θ_{H_2O} reaches a value of ~0.2, practical constancy is observed in which the dotted curves intersect solid curves in the region of large θ_{H_2O}. The reason for the difference in the course of the curves for fast and slow reactions is that for these two approximations the weights of the configurations

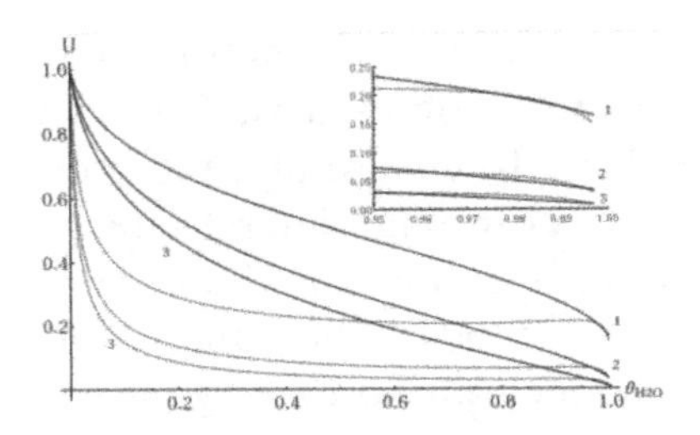

Fig. 64.2. The rate U_A of the monomolecular reaction (in arbitrary units) as a function of the concentration of the solvent θ_{H_2O}; $T = 300$ K, $\alpha = 0.25$ (*1*), 0.5 (*2*), 0.75 (*3*). On the inset – the region of large densities.

of the solvate shells of AC differ in the expression for the reaction rate differently. The limiting values of both expressions for $\theta_{H2O} \to 0$ and $\theta_{H2O} \to 1$ coincide.

The results of the calculation confirm the previous conclusion [87] that there is a significant difference in the concentration dependences of the rates, depending on the nature of the relaxation of the environment. Recall that the difference in reaction rates can reach up to 4–5 times in the region of large and / or small fillings of the solvent. The displacement of the position of θ_{H_2O} for the points of intersection of the curves from the value $\theta_{H2O} \sim 0.5$ [87] to high densities is related to the nature of the dipole–dipole interaction: depending on the mutual orientations of neighbouring molecules, they can both be attracted and repelled. (Whereas in the case of a spherically symmetric potential, the nature of the interaction between the molecules does not depend on their orientations.) The position of the point of intersection is also affected by the interaction energies of the solvent molecules with the reagents.

Bimolecular reaction. Figure 64.3 shows how the rate of the bimolecular reaction depends on the concentration of the solvent. This figure shows the total velocity of U_{AC}, which is the sum $\sigma_A\sigma_C = 64$ of different partial velocities from different locations of reagents *A* and *C*. The calculation in Fig. 64.3 is carried out for $\alpha = 0.5$ (1), 0.75 (2) and 1.5 (3) . The results of comparisons of the normalized values of the rates of the bimolecular reaction *U* are given at $T = 300$ K for two limiting cases of environment relaxation: fast (solid lines) and slow (points) reactions.

As in the case of a monomolecular reaction, in the case of a fast bimolecular reaction we observe a practically monotonic change in the value of the rate, depending on the concentration of the solvent, which reaches its lowest value when the solvent concentration reaches its maximum value. In this case, the solvent inhibits the course of the reaction, making it difficult to meet reagents and increasing the value of the effective activation energy of this stage due to lateral solvent–reagent interactions. For a slow reaction, as the fraction of the solvent increases, the reaction rate decreases much more rapidly (curve 2), then there is a region of a slight change in the velocity, but when θ_{H2O} reaches ~0.8, an increase in the reaction rate occurs, passing through a maximum, reaches a common point at $\theta_{H2O} \rightarrow 1$. In the region of large densities, curves intersect for two types of relaxation of the environment. As above, the reason for the difference in the course of these two types of curves is that for these two approximations the weights of the configurations of the solvate shells of AC differ in the expression for the reaction rate differently.

To illustrate the existence of a point of intersection of curves for fast and slow reactions, a calculation is given for the situation when $\alpha = 1.5$ (curve *3*). In this case, an increase in the concentration of the solvent increases the reaction rate in both cases, however, even here, at high densities, there is an obligatory intersection of the curves. (Both curves 3 are given on a scale reduced by three orders of magnitude to put all the curves in one field.)

The obtained results testify to the generality of the previous conclusions [87] about the significant difference in the concentration dependences of the reaction rates depending on the nature of the relaxation of the environment and the mandatory intersection of curves corresponding to different types of relaxation of the nearest environment. The position of the point of intersection θ_{H2O} is due to the nature of the dipole–dipole interaction and the coupling of reagents with the solvent. According to the general analysis of the lattice gas model (LGM) [81,98], the presence of attraction between the laterally interacting particles can lead to a non-linear

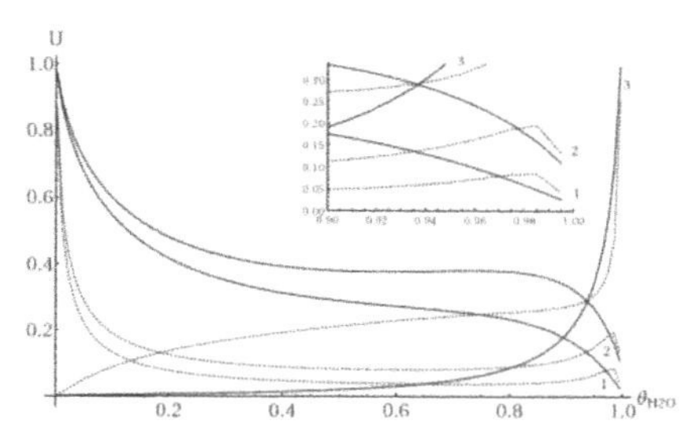

Fig. 64.3. The rate U_A of the bimolecular reaction (in relative units) as a function of the concentration of the solvent θ_{H2O}; $T = 300$ K, $\alpha = 0.25$ (*1*), 0.5 (*2*), 1.5 (*3*). The inset – the region of large densities.

concentration dependence of the rate of the elementary reaction. The average values of the rate shown in Fig. 64.3 are obtained by weighting over all orientations of neighbouring molecules. The nature of the non-linearity of curves *1* and *2* for a fast reaction is small. In contrast, the course of the corresponding curves for the slow relaxation of the environment is much more nonlinear, and gives maxima in the region of large θ_{H2O}. This type of non-linearity is a consequence of phase transitions that are absent in the given system (and it is not present when considering the reaction as a fast process) [81,98]. This indicates that the assumption of an equilibrium distribution of surrounding molecules leads to an non-physical result, i.e. to the distortion of the properties of a real system.

Self-diffusion coefficient. The process of self-diffusion can be formally written in the form of a bimolecular reaction $(H_2O)_f + V_g = V_f + (H_2O)_g$, in which the indexes of adjacent sites f and g appear in the form of indices, and the symbol V_g denotes a free site with the number g. Thus, the hopping of the water molecule is a particular case of the bimolecular reaction proceeding at the sites f and g. The theory makes it possible to find the value of the coefficient of self-diffusion of water and to study its dependence on temperature (Fig. 64.4) [90].

The calculation is carried out under the condition that the considered hopping of a molecule into an adjacent vacancy occurs much faster than a local restructuring of the whole environment. The transition to the processes of thermal motion of molecules is associated with a graphic illustration of the correctness of the hypothesis of a weak change in the state of neighbours during the jump of the molecule under consideration, according to the Frenkel hypothesis [14]. The assumption [78] about the equilibrium distribution of neighbours would be in contradiction with the logic of the process: that the selected molecule should be moved so that similar neighbours quickly adapt to the transition state of this stage. That is, the set of neighbour jumps is necessary to make one hop of the molecule in question.

For the energy parameters of water molecules found in [93], the optimal value of the parameter $\alpha = 0.65$ is determined from the comparison of the constructed theoretical curve with the experiment [99]. (It should be noted that the same experiment can also be satisfactorily approximated by the molecular dynamics method [100].) The good agreement obtained indicates the adequacy of the lattice model and the validity of its assumptions. The found value

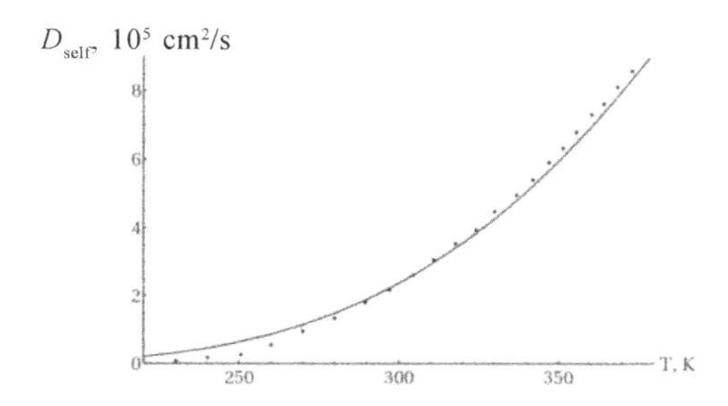

Fig. 64.4. Comparison of the temperature dependence of the coefficient of self-diffusion of water molecules: experiment (points) [99] and calculation of this work (line).

of α indicates a relatively small activation energy, which must be overcome by the water molecule during the jump in the course of this 'reaction'.

Thus, the comparison of the rates of relaxation of the medium to the concentration dependences of the rates of mono- and bimolecular reactions (in the kinetic regime of their course) leads to differences in reaction rates up to 4–5 times and they have a different course of concentration dependences. The considered equations of reaction rates refer to two limiting cases: $\tau_{rel} \ll \tau_{reaction}$ (slow reaction) and $\tau_{rel} \gg \tau_{reactions}$ (fast reaction), which are determined independently by the characteristic migration times of τ_{migr} and rotations τ_{rot} of the solvent molecules in the pure solvent $\tau_{rel} = \max(\tau_{migr}, \tau_{rot})$ and reactions $\tau_{reaction}$ in an inert solvent (or vacuum). Analysis of the concentration dependences of the reaction rates in a wide range of solvent concentrations θ_R (0.3–$0.7 < \theta_R < 1$) allows one to discriminate which of the two situations is realized in the experiment. If there are less strict ratios $\tau_{rel} < \tau_{reaction}$ and $\tau_{rel} > \tau_{reaction}$, then two situations are possible, caused by the imperfection of the reaction system. The first situation is that when the change in ΔE^{ef} preserves these relations. Then, for the same characteristic concentration dependences of reaction rates on θ_R, these relations can be discriminated. The second situation is such that when the times τ_{rel} and $\tau_{reactions}$ become commensurable, and to take this fact into account it is necessary to use more exact equations [56, 88], which take into account the equilibrium with respect to the degrees of freedom and the disequilibrium with respect to the remaining degrees of freedom.

The traditional idea of the existence of the chemical potential of AC in dense phases is in contradiction with the real times of molecular motions of molecules and with the procedure of statistical averaging of molecular distributions. Typical times of translational displacements and changes in molecular orientations in dense phases are, as a rule, larger than the characteristic time of chemical transformation, estimated on the order of 10^{-13} s [82,83]. Actually, the chemical reaction associated with the passage of the top of the activation barrier occurs in the unchanged field of neighboring molecules that form a local barrier. The notions of equilibrium relaxation of the molecular environment are unlikely, since the relaxation rate would have to be on the order of 10^{-14} s and less, which is physically unrealistic. Analysis of the molecular model does not exclude the possibility of realizing the equilibrium distribution of atomic and molecular polarization, which has characteristic relaxation times of less than 10^{-13} s. In this case, the model must be supplemented by taking into account the polarization contributions, which are discussed in [101, 102].

Concluding the analysis of thermodynamic interpretations by discussing the question of the inadmissibility of introducing parameters into any thermodynamic equations, it should be emphasized that the plausibility of the expression for the reaction rate constant through the internal states of the reagents and the transition state in the theory of absolute reaction rates (k'_0) is associated with the influence of environments through activity coefficients. For equilibrium characteristics, activity coefficients are a formal way of expressing experimental data. But for a transitional state such an interpretation gives a qualitatively wrong result. The relaxation times of the transition state are characterized by τ_{temp}. The calculation of the activity coefficient for an activated complex implies the realization of averagings over all configurations with a characteristic relaxation time τ_{den}. To use a_M^* we required the time τ when the activated complex repeatedly terminates the reaction, $\tau > \tau_{den} \gg \tau_{temp}$. This leads to qualitative distortions in the magnitude of the reaction rate since the introduction of γ^{**} violates the meaning of the activated complex as a transition state [78,82,83]. Calculations [87,89,90] illustrate this discrepancy. This factor is principally important for the averaging of the kinetic equations in obtaining the mean values $< S_{fg}^{(r,n)} \theta_{fg} >$ of the transported moments that enter into the expressions for the kinetic equations of Chapter 5.

References

1. Thomson W.T., Phil. Mag. 1971, V. 42, P. 448.
2. Yang C.N., Lee T.D., Phys. Rev. 1952. V. 87. P. 404.
3. Lee T.D., Yang C.N., Phys. Rev. 1952. V. 87. P. 410.

4. Hill T.L., Thermodynamics of Small Systems. Part 1. New York Amsterdam: W. A. Benjamin, Inc., Publ., 1963. Part 2. 1964.
5. Huang K.. Statistical Mechanics. Moscow, Mir, 1966. 520 p.
6. Tovbin Y.K., Zh, fiz. khimii. 2010. V. 84. No. 10. P. 1882. [Rus. J. Phys. Chem. A, 2010, V. 84. No. 10. C. 1717].
7. Tovbin Y.K., Fiz.-khim. poverkhnosti i zashchita materialov. 2010. V. 46. No. 3. P. 261. [Protection of Metals and Physical Chemistry of Surfaces, 2010, Vol. 46, No. 3, P. 309].
8. Gibbs J.W., Thermodynamics. Statistical Mechanics. Moscow, Nauka, 1982.
9. Rice O.K., J. Phys. Chem. 1927. V. 31. P. 207.
10. Prigogine I., Defay R., J. Chem. Phys. 1949. V. 46, p. 367.
11. Ono S., Kondo C. Molecular theory of surface tension. Moscow, IL, 1963. [Handbuch der Physik, Vol X (Springer) 1960].
12. Tovbin Yu.K., The Molecular Theory of Adsorption in Porous Solids, Moscow, Fizmatlit, 2012. [CRC Press, Taylor & Francis Group, 2017].
13. Anisimov M.P., Usp. khimii. 2003. V. 72. No. 7. P. 664.
14. Frenkel J., Kinetic theory of liquids., Publishing House of the USSR Academy of Sciences, 1945.
15. Timoshenko S.P., Goodier J., Theory of elasticity. Moscow, Nauka, 1979.
16. Lame G., Lecons sur la Theorie ... d el'Elasticite, Paris, 1852.
17. Landau L.D., Lifshits E.M. Theory of elasticity. T.7. (page 11, problem 2).
18. Tovbin Yu.K., Zh. fiz. khimii. 2017. V. 91. No. 9. P. 1453. [Rus. J. Phys. Chem. A, 2017, V. 91, No. 9, P. 1621].
19. Rowlinson, J., Widom B., Molecular theory of capillarity. Moscow, Mir, 1986. p. [Oxford: ClarendonPress, 1982].
20. Adamson A. W., The Physical Chemistry of Surfaces, Mir, Moscow, 1979. [Wiley, New York, 1976].
21. Buff F.P., J. Chem. Phys. 1955. V. 23. P. 419.
22. Kondo S.J., Chem. Phys. 1956, V. 25, p. 662.
23. 23. Rusanov A.I.. Phase equilibria and surface pnehomena. Leningrad, Khimiya, 1967.
24. Tovbin Yu.K., Rabinovich A.B., Izv. AN, Ser. khim., 2009. No. 11. P. 2127. [Russ. Chem. Bull. 58, 2193 (2009)].
25. Tovbin Yu.K., Zh. fiz. khimii. 2010. V. 84. No. 2. P. 231. [Rus. J. Phys. Chem. A, 2010, V. 84. No. 2. P. 180].
26. Tovbin Yu.K., Rabinovich A.B., Zh. fiz. khimii. 2013. V. 87. No. 2. P. 337. [Russ. J. Phys. Chem. A, 2013, V. 87, No. 2, P. 329].
27. Tovbin Yu.K., Rabinovich A.B., Izv. AN. Ser. khim. 2010. No. 4. P. 663. [Rus. Chem. Bull. 2010. T. 59. No. 4. P. 677].
28. Bykov T.V., Shchekin A.K., Kolloid. zh. 1999, V. 61, P. 164.
29. Bykov T.V., Shchekin A.K., Neorgan. mater, 1999, V. 35, P. 759.
30. Bykov T.V., Zeng X.C., J. Chem. Phys., 1999, V. 111, P. 3705.
31. Bykov T.V., Zeng X.C., J. Chem. Phys., 1999, V. 111, P. 10602.
32. Tovbin Yu.K., Rabinovich A.B., Izv. AN. Ser. khim. 2010. No. 4. P. 839. [Russ. Chem. Bull. 59, 857 (2010)].
33. Hill T.L., Thermodynamics of Small Systems. Part 2. New York Amsterdam: W. A. Benjamin, Inc., Publ., 1964.
34. Van der-Waals I.D., Konstamm F., Courses in thermostatics. Moscow, ONTI, 1936.
35. Bakker G., Kapillaritat und Oberflachenspannung, Handbuch der Experim-ental physik, Bd. VI, Leipzig, 1928.

36. Tolman R.C., Journ. Chem. Phys. 1948, V. 16, P. 758.
37. Mason E.A., Spurling T.H., The Virial Equation of State. Moscow, Mir, 1972. [The International Encyclopedia of Physical Chemistry and Chemical Physics. Topic 10. The Fluid state. V. 2].
38. Hirschfelder J. O., Curtiss Ch. F., Bird R. B., Molecular theory of gases and liquids. – Moscow: Inostr. Lit., 1961. – 929 p. [Wiley, New York, 1954].
39. Hill T.L.. J. Chem. Phys. 1951. V. 19. P. 261.
40. Hill T.L.. J. Chem. Phys. 1952. V. 20, P. 141.
41. Tonks L., Phys. Rev. 1936, V. 50, P. 955.
42. Pawlow P. Z., Phys. Chem. (Munich) 1909, V. 65, P. 548.
43. Couchman P.R., Jesser W.A., Nature. 1977, V. 236, P. 481.
44. Sirotin Yu.I., Shaskol'skaya M.P., Fundamentals of crystal physics. Moscow, Nauka, 1979.
45. Landau L.D., Zh. Eksp. Teor. Fiz. 1937. V. 5. P.627.
46. Tolman R.C., J. Chem. Phys. 1949, V. 17, P. 333.
47. Kuni F.M., Vestnik LGU. 1964. N. 22. P. 3.
48. Kuni F.M., Phys. Lett. A. 1968, V. 26, p. 305.
49. Kuni F.M., Rusanov A.I., Dokl. AN SSSR. 1967. V. 174. No. 2. P. 406.
50. Landau L.D., Lifshitz E.M., Statistical Physics. V. 5. Moscow, Nauka, 1964.
51. Tovbin Yu.K., Komarov V.N., Zaitseva E.S., Zh. fiz. khim. 2016. V. 90. No. 10. P. 1570. [Rus. J. Phys. Chem. A, 2016, V. 90, No. 10, P. 2096].
52. Naghizadeh J., Rice S.A. J. Chem. Phys. 1962, V. 36, P. 2710.
53. Borovskii, I.B., Gurov, K.P., Machukova, I.D., Ugaste, Yu.E., Processes of mutual diffusion in alloys. Moscow, Nauka, 1973.
54. Bokshtein, B.S., Bokshtein, S.Z., Zhukhovitskii, A.A., Thermodynamic and Kinetic of Diffusion in Solids, Moscow: Metallurgiya, 1974.
55. Gurov B.A., Kartashkin B.A., Ugaste Yu.E., Mutual diffusion in a multiphase metallic system. Moscow, Nauka, 1981.
56. Tovbin Yu. K., Theory of physical and chemical processes at the gas-solid interface. – Moscow: Nauka, 1990. – 288 p. [CRC, Boca Raton, Florida, 1991].
57. Hill T.L., Statistical Mechanics. Principles and Selected Applications. – Moscow: Izd. Inostr. lit., 1960. – 486 p. [N.Y.: McGraw–Hill Book Comp. Inc., 1956].
58. Properties of elements. Directory, ed.. M.E. Drits. Moscow, Metallurgiya, 1985.
59. 59. Fisher I.Z., The problem of many bodies and plasma physics. Moscow, Nauka, 1967.
60. Raist P. Aerosols. Introduction to the theory. Moscow, Mir, 1987.
61. Chandrasekar S., Liquid crystals. Moscow, Mir, 1980.
62. Tager A.A.. Polymer chemistry. Moscow, Khimiya, 1978. 544 p.
63. Kireev V.A., Physical chemistry course. Moscow, Khimiya, 1975.
64. Kubo R. Thermodynamics. Moscow, Mir, 1970.
65. Panin V.E., Egorushkin B.E. Fiz. mezomekhanika. 2008. V. 11. No. 2. P. 9.
66. Leontovich A.M., Zh. Eksper. Teor. Fiz., 1938. V. 8. V. 844.
67. Bazarov I.P., Thermodynamics. Moscow, Vysshaya shkola. 1991.
68. Leontovich A.M. Introduction to thermodynamics. Statistical Physics. Moscow, Nauka, 1983.
69. Ionov V.N., Selivanov V.V. Dynamics of destruction of a deformed body. Moscow, Mashinostroenie, 1987.
70. Tovbin Yu.K., Titov S.V., Komarov V.N., Fiz. Tverd. Tela. 2015. V. 57, No. 2. P. 342. [Physics Solid State, 2015, V. 57, No. 2, P. 360].
71. 71. Fuchs N.A. Mechanics of Aerosols. Moscow, Khimiya, 1959. 500 p.

72. Lushnikov A.A., Sytygin A.G., Usp. khimii. 1976. V. 45. P. 385.
73. Lushnikov A.A., Phys. Rev. E. 2007. V. 76. P. 011120.
74. New materials. ed. Yu.S. Karabasov. Moscow, MISA, 2002.
75. Zhilyaev A.P., Pshenichnyuk A.I., Superplasticity and grain boundaries in ultrafine-grained materials. Moscow, Fizmatlit, 2008.
76. Chuvil'deev V.N. Non-equilibrium grain boundaries in metals. Theory and applica¬tions. Moscow, Fizmatlit, 2004.
77. Aaronson H.I., et al., Mechanisms of Diffusional Phase Transformations in Metals and Alloys. CRC Press, Taylor & Francis Group, Boca Raton, FL. 2010.
78. Glasston S., Laidler K.J., Eyring H., Theory of absolute reaction rates. Moscow, IL, 1948 [Princeton Univ. Press, New York, London, 1941].
79. Temkin M.I., Zh. fiz. khim. 1941. V. 15. P. 296.
80. Temkin M.I., Zh. fiz. khim. 1950. V. 24. P. 1312.
81. Tovbin Yu.K., Progress in Surface Sci. 1990. V. 34. No. 1-4. P. 1-236.
82. Entelis S.G., Tiger R.L. Kinetics of reactions in the liquid phase. Moscow, Khimiya, 1973.
83. Melvin-Hughes E.A., Equilibrium and kinetics of reactions in solutions. Moscow, Khimiya, 1975.
84. Marcus R.A., Ann. Rev. Phys. Chem. 1964. V. 15. P. 1.
85. Levich V.G., Itogi nauki. Elektrokhimiya. Moscow: VINITI, 1967. P. 5.
86. Dogonadze R.R., Kuznetsov A.M., Itogi nauki i tekhniki. Kinetika i kataliz. V. 5. Moscow, VINITI, 1978. P. 5.
87. Tovbin Yu.K., Votyakov E.V., Zh. fiz. khimii. 1997. V. 71. No.1. P. 271. [Russ. J. Phys. Chem. 1997. V. 71. No. 2. P. 214].
88. Tovbin Yu.K., Zh. fiz. khim. 1996. V. 70. No. 10. P. 1783. [Russ. J. Phys. Chem. 1996. V.70. No. 10, C. 1655].
89. Tovbin Yu.K., Titov S.V., Sverkhkriticheskie flyuida: teoriya i praktika. 2011, V. 6, No. 2. P. 35. [Rus. J. Phys. Chem. C, 2011. V. 5. No. 7. P. 1135].
90. Tovbin Yu.K., Titov S.V., Zh. fiz. khim. 2013. V. 87. No. 2. P. 205. [Rus. J. Phys. Chem. A, 2013, V. 87, No. 2, P. 185].
91. Bell G.M.. J. Phys. S. 1972. V. 5. No. 9. P. 889.
92. Bell G.M., Salt D.W., J. Chem. Soc: Faraday Trans. Pt. 2. 1976. V. 72. No. 1. P. 76.
93. Titov S.V., Tovbin Yu.K., Izv. AN. Ser. khimii. 2011. No. 1. P. 12. [Rus. Chem. Bull. 2011. V. 60, No. 1, P. 11].
94. Smirnova N.. Molecular theory of solutions. Leningrad. Khimiya, 1987.
95. Prausnitz J.M., et al., Molecular thermodynamics of fluid-phase equlibria. 2nd Ed. New Jersey: Prentice-Hall Inc. Englewood Cliffs, 1986.
96. Water. A comprehensive treatise. Ed. F. Franks. New York - London: Plenum, 1972. V. 1.
97. Eisenberg, D., Kautsman V. Structure and properties of water. Leningrad, Gidrometeoizdat, 1975.
98. Tovbin Yu.K., Dynamics of gas adsorption on non-uniform solid surfaces. Eds. W.Rudzinski, W.A. Steele, G. Zgrablich. Amsterdam: Elsevier, 1996. P. 240.
99. Angell C.A., Water: A Comprehensive Treatise, F. Franks, ed. 1978. V. 7. P. 23.
100. Malenkov G.G., et al., J. Molec. Liquids. 2003. V. 106. No. 2–3. P. 179.
101. Misurkin I.A., Titov S.V., Zh. fiz. khimii. 2008. V. 81. P. 1781. [Russ. J. Phys. Chem. A 82, 1672 (2008)].
102. Tovbin Yu.K., Zh. fiz. khim. 2014. V. 88. No. 11. P. 1752. [Rus. J. Phys. Chem. A, 2014, V. 88, No. 11, P. 1932].

Conclusion

Statistical thermodynamics answered the questions posed in the Preface about the limitations of the use of the equations of macroscopic thermodynamics in the description of small systems. The molecular theory singled out three characteristic sizes of the radius of spherical drops (as the simplest example of a small body), corresponding to:

(1) the beginning of the process of appearance of a dense phase with a surface tension $\sigma(R_0) \geq 0$, $R_0 \sim 10\lambda$, where λ is the average distance between adjacent molecules of the drop. The minimum phase size R_0 allowed to unambiguously separate the notions of clusters and droplets (as small phases). The constructed theory of small systems made it possible to introduce a strict statistical definition of the 'embryo' of the new phase (by the condition $\sigma = 0$), which is a criterion distinguishing between the concepts 'cluster/associate' or 'small phase'.

(2) the region of applicability of the thermodynamic description of the surface tension of a droplet, when for $R > R_{t2} \sim 90\lambda$ the discreteness of the matter and the contributions of spontaneous fluctuations can be completely neglected. When $R_{t1} < 41\lambda$, density fluctuations must be taken into account. The region of thermodynamic description for bulk phases without an allowance for boundary contributions is given by analogous quantities $R_{t1}^{(V)} = 17\lambda$ and $R_{t2}^{(V)} = 29\lambda$. The theory has shown that thermodynamics can not be applied to small systems and to the calculation of the surface tension of any interfaces. Although statistical mechanics limits the applicability of thermodynamics beyond the phase approximation (for $R < R_t$), but extends this approximation to small systems for all $R > R_0$.

(3) large droplet size regions, in which surface tension values are close to the bulk value, $R_b \sim 10^2 \div 10^3\lambda$; at $R > R_b$, the dimensional dependence of the surface tension $\sigma(R) = \sigma_{\text{bulk}}$ can be neglected.

Statistical thermodynamics clarifies and restricts various thermodynamic constructions in the field of non-equilibrium processes. From the standpoint of the non-equilibrium theory, the artificiality of the concept of 'passive forces' laid down by Gibbs in the thermodynamics for bulk phases, which reflects the level of knowledge of those years with respect to the chemical kinetics and properties of solids, but not the violation of any laws of thermodynamics, was shown. Also, the principle of deriving criteria for estimating the yield of system states from the condition of local equilibrium was developed. This makes it possible to correctly relate the description of equilibrium and non-equilibrium processes, which is especially important for ensembles of small solids, which have been actively investigated recently (and which currently have virtually no thermodynamic interpretation). The principal importance of taking into account the relaxation times of various properties and the incorrectness of using the concept of the coefficient of activity of the activated complex in the theory of absolute reaction rates are noted.

The formulation of the list of constraints completes the construction of classical thermodynamics, since it specifies the field of its application and limits its use to small systems.

We should separately dwell on the problem of the Kelvin equation (KE). The analysis showed that this equation can not be derived either from the equilibrium or from the kinetic molecular theory. Historically, it has become widespread in all problems with a curved interface: it is used for both equilibrium and dynamic conditions. When discussing the conditions for its application, one can only speak of deviations between the calculated values of the KE and the molecular theory under static conditions. Today, KE is used for all droplet sizes, ranging from the micron range to the nanometer range. The natural criterion for the use of KE is the difference between $P(R)/P_s$ from a similar value in the molecular theory. Such a criterion is the dimensionless parameter $\psi = 2\beta\sigma V_0/R$. The smaller the deviation of the parameter ψ from zero, the more 'justified' is the use of KE the less it distorts the molecular characteristics.

Here it is necessary to separate two situations: the problem of forming an isolated new phase in which the main role is played by intermolecular interactions and the problem of the form of a liquid film near the surface of a solid body in which the surface potential plays the main role, like any other external field (Gibbs considered the action of the gravitational field on surface tension).

In the first problem, the deviation from the KE appears only due to the size dependence of the surface tension $\sigma(R)$ (provided there are no external fields).

In the second problem, the distribution of the film and the formation of its meniscus is the result of the combined action of the surface and intermolecular potentials, and the difference in the shape of the meniscus from the circular one, embedded in the KE, largely depends on the shape of the surface. In the case of cylindrical pores, the form of the vapour–liquid interface varies greatly from the pore size: at small nanometer diameters, the meniscus can be approximated by a spherical shape, whereas as the diameter increases, the meniscus flattens and tends at the centre of the pore to a flat shape, partially retaining a curved shape (convex or concave, depending on how the interaction potential of the liquid molecules with the walls is) only near the walls (reference [12] of Chapter 7). Therefore, the analysis of capillary condensation in the pores with the help of KE is conditional – for the use of KE the shape of the meniscus should remain spherical.

Parameter ψ corresponds to both situations. The increase in R equally flattens the phase boundary of an isolated small body in the bulk phase and, as indicated above, the meniscus in the pores. In the first case, deviations of $P(R)/P_s$ in terms of KE due to differences in surface tension correspond to a large range of sizes from 20 nm at temperatures near the melting point to 50 nm at elevated temperatures, but outside the critical region.

In the second case, the specific results of the comparisons of the KE and the theory in terms of the $P(R)/P_s$ values for the simplest gases (nitrogen, argon and others, which obey the so-called law of the corresponding states) show: at a diameter of 50 nm, the deviations from the molecular theory are 1%, at 12 nm ~ 10%, and at 4.5 nm the difference is ~100% (*and in general is not applicable* at sizes less than 4.1 nm!) (Ref. [12] Chapters 7). Thus, the application of the KE here is limited from below by values of the order of 20 nm.

The above numbers refer to weakly interacting molecules. As σ increases, the numerator of the parameter y increases accordingly, and similar deviations refer to large values of the radius R. So, to compare the characteristic values, it can be pointed out that the above characteristic values of the deviations of the KE and the theory for argon should be attributed to the large linear size of the aluminum droplets in 2.6, in 3.8, mercury in 7.9 times. Thus, the lower limit of the applicability of the KE for atoms with a strong attraction can exceed 100 nm and enter the submicron range. In these qualitative estimates, of course, temperature effects affecting the values of the parameter ψ and the magnitude of the surface tension depending on T and R are not taken into account, of course. Therefore, the value of the parameter, permissible for using the approximation with the help of KE, must be specified in concrete conditions.

The analysis of the foundations of classical thermodynamics from the point of view of statistical thermodynamics has often confirmed two opinions in the literature on the place of thermodynamics among other disciplines. Most often quoted are the words of A. Einstein, 1949, "A theory is the more impressive the greater the simplicity of its premises, the more different kinds of things it relates, and the more extended its area of applicability. Therefore the deep impression that classical thermodynamics made upon me. It is the only physical theory of universal content which I am convinced will never be overthrown, within the framework of applicability of its basic concepts"

The analysis showed that the first three stages (out of four) of thermodynamics introduced in the Preface are not subject to doubt – when moving from small bodies to macroscopic dimensions, all the positions introduced by Gibbs in the formulation of the phase approximation for describing non-uniform systems are completely satisfied. However, the fourth stage of classical thermodynamics, which pertains to small systems, can not be attributed to its basic propositions.

Also popular is the opinion of J.W. Gibbs, 1902: "But although, as a matter of history, statistical mechanics owes its origin to investigations in thermodynamics, it seems eminently worthy of an independent development, both an account of the elegance and simplicity of its principles, and because it yields new results and places old truths in a new light in departments quite outside of thermodynamics". The words about the presentation of 'old truths in a new light' proved to be prophetic not only in relation to the numerous applications of statistical physics at the present time, but also in relation to the analysis of the very foundations of thermodynamics for small systems from the positions developed by Gibbs to the methods of statistical description of the thermodynamic properties of systems. The fourth stage of thermodynamics, connected with the approach to small systems, turned out to be incorrect. The problem of the thermodynamic description of curved surfaces, which began with the Kelvin equation (1871), arose from the formal transfer of mechanical representations to thermodynamics. This is a purely historical circumstance, because then mechanical ideas prevailed about the curved boundaries of the phase separation, and they were in no way connected with the notion of chemical potential, which was not yet introduced. During the development of Gibbs' thermodynamics (1878), it was impossible to foresee those mathematically rigorous results of statistical mechanics that would indicate the incorrectness of the thermodynamic approach to small systems. We are talking about the Yang–Lee condensation theory (1952) and the existence of stable solutions to the droplet concentration profile without taking into account, and taking into account the influence of the Laplace equation (2010) (both types of solutions satisfy the Yang–Lee theories). This radically changes the situation with the treatment of small systems – their description is possible only within the framework of statistical thermodynamics.

The molecular theory made it possible to reveal the main reason for the difference between the molecular statistical and thermodynamic approaches in describing small systems. The reason for the incorrectness of the application of thermodynamics is based on neglecting the consideration of the real properties of systems connected under isothermal conditions with the relaxation times of impulse and mass transfer. This is a non-trivial reason, caused by the essence of thermodynamics itself: its bases are based solely on experimental data. If they are violated, then thermodynamics can not give an adequate description of the properties of systems. Equilibrium thermodynamics ignores the well-known relationships between the relaxation times of impulse transfer and mass, assuming that for large times this does not play a role. However, the molecular theory shows that the trajectory of the system's output to the limiting states (at large times) during phase transformations is ambiguous, and the difference in the relaxation times for impulse and mass transfer plays a fundamental role, since it leads to two different final states.

The experiment shows that the impulse relaxation times are, as a rule, much less than the relaxation times of the mass. This automatically excludes metastable states from the phase equilibrium of the vapour–liquid system, which, in contradiction to the Yang–Lee condensation theory, is present in the Kelvin equation, and leads to the appearance of equilibrium drops. This once again confirmed the basic position of thermodynamics that the violation of the conditions relating to the properties of experimental systems should lead to a violation of the conditions for the applicability of thermodynamic relations. Formally, for macroscopic systems, ignoring the role of relaxation times in the transfer of impulse and mass, does not affect the conditions for complete phase equilibrium at asymptotically large times. Here the term 'asymptotically' large times plays a key role. (For simple liquid systems, they are realized, and for solid ones, not always.) The relaxation times of the dynamic variables of the system form a hierarchy of relaxation times of different dynamic variables with respect to a given time scale in the experimental system under study. 'Automatically' for this time scale, the size scale of the subsystem and the type of dynamic variables that corresponds to the condition of internal complete equilibrium for these dynamic variables are formed, and the sizes of subsystems in which complete equilibrium is not realized for each of the remaining dynamic variables of the entire system. Therefore, the concept of complete equilibrium must always be strictly defined in relation to a fixed time scale. The discussed choice of the time scale was omitted by Gibbs (1875) from an analysis of the conditions for generalizing the principle of mechanical equilibrium to thermodynamic variables. Small systems were considered by Gibbs later in 1878 without revision of the previously formulated provisions.

At one time the development of classical thermodynamics served as a catalyst for the development of science. Classical thermodynamics remains today the foundation for all natural sciences, in which it is necessary to take into account the conservation of energy and the study of thermal effects. The internal energy of matter is associated with different types of interactions on different spatial scales and questions related to thermodynamics will arise in the case of macroscopic sizes of the systems under study. The properties of particles themselves can be related to their quantum nature and through intermolecular interactions in ensembles lead to the corresponding dependences of thermodynamic functions on their quantum nature. With a decrease in the size of small systems classical thermodynamics loses its validity and it is necessary to take into account the size effects indicated in this book. The found size criteria are characteristic values that reflect the properties of short-range potentials. The specifics of Coulomb charges were not discussed in the book. According to the most rough estimates, all the found size criteria will be increased approximately two times, because in systems with charge components they are neutral particles.

Thermodynamics is not an interpolation tool because of its model-free field of science. The introduction of any molecular specificity into it distorts its essence and leads to quasi-thermodynamic approaches, more precisely to 'hybrids', creating the illusion of scientific consideration. The term thermodynamic model is incorrect because the thermodynamics reflects the experiment to the full extent of simultaneous manifestation in it of all molecular effects. Any molecular interpretation is possible only on the basis of a molecular model, and its applicability should be tested by comparison with experiment and have a sufficient correct basis (that is, do not be too crude) to allow some extrapolation of results outside the parameter range (used in the comparison with experiment).

Today, the important role of thermodynamics for the organization of the experiment and its interpretation remains: it is control over the correctness of the use of the law of conservation of energy in real systems and the 'storage' of fundamental information on thermodynamic characteristics and the organization of experimental measurements to measure new and refine old thermodynamic characteristics.

For many solid-phase systems, the information extracted from the experiment today is approximate, the degree of approximation is difficult to control – there is not always evidence of achieving complete equilibrium. Achieving equilibrium should reflect the dynamic nature of processes that ensure the constancy of the values of the parameters in time and the independence of the values of the characteristics from the path of the transition. At low T, the system necessarily goes into a solid state, and for monitoring in them, measurements of the relaxation times of thermophysical characteristics, which are very few, are needed. The binding to solid state states follows from Nernst's heat theorem, which establishes a common reference point for all thermodynamic characteristics.

The recent development of the statistical mechanics of 'three-aggregate' state equations and their boundaries has made it possible to reach the same level of generality in describing three aggregate states and different phases, both in equilibrium and in non-equilibrium conditions, which were formerly characteristic only of thermodynamics. This opens the possibility of transition to a correct description of processes involving small systems, providing their interpretation of the thermophysical and thermodynamic characteristics, provided it is supplemented by modern statistical thermodynamics.

Appendix 1

Metastable drops

Chapter 3 shows that the molecular theory leads to the existence of equilibrium drops, which are not found in thermodynamics: they do not have a pressure jump inside the transition region and the Laplace equation does not need to be calculated for the concentration profile of the vapour–liquid interface. The theory makes it possible to obtain the equation of state both within the coexisting phases and the transition region between the phases. This eliminates the need to draw information about the state of the boundary from an independent condition of mechanical equilibrium (not related to the current chemical potential values). The mechanical equilibrium at the boundary of the equilibrium drop is ensured by the equality of the chemical potential and the equation of state in each transition monolayer. This kind of detailed information is impossible in thermodynamics.

Today, all the work on drops uses the Laplace equation, as in the derivation of the Kelvin equation, so the goal of this application is to show that the theory allows us to include in our consideration the traditional condition of mechanical equilibrium (via the Laplace equation). This allows us to compare the thermodynamic functions of equilibrium and metastable drops. This appendix gives an account of the traditional thermodynamics of metastable drops and its microscopic version in a discrete approach (including the 4th method of determining the surface of tension), which allows one to simultaneously discuss the differences between continual and discrete descriptions. In Chapter 7, both types of drops are compared with thermodynamic metastable drops, which follow from the Kelvin equation.

First, we consider the traditional thermodynamic description of a drop. In order not to complicate the formulas, we follow the exposition [1, 2], which is oriented only to a spherical boundary.

Determination of surface tension. A spherical surface (bubble or drop) is the only stable in the absence of an external field. We consider the situation with curved surfaces by the example of a drop of liquid in the surrounding vapour. We consider a closed system consisting of two phases α (internal) and β (external), separated by a spherical dividing layer, that is, the case of a transition zone of constant curvature. Consider a limited part of the system enclosed in a conical vessel, such as shown in Fig. A1.1, where ω and r denote the solid angle of the cone and the radial coordinate, measured from the vertex of the cone 0 [1], respectively.

The vessel consists of the part of the cone, enclosed between $r = R_\alpha$ and $r = R_\beta$, where $R_\alpha < R_\beta$. We assume that the system consists of (s–1) independent components, there are no external fields, and the centre of curvature of the spherical surface layer lies at the point 0. Then any intensive property of the system can depend only on r. We choose the sphere $r = a$ (dotted line in Fig. A1.1) as the dividing surface. It will divide the total volume V into two volumes V_α and V_β, equal to, respectively

$$V_\alpha = \omega\left[a^3 - (R_\alpha)^3\right]/3, \quad V_\beta = \omega\left[(R_\beta)^3 - a^3\right]/3, \qquad \text{(A1.1)}$$

where the area of the dividing surface will be equal to: $A = \omega a^2$.

It is assumed that while the system is in mechanical equilibrium the pressure P_α in the internal phase α can not equal the pressure P_β in the outer phase β. Inside the bulk phases α and β, the pressures P_α and P_β are constant. However, the method can not be applied to the case of spherical drops so small that it is impossible to achieve uniform properties even at the centre of the phase α.

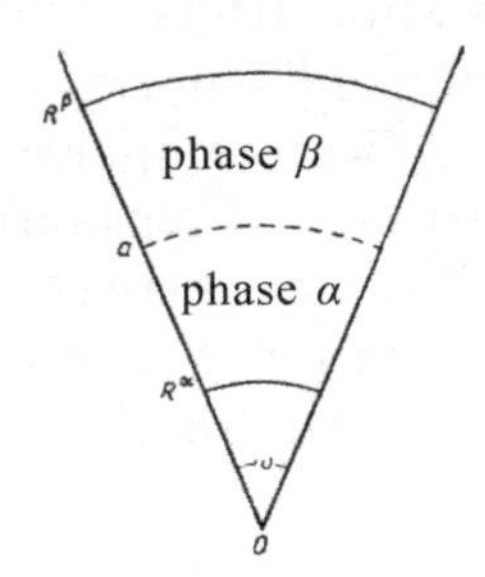

Fig. A1.1. Scheme for determining the surface tension of a drop.

Imagine that the solid angle ω increases by $d\omega$ by reversible isothermal displacement of the side wall, with all other variables remaining unchanged. The work done by the system in this process is proportional to $d\omega$ and can be written as $-\eta d\omega$. This work can be considered as consisting of two parts: the work $P_\alpha dV_\alpha + P_\beta dV_\beta$, connected with the change in the volumes of two uniform bulk phases, and the excess work, which we denote by $-\sigma dA$. Thus, it can be written that

$$dW = -\eta d\omega = P_\alpha dV_\alpha + P_\beta dV_\beta - \sigma dA. \qquad \text{(A1.2)}$$

From a purely mathematical point of view, the term $-\sigma dA$ is simply the correction term introduced for the sum to give the complete work dW, which is a completely determined quantity containing (in addition to contributions from uniform parts in the phases α and β) the complex effects arising as a result of the presence of the surface layer. The coefficient σ can be considered, however, as a work done on the system associated with a unit increment of the area of the dividing surface for a given curvature. This allows us to consider the real system as if it consisted of two uniform phases, α and β, separated by a spherical film of zero thickness, with a radius a and a uniform tension σ in all directions.

Then it is natural to take σ, determined according to (A1.2), for the surface tension of the spherical boundary.

Using (A1.1), we obtain from (A1.2) [3]

$$\sigma = \frac{1}{3}a(P_\alpha - P_\beta) + \frac{K}{a^2}, \quad K = \eta + \frac{1}{3}(R_\beta)^3 - \frac{1}{3}(R_\alpha)^3, \qquad \text{(A1.3)}$$

The quantity K does not depend on the choice of the dividing surface. Therefore, according to (A1.3), the surface tension determined by (A1.2), unlike the planar case, depends on the choice of the dividing surface.

Taking into account the definition of K and the expressions (A1.1), equality (A1.3) can be rewritten in an alternative form:

$$\eta\omega = \sigma A - P_\alpha V_\alpha - P_\beta V_\beta. \qquad \text{(A1.4)}$$

Differentiating (A1.3) with respect to a, we obtain [3]

$$\sigma + \frac{1}{2}a\left[\frac{\partial\sigma}{\partial a}\right] = \frac{1}{2}a(P_\alpha - P_\beta). \tag{A1.5}$$

Here $[\partial\sigma/\partial a]da$ is the change in the surface tension value associated with the mathematical displacement of the dividing surface by da, provided that all physical quantities within the system and external conditions remain unchanged (the notation $[\partial\sigma/\partial a]$ should be distinguished from the derivative by the real dependence on the radius, which will be recorded as $\partial\sigma/\partial a$). This change in the magnitude of surface tension is due simply to the arbitrariness in the choice of the dividing surface, and it should not be confused with the variation of σ from the increase in the radius of the physical surface of the discontinuity. The derivative $[\partial\sigma/\partial a]$ plays an important role in the thermodynamics of a spherical interface.

Since the work done by the system associated with the change in dR_α and dR_β is $-P_\alpha\omega(R_\alpha)^2dR_\alpha + P_\beta\omega(R_\beta)^2dR_\beta$, the general expression for the elementary work dW instead of (A1.2) should be written in the form

$$dW = -\eta\omega + P_\beta\omega(R_\beta)^2 dR_\beta - P_\alpha\omega(R_\alpha)^2 dR_\alpha. \tag{A1.6}$$

Each of the terms on the right-hand side of (A1.6) does not depend on the choice of the dividing surface. From the formulas (A1.1) we obtain

$$dV_\alpha = \frac{d\omega}{3}[a^3 - (R_\alpha)^3] + \omega a^2 da - \omega(R_\alpha)^3 dR_\alpha \tag{A1.7}$$

$$dV_\beta = \frac{d\omega}{3}[(R_\beta)^3 - a^3] + \omega(R_\beta)^3 dR_\beta - \omega a^2 da \tag{A1.8}$$

$$dA = a^2 d\omega + 2\omega a da. \tag{A1.8}$$

Then, solving these equations for dR_α, dR_β and $d\omega$, substitute the result in (A1.6) and using (A1.1) and (A1.4), we obtain

$$dW = P_\beta dV_\beta + P_\alpha dV_\alpha - \sigma dA - \{P_\alpha - P_\beta - 2\sigma / a\}Ada. \tag{A1.9}$$

The last term on the right-hand side of this equation determines the differences in different ways of describing the thermodynamics of a spherical interface.

Using (A1.5), we can rewrite (A.1.9) in an equivalent form:

$$dW = P_\beta dV_\beta + P_\alpha dV_\alpha - \sigma dA - \left[\frac{\partial\sigma}{\partial a}\right] Ada \qquad \text{(A1.10)}$$

It is not difficult to show that (A1.10) remains unchanged with an infinitesimal unphysical change

$$dV_\alpha = -dV_\beta = Ada, \quad dA = 2Ada / a, \qquad \text{(A1.11)}$$

This is due to a purely mathematical displacement of the position of the dividing surface.

We now consider a change not only of ω, R_α and R_β, but also of other state variables. Equation (A1.10) remains valid even in this general case. Therefore, if we denote by dQ the heat received by the system, and by dU the change in its internal energy during this process, then the expression for the First Law of thermodynamics takes the following form:

$$dU = dQ - P_\beta dV_\beta - P_\alpha dV_\alpha + \sigma dA + \left\{P_\alpha - P_\beta - 2\sigma / a\right\} Ada. \qquad \text{(A1.12)}$$

It should be noted that this equation is also valid for systems that are not in thermodynamic equilibrium if the values of P_α and P_β are completely determined for them.

Surface tension. The mechanical equilibrium condition for a system consisting of two phases α and β separated by a spherical membrane of radius a and having a uniform tension σ leads to the well-known relation

$$P_\alpha - P_\beta = 2\sigma / a, \qquad \text{(A1.13)}$$

where $P_\alpha > P_\beta$, the pressure at the liquid boundary from the inside differs from the pressure outside.

On the other hand, we have the relation (A1.5), which is applicable to any system with a spherical interphase boundary. We now choose the dividing surface in such a way that the derivative $[\partial\sigma/\partial a]$ disappears, noting the quantities associated with this choice by the index s. The radius of this dividing surface a_s is determined from the relation [3]

$$\left[\frac{\partial\sigma}{\partial a}\right]_{a=a_s} = 0. \qquad \text{(A1.14)}$$

Thus, the chosen dividing surface turns out to be one of the most important, since only for such a surface from (A1.5) follows a simple form of the relation (A1.13):

$$P_\alpha - P_\beta = 2\sigma_s / a_s, \tag{A1.15}$$

According to Gibbs [4], this particular dividing surface is called the *surface of tension*. The agreement between (A.1.15) and the relation (A1.13) means that the *mechanical action* of a real spherical boundary having a complex structure can be replaced by the action of a simple flexible film whose location coincides with the position of the surface of tension having zero thickness and tension σ, the same in all directions [4,5].

The substitution of (A1.3) into (A.1.15) leads to the relation

$$a_s = \left(\frac{6K}{P_\alpha - P_\beta} \right)^{1/3} \tag{A1.16}$$

Using it, we can exclude K from (A1.3) and obtain [3]

$$\sigma = \frac{a_s^2 \sigma_s}{3a^2} + \frac{2\sigma_s a}{3a_s} \tag{A1.17}$$

Since the surface tension values are always positive, it is easy to see from (A1.17) that σ reaches its minimum value σ_s for $a = a_s$. Therefore, we can conclude that as long as a_s is large in comparison with the thickness of the transition layer, all the values of σ related to those dividing surfaces that are located inside or near this layer (we can expect that there is also a surface of tension in it), are almost equal to the minimum value of σ_s [3] (more precisely $\sigma = \sigma_s \left\{ 1 + \frac{(a - a_s)^2}{a_s^2} + o\left(\left(\frac{(a - a_s)}{a_s} \right)^3 \right) \right\}$). This means that from a macroscopic point of view, the surface of tension can be considered as *practically independent* of the position of the dividing surface, as long as the latter lies inside the surface layer (except for the case of drops and bubbles so small that their radii are comparable to the thickness of the surface layer).

Thus, the physical surface tension in the case of a spherical boundary can be completely determined by means of such σ. Finally, we must emphasize that the relations obtained in this subsection are also valid for systems that are not in equilibrium.

The considered difference of spherical surfaces in comparison with flat surfaces allows to repeat word for word and rewrite expressions for fundamental thermodynamic equations and expressions between excess values. As an example, we give only the expression for the excess free energy

$$F_b = \sum_{i=1}^{s-1} \mu_i N_b^i + \sigma A. \tag{A1.18}$$

If the dividing surface is chosen so that the sum $\sum_{i=1}^{s-1} \mu_i N_b^i$ vanishes, then (A1.18) reduces to equality

$$F_b = \sigma A. \tag{A1.19}$$

It follows that the surface tension on the spherical boundary is equal to the excess free energy per unit area of the dividing surface, if the latter is chosen so as $\sum_{i=1}^{s-1} \mu_i N_b^i$ vanishes. However, in the case of any other dividing surface, σ is no longer equal to excess free energy.

Discrete version of the theory [6]. We use the developments of [3,7–9], in which we explicitly describe how the surface tension depends on the position of the dividing surface to which it relates, and give a discrete exposition of the mechanical determination of the surface tension to show that it is fully equivalent continuum description, and to show what the refinements are when accounting for fluctuations in small systems. The symmetry of the spherical interface is considered in Section 24.

The presence of normal stresses in deformed cells in the form of curved monolayers should cause the appearance of tangential stresses. This situation is analogous to the relationship between normal and tangential stresses in macrosystems [10]. If we denote by S_q the surface area in the monolayer q ($S_q = 4\pi q^2$), then the mechanical equilibrium condition is formally written as

$$\left(\pi^N S\right)_q = \left(\pi^{N+1} S\right)_{q+1} + \left(\pi^T \Delta S\right)_q, \tag{A1.20}$$

where the subscript of the layer q refers to both factors in the parentheses. Here π_q^N is the normal component of the pressure tensor in the layer q, related to the surface of this monolayer S_q. The difference in the normal components of the pressure tensor in the layers q and $q + 1$ is balanced by the tangential component of the pressure tensor π^T_q in the layer q. The component π^T_q in a curved coordinate system is directed at an angle to the radius of curvature of the local section along which the normal component acts. The slope of the component π^T_q depends on the curvature of the surface under consideration. Therefore, the action of the tangential component consists in stretching the lower surface of the monolayer q from $S_{q-0.5}$ to its upper surface $S_{q+0.5}$ (or vice versa in compression from the surface $S_{q+0.5}$ to the surface $S_{q-0.5}$).

Equation (A1.20), constructed on a discrete set of monolayers of the boundary of a drop, can be written in two ways, using finite differences [11,12]: (a) to expand the increments of $(\pi^N S)_q$ and go to the difference derivatives for π^N_q and S_q, or (b) consider the difference derivative of the product $(\pi^N S)_q$. The difference increment of the function α between the layers q and $q + 1$ is denoted by $\Delta(\alpha)_q = \alpha_{q+1} - \alpha_q$.

(a) In the first case we rewrite (A1.20) as

$$\pi_q^N S_q = \pi_q^{N+1} S_{q+1} + \pi_q^T (S_{q-0.5} - S_{q+0.5}). \tag{A1.21}$$

Representing π^{N+1}_q in the form of the difference derivative as $\pi^{N+1}_q = \pi^N_q + \Delta\pi^N_q/\Delta q$ it follows that $\Delta\pi^N_q/\Delta q = \pi^N_q(S_q - S_{q+1})/S_{q+1} - \pi^T_q(S_{q-0.5} - S_{q+0.5})/S_{q+1}$. Neglecting the small difference between $(S_{q-0.5} - S_{q+0.5})$ and $(S_q - S_{q+1})$, which is equal to $1/q^2$, we obtain: $\Delta\pi_q^N / \Delta q = (\pi_q^T - \pi_q^N)(1 - S_q / S_{q+1})$. In the last bracket, the area ratio can be represented in the form $S_q/S_{q+1} = (q/q+1)^m = 1 - m/q$, where $m = 2$ for the sphere and $m = 1$ for the cylinder, which gives

$$\Delta\pi_q^N / \Delta q = m(\pi_q^T - \pi_q^N) / q. \tag{A1.22}$$

The discrete equation (A1.22) is equivalent to the first form of writing a continual differential equation – it is a microscopic analog of the macroscopic equation for the normal component of the pressure tensor. For comparison, we write out the equivalent forms of writing equations for the variation along the radius of the normal pressure component [1,9]: $\dfrac{dP_N(r)}{dr} = \dfrac{2[P_T(r) - P_N(r)]}{r}$, (1st form), $\dfrac{d(r^3 P_N(r))}{dr} = r^2[2P_T(r) + P_N(r)]$, (2nd form), and $\dfrac{d(r^2 P_N(r))}{d(r^2)} = P_T(r)$ (3rd form).

(b) In the second case, equation (A1.20) is written as

$$\Delta(\pi^N S)_q = \pi_q^T \Delta S_q, \quad \text{(A1.23)}$$

Here, the difference of the order of $1/q^2$ between the areas $(S_{q-0.5}-S_{q+0.5})$ and (S_q-S_{q+1}) is also neglected. To the right is a discrete increment of the surface area $\Delta S_q = 2q\lambda$, which is proportional to $dr^2 = 2rdr$ ($dr = \lambda$ and $r = q$) in the continuum calculus. Thus, the discrete equation (A1.23) is equivalent to the third form of the continuous record of the mechanical equilibrium equation in the drop.

The boundary conditions of both discrete equations are the pressure valurd in the liquid drop at $q = 1$ and in the vapour phase at $q = \kappa$. The quantity $\kappa = q_{\text{vap}} - q_{\text{liq}}$, $r \gg \kappa$, here q_{vap} and q_{liq} are the numbers of monolayers that determine the radii of the vapour and liquid phases, bounding the transition region.

We emphasize that the discrete analogues (A1.22) and (A1.23) of the first and third equivalent forms of equations [9] follow from the same discrete equation (A1.20). The difference in these equations consists in the method of expressing the difference between the pressures (the first form) or in the difference of the product of pressure on the area of the drop (the third form). Formally, the differences consist only in the replacement of the difference derivative by the differential derivative [11, 12]. However, there is a fundamental difference between the expressions for the components of the pressure tensor themselves. At the macroscopic level, we are dealing with stresses in local volumes containing a large number of particles. Here, the monolayer size is equal to the diameter of the molecule and the expressions for π_q^T and π_q^N reflect the curvature on the characteristic cell size $\Delta q = \lambda$.

In principle, the microscopic theory allows us to abandon the use of the concept of surface tension for the calculation of quasi-equilibrium distributions of molecules at the vapour–liquid interface in metastable conditions. For this, using the first form of equations for the normal components of the pressure tensor, an alternative description of the properties of the transition region of the boundary can be obtained. The difference equation (A1.22) determines the way the normal component π_q^N varies from layer to layer. For each layer q there exists a unique connection between π_q^N and the local section of the profile $\{\theta_q\}$ and $\pi_q^T(\{\theta_q\})$, which constitutes the system of equations on $\{\theta_q\}$. The equation as a whole from one boundary (with a given vapour pressure) to the other (up to the inside of the drop) determines the pressure inside the drop. This description, in principle, is suitable for any curved surface. After determining the profile $\{\theta_q\}$, local values $(aP)_q$ – the variable profile of the chemical potential in the metastable system, can be calculated.

The first mechanical determination of surface tension. The microscopic theory makes it possible to obtain a molecular interpretation of surface tension. To this end, consider equation (A1.23) and obtain by summation over the layers an integral relationship between the vapour pressure and the liquid, which describes the hydrostatic equilibrium of the drop as a whole. By transforming this connection, by analogy with the derivation of [1,7,9], we can construct a discrete analogue of the Laplace equation, from which the first definition of the surface tension σ is introduced.

Summation over the layers on the left side of the expression (A1.23) leads to the integral form of the notation $\Sigma_{1\kappa} = S_{\kappa+1}\pi_{\kappa+1} - S_0\pi_0$, where $S_{\kappa+1}$ and S_0 are the areas of the boundary circles of the drop from the vapour side and the liquid, $\pi_{\kappa+1}$ and π_0 are the pressures from the side of the vapour and the liquid. The sum on the right side (A1.23) gives $\sum_{1\kappa} = \sum_{q=1}^{\kappa} \pi_q^T \Delta S_q = \sum_{q=1+R}^{\kappa+R} \pi_q^T 2q\Delta q = 2\lambda \sum_{q=1+R}^{\kappa+R} \pi_q^T q$.

Thus, $S_{\kappa+1}\pi_{\kappa+1} - S_0\pi_0 = 2\lambda \sum_{q=1+R}^{\kappa+R} \pi_q^T\ q$, or taking into account the definitions of S_q and ΔS_q indicated above, we have

$$R_{\kappa+1}^2 \pi_{\text{vap}} - R_0^2\, \pi_{\text{liq}} = 2\lambda \sum_{q=1+R}^{\kappa+R} \pi_q^T q \qquad \text{(A1.24)}$$

Equation (A1.24) is identical to the continual expression for hydrostatic pressure around a drop of radius *R*. The derivation of the equation for the first mechanical determination of σ is largely identical to the derivation of σ for the continuum calculus. The essence of the derivation is the transition from the formula (A1.25) to the equation for mechanical equilibrium with respect to the reference dividing surface ρ_s, relative to which the surface tension σ_s determined from the condition of mechanical equilibrium of the moments of forces in the transition region is determined. We note that without determining the reference surface, it is impossible to uniquely introduce the surface tension and it is necessary to use the first form of the equations for the normal components of the pressure tensor, as indicated above.

We take the area in the form of a sectorial strip in the *yz* plane rotating around the axis of the cone with the centre coinciding with the centre of the drop and having the shape of a sector with an angle $d\theta$ and bounded by the circles $q = R_\alpha \equiv R_0$ and $q = R_\beta \equiv R_{\kappa+1}$ (Fig. A1.2). In our case, for a discrete number of monolayers of the interface, the end of the sector strip with the angle $d\theta$ has a discrete broken line passing along the faces of the cells (instead of a strict line along the radius). To obtain a mechanical definition of the surface tension for a spherical boundary, let us imagine a hypothetical system consisting of the phases α and β, uniform up to a spherical film of radius ρ_s separating them [1]. The values of ρ_s and σ_s are determined from the condition that the hypothetical system is mechanically equivalent to the real system both by the resultant force, and by the resultant moment acting on the selected sectoral area. The subscript s reflects the position of the dividing surface $\rho = \rho_s$, referring to the hypothetical phase boundary having zero thickness, and the uniform tension σ_s related to the Laplace equation.

The resultant stress $d\sum_x$, acting in the x-direction normal to the selected area in the real system, is given by

$$d\sum_x = -d\theta \sum_{q=R+1}^{R+\kappa} \pi_q^T q\, dq, \quad (dq = \Delta q = \lambda). \tag{A1.25}$$

On the other hand, in a hypothetical system, the stress $d\sum^{\alpha\beta}$, acting in the x-direction on our sectoral strip can be represented in the form

$$d\sum^{\alpha\beta} = -d\theta \sum_{q=R+1}^{R+\kappa} \pi_q^{\alpha\beta} q\, dq + \sigma_s \rho_s d\theta, \tag{A1.26}$$

where we use the notation

$$\pi_q^{\alpha\beta} = \pi_q^\alpha \text{ for } 1 \le q \le \rho_s \text{ and } \pi_q^{\alpha\beta} = \pi_q^\beta \text{ for } \rho_s < q \le \kappa. \tag{A1.27}$$

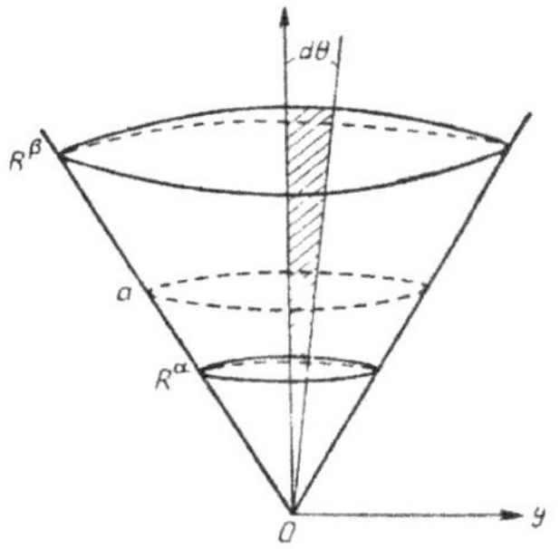

Fig. A1.2. The shaded part represents an area in the form of a sectorial strip. A rectangular coordinate system x, y, z with the origin at the vertex of the cone 0 and the z axis directed along the axis of the cone is introduced.

Equating (A1.25) and (A1.26), we obtain a formula that defines the surface tension for a spherical interface:

$$\sigma_s = \sum_{q=R+1}^{R+\kappa} (\pi_q^{\alpha\beta} - \pi_q^T) q\, dq / \rho_s, \tag{A1.28}$$

where the value of ρ_s is determined by the equilibrium condition with respect to the resultant moment of forces inside the transition region of the boundary:

$$\sum_{q=R+1}^{R+\kappa} (\pi_q^{\alpha\beta} - \pi_q^T) q (q - \rho_s) = 0 \tag{A1.29}$$

where it is taken into account that $dq = \lambda$ is the same for all layers q and it can be shortened. In the expression (A1.29) there are both discrete (number of layers) and continual values (π_q), therefore the quantity ρ_s has in the general case a continual value, which is taken into account by the relations between q and ρ_s in (A1.27). For the vapour region ($q > R + \kappa$) and drops ($q < R + 1$), the pressures and fluid densities remain constant, and in principle the summation limits in formulas (A1.28) and (A1.29) can be formally extended to a larger (infinite) range of values of q, as is done in the continuum calculus, in order to emphasize the independence of the introduced definitions from the width of the transition region. However, in the above forms of writing equations (A1.28) and (A1.29), the inclusion of the boundedness of the quantity κ is visibly reflected, which is of fundamental importance in the interpretation of the temperature dependences of the surface tension [13]. In the case of $R \gg \kappa$, the expressions (A1.28) and (A1.29) automatically become expressions for the surface tension and the positions of the dividing surface of the flat boundary, which correspond to the conditions of mechanical equilibrium.

Let us verify the correspondence of expression (A1.28) and the condition of mechanical equilibrium (A1.24). To do this, we rewrite the last formula in the form

$$\rho_s^2 (\pi_{\text{vap}} - \pi_{\text{liq}}) = 2\lambda \sum_{q=1+R}^{\kappa+R} (\pi_q^{\alpha\beta} - \pi_q^T) q, \tag{A1.30}$$

where ρ_s is introduced via relation

$$2\lambda \sum_{q+1+R}^{\kappa+R} \pi_q^{\alpha\beta} q = R_{\kappa+1}^2 \pi_{\text{vap}} - R_0^2 \pi_{\text{liq}} - \rho_s^2 (\pi_{\text{vap}} - \pi_{\text{liq}}), \tag{A1.31}$$

which is also obtained from (A1.24) by substituting in it a hypothetical pressure profile $\pi_q^{\alpha\beta}$ given by the expression (A1.27) instead of the real profile for the tangential components of the pressure tensor π_q^T.

As a result, we have the Laplace equation for a spherical drop in the lattice gas model

$$\pi_{\text{vap}} - \pi_{\text{liq}} = \left(2\lambda / \rho_s^2\right) \sum_{q=1+R}^{\kappa+R} (\pi_q^{\alpha\beta} - \pi_q^T) q = 2\sigma_s / \rho_s, \qquad \text{(A1.32)}$$

in which σ_s is given by the expression (A1.28), which proves the self-consistency of the definition of both σ_s and ρ_s by the formula (A1.29). For any violation of equality (A1.29), the moment of forces is violated in the transition region, and the value of ρ_s remains uncertain, since the transformation procedure (A1.29) providing the recalculation of the mechanical equilibrium in the transition region to the reference dividing surface is not defined.

Mechanical equilibrium in the continuous and discrete calculus. Before proceeding to the second mechanical definition of surface tension, we discuss the relationship between the three forms of mechanical equilibrium recording in the continuous and discrete calculi.

Equations for a drop in the continuous calculus are equivalent in three forms of writing. The discrete analogues of the first (A1.22) and the third (A1.23) form, as shown above, follow directly from one equation (A1.20). The difference in these equations consists in the way of expressing the difference between the pressures (the first form) or in the difference in the product of the pressure on the area of the drop (the third form). From the point of view of continuous analysis, the transition from the third form to the first is carried out by simple differentiation of r^2 in the denominator and in the form of a factor in the numerator.

However, the final formulas for the discrete notation for the third form $\Delta(\pi^N S)q = \pi_q^T \Delta S_q$ (A1.23) and for the first form $\Delta\pi_q^N / \Delta q = m\left(\pi_q^T - \pi_q^N\right) / q$ (A1.22) differ from each other due to differences in the rules for calculating difference derivatives. For the first form, the increment in the layer $\Delta\pi_q^N \Delta q = m\lambda\left(\pi_q^T - \pi_q^N\right) / q$ leads to the coincidence with the differential form of the record $dP_N(r) = m\lambda(P_T(r) - P_N(r))/r$, where $\lambda = dr$.

The third forms of writing derivatives in discrete and continuous analysis of the product of functions lead to similar expressions:

discrete calculus: $\Delta(q^2\pi_q^N)_q / \Delta(q^2) = \left[(q+\lambda)^2 \pi_{q+1}^N - q^2\pi_q^N\right] / \Delta(q^2) =$
$= q^2\Delta(\pi_q^N) / \Delta(q^2) + (2q+\lambda)\pi_{q+1}^N / (2q+\lambda) = q^2\Delta(\pi_q^N) / \Delta(q^2) + \pi_{q+1}^N.$

continuous analysis: $d\left(r^2\pi_q^N\right)_q / d(r^2) = r^2 d\pi_q^N / d(r^2) + \pi_q^N.$

That is, the differences in the second summand consist in the value of the function π_{q+1}^N at another point $q+1$ instead of π_q^N. On the macroscale at $\Delta r = \lambda \to 0$ both methods of recording the third form of the equations of mechanical equilibrium coincide.

For the second differential continuum form of recording, the corresponding discrete record is constructed as

$$\Delta(q^3\pi_q^N)_q / \Delta q = \left[(q+\lambda)^3 \pi_{q+1}^N - q^3\pi_q^N\right] / \lambda = q^2\left[2\pi_q^T + \pi_q^N\right]. \quad (A1.33)$$

The differences form both the additional terms $(3r\lambda + \lambda^2)\pi_{q+1}^N$, and the values of the function π_{q+1}^N at the other point $q + 1$. The transition from the formula (A1.33) to the first form of writing leads to the following expression

$$\Delta\pi_q^N / \Delta q = q^{-1}\left\{2\pi_q^T + \pi_q^N - q^2\pi_{q+1}^T - (3q\lambda + \lambda^2)\pi_{q+1}^N\right\},$$

which differs from both types of records, both in continuous and discrete analysis. However, for $\Delta r = \lambda \to 0$, both ways of recording the second form of the equations of mechanical equilibrium coincide, and the form two goes into the first form. For a finite value of λ this is not so.

This difference creates the first problem using the second form of the record. Thus the transition to the first discrete form of writing from the second discrete form of writing leads to another equation with respect to $\Delta\pi_q^N/\Delta q$ (additional terms and coordinates are obtained). This situation is not unexpected. Differences between discrete and continuous calculi are well known [12]. At the same time, the use of the lattice gas model (LGM) in contrast to the general problems of discrete calculus [12] consists in the fact that for all functions (here, first of all, this refers to $\pi_q^{T,N}$) no approximation by polynomials is necessary, since the functions themselves in the LGM. This simplifies their practical use.

In the second continuous record form on the left side there is a derivative with respect to the radius of the product of the normal component of the tensor by volume, which reflects the derivative of the work associated with deformation of the layer along the normal. Its right-hand side can be represented as $3q^2\pi_q$, where $\pi_q = (2\pi_q^T + \pi_q^N)/3$ is the average pressure in the cell of the layer q. That is, unlike traditional thermodynamic constructions [1–4], which use only tangential components of the pressure tensor to construct the surface tension, the second form of mechanical equilibrium leads to a different relationship between the work and the components of the pressure tensor. This defines the second problem with the second form of recording directly related to the second mechanical definition of surface tension.

Second mechanical determination of surface tension. The second mechanical definition is based on the displacement of the surface element of the conical wall of a two-phase system with a spherical boundary, bounding the same cone-shaped sector as for the first determination. The force acting across the surface element of the conical wall, enclosed between r and $r + dr$ and between φ and $\varphi + d\varphi$, can be written as $\pi^T(r)r \sin\theta \, dr \, d\varphi$. Let us give the solid angle of the cone an infinitesimal increment $d\omega$, then the work on moving the surface element of the conical wall will be expressed as

$$dW = d\omega \sum_{q=R+1}^{R+\kappa} \pi_q^T q^2 dq, \tag{A1.34}$$

so that according to the general thermodynamic relation $dW = \pi^\alpha dV^\alpha + \pi^\beta dV^\beta - \sigma dA$ [1–4], we have

$$d\omega \sum_{q=R+1}^{R+\kappa} \pi_q^T q^2 dq = \pi^\alpha dV^\alpha + \pi^\beta dV^\beta - \sigma dA, \tag{A1.35}$$

where $dA = \rho^2 d\omega$ is the change in the surface area of the interface, for a given displacement. Since $dV^\alpha = d\omega \sum_{q=R+1}^{\rho} q^2 dq$ and $dV^\beta = d\omega \sum_{q=\rho}^{R+\kappa} q^2 dq$, then taking into account (A1.27) we obtain the second mechanical definition of the surface tension

$$\sigma_2 = \sum_{q=R+1}^{R+\kappa} (\pi^{\alpha\beta} - \pi_q^T) d^2 dq / \rho^2. \tag{A1.36}$$

which must be satisfied for any positions of the dividing surface ρ.

The formal differentiation of the given expression σ_2 by ρ yields the formula

$$\left[\frac{d\sigma}{d\rho}\right] = -\frac{2}{\rho^3} \sum_{q=R+1}^{R+\kappa} \left(\pi^{\alpha,\beta} - \pi_q^T\right) q^2 dq + (\pi^\alpha - \pi^\beta), \tag{A1.37}$$

coinciding with the traditional thermodynamic definition, which allows a change in the value of surface tension σ_2 associated with a mathematical change in the position of the dividing surface by an amount $d\rho$ (here, in the discrete calculus of $\Delta\rho$), provided that all physical quantities within the transition region and , in particular, the physical radius of the drop R, remained unchanged [1–4].

Thermodynamic treatments lead to different definitions of surface tension. There are two thermodynamic approaches in determining σ through the mathematical (Ono) and physical (Gibbs) displacements of the dividing surface and the two above-mentioned mechanical approaches [1–4]. It is believed that both thermodynamic approaches and the second mechanical definition lead to equivalent formulations in which the position of the dividing surface is found from the minimum (in the general case of the extremum) of the function $\sigma(\rho)$. This procedure involves fixing the state of the transition layer when ρ changes. This hypothesis is necessary for thermodynamic constructions, since it is necessary to determine the position of the dividing surface on which the surface tension is introduced. The state of the system at the microlevel is described by a much larger number of variables than at the macrolevel, so the requirement of the constancy of any physical property within the transition region with a change in the value of ρ is some *a priori* assumption (hypothesis), which need not necessarily be satisfied at the molecular level. This was clearly demonstrated by the analysis of Kondo drops [3] (see [14] and Section 59).

References

1. Ono S., Kondo S., Molecular theory of surface tension. Moscow: IL, 1963. [Springer, Berlin, Gottinhen, Heidelberg, 1960].
2. Rowlinson G., Widom B. Molecular theory of capillarity. Moscow, Mir, 1986. [Clarendon, Oxford, 1982].
3. Kondo S., Journ. Chem. Phys., 25, 662 (1956).
4. Gibbs J.W., Thermodynamics. Statistical mechanics. Moscow, Nauka, 1982.
5. Tolman R.C., J. Chem. Phys., 1948. V. 16. P. 758.
6. Tovbin Yu.K., Zh. fiz. khimii. 2010. V. 84. No. 2. P. 231. [Russ. J. Phys. Chem. A. 2010. V. 84. No. 2. P. 180].
7. Kirkwood J. G., Buff F. P., Journ. Chem. Phys. 1949. V. 17. P. 338.
8. Hill T.L., J. Phys. Chem. 1952. V. 56. P. 526.
9. Buff F.P., J. Chem. Phys. 1955. V. 23. P. 419.
10. Timoshenko S.P., Goodier J.N., Theory of Elasticity. Moscow, Nauka, 1979.
11. Godunov S.K., Ryaben'kii V.S., Difference schemes (introduction to theory). Moscow, Nauka, 1973.
12. Gelfond A.O., Calculus of finite differences. Moscow, Nauka. 1967.

13. 13. Tovbin Yu.K., Rabinovich A.B., Izv. AN, ser. khim. 2009. No. 11. P. 2127. [Russ. Chem. Bull. 58, 2193 (2009)]
14. 14. Tovbin Yu.K., Rabinovich A.B., Zh. fiz. khimii. 2013. V. 87. No. 2. P. 337. [Russ. J. Phys. Chem. A 87, 329 (2013)]

Appendix 2

Transfer equations and dissipative coefficients

The following subjects are discussed in this appendix: what are the transfer equations for the properties contained in the system (38.6) and (38.7), the principle of constructing the dissipative coefficients and applying the first equation from the system of equations (38.7) to describe the dynamics of solid-phase processes within the two-level model.

Transfer equations. Let us write for specificity the structure of equations (38.6) and (38.7) of a one-component system in the absence of contributions from external fields. The transport equations reflect the evolution of the mass, impulse and energy of the system. Substituting S_f ($S_f = m_f$, $m_f v_{fi}$, $mv_f^2/2$) into equation (38.6), we obtain an analog of the well-known system of hydrodynamic equations [1–4]:

The continuity equation (only molecular mechanisms of molecular transport are considered)

$$\frac{\partial\langle\theta_f\rangle}{\partial t}+\sum_{j=1}^{3}\frac{\partial\langle\theta_f v_{fj}\rangle}{\partial r_{fj}}=I_f^{(m)}=0, \tag{A2.1}$$

since there are no direct (collisionless) displacements of molecules over long distances, and their collisions do not lead to mass transfer (however, in this case $I_f^{(ik)}\neq 0$ and $I_f^{(e)}\neq 0$).

The equation of the viscous flow

$$\theta_f\left(\frac{\partial u_{fk}}{\partial t}+\sum_{j=1}^{3}u_{fj}\frac{\partial u_{fk}}{\partial r_{fj}}\right)+\sum_{j=1}^{3}\left[\frac{\partial\left[P_f^{ne}\delta_{jk}-\Pi_{ik}^{ne}(f)\right]}{\partial r_{fj}}\right]=I_f^{(ik)}. \tag{A2.2}$$

where u_{fk} is the microhydrodynamic velocity of molecules at site f in the direction $k = x, y, z$; P_f^{ne} is the non-equilibrium pressure at the given local parameters of the system at site f (see Section 43) [5]. The principle of constructing the components of the impulse flux density tensor $\Pi_{ik}^{ne}(f)$ is discussed below. A characteristic feature of this approach is that all characteristics (P_f^{ne} and $\Pi_{ik}^{ne}(f)$) are expressed through unary and pairwise distribution functions (DF) with identical functional connections for any degree of non-equilibrium. The symbol of nonequilibrium (ne) means the use of non-equilibrium paired DFs. In the case of local equilibrium, formula (A2.2) becomes the hydrodynamic equation. In them, there are no exchange terms, $I_f^{(ik)} = 0$, and the quantities P_f^{ne} and $\Pi_{ik}^{ne}(f)$ are calculated by means of equilibrium paired DFs.

The energy transfer equation

$$\frac{\partial\left(\theta_f(m_f u_f^2/2 + U_f^{full})\right)}{\partial t} + \\ + \sum_{i=1}^{3} \frac{\partial}{\partial r_{fi}} \left\{\left(\theta_f u_f^2 u_{fi}/2\right) + \left[\theta_f U_f^{full} + P_f^{ne}\right] u_{fi}\right\} + \\ + \sum_{i=1}^{3} \frac{\partial}{\partial r_{fi}} \left[L_{fi}^{ne} - \sum_k u_{fk} \Pi_{fi,fk}^{ne}\right] = I_f^{(e)}. \tag{A2.3}$$

where U_f^{full} is the total internal energy per unit volume, L_{fi}^{ne} is the vector of the heat flux density under non-equilibrium conditions [5]. This expression generalizes the well-known expression for energy transfer under the condition that local equilibrium is fulfilled for the case of local disequilibrium. The rearrangement $\partial P_f^{ne} u_{fi} / \partial r_{fi}$ of the term to the internal energy $\theta_f U_f^{full} u_{fi}$ reflects the enthalpy flow for the unit volume of the system.

Equations of transfer of paired properties. The first group for transferring the pair properties of the equations (38.7) consists of 5 functions $\left\langle m_f S_\Psi^m \right\rangle$ (and vice versa $\left\langle S_f^m m_\Psi \right\rangle$) for different S_ψ^m (below the mass in all terms are canceled).

For $S_\psi^m = m_\psi$ and $S_{f\psi}^{m,m} = m_f m_\psi$ we have an equation for the evolution of paired DFs, taking into account the separation of spatial and concentration variables,

$$\frac{\partial\langle\theta_{f\psi}\rangle}{\partial t}+\sum_{j=1}^{3}\left\{\frac{\partial\langle\theta_{f\psi}v_{fj}(\psi)\rangle}{\partial r_{fj}}+\frac{\partial\langle\theta_{f\psi}v_{\psi j}(f)\rangle}{\partial r_{\psi j}}\right\}=$$
$$=\frac{\partial\theta_{f\psi}}{\partial t}+\sum_{j=1}^{3}\left\{\frac{\partial\theta_{f\psi}u_{fj}(\psi)}{\partial r_{fj}}+\frac{\partial\theta_{f\psi}u_{\psi j}(f)}{\partial r_{\psi j}}\right\}=I_{f\psi}^{(m,m)} \quad \text{(A2.4)}$$

In the second term, new unknown functions of the type $\theta_{f\psi}\langle v_{fj}(\psi)\rangle$ appeared. The symbol (ψ) indicates that the value is determined for a fixed occupied state of the neighbouring site ψ by the molecule A.

Kinetic equations for new variables $\theta_{f\psi}u_{fi}(\psi)$ are written out as equations for the conservation of the mass and impulse of particles of the pair $f\psi$: $S_{f\psi}^{m,ik}=m_f i_{\psi k}(f)$, k = 1–3 (here $i_{\psi k}(f)$ is the total impulse of the molecule, reflecting the translational motion and the influence of the potential fields of neighbours on this motion [5]):

$$\frac{\partial\theta_{f\psi}u_{\psi k}(f)}{\partial t}+\sum_{j=1}^{3}\left\{\frac{\partial\theta_{\psi f}\langle v_{fj}(\psi)i_{\psi k}(f)\rangle}{\partial r_{fj}}+\right.$$
$$\left.+\left[\frac{\partial\theta_{f\psi}u_{\psi j}(f)u_{\psi k}(f)}{\partial r_{\psi j}}+\frac{\partial t_{\psi f}[P_{\psi}^{ne}(f)\delta_{jk}-\Pi_{ik}^{ne}(\psi f)]}{\partial r_{\psi j}}\right]\right\}=I_{f\psi}^{(m,ik)}, \quad \text{(A2.5)}$$

Note that in the square bracket there is a conditional probability $t_{\psi f}$ instead of a pair DF, associated with the method of calculating the components of the impulse transfer tensor [5]. Here the term in the second part of the formula $\Pi_{ik}^{ne}(\psi f)$ in the square brackets is a modified expression for the dissipative coefficient constructed on the site ψ – here the second index f points to the site containing the fixed particle A. In equation (A2.5), the first summand of the sum is a new unknown $\theta_{f\psi}\langle v_{fj}(\psi)i_{\psi k}(f)\rangle$ dynamic variable, to describe the evolution of which we need a microscopic analogue of the first equation of the Keller–Friedman chain [6] inside the cell f, ($\psi \in V(f)$).

To describe the evolution $S_{f\psi}^{m,e}=m_f e_{\psi}(f)$ (mass$_f$ – energy$_\psi$) we have an equation for $\theta_{f\psi}\langle e_{\psi}(f)\rangle$ (here $e_{\psi}(f)$ is the total energy of the particle, taking into account the kinetic and potential contributions [5]), in which taking into account the fixation of the neighbouring molecule at the site f, we obtain

$$\frac{\partial\theta_{f\psi}\left\langle e_{\psi}(f)\right\rangle}{\partial t}+\sum_{j=1}^{3}\left\{\frac{\partial\theta_{f\psi}\left\langle v_{fj}(\psi)e_{\psi}(f)\right\rangle}{\partial r_{fj}}+\right.$$

$$\left.+\sum_{i=1}^{3}\frac{\partial}{\partial r_{\psi i}}\theta_{f\psi}\left\{u_{\psi}^{2}(f)u_{\psi i}(f)/2+U_{\psi}^{full}(f)u_{\psi i}(f)\right\}\right\}+ \tag{A2.6}$$

$$+\left\{\sum_{i=1}^{3}\frac{\partial}{\partial r_{\psi i}}\left\{t_{\psi f}[P_{\psi}^{ne}(f)u_{\psi i}(f)+L_{\psi i}^{ne}(f)-\sum_{k}u_{\psi k}(f)\Pi_{ik}^{ne}(\psi f)]\right\}\right\}=I_{f\psi}^{(m,e)}$$

In the second term (A2.6), a new unknown function appeared $\theta_{f\psi}\langle v_{fi}(\psi)e_{\psi}(f)\rangle$, built on two internal sites f and ψ of the region $V(f)$, requiring its new kinetic equation. The third term takes into account the flux of internal energy $U_{\psi}^{full}(f)$ pertaining to one site ψ, but in the presence of a particle at site f. In the fourth term, the modified expressions for the pressure and dissipative coefficients of the heat $L_{\psi i}^{ne}(f)$ and impulse $\Pi_{ik}^{ne}(\psi f)$ appear.

The evolution of the pair property impulse$_{fn}$–impulse$_{\psi k}$ $S_{f\psi}^{(in,ik)}=i_{fn}(\psi)i_{\psi k}(f)$, n, k = 1–3 is described by the following equation

$$\frac{\partial\theta_{f\psi}\left\langle i_{fn}(\psi)i_{\psi k}(f)\right\rangle}{\partial t}+$$
$$+\sum_{j=1}^{3}\left\{\frac{\partial\theta_{f\psi}\left\langle v_{fj}(\psi)i_{fn}(\psi)i_{\psi k}(f)\right\rangle}{\partial r_{fj}}+\right. \tag{A2.7}$$
$$\left.+\frac{\partial\theta_{f\psi}\left\langle\dfrac{\partial\theta_{f\psi}< i_{fn}(\psi)i_{\psi k}(f)v_{\psi j}(f)>}{\partial r_{\psi j}}\right\rangle}{\partial r_{\psi j}}\right\}=I_{f\psi}^{(in,ik)},$$

Under the sign of the time derivative (A2.7), a correlator $\theta_{f\psi}\langle i_{fn}(\psi)i_{\psi k}(f)\rangle$ is associated with the new dynamical variable that appeared in (A2.5) $\theta_{f\psi}\langle v_{fj}(\psi)i_{\psi k}(f)\rangle$.

In the flux terms there are the same triple correlators of hydrodynamic velocities. The most consistent procedure from the statistical point of view of the closure of triple correlators is the Kirkwood superposition approximation [7] (see also [8]). In each of the subspaces, the triple correlators are expressed in terms of unary and paired DFs:

$$\left\langle v_{fj} v_{fk} v_{\psi i} \right\rangle = \frac{\left\langle v_{fj} v_{fk} \right\rangle \left\langle v_{fj} v_{\psi i} \right\rangle \left\langle v_{fk} v_{\psi i} \right\rangle}{\left\langle v_{fj} \right\rangle \left\langle v_{fk} \right\rangle \left\langle v_{\psi i} \right\rangle},$$

$$\theta_{f\psi h}^{ikj}(r_f r_\psi r_h) = \frac{\theta_{fh}^{ij}(r_f r_h)\theta_{f\psi}^{ik}(r_f r_\psi)\theta_{\psi h}^{kj}(r_\psi r_h)}{\theta_f^i \theta_\psi^k \theta_h^j}. \tag{A2.8}$$

This approximation was first introduced to describe the spatial equilibrium distribution of triplets of molecules $\theta_{f\psi h}^{ikj}(r_f r_\psi r_h)$ [7]. Concentration triple correlators are present in all exchange terms (38.7).

All subsequent equations of the complete system after formula (A2.7) contain third correlators with respect to hydrodynamic velocities, and expressions (A2.8) should be used to close them. The remaining equations of system (38.7) related to energy transfer are written out in the form: three functions 'impulse$_{fk}$–energy$_\psi$' (k = 1–3)

$$\frac{\partial \theta_{f\psi} \left\langle i_{fk}(\psi) e_\psi(f) \right\rangle}{\partial t} + \sum_{j=1}^{3} \left\{ \frac{\partial \theta_{f\psi} \left\langle v_{fj}(\psi) i_{fk}(\psi) e_\psi(f) \right\rangle}{\partial r_{fj}} + \right.$$
$$\left. + \frac{\partial \theta_{f\psi} \left\langle v_{\psi j}(f) i_{fk}(\psi) e_\psi(f) \right\rangle}{\partial r_{\psi j}} \right\} = I_{f\psi}^{(ik,e)}, \tag{A2.9}$$

one function 'energy$_f$–energy$_\psi$'

$$\frac{\partial \theta_{f\psi} \left\langle e_f(\psi) e_\psi(f) \right\rangle}{\partial t} + \sum_{j=1}^{3} \left\{ \frac{\partial \theta_{f\psi} \left\langle v_{fj}(\psi) e_f(\psi) e_\psi(f) \right\rangle}{\partial r_{fj}} + \right.$$
$$\left. + \frac{\partial \theta_{f\psi} \left\langle e_f(\psi) e_\psi(f) v_{\psi j}(f) \right\rangle}{\partial r_{\psi j}} \right\} = I_{f\psi}^{(e,e)}. \tag{A2.10}$$

The use of closure (A2.8) leads to the appearance of a non-linearity for pair correlators in equations (A2.7), (A2.9) and (A2.10), describing the mean-square fluctuations of the pair properties 'impulse–energy' (by the construction of means $u_{fi} = u_{\psi i}$ within the domain f).

Thus, although the general structure of the transport equations for the paired properties is that for each unknown function new unknown functions of higher dimension appear at the hydrodynamic velocities: either for both sites of the pair due to the displacement of this property by the hydrodynamic flow, or for one of the sites of the pair. However, the closure of the system of equations (A2.8) reduces all the flow terms to the mean of the thermal velocities. The above modified dissipative coefficients differ from the usual dissipative coefficients by the presence of one of the neighbours for which the averaging is not carried out.

To work with the constructed system of equations, it is necessary to introduce a procedure for constructing dissipative coefficients and take into account the dimensions of the obtained dynamic variables at the microscopic and hydrodynamic levels.

Closure of hydrodynamic flows. To close the transport equations in hydrodynamics it is necessary to have the dissipative coefficients. The flows of the properties (the impulse flux density tensor $\Pi_{ik}(f)$ and the energy flux density vector L_{fi}) in the presence of local equilibrium are closed within the framework of the phenomenological laws of non-equilibrium thermodynamics and for each dynamic variable the property flow is considered with respect to a particular variable [3]. It follows from (38.6) and (38.7) that in the case of local equilibrium $U_{fg}^{VA}=U_{fg}^{AV}$ and $U_{fg}^{AA^*}=U_{fg}^{A^*A}$, and the right-hand sides of the transfer equations vanish. They go over into the well-known equations of hydrodynamics [1–4]; the system establishes a dynamic equilibrium of the transport flows of molecules in the forward and backward directions.

The general idea of constructing dissipative coefficients is connected with calculating the amount of transfer of a property through an isolated plane used earlier in microscopic hydrodynamics [9,10]. It does not change when three-aggregate states are considered. Here we shall only outline the principle of constructing dissipative coefficients by the example of spherically symmetric particles (more precisely, a monatomic fluid) in order to focus our attention only on the concentration dependences of the self-diffusion transfer coefficients (D^*), shear (η) and bulk (ξ) viscosities, and also the thermal conductivity (κ) to exclude the effect of internal degrees of freedom on the coefficients of bulk viscosity and thermal conductivity. It follows from the kinetic theory [1, 2] that the transfer coefficients characterize the flows for small deviations of the state of the system from the equilibrium state. We will characterize the state of the fluid far from the pore walls by the concentration θ and the temperature T. We will calculate the equilibrium particle distribution with respect to each other in a quasi-chemical approximation taking into account direct correlations between the interacting particles.

To calculate the dissipative coefficients, we select in space some plane 0 and consider the particle fluxes and the impulses and energy transferred by them. We will use the notion of the mean velocity of moving particles w. We draw two planes parallel to the plane 0 (with $x = 0$) at distances $x = \pm\rho$, where ρ is the mean free path of the particle, then the properties of the particles in these planes are written as $S(x = \rho) = S(x = 0) \pm \rho dS/dx$, where the symbol S denotes the concentration, the impulse (in the direction of y, for example) or the energy of the particles moving along the X axis. The flow of the quantity S through the plane 0 is composed of two oppositely directed motions of the particles from the planes $x = \pm\rho$.

Two channels of impulse and energy transfer are realized in dense fluids (Fig. A2.1). The first is connected with the displacement of particles, as in the rarefied phase, and the second is determined by collisions between the particles. The particle under consideration can not intersect this plane 0 if its trajectory has been blocked by other particles located at the sites up to the plane 0 or by a particle located in the immediate vicinity of the other side of the plane 0 and preventing its crossing of the given plane. Both these cases are not taken into account by the elementary kinetic theory in gas, and it is necessary to use the kinetic theory of condensed systems [10–12].

The transfer of the property S through the distinguished plane, where S is 1) the number of molecules – to calculate the self-diffusion coefficient D_i^* and the mass transfer coefficients D_{ij}, 2) the impulse amount – to calculate the shear coefficients η and the bulk viscosity ξ, and 3) the amount of energy – to calculate coefficient of thermal conductivity λ. There are two channels for transferring the property S: $\eta_{fg} = \eta_{fg}(1) + \eta_{fg}(2)$, $\lambda_{fg} = \lambda_{fg}(1)+\lambda_{fg}(2)$, where the number (1) means the transfer of molecules through the selected plane – the calculation of the coefficients D_i^*, D_{ij}, $\eta_{fg}(1)$ and $\lambda_{fg}(1)$; and (2) the transfer of the impulse and energy properties through collisions – the calculation of the coefficients $\eta_{fg}(2)$, ξ and $\lambda_{fg}(2)$.

This principle of constructing molecular transport models in a uniform bulk phase has been generalized to non-uniform systems at the phase boundary, taking into account the influence of the surface potential [10].

In the absence of local equilibrium, the flow of properties of any dynamic variable $\Pi_{ik}^{ne*}(f)$, L_{fi}^{ne*} will be a function of all dynamic variables of the system. When varying any property, the derivatives

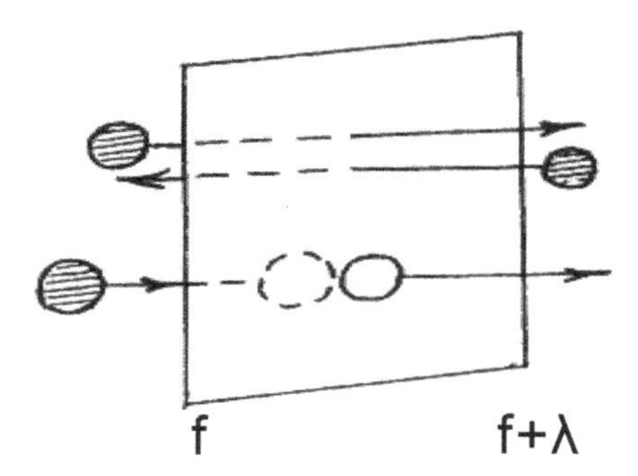

Fig. A2.1. A scheme for transferring properties across a selected plane.

must be used for all independent dynamic variables $\frac{\partial J_f}{\partial x} = \sum_{i=1}^{b} \frac{\partial J_f}{\partial y_i} \frac{\partial y_i}{\partial x}$, where $b = b_1 + b_2$ is the list of dynamic variables y_i. Obviously, in the case of local equilibrium, the number $b = b_1$. Far from equilibrium, the list of independent variables is further extended by $b_2 = 15$ variables on the scale considered associated with the transfer of paired properties [13, 14].

Consider the local flow of properties $J_f^{(r)} = \left\langle \theta_f v_{fj} S_f^{(r)} \right\rangle$, where $r \in b_1$, and $J_{f\psi}^{(r,n)} = \left\langle \theta_{f\psi} v_{fj} S_{f\psi}^{(r,n)} \right\rangle$, $r, n \in b_2$. We select the plane through which the flow of the property passes $S_f^{(r)}, S_{f\psi}^{(r,n)}$, situated between the neighbouring sites f and ψ with the coordinates $x = x_0 - l/2$ and $x = x_0 + l/2$, l is the distance between the sites f and ψ inside the region f. In the continual approximation, the expressions for the $J_f^{(r)}$ fluxes can be written in terms of the thermal velocities $U_{f\psi}$ of the molecules moving by two molecular mechanisms (direct jumps and their collisions) between the sites f and ψ, and likewise $J_{f\psi}^{(r,n)}$ – the thermal velocities of the molecules $U_{(hf)\psi}$ moving between the sites f and h, the state of the site ψ does not change in the course of this process, as [10]

$$J_f^{(r)} = U_{f\psi} S_f^{(r)}{}_{|x=x_0-l/2} - U_{f\psi} S_f^{(r)}{}_{|x=x_0+l/2} = \sum_{j=1}^{b} K_{f\psi}^{rj} \frac{\partial y_j}{\partial x}, \quad r \in b_1, \qquad \text{(A2.11)}$$

$$J_{f\psi}^{(r,n)} = U_{(hf)\psi} S_{f\psi}^{(r,n)}{}_{|x=x_0-l/2} - U_{(hf)\psi} S_{f\psi}^{(r,n)}{}_{|x=x_0+l/2} = \sum_{j=1}^{b} K_{(hf)\psi}^{r,n,j} \frac{\partial y_j}{\partial x}, \quad r,n \in b_2, \qquad \text{(A2.12)}$$

The coefficients of the expansion of the flows with respect to the spatial variations of the dynamic variables are dissipative coefficients that characterize the effect of the change in the dynamic variable y_j on the transfer of the local property r for $j \in b_1$ and on the transfer of the local property r, n for $j \in b_2$ in the domain f. The coefficients are calculated from the current solutions of the kinetic equations to the evolution of unary and pair properties.

Imagine $S_f^{(r)}$ as $\langle S_f^{(r)}\rangle = d_r y_r$, for $j=r\in b_1$, and similarly $\langle S_{f\psi}^{(r,n)}\rangle = d_{r,n} y_{r,n}$ for $j = r,\ n \in b_2$, then

$$\begin{aligned} K_{f\psi}^{rj} &= \partial\left(U_{f\psi} S_f^{(r)}\right)/\partial y_j = \\ &= \partial\left(U_{f\psi} S_f^{(r)}\right)/\partial y_j = K_{f\psi}^{r=j}\delta_{rj} + K_{f\psi}^{r*}, \\ K_{(hf)\psi}^{r,n,j} &= \partial(U_{(hf)\psi} S_{f\psi}^{(r,n)})/\partial y_j = K_{(hf)\psi}^{r,n=j}\delta_{(r,n)j} + K_{(hf)\psi}^{r,n*j}, \end{aligned} \tag{A2.13}$$

where δ_{rj} is the delta function and

$$\begin{aligned} &K_{f\psi}^{r=j} = U_{f\psi} d_r, \quad K_{f\psi}^{r*j} = y_r \partial(U_{f\psi} d_r)/\partial y_j, \\ &K_{(hf)\psi}^{r,n=j} = U_{(hf)\psi} d_{r,n} \quad K_{(hf)\psi}^{r,n*j} = y_{r,n}\partial(U_{(hf)\psi} d_{r,n})/\partial y_j. \end{aligned} \tag{A2.14}$$

Equations (A2.13) show that among the complete set of dynamic variables j there is a situation when the symbol r corresponds to the symbol j, which is the 'ordinary' dissipation coefficient when this property is transferred ($r = j$) under the condition of local equilibrium. In the absence of local equilibrium, this coefficient $K_{f\psi}^{r=j}$ forms only part of the flow of property r. The coefficient $K_{f\psi}^{r*j}$ forms an additional part of the flow for the same dynamic variable. The remaining coefficients of the type $K_{f\psi}^{r*j}$, for $r \neq j$, correspond to cross contributions from the influence of other dynamic variables. All the same applies to the coefficients for the transfer of pair properties (r, $n = j$).

Taking into account (A2.13), the expressions for the flows (A2.11) and (A2.12) will be rewritten as

$$J_f^{(r)} = K_{f\psi}^{r=j}\frac{\partial y_r}{\partial x} + \sum_{j=1}^{b} K_{f\psi}^{r*j}\frac{\partial y_j}{\partial x}, \quad r\in b_1, \tag{A2.15}$$

$$J_{f\psi}^{(r,n)} = K_{(hf)\psi}^{r,n=j}\frac{\partial y_{r,n}}{\partial x} + \sum_{j=1}^{b} K_{(hf)\psi}^{r,n*j}\frac{\partial y_j}{\partial x}, \quad r\in b_2, \tag{A2.16}$$

In the case of local equilibrium, by virtue of the equilibrium relationship between unary $K_{f\psi}^{rj}$ and paired DFs, only one nonzero coefficient remains among the coefficients $K_{f\psi}^{r=j}$, and all the $K_{(hf)\psi}^{r,n,j}$ coefficients vanish.

In the general case of the absence of local equilibrium, the system (38.6), (38.7) can not be divided into subsystems and a complete system of equations must be solved in order to obtain all the current values of the dynamic variables. The solution of the system of equations (38.6), (38.7) allows one to obtain the average values of five dynamic variables $< S_f^{(r)} >$ and their 15 pair combinations $< S_{f\psi}^{(r,n)} >$ as a function of time. Introducing in the usual way [15], the deviations of dynamic variables $\Delta S_f^{(r)} = S_f^{(r)} - < S_f^{(r)} >$, and their root-mean-square deviations as $\Delta S_{f\psi}^{(r,n)} =< S_{f\psi}^{(r,n)} > - < S_f^{(r)} >< S_\psi^n >$ as a result of the solution of this system of equations, the time dependences of root-mean-square fluctuations of dynamic variables will be obtained $S_f^{(r)}$. In particular, this refers to a function $<u_{fi}u_{\psi k}>$ that is directly obtained from the solution of the system (38.6), (38.7) (recall that $u_{fi} = u_{\psi i}$ inside the region *f*) and the expressions (A2.15) and (A2.16) determine all the dissipative coefficients in the absence of local equilibrium.

Unlike the higher moments used in the analysis of turbulent flows, which can be introduced for one or two times [3,4,6,8], all the new 15 dynamic variables $< S_{f\psi}^{(r,n)} >$ refer to one point in time, for which defines the variables $< S_f^{(r)} >$.

As an example, we give the expressions for the coefficients of shear (η_{zx}^{turb}) and bulk (ξ_{zz}^{tur}) viscosities and thermal conductivity (κ_{fi}^{ne}) of the turbulent flow.

$$\eta_{zx}^{turb} = u_y \sum_{j=1}^{b} \frac{\partial (U_{f\psi} d_z)}{\partial y_j} \frac{\partial y_j}{\partial u_z}. \qquad \text{(A2.17)}$$

$$\xi_{zz}^{tur} = u_z \sum_{j=1}^{b} \frac{\partial (U_{f\psi} d_z)}{\partial y_j} \frac{\partial y_j}{\partial u_z} \qquad \text{(A2.18)}$$

where all the quantities refer to the coordinate *f*. In (A2.17), the first derivative under the summation sign is ordinary, while the second derivative is a functional derivative (otherwise it would be zero). This functional derivative is also taken over the complete system of equations of dimension *b*. It follows from the structure of (A2.17) that η_{yx}^{turb} is a function of the local flow velocity and all cross dissipative coefficients.

$$\kappa_z^{turb} = T\sum_{j=1}^{b}\frac{\partial(U_{f\psi}d_{ez})}{\partial y_j}\frac{\partial y_j}{\partial T}. \qquad \text{(A2.19)}$$

where κ_{fi}^{ne} is the local coefficient of thermal conductivity in the direction i (here $i = z$) in the framework of the Kolmogorov equations [8] for the transfer $S_f^{(e)} = \langle e_f \rangle = C_v(f)T_f$, T_f is the local temperature of the region $V(f)$ and $C_v(f)$ is its heat capacity at constant volume per molecule; that is $d_e = C_v$, which depends on paired DFs. The expressions for $C_v(f)$ in terms of unary and paired DFs are discussed in detail in [10]. Expression (A2.19) is written out using the same functional derivative.

Two-level models. The constructed equations have a wide potential range of applications. We confine ourselves to discussing only the range of problems associated with small solid particles. Hydrodynamic variables are usually determined in elementary macroscopic volumes containing a large number of molecules [3]. In the microscopic theory, a microscopic volume of the order of the volume of one molecule is considered as an elementary volume (cell) [10]. This is the minimum size at which one can talk about the existence of average flow characteristics (section 37). Small particles form the second level of the size of the system. An important role in non-uniform systems is played by ensembles of small particles, the size of which does not fit into the Gibbs phase approximation, since each of them can no longer be considered macroscopic. This specificity should be taken into account in solid-phase systems consisting of ensembles of small bodies, with the details of models used for modelling.

As one of the most important areas in the field of applications of new equations, one should point to the theoretical justification of the 'discrete element method' approach, widely known in mechanics [16–20], which is a family of numerical methods for calculating the motion of a large number of particles, such as molecules or grains of sand. Such situations are very common in technologies: in the sintering processes [21], in the flow of bi-dispersed bulk material [22,23], if the granular flow is similar to gas or liquid, and many others. The behaviour of individual particles is described in terms of a continuous medium using computational fluid dynamics. Considering in an explicit form the forces of different nature acting between the particles, it should be noted that the forces can have a macroscopic character (such as friction, bounce, etc.) along with the Coulomb and van der Waals molecular forces. In this connection, the treatment of the discrete element method as a generalization of the finite element method, on the one hand, and as an analog of the MD method, although the particles are not molecules, on the other hand, points to its intermediate position and the predominance of the practical value of the received qualitative information over the rigorous statistical substantiation of results. The question of reconciling both points of view in such situations is still open [24].

The works on investigation of the shear flow of suspensions, as well as the results of computer simulation of the properties of disperse systems under dynamic conditions [25–27], which show the determining role of the microstructure of disperse systems in the formation of their volume–rheological properties, are similar in style and the causes and conditions for the appearance of a macroscopic non-uniformity – the discontinuity of continuity in highly concentrated systems. Here we have the same problem of correct determination of the forces having a complex nature and depending on the state of the medium, the velocity of the particles and the state of the surface.

To simulate such processes, it is necessary to describe the system at different levels: as a molecular level problem, in order to take into account the specific local non-equilibrium distributions of reagents for the creation of specific nanoparticles (microcrystals) and as a task to create in such a macroscopic system such flows that provide for the spatial displacement of nanoparticles. These estimates for the minimum size of the field of application of thermodynamic approaches make it possible to divide the use of theoretical methods for local processes and for supermolecular flows. These estimates apply to all methods of a continuous medium. In fact, they determine the boundary of the application of the very equations of thermodynamics of irreversible processes and the calculation of dissipative coefficients for them.

Dense solid-phase systems. The systems specified in Section 1 are generally complex non-uniform systems. In the general case, four types of characteristic time-of-flight relaxation time scales can be distinguished, corresponding to the complete and partial equilibrium distribution of the components [28]. A fully equilibrium state refers to the state of all components of the system with commensurate mobilities of the components. First, we single out two relaxation time scales associated with (1) the transfer of vacancies with a weak relative redistribution of the components of the mixture, and (2) and with the relaxation of the relative redistribution of the components of the mixture for quasi-equilibrium vacancies. For large differences in the mobilities of the components, the total number s can be divided into 'fast' (mobile) and 'slow' (partially frozen): $s = s_1 + s_2$, s_1 – the number of fast components, and s_2 – the number of slow components. (3) On the third time scale, the transfer of slow components is practically absent, and the migration of mobile components occurs by the vacancy mechanism in the field of action of the potentials of the slow components, and their relaxation is studied in partial equilibrium. (4) On the fourth time scale, the fast components are distributed in equilibrium, and the migration of slow components by the vacancy mechanism occurs in the field of action of the fast components.

The last two situations correspond to the processes of adsorption, absorption, desorption, drying and wetting, etc., when slow components form a 'fixed' matrix, but the latter is usually rearranged, deformed and relaxed under the influence of mobile vapour and liquid components (in existing models this factor is taken into account roughly). Usually, such matrices are ensembles of small phases connected in a single whole macrobody. In these bodies, there can exist a developed surface, whose role increases with the decrease in the size of non-uniform phases – changing to micro-non-uniform phases uses different terms: microcrystals, grains, granules, globules, etc. Below, the term 'grain' will be used.

In practice, in mechanics of continuous media, there is a reliably established upper bound for the grain size in macroscopic systems [29]. It follows from the fact that 1 cm^3 contains several million grains. This gives the size of the side of the cube L of one grain about $\sim 10^5$ nm (and provides good statistics on the macroscale at $L \sim 10^7$ nm). It is natural to take this value of $L \sim 10^5$ nm as the upper value of the mesoscale. On it, each macrograin should be regarded as a separate body with its boundary. When moving to micro-non-uniform systems, the average size L decreases. It is advisable to introduce the following classification of grain sizes, based on the results obtained earlier on the size dependences of the surface tension of vapour–liquid in droplets [30, 31]. Note that the sizes of spherical droplets and cubic grains have close values: the ratio between the volume of the cube and the sphere is connected as $L = 2^{-1/3}D \sim 0.8D$, where L is the side of the cube, D is the diameter of the sphere, and the surface-to-volume ratio for the sphere ($S/V = 6/D$) and for the cube ($S/V = 6/L$) is the same. We denote $R_{mac} \sim L \sim 10^5$ nm.

In Section 27, three characteristic droplet sizes R were singled out: $R_0 \sim 10\lambda$, $R_t \sim 60\lambda$, and $R_b \sim 10^2 \div 10^3\lambda$. Section 34 lists the grain size ranges for which thermodynamics can be used. Thus, the general range of grain sizes to which it is possible to correctly use the thermodynamic description of the diffusion process within the framework of non-equilibrium thermodynamics goes from R_{t2} (36 nm) to R_{mac} (10^5 nm). However, moving from R_{mac} (one can neglect the role of grain surfaces in thermodynamic functions) to micro-non-uniform systems with a characteristic grain size $R < 10^2 \div 10^4$ nm, the question arises of the commensurability of R and R_{t2} and the need for different averaging procedures for substance flows. This range is naturally divided by the value of R_b found by the dependence of the surface tension on the size of the small droplet on two ranges: 1) $R_{t2} < R < R_b$ (the surface tension depends on the grain radius) and 2) $R_b < R < R_{mac}$ (the surface tension does not depend on the grain radius). In the second range, the grain size has practically no effect on the surface tension, whereas in the first range the grain sizes affect the surface (equilibrium and dynamic) tension values, which can also be manifested in diffusion processes.

Types of grains and boundaries. The main property of micro-non-uniform systems is the developed surface of grains with 'interphase' boundaries. In micro-non-uniform solids, it is necessary to isolate the regions inside the grains and their near-surface transition regions, and also to differentiate the state of the *intermediate regions* between neighbouring grains. Transitional sections are analogues of the interface between phases, which are characterized in thermodynamics by surface tension between coexisting or immiscible phases. Intermediate areas exceed the size of the transition areas. The size of the intermediate regions can be commensurable with the grain size, then they also need to allocate their *transition areas*. These boundaries differ in the nature of the transition regions between grains according to the type of coherence of the crystal structure [32] and the type of intermediate regions [10,33].

The grain boundaries are coherent, partially coherent and incoherent. The coupling of phases with different crystal lattices should provide for mutual accommodation of these lattices due to elastic displacements of atoms from their equilibrium positions (coherent boundaries), and also due to inelastic displacements associated with discontinuities in the continuity of the material, caused by misfit dislocations and vacancies condensing at the boundaries coherent boundaries). The existence of completely incoherent conjugation is defined as the absence of tangential shear stresses when a particle of a new phase is inserted into the corresponding cavity in the matrix in which there is no friction between the surface of this particle and the inner surface of the cavity. Under these conditions, the boundaries can freely slip relative to each other. This type of boundaries, however, is interpreted somewhat simplistically, both because of the impossibility of creating an ideally spherical boundary for the new phase (which eliminates the possibility of slippage) and because of the appearance of intermediate transition regions of amorphous and/or altered crystal structure.

An important role in solids is played by the voids that arise in various processes of redistribution of atoms in the course of mutual diffusion or mechanical disturbances. In this case, in addition to the various crystal conjugations of neighbouring grains, the character of the distribution of voids (isolated or forming a porous subsystem of a solid body) plays an important role. The type of incoherent boundaries should be classified into three classes of contacts (considering the possibility of free slippage as an idealized exception). The first class is the presence of intermediate regions and regions with an amorphous and/or distorted crystalline structure with a predominant normal stress when the majority of the bonds between the grains are ruptured. Such regions take on a mechanical load from outside the body and participate in the transfer of impulse throughout the volume of the solid. The second class is areas with an amorphous and/or distorted crystalline structure outside the zones of mechanical stress. By mechanical behaviour the microporous region also refers to them – opposite walls of micropores influence each other through direct potential interaction, but mechanical contact (as an elastic body) is absent in the absence of a mobile phase and is strongly attenuated when the micropores completely fill the mobile phase. The third class is the regions with a meso- and macroporous structure. They do not directly interact with opposite walls of the pores in the absence of the mobile phase, but there is indirect influence through the mobile phase with complete filling of the pores. It should be noted that the deformation component is taken into account when classifying boundaries. A subsystem of slow components and a subsystem of fast components are considered together.

Equations of diffusion transfer. Equations [34] take into account the existence of a non-uniform system in which internal local redistributions of components along different sublattices occur in the volume dV. These equations reflect these two-level non-uniform grain distributions and their boundaries. We denote by x the coordinate of the system along which diffusion is considered. The processes on different sublattices have different characteristic times, which are determined by the constants of elementary velocities of component jumps in the right-hand sides of the kinetic equations. Important is the non-uniformity of the spatial structure of the system, given by the volume shares and the method of locating sites of different types [35–39].

The diffusion description of the transport process is associated with the rejection of the site-by-site description of the concentrations in different sites of the system. Since the hopping velocities (Section 55) depend on local concentrations and on pair functions, in the general case of the absence of local equilibrium, the diffusion equations must take into account the evolution of these independent variables. Therefore, migration processes are characterized by four types of transfer coefficients of components and their pairs. For the flow along the x-axis, the equations [34] were obtained in the form

$$\begin{aligned} \dot{\theta}_q^i(x) &= \sum_p \sum_k \frac{\partial}{\partial x} D_{qp}^{ik} \frac{\partial \theta_q^k}{\partial x} + \sum_{p\eta} \sum_{kl} \frac{\partial}{\partial x} D_{q(p\eta)}^{i(kl)} \frac{\partial \theta_{p\eta}^{kl}}{\partial x} \\ \dot{\theta}_{q\xi}^{ij}(x) &= \sum_p \sum_k \frac{\partial}{\partial x} D_{q\xi(p)}^{ij(k)}(r) \frac{\partial \theta_q^k}{\partial x} + \sum_{p\eta} \sum_{kl} \frac{\partial}{\partial x} D_{q\xi(p\eta)}^{ij(kl)} \frac{\partial \theta_{p\eta}^{kl}}{\partial x} \end{aligned} \tag{A2.20}$$

The coefficients D_{qp}^{ik}, and $D_{qp\eta}^{i(kl)}$, $D_{q\xi(p)}^{ij(k)}$ and $D_{q\xi(p\eta)}^{ij(kl)}$ characterize the diffusion fluxes of the component i from the sites of type q and the pairs of components ij on pairs of sites $q\xi$. The second terms characterize the flows under the influence of the gradients of the pairs. Summation over the pair functions in (A2.20) is carried out over linearly independent pairs kl. The choice of the latter is related to the normalization conditions. All the coefficients D are obtained by expanding in series in gradθ_p^k and grad$\theta_{p\eta}^{kl}$ the flows of particles and their pairs through jumps $J_{mm+1}^i(qp)$, where m is the number of the layer perpendicular to the x axis, between the layers m and $m + 1$ there is a secant plane with coordinate x. In a one-step approximation in time, the flow between adjacent layers is written as $J_{mm+1}^i(qp) = U_{mm+1}^{Vi}(qp) - U_{mm+1}^{iV}(qp)$, where $U_{mm+1}^{iV}(qp)$ is the migration rate of particle i from site q of layer m to site p of layer $m + 1$; V is the vacancy symbol. The total flow of pairs of particles ij is composed of the corresponding migration contributions by the vacancy mechanism of both components of the pair i and j. In the second equation (A2.20), the hopping rates of particles i appear in the presence of neighbours j: $U_{pq\xi}^{(Vi)j} = U_{qp}^{iV}\psi_{q\xi}^{ij}$, where $\psi_{q\xi}^{ij} = t_{q\xi}^{ij}\exp\left[\beta(\varepsilon_{ij}^{*} - \varepsilon_{ij})\right] / S_{q\xi}^{i}$.

Equations (A2.20) are an example of equation (A2.1) for a strongly non-equilibrium solid-phase system. Its reduction with the transition to the ordinary diffusion equations in the case of local equilibrium is discussed in [34], and the transition to the problem of estimating the relaxation times near the local equilibrium for a two-scale grain structure (the supra-atomic structure is modeled with the grains of a dispersed body of characteristic size $L \gg \lambda$) are discussed in [28].

Other situations discussed in Section 41, in particular analogues of turbulence in solids, require the analysis of the following equations of the system (38.7), except (A2.1). For them, the coefficients (A2.17)–(A2.19) obtained above must be specified with the properties of the two-level model under discussion.

References

1. Huang K. Statistical mechanics. Moscow, Mir, 1966. [Wiley, New York, 1963].
2. Rumer Yu.B., Ryvkin M.Sh., Thermodynamics, statistical physics and kinetics. Moscow, Nauka, 1977.
3. Landau L.D., Lifshitz E.M. Theoretical physics.VI. Hydrodynamics. Moscow, Nauka, 1986. 734 p.[Pergamon, New York, 1987].
4. Bird R. B., Stuart W. E., Lightfoot E. N., Transport phenomena. Moscow, Khimiya, 1974. [Wiley, New York, London, 1965].
5. Tovbin Yu.K., Zh. Fiz. Khimi. 2017. Vol. 91. No. 3. P. 381. [J. Phys. Chem. A 91, 403 (2017)].
6. Keller L., Fridman A.D., Proc. 1st Intern. Congr. Appl. Mech., Delft, 1924. P. 395.
7. Kirkwood, J., J. Chem. Phys. 1935. V. 3. P. 300.
8. Monin A.S., Yaglom A.M. Statistical hydrodynamics. Moscow: Nauka, 1965 Part 1. 639 p; 1967 Part 2. 720 pp.
9. Tovbin Yu.K., Zh. Fiz. Khimii. 1998. V. 72. No. 8. P.1446. [Russ. J. Phys. Chem. 1998. V. 72. No. 8. P. 1298].
10. Tovbin Yu.K., The Molecular Theory of Adsorption in Porous Solids, Moscow, Fizmatlit, 2012. [CRC Press, Taylor & Francis Group, 2017].
11. Tovbin Yu.K., Khim. Fiz., 2002. V .21. No. 1. P. 83.
12. Tovbin Yu.K., Zh. Fiz. Khimii. 2002. T. 74. No. 1. C. 76. [Russ. J. Phys. Chem. 2002. V. 76. № 1. P. 64].
13. Tovbin Yu.K., Zh. Fiz. Khimii. 2014. T. 88. No. 2. P. 261. [Russ. J. Phys. Chem. A 88, 213 (2014)].
14. Tovbin Yu.K., Teoret. Fundamentals of chemical. technol. 2013. V. 47. No. 6. P. 734.
15. Landau L.D., Lifshitz E.M. Theoretical physics. V. Statistical Physics. Moscow: Nauka, 1964.

16. Bicanic N., Discrete Element Methods, in Encyclopedia of Computational Mechanics, Ed. by E. Stein, R. de Borst, and T. J. R. Hughes (Wiley, New York, 2004), Vol. 1.
17. 2nd International Conference on Discrete Element Methods, Editors Williams, J.R. and Mustoe, G.G.W., IESL Press. 1992.
18. Williams, J. R., O'Connor R., Discrete Element Simulation and the Contact Problem, Archives of Computational Methods in Engineering. 1999. V. 6. No. 4. P. 279.
19. Herrmann H.J.P., Statistical Physics. Invited Papers from STATPHYS 20. P. 51 / Reprinted from Physica A. 263 No. 1-4. Elsevier. North-Holland 1999.
20. Munjiza A., The Combined Finite-Discrete Element Method Wiley. 2004.
21. Galkin V.A., Smoluchowski's equation. Moscow, Fizmatlit. 2001.
22. Dorofeenko S.O., Polyanchik E.V., Manelis G.B., DAN. 2008. V. 422. P. 1.
23. Dorofeenko S.O., Teoret. Fundamentals of chemical. technol. 2007. V. 41. P. 1.
24. Tovbin Yu.K., Ross. nanotekhnologii. 2010, Vol. 5, No. 11–12, P. 715.
25. Uryev N.B., Potanin A.A., Flowability of suspensions and powders. Moscow, Khimiya. 1992.
26. Potanin A.A., Muller V.M., Kolloid. Zh. 1995. V. 57. No. 4. P. 533.
27. Uryev N.B., Kuchin I.V., Usp. khimii. 2006. V. 75. P. 36.
28. Tovbin Yu.K., Zh. Fiz. Khimii. 2017. V. 91. No. 8. P. 1243. [Rus. J. Phys. Chem. A, 91, No. 8, P. 1357 (2017)].
29. Timoshenko S.P., Goodier, J. Theory of Elasticity. Moscow, Nauka, 1979.
30. Tovbin Yu.K., Zh. Fiz. Khimii. 2010. V. 84. No. 4. P. 797. [Russ. J. Phys. Chem. A 84 (4), 705 (2010)].
31. Tovbin Yu.K., Zh. Fiz. Khimii. 2010. V. 84. No. 10. P. 1882. [Russ. J. Phys. Chem. A 84 (10) 1717 (2010)].
32. Khachaturyan A.G., The theory of phase transitions and the structure of solids. Moscow: Nauka, 1974.
33. Tovbin Yu.K., Izv. AN. Ser. khim. 2003. No. 4. P. 827. [Russ. Chem. Bull. 2003. V. 52. № 4. P. 869].
34. Tovbin Yu.K., Dokl. AN SSSR. 1988. V. 302. No. 2. P. 385.
35. Ovchinnikov A.A., Timashev S.F., Belyi A.A., Kinetics of diffusion-controlled chemical processes. Moscow, Khimiya. 1986.
36. Chalykh A.E., Diffusion in polymer systems. Moscow, Khimiya. 1987.
37. Timashev S.F., Physicochemistry of membrane processes. Moscow, Khimiya.1988.
38. Tovbin Yu.K., Usp. khimii. 1988. Vol. 57. P. 929.
39. Tovbin Yu.K., Zh. Fiz. Khimii. 1997. 71. No. 8. P. 1454. [Russ. J. Phys. Chem. 1997. V. 71. No. 8. P. 1304]

Appendix 3

Coefficients of activity in associated solutions

The notion of 'associated solution' reflects the existence in the mixture of new stable particles – associates, which consist of the original components of the system. Depending on the number of initial components of different sorts that are included in the associates, their appearance may differ greatly in their composition and structure. In this case, the principal property of associates is the possibility of their decay under the conditions (temperature and pressure) considered for the initial components, so that the chemical equilibrium between the associates and the original components is established in the system. The same condition allows the possibility of the transition of the original components between different associates, than a wide range of associates is formed. For liquid solutions, the concept of associated solutions focuses on the relationship between internal and external molecular interactions for all molecules present in the mixture.

The use of activity coefficients simplifies the formal recording of the thermodynamic functions of real systems by analogy with ideal systems. In the concept of associated solutions, the dominance of the energy of internal bonds between the constituent components of associates over the lateral interactions with the remaining components of solutions is fundamental.

Intermolecular interactions in condensed phases are approximated by different potential functions [1–4]. Most often, intermolecular potential functions are modeled by the Lennard-Jones potential. Even in such simple systems, the interaction of the nearest neighbouring molecules is about 70% of the total binding energy, which is formed as a result of taking into account all neighbours at different distances. In polar media in the absence of charged ions, it is necessary to take into account dipole–dipole interactions, which depend on the mutual orientation of the molecules. Such interactions are formed by local charges on atoms and groups of atoms belonging to different molecules. This type of interaction is characterized by its length over distances of the order of up to ten molecular diameters. It is also possible for the induction effect that occurs when molecules are deformed under the influence of the electrostatic forces of neighbouring molecules, which grabs a sufficiently extended region around the components under consideration. This effect is of greatest importance in the interaction between molecules with a large dipole moment and easily polarizable molecules. Thus, in real solutions, intermolecular interactions are of considerable length and in the analysis of the thermodynamic characteristics it is necessary to take this circumstance into account.

The original idea of associated solutions was reduced to the formulation of ideal associated solutions [5]. Physically, it is quite difficult to imagine strict compensation of local charges on different molecules, so that the associates practically do not interact with the other components of the solution and with each other, so later the notions of non-ideal associated solutions appeared.

In [6] the concept of associated solutions is analyzed from the point of view of correspondence to real potential functions, which are used in describing the equilibrium distribution of solution components. For the analysis, the molecular theory of solutions within the framework of the lattice gas model (LGM) [3,4,7,8] is used, which makes it possible to calculate molecular distributions and thermodynamic functions from intermolecular forces. In [9,10], the main limitations on lattice models were removed: the lattice stiffness, the internal motions of the centres of mass of molecules inside the cells, the possibility of their rotation and orientation interactions. This allowed us to consider a wide range of potentials of intermolecular interactions that simultaneously take into account all the intermolecular contributions to the overall interaction potential that reach this r-th coordination sphere:

$$\xi_{ij}(r) = u_{\text{rep}}(ij\,|\,r) + u_{\text{el}}(ij\,|\,r) + u_{\text{disp}}(ij\,|\,r) + u_{\text{ch.tr.}}(ij\,|\,r) \qquad \text{(A3.1)}$$

where there are contributions to the total interaction between a pair of molecules ij: 1) exchange (repulsive) interaction (u_{rep}); 2) direct electrostatic interaction (u_{el}); 3) dispersion interaction (u_{disp}); 4) the interaction due to charge transfer ($u_{ch.tr.}$). As the distance r between the molecules i and j increases, the terms on the right-hand side of equation (A3.1) decrease or may be zero. (An inductive interaction leading to multiparticle interactions is excluded from the 'usual' list [1–4] in (A3.1)).

To specify the orientations of polyatomic molecules of commensurable sizes, we represent a vector from the centre of the molecule mass directed to one of its atoms, and we will characterize the orientation of the rigid molecule by the angles between the given vector and the external (fictitious) field vector. In a spherical coordinate system, the orientation is given by the angle η_x, $0 \le \eta_x \le 2\pi$, which determines the projection of the vector onto the X0Y plane reckoned from the X axis (the external field vector coincides with the z axis) and the angle η_z, $0 \le \eta_z \le 2\pi$, between the vector and the Z axis (for linear molecules), and also by the additional angle η, $0 \le \eta \le 2\pi$, which describes the rotation with respect to a given vector for non-linear molecules. For brevity, we shall denote the set of values of the angles η_x, η_z, η of the molecule i by the single symbol ϕ_i.

By analogy with the transition from a continuous set of values of the coordinates of the centre of mass of molecules to the discrete one used in lattice models, we introduce a set of discrete molecular orientations relative to their mass centres instead of continuous orientations. In this case, the values of ϕ_i correspond to the discrete values of the angles η_x, η_z, η [11]. The number of different orientations of the molecule i will be denoted by σ_i. This model made it possible to explain the existence of the upper and lower critical points on the phase diagrams of the stratification of binary solutions [11]. In later works [12–15], the number of orientations was considered as a parameter of the model, not related to the lattice structure z.

To describe the molecular orientations, we introduce the random variable γ_i^ϕ, which characterizes the orientation ϕ_i of the molecule i: $\gamma_i^\phi = 1$ if the molecule i has the orientation ϕ_i, and $\gamma_i^\phi = 0$ otherwise. For the value γ_i^ϕ, the equality $\sum_{\varphi_i=1}^{\Phi_i} \gamma_i^\varphi = 1$ is fulfilled. For vacancies, there are no orientations, $\sigma_s = 1$. Then we can introduce the values $\gamma_i^\phi(f) = \gamma_f^i / \gamma_i^\phi$ characterizing the complex event: at site f there is a molecule i and it has the orientation ϕ_i. This allows us to introduce a new notion of the kind of particle ℓ at site f if the sort of molecule i and its orientation ϕ_i are related to one index ℓ ($\ell \leftrightarrow i$, ϕ_i), then $\gamma_i^\ell \equiv \gamma_i^\phi(f)$, where the values of the index ℓ vary from unity to $\Phi = \Sigma_{i=1}^s \sigma_i$. (Let us agree to refer the last index Φ to vacancies.)

It should also be taken into account that for fixed indices r, ϕ_i and ψ_i, the change in the position of the neighbouring molecule (variation of g_r) changes the values of the energy parameters. This is the main difference of the potentials that reflects the directivity of the bonds from the spherical potentials. Although it is obvious that the sequence of indices in itself does not change the values of the energy parameters $\varepsilon_{ij}^{\phi\psi}(f,g|r) = \varepsilon_{ji}^{\psi\phi}(g,f|r)$. Equations for the molecular distribution are written as

$$\theta_f^\Phi = \theta_f^l \exp\beta(v_f^l - v_f^\Phi)\prod_{r=1}^{R}\prod_{g\in z_f(r)} S_{fg}^l(r), \quad \sum_{l=1}^{\Phi}\theta_f^l = 1, \tag{A3.2}$$

$$\begin{aligned}
&S_{fg}^l(r) = 1 + \sum_{\lambda=1}^{\Phi-1} t_{fg}^{l\lambda}(r)x_{fg}^{l\lambda}(r), \quad x_{fg}^{lm}(r) = \exp[-\beta\varepsilon_{fg}^{lm}(r)] - 1,\\
&\theta_{fg}^{lm}(r)\theta_{fg}^{\Phi\xi}(r) = \theta_{fg}^{l\xi}(r)\theta_{fg}^{\Phi m}(r)\exp[\beta\omega_{fg}^{lm\Phi\xi}(r)],\\
&\sum_{\lambda=1}^{\Phi}\theta_{fg}^{l\lambda}(r) = \theta_f^l, \quad \sum_{l=1}^{\Phi}\theta_{fg}^{l\lambda}(r) = \theta_g^\lambda,\\
&\omega_{fg}^{lmb\Phi}(r) = \varepsilon_{fg}^{lm}(r) + \varepsilon_{fg}^{b\Phi}(r) - \varepsilon_{fg}^{l\Phi}(r) - \varepsilon_{fg}^{bm}(r).
\end{aligned} \tag{A3.3}$$

Equations (A3.2) are the equations for the molar fractions of the particles l $\left(\theta_l = \left\langle \gamma_f^1 \right\rangle\right)$. They describe the vapour–liquid equilibrium when substituted $v_f^l = v_l$ into the expression for the values of chemical potentials v_l of molecules in the gas phase. In the theory of liquid solutions, it is usually assumed that the mole fraction θ_l of the component i is known, $\theta_i = \sum_{\varphi_i=1}^{\Phi_i} \theta_i^\varphi$. In this case, equations (A3.2) allow us to determine the molar fraction of molecules i with a specific orientation ϕ_i, and also allow us to use experimental data on vapour–liquid equilibrium or other thermodynamic characteristics of solutions for determining the parameters of intermolecular interaction. For molecules in the dense phase, the values v_f^l take into account the internal motions of molecules, modified by the influence of neighbouring molecules [10].

Equations (A3.3) is a system of equations for paired distribution functions $\theta_{fg}^{l\lambda}(r) = <\gamma^l{}_f \gamma_{g_r}^{\lambda}>$ characterizing the probability of finding particles l and λ (i.e., molecules i with orientation ϕ_i at site f and molecule j with orientation ψ_j at site g_r) at a relative distance r. The size of the algebraic system of equations (A3.3) can also be reduced by introducing new variables, as explained in [7].

Equations (A3.2), (A3.3) are a generalization of the equations given in [11–15], used earlier to study the phase diagrams of binary solutions of molecules with specific interactions. They reflect the contributions of all possible molecular orientations of each variety. 'Selection' of orientations is carried out through specific values of energy parameters, which in the case of specific interactions 'cut out' the locally oriented locations of neighbouring molecules, taking into account the tendency of molecules to organize specific solvation at which the internal motions of molecules change, which changes the values of the parameters v_l (for more details, [10,16]).

Equations (A3.2) and (A3.3) follow a modified expression for the chemical potential of the molecule i, which is an orientation-averaged expression for the 'quasiparticles' k_i, $\theta_i = \sum_{k_i=1}^{\sigma_i} \theta_f^{k_i} / \sigma_i$ also containing contributions from different coordination spheres

$$\mu_i = v_i + kT\ln(\theta_f^i) + \frac{kT}{2}\sum_{r=1}^{R} z(r) \sum_{\ell=1}^{\Phi} t_{fg}^{k_i\ell}(r) \ln\frac{\hat{\theta}_{fg}^{k_i\ell_j}(r)}{(\theta_f^i)^2 \sigma_j}, \qquad \text{(A3.4)}$$

where $\hat{\theta}_{fg}^{k_i\ell}(r) = \theta_{fg}^{k_i\ell}(r)\exp(-\beta\varepsilon_{fg}^{k_i\ell})$, here the type of contact k_i of the molecule i at its fixed position depends on the site number g in the r-th coordination sphere $z(r)$ around the site f. In (A3.4) for simplicity of recording it is assumed that each contact is different from the others and it is possible not to introduce a weighting factor. The co-factor σ_j provides the transition (A3.4) to the formula for point particles in the case of a spherical potential. Here $\mu_i^{res} = \mu_i - \mu_i^{id} = kT\sum_{r=1}^{R} \ln\gamma_i(r)$ consists of contributions from different coordination spheres, $\mu_i^{id} = v_i + kT\ln\theta_i$, where $\ln\gamma_i(r) = \frac{z(r)}{2}\sum_{\ell=1}^{\Phi} t_{fg}^{k_i\ell}(r)\ln\frac{\hat{\theta}_{fg}^{k_i\ell_j}(r)}{(\theta_f^i)^2\sigma_j}$.

For point particles from formulas (A3.4), an expression for the chemical potential follows $\mu_i = v_i + kT\ln\theta_i + \frac{kT}{2}\sum_{r=1}^{R} z(r)\ln\frac{\hat{\theta}_{ii}}{\theta_i^2}$.

In the particular case of $R = 1$, this formula becomes the well-known expression [4,17]. Here and below, we use equations that explicitly contain paired functions that more clearly reflect the physical meaning. Thus, taking into account the extent of potential functions leads to the fact that the residual contribution to the activity coefficient consists of the sum of the terms of contributions from individual coordination spheres: $\mu_i^{res} = \frac{kT}{2}\sum_{r=1}^{R} z(r)\ln\frac{\hat{\theta}_{ii}(r)}{\theta_i^2}$. We note that the expansion of the excess chemical potential over different distances r in the formula $\mu_i^{res} = kT\ln\gamma_i^{res} = kT\sum_{r=1}^{R}\ln\gamma_i(r)$ refers to a spherically symmetric potential of radius $R > 1$. To relate to the expressions given below, we give an equivalent notation for $\gamma_i(r)$:

$$\ln\gamma_i(r) = \frac{z(r)}{2}\ln\frac{\hat{\theta}_{ii}(r)}{\theta_i^2} = \frac{z(r)}{2}\sum_{j=1}^{s} t_{ij}(r)\ln\frac{\hat{\theta}_{ij}(r)}{\theta_i^2}. \tag{A3.5}$$

If we use more accurate methods for describing the distribution of neighboring molecules than the quasi-chemical approximation, then the only term remains μ_i^{res} in the expression for, and the expression itself γ_i^{res} becomes more cumbersome [18].

Mean-field approximation (MFA). In the early works, the so-called mean-field approximation was actively used, with the help of which the presence of the first term in the formal expansion of Margules for the chemical potential of the components of the binary mixture was explained [5.19] (see below). In the framework of this approximation, the effect of direct correlations, which is present in the quasi-chemical approximation, is neglected. As a result, equations

$$\begin{aligned} a_i P_i &= \frac{\theta_i}{\theta_s}\exp[\sum_{r=1}^{R} z(r)\beta\sum_{j=1}^{s}(\varepsilon_{sj}(r) - \varepsilon_{ij}(r))\theta_j] = \\ &\frac{\theta_i}{\theta_s}\exp(z\beta\sum_{j=1}^{s}(\hat{\varepsilon}_{sj} - \hat{\varepsilon}_{ij})\theta_j), \end{aligned} \tag{A3.6}$$

where new effective interaction parameters are introduced $\hat{\varepsilon}_{ij} = \sum_{r=1}^{R} z(r)\varepsilon_{ij}(r)/z$, which include contributions from extended potential regions in all coordination spheres [20]. The main conclusion from the expression (A3.6) is that taking into account the long-term contributions of potential functions in the MFA does not lead to a change in the concentration dependence in comparison with taking into account only the nearest neighbours, therefore any variants of potential functions are reduced only to renaming the short-range parameters $\varepsilon_{ij} \equiv \varepsilon_{ij}(r=1)$. Therefore, in the expression for the chemical potential (for the residual contribution), there are only α_{ij} terms like for $R = 1$. As an example, we give a well-known expression for the chemical potential in a three-component strictly regular solution [5] $\mu_1 = kT \ln \theta_1 + \alpha_{12}\theta_2^2 + \alpha_{13}\theta_3^2 + (\alpha_{12} - \alpha_{23} + \alpha_{13})\theta_2\theta_3$, (expressions for μ_2 and μ_3 are obtained by cyclic replacement of the indices 1, 2 and 3) here, which excludes $\alpha_{ij} = z\varepsilon_{ij}/2$ the possibility of a difference in the concentration dependences of the activity coefficient for interaction potentials of different lengths.

Similar effective short-range interaction parameters are obtained by using the potential (A3.1) if, in addition to averaging over different distances r, averaging over different orientations of neighboring molecules is performed ϕ_i, which illustrates the inapplicability of this approximation to the molecular interpretation of thermodynamic characteristics.

Associates. The transition to associates in non-ideal mixtures within the molecular theory is carried out in a similar way, as for ideal systems. Specific intermolecular interactions are intermediate between van der Waals interactions and a chemical bond, and in many respects they are associated with a partial charge redistribution, which increases the energy of interaction of neighbouring components in comparison with non-specific interactions. The formation of any associate increases the size of the original component, and as the degree of association grows, larger molecules grow in size.

Assume that in the associated solution of their original components A and B there are complexes A_i formed from i monomolecules A, complexes B_j and complexes A_iB_j formed as a result of the association of i molecules A and j molecules B [5]. The complexes are in equilibrium with each other and with monomers A and B. Possible reactions between these particles can be represented by the equations $A_i = iA_i$; $B_j = jB_j$, $A_iB_j = iA_1 + jB_1$. If the total numbers of moles A and B in the solution are equal to n_A and n_B, and the number of moles of the various complexes present in the solution are $n_{A(i)}$, $n_{B(j)}$, $n_{A(i)B(j)}$, then

$$n_A = \sum_i in_{A(i)} + \sum_i \sum_j in_{A(i)B(j)}$$
$$n_B = \sum_j jn_{B(j)} + \sum_j \sum_i jn_{A(i)B(j)} \qquad \text{(A3.7)}$$

Denoting the chemical potentials of complexes in a solution through $\mu_{A(i)}$, $\mu_{B(j)}$, $\mu_{A(i)B(j)}$, we have [5] that $\mu_{A(i)} = i\mu_{A(1)}$; $\mu_{B(i)} = i\mu_{B(1)}$; $\mu_{A(i)B(j)} = i\mu_{A(1)} + j\mu_{B(1)}$, where the macroscopic chemical potentials of the components A and B are denoted by μ_A, μ_B and $\mu_A = \mu_{A(1)}$ and $\mu_B = \mu_{B(1)}$. Thus, the macroscopic chemical potentials μ_A and μ_B are equal to the chemical potentials of monomeric molecules.

Taking the non-ideality of the solution into account retains the balance expressions (A3.7) unchanged. The above formulas for chemical potentials (A3.6) could be used if we neglect the differences in the sizes of the associates and the original components. However, it is not. The question of the role of differences in the size of the components of mixtures on lattice structures has been devoted to a large number of papers [3–5,21–37]. A certain number of places are assigned to the molecule in the lattice, the arrangement of which agrees with the shape of the molecule and its flexibility. Much attention was paid to chain molecules, which are represented by a linear sequence of segments, each of which occupies one site. The forces of attraction are taken into account in the same way as in theories for small spherical molecules. All available works are limited to taking into account only the lateral interactions of the nearest neighbors. This is reflected in the Barker–Guggenheim model, which takes into account both the size of the associate and the presence of a multitude of different functional groups on the polyatomic molecule, which are approximated in terms of the 'contact model'. According to this model, a certain characteristic size of the surface area of a polyatomic molecule with its energy parameter is singled out in the system.

The expression for the activity coefficient of component i is expressed as two terms

$$\ln \gamma_i = \ln \gamma_i^{comb} + \ln \gamma_i^{res}, \qquad \text{(A3.8)}$$

where the first term γ_i^{comb} term reflects the size (volume) and shape of the large particle i, due to the contribution of the surface contacts of the given molecule to the total energy of the system. The second term γ_i^{res} reflects the lateral interactions of a large molecule with its neighbours.

The appearance of two terms in the formulas for $\ln \gamma_i$ is connected with the choice as the point of reference of an ideal mixture consisting of molecules with approximately the same size: $\mu_i^{id} = \nu_i + kT \ln \theta_i$. If the sample for an ideal system were calculated taking into account the different sizes γ_i of the components, then only the non-ideal part of the contributions, associated γ_i^{res} with lateral interactions, should be present in the expression. In turn, it can be represented as one γ_i^{res} or two terms reflecting the structure of molecules or free volume when using group energy parameters (see, for example, [4]).

The above equations for large particles on the basis of the molecular theory for non-ideal systems were generalized: in [21-37] molecules were considered that were in the form of three-dimensional parallelepipeds and their simple modifications. As a result, the presence of associates leads only to an increase in the number of components in the solution and additionally it is necessary to take into account the differences in the sizes of the associates and the initial components in the molecular theory.

Since our task is to discuss the relationship between the above molecular equations that reflect the extent of intermolecular interactions and the structure of the expression for the activity coefficient, then taking into account the different sizes of the associates, we have the formula (A3.8), in γ_i^{res} which

$$\ln \gamma_i^{\mathrm{res}} = \frac{1}{2} \sum_{r=1}^{R} \sum_{h \in S_i(f)} z_h(r) \sum_{\ell=1}^{\Phi} t_{hg}^{k_i \ell}(r) \ln \frac{\hat{\theta}_{hg}^{k_i \ell_j}(r)}{(\theta_h^i)^2 \sigma_j} \tag{A3.9}$$

where intermolecular interactions are taken into account in the quasi-chemical approximation. The formula (A3.9) differs from the expression (A3.4) in that it contains sets of numbers $z_h(r)$ that are constructed with respect to the contact number h (the energy properties of the contact h are fixed by the symbol k) on the surface of the large molecule instead of the coordination spheres with respect to central site f containing the center of mass of the molecule i. Here the list of contacts h refers to the entire surface $S_i(f)$ of a molecule i having a centre of mass at site f. The coordinate recalculation from the center of the molecule f to the contact h is given by the geometry of the molecule, which is considered known. As a consequence, the dependence of the sets of numbers $z_h(r)$ on the presence of neighboring particles inside the region of the interaction radius R appears in the formula (9), in contrast to the formula (A3.4).

The generalization of the system of equations (A3.2) and (A3.3) for associates leads to a system that is more cumbersome and larger in dimension, but the physical $\mu_i^{res} = kT\sum_{r=1}^{R} \ln\gamma_i(r)$ meaning of each contribution is completely preserved, including the expansion in terms of contributions from different distances. Note that the formula (A3.9) refers only to the disorganized mutual arrangement of molecules of different sizes. In the case of the appearance of ordered structures that are characteristic of liquid-crystal substances, this expression should reflect the disparity of the distributions of neighboring molecules.

Molecular interpretation. The obtained results indicate the importance of taking into account the real properties of potential functions and the need to take into account the correlation between interacting molecules in the solution. Neglecting the effects of correlation in the mean-field approximation makes molecular models devoid of physical meaning. For the molecular interpretation of experimental data, the quality of the molecular models used and the accuracy of the description of the actual factors involved are important. It is well known that any description of experimental data by selecting model parameters is a poorly-specified problem that allows an ambiguous solution. Therefore, depending on the physical validity of the functional relationships of the equations used, the reliability of the molecular interpretation of the experiment largely depends.

In this connection, it is necessary to emphasize the qualitative difference between the thermodynamic approach and molecular models in the processing of experimental data. The thermodynamic problem consists in the development of functional connections only between the measured parameters of the state of the system, the number of which is always small. If we neglect the influence of external fields, then the parameters of the solution state are: temperature (T), pressure (P) and concentration of the components of mixtures θ_i (molar fractions). The mixing of the thermodynamic characteristics and molecular properties (the size and orientation of the molecule, intermolecular potentials, the structure of the solution, etc.) is unacceptable. Molecular properties can be used only in molecular models that operate with a large (in general, macroscopic) number of variables describing the states of all molecules of the mixture. When constructing models, naturally, the number of variables is reduced by the simplifications introduced. To take into account the local structure of the solution and its effect on the thermodynamic characteristics, molecular models use, in addition to concentrations, additional pairs of higher and higher order distribution functions.

The introduction of the concepts of chemical potential and the coefficient of activity closes classical thermodynamics and does not allow any molecular interpretation of these characteristics in them without involving molecular models. Additional binding to the concept of ideal solutions when separating excess contributions to thermodynamic characteristics makes it possible to characterize the degree of non-ideality of the system and make a number of conclusions from the form of the concentration dependences of excess functions only about the intensity of deviations from the behaviour of ideal systems.

In the case of significant deviations from the properties of ideal systems, in practice, mathematical methods for describing the redundant functions are introduced. For the first time such constructions were introduced by Margules [5,19], and now they are used in different versions, the so-called thermodynamic models. Thus, the logarithm $\ln\gamma_1 = \sum_{k\geq 2}\alpha_k x_2$ of the activity coefficient in a binary mixture can be represented as a series with integer exponents. If, for example $\ln\gamma_1 = \alpha_2 x_2^2 + \alpha_3 x_2^3 + \alpha_4 x_2^4$, then, taking into account the Gibbs-Dugem equation, the activity coefficient of the second component is defined as $\ln\gamma_2 = (\alpha_2 + 3\alpha_3/2 + 2\alpha_4)x_1^2 - (\alpha_3 + 8\alpha_4/3)x_1^3 + \alpha_4 x_1^4$. The same constructions create formal possibilities for the quantitative description of the experiment also in the case of a larger number of variables, although the written out series for ln γ_1 is a conventional mathematical method of expanding a series of functions of one variable that is devoid of physical premises.

The molecular theory provides ‘decoding’ the coefficient of activity through molecular parameters. When deriving equations for the thermodynamic characteristics by the methods of statistical thermodynamics, the simplifications and the field of their application are always formulated. Therefore, all molecular parameters are always limited in magnitude by the physical meaning and/or the conditions of mathematical transformations, as well as the domain of applicability of the constructed equations. This accompanying information obtained in the construction of molecular models and the derivation of expressions for the chemical potential can not be obtained in any 'thermodynamic models', and therefore the latter create the illusion of interpretation. Thus, thermodynamic models are not controlled by the concentration range of application, as well as by the number of phases and by the concentration regions of existence of the phases.

In essence, the thermodynamic description is always oriented to 'accurate' experimental information, and not to give 'model' equations for its approximate description. As noted above, there is always a multiplicity (i.e. arbitrariness) of variants of thermodynamic models, therefore, in principle, there can not be 'model thermodynamic' theories. This position illustrates the connections in the binary mixtures $\ln \gamma_1 = ax_2^2$ and $\ln \gamma_2 = ax_1^2$, which can improve the quantitative agreement with the experimental curve, but from the physical point of view the coefficient α does not have a strictly molecular meaning. The same arbitrariness is represented by proposals to represent the residual contribution of $\ln \gamma_i^{res}$, caused by intermolecular interactions, in the form of terms from different types of potential functions for specific and non-specific constituents [38]. It was shown above that this procedure is impossible in principle by the very nature of statistical averaging procedures relating to different spatial coordinates. From the additivity of the dispersion and dipole potential contributions to the total energy of the system (A3.1), it is not possible to represent these contributions in the activity coefficients as separate terms. Different potential contributions act simultaneously in space and can not be separated as independent. From a physical point of view, it is the joint action of different potentials that is important for the formation of associates, and also allows us to approach the molecular interpretation of the dielectric permittivity of polar liquids that is absent at present, including water, whose solutions are often interpreted from the position of associates formation.

The molecular theory shows that in the general case the structure of the expression for the activity coefficient (A3.8) is approximate. The first term γ_i^{comb} is a consequence of the historical choice for determining redundant functions through expressions for ideal mixtures of the approximately commensurable size. Large differences in the size of the components of the solution with this choice of 'excess properties' lead to the appearance of a contribution γ_i^{comb} even in the absence of intermolecular interactions. Depending on the accuracy of the molecular model used for describing the differences in the size of components in γ_i^{comb} there may be one or two terms [4], therefore, as explained above, one can not use γ_i^{comb} in 'thermodynamic models'.

Intermolecular interactions between all components of the mixture, including associates as their components, are taken into account by the second term in (8). The number of terms in γ_i^{res} depends on the accuracy of the correlation effects and does not depend on the method of constructing the potential functions: for the molecule as a whole or in the contact approximation in the framework of atom–atom potentials [1–4] or quantum chemical methods [39]. The structure of the expression γ_i^{res} is determined by the extent of the potential functions. When the accuracy of the quasi-chemical approximation is sufficient to consider the experiment, the maximum of the number of terms in γ_i^{res} is equal to the radius of the interaction potential, measured in the coordination spheres, which is determined by the total action of all the potentials, rather than individual contributions of different potential functions. With increasing accuracy of accounting for lateral interactions $\ln \gamma_i^{res}$ consists of a single term for any number of contributions in the formula (A3.1), but each contribution affects the concentration behaviour $\ln \gamma_i^{res}$.

The increase in the accuracy of the molecular model inevitably links the configurations of large molecules and their intermolecular interactions. This leads to the impossibility of separation of $\ln \gamma_i$ according to the formula (A3.8). It is necessary to calculate $\ln \gamma_i$ as a single characteristic of the mixture – this is the essence why the thermodynamic models can not exist, as noted above.

References

1. Hirschfelder J.O., Curtiss C., Bird R.B. Molecular theory of gases and liquids. Moscow: IL, 1961. 930 p (Wiley, New York, 1954).
2. Kaplan I.G., Introduction to the Theory of Intermolecular Interactions. Moscow, Nauka, 1982. 312 p.
3. Prigogine I.P., The Molecular Theory of Solutions. Amsterdam; New York, Amster¬dams Interscience Publishers Inc., 1957.
4. Smirnova N.A., Molecular theory of solutions. Leningrad, Khimiya, 1987.
5. Prigogine I., Defay R., Chemical Thermodynamics. Novosibirsk, Nauka, 1966. [Longmans Green, London, 1954].
6. Tovbin Yu.K., Zh. Fiz. Khimii. 2012. V. 86. No. 8. P. 1355.
7. Hill T.L., Statistical Mechanics, New York, McGraw-Hill, 1956.
8. Tovbin Yu.K., Theory of physico-chemical processes at the gas–solid interface, Moscow, Nauka, 1990. [CRC, Boca Raton, Florida, 1991].
9. Tovbin Yu.K., Zh. Fiz. Khimii. 1995. V. 69. No. 2. P. 214.
10. Tovbin Yu.K., Zh. Fiz. Khimii. 1995. V.69. No. 2. P. 220.

11. Barker J.A., Fock W., Disc. Farad. Soc. 1953. No. 15. P. 188.
12. Anderson G.A., Wheeler, J.C., J. Chem. Phys. 1978. V. 69. P. 2082.
13. Walker J.C., Vause S.A., Phys. Lett. 1980. V. A79. P. 421.
14. Goldstein R.E., Walker J.C., J. Chem. Phys. 1983. V. 78. P. 1492.
15. Walker J.C., Vause S.A., Ibid. 1983. V. 78. P. 2660.
16. Tovbin Yu.K., Zh. Fiz. Khimii. 2015. V. 89. No. 11. P. 1752.
17. Fowler R.H., Guggenheim E.A. Statistical Thermodynamics. Cambridge Univer. Press, 1939.
18. Tovbin Yu.K., Kinetika i kataliz. 1982. V.23. P. 1231.
19. Durov V.A., Ageev E.P., Thermodynamic theory of solutions. Moscow, Editorial URSS. 2002.
20. Tovbin Yu.K., Teoret. eksper. khimiya. 1982. V. 18. No. 4. P. 419.
21. Guggenheim E.A., Mixtures. Oxford Univer. Press, 1952.
22. Miller A.R., Theory of Solutions of High Polymers. Oxford, Clarendon Press, 1948.
23. Barker J.A., J. Chem. Phys. 1952. V.20. P. 1526.
24. Flory R.J., Principles of Polymer chemistry. New York, Cornell Univ. Press, 1953.
25. Shakhparonov M.I., Introduction to the molecular theory of solutions. Moscow: GITTL, 1956. 507 pp.
26. Sanchez I.S., Lacombe R.H., J. Phys. Chem. 1976. V.80. P. 2352.
27. Lacombe R.N., Sanchez I.C., Ibid. P. 2568.
28. Sanchez I.S., Lacombé R.H., Macromolecules. 1978. V. 11. P. 1145.
29. Tovbin Yu.K., Khim. Fizika. 2011. V.30. No.4. P.27. [Russ. J. Phys. Chem. B. 5, 256 (2011).
30. DiMarzio E.A., J. Chem. Phys. 1961. V. 35. P. 658.
31. Mitra S.K., Alnatt A.R., J. Phys. Ser. C. Solid State Phys. 1979. V.12. P. 2261.
32. Tumanyan N.P., Shahatuni A.G., Arm. Khim. Zh. 1982. V. 35. P. 103.
33. Tumanyan N.P. , Sokolova E.P., Zh. fiz. khimii. 1984. V.58. P. 2488.
34. Tovbin Yu.K., Izv. AN SSSR. ser. khim. 1997. No.3. P. 458. [Russ. Chem. Bull. 46, 437 (1997)].
35. Tovbin Yu.K., Khim. Fizika. 1997. V.16. No.6. P. 96.
36. Tovbin Yu.K., Zhidkova L.K., Komarov V.N., Izv. AN SSSR. ser. khim.. 2001. No. 5. P. 752. [Russ. Chem. Bull. 50, 786 (2001)].
37. Tovbin Yu.K., Senyavin M.M., Sokolova E.P., Zh. fiz. khimii. 2003. V. 77. No. 4. P. 714 [Russ. J. Phys. Chem. A 77, 636 (2003)].
38. Durov V.A., Shilov I.Yu., Zh fiz. khimii. 2012. V. 86. № 2. P. 216 [Russ. J. Phys. Chem. A 86, 162 (2012)].
39. Simkin B.Ya., Sheikhet I.I., Quantum-chemical and statistical theory of solutions. Moscow, Khimiya. 1989 [Prentice Hall, Englewood Cliffs, NJ, 1995].

Index

A

C

D

E

N

O

P

Q

R

S

T

V